ORBAN'S
ORAL HISTOLOGY
AND EMBRYOLOGY

ORBAN'S
ORAL HISTOLOGY AND EMBRYOLOGY

EDITED BY

S.N. Bhaskar, B.D.S., D.D.S., M.S., Ph.D.

Major General, U.S. Army (Retired); formerly Chief, U.S. Army Dental Corps;
Diplomate, American Board of Oral Pathology; Diplomate, American Board of Oral Medicine

TENTH EDITION

with 650 illustrations and four color plates

The C. V. Mosby Company

ST. LOUIS · TORONTO · PRINCETON   1986

MOSBY

A TRADITION OF PUBLISHING EXCELLENCE

Editor: Darlene Barela Warfel
Assistant editor: Donna Saya Sokolowski
Editing supervisor: Lin Dempsey
Manuscript editor: Liz Williams
Designer: Jeanne Genz
Production: Suzanne C. Glazer, Jeanne Gulledge

TENTH EDITION

The C.V. Mosby Company
11830 Westline Industrial Drive, St. Louis, Missouri 63146

Library of Congress Cataloging in Publication Data

Oral histology and embryology.
   Orban's Oral histology and embryology.

   Includes bibliographies and index.
   1. Mouth.   2. Teeth.   3. Histology.   4. Embryology,
Human.   I. Orban, Balint J. (Balint Joseph), 1899-1960.
II. Bhaskar, S.N. (Surindar Nath), 1923-
III. Title.   [DNLM: 1. Mouth—anatomy & histology.
2. Mouth—embryology.   3. Tooth—anatomy & histology.
4. Tooth—embryology.   WU 101 064]
QM306.07   1985   611'.018931   85-4904
ISBN 0-8016-0812-0

C/VH/VH   9   8   7   6   5   4   3   2   1   03/C/369

CONTRIBUTORS

GARY C. ARMITAGE, D.D.S., M.S.

Professor and Chairman, Division of Periodontology, University of California, San Francisco School of Dentistry, San Francisco, California

JAMES K. AVERY, D.D.S., Ph.D.

Professor and Chairman, Department of Oral Biology, Director Dental Research Institute, The University of Michigan School of Dentistry; Professor of Anatomy and Cell Biology, The University of Michigan School of Medicine, Ann Arbor, Michigan

S.N. BHASKAR, B.D.S., D.D.S., M.S., Ph.D.

Major General, U.S. Army (Retired); formerly Chief, U.S. Army Dental Corps; Diplomate, American Board of Oral Pathology; Diplomate, American Board of Oral Medicine

†BALDEV RAJ BHUSSRY, B.D.S., D.D.S., M.S., Ph.D.

Formerly Associate Professor and Chairman, Department of Anatomy, Georgetown University Schools of Medicine and Dentistry, Washington, D.C.

ARTHUR R. HAND, D.D.S.

Laboratory of Oral Biology and Physiology, National Institute of Dental Research, National Institutes of Health, Bethesda, Maryland

MALCOLM C. JOHNSTON, D.D.S., M.Sc.D., Ph.D.

Professor of Orthodontics and Anatomy, Schools of Dentistry and Medicine; Sr. Scientist, Dental Research Center, University of North Carolina at Chapel Hill, Chapel Hill, North Carolina

SHAKTI P. KAPUR, M.S., Ph.D.

Associate Professor, Department of Anatomy, Georgetown University Schools of Medicine and Dentistry, Washington, D.C.

ANTONY H. MELCHER, M.D.S., H.D.D., Ph.D., D.Sc.

Professor of Dentistry, Faculty of Dentistry; Associate Dean (Life Sciences), School of Graduate Studies; University of Toronto, Toronto, Ontario, Canada

MOHAMED SHARAWY, B.D.S., Ph.D.

Professor and Chairman, Department of Oral Biology/Anatomy, School of Dentistry; Professor of Anatomy, School of Medicine, Medical College of Georgia, Augusta, Georgia

IRVING B. STERN, D.D.S.

Formerly Professor and Chairman, Department of Periodontology, Tufts University School of Dental Medicine, Boston, Massachusetts; formerly Professor of Periodontics, Department of Periodontics; and currently Affiliate Professor of Oral Biology, University of Washington, School of Dentistry, Seattle, Washington

†FAUSTINO R. SUAREZ, M.D.

Formerly Assistant Professor, Department of Anatomy, Georgetown University Schools of Medicine and Dentistry, Washington, D.C.

KATHLEEN K. SULIK, Ph.D.

Assistant Professor of Anatomy, Department of Anatomy, School of Medicine, University of North Carolina at Chapel Hill, Chapel Hill, North Carolina

A. RICHARD TEN CATE, B.Sc., B.D.S., Ph.D.

Faculty of Dentistry, University of Toronto, Toronto, Ontario, Canada

BRANISLAV VIDIĆ, S.D.

Professor, Department of Anatomy, Georgetown University Schools of Medicine and Dentistry, Washington, D.C.

JAMES A. YAEGER, D.D.S., Ph.D.

Professor, Department of Oral Biology, University of Connecticut Health Center, School of Dental Medicine, Farmington, Connecticut

---

†Deceased.

---

†Deceased.

TO

Balint J. Orban

Joseph P. Weinmann

Harry Sicher

Baldev Raj Bhussry

and

Faustino R. Suarez

PREFACE

The ultimate test of all dental education is to see how well it prepares the practitioner to serve the patient. No aspect of a dental school curriculum can give greater meaning to a clinical procedure or put it on a more rational foundation than a thorough understanding of basic sciences. An understanding of basic sciences can be the difference between an excellent clinician and one who can treat a patient only as a technician, between a dentist who can lead and one who can only follow, between an innovator and one whose clinical resources are limited and dated.

In the tenth edition of *Orban's Oral Histology and Embryology,* a group of scientists and dentists, varyingly engaged in basic science and clinical research, in teaching dental students and dental practitioners, and in clinical practice of dentistry, have joined together to present the subject with the hope that it will better prepare the student to practice the profession wisely and with confidence.

I record with deep regret that since the publication of the last edition two of my friends and collegues have passed away. Professors Baldev Raj Bhussry and Faustino R. Suarez, both from Georgetown University School of Dentistry, were teachers and researchers of exceptional talents. Their contributions to science and to the development of future generations of dentists will be deeply missed. It is with great respect and gratitude that I add their names to those to whom this edition of the book is dedicated.

S.N. Bhaskar

CONTENTS

†Deceased.

14 MAXILLARY SINUS, 405
Branislav Vidić

15 HISTOCHEMISTRY OF ORAL TISSUES, 422
Shakti P. Kapur

Appendix

Preparation of Specimens for Histologic Study, 455

COLOR PLATES

†Deceased.

ORBAN'S
ORAL HISTOLOGY
AND EMBRYOLOGY

1

DEVELOPMENT OF FACE AND ORAL CAVITY

This chapter deals primarily with the development of the human face and oral cavity. Consideration is also given to information about underlying mechanisms that is derived from experimental studies conducted on developing subhuman embryos. Much of the experimental work has been conducted on amphibian and avian embryos. Evidence derived from these and more limited studies on other vertebrates including mammals indicates that the early facial development of all vertebrate embryos is similar. Many events occur, including cell migrations, interactions, differential growth, and differentiation, all of which lead to progressively maturing structures (Fig. 1-1). Progress has also been made with respect to abnormal developmental alterations that give rise to some of the most common human malformations (Fig. 1-15). Further information on the topics discussed can be obtained by consulting the references at the end of the chapter.

ORIGIN OF FACIAL TISSUES

After fertilization of the ovum, a series of cell divisions gives rise to an egg cell mass known as the *morula* in mammals. In most vertebrates, including humans, the major portion of the egg cell mass forms the extraembryonic membranes and other supportive structures such as the placenta. Less than one fourth of the cells of the egg cell mass eventually assemble to form a single layer, which will form the embryo. Cell movements then convert this *embryonic disc* into two layers with an intervening space, or potential space. An additional, well-integrated movement of actively migrating cells from the upper of the two layers leads to the formation of a third layer of cells, which occupies the intervening space. The uppermost of these three "germ" layers is called the *ectoderm;* the middle layer, the *mesoderm;* and the lowest layer, the *endoderm* (Fig. 1-2, *A*). Thus, at this stage, three distinct populations of embryonic cells have arisen largely through division and migration. They follow distinctly separate courses during later development.

Migrations, such as those described above, create new associations between cells, which, in turn, allow unique possibilities for subsequent development through interactions be-

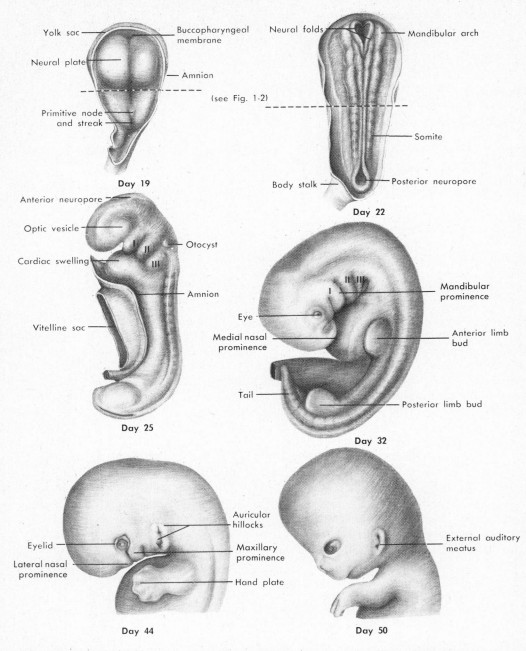

Yolk sac
Neural plate
Primitive node and streak
Buccopharyngeal membrane
Amnion
(see Fig. 1-2)
Day 19

Neural folds
Mandibular arch
Somite
Body stalk
Posterior neuropore
Day 22

Anterior neuropore
Optic vesicle
Cardiac swelling
Vitelline sac
Otocyst
II
III
Amnion
Day 25

Mandibular prominence
Anterior limb bud
I
II III
Eye
Medial nasal prominence
Tail
Posterior limb bud
Day 32

Eyelid
Lateral nasal prominence
Auricular hillocks
Maxillary prominence
Hand plate
Day 44

External auditory meatus
Day 50

Fig. 1-1. Emergence of facial structures during development of human embryos. At gestational days 19 and 22, dorsal views of earlier stages are illustrated. At days 25 and 32, visceral arches are designated by Roman numerals. Gradually, the embryos become recognizable as "human" by gestational day 50. Section planes for Fig. 1-2 are illustrated in the upper (days 19 and 22) diagrams.

tween the cell populations. Such interactions have been studied experimentally by isolating the different cell populations or tissues and recombining them in different ways in culture or in transplants. From such studies it is known, for example, that a median strip of mesoderm cells (the chordamesoderm) extending throughout the length of the embryo induces *neural plate* formation within the overlying ectoderm (Fig. 1-2). The nature of such inductive stimuli is presently unknown. Sometimes cell-to-cell contact appears to be necessary, whereas in

other cases (as in neural plate induction) the inductive influences appear to be able to act between cells separated by considerable distances and to consist of diffusible substances. It is known that inductive influences need only be present for a short time, after which the responding tissue is capable of independent development. For example, an induced neural plate isolated in culture will roll up into a tube, which then differentiates into the brain, spinal cord, and other structures.

A unique population of cells develops from

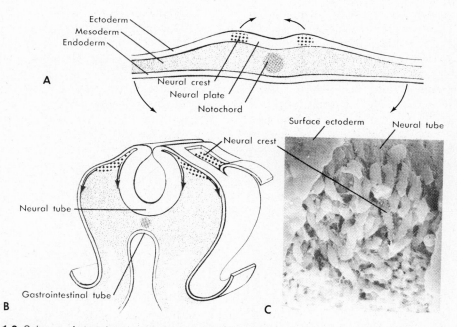

Fig 1-2. Scheme of neural and gastrointestinal tube formation in higher vertebrate embryos (section planes illustrated in Fig. 1-1). **A,** Cross section through three-germ layer embryo. Similar structures are seen in both head and trunk regions. Neural crest cells (diamond pattern) are initially located between neural plate and surface ectoderm. Arrows indicate directions of folding processes. **B,** Neural tube, which later forms major components of brain and spinal cord, and gastrointestinal tube will separate from embryo surface after fusions are completed. Arrows indicate directions of migration of crest cells, which are initiated at about fourth week in human embryo. **C,** Scanning electron micrograph (SEM) of mouse embryo neural crest cells migrating over neural tube and under surface ectoderm near junction of brain and spinal cord following removal of piece of surface ectoderm as indicated in **B.** Such migrating cells are frequently bipolar (e.g., outlined cell at end of leader) and oriented in path of migration *(arrow)*.

the ectoderm along the lateral margins of the neural plate. These are the neural crest cells. They undergo extensive migrations, usually beginning at about the time of tube closure (Fig. 1-2), and give rise to a variety of different cells that form components of many tissues. The crest cells that migrate in the trunk region form mostly neural, endocrine, and pigment cells, whereas those that migrate in the head and neck also contribute extensively to skeletal and connective tissues (i.e., cartilage, bone, dentin, dermis, etc.). In the trunk, all skeletal and connective tissues are formed by mesoderm. Of the skeletal or connective tissue of the facial region, it appears that tooth enamel (an acellular skeletal tissue) is the only one not formed by crest cells. The enamel-forming cells are derived from ectoderm lining the oral cavity.

The migration routes that cephalic (head) neural crest cells follow are illustrated in Fig. 1-3. They move around the sides of the head beneath the surface ectoderm, en masse, as a sheet of cells. They form all the mesenchyme* in the upper facial region, whereas in the lower facial region they surround mesodermal cores already present in the visceral arches. The pharyngeal region is then characterized by grooves (clefts) in the lateral pharyngeal wall endoderm and ectoderm that approach each other and appear to effectively segment the mesoderm into a number of bars that become surrounded by crest mesenchyme (Fig. 1-6, A).

Toward the completion of migration, the trailing edge of the crest cell mass appears to attach itself to the neural tube at locations where sensory ganglia of the fifth, seventh, ninth, and tenth cranial nerves will form. In the trunk sensory ganglia, supporting (e.g., Schwann) cells and all neurons are derived

*Mesenchyme is defined here as the loosely organized embryonic tissue in contrast to epithelium, which is compactly arranged.

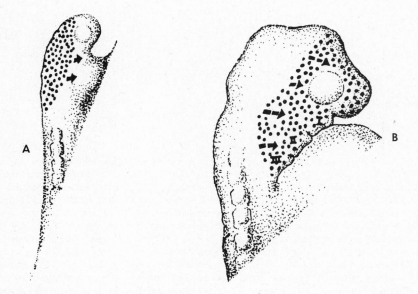

Fig. 1-3. Scheme of subectodermal distribution of neural crest cells (large stipples) during, **A,** and toward the completion, **B,** of migration. Arrows, Direction of migration. First three visceral arches are indicated by Roman numerals.

from neural crest cells. On the other hand, many of the sensory neurons of the cranial sensory ganglia originate from surface ectoderm.

Eventually, capillary endothelial cells derived from mesoderm cells invade the crest cell mesenchyme, and it is from this mesenchyme that the supporting cells of the developing blood vessels are derived. Initially, these supporting cells include only pericytes, which are closely apposed to the outer surfaces of endothelial cells. Later, additional crest cells differ-

entiate into the fibroblasts and smooth muscle cells that will form the vessel wall. The developing blood vessels become interconnected to form vascular networks. These networks undergo a series of modifications, examples of which are illustrated in Fig. 1-4, before they eventually form the mature vascular system. The underlying mechanisms are not clearly understood.

Almost all the myoblasts that subsequently fuse with each other to form the multinu-

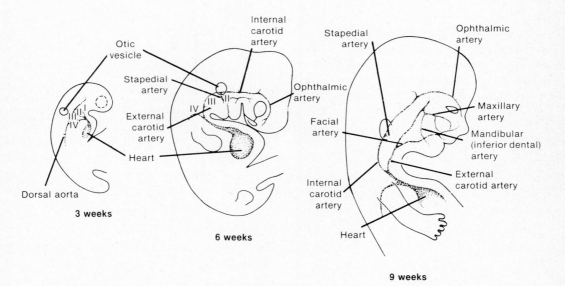

Fig. 1-4. Development of the arterial system serving the facial region with emphasis on its relation to the visceral arches. In the 3-week human embryo the visceral arches are little more than conduits for blood traveling through aortic arch vessels (indicated by Roman numerals according to the visceral arch containing them) from the heart to the dorsal aorta. Other structures indicated are the eye *(broken circle)* and opthalmic artery. In the 6-week embryo the first two aortic arch vessels have regressed almost entirely, and the distal portions of the arches have separated from the heart. The portion of the third aortic arch vessel adjacent to the dorsal aorta persists and eventually forms the stem of the external carotid artery by fusing with the stapedial artery. The stapedial artery, which develops from the second aortic arch vessel, temporarily (in humans) provides the arterial supply for the embryonic face. After fusion with the external carotid artery, the proximal portion of the stapedial artery regresses. The aortic arch vessel of the fourth visceral arch persists as the arch of the aorta. By 9 weeks the primordium of the definitive vascular system of the face has been laid down. (From Ross, R.B., and Johnston, M.C.: Cleft lip and palate, Baltimore, 1972, The Williams & Wilkins Co.)

cleated striated muscle fibers are derived from mesoderm. The myoblasts that form the hypoglossal (tongue) muscles are derived from somites located beside the developing hindbrain. Somites are condensed masses of cells derived from mesoderm located adjacent to the neural tube. The extrinsic ocular muscles originate from similar, more anterior mesoderm (Fig. 1-5), which fails to condense in higher vertebrates. However, direct visualization by scanning electron microscopy shows enough organization into "somitomeres" to distinguish them as the equivalent of more caudal somites. These prospective myoblasts (still not recognizable as premuscle cells) must undergo extensive migrations (Fig. 1-5). The supporting connective tissue found in facial muscles is derived from neural crest cells. Much of the development of the masticatory and other facial musculature is closely related to the final stages of visceral arch development and will be described later.

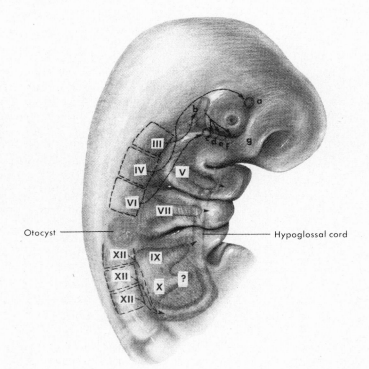

Fig. 1-5. Migration paths followed by prospective skeletal muscle cells. Somites, or comparable structures from which muscle cells are derived, give rise to most skeletal (voluntary) myoblasts (differentiating muscle cells). Condensed somites tend not to form in head region of higher vertebrates, and their position in lower forms is indicated by broken lines. It is from these locations that extrinsic ocular and "tongue" (hypoglossal cord) muscle contractile cells are derived—their regions of origin are indicated by Roman numerals of cranial nerves that innervate them. Recent studies indicate that myoblasts which contribute to the visceral arch musculature have similar origins and originate as indicated by Roman numerals according to their nerves of innervation. At this stage of development (approximately day 34) they are still migrating *(arrowheads)* into cores of each visceral arch. Information about fourth visceral arch is still inadequate, as indicated by question mark *(?)*.

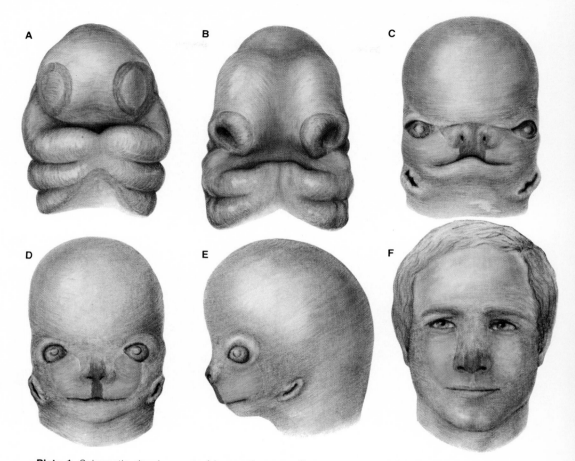

Plate 1. Schematic development of human face, maxillary process *(yellow)*, lateral nasal process *(blue),* and medial nasal process *(red).* **A,** Embryo 4 to 6 mm in length, approximately 28 days. Prospective medial nasal and lateral nasal processes are just beginning to form from mesenchyme surrounding olfactory placode. Maxillary process is undergoing its initial separation from remainder of first mandibular arch (compare to Fig. 1-12) under eye. **B,** Embryo 8 to 11 mm in length, approximately 37 days. Medial nasal process is beginning to make contact with lateral nasal and maxillary processes. **C,** Embryo 16 to 18 mm in length, approximately 47 days. Maxillary process is beginning to overlap lateral nasal process mesenchyme to some extent *(stippled area)* and medial nasal process is firmly in contact with lateral nasal and maxillary process. **D** and **E,** Embryo 23 to 28 mm in length, approximately 54 days. Maxillary process further overlaps lateral nasal process and may overlap much or all of medial nasal process mesenchyme in upper lip *(red stipple).* **F,** Adult face. Approximate derivatives of medial nasal process, lateral nasal process, and maxillary process are indicated.

A number of other structures in the facial region, such as glands and the enamel organ of the tooth bud, are derived from epithelium that grows (invaginates) into underlying mesenchyme. Again, the connective tissue components in these structures (e.g., fibroblasts, odontoblasts, and the cells of tooth-supporting tissues) are derived from neural crest cells.

DEVELOPMENT OF FACIAL PROMINENCES

On the completion of the initial crest cell migration and the vascularization of the derived mesenchyme, a series of outgrowths or swellings termed "facial prominences" initiates the next stages of facial development (Fig. 1-6). The growth and fusion of upper facial prominences produce the primary and secondary palates. As will be described below, other prominences developing from the first two visceral arches considerably alter the nature of these arches.

Development of nasal placodes, frontonasal region, primary palate, and nose. Before crest cell migration, the surface ectoderm lies in apposition to portions of the developing forebrain. Inductive influences originating from the forebrain initiate the formation of the nasal placodes in the apposed ectoderm. Placodes are recognizable as ectodermal thickenings, and they give rise to a variety of structures such as the lens of the eye and the inner ear epithelium. The nasal placodes will later form the sensory epithelium for olfaction. After induction, mesenchymal cells separate the nasal placodal ectoderm from the underlying forebrain. The thickening nasal placodes appear to interact with mesenchymal cells that aggregate along their undersurfaces. It is not clear whether this interaction is related to placodal invagination.

Up to the present time, the term "frontonasal region" has been ill defined. In view of the following material, it would clarify the problem to refer to the crest mesenchyme that underlies the "prospective" olfactory placodes together with the overlying epithelium located between the eyes as the "frontonasal region." At least partly under the direction of the olfactory placode, new outgrowths (the medial and lateral nasal prominences, Fig. 1-6) later appear on either side of each olfactory placode. Their growth is associated with an extensive meshwork of mesenchymal cell processes (CPM; Fig. 1-6, *C*), which interacts with the overlying epithelium much as it does in limb bud formation and other structures undergoing rapid growth and other morphogenetic changes.

Eventually, the medial and lateral nasal prominences contact each other below the developing nasal pit (Figs. 1-6 to 1-8). During the fifth week of human pregnancy a portion of the adhering epithelium breaks down so that the mesenchyme of the two prominences becomes continuous (Fig. 1-7). Fluid accumulates between the cells of the persisting epithelium behind the point of epithelial breakdown. Eventually, these fluid-filled spaces coalesce to form the initial nasal passageway connecting the olfactory pit with the roof of the primitive oral cavity (Fig. 1-8). The tissue resulting from development and fusion of these prominences is termed the *primary palate* (outlined by broken lines in Fig. 1-8). It forms the roof of the anterior portion of the primitive oral cavity, as well as forming the initial separation between the oral and nasal cavities. In later development, derivatives of the primary palate form portions of the upper lip, anterior maxilla, and upper incisor teeth.

The outlines of the developing external nose can be seen in Fig. 1-7, *D*. Although the nose is disproportionately large, the basic form is easily recognizable. Subsequent alterations in form lead to progressively more mature structure (Fig. 1-1, day 50 specimen). Plate 1 is a schematic illustration of the contribution of various facial prominences to the development of the external face.

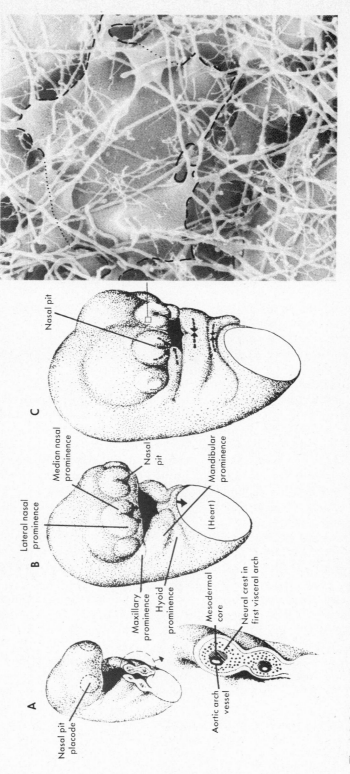

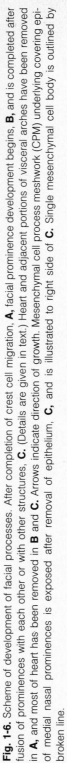

Fig. 1-6. Scheme of development of facial processes. After completion of crest cell migration, **A,** facial prominence development begins, **B,** and is completed after fusion of prominences with each other or with other structures, **C.** (Details are given in text.) Heart and adjacent portions of visceral arches have been removed in **A,** and most of heart has been removed in **B** and **C.** Arrows indicate direction of growth. Mesenchymal cell process meshwork (CPM) underlying covering epithelium, **C,** and is illustrated to right side of **C.** Single mesenchymal cell body is outlined by broken line.

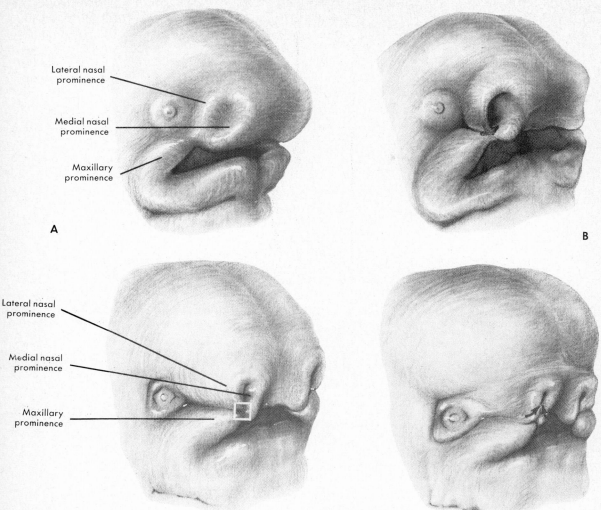

Fig. 1-7. Formation of primary palate, embryonic structure initially separating oral and nasal cavities (see Fig. 1-8). Recent investigations on higher vertebrate embryos, including humans, have contributed a great deal to clarification of this complex and important aspect of development. **A,** Three outgrowths of tissue, lateral nasal, medial nasal, and maxillary prominences, are apparent. **B,** Coalescence or "merging" of three facial prominences results in formation of considerable rim of tissue at base of olfactory pit, which grows outward from underlying brain and eye in direction indicated by arrow. Much confusion has arisen because of inadequate understanding of this and succeeding events in primary palate formation. **C,** Further outgrowth of medial nasal prominence on one side and of lateral nasal and maxillary prominences on other results in formation of an epithelially lined groove (not shown). Although there are minor species variations, initial contacts and epithelial fusion and breakdown occur at region indicated (☐). Epithelial cell death initially occurs at superficial location close to junction between lateral nasal and maxillary prominences with subsequent cell death in adjacent epithelium of medial nasal prominence permitting underlying mesenchymal cells to make contact with one another. **D,** Consolidation of this initial mesenchymal contact is achieved through "zipping up" of outer margin of olfactory pit *(arrows)* together with subsequent further widening of mesenchymal contact through epithelial regression and "merging" phenomenon. Epithelial continuity between base of pit and roof of primitive oral cavity is always maintained behind this mesenchymal contact. Hollowing out of this epithelial connection gives rise to posterior position of primitive nasal cavities (see Fig. 1-8).

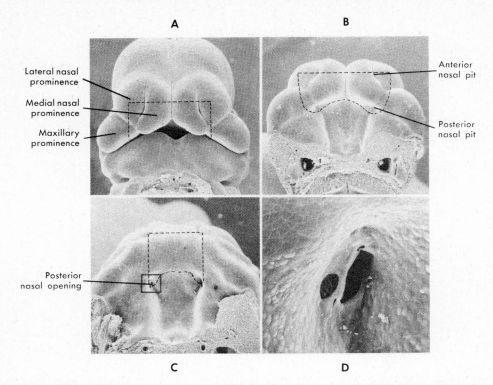

Fig. 1-8. Some of the details of primary palate formation, here shown in mouse, are conveniently demonstrated by scanning electron micrographs (SEMs). Area encompassed by developing primary palate is outlined by broken lines. **A** and **B**, Frontal and palatal views showing moderately advanced stage of primary palate formation. **C** and **D**, In this more advanced stage, elimination of epithelial connection between anterior and posterior nasal pits is nearing completion. Area outlined by solid lines in **C** is given in **D**, showing that the last epithelial elements are regressing as the nasal passage is now almost completely opened.

Development of maxillary prominences and secondary palate. New outgrowths from the medial edges of the maxillary prominences form the shelves of the secondary palate. These palatal shelves grow downward beside the tongue (Figs. 1-9 and 1-10), at which time the tongue partially fills the nasal cavities. At about the ninth gestational week, the shelves elevate, make contact, and fuse with each other above the tongue (Fig. 1-10). In the anterior region, the shelves are brought to the horizontal position by a rotational (hingelike) movement. In the more posterior regions, the shelves appear to alter their position by changing shape (remodeling) as well as by rotation. Available evidence indicates that the shelves are incapable of elevation until the tongue is first withdrawn from between them. Although the motivating force for shelf elevation is not clearly defined, contractile elements may be involved.

Fusion of palatal shelves requires alterations in the epithelium of the medial edges that begin prior to elevation. These alterations consist of cessation of cell division, which appears to be mediated through distinct underlying biochemical pathways, including a rise in cyclic

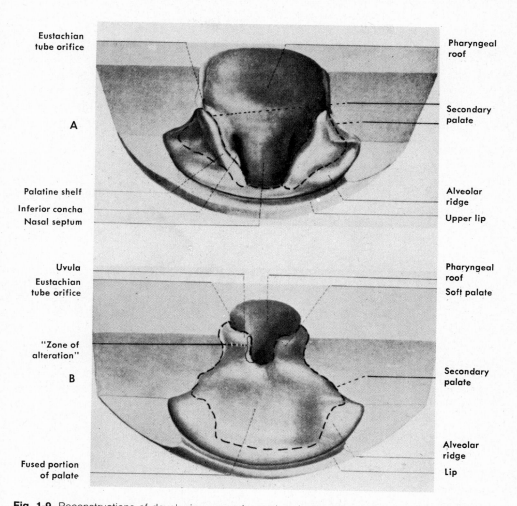

Eustachian tube orifice

Pharyngeal roof

A

Secondary palate

Palatine shelf

Inferior concha

Nasal septum

Alveolar ridge

Upper lip

Uvula

Eustachian tube orifice

Pharyngeal roof

Soft palate

"Zone of alteration"

B

Secondary palate

Alveolar ridge

Fused portion of palate

Lip

Fig. 1-9. Reconstructions of developing secondary palate (outlined by broken lines) of human embryos viewed from below and behind. **A,** Reconstruction of roof of primitive oral and pharyngeal cavities of 8-week human embryo. Primary palate and internal surface of maxillary prominence form horseshoe-shaped incomplete roof of oral cavity. In center, oral cavity communicates with nasal cavities. At edges of maxillary prominences, palatal shelves develop. **B,** Reconstruction of palate of slightly older human embryo. Palatal shelves are fused in area of hard palate. Fusion has not reached soft palate and uvula. Nature of "zone of alteration" is described in Fig. 1-11. (From Sicher, H., and Tandler, J.: Anatomie für Zahnärzte [Anatomy for dentists], Berlin, 1928, Springer Verlag.)

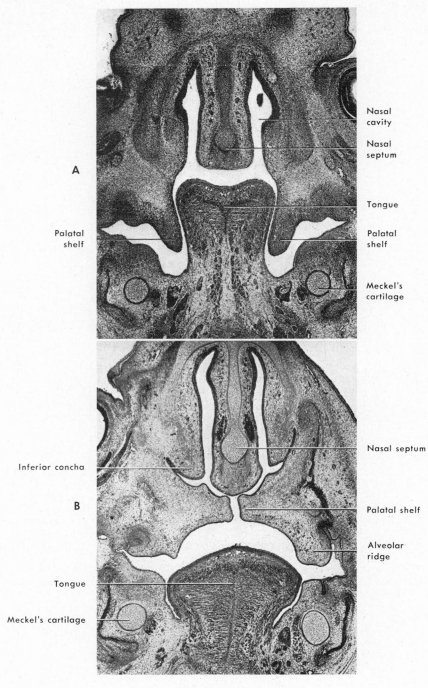

A

Nasal cavity

Nasal septum

Tongue

Palatal shelf

Palatal shelf

Meckel's cartilage

B

Inferior concha

Nasal septum

Palatal shelf

Alveolar ridge

Tongue

Meckel's cartilage

Fig. 1-10. For legend see opposite page.

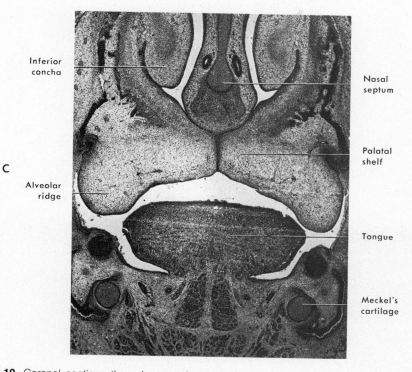

Inferior
concha

Nasal
septum

Palatal
shelf

C

Alveolar
ridge

Tongue

Meckel's
cartilage

Fig. 1-10. Coronal sections through secondary palates of human embryos showing progressive stages of development. **A,** Frontal section through head of 8-week embryo. Tongue is high and narrow between vertical palatal shelves. Meckel's cartilage is first visceral arch cartilage. **B,** Frontal section through head of slightly more advanced embryo. Tongue has left space between palatal shelves and lies flat and wide within mandibular arch. Palatal shelves have assumed horizontal positon. **C,** Frontal section through head of embryo slightly older than that in **B.** Horizontal palatal shelves are fusing with each other and with nasal septum. Secondary palate separates nasal cavities from oral cavity. (**A** and **B** courtesy P. Gruenwald, Richmond, Va.)

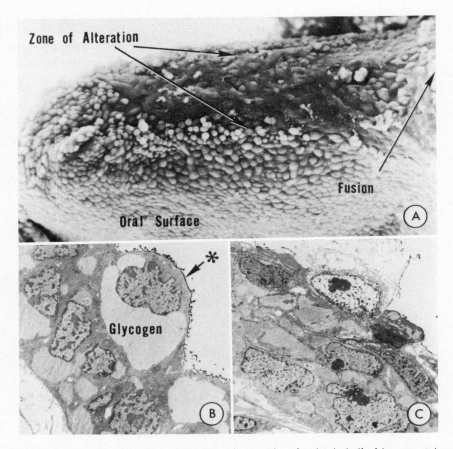

Fig. 1-11. Scanning and transmission electron micrographs of palatal shelf of human embryo at same stage of development as reconstruction in Fig. 1-9, *B*. **A,** Posterior region of palatal shelf viewed from below and from opposite side. Fusion will occur in "zone of alteration," location of which is indicated in Fig. 1-9, *B*. **B** and **C,** Transmission electron micrographs of specimen in **A.** Surface cells of oral epithelium in **B** contain large amounts of glycogen, whereas those of zone of alteration in **C** are undergoing degenerative changes and many of them are presumably desquamated into oral cavity fluids. Asterisk in **B** indicates heavy metal deposited on embryo surfaces for scanning electron microscopy. (**A** to **C** from Waterman, R.E., and Meller, S.M.: Anat. Rec. **180:**11, 1974.)

AMP levels. There is also loss of some surface epithelial cells (Fig. 1-11) and production of extracellular surface substances, particularly glycoproteins, that appear to enhance adhesion between the shelf edges as well as between the shelves and inferior margin of the nasal septum (Fig. 1-10). Finally, the adhering epithelia, together with their basement membranes, break down and are replaced by mesenchyme. Epithelial cell debris is phagocytosed by mesenchymal cells. Not all the epithelial cells are lost in this process; some remain indefinitely in clusters (cell rests) along the fusion line. Eventually, most of the hard palate and all of the soft palate form from the secondary palate (see Chapter 8).

Development of pituitary gland, visceral arches, and tongue. The pituitary gland develops as a result of inductive interactions be-

tween the ventral forebrain and oral ectoderm and is derived in part from both tissues (Figs. 1-12 and 1-13). Following initial crest cell migration (Fig. 1-6, *A*), these cells invade the area of the developing pituitary gland and are continuous with cells that will later form the maxillary prominence. Eventually, crest cells form the connective tissue components of the gland.

In humans there is a total of six visceral arches, of which the fifth is rudimentary. The proximal portion of the first (mandibular) arch becomes the maxillary prominence (Figs. 1-1 and 1-12). As the heart recedes caudally, the mandibular and hyoid arches develop further at their distal portions to become consolidated in the ventral midline (Figs. 1-6, 1-12, and 1-13). As noted previously, the mesodermal core of each visceral arch (Fig. 1-6, *A*) is concerned primarily with the formation of vascular endo-

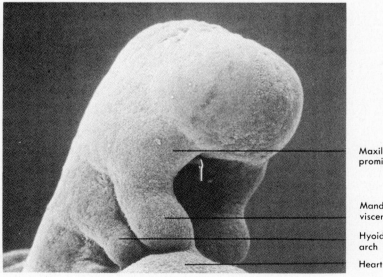

Maxillary prominence

Mandibular visceral arch

Hyoid viscera arch

Heart

Fig. 1-12. Scanning electron micrograph of ferret embryo showing intermediate stage of visceral arch development. Eventually both mandibular and hyoid visceral arches come together in ventral midline as heart recedes caudally. *Arrow,* Opening to Rathke's pouch is located at medial edge of maxillary prominence, which is just beginning to form recognizable structure. (Courtesy A.J. Steffek and D. Mujwid, Chicago.)

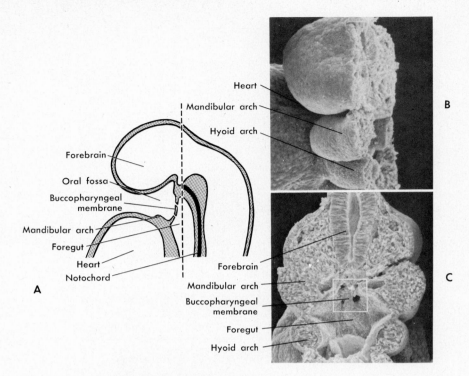

Fig. 1-13. Oropharyngeal development. **A,** Diagram of sagittal section through head of 3½- to 4-week-old human embryo. Oral fossa is separated from foregut by double layer of epithelium (buccopharyngeal membrane), which is in its early stages of breakdown. **B** and **C,** Scanning electron micrographs (SEMs) of mouse head sectioned in plane indicated by broken line in **A. B** represents a more lateral view of specimen while in **C** it is viewed from its posterior aspect. Rupturing buccopharyngeal membrane is outlined by rectangle in this figure.

thelial cells. As noted below, these cells appear to be later replaced by cells that eventually form visceral arch myoblasts.

The first (mandibular) and second (hyoid) visceral arches undergo further developmental changes. As the heart recedes caudally, both arches send out bilateral processes that fuse with their opposite members in ventral midline (Figs. 1-6 and 1-12).

Nerve fibers from the fifth, seventh, ninth, and tenth cranial nerves extend into the mesoderm of the first four visceral arches. The mesoderm of the definitive mandibular and

hyoid arches gives rise to the fifth and seventh nerve musculature, while mesoderm associated with the less well developed third and fourth arches forms the ninth and tenth nerve musculature. Recent studies show that myoblast cells in the visceral arches actually originate from mesoderm more closely associated with the neural tube (as do the cells that form the hypoglossal and extrinsic eye musculature; Fig. 1-5). They would then migrate into the visceral arches and replace the mesodermal cells that initiated blood vessel formation earlier (see p. 5). It therefore appears that myoblasts forming

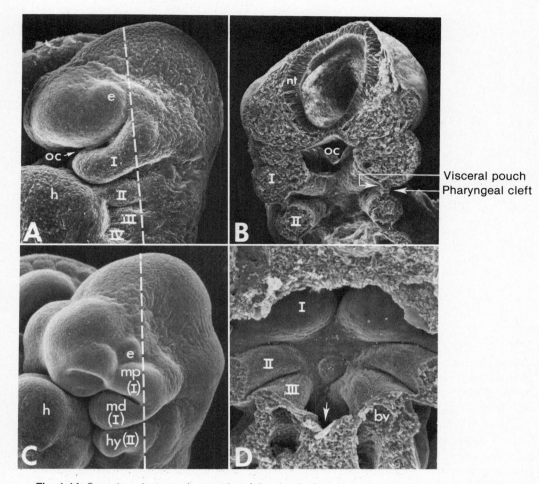

Visceral pouch
Pharyngeal cleft

Fig. 1-14. Scanning electron micrographs of the developing visceral arches and tongue of mouse embryos. The planes of section illustrated in **B** and **D** (dorsal views of floor of pharynx) are shown in **A** and **C. A** and **B,** Embryos whose developmental age is approximately equivalent to that of human 30-day-old embryos. (See Fig. 1-1) Development of the medial and lateral nasal prominences has yet to be initiated. The visceral arches are indicated by Roman numerals. The first (mandibular) arch is almost separated from the heart (h). Other structures indicated are the eye (e), oral cavity (oc; compare to buccopharyngeal membrane in Fig. 1-13, C), and neural tube (nt). **C** and **D,** These are comparable to 35-day-old human embryos. The mandibular arch now has two distinct prominences, the maxillary prominence (mp) and the mandibular prominence (md). The second arch is called the hyoid arch (hy). In **D** the blood vessel exiting from the third arch is labeled bv. The arrow indicates entry into the lower pharynx.   *Continued.*

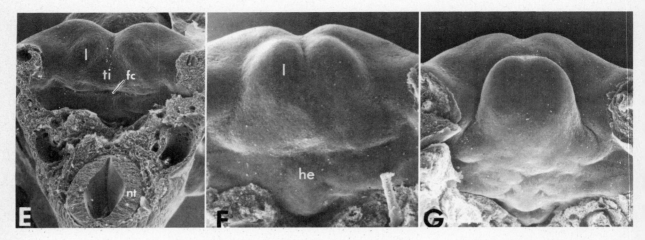

Fig. 1-14, cont'd. E to **G,** Older specimens, prepared in a manner similar to **B** and **D,** illustrate development of the tongue. The lingual swellings *(l)* presumably represent accumulations of myoblasts derived from the hypoglossal cord. The tuberculum impar *(ti)* also contributes to the anterior two thirds of the tongue. The foramen cecum *(fc)* is the site of endodormal invagination that gives rise to epithelial components of the thyroid gland. It lies at the junction between the anterior two thirds and posterior one third of the tongue. The hypobranchial eminence *(he)* is the primordium of the epiglottis. (From Johnston, M.C., and Sulik, K.K.: Embryology of the head and neck. In Serafin, D., and Georgiade, N.G., editors: Pediatric plastic surgery, vol. I, St. Louis, 1984, The C.V. Mosby Co.)

voluntary striated muscle fibers of the facial region would then originate from mesoderm adjacent to the neural tube.

Groups of visceral arch myoblasts that are destined to form individual muscles each take a branch of the appropriate visceral arch nerve. Myoblasts from the second visceral arch, for example, take branches of the seventh cranial nerve and migrate very extensively throughout the head and neck to form the contractile components of the "muscles of facial expression." Myoblasts from the first arch contribute mostly to the muscles of mastication, while those from the third and fourth arches contribute to the pharyngeal and soft palate musculature. As noted earlier, connective tissue components of each muscle in the facial region are provided by mesenchymal cells of crest origin.

The crest mesenchymal cells of the visceral arches give rise to skeletal components such as the temporary visceral arch cartilages (e.g., Meckel's cartilage; Fig. 1-10), middle ear cartilages, and mandibular bones. Also visceral arch crest cells form connective tissues such as dermis and the connective tissue components of the tongue.

The tongue forms in the ventral floor of the pharynx after arrival of the hypoglossal muscle cells. The significance of the lateral lingual tubercles (Fig. 1-14) and other swellings in the forming tongue has not been carefully documented. It is known that the anterior two thirds of the tongue is covered by ectoderm whereas endoderm covers the posterior one third. The thyroid gland forms by invagination of the most anterior endoderm (thyroglossal duct). A residual pit (the foramen cecum; Fig. 1-14, *C*) left in the epithelium at the site of invagination marks the junction between the anterior two thirds and posterior one third of the tongue, which are, respectively, covered by epithelia of ectodermal and endodermal origin. It is also known that the connective tissue components of the anterior two thirds of the tongue are derived from first-arch mesenchyme, whereas those of the posterior one third appear to be primarily derived from the third-arch mesenchyme.

The epithelial components of a number of glands are derived from the endodermal lining of the pharynx. In addition to the thyroid, these include the parathyroid and thymus. The epithelial components of the salivary and anterior pituitary glands are derived from oral ectoderm.

Finally, a lateral extension from the inner groove between the first and second arch gives rise to the eustachian tube, which connects the pharynx with the ear. The external ear, or pinna, is formed at least partially from tissues of the first and second arches (Fig. 1-1, day 44).

FINAL DIFFERENTIATION OF FACIAL TISSUES

The extensive cell migrations referred to above bring cell populations into new relationships and lead to further inductive interactions, which, in turn, lead to progressively more differentiated cell types. For example, some of the crest cells coming into contact with pharyngeal endoderm are induced by the endoderm to form visceral arch cartilages (see Chapter 8). Other crest cells that have migrated in the vicinity of the pharyngeal endoderm on their way to the oral cavity are preconditioned by the endoderm and will react with oral ectoderm and differentiate into tooth papilla mesenchyme.

In many instances, such as those cited above, only crest mesenchymal cells and not mesodermal mesenchymal cells will respond to inducing tissues such as pharyngeal endoderm. In other cases, as in the differentiation of dermis and meninges, it appears that the origin of the mesenchyme is of no consequence. In any case it is clear that one function, the formation

of skeletal and connective tissues, ordinarily performed by mesodermal cells in other regions, has been usurped by neural crest cells in the facial region. The crest cells therefore play a very dominant role in facial development, since they form all nonepithelial components except endothelial cells and the contractile elements of skeletal (voluntary) muscle.

The onset of bone formation or the establishment of all the organ systems (about the eighth week of development) is considered the termination of the embryonic period. Bone formation and other aspects of the final differentiation of facial tissues will be considered in detail elsewhere in this text.

CLINICAL CONSIDERATIONS

Aberrations in embryonic facial development lead to a wide variety of defects. Although any step may be impaired, defects of primary and secondary palate development are most common. There is evidence that other developmental defects may be even more common but they are not compatible with completion of intrauterine life and are therefore not as well documented.

Facial clefts. Most cases of clefts of the lip with or without associated cleft palate (Fig. 1-15) appear to form a group etiologically different from clefts involving only the secondary palate. For example, when more than one child in a family has facial clefts, the clefts are almost always found to belong only to one group. There is some evidence that underdevelopment (small size) of the medial or lateral nasal prominences is involved in primary palate clefting in humans so that contact at the site of fusion is either prevented or inadequate. Spontaneous clefting in one animal model obviously results from small prominences that fail to contact, whereas in another it appears that the direction of growth of the median nasal prominence is such that it does not make adequate contact with the lateral nasal prominence. Clefts of the primary palate can be produced in

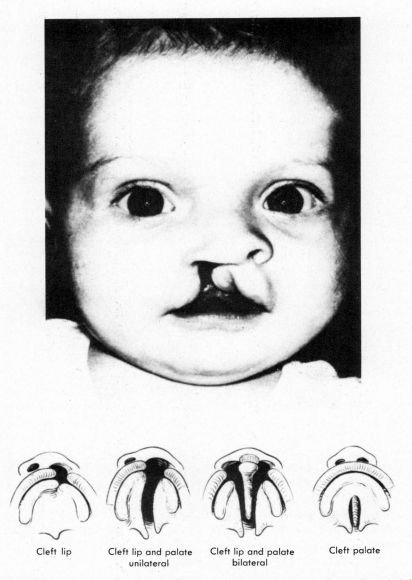

Cleft lip   Cleft lip and palate   Cleft lip and palate   Cleft palate
unilateral   bilateral

Fig. 1-15. Clefts of lip and palate in infants. Infant in photograph has complete unilateral cleft of lip and palate. (From Ross, R. B., and Johnston, M. C.: Cleft lip and palate, Baltimore, 1972, The Williams & Wilkins Co.)

experimental animals by several procedures that reduce the number of crest cells prior to migration and consequently reduce the size of the facial prominences. Also, clefts of the primary palate can be produced experimentally by agents that cause cell death in the facial prominences or interference with their growth in other ways prior to fusion. There is some evidence that many cases of cleft lip in humans result from a combination of developmental abnormalities that together reduce the mesenchyme in the primary palate at the point of fusion by an amount sufficient to prevent normal contact and fusion.

About two thirds of patients with clefts of the primary palate also have clefts of the secondary palate. Studies of experimental animals suggest that excessive separation of jaw segments as a result of the primary palate cleft prevents the palatal shelves from contacting after elevation. The degree of clefting is highly variable. Clefts may be either bilateral or unilateral (Fig. 1-15) and complete or incomplete. Most of this variation results from differing degrees of fusion and may be explained by variable degrees of mesenchyme in the facial prominences. Some of the variation may represent different initiating events.

Clefts involving only the secondary palate (cleft palate, Fig. 1-15) constitute, after clefts involving the primary palate, the second most frequent facial malformation in humans. Cleft palate can also be produced in experimental animals with a wide variety of chemical agents or other manipulations affecting the embryo. Usually, such agents retard or prevent shelf elevation. In other cases, however, it is shelf growth that is retarded so that, although elevation occurs, the shelves are too small to make contact. There is also some evidence that indicates that failure of the epithelial seam or failure of it to be replaced by mesenchyme occurs after the application of some environmental agents. Cleft formation could then result from rupture of the persisting seam, which would not have sufficient strength to prevent such rupture indefinitely.

Less frequently, other types of facial clefting are observed. In most instances they can be explained by failure of fusion between facial prominences of reduced size, and similar clefts can be produced experimentally. Examples include failure of merging and fusion between the maxillary prominence and the lateral nasal prominence, leading to oblique facial clefts, or failure of merging of the maxillary prominence and mandibular arch, leading to lateral facial clefts (macrostomia). Many of the variations in the position or degree of these rare facial clefts may depend on the timing or position of arrest of growth of the maxillary prominence that normally merges and fuses with adjacent structures (Plate 1). Other rare facial malformations (including oblique facial clefts) may also result from abnormal pressures or fusions with folds in the fetal (e.g., amniotic) membranes.

Also new evidence regarding the apparent role of epithelial-mesenchymal interactions via the mesenchymal cell process meshwork (CPM) may help to explain the frequent association between facial abnormalities, especially clefts, and limb defects. Genetic and/or environmental influences on this interaction might well affect both areas in the same individual.

Hemifacial microsomia. Hemifacial microsomia is the third most common facial malformation. Affected individuals have underdevelopment and other abnormalities of the temporomandibular joint, middle and external ear, and other structures in this region, such as the parotid gland and muscles of mastication. The defect is almost invariably unilateral.

Recent studies strongly suggest that at least some cases of hemifacial microsomia result from hemorrhage at the point of fusion between the external carotid and stapedial arteries. The stapedial artery supplies much of the facial region during an early embryonic stage. The newly developing external carotid takes over most of this function by fusing with the stapedial, after which the proximal portion of the latter artery regresses (Fig. 1-4). In experimental animals thalidomide and other chemical agents cause hemorrhage at the point of fusion and later result in malformations similar to hemifacial microsomia. Malformations similar to hemifacial microsomia also occur as part of the thalidomide syndrome in humans.

Treacher Collins' syndrome. Treacher Collins' syndrome (mandibulofacial dysostosis) is an inherited disorder that results from the action of a dominant gene and may be almost as common as hemifacial microsomia. The syndrome consists of underdevelopment of the tissues derived from the maxillary, mandibular, and hyoid prominences. The external and middle ear is often defective, and clefts of the secondary palate sometimes are found. Defects of a similar nature result from the action of an abnormal gene in mice and can also be produced experimentally with excessive doses of vitamin A.

Labial pits. Small pits may persist on either side of the midline of the lower lip. They are caused by the failure of the embryonic labial pits to disappear.

Lingual anomalies. Median rhomboid glossitis, an innocuous, red, rhomboidal smooth zone of the tongue in the midline in front of the foramen cecum, is considered the result of persistence of the tuberculum impar. Lack of fusion between the two lateral lingual prominences may produce a bifid tongue. Thyroid tissue may be present in the base of the tongue. Part of the thyroglossal duct may persist and form cysts at the base of the tongue.

Developmental cysts. Epithelial rests in lines of union, of facial or oral prominences or from epithelial organs, (e.g., vestigial nasopalatine ducts) may give rise to cysts lined with epithelium.

Branchial cleft (cervical) cysts or fistulas may arise from the rests of epithelium in the visceral arch area. They usually are laterally disposed on the neck. Thyroglossal duct cysts may occur at any place along the course of the duct, usually at or near the midline.

Cysts may arise from epithelial rests after the fusion of medial and lateral nasal prominences. They are called globulomaxillary cysts and are lined with pseudostratified columnar epithelium and squamous epithelium. They may, however, develop as primordial cysts from a supernumerary tooth germ.

Anterior palatine cysts are situated in the midline of the maxillary alveolar prominence. Once believed to be from remnants of the fusion of two prominences, they may be primordial cysts of odontogenic origin; their true nature is a subject of discussion.

Nasolabial cysts, originating in the base of the wing of the nose and bulging into the nasal and oral vestibule and the root of the upper lip, sometimes causing a flat depression on the anterior surface of the alveolar prominence, are also explained as originating from epithelial remnants in the cleft-lip line. It is, however, more probable that they derive from excessive epithelial proliferations that normally, for some time in embryonic life, plug the nostrils. It is also possible that they are retention cysts of vestibular nasal glands or that they develop from the epithelium of the nasolacrimal duct.

REFERENCES

Balinsky, B.I.: An introduction to embryology, ed. 3, Philadelphia, 1970, W.B. Saunders Co.

Bartelmez, G.W.: Neural crest in the forebrain of mammals, Anat. Rec. **138:**269, 1960.

Gasser, R.F.: The development of the facial muscles in man, Am. J. Anat. **120:**357, 1967.

Hamilton, W.J., and Mossman, H.: Human embryology, ed. 4, Cambridge, 1972, W. Heffer & Sons. Ltd.

Hay, E.D., and Meier, S.: Tissue interactions in development. In Shaw, J.H., et al., editors: Textbook of oral biology, Philadelphia, 1978, W.B. Saunders Co.

Hazelton, R.B.: A radioautographic analysis of the migration and fate of cells derived from the occipital somites of the chick embryo with specific reference to the hypoglossal musculature, J. Embryol. Exp. Morphol. **24:**455, 1970.

Jirásek, J.E.: Atlas of human prenatal morphogenesis, Hingham, Mass. 1983, Martinus Nijhoff Publishers.

Johnston, M.C., and Listgarten, M.A.: The migration interaction and early differentiation of oral-facial tissues. In Slavkin, H.S., and Bavetta L.A., editors: Developmental aspects of oral biology, New York, 1972, Academic Press, Inc.

Johnston, M.C., and Sulik, K.K.: Embryology of the head and neck. In Serafin, D., and Georgiade, N.G., editors: Pediatric plastic surgery, vol. 1, St. Louis, 1984, The C.V. Mosby Co.

Kawamato, H.K.: The kaleidoscopic world of rare craniofacial clefts: order out of chaos (Tessier classification), Clin. Plast. Surg. **3**:529, 1976.

Kraus, B.S., Kitamura, H., and Latham, R.A.: Atlas of the developmental anatomy of the face, New York, 1966, Harper & Row, Publishers.

Langman, J.: Medical embryology, ed. 2, Baltimore, 1969, The Williams & Wilkins Co.

LeLievre, C., and LeDouarin, N.M.: Mesenchymal derivatives of the neural crest: analysis of chimeric quail and chick embryos, J. Embryol. Exp. Morphol. **34**:125, 1975.

Minkoff, R., and Kuntz, A.J.: Cell proliferation and cell density of mesenchyme in the maxillary process on adjacent regions during facial development in the chick embryo, J. Embryol. Exp. Morphol. **46**:65, 1978.

Minkoff, R., and Kuntz, A.J.: Cell proliferation during morphogenetic changes: analysis of frontonasal morphogenesis in the chick embryo employing DNA labelling indices, J. Embryol. Exp. Morphol. **40**:101, 1977.

Nishimura, H.: Incidence of malformations in abortions. In Fraser, F.C., and McKusick, V.A., editors: Congenital malformations, Amsterdam, 1969, Excerpta Medica Press.

Nishimura, H., Semba, R., Tanimura, P., and Tanaka, O.: Prenatal development of humans with special reference to craniofacial structures: an atlas, Washington, D.C., 1977, U.S. Government Printing Office.

Noden, D.M.: Embryonic origins of avain cephalic and cervial muscles and associated connective tissue, Am. J. Anat. **168**:257, 1983.

Noden, D.M.: Interactions directing the migration and cytodifferentiation of avian neural crest cells. In Garrod, D.R., editor: Specificity of embryological interactions, vol. 5, London, 1978, Chapman & Hall Ltd.

Patterson, S., Minkoff, R., and Johnston, M.C.: Autoradiographic studies of cell migration during primary palate formation, J. Dent. Res. **58**:113, 1979. (Abstract.)

Poswillo, D.: The pathogenesis of the first and second branchial arch syndrome, Oral Surg. **35**:302, 1973.

Pourtois, M.: Morphogenesis of the primary and secondary palate. In Slavkin, H.S., and Bavetta, L.A., editors: Developmental aspects of oral biology, New York, 1972, Academic Press.

Pratt, R.M., and Martin, G.R.: Epithelial cell death and elevated cyclic AMP during palatal development, Proc. Natl. Acad. Sci. U.S.A. **72**:814, 1975.

Ross, R.B., and Johnston, M.C.: Cleft lip and palate, Baltimore, 1972, The Williams & Wilkins Co.

Sicher, H., and Tandler, J.: Anatomie fur Zahnarzte (Anatomy for dentists), Berlin, 1928, Springer Verlag.

Sperberg, G.H.: Craniofacial embryology, Bristol, England, 1976, John Wright & Sons, Ltd.

Streeter, G.L.: Developmental horizons in human embryos, Contrib. Embryol. **32**:133, 1948.

Sulik, K.K., Johnston, M.C., Ambrose, J.L.H., and Dorgan, D.R.: Phenytoin (Dilantin)-induced cleft lip, a scanning and transmission electron microscopic study, Anat. Rec. **195**:243, 1979.

Tam, P.P.L., and Meier, S.: The establishment of a somitomeric pattern in the mesoderm of the gastrulating mouse embryo, J. Anat. **164**:209, 1982.

Tamarin, A., and Boyde, A.: Facial and visceral arch development in the mouse embryo: a study by scanning electron microscopy, J. Anat. **124**:563, 1977.

Trasler, D.G.: Pathogenesis of cleft lip and its relation to embryonic face shape in A/Jax and C57BL mice, Teratology **1**:33, 1968.

Trasler, D.G., and Fraser, F.C.: Time-position relationships with particular references to cleft lip and cleft palate. In Wilson, J.C., and Fraser, F.C., editors: Handbook of teratology, vol 2, New York, 1977, Plenum Press.

Waterman, R.E., and Meller, S.M.: A scanning electron microscope study of secondary palate formation in the human, Anat. Rec. **175**:464, 1973.

Waterman, R.E., and Meller, S.M.: Normal facial development in the human embryo. In Shaw, J.H., et al., editors: Textbook of oral biology, Philadelphia, 1978, W.B. Saunders Co.

Weston, J.A.: The migration and differentiation of neural crest cells, Adv. Morphol. **8**:41, 1970.

2
DEVELOPMENT AND GROWTH OF TEETH

The primitive oral cavity or stomodeum, is lined by stratified squamous epithelium called the oral ectoderm. The oral ectoderm contacts the endoderm of the foregut to form the buccopharyngeal membrane (Fig. 1-9). At about the twenty-seventh day of gestation this membrane ruptures and the primitive oral cavity establishes a connection with the foregut. Most of the connective tissue cells underlying the oral ectoderm are neural crest or ectomesenchyme in origin. These cells are thought to instruct or induce the overlying ectoderm to start tooth development, which begins in the anterior portion of what will be the future maxilla and mandible and proceeds posteriorly.

DENTAL LAMINA

Two or 3 weeks after the rupture of the buccopharyngeal membrane, when the embryo is about 6 weeks old, certain areas of basal cells of the oral ectoderm proliferate at more rapid rate than do the cells of the adjacent areas. This leads to the formation of the dental lamina, which is a band of epithelium that has invaded the underlying ectomesenchyme along each of the horseshoe-shaped future dental arches (Figs. 2-1, A, and 2-3). The dental laminae serve as the primordium for the ectodermal portion of the deciduous teeth. Later, during the development of the jaws, the permanent molars arise directly from a distal extension of the dental lamina.

The development of the first permanent molar is initiated at the fourth month in utero. The second molar is initiated at about the first year after birth, the third molar at the fourth or fifth years. The distal proliferation of the dental lamina is responsible for the location of the germs of the permanent molars in the ramus of the mandible and the tuberosity of the

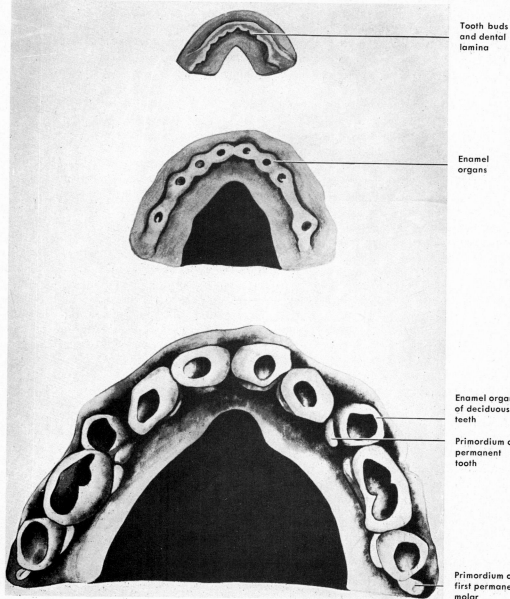

A

B

C

Tooth buds
and dental
lamina

Enamel
organs

Enamel organs
of deciduous
teeth

Primordium of
permanent
tooth

Primordium of
first permanent
molar

Fig. 2-1. Diagrammatic reconstruction of dental lamina and enamel organs of mandible. **A,** 22 mm embryo, bud stage (eighth week). **B,** 43 mm embryo, cap stage (tenth week). **C,** 163 mm embryo, bell stage (about 4 months). Primordia of permanent teeth are seen as thickenings of dental lamina on lingual side of each tooth germ. Distal extension of dental lamina with primordium of first molar.

maxilla. The successors of the deciduous teeth develop from a lingual extension of the free end of the dental lamina opposite to the enamel organ of each deciduous tooth (Fig. 2-2, *C*). The lingual extension of the dental lamina is named the successional lamina and develops from the fifth month in utero (permanent central incisor) to the tenth month of age (second premolar).

Fate of dental lamina. It is evident that the total activity of the dental lamina extends over a period of at least 5 years. Any particular portion of the dental lamina functions for a much briefer period since only a relatively short time elapses after initiation of tooth development before the dental lamina begins to degenerate at that particular location. However, the dental lamina may still be active in the third molar region after it has disappeared elsewhere, except for occasional epithelial remnants. As the teeth continue to develop, they lose their connection with the dental lamina. They later break up by mesenchymal invasion, which is at first incomplete and does not perforate the total thickness of the lamina (Fig. 2-8). Remnants of the dental lamina persist as epithelial pearls or islands within the jaw as well as in the gingiva.

Vestibular lamina. Labial and buccal to the dental lamina in each dental arch, another epithelial thickening develops independently and somewhat later. It is the vestibular lamina, also termed the lip furrow band (Figs. 2-6 and 2-7). It subsequently hollows and forms the oral vestibule between the alveolar portion of the jaws and the lips and cheeks (Figs. 2-10 and 2-11).

TOOTH DEVELOPMENT

At certain points along the dental lamina, each representing the location of one of the 10 mandibular and 10 maxillary deciduous teeth, the ectodermal cells multiply still more rapidly and form little knobs that grow into the underlying mesenchyme (Figs. 2-2 and 2-4). Each of these little downgrowths from the dental lamina represents the beginning of the *enamel or-*

gan of the tooth bud of a deciduous tooth. Not all of these enamel organs start to develop at the same time, and the first to appear are those of the anterior mandibular region.

As cell proliferation continues, each enamel organ increases in size and changes in shape. As it develops, it takes on a shape that resembles a cap, with the outside of the cap directed toward the oral surface (Figs. 2-5 and 2-7).

On the inside of the cap (i.e., inside the depression of the enamel organ), the ectomesenchymal cells increase in number. The tissue appears more dense than the surrounding mesenchyme and represents the beginning of the *dental papilla.* Surrounding the combined enamel organ and dental papilla, the third part of the tooth bud forms. It is the *dental sac,* and it consists of ectomesenchymal cells and fibers that surround the dental papilla and the enamel organ (Fig. 2-8).

During and after these developments the shape of the enamel organ continues to change. The depression occupied by the dental papilla deepens until the enamel organ assumes a shape resembling a bell. As this development takes place, the dental lamina, which had thus far connected the enamel organ to the oral epithelium, breaks up and the tooth bud loses its connection with the epithelium of the primitive oral cavity.

DEVELOPMENTAL STAGES

Although tooth development is a continuous process, the developmental history of a tooth is divided into several morphologic "stages" for descriptive purposes. While the size and shape of individual teeth are different, they pass through similar stages of development. They are named after the shape of the epithelial part of the tooth germ and are called the bud, cap, and bell stages (Fig. 2-2, *A* to *C*).

Bud stage

The epithelium of the dental laminae is separated from the underlying ectomesenchyme by a basement membrane (Fig. 2-3). Simulta-

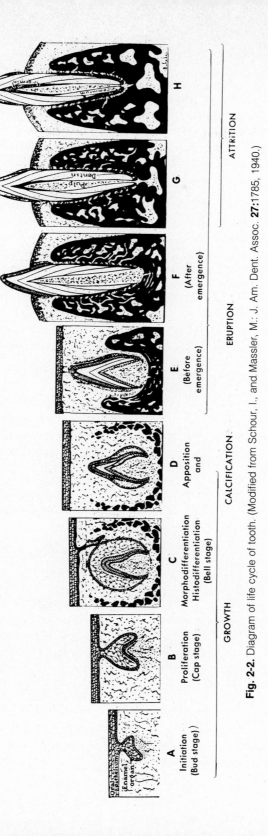

Fig. 2-2. Diagram of life cycle of tooth. (Modified from Schour, I., and Massler, M.: J. Am. Dent. Assoc. **27**:1785, 1940.)

A
Initiation
(Bud stage)

Oral epithelium
Enamel organ

B
Proliferation
(Cap stage)

C
Morphodifferentiation
Histodifferentiation
(Bell stage)

D
Apposition
and

E
(Before
emergence)

F
(After
emergence)

Dentin
Pulp

G

H

GROWTH · · · · CALCIFICATION · · · · ERUPTION · · · · ATTRITION

neous with the differentiation of each dental lamina, round or ovoid swellings arise from the basement membrane at 10 different points, corresponding to the future positions of the deciduous teeth. These are the primordia of the enamel organs, the tooth buds (Fig. 2-4). Thus the development of tooth germs is initiated, and the cells continue to proliferate faster than adjacent cells. The dental lamina is shallow, and microscopic sections often show tooth buds close to the oral epithelium. Since the main function of certain epithelial cells of the tooth bud is to form the tooth enamel, these cells constitute the enamel organ, which is critical to normal tooth development. In the bud stage, the enamel organ consists of peripherally located low columnar cells and centrally located polygonal cells (Fig. 2-4). Many cells of the tooth bud and the surrounding mesenchyme undergo mitosis (Fig. 2-4). As a result of the increased mitotic activity and the migration of neural crest cells into the area the ectomesenchymal cells surrounding the tooth bud condense. The area of ectomesenchymal condensation immediately subjacent to the enamel organ is the dental papilla. The condensed ectomesenchyme that surrounds the tooth bud and the dental papilla is the dental sac (Figs. 2-

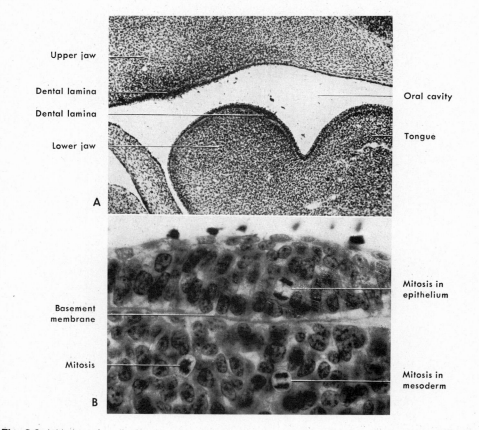

Fig. 2-3. Initiation of tooth development. Human embryo 13.5 mm in length, fifth week. **A,** Sagittal section through upper and lower jaws. **B,** High magnification of thickened oral epithelium. (From Orban, B.: Dental histology and embryology, Philadelphia, 1929, P. Blakiston's Son & Co.)

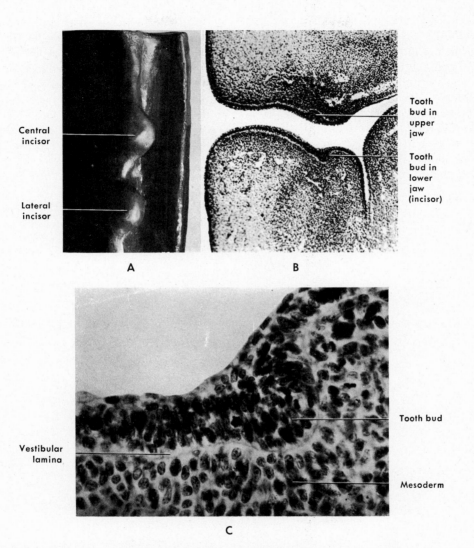

Central
incisor

Lateral
incisor

Tooth
bud in
upper
jaw

Tooth
bud in
lower
jaw
(incisor)

A

B

Vestibular
lamina

Tooth bud

Mesoderm

C

Fig. 2-4. Bud stage of tooth development, proliferation stage. Human embryo 16 mm in length, sixth week. **A,** Wax reconstruction of germs of lower central and lateral incisors. **B,** Sagittal section through upper and lower jaws. **C,** High magnification of tooth germ of lower incisor in bud stage. (From Orban, B.: Dental histology and embryology, Philadelphia, 1929, P. Blakiston's Son & Co.)

6, 2-7, and 2-8). Both the dental papilla and the dental sac become more well defined as the enamel organ grows into the cap and bell shapes (Fig. 2-8). The cells of the dental papilla will form tooth pulp and dentin. The cells in the dental sac will form cementum and the periodontal ligament.

Cap stage

As the tooth bud continues to proliferate, it does not expand uniformly into a larger sphere. Instead, unequal growth in different parts of the tooth bud leads to the cap stage, which is characterized by a shallow invagination on the deep surface of the bud (Figs. 2-2, *B,* and 2-5).

Outer and inner enamel epithelium. The peripheral cells of the cap stage are cuboidal, cover the convexity of the "cap," and are called the outer enamel (dental) epithelium. The cells in the concavity of the "cap" become tall, columnar cells and represent the inner enamel (dental) epithelium (Figs. 2-6 and 2-7). The outer enamel epithelium is separated from the dental sac, and the inner enamel epithelium

A

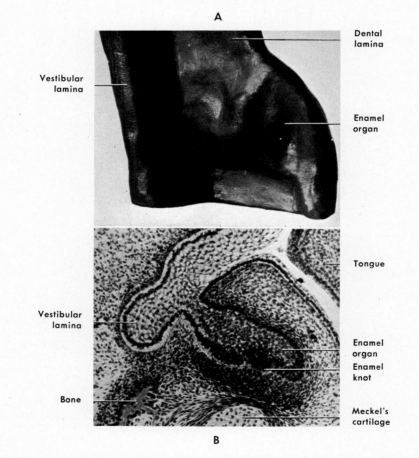

B

Fig. 2-5. Cap stage of tooth development. Human embryo 31.5 mm in length, ninth week. **A,** Wax reconstruction of enamel organ of lower lateral incisor. **B,** Labiolingual section through same tooth. (From Orban, B.: Dental histology and embryology, Philadelphia, 1929, P. Blakiston's Son & Co.)

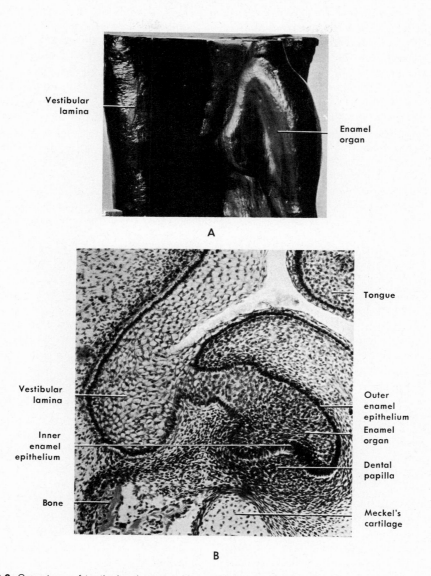

Fig. 2-6. Cap stage of tooth development. Human embryo 41.5 mm in length, tenth week. **A,** Wax reconstruction of enamel organ of lower central incisor. **B,** Labiolingual section through same tooth. (From Orban, B.: Dental histology and embryology, Philadelphia, 1929, P. Blakiston's Son & Co.)

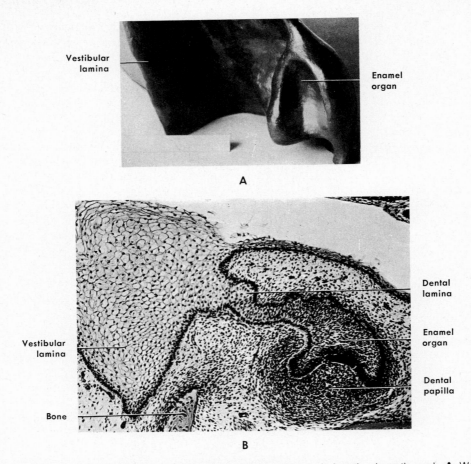

Vestibular
lamina

Enamel
organ

A

Dental
lamina

Enamel
organ

Dental
papilla

Vestibular
lamina

Bone

B

Fig 2-7. Cap stage of tooth development. Human embryo 60 mm in length, eleventh week. **A,** Wax reconstruction of enamel organ of lower lateral incisor. **B,** Labiolingual section through same tooth. (From Orban, B.: Dental histology and embryology, Philadelphia, 1929, P. Blakiston's Son & Co.)

from the dental papilla, by a delicate basement membrane. Hemidesmosomes anchor the cells to the basal lamina.

Stellate reticulum (enamel pulp). Polygonal cells located in the center of the epithelial enamel organ, between the outer and inner enamel epithelia, begin to separate as more intercellular fluid is produced and form a cellular network called the stellate reticulum (Figs. 2-8 and 2-9). The cells assume a branched reticular form. The spaces in this reticular network are filled with a mucoid fluid that is rich in albu-

min, which gives the stellate reticulum a cushionlike consistency that may support and protect the delicate enamel-forming cells.

The cells in the center of the enamel organ are densely packed and form the *enamel knot* (Fig. 2-5). This knot projects in part toward the underlying dental papilla, so that the center of the epithelial invagination shows a slightly knoblike enlargement that is bordered by the labial and lingual enamel grooves (Fig. 2-5). At the same time there arises in the increasingly high enamel organ a vertical extension of the

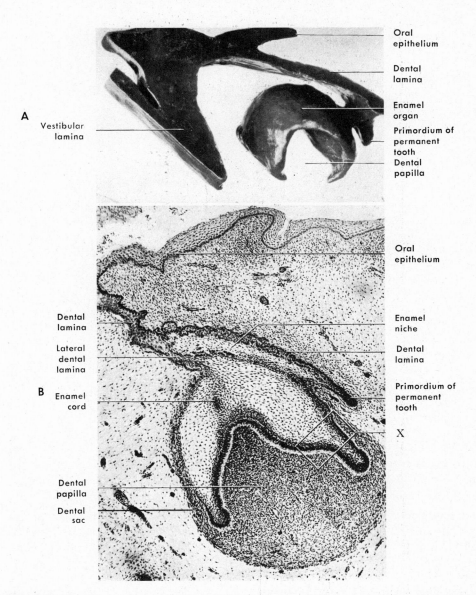

A

Vestibular lamina

Oral epithelium

Dental lamina

Enamel organ

Primordium of permanent tooth

Dental papilla

Oral epithelium

Dental lamina

Lateral dental lamina

B Enamel cord

Enamel niche

Dental lamina

Primordium of permanent tooth

X

Dental papilla

Dental sac

Fig. 2-8. Bell stage of tooth development. Human embryo 105 mm in length, fourteenth week. **A,** Wax reconstruction of lower central incisor. **B,** Labiolingual section of the same tooth. X, See Fig. 2-9. (From Orban, B.: Dental histology and embryology, Philadelphia, 1929, P. Blakiston's Son & Co.)

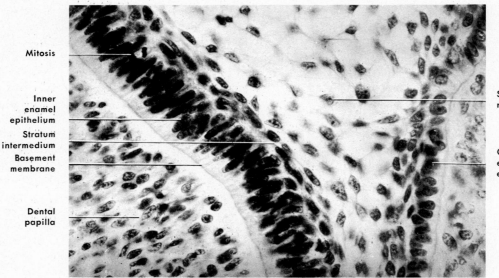

Mitosis

Inner enamel epithelium

Stratum intermedium

Basement membrane

Dental papilla

Stellate reticulum

Outer enamel epithelium

Fig. 2-9. Layers of epithelial enamel organ at high magnification. Area X of Fig. 2-8.

enamel knot, called the *enamel cord* (Fig. 2-8). Both are temporary structures that disappear before enamel formation begins. The function of the enamel knot and cord may be to act as a reservoir of dividing cells for the growing enamel organ.

Dental papilla. Under the organizing influence of the proliferating epithelium of the enamel organ, the ectomesenchyme (neural crest cells) that is partially enclosed by the invaginated portion of the inner enamel epithelium proliferates. It condenses to form the dental papilla, which is the formative organ of the dentin and the primordium of the pulp (Figs. 2-5 and 2-6). The changes in the dental papilla occur concomitantly with the development of the epithelial enamel organ. Although the epithelium exerts a dominating influence over the adjacent connective tissue, the condensation of the latter is not a passive crowding by the proliferating epithelium. The dental papilla shows active budding of capillaries and mitotic fig-

ures, and its peripheral cells adjacent to the inner enamel epithelium enlarge and later differentiate into the odontoblasts.

Dental sac. Concomitant with the development of the enamel organ and the dental papilla, there is a marginal condensation in the ectomesenchyme surrounding the enamel organ and dental papilla. Gradually, in this zone, a denser and more fibrous layer develops, which is the primitive dental sac. The cells of the dental sac are important for the formation of cementum and the periodontal ligament.

The epithelial enamel organ, the dental papilla, and the dental sac are the formative tissues for an entire tooth and its supporting structures.

Bell stage

As the invagination of the epithelium deepens and its margins continue to grow, the enamel organ assumes a bell shape (Figs. 2-2, C, and 2-8). Four different types of epithelial cells can be distinguished on light microscopic examination of the bell stage of the enamel organ. The cells form the inner enamel epithelium, the stratum intermedium, the stellate reticulum, and the outer enamel epithelium.

Inner enamel epithelium. The inner enamel epithelium consists of a single layer of cells that differentiate prior to amelogenesis into tall columnar cells called ameloblasts (Figs. 2-8 and 2-9). These cells are 4 to 5 micrometers (μm) in diameter and about 40 μm high. These elongated cells are attached to one another by junctional complexes laterally and to cells in the stratum intermedium by desmosomes (Fig. 2-9). The fine structure of inner enamel epithelium and ameloblasts is described in Chapter 3.

The cells of the inner enamel epithelium exert an organizing influence on the underlying mesenchymal cells in the dental papilla, which later differentiate into odontoblasts.

Stratum intermedium. A few layers of squamous cells form the stratum intermedium, between the inner enamel epithelium and the stellate reticulum (Fig. 2-9). These cells are closely attached by desmosomes and gap junctions. The well-developed cytoplasmic organelles, acid mucopolysaccharides, and glycogen deposits indicate a high degree of metabolic activity. This layer seems to be essential to enamel formation. It is absent in the part of the tooth germ that outlines the root portions of the tooth but does not form enamel.

Stellate reticulum. The stellate reticulum expands further, mainly by an increase in the amount of intercellular fluid. The cells are star shaped, with long processes that anastomose with those of adjacent cells (Fig. 2-9). Before enamel formation begins, the stellate reticulum collapses, reducing the distance between the centrally situated ameloblasts and the nutrient capillaries near the outer enamel epithelium. Its cells then are hardly distinguishable from those of the stratum intermedium. This change

begins at the height of the cusp or the incisal edge and progresses cervically (see Fig. 3-37).

Outer enamel epithelium. The cells of the outer enamel epithelium flatten to a low cuboidal form. At the end of the bell stage, preparatory to and during the formation of enamel, the formerly smooth surface of the outer enamel epithelium is laid in folds. Between the folds the adjacent mesenchyme of the dental sac forms papillae that contain capillary loops and thus provide a rich nutritional supply for the intense metabolic activity of the avascular enamel organ.

Dental lamina. In all of the teeth, except the permanent molars, the dental lamina proliferates at its deep end to give rise to the enamel organs of the permanent teeth (Figs. 2-10 and 2-11).

Dental papilla. The dental papilla is enclosed in the invaginated portion of the enamel organ. Before the inner enamel epithelium begins to produce enamel, the peripheral cells of the mesenchymal dental papilla differentiate into odontoblasts under the organizing influence of the epithelium. First, they assume a cuboidal form; later they assume a columnar form and acquire the specific potential to produce dentin.

The basement membrane that separates the enamel organ and the dental papilla just prior to dentin formation is called the *membrana preformativa.*

Dental sac. Before formation of dental tissues begins, the dental sac shows a circular arrangement of its fibers and resembles a capsular structure. With the development of the root, the fibers of the dental sac differentiate into the periodontal fibers that become embedded in the developing cementum and alveolar bone.

Advanced bell stage. During the advanced bell stage, the boundary between inner enamel epithelium and odontoblasts outlines the future dentinoenamel junction (Figs. 2-8 and 2-10). In addition, the cervical portion of the enamel or-

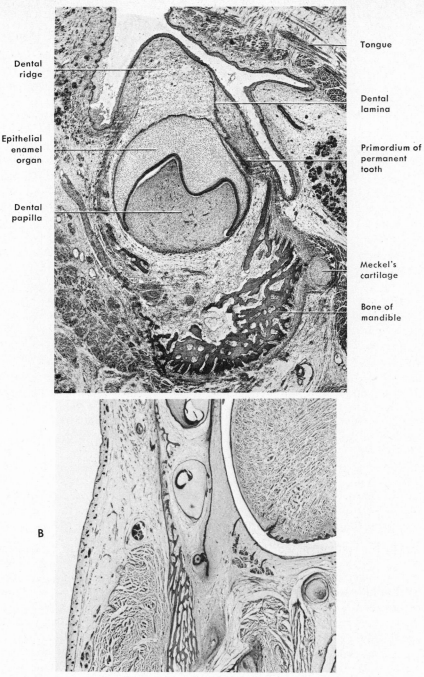

Dental
ridge

Epithelial
enamel
organ

A

Dental
papilla

Tongue

Dental
lamina

Primordium of
permanent
tooth

Meckel's
cartilage

Bone of
mandible

B

Fig. 2-10. A, Advanced bell stage of tooth development. Human embryo 200 mm in length; about 18 weeks. Labiolingual section through deciduous lower first molar. **B,** Horizontal section through human embryo about 20 mm in length showing extension of dental lamina distal to second deciduous molar and formation of permanent first molar tooth germ. (**B** from Bhaskar, S.N.: Synopsis of oral histology, ed. 5, St. Louis, 1977, The C.V. Mosby Co.)

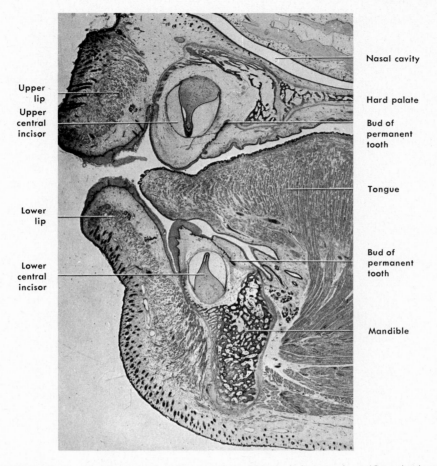

Nasal cavity

Hard palate

Bud of
permanent
tooth

Tongue

Bud of
permanent
tooth

Mandible

Upper
lip

Upper
central
incisor

Lower
lip

Lower
central
incisor

Fig. 2-11. Sagittal section through head of human fetus 200 mm in length, about 18 weeks, in region of central incisors.

gan gives rise to the epithelial root sheath of Hertwig.

Hertwig's epithelial root sheath and root formation

The development of the roots begins after enamel and dentin formation has reached the future cementoenamel junction. The enamel organ plays an important part in root development by forming Hertwig's epithelial root sheath, which molds the shape of the roots and initiates radicular dentin formation. Hertwig's

root sheath consists of the outer and inner enamel epithelia only, and therefore it does not include the stratum intermedium and stellate reticulum. The cells of the inner layer remain short and normally do not produce enamel. When these cells have induced the differentiation of radicular cells into odontoblasts and the first layer of dentin has been laid down, the epithelial root sheath loses its structural continuity and its close relation to the surface of the root. Its remnants persist as an epithelial network of strands or tubules near the external surface of the root. These epithelial remnants are found in the periodontal ligament of erupted teeth and are called *rests of Malassez* (see Chapter 7).

There is a pronounced difference in the development of Hertwig's epithelial root sheath in teeth with one root and in those with two or more roots. Prior to the beginning of root formation, the root sheath forms the epithelial diaphragm (Fig. 2-12). The outer and inner enamel epithelia bend at the future cementoenamel junction into a horizontal plane, narrowing the wide cervical opening of the tooth germ. The plane of the diaphragm remains relatively fixed during the development and growth of the root (see Chapter 11). The proliferation of the cells of the epithelial diaphragm is accompanied by proliferation of the cells of the connective tissue of the pulp, which occurs in the area adjacent to the diaphragm. The free

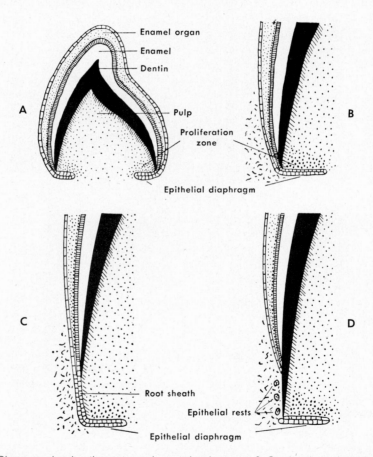

Fig. 2-12. Diagrams showing three stages in root development. **A,** Section through tooth germ. Note epithelial diaphragm and proliferation zone of pulp. **B,** Higher magnification of cervical region of **A. C,** "Imaginary" stage showing elongation of Hertwig's epithelial sheath coronal to diaphragm. Differentiation of odontoblasts in elongated pulp. **D,** In area of proliferation, dentin has been formed. Root sheath is broken up into epithelial rest and is separated from dentinal surface by connective tissue. Differentiation of cementoblasts.

end of the diaphragm does not grow into the connective tissue, but the epithelium proliferates coronally to the epithelial diaphragm (Fig. 2-12, *B*). The differentiation of odontoblasts and the formation of dentin follow the lengthening of the root sheath. At the same time the connective tissue of the dental sac surrounding the root sheath proliferates and divides the continuous double epithelial layer (Fig. 2-12, *C*) into a network of epithelial strands (Fig. 2-12, *D*). The epithelium is moved away from the surface of the dentin so that connective tissue cells come into contact with the outer surface of the dentin and differentiate into cementoblasts that deposit a layer of cementum onto the surface of the dentin. The rapid sequence of proliferation and destruction of Hertwig's

root sheath explains the fact that it cannot be seen as a continuous layer on the surface of the developing root (Fig. 2-12, *D*, and Fig. 7-7). In the last stages of root development, the proliferation of the epithelium in the diaphragm lags behind that of the pulpal connective tissue. The wide apical foramen is reduced first to the width of the diaphragmatic opening itself and later is further narrowed by apposition of dentin and cementum to the apex of the root.

Differential growth of the epithelial diaphragm in multirooted teeth causes the division of the root trunk into two or three roots. During the general growth of the enamel organ the expansion of its cervical opening occurs in such a way that long tonguelike extensions of the horizontal diaphragm develop (Fig. 2-13).

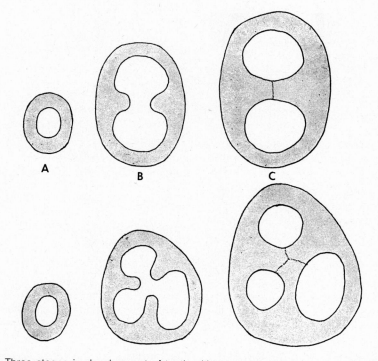

Fig. 2-13. Three stages in development of tooth with two roots and one with three roots. Surface view of epithelial diaphragm. During growth of tooth germ, simple diaphragm, **A,** expands eccentrically so that horizontal epithelial flaps are formed. **B,** Later these flaps proliferate and unite (dotted lines in **C**) and divide single cervical opening into two or three openings.

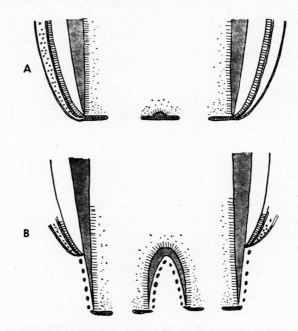

Fig. 2-14. Two stages in development of two-rooted tooth. Diagrammatic mesiodistal sections of lower molar. **A,** Beginning of dentin formation at bifurcation. **B,** Formation of two roots in progress. (Details as shown in Fig. 2-12.)

Two such extensions are found in the germs of lower molars and three in the germs of upper molars. Before division of the root trunk occurs, the free ends of these horizontal epithelial flaps grow toward each other and fuse. The single cervical opening of the coronal enamel organ is then divided into two or three openings. On the pulpal surface of the dividing epithelial bridges, dentin formation starts (Fig. 2-14, *A*), and on the periphery of each opening, root development follows in the same way as described for single-rooted teeth (Fig. 2-14, *B*).

If cells of the epithelial root sheath remain adherent to the dentin surface, they may differentiate into fully functioning ameloblasts and produce enamel. Such droplets of enamel, called *enamel pearls,* are sometimes found in the area of furcation of the roots of permanent molars. If the continuity of Hertwig's root sheath is broken or is not established prior to dentin formation, a defect in the dentinal wall of the pulp ensues. Such defects are found in the pulpal floor corresponding to the furcation or on any point of the root itself if the fusion of the horizontal extensions of the diaphragm remains incomplete. This accounts for the development of accessory root canals opening on the periodontal surface of the root (see Chapter 5).

HISTOPHYSIOLOGY AND CLINICAL CONSIDERATIONS

A number of physiologic growth processes participate in the progressive development of the teeth (Table 1). Except for their initiation, which is a momentary event, these processes overlap considerably, and many are continuous throughout the various morphologic stages of odontogenesis. Nevertheless, each physiologic process tends to predominate in one stage more than in another.

For example, the process of histodifferentiation characterizes the bell stage, in which the

Table 1. Stages in tooth growth

Morphologic stages	Physiologic processes
Dental lamina ⟷	Initiation
Bud stage	
Cap stage	Proliferation
Bell stage (early)	Histodifferentiation
Bell stage (advanced)	Morphodifferentiation
Formation of enamel and dentin matrix	Apposition

cells of the inner enamel epithelium differentiate into functional ameloblasts. However, proliferation still progresses at the deeper portion of the enamel organ.

Initiation. The dental laminae and associated tooth buds represent those parts of the oral epithelium that have the potential for tooth formation. Specific cells within the horseshoe-shaped dental laminae have the potential to form the enamel organ of certain teeth by responding to those factors that initiate or induce tooth development. Different teeth are initiated at definite times. Initiation induction requires ectomesenchymal-epithelial interaction. The mechanism of such interaction is not clearly understood. However, it has been demonstrated that dental papilla mesenchyme can induce or instruct tooth epithelium and even nontooth epithelium to form enamel.

Teeth may develop in abnormal locations, for example, in the ovary (dermoid tumors or cysts) or in the hypophysis. In such instances the tooth undergoes stages of development similar to those in the jaws.

A lack of initiation results in the absence of either a single tooth or multiple teeth, most frequently the permanent upper lateral incisors, third molars, and lower second premolars. There also may be a complete lack of teeth (anodontia). On the other hand, abnormal initiation may result in the development of single or multiple supernumerary teeth.

Proliferation. Enhanced proliferative activity ensues at the points of initiation and results successively in the bud, cap, and bell stages of the odontogenic organ. Proliferative growth causes regular changes in the size and proportions of the growing tooth germ (Figs. 2-3 and 2-7).

Even during the stage of proliferation, the tooth germ already has the potential to become more highly developed. This is illustrated by the fact that explants of these early stages continue to develop in tissue culture through the subsequent stages of histodifferentiation and appositional growth. A disturbance or experimental interference has entirely different effects, according to the time of occurrence and the stage of development that it affects.

Histodifferentiation. Histodifferentiation succeeds the proliferative stage. The formative cells of the tooth germs developing during the proliferative stage undergo definite morphologic as well as functional changes and acquire their functional assignment (the appositional growth potential). The cells become restricted in their functions. They differentiate and give up their capacity to multiply as they assume their new function; this law governs all differentiating cells. This phase reaches its highest development in the bell stage of the enamel organ, just preceding the beginning of formation and apposition of dentin and enamel (Fig. 2-8).

The organizing influence of the inner enamel epithelium on the mesenchyme is evident in the bell stage and causes the differentiation of the adjacent cells of the dental papilla into odontoblasts. With the formation of dentin, the

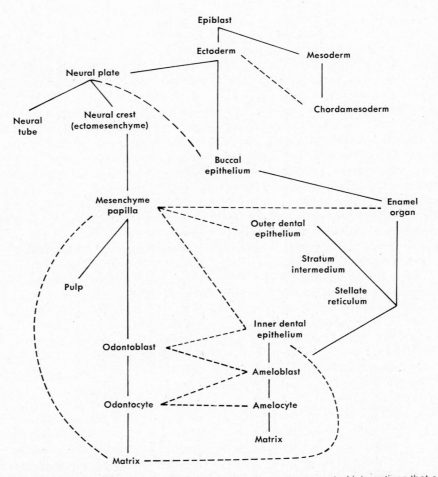

Fig. 2-15. Outline of development of tooth. *Broken lines,* Known or suspected interactions that occur between tissues. Data suggesting placement of these lines derive from transplantations and in vitro studies. Words "amelocyte" and "odontocyte" are employed only to indicate that these cells may possess different capabilities for interaction with other tissues after their overt differentiation. (Courtesy Dr. William E. Koch., Chapel Hill, N.C.)

cells of the inner enamel epithelium differentiate into ameloblasts and enamel matrix is formed opposite the dentin. Enamel does not form in the absence of dentin, as demonstrated by the failure of transplanted ameloblasts to form enamel when dentin is not present. Dentin formation therefore precedes and is essential to enamel formation. The differentiation of the epithelial cells precedes and is essential to the differentiation of the odontoblasts and the initiation of dentin formation.

In vitro studies on tooth development have provided vital information concerning the interaction of dermal-epidermal components of tooth tissues on differentiation of odontoblasts and ameloblasts. The importance of the basement membrane of this interface has been recognized. However, the criteria for the development of this complex organ system will have to await the delineation of the precise roles of the stellate reticulum, the stratum intermedium, and the outer dental epithelial components. One of the models that has been suggested for the interactions that may occur between tissues during the development of a tooth is presented in Fig. 2-15.

In vitamin A deficiency the ameloblasts fail to differentiate properly. Consequently, their organizing influence on the adjacent mesenchymal cells is disturbed, and atypical dentin, known as osteodentin, is formed.

Morphodifferentiation. The morphologic pattern, or basic form and relative size of the future tooth, is established by morphodifferentiation, that is, by differential growth. Morphodifferentiation therefore is impossible without proliferation. The advanced bell stage marks not only active histodifferentiation but also an important stage of morphodifferentiation in the crown, outlining the future dentinoenamel junction (Figs. 2-8 and 2-10).

The dentinoenamel and dentinocemental junctions, which are different and characteristic for each type of tooth, act as a blueprint pattern. In conformity with this pattern the ameloblasts, odontoblasts, and cementoblasts deposit enamel, dentin, and cementum, respectively, and thus give the completed tooth its characteristic form and size. For example, the size and form of the cuspal portion of the crown of the first permanent molar are established at birth long before the formation of hard tissues begin.

The frequent statement in the literature that endocrine disturbances affect the size or form of the crown of teeth is not tenable unless such effects occur during morphodifferentiation, that is, in utero or in the first year of life. Size and shape of the root, however, may be altered by disturbances in later periods. Clinical examinations show that the retarded eruption that occurs in persons with hypopituitarism and hypothyroidism results in a small clinical crown that is often mistaken for a small anatomic crown.

Disturbances in morphodifferentiation may affect the form and size of the tooth without impairing the function of the ameloblasts or odontoblasts. New parts may be differentiated (supernumerary cusps or roots), twinning may result, a suppression of parts may occur (loss of cusps or roots), or the result may be a peg or malformed tooth (e.g., Hutchinson's incisor) with enamel and dentin that may be normal in structure.

Apposition. Apposition is the deposition of the matrix of the hard dental structures. It will be described in separate chapters on enamel, dentin, and cementum. This chapter deals with certain aspects of apposition in order to complete the discussion of the physiologic processes concerned in the growth of teeth.

Appositional growth of enamel and dentin is a layerlike deposition of an extracellular matrix. This type of growth is therefore additive. It is the fulfillment of the plans outlined at the stages of histodifferentiation and morphodifferentiation. Appositional growth is characterized by regular and rhythmic deposition of the extracellular matrix, which is of itself incapable of further growth. Periods of activity and rest alternate at definite intervals during tooth formation.

Genetic and environmental factors may disturb the normal synthesis and secretion of the organic matrix of enamel leading to a condition called *enamel hypoplasia.*

If the organic matrix is normal but its mineralization is defective, then the enamel or dentin is said to be hypocalcified or hypomineralized. Both hypoplasia and hypocalcification can occur as a result of an insult to the cells responsible for the apposition stage of tooth development (see Chapter 3, Clinical Considerations).

REFERENCES

Avery, J.K.: Embryology of the teeth, J. Dent. Res. **30:**490, 1951.

Avery, J.K.: Primary induction of tooth formation, J. Dent. Res. **33:**702, 1954. (Abstract.)

Bhaskar, S.N.: Synopsis of oral pathology, ed. 5, St. Louis, 1977, The C.V. Mosby Co.

Diamond, M., and Applebaum, E.: The epithelial sheath, J. Dent. Res. **21:**403, 1942.

Fisher, A.R.: The differentiation of the molar tooth germ of the mouse in vivo and in vitro with special reference to cusp development, Ph.D. thesis, 1957, University of Bristol.

Fleming, H.S.: Homologous and heterologous intraocular growth of transplanted tooth germs, J. Dent. Res. **31:**166, 1952.

Gaunt, W.A.: The vascular supply to the dental lamina during early development, Acta Anat. (Basel) **37:**232, 1959.

Glasstone, S.: Regulative changes in tooth germs grown in tissue culture, J. Dent. Res. **42:**1364, 1963.

Hoffman, R., and Gillete, R.: Mitotic patterns in pulpal and periodontal tissue in developing teeth, Fortieth General Meeting of the International Association of Dental Research, St. Louis, 1962.

Johnson, P.L., and Bevelander, G.: The role of the stratum intermedium in tooth development, Oral Surg. **10:**437, 1957.

Koch, W.E.: Tissue interaction during in vitro odontogenesis. In Slavkin, H.S., and Bavetta, L.A., editors: Developmental aspects of oral biology, New York, 1972, Academic Press, Inc.

Kollar, E.J.: Histogenetics of dermal-epidermal interactions. In Slavkin, H.S., and Bavetta, L.A., editors: Developmental aspects of oral biology, New York, 1972, Academic Press, Inc.

Kraus, B.S.: Calcification of the human deciduous teeth, J. Am. Dent. Assoc. **59:**1128, 1959.

Lefkowitz, W., and Swayne, P.: Normal development of tooth buds cultured in vitro, J. Dent. Res. **37:**1100, 1958.

Marsland, E.A.: Histological investigation of amelogenesis in rats, Br. Dent. J. **91:**251, 1951.

Marsland, E.A.: Histological investigation of amelogenesis in rats, Br. Dent. J. **92:**109, 1952.

Orban, B.: Growth and movement of the tooth germs and teeth, J. Am. Dent. Assoc. **15:**1004, 1928.

Orban, B.: Dental histology and embryology, Philadelphia, 1929, P. Blakiston's Son & Co.

Orban, B., and Mueller, E.: The development of the bifurcation of multirooted teeth, J. Am. Dent. Assoc. **16:**297, 1929.

Schour, I., and Massler, M.: Studies in tooth development: the growth pattern of human teeth, J. Am. Dent. Assoc. **27:**1778, 1940.

Sicher, H.: Tooth eruption: axial movement of teeth with limited growth, J. Dent. Res. **21:**395, 1942.

Slavkin, H.C.: Embryonic tooth formation. In: Melcher, A.H., and Zarb, G.A., editors: Oral sciences reviews 4, Copenhagen, 1974, Munksgaard, International Booksellers and Publishers, Ltd.

3

ENAMEL

HISTOLOGY
Physical characteristics

Enamel forms a protective covering of variable thickness over the entire surface of the crown. On the cusps of human molars and premolars the enamel attains a maximum thickness of about 2 to 2.5 mm, thinning down to almost a knife edge at the neck of the tooth. The shape and contour of the cusps receive their final modeling in the enamel.

Because of its high content of mineral salts and their crystalline arrangement, enamel is the hardest calcified tissue in the human body. The function of the enamel is to form a resistant covering of the teeth, rendering them suitable for mastication. The structure and hardness of the enamel render it brittle, which is particularly apparent when the enamel loses its foundation of sound dentin. The specific gravity of enamel is 2.8.

Another physical property of enamel is its permeability. It has been found with radioactive tracers that the enamel can act in a sense like a semipermeable membrane, permitting complete or partial passage of certain molecules: ^{14}C-labeled urea, I, etc. The same phenomenon has also been demonstrated by means of dyes.

The color of the enamel-covered crown ranges from yellowish white to grayish white, It has been suggested that the color is determined by differences in the translucency of

45

enamel, yellowish teeth having a thin, translucent enamel through which the yellow color of the dentin is visible and grayish teeth having a more opaque enamel. The translucency may be attributable to variations in the degree of calcification and homogeneity of the enamel. Grayish teeth frequently show a slightly yellowish color at the cervical areas, presumably because the thinness of the enamel permits the light to strike the underlying yellow dentin and be reflected. Incisal areas may have a bluish tinge where the thin edge consists only of a double layer of enamel.

Chemical properties

The enamel consists mainly of inorganic material (96%) and only a small amount of organic substance and water (4%). The inorganic material of the enamel is similar to apatite. The bar graph in Fig. 3-1 indicates the composition by volume of mineralized tissues in which odontoblast processes have been replaced with peritubular dentin (sclerotic dentin) and the equivalent situation in bone in which osteocyte lacunae are filled with mineral.

The origins shown at the left of Fig. 3-1 reflect the facts that enamel matrix mineralization

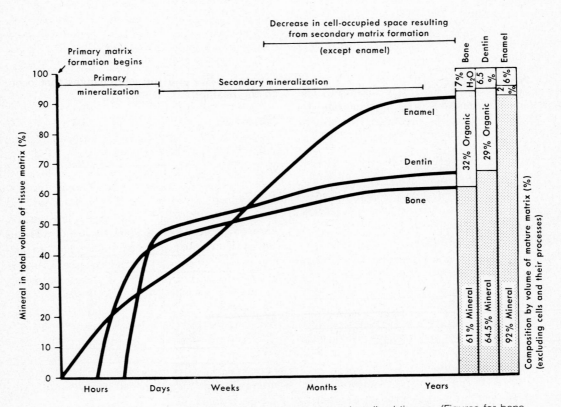

Fig. 3-1. Formation, mineralization, and maturation of some mineralized tissues. (Figures for bone from Robinson, R.A.: In Rodahl, K., Nicholson, J.T., and Brown, E.M., editors: Bone as a tissue, New York, 1960, The Blakiston Division, McGraw-Hill Book Co., pp. 186-250. Figures for dentin and enamel from Brudevold, F.: In Sognnaes, R.F., editor: Chemistry and prevention of dental caries, Springfield, Ill., 1962, Charles C Thomas, Publisher, pp. 32-88.)

begins immediately after it is secreted and that the lag in mineralization after matrix formation is greater in dentin than in bone. Enamel primary mineralization and secondary mineralization (maturation) increase mineral content in a relatively smooth curve. In both bone and dentin, well over one half of the mineral accumulates rapidly (primary mineralization). The curves then flatten as secondary mineralization occurs. The curves continue to rise slowly as cell-occupied space is filled with mineralized matrix (secondary matrix formation) in bone and dentin.

The relative *space occupied* by the organic framework and the entire enamel is almost equal. Fig. 3-2 illustrates this by comparing a stone and a sponge of approximately *equal size*. The stone represents the mineral content, and the sponge represents the organic framework of the enamel. Although their sizes are almost equal, their weights are vastly different. The stone is more than 100 times heavier than the sponge, or expressed in percentage, the weight of the sponge is less than 1% of that of the stone.

The nature of the organic elements of enamel is incompletely understood. In development and histologic staining reactions the enamel matrix resembles keratinizing epidermis. More specific methods have revealed sulfhydryl groups and other reactions suggestive of keratin. However, chemical analyses of the matrix of mature enamel indicate that the amino acid composition is not closely related to keratin and is distinctly different from collagen. Proteins can be isolated in several different fractions, and they generally contain high percentages of serine, glutamic acid, and glycine. Roentgen-ray diffraction studies reveal that the molecular structure is typical of the group of proteins called cross-β-proteins. In addition, histochemical reactions have suggested that the enamel-forming cells of developing teeth also contain a polysaccharide-protein complex and that an acid mucopolysaccharide enters the enamel itself at the time when calcification becomes a prominent feature. Tracer studies have indicated that the enamel of erupted teeth of rhesus monkeys can transmit and exchange radioactive isotopes originating from the saliva and the pulp. Considerable investigation is still required to determine the normal physiologic characteristics and the age changes that occur in the enamel.

Structure

Rods. The enamel is composed of enamel rods or prisms, rod sheaths, and in some regions a cementing interprismatic substance. The number of enamel rods has been estimated as ranging from 5 million in the lower lateral incisors to 12 million in the upper first molars. From the dentinoenamel junction the rods run somewhat tortuous courses outward to the surface of the tooth. The length of most rods is greater than the thickness of the enamel be-

Fig. 3-2. A sponge, **A,** and a stone, **B,** are comparable to organic and mineral elements of enamel. Their sizes are approximately equal, but their weights differ greatly. (From Bodecker, C.F.: Dent. Rev. **20:**317, 1906.)

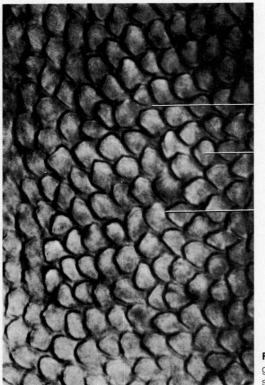

Interrod substance
(rod "tail")

Rod

Rod sheath

Fig. 3-3. Decalcified section of enamel of human tooth germ. Rods cut transversely have appearance of fish scales.

cause of the oblique direction and the wavy course of the rods. The rods located in the cusps, the thickest part of the enamel, are longer than those at the cervical areas of the teeth. It is stated generally that, as observed with the light microscope, the diameter of the rods averages 4 μm, but this measurement necessarily varies, since the outer surface of the enamel is greater than the dentinal surface where the rods originate. It is claimed that the diameter of the rods increases from the dentinoenamel junction toward the surface of the enamel at a ratio of about 1:2.

The enamel rods normally have a clear crystalline appearance, permitting light to pass through them. In cross section under the light microscope they occasionally appear hexagonal. Sometimes they appear round or oval. In cross

sections of human enamel, many rods resemble fish scales (Fig. 3-3).

Submicroscopic structure. Since many features of enamel rods are below the limit of resolution of the light microscope, many questions concerning their morphology can only be answered by electron microscopy. Although many areas of human enamel seem to contain rods surrounded by rod sheaths and separated by interrod substance (Fig. 3-4), a more common pattern is a keyhole- or paddle-shaped prism in human enamel (Fig. 3-5). When cut longitudinally (Fig. 3-6), sections pass through the "heads" or "bodies" of one row of rods and the "tails" of an adjacent row. This produces an appearance of rods separated by interrod substance. These rods measure about 5 μm in breadth and 9 μm in length. Rods of this shape

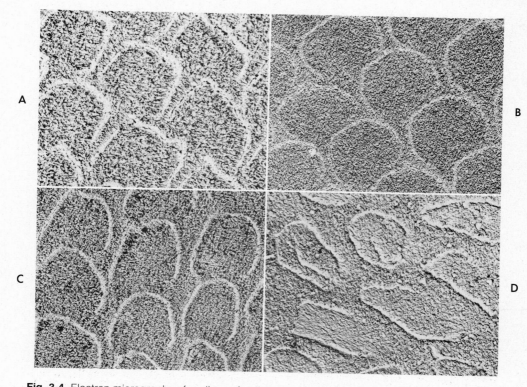

Fig. 3-4. Electron micrographs of replicas of polished and etched human subsurface enamel. Rods are cut in cross section. Various patterns are apparent. **A,** "Keyholes." **B,** "Staggered arches." **C,** "Stacked arches." **D,** Irregular rods near dentinoenamel junction. (Approximately ×3000.) (From Swancar, V.R., Scott, D.B., and Njemirovskij, Z.: J. Dent. Res. **49:**1025, 1970. Copyright by the American Dental Association. Reprinted by permission.)

can be packed tightly together (Fig. 3-7), and enamel with this structure explains many bizarre patterns seen with the electron microscope. The "bodies" of the rods are nearer occlusal and incisal surfaces, whereas the "tails" point cervically.

Studies with polarized light and roentgen-ray diffraction have indicated that the apatite crystals are arranged approximately parallel to the long axis of the prisms, although deviations of up to 40 degrees have been reported. Careful electron microscope studies have made it possible to describe more precisely the orientation of these crystals. They are approximately par-

allel to the long axes of the rods in their "bodies" or "heads" and deviate about 65 degrees from this axis as they fan out into the "tails" of the prisms (Fig. 3-8). Since it is extremely difficult to prepare a section that is exactly parallel to the long axes of the crystals, there is some question about their length, but they are estimated to vary between 0.05 and 1 μm. When cut in cross section, the crystals of human enamel are somewhat irregular in shape (Fig. 3-9) and have an average thickness of about 30 nanometers* (nm; 300 angstrom units

---

*1 nanometer (new terminology) = 10 Å.

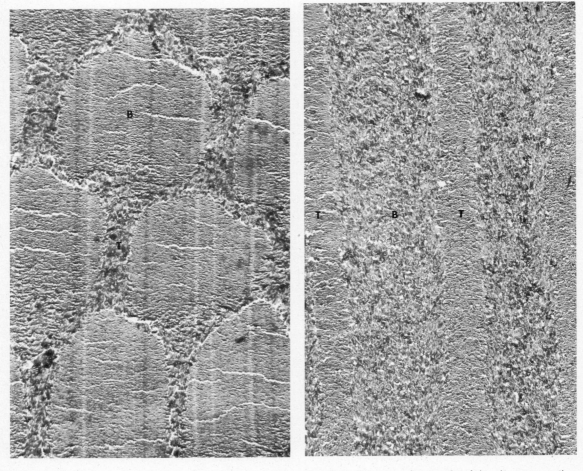

Fig. 3-5. Electron micrograph of cross sections of rods in mature human enamel. Rods are keyhole shaped, and crystal orientation is different in "bodies," *B,* than in "tails," *T.* (Approximately ×5000.) (From Meckel, A.H., Griebstein, W. J., and Neal, R.J.: Arch. Oral Biol. **10**:775, 1965.)

Fig. 3-6. Electron micrograph of longitudinal section through mature human enamel. Alternating "tails," *T,* and "bodies," *B,* of rods are defined by abrupt changes in crystal direction where they meet. (Approximately ×5000.) (From Meckel, A.H., Griebstein, W.J., and Neal, R.J.: Arch. Oral Biol. **10**:775, 1965.)

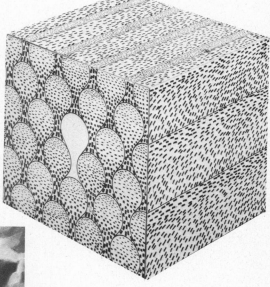

Fig. 3-7. Model indicating packing of keyhole-shaped rods in human enamel. Various patterns can be produced by changing plane of sectioning. (From Meckel, A.H., Griebstein, W.J., and Neal, R.J.: Arch. Oral Biol. **10:**775, 1965.)

Fig. 3-8. Drawing of keyhole pattern of human enamel indicating orientation of apatite crystals within individual rods. Crystals are oriented parallel to long axes of "bodies" of rods and fan out at an angle of approximately 65 degrees in "tails" of rods. (From Griebstein, W.J.: In Stack, M.V., and Fearnhead, R.W., editors: Tooth enamel, Bristol, 1965, John Wright & Sons, Ltd., p. 190.)

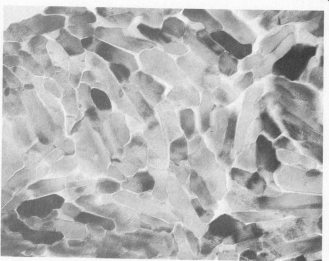

Fig. 3-9. Cross section of apatite crystals within enamel rod in human enamel. Crystals are tightly packed and irregular in shape. (Approximately ×168,000.) (From Frazier, P.D.: J. Ultrastruct. Res. **22:**1, 1968.)

Fig. 3-10. Electron micrograph of decalcified section of immature bovine enamel. Although shape of rods in bovine enamel is not clearly established, this electron micrograph reproduces pattern one would expect in longitudinal sections through human enamel. Organic sheaths around individual apatite crystals are oriented parallel to long axes of rods in their "bodies," *B,* and more nearly perpendicular to long axes in their "tails," *T.* (Approximately ×38,000.) (From Travis, D.F., and Glimcher, M.J.: J. Cell Biol. **23:**447, 1964.)

[Å]) and an average width of about 90 μm (900 Å).

Early investigators using electron microscopy described a network of fine organic fibrils running throughout the rods and interrod substance. Recent improvements in preparative methods have disclosed that the organic matrix probably forms an envelope surrounding each apatite crystal (Fig. 3-10). In electron micrographs the surfaces of rods are visible because of abrupt changes in crystal orientation from one rod to another. For this reason the crystals are not as tightly packed and there may be more space for organic matrix at these surfaces. This accounts for the rod sheath visible in the light microscope (Fig. 3-3).

Striations. Each enamel rod is built up of segments separated by dark lines that give it a striated appearance (Fig. 3-11). These transverse striations demarcate rod segments and become more visible by the action of mild acids. The striations are more pronounced in enamel that is insufficiently calcified. The rods are segmented because the enamel matrix is formed in a rhythmic manner. In humans these segments seem to be a uniform length of about 4 μm.

Direction of rods. Generally the rods are oriented at right angles to the dentin surface. In the cervical and central parts of the crown of a deciduous tooth they are approximately horizontal (Fig. 3-12, A). Near the incisal edge or tip of the cusps they change gradually to an increasingly oblique direction until they are almost vertical in the region of the edge or tip of the cusps. The arrangement of the rods in permanent teeth is similar in the occlusal two thirds of the crown. In the cervical region, however, the rods deviate from the horizontal in an apical direction (Fig. 3-12, B).

The rods are rarely, if ever, straight throughout. They follow a wavy course from the dentin to the enamel surface. The most significant deviations from a straight radial course can be described as follows. If the middle part of the crown is divided into thin horizontal discs, the rods in the adjacent discs bend in opposite directions. For instance, in one disc the rods start from the dentin in an oblique direction and bend more or less sharply to the left side (Fig. 3-13, A), whereas in the adjacent disc the rods bend toward the right (Fig. 3-13, B). This

Fig. 3-11. Ground section through enamel. Rods cut longitudinally. Cross-striation of rods.

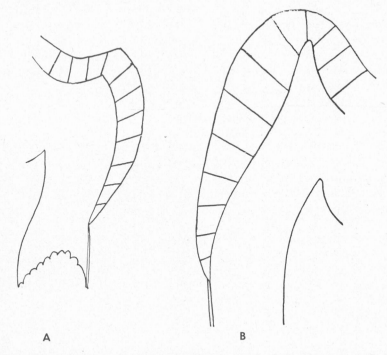

A   B

Fig. 3-12. Diagrams indicating general direction of enamel rods. **A,** Deciduous tooth. **B,** Permanent tooth.

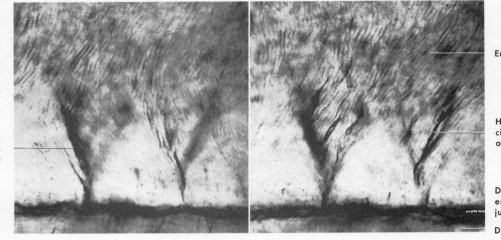

Fig. 3-13. Horizontal ground section through enamel near dentinoenamel junction. **A** and **B** show change in direction of rods in two adjacent layers of enamel, which is made visible by change in focus of microscope.

alternating clockwise and counter-clockwise deviation of the rods from the radial direction can be observed at all levels of the crown if the discs are cut in the planes of the general rod direction.

If the discs are cut in an oblique plane, especially near the dentin in the region of the cusps or incisal edges, the rod arrangement appears to be further complicated—the bundles of rods seem to intertwine more irregularly. This optical appearance of enamel is called *gnarled enamel*.

The enamel rods forming the developmental fissures and pits, as on the occlusal surface of molar and premolars, converge in their outward course.

Hunter-Schreger bands. The more or less regular change in the direction of rods may be regarded as a functional adaptation, minimizing the risk of cleavage in the axial direction under the influence of occlusal masticatory forces. The change in the direction of rods is responsible for the appearance of the Hunter-Schreger bands. These are alternating dark and light strips of varying widths (Fig. 3-14, *A*) than can best be seen in a longitudinal ground section under oblique reflected light. They originate at the dentinoenamel border and pass outward,

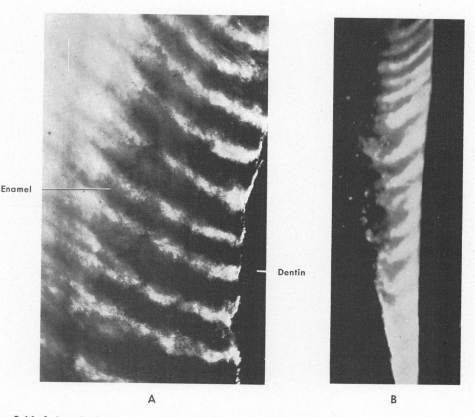

Enamel

Dentin

A   B

Fig. 3-14. A, Longitudinal ground section through enamel photographed by reflected light. Hunter-Schreger bands. **B,** Decalcified enamel, photographed by reflected light, showing Hunter-Schreger bands.

ending at some distance from the outer enamel surface. Some investigators claim that there are variations in calcification of the enamel that coincide with the distribution of the bands of Hunter-Schreger. Careful decalcification and staining of the enamel have provided further evidence that these structures may not be the result solely of an optical phenomenon but that they are composed of alternate zones having a slightly different permeability and a different content of organic material (Fig. 3-14, *B*).

Incremental lines of Retzius. The incremental lines of Retzius appear as brownish bands in ground sections of the enamel. They illustrate

the incremental pattern of the enamel, that is, the successive apposition of layers of enamel during formation of the crown. In longitudinal sections they surround the tip of the dentin (Fig. 3-15, *A*). In the cervical parts of the crown they run obliquely. From the dentino-enamel junction to the surface they deviate occlusally (Fig. 3-15, *B*). In transverse sections of a tooth the incremental lines of Retzius appear as concentric circles (Fig. 3-16). They may be compared to the growth rings in the cross section of a tree. The term "incremental lines" designates these structures appropriately, for they do, in fact, reflect variations in structure

Fig. 3-15. Incremental lines of Retzius in longitudinal ground sections. **A,** Cuspal region. **B,** Cervical region, X.

and mineralization, either hypomineralized or hypermineralized, that occur during growth of the enamel. The exact nature of these developmental changes is not known. The incremental lines have been attributed to periodic bending of the enamel rods, to variations in the basic organic structure (Fig. 3-17), or to a physiologic calcification rhythm.

The incremental lines of Retzius, if present in moderate intensity, are considered normal. However, the rhythmic alteration of periods of enamel matrix formation and of rest can be upset by metabolic disturbances, causing the rest periods to be unduly prolonged and close together. Such an abnormal condition is responsible for the broadening of the incremental lines of Retzius, rendering them more prominent.

Surface structures. A relatively structureless layer of enamel, approximately 30 μm thick, has been described in 70% of permanent teeth and all deciduous teeth. This structureless enamel is found least often over the cusp tips and most commonly toward the cervical areas of the enamel surface. In this surface layer no prism outlines are visible, and all of the apatite crystals are parallel to one another and perpendicular to the striae of Retzius. It is also somewhat more heavily mineralized than the bulk of enamel beneath it (Fig. 3-18). Other microscopic details that have been observed on outer enamel surfaces of newly erupted teeth are perikymata, rod ends, and cracks (lamellae).

Perikymata are transverse, wavelike grooves, believed to be the external manifestations of the striae of Retzius. They are continuous around a tooth and usually lie parallel to each other and to the cementoenamel junction (Figs. 3-19 and 3-20). Ordinarily there are about 30 perikymata per millimeter in the region of the cementoenamel junction, and their concentration gradually decreases to about 10 per millimeter near the occlusal or incisal edge of a surface. Their course usually is fairly regular, but in the cervical region it may be quite irregular.

The enamel rod ends are concave and vary in depth and shape. They are shallowest in the cervical regions of surfaces and deepest near the incisal or occlusal edges (Fig. 3-19, *B*).

The term "cracks" originally was used to describe the narrow, fissurelike structures that

Perikymat [margin annotation]

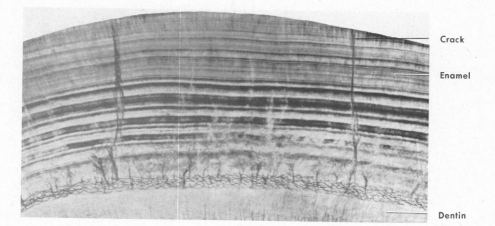

Crack

Enamel

Dentin

Fig. 3-16. Incremental lines of Retzius in transverse ground section, arranged concentrically.

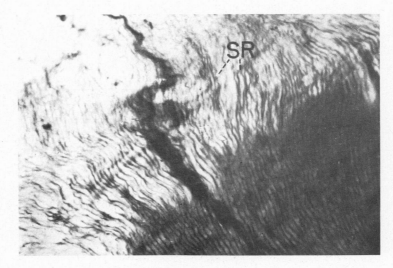

Fig. 3-17. Carefully decalcified section through enamel. Thickening of sheath substance, *SR,* in Retzius lines. (From Bodecker, C.F.: Dent. Rev. **20:**317, 1906.)

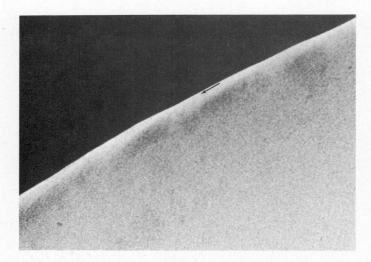

Fig. 3-18. Microradiograph of ground section of sound human enamel. Relatively structureless surface layer *(arrow)* is more radiopaque than bulk of enamel below it. (Approximately ×200.) (Courtesy Dr. A.J. Gwinnett, Stony Brook, N.Y.)

are seen on almost all surfaces (Fig. 3-20, *D*). It has since been demonstrated that they are actually the outer edges of lamellae (see discussion of enamel lamellae). They extend for varying distances along the surface, at right angles to the dentinoenamel junction, from which they originate. Most of them are less than a millimeter in length, but some are longer, and a few reach the occlusal or incisal edge of a surface. They are fairly evenly spaced, but long lamellae appear thicker than short ones.

The enamel of the deciduous teeth develops partly before and partly after birth. The boundary between the two portions of enamel in the deciduous teeth is marked by an accentuated incremental line of Retizius, the *neonatal line* or *neonatal ring* (Fig. 3-21). It appears to be the result of the abrupt change in the environment and nutrition of the newborn infant. The prenatal enamel usually is better developed than the postnatal enamel. This is explained by the fact that the fetus develops in a well-protected environment with an adequate supply of all the essential materials, even at the expense of the mother. Because of the undisturbed and even development of the enamel prior to birth, perikymata are absent in the occlusal parts of the deciduous teeth, whereas they are present in the postnatal cervical parts.

Enamel cuticle. A delicate membrane called *Nasmyth's membrane,* after its first investigator, or the *primary enamel cuticle* covers the entire crown of the newly erupted tooth but is probably soon removed by mastication. Electron microscope studies have indicated that this membrane is a typical basal lamina found beneath most epithelia (Fig. 3-22). It is probably visible with the light microscope because of

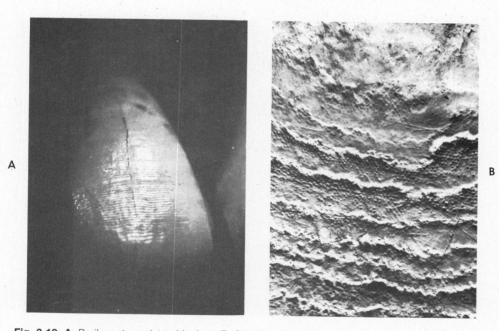

Fig. 3-19, A, Perikymata on lateral incisor. **B,** Shadowed replica of surface of intact enamel (buccal surface of upper left second molar showing perikymata). (×1500.) (**B** from Scott, D.B., and Wyckoff, R.W.G.: Public Health Rep. **61:**1397, 1946.)

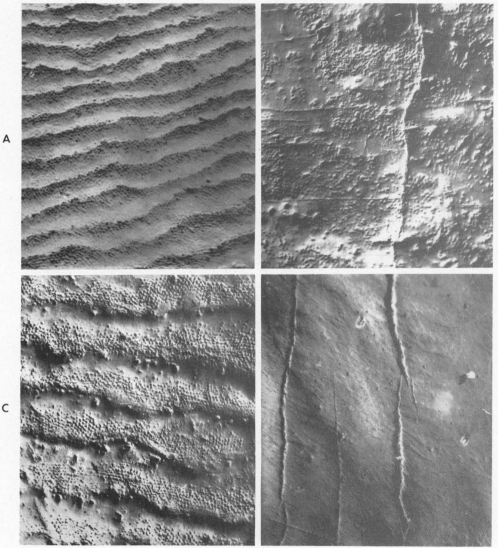

Fig. 3-20. Progressive loss of surface structure with advancing age. **A,** Surface of recently erupted tooth showing pronounced enamel prism ends and perikymata. Patient is 12 years of age. **B,** Early stage of structural loss that occurs during first few years (wear is more rapid on anterior teeth than on posterior teeth and more rapid on facial or lingual surfaces than on proximal surfaces). Note small regions where prism ends are worn away. Patient is 25 years of age. **C,** Later stage. Here elevated parts between perikymata are worn smooth, while structural detail in depths of grooves is still more or less intact. Eventually wearing proceeds to point where all prism ends and perikymata disappear. Patient is 52 years of age. (Since these are negative replicas, surface details appear inverted. Raised structures represent depressions in actual surface.) **D,** Surface worn completely smooth and showing only "cracks," which actually represent outer edges of lamellae. Patient is 50 years of age. (All magnifications ×105.) (From Scott, D.B., and Wyckoff, R.W.G.: J. Am. Dent. Assoc. **39:**275, 1949.)

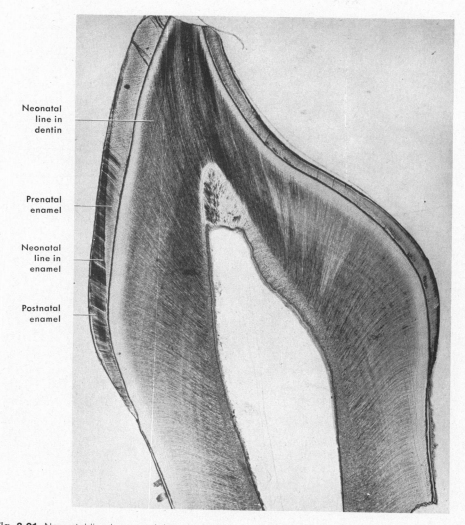

Neonatal
line in
dentin

Prenatal
enamel

Neonatal
line in
enamel

Postnatal
enamel

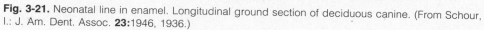

Fig. 3-21. Neonatal line in enamel. Longitudinal ground section of deciduous canine. (From Schour, I.: J. Am. Dent. Assoc. **23:**1946, 1936.)

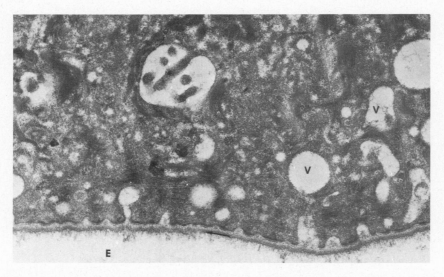

Fig. 3-22. Electron micrograph of reduced enamel epithelium covering surface of unerupted human tooth. Enamel has been removed by demineralization, *E*. Typical basal lamina separates enamel space from epithelium *(arrow)*. Epithelial cells contain a number of intracytoplasmic vacuoles, *V*. (Approximately ×24,000.) (From Listgarten, M.A.: Arch. Oral Biol. **11:**999, 1966.)

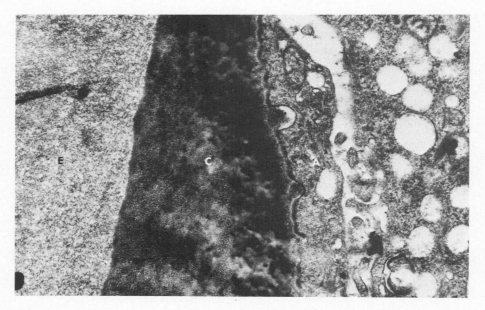

Fig. 3-23. Electron micrograph of gingival area of erupted human tooth. Remnants of enamel matrix appear at left, *E*. Cuticle, *C,* separates enamel matrix from epithelial cells of attached epithelial cuff, *A*. Inner layers of cuticle (afibrillar cementum) are deposited before eruption; origin of outer layers is not known. (Approximately ×37,000.) (From Listgarten, M.A.: Am. J. Anat. **119:**147, 1966.)

Fig. 3-24. Electron micrograph of surface of undemineralized human enamel. Enamel surface, *E,* is covered by pellicle, *P.* Individual crystals can be seen in enamel. (Approximately ×58,000.) (From Houver, G., and Frank, R.M.: Arch. Oral Biol. **12:**1209, 1967.)

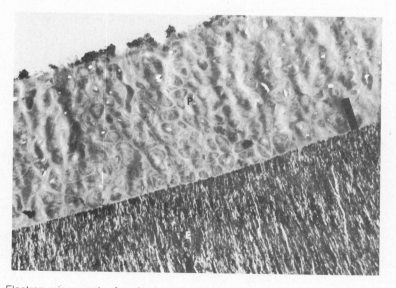

Fig. 3-25. Electron micrograph of undemineralized human enamel surface. Enamel, *E,* is covered by a bacterial plaque, *P. Black bar at right,* Thickness of pellicle seen in Fig. 3-24. (Approximately ×12,000.) (From Frank, R.M., and Brendel, A.: Arch. Oral Biol. **11:**883, 1966.)

its wavy course. This basal lamina is apparently secreted by the ameloblasts when enamel formation is completed. It has also been reported that the cervical area of the enamel is covered by afibrillar cementum, continuous with the cementum and probably of mesodermal origin (Fig. 3-23). This cuticle is apparently secreted after the epithelial enamel organ retracts from the cervical region during tooth development.

Finally, erupted enamel is normally covered by a *pellicle,* which is apparently a precipitate of salivary proteins (Fig. 3-24). This pellicle reforms within hours after an enamel surface is mechanically cleaned. Within a day or two after the pellicle has formed, it becomes colonized by microorganisms to form a bacterial plaque (Fig. 3-25).

Enamel lamellae. Enamel lamellae are thin, leaflike structures that extend from the enamel surface toward the dentinoenamel junction (Fig. 3-26). They may extend to, and sometimes penetrate into, the dentin. They consist of organic material, with but little mineral content. In ground sections these structures may

be confused with cracks caused by grinding of the specimen (Fig. 3-16). Careful decalcification of ground sections of enamel makes possible the distinction between cracks and enamel lamellae. The former disappear, whereas the latter persist (Figs. 3-26, *A,* and 3-27).

Lamellae may develop in planes of tension. Where rods cross such a plane, a short segment of the rod may not fully calcify. If the disturbance is more severe, a crack may develop that is filled either by surrounding cells, if the crack occurred in the unerupted tooth, or by organic substances from the oral cavity, if the crack developed after eruption. Three types of lamellae can thus be differentiated: type A, lamellae composed of poorly calcified rod segments (Fig. 3-27, *B*); type B, lamellae consisting of degenerated cells; and type C, lamellae arising in erupted teeth where the cracks are filled with organic matter, presumably originating from saliva. The last type may be more common than formerly believed. Although lamellae of type A are restricted to the enamel, those of types B and C may reach into the dentin (Fig.

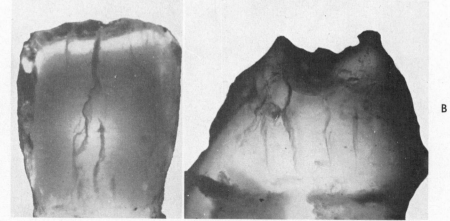

Fig. 3-26. A, Decalcified incisor with moderately severe mottled enamel. Numerous lamellae can be observed. (×8.) **B,** Maxillary first permanent molar of caries-free-2-year-old rhesus monkey. Numerous bands of organic matter, lamellae, can be seen after decalcification. (×8.) (**B** from Sognnaes, R.F.: J. Dent. Res. **29:**260, 1950.)

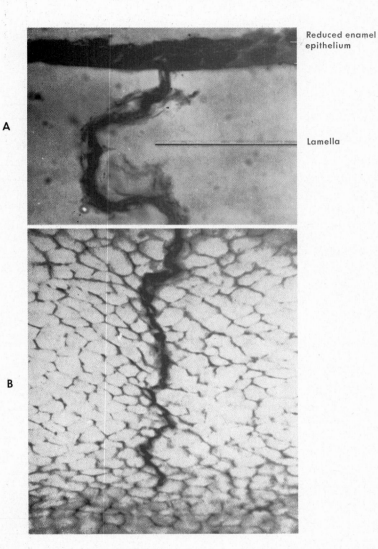

Reduced enamel
epithelium

Lamella

Fig. 3-27. A, Paraffin section through reduced enamel epithelium, enamel cuticle, and lamella, isolated together by acid flotation from surface of unerupted human tooth. Note intimate relationship
between three elements. (Hematoxylin and eosin; ×1300.) **B,** Paraffin section of decalcified enamel
of human molar showing relation between lamella and surrounding organic sheath substance. (Hematoxylin and eosin; ×1000.) (**A** from Ussing, M.J.: Acta Odontal. Scand. **13:**23, 1955; reprinted in
J. West. Soc. Periodont. **3:**71, 1955; **B** courtesy Dr. R.F. Sognnaes, Los Angeles.)

3-28). If cells from the enamel organ fill a crack in the enamel, those in the depth degenerate, whereas those close to the surface may remain vital for a time and produce a hornified cuticle in the cleft. In such cases the inner parts of the lamella consist of an organic cell detritus, the outer parts of a double layer of the cuticle. If connective tissue invades a crack in the enamel, cementum may be formed. In such

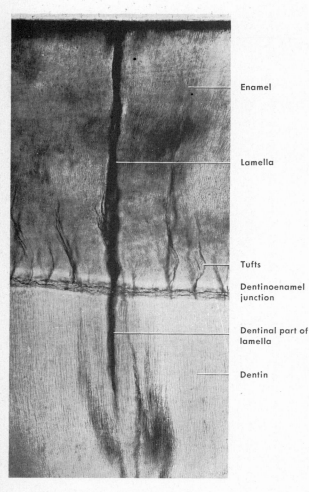

Fig. 3-28. Transverse ground section through lamella reaching from surface into dentin.

[image labels: Enamel; Lamella; Tufts; Dentinoenamel junction; Dentinal part of lamella; Dentin]

cases lamellae consist entirely or partly of cementum.

Lamellae extend in the longitudinal and radial direction of the tooth, from the tip of the crown toward the cervical region (Fig. 3-26). This arrangement explains why they can be observed better in horizontal sections. It has been suggested that enamel lamellae may be a site of weakness in a tooth and may form a road of entry for bacteria that initiate caries.

Enamel tufts. Enamel tufts (Fig. 3-29) arise at the dentinoenamel junction and reach into the enamel to about one fifth to one third of its thickness. They were so termed because they resemble tufts of grass when viewed in ground sections. This picture is erroneous. An enamel tuft does not spring from a single small area but is a narrow, ribbonlike structure, the inner end of which arises at the dentin. The impression of a tuft of grass is created by examining such structures in thick sections under low magnification. Under these circumstances the imperfections, lying in different planes and curving in different directions (Fig. 3-13), are projected into one plane (Fig. 3-29).

Tufts consist of hypocalcified enamel rods and interprismatic substance. Like the lamellae, they extend in the direction of the long axis of the crown. Therefore they are seen abundantly in horizontal, and rarely in longitudinal, sections. Their presence and their development are a consequence of, or an adaptation to, the spatial conditions in the enamel.

Dentinoenamel junction. The surface of the dentin at the dentinoenamel junctions is pitted. Into the shallow depressions of the dentin fit rounded projections of the enamel. This relation assures the firm hold of the enamel cap on the dentin. In sections, therefore, the dentinoenamel junction appears not as straight but as a scalloped line (Figs. 3-29 and 3-30). The convexities of the scallops are directed toward the dentin. The pitted dentinoenamel junction is preformed even before the development of hard tissues and is evident in the arrangement

Enamel

Tufts

Dentinoenamel
junction

Dentin

Fig. 3-29. Transverse ground section through tooth under low magnification. Numerous tufts extend from dentinoenamel junction into enamel.

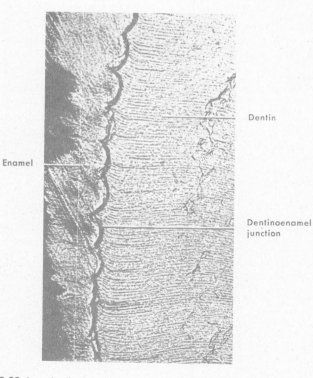

Enamel

Dentin

Dentinoenamel
junction

Fig. 3-30. Longitudinal ground section. Scalloped dentinoenamel junction.

of the ameloblasts and the basement membrane of the dental papilla (Fig. 3-43).

In microradiographs of ground sections a hypermineralized zone about 30 μm thick can sometimes be demonstrated at the dentinoenamel junction. It is most prominent before mineralization is complete.

Odontoblast processes and enamel spindles. Occasionally odontoblast processes pass across the dentinoenamel junction into the enamel. Since many are thickened at their end (Fig. 3-31), they have been termed *enamel spindles*. They seem to originate from processes of odontoblasts that extended into the enamel epithelium before hard substances were formed. The direction of the odontoblast processes and spindles in the enamel corresponds to the orignal direction of the ameloblasts—at right angles to the surface of the dentin. Since the enamel rods are formed at an angle to the axis of the ameloblasts, the direction of spindles and rods is divergent. In ground sections of dried teeth the organic content of the spindles disintegrates and is replaced by air, and the spaces appear dark in transmitted light.

Age changes

The most apparent age change in enamel is attrition or wear of the occlusal surfaces and proximal contact points as a result of mastication. This is evidenced by a loss of vertical dimension of the crown and by a flattening of the proximal contour. In addition to these gross changes, the outer enamel surfaces themselves undergo posteruptive alterations in structure at the microscopic level. These result from environmental influences and occur with a regularity that can be related to age (Fig. 3-20).

The surfaces of unerupted and recently erupted teeth are covered completely with pronounced rod ends and perikymata. At the points of highest contour of the surfaces these structures soon begin to disappear. This is followed by a generalized loss of the rod ends and a much slower flattening of the perikymata. Finally, the perikymata disappear completely. The rate at which structure is lost depends on the location of the surface of the tooth and on the location of the tooth in the mouth. Facial and lingual surfaces lose their structure much more rapidly than do proximal surfaces, and

Fig. 3-31. Ground section. Odontoblast processes extend into enamel as enamel spindles.

anterior teeth lose their structure more rapidly than do posterior teeth.

Age changes within the enamel proper have been difficult to discern microscopically. The fact that alterations do occur has been demonstrated by chemical analysis, but the changes are not well understood. For example, the total amount of organic matrix is said by some to increase, by others to remain unchanged, and by still others to decrease. Localized increases of certain elements such as nitrogen and fluorine, however, have been found in the superficial enamel layers of older teeth. This suggests a continuous uptake, probably from the oral environment, during aging. As a result of age changes in the organic portion of enamel, presumably near the surface, the teeth may become darker, and their resistance to decay may be increased. Suggestive of an aging change is the greatly reduced permeability of older teeth to fluids. There is insufficient evidence to show that enamel becomes harder with age.

Clinical considerations

The course of the enamel rods is of importance in cavity preparations. The choice of instruments depends on the location of the cavity in the tooth. Generally the rods run at a right angle to the underlying dentin or tooth surface. Close to the cementoenamel junction the rods run in a more horizontal direction (Fig. 3-12, *B*). In preparing cavities, it is important that unsupported enamel rods are not left at the cavity margins because they would soon break and produce leakage. Bacteria would lodge in these spaces, inducing secondary dental caries. Enamel is brittle and does not withstand forces in thin layers or in areas where it is not supported by the underlying dentin (Fig. 3-32, *A*). Deep enamel fissures predispose teeth to caries. Although these deep clefts between adjoining cusps cannot be regarded as pathologic, they afford areas for retention of caries-producing agents. Caries penetrate the floor of fissures rapidly because the enamel in

these areas is very thin (Fig. 3-32, *B*). As the destructive process reaches the dentin, it spreads along the dentinoenamel junction, undermining the enamel. An extensive area of dentin becomes carious without giving any warning to the patient because the entrance to the cavity is minute. Careful examination is necessary to discover such cavities because most enamel fissures are more minute than a single toothbrush bristle and cannot be detected with the dental probe.

Dental lamellae may also be predisposing locations for caries because they contain much organic material. Primarily from the standpoint of protection against caries, the structure and reactions of the outer enamel surface are subject to much current research. In vitro tests have shown that the acid solubility of enamel can be greatly reduced by treatment with fluoride compounds. Clinical trials based on these studies have demonstrated reductions of 40% or more in the incidence of caries in children after topical applications of sodium or stannous fluoride. Incorporation of fluorides in dentifrices is now a well-accepted means of caries prevention. Fluoride-containing mixtures such as stannous fluoride pastes, sodium fluoride rinses, and acidulated phosphate fluoride are also used by the dentist to alter the outer surface of the enamel in such a manner that it becomes more resistant to decay.

The most effective means for mass control of dental caries to date has been adjustment of the fluoride level in communal water supplies to 1 part per million. Epidemiologic studies in areas in which the drinking water contained natural fluoride revealed that the caries prevalence in both children and adults was about 65% lower than in nonfluoride areas, and long-term studies have demonstrated that the same order of protection is afforded through water fluoridation programs. The mechanisms of action are believed to be primarily a combination of changes in enamel resistance, brought about by incorporation of fluoride during calcification, and alterations in the environment of the teeth, particularly with respect to the oral bacterial flora.

The surface of the enamel in the cervical region should be kept smooth and well polished by proper home care and by regular cleansing by the dentist. If the surface of the cervical enamel becomes decalcified or otherwise roughened, food debris, bacterial plaques, and so on accumulate on this surface. The gingiva in contact with this roughened, debris-covered enamel surface undergoes inflammatory changes. The ensuing gingivitis, unless promptly treated, may lead to more serious periodontal disease.

One of the more recently developed techniques in operative dentistry consists of the use of composite resins. These materials can be mechanically "bonded" directly to the enamel surface. In this procedure the enamel surface is first etched with an acid (phosphoric acid 50%). This produces an uneven dissolution of the enamel rods and their "sheaths" or enamel "heads" and their "tails" so that a relatively smooth enamel surface becomes pitted and irregular. When a composite resin is put on this irregular surface, it can achieve mechanical bonding with the enamel. The same principle is used in coating the susceptible areas of the enamel with the so-called pit fissure sealants.

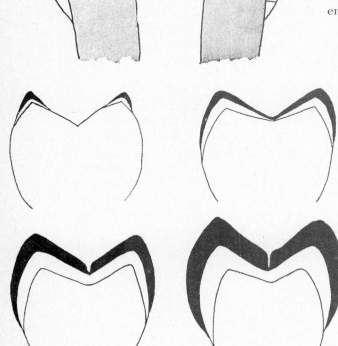

Fig. 3-32. **A,** Diagram of course of enamel rods in molar in relation to cavity preparation. *1* and *2* indicate wrong preparation of cavity margins. *3* and *4* indicate correct preparation. **B,** Diagram of development of deep enamel fissure. Note thin enamel layer forming floor of fissure. (**B** from Kronfeld, R.: J. Am. Dent. Assoc. **22:**1131, 1935.)

DEVELOPMENT
Epithelial enamel organ

The early development of the enamel organ and its differentiation have been discussed in Chapter 2. At the stage preceding the formation of hard structures (dentin and enamel) the enamel organ, originating from the stratified epithelium of the primitive oral cavity, consists of four distinct layers: outer enamel epithelium, stellate reticulum, stratum intermedium, and inner enamel epithelium (ameloblastic layer) (Fig. 3-33). The borderline between the inner enamel epithelium and the connective tissue of the dental papilla is the subsequent dentinoenamel junction. Thus its outline determines the pattern of the occlusal or incisal part of the crown. At the border of the wide basal opening of the enamel organ, the inner enamel epithelium reflects onto the outer enamel epithelium. This is the *cervical loop.* The inner and outer enamel epithelia are elsewhere separated from each other by a large mass of cells differentiated into two distinct layers. The layer that is close to the inner enamel epithelium consists of two or three rows of flat polyhedral cells—the stratum intermedium. The other layer, which is more loosely arranged, constitutes the stellate reticulum.

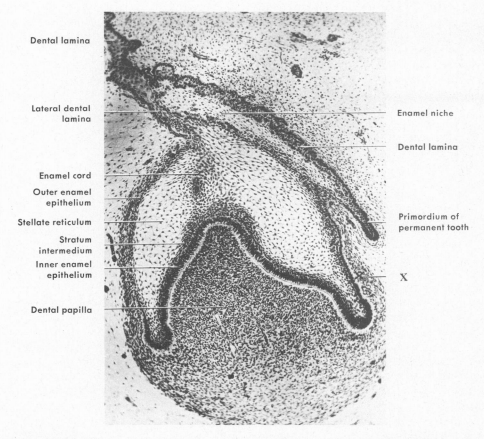

Fig. 3-33. Tooth germ (deciduous lower incisor) of human embryo 105 mm, fourth month. Four layers of enamel organ. Area at X is shown at a higher magnification in Fig. 3-35.

The different layers of epithelial cells of the enamel organ are named according to their morphology, function, or location. The stellate reticulum derives its name from the morphology of its cells. The outer enamel epithelium and the stratum intermedium are so named because of their location. The inner enamel epithelium is so named on the basis of its position. On the basis of function it is called the ameloblastic layer.

Outer enamel epithelium. In the early stages of development of the enamel organ the outer enamel epithelium consists of a single layer of cuboid cells, separated from the surrounding connective tissue of the dental sac by a delicate basement membrane (Fig. 3-34). Prior to the formation of hard structures, this regular arrangement of the outer enamel epithelium is maintained only in the cervical parts of the enamel organ. At the highest convexity of the organ (Fig. 3-33) the cells of the outer enamel epithelium become irregular in shape and cannot be distinguished easily from the outer portion of the stellate reticulum. The capillaries in the connective tissue surrounding the epithelial enamel organ proliferate and protrude toward it (Fig. 3-34). Immediately before enamel formation commences, capillaries may even indent the stellate reticulum. This increased vascularity ensures a rich metabolism when a plentiful supply of substances from the bloodstream to the inner enamel epithelium is required (Fig. 3-35).

During enamel formation, cells of the outer enamel epithelium develop villi and cytoplasmic vesicles and large numbers of mito-

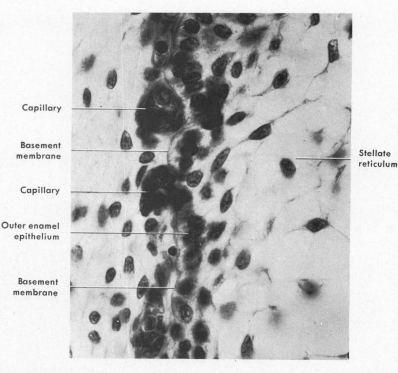

Capillary

Basement membrane

Capillary

Outer enamel epithelium

Basement membrane

Stellate reticulum

Fig. 3-34. Capillaries in contact with outer enamel epithelium. Basement membrane separates outer enamel epithelium from connective tissue.

chondria, all indicating cell specialization for the active transport of materials. The capillaries in contact with the outer enamel epithelium show areas with very thin walls, a structural modification also commonly found in areas of active transport.

Stellate reticulum. In the stellate reticulum, which forms the middle part of the enamel organ, the neighboring cells are separated by wide intercellular spaces filled by a large amount of intercellular substance. The cells are star shaped, with long processes reaching in all directions from a central body (Figs. 3-34 and 3-36). They are connected with each other and with the cells of the outer enamel epithelium and the stratum intermedium by desmosomes.

The structure of the stellate reticulum renders it resistant and elastic. Therefore it seems probable that it acts as a buffer against physical forces that might distort the conformation of

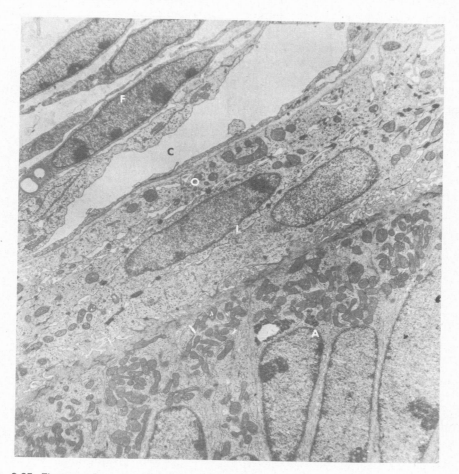

Fig. 3-35. Electron micrograph of epithelial enamel organ over area of rodent incisor in which enamel secretion is underway. From above downward are fibroblasts of dental sac, *F;* capillary, *C;* cells of outer enamel epithelium, *O;* cells of stratum intermedium, *I;* and proximal ends of ameloblasts, *A.* (Approximately ×7000.) (Courtesy Dr. P.R. Garant, Stony Brook, N.Y.)

the developing dentinoenamel junction, giving rise to gross morphologic changes. It seems to permit only a limited flow of nutritional elements from the outlying blood vessels to the formative cells. Indicative of this is the fact that the stellate reticulum is noticeably reduced in thickness when the first layers of dentin are laid down, and the inner enamel epithelium is thereby cut off from the dental papilla, its original source of supply (Fig. 3-37).

Stratum intermedium. The cells of the stratum intermedium are situated between the stellate reticulum and the inner enamel epithe-

lium. They are flat to cuboid in shape and are arranged in one to three layers. They are connected with each other and with the neighboring cells of the stellate reticulum and the inner enamel epithelium by desmosomes. Tonofibrils, with an orientation parallel to the surface of the developing enamel, are found in the cytoplasm. The function of the stratum intermedium is not understood, but it is believed to play a role in production of the enamel itself, either through control of fluid diffusion into and out of the ameloblasts or by the actual contribution of necessary formative elements or

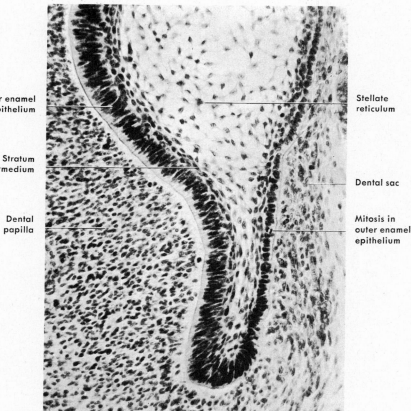

Inner enamel epithelium

Stratum intermedium

Dental papilla

Stellate reticulum

Dental sac

Mitosis in outer enamel epithelium

Fig. 3-36. Region of cervical loop (higher magnification of area X in Fig. 3-33). Transition of outer into inner enamel epithelium.

enzymes. The cells of the stratum intermedium show mitotic division even after the cells of the inner enamel epithelium cease to divide.

Inner enamel epithelium. The cells of the inner enamel epithelium are derived from the basal cell layer of the oral epithelium. Before enamel formation begins, these cells assume a columnar form and differentiate into ameloblasts that produce the enamel matrix. The changes in shape and structure that the cells of the inner enamel epithelium undergo will be described in detail in the discussion of the life cycle of the ameloblasts. It should be mentioned, however, that cell differentiation occurs earlier in the region of the incisal edge or cusps than in the area of the cervical loop.

Cervical loop. At the free border of the enamel organ the outer and inner enamel epi-

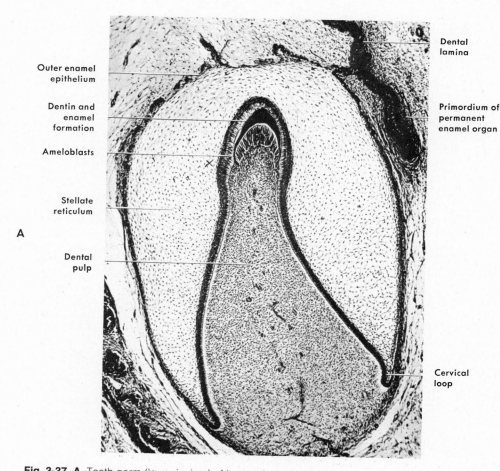

Outer enamel epithelium

Dentin and enamel formation

Ameloblasts

Stellate reticulum

Dental pulp

Dental lamina

Primordium of permanent enamel organ

Cervical loop

A

Fig. 3-37. A, Tooth germ (lower incisor) of human fetus (fifth month). Beginning of dentin and enamel formation. Stellate reticulum at tip of crown is reduced in thickness.

Continued.

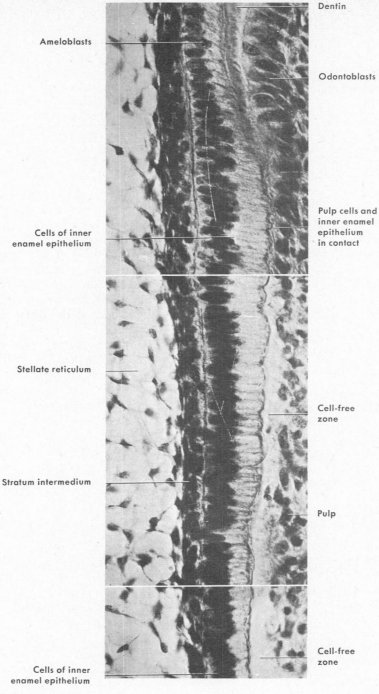

Ameloblasts

Dentin

Odontoblasts

Cells of inner
enamel epithelium

Pulp cells and
inner enamel
epithelium
in contact

B

Stellate reticulum

Cell-free
zone

Stratum intermedium

Pulp

Cells of inner
enamel epithelium

Cell-free
zone

Fig. 3-37, cont'd. B, High magnification of inner enamel epithelium from area X in **A.** In cervical region, cells are short, and outermost layer of pulp is cell free. Occlusally cells are long, and cell-free zone of pulp has disappeared. Ameloblasts are again shorter were dentin formation has begun and enamel formation is imminent. (**B** from Diamond, M., and Weinmann, J.P.: J. Dent. Res. **21:**403, 1942.)

thelial layers are continuous and reflected into one another as the cervical loop (Figs. 3-33 and 3-35). In this zone of transition between the outer enamel epithelium and the inner enamel epithelium the cuboid cells gradually gain in length. When the crown has been formed, the cells of this portion give rise to Hertwig's epithelial root sheath (see Chapter 2).

Life cycle of the ameloblasts

According to their function, the life span of the cells of the inner enamel epithelium can be divided into six stages: (1) morphogenic, (2) organizing, (3) formative, (4) maturative, (5) protective, and (6) desmolytic. Since the differentiation of ameloblasts is most advanced in the region of the incisal edge or tips of the cusps and least advanced in the region of the cervical loop, all or some stages of the developing ameloblasts can be observed in one tooth germ.

Morphogenic stage. Before the ameloblasts are fully differentiated and produce enamel, they interact with the adjacent mesenchymal cells, determining the shape of the dentinoenamel junction and the crown (Fig. 3-37, A). During this morphogenic stage the cells are short and columnar, with large oval nuclei that almost fill the cell body.

The Golgi apparatus and the centrioles are located in the proximal end of the cell,* whereas the mitochondria are evenly dispersed throughout the cytoplasm. During ameloblast differentiation, terminal bars appear concomitantly with the migration of the mitochondria to the basal region of the cell (Fig. 3-44). The terminal bars represent points of close contact between cells. They were previously believed to consist of dense intercellular substance, but

---

*In modern usage, to conform with the terminology applied to other secretory cells, the dentinal end of the ameloblast, at which enamel is formed, is called *distal*, and the end facing the stratum intermedium is called *basal* or *proximal*.

under the electron microscope it has been found that they comprise thickening of the opposing cell membranes, associated with condensations of the underlying cytoplasm.

The inner enamel epithelium is separated from the connective tissue of the dental papilla by a delicate basal lamina. The adjacent pulpal layer is a cell-free, narrow, light zone containing fine argyrophil fibers and the cytoplasmic processes of the superficial cells of the pulp (Figs. 3-37, B, and 3-38).

Organizing stage. In the organizing stage of development the inner enamel epithelium interacts with the adjacent connective tissue cells, which differentiate into odontoblasts. This stage is characterized by a change in the appearance of the cells of the inner enamel epithelium. They become longer, and the nucleus-free zones at the distal ends of the cells become almost as long as the proximal parts containing the nuclei (Fig. 3-37, B). In preparation for this development a reversal of the functional polarity of these cells takes place by the migration of the centrioles and Golgi regions from the proximal ends of the cells into their distal ends (Fig. 3-39).

Special staining methods reveal the presence of fine acidophil granules in the proximal part of the cell. Electron microscope studies have shown that these granules are actually the mitochondria, which have become concentrated in this part of the cell. At the same time the clear cell-free zone between the inner enamel epithelium and the dental papilla disappears (Fig. 3-37, B), probably because of elongation of the epithelial cells toward the papilla. Thus the epithelial cells come into close contact with the connective tissue cells of the pulp, which differentiate into odontoblasts. During the terminal phase of the organizing stage the formation of the dentin by the odontoblasts begins (Fig. 3-37, B).

The first appearance of dentin seems to be a critical phase in the life cycle of the inner enamel epithelium. As long as it is in contact with the connective tissue of the dental papilla, it receives nutrient material from the blood vessels of this tissue. When dentin forms, however, it cuts off the ameloblasts from their original source of nourishment, and from then on they are supplied by the capillaries that surround and may even penetrate the outer enamel epithelium. This reversal of nutritional source is characterized by proliferation of capillaries of the dental sac and by reduction and gradual disappearance of the stellate reticulum (Figs. 3-35 and 3-37, *A*). Thus the distance between the capillaries and the stratum intermedium and the ameloblast layer is shortened. Experiments with vital stains demonstrate this reversal of the nutritional stream.

Formative stage. The ameloblasts enter their formative stage (Fig. 3-39) after the first layer of dentin has been formed. The presence of dentin seems to be necessary for the beginning of enamel matrix formation just as it was necessary for the epithelial cells to come into close

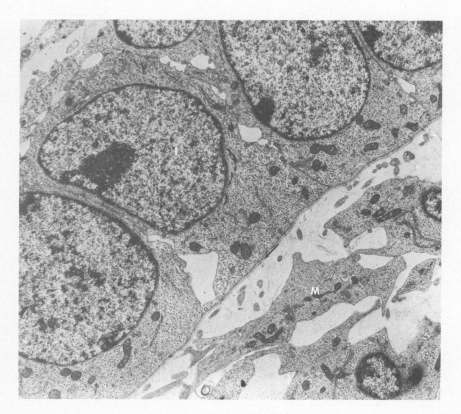

Fig. 3-38. Electron micrograph of inner enamel epithelium, *I,* and adjacent mesenchymal cells of dental papilla, *M,* at early stage of tooth formation. Cytoplasm of cells of inner enamel epithelium is filled with mitochondria and free ribosomes. Typical basement membrane separates epithelium from mesenchyme *(arrow).* Reticular fibers and cytoplasmic processes of mesenchymal cells appear between inner enamel epithelium and cells of dental papilla. (Approximately ×9000.) (Courtesy Dr. P.R. Garant, Stony Brook, N.Y.)

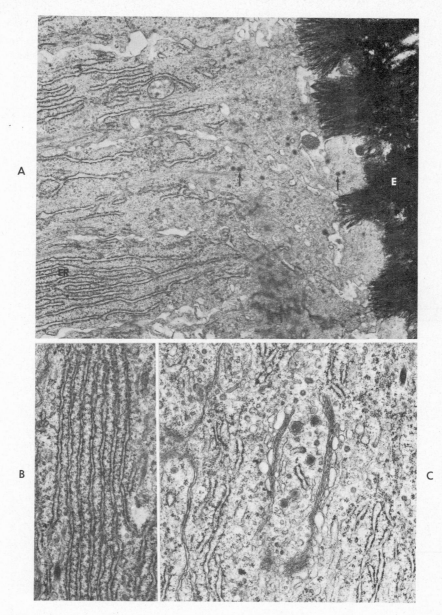

Fig. 3-39. A, Electron micrograph of secreting ends of ameloblasts. Electron-opaque, partially mineralized enamel matrix is at right, *E.* Ameloblasts contain abundant endoplasmic reticulum at left, *ER,* and number of secretory granules at right *(arrows).* (Approximately ×15,000.) **B,** Electron micrograph of area of ameloblast cytoplasm between nucleus and secreting end of cell. Cytoplasm in this region is packed with rough-surfaced endoplasmic reticulum. (Approximately ×25,000.) **C,** Electron micrograph of region of ameloblast cytoplasm approximately halfway between nucleus and secreting end. In this region, rough-surfaced endoplasmic reticulum is displaced by Golgi apparatus, which can be seen in center of this figure. (Approximately ×25,000.) (**A** from Garant, P.R., and Nalbandian, J.: J. Ultrastruct. Res. **23:**427, 1968; **B** and **C** courtesy Dr. P.R. Garant, Stony Brook, N.Y.)

contact with the connective tissue of the pulp during differentiation of the odontoblasts and the beginning of dentin formation. This mutual interaction between one group of cells and another is one of the fundamental laws of organogenesis and histodifferentiation.

During formation of the enamel matrix the ameloblasts retain approximately the same length and arrangement. Changes in the organization and number of cytoplasmic organelles and inclusions are related to the initiation of secretion of enamel matrix.

The earliest apparent change is the development of blunt cell processes on the ameloblast surfaces, which penetrate the basal lamina and enter the predentin (Fig. 3-40).

Maturative stage. Enamel maturation (full mineralization) occurs after most of the thickness of the enamel matrix has been formed in the occlusal or incisal area. In the cervical parts of the crown, enamel matrix formation is still progressing at this time. During enamel maturation the ameloblasts are slightly reduced in length and are closely attached to enamel matrix. The cells of the stratum intermedium lose their cuboidal shape and regular arrangement and assume a spindle shape. It is certain that the ameloblasts also play a part in the maturation of the enamel. During maturation, ameloblasts display microvilli at their distal extremities, and cytoplasmic vacuoles containing material resembling enamel matrix are present (Figs. 3-50 and 3-52). These structures indicate an absorptive function of these cells.

Protective stage. When the enamel has completely developed and has fully calcified, the ameloblasts cease to be arranged in a well-defined layer and can no longer be differentiated from the cells of the stratum intermedium and outer enamel epithelium (Fig. 3-50). These cell layers then form a stratified epithelial covering of the enamel, the so-called reduced enamel epithelium. The function of the reduced enamel epithelium is that of protecting the mature enamel by separating it from the connective tissue until the tooth erupts. If connective tissue comes in contact with the enamel, anomalies may develop. Under such conditions the enamel may be either resorbed or covered by a layer of cementum.

During this phase of the life cycle of ameloblasts the epithelial enamel organ may retract from the cervical edge of the enamel. The adjacent mesenchymal cells may then deposit afibrillar cementum on the enamel surface (Fig. 3-41).

Desmolytic stage. The reduced enamel epithelium proliferates and seems to induce atrophy of the connective tissue separating it from the oral epithelium, so that fusion of the two epithelia can occur (see Chapter 9). It is probable that the epithelial cells elaborate enzymes that are able to destroy connective tissue fibers by desmolysis. Premature degeneration of the reduced enamel epithelium may prevent the eruption of a tooth.

Amelogenesis

On the basis of ultrastructure and composition, two processes are involved in the development of enamel: organic matrix formation and mineralization. Although the inception of mineralization does not await the completion of matrix formation, the two processes will be treated separately.

Formation of the enamel matrix

The ameloblasts begin their secretory activity when a small amount of dentin has been laid down. The ameloblasts lose the projections that had penetrated the basal lamina separating them from the predentin (compare Figs. 3-40, B, and 3-42, A), and islands of enamel matrix are deposited along the predentin (Fig. 3-42, B,). As enamel deposition proceeds, a thin, continuous layer of enamel is formed along the dentin (Figs. 3-37, B, and 3-43). This has been termed the dentinoenamel membrane. Its

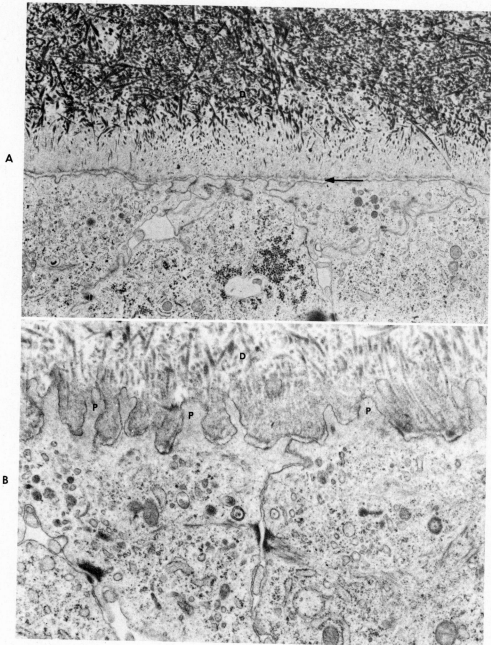

Fig. 3-40. Electron micrographs of distal (secretory) ends of ameloblasts in stage of differentiation shortly before enamel formation begins. **A,** Relatively smooth ameloblast surfaces are separated from predentin *(D)* by basal lamina *(arrow)*. (Approximately ×11,000.) **B,** At slightly later stage, ameloblast cell processes *(P)* have penetrated basal lamina and protrude into predentin *(D)*. (Approximately ×16,000.) (From Kallenbach, E.: Am. J. Anat. **145:**283-317, 1976.)

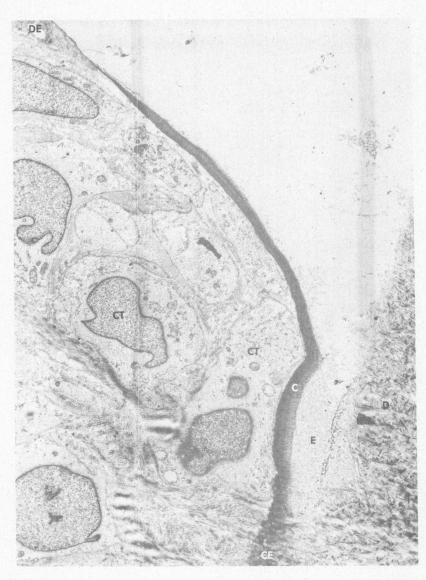

Fig. 3-41. Electron micrograph of cervical region of unerupted human tooth. Dentin matrix, *D,* and remnants of demineralized enamel matrix, *E,* are at right. Afibrillar cementum, apparently of meso-dermal origin, runs through center of figure, *C,* and is continuous with cementum, *CE.* Cells of adjacent connective tissue, *CT,* and retracted end of enamel organ, *DE,* are at left. (Approximately ×6500.) (From Listgarten, M.A.: Arch. Oral Biol. **11:**999, 1966.)

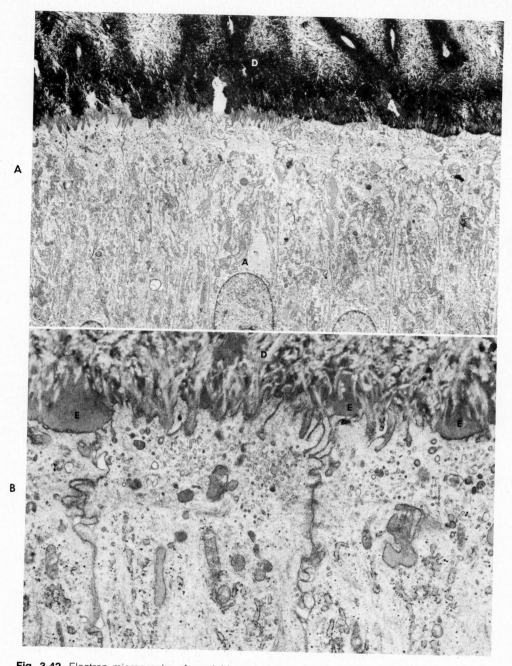

Fig. 3-42. Electron micrographs of ameloblasts in later stage of differentiation than those in Fig. 3-40. **A,** Ameloblasts *(A)* at left still retain their processes, while those at right, at a slightly later stage of differentiation, have smooth surfaces facing dentin *(D)*. (Approximately ×2700.) **B,** Higher magnification of region similar to that in **A.** In this decalcified section, dentin *(D)* is pale and islands of enamel matrix *(E)* are more darkly stained. (Approximately ×12,000.) (From Kallenbach, E.: Am. J. Anat. **145:**283-317, 1976.)

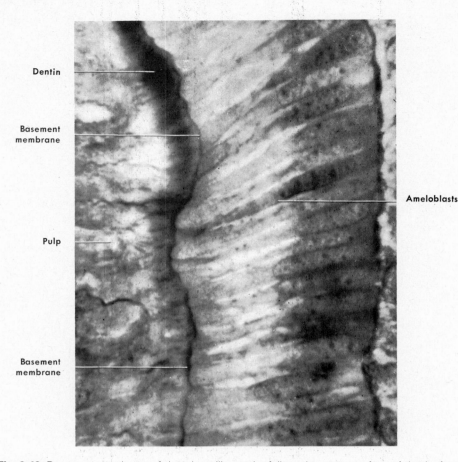

Dentin

Basement membrane

Pulp

Basement membrane

Ameloblasts

Fig. 3-43. Basement membrane of dental papilla can be followed on outer surface of dentin, forming dentinoenamel membrane. (From Orban, B., Sicher, H., and Weinmann, J.P.: J. Am. Coll. Dent. **10:**13, 1943.)

presence accounts for the fact that the distal ends of the enamel rods are not in direct contact with the dentin.

Development of Tomes' processes. The surfaces of the ameloblasts facing the developing enamel are not smooth. There is an interdigitation of the cells and the enamel rods that they produce (Fig. 3-44). This interdigitation is partly a result of the fact that the long axes of the ameloblasts are not parallel to the long axes of the rods (Figs. 3-45 and 3-46). The projections of the ameloblasts into the enamel matrix

have been named Tomes' processes. It was once believed that these processes were transformed into enamel matrix, but more recent electron microscope studies have demonstrated that matrix synthesis and secretion by ameloblasts are very similar to the same processes occurring in other protein-secreting cells. Although Tomes' processes are partly delineated by incomplete septa (Fig. 3-45), they also contain typical secretion granules as well as rough endoplasmic reticulum and mitochondria (Figs. 3-39, 3-47, *B*, and 3-53, *B*).

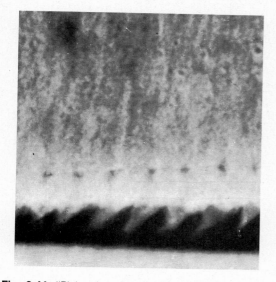

Fig. 3-44. "Picket fence" arrangement of Tomes' processes. Rods are at angle to ameloblasts and Tomes' processes. (From Orban, B., Sicher, H., and Weinmann, J.P.: J. Am. Coll. Dent. **10:**13, 1943.)

Fig. 3-46 is a drawing derived from the electron micrograph in Fig. 3-45. It is clear from this sketch that at least two ameloblasts are involved in the synthesis of each enamel rod. If the surface of developing enamel is examined in the scanning electron microscope, which permits a three-dimensional visualization of the surface, the depressions resulting from the presence of Tomes' processes are quite obvious (Fig. 3-48). One interpretation of the relationships between the keyhole-shaped enamel rods and the roughly hexagonal ameloblasts is indicated in Fig. 3-49. The bulk of the "head" of each rod is formed by one ameloblast, whereas three others contribute components to the "tail" of each rod. According to this interpretation, each rod is formed by four ameloblasts, and each ameloblast contributes to four different rods.

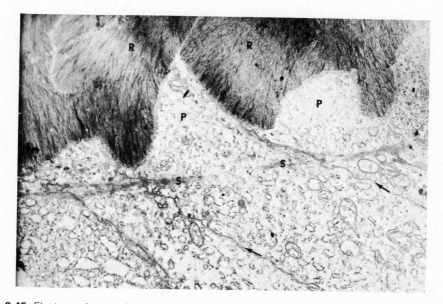

Fig. 3-45. Electron micrograph of ends of ameloblasts and adjacent enamel in developing human deciduous tooth. Positions of ameloblast cell membranes *(arrows)* indicate that cells are nearly perpendicular to long axes of rods, *R.* An incomplete septum, *S,* can be seen, indicating approximate position of Tomes' processes, *P.* (Approximately ×16,000.) (From Rönnholm, E.: J. Ultrastruct. Res. **6:**249, 1962.)

Distal terminal bars. At the time Tomes' processes begin to form, terminal bars appear at the distal ends of the ameloblasts, separating the Tomes' processes from the cell proper (Fig. 3-47, A). Structurally, they are localized condensations of cytoplasmic substance closely associated with thickened cell membranes. They are observed during the enamel-producing stage of the ameloblasts, but their exact function is not known.

Ameloblasts covering maturing enamel. At the light microscope level one can see that the ameloblasts over maturing enamel are considerably shorter than the ameloblasts over incompletely formed enamel (Fig. 3-50). These short ameloblasts have a villous surface near the enamel, and the ends of the cells are packed with mitochondria (Figs. 3-51 and 3-52). This morphology is typical of absorptive cells, and it has been demonstrated that ameloblasts are apparently transporting organic components from the matrix. The fact that organic components as well as water are lost in mineralization is a striking difference between enamel and other mineralized tissues. Over 90% of the initially secreted protein is lost during enamel maturation, and that which remains forms envelopes around individual crystals (Fig. 3-10), although there may be a higher content of organic matter in the area of the prism sheath where the abrupt change in crystal orientation occurs. In the electron micro-

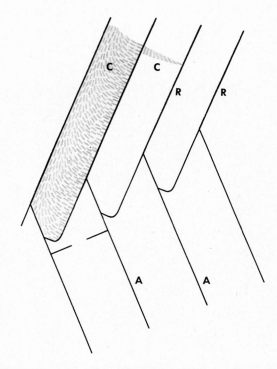

Fig. 3-46. Drawing derived from Fig. 3-45. Dark lines indicate rod boundaries, *R*, and ameloblast cell surfaces, *A*, as well as incomplete septum near distal end of ameloblast at left. Gray lines indicate approximate orientation of apatite crystals, *C*.

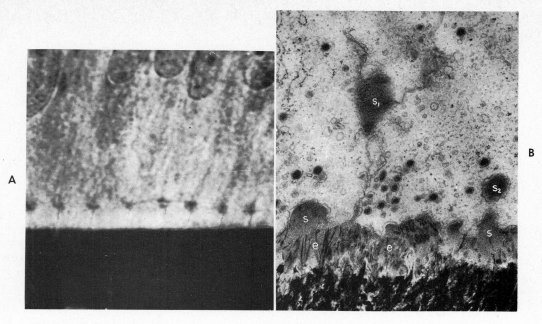

A

B

Fig. 3-47. A, Formation of Tomes' processes and terminal bars as first step in enamel rod formation. Rat incisor. **B,** Electron photomicrograph showing an early stage in formation of enamel in lower incisor of rat. At this stage, dentin (at bottom of photomicrograph) is well developed. Enamel, *e,* appears as a less dense layer on surface of dentin and consists of thin, ribbon-shaped elements running more or less perpendicular to dentinoenamel junction and masses of a less dense stippled material, *s.* Separating enamel from cytoplasm of ameloblasts, which occupies most of upper part of photomicrograph, is ameloblast plasma membrane. Parts of three ameloblasts are shown. In middle of photomicrograph in region bounded by membranes of three ameloblasts lies another mass of stippled material, s_1, while a second mass, s_2, lies at right, surrounded by membrane, but within bounds of ameloblast. Numerous small, membrane-bound granules lie within cytoplasm. Contents of these have same general consistency as stippled material, but rather higher density. It is possible that these represent unsecreted granules of stippled material, which in turn is a precursor of enamel matrix. ($\times 24,000$.) (**A** from Orban, B., Sicher, H., and Weinmann, J.P.: J. Am. Coll. Dent. **10:**13, 1943; **B** from Watson, M.L.: J. Biophys. Biochem. Cytol. **7:**489, 1960.)

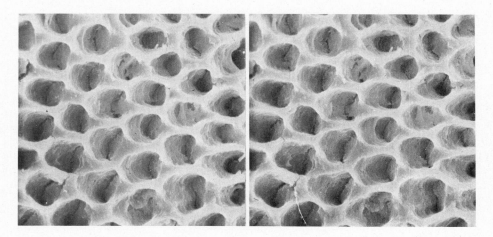

Fig. 3-48. Pair of stereographic scanning electron micrographs of surface of developing human enamel. Great depth of focus of this instrument permits visualization of interdigitated nature of this surface. Depressions were occupied by Tomes' processes, which were stripped away with epithelial enamel organ. (Courtesy Dr. A.R. Boyde, London.)

Fig. 3-49. Drawing illustrating one interpretation of relationships between enamel rods and ameloblasts. Cross sections of ameloblasts are indicated by thin lines arranged in regular hexagonal array. Enamel rods are indicated by thicker curved black lines, outlining keyhole- or paddle-shaped rods. Gray lines indicate approximate orientation of enamel crystals, which are parallel to long axes of rods in their "bodies" and approach a position perpendicular to long axes in "tails." One can see that each rod is formed by four ameloblasts and that each ameloblast contributes to four different rods. (Modified from Boyde, A.: In Stack, M.V., and Fearnhead, R.W., editors: Tooth enamel, Bristol, 1965, John Wright & Sons, Ltd.)

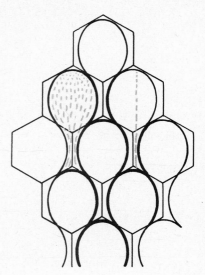

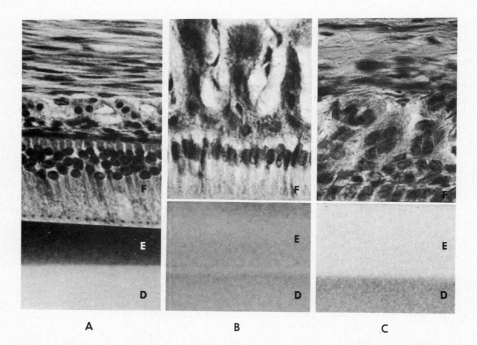

A   B   C

Fig. 3-50. Light micrographs of various stages in life cycle of ameloblasts, *F,* in rat incisor matched with microradiographs of corresponding adjacent enamel, *E,* and dentin, *D.* **A,** Ameloblasts are secreting enamel, which is incompletely formed. Enamel is less radiopaque than dentin, indicating that it is less mineralized. **B,** In area of enamel maturation, ameloblasts are shorter, and enamel matrix is about as heavily mineralized as dentin. **C,** In area in which ameloblasts are in protective stage, enamel is fully mineralized and is much more radiopaque than underlying dentin. (All approximately ×260.)

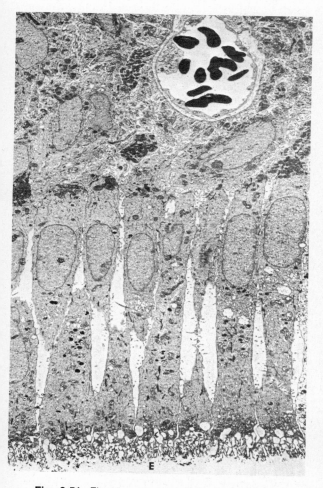

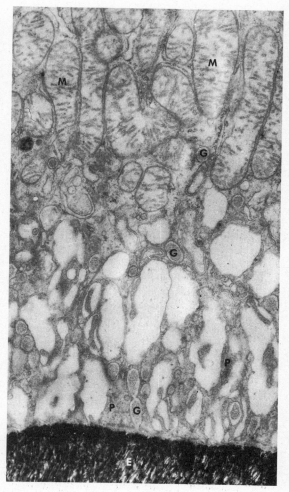

Fig. 3-51. Electron micrograph of ameloblasts during stage of enamel maturation. Enamel has been lost during demineralization. Cells are covered at their surfaces adjoining enamel, *E,* and on their lateral surfaces by numerous microvilli. (Approximately ×3400.) (From Reith, E.J.: J. Biophys. Biochem. Cytol. **9:**825, 1961.)

Fig. 3-52. Higher magnification of electron micrograph of ends of ameloblasts during stage of enamel maturation. Adjacent to enamel, *E,* are elaborate cell processes of ameloblasts, *P,* as well as numerous mitochondria within ameloblast cytoplasm, *M.* Granular material, possibly being resorbed, is seen both between cell processes and within ameloblast cytoplasm, *G.* This structure is typical of resorptive cells. (Approximately ×36,000.) (From Reith, E.J.: J. Cell Biol. **18:**691, 1963.)

scope, several substages can be identified in the transition of ameloblasts from the formative stage through the maturative stage (Fig. 3-53). Shifts are apparent in the cellular organelles from those associated with protein synthesis and secretion to those related to absorption. In addition, a sequence of changes in cell-to-cell contacts and communications between cell layers occurs.

Mineralization and maturation of the enamel matrix

Mineralization of the enamel matrix takes place in two stages, although the time interval between the two appears to be very small. In the first stage an immediate partial mineralization occurs in the matrix segments and the interprismatic substance as they are laid down.

Chemical analyses indicate that the initial influx may amount to 25% to 30% of the eventual total mineral content. It has been shown recently by electron microscopy and diffraction that this first mineral actually is in the form of crystalline apatite (Fig. 3-56, *A*).

The second stage, or *maturation,* is characterized by the gradual completion of mineralization (Fig. 3-50). The process of maturation starts from the height of the crown and progresses cervically (Fig. 3-54). However, at each level, maturation seems to begin at the dentinal end of the rods. Thus there is an integration of two processes: each rod matures from the depth to the surface, and the sequence of maturing rods is from cusps or incisal edge toward the cervical line.

Maturation begins before the matrix has

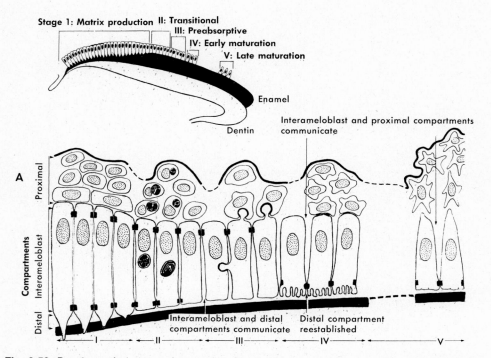

Fig. 3-53. Drawings of electron micrographs of enamel organ of rat incisor. Five substages have been identified from formative to maturative. **A,** Overview of enamel organ.

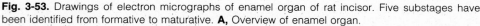

Continued.

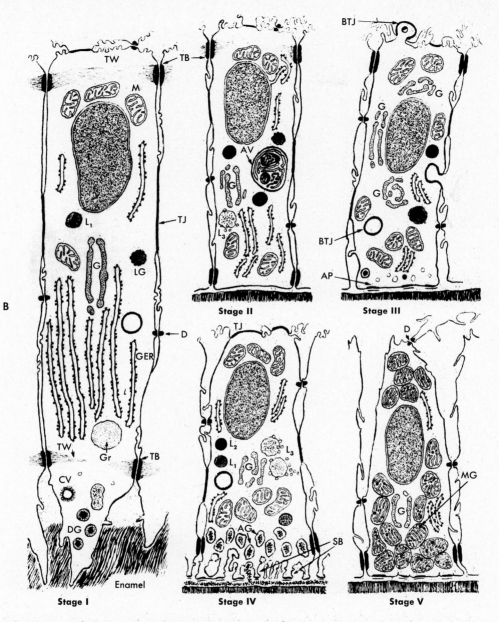

Fig. 3-53, cont'd. B, Individual ameloblasts from five substages. Organelles: *AG,* absorption granules; *AP,* apical contact specialization (hemidesmosomes); *AV,* autophagic vacuoles (lysosomes); *BTJ,* bulb type of contacts; *CV,* coated (absorptive?) vesicles; *D,* desmosomes; *DG,* dense (secretory) granules; *G,* Golgi apparatus; *GER,* granular (rough) endoplasmic reticulum; *Gr,* pale (secretory?) granules; L_1, L_2, L_3, lysosomes; *LG,* lipid granules; *M,* mitochondria; *MG,* mitochondrial granules; *SB,* striated border; *TB,* terminal bars; *TJ,* tight junctions; *TW,* terminal web. (From Reith, E.J.: J. Ultrastruct. Res. **30:**111, 1970.)

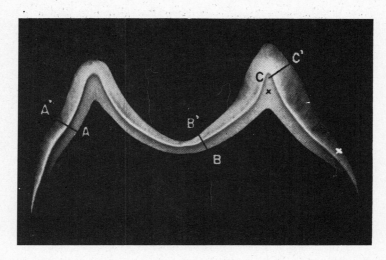

Fig. 3-54. Microradiograph of ground section through developing deciduous molar. From gradation in radiopacity, maturation can be seen to progress from dentinoenamel junction toward enamel surface. Mineralization is more advanced occlusally than in cervical region. Lines *A, B,* and *C* indicate planes in which actual microdensitometric tracings were made. *Black* X, Cusp area. *White* X, Cervical area. (×15.) (From Hammarlund-Essler, E.: Trans. R. Schools Dent., Stockholm and Umea **4:**15, 1958.)

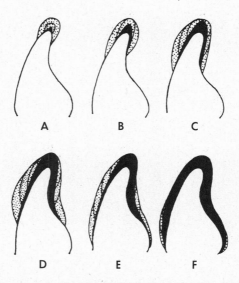

Fig. 3-55. Diagram showing pattern of mineralization of incisor tooth. *Stippled zones,* Consecutive layers of partly mineralized enamel matrix. *Black areas,* Advance of final mineralization during maturation. (From Crabb, H.S.M.: Proc. R. Soc. Med. **52:**118, 1959; and Crabb, H.S.M., and Darling, A.I.: Arch. Oral Biol. **2:**308, 1960.)

reached its full thickness. Thus it is going on in the inner, first-formed matrix at the same time as initial mineralization is taking place in the outer, recently formed matrix. The advancing front is at first parallel to the dentinoenamel junction and later to the outer enamel surface. Following this basic pattern, the incisal and occlusal regions reach maturity ahead of the cervical regions (Fig. 3-55).

At the ultrastructural level, maturation is characterized by growth of the crystals seen in the primary phase (Fig. 3-56, A). The original ribbon-shaped crystals increase in thickness more rapidly than in width (Fig. 3-57). Concomitantly the organic matrix gradually becomes thinned and more widely spaced to make room for the growing crystals. Chemical analysis shows that the loss in volume of the organic matrix is caused by withdrawal of a substantial amount of protein as well as water.

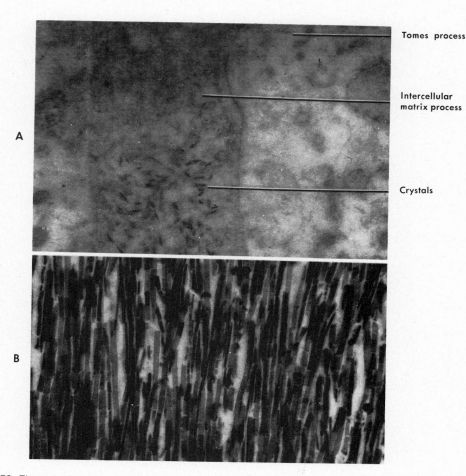

Fig. 3-56. Electron photomicrographs illustrating difference between short, needlelike crystals laid down in newly deposited enamel matrix **(A)**, and long, ribbonlike crystals seen in mature enamel **(B).** (×70,000.)

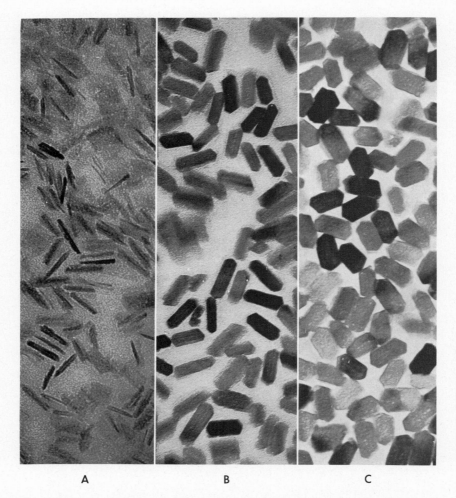

A B C

Fig. 3-57. Electron photomicrographs of transverse sections through enamel rods in rat incisor showing three stages in growth of apatite crystals during enamel maturation. From **A** (recently formed enamel) through **C** (more mature enamel) crystals increase in thickness more rapidly than in width. Spaces between crystals will become even smaller as maturation is completed. ($\times$240,000.) (Modified from Nylen, M.U., Eanes, E.D., and Omnell, K.-Å.: J. Cell Biol. **18:**109, 1963.)

Clinical considerations

Clinical interest in amelogenesis is centered primarily on the perfection of enamel formation. Although there is relatively little the dentist can do directly to alter the course of events in amelogenesis, it may be possible to minimize certain factors believed to be associated with the etiology of defective enamel structure. The principal expressions of pathologic amelogenesis are hypoplasia, which is manifested by pitting, furrowing, or even total absence of the enamel, and hypocalcification, in the form of opaque or chalky areas on normally contoured enamel surfaces. The causes of such defective enamel formation can be generally classified as systemic, local, or genetic. The most common systemic influences are nutritional deficiencies, endocrinopathies, febrile diseases, and certain chemical intoxications. It thus stands to reason that the dentist should exert his or her influence to ensure sound nutritional practices and recommended immunization procedures during periods of gestation and postnatal amelogenesis. Chemical intoxication of the ameloblasts is not prevalent and is limited essentially to the ingestion of excessive amounts of water-borne fluoride. Where the drinking water contains fluoride in excess of 1.5 parts per million, chronic endemic fluorosis may occur as a result of continuous use throughout the period of amelogenesis. In such areas it is important to urge substitution of a water with levels of fluoride (about 1 part per million) well below the threshold for fluorosis, yet optimal with regard to protection against dental caries (see discussion of clinical considerations in section on histology).

Since it has been realized that enamel development occurs in two phases, that is, matrix formation and maturation, developmental disturbances of the enamel can be understood more fully. If matrix formation is affected, enamel hypoplasia will ensue. If maturation is lacking or incomplete, hypocalcification of the enamel results. In the case of hypoplasia a defect of the enamel is found. In the case of hypocalcification a deficiency in the mineral content of the enamel is found. In the latter the enamel persists as enamel matrix and is therefore soft and acid insoluble in routine preparation after formalin fixation.

Hypoplasia as well as hypocalcification may be caused by systemic, local, or hereditary factors. Hypoplasia of systemic origin is termed "chronologic hypoplasia" because the lesion is found in the areas of those teeth where the enamel was formed during the systemic (metabolic) disturbance. Since the formation of enamel extends over a longer period and the systemic disturbance is, in most cases, of short duration, the defect is limited to a circumscribed area of the affected teeth. A single narrow zone of hypoplasia (smooth or pitted) may be indicative of a disturbance of enamel formation during a short period in which only those ameloblasts that at that time had just started enamel formation were affected. Multiple hypoplasia develops if enamel formation is interrupted on more than one occasion.

No specific cause of chronologic hypoplasia has been established as yet. Recent investigations have demonstrated that exanthematous diseases are not so frequently a cause of enamel hypoplasia as was heretofore commonly believed. The more frequent causes are said to be rickets and hypoparathyroidism, but hypoplasia cannot be predicted with any reliability even in the most severe forms of those diseases.

The systemic influences causing enamel hypoplasia are, in the majority of cases, active during the first year of life. Therefore the teeth most frequently affected are the incisors, canines, and first molars. The upper lateral incisor is sometimes found to be unaffected because its development starts later than that of the other teeth mentioned.

Local factors affect single teeth, in most cases only one tooth. If more than one tooth is affected by local hypoplasia, the location of the defects shows no relation to chronology of development. The cause of local hypoplasia may be an infection of the pulp with subsequent infection of the periapical tissues of a deciduous tooth if the irritation occurred during the period of enamel formation of its permanent successor.

The hereditary type of enamel hypoplasia is probably a generalized disturbance of the ameloblasts. Therefore the entire enamel of all the teeth, deciduous as well as permanent, is affected rather than merely a beltlike zone of the enamel of a group of teeth, as in systemic cases. The anomaly is transmitted as a mendelian dominant character. The enamel of such teeth is so thin that is cannot be noticed clinically or in radiographs. The crowns of the teeth of affected family members are yellow-brown, smooth, glossy, and hard, and their shape resembles teeth prepared for jacket crowns.

An example of systemic hypocalcification of the enamel is the so-called mottled enamel. A high fluoride content in the water is the cause of the deficiency in calcification. Fluoride hypocalcification is endemic; that is, it is limited in its distribution to definite areas in which the drinking water contains more than 1 part of fluoride per 1 million parts of water. It has been demonstrated that a small amount of fluoride (about 1 to 1.2 parts per million) reduces susceptibility to dental caries without causing mottling. For this reason many communities are adding small quantities of fluoride to the community water supplies.

The same local causes that might affect the formation of the enamel can disturb maturation. If the injury occurs in the formative stage of enamel development, hypoplasia of the enamel will result. An injury during the maturation stage will cause a deficiency in calcification.

The hereditary type of hypocalcification is characterized by the formation of a normal amount of enamel matrix that, however, does not fully mature. Such teeth, if investigated before or shortly after eruption, show a normal shape. Their surfaces do not have the luster of normal enamel but appear dull. The enamel is opaque. The hypocalcified soft enamel matrix is soon discolored, abraded by mastication, or peeled off in layers. When parts of the soft enamel are lost, the teeth show an irregular, rough surface. When the enamel is altogether lost, the teeth are small and brown, and the exposed dentin is extremely sensitive. In a rare hereditary disturbance of the enamel organ called odontodysplasia, both the apposition and maturation of the enamel are disturbed. Such teeth have irregular, "moth-eaten," poorly calcified enamel.

The discoloration of teeth from administration of tetracyclines during childhood is a very common clinical problem. Whereas usually this discoloration is because of deposition of tetracycline in the dentin, a small amount of the drug may be deposited in the enamel. In mild cases, the use of some of the newly developed surface-binding restorative materials can produce good esthetic results.

REFERENCES
Structure

Arnold, F.A., Jr.: Grand Rapids fluoridation study—results pertaining to the eleventh year of fluoridation, Am. J. Public Health **47:**539, 1957.

Bartelstone, H.J., Mandel, I.D., Oshry, E., and Seidlin, S.M.: Use of radioactive iodine as a tracer in the study of the physiology of the teeth, Science **106:**132, 1947.

Bergman, G., Hammerlund-Essler, E., and Lysell, L.: Studies on mineralized dental tissues. XII. Microradiographic study of caries in deciduous teeth, Acta Odontol. Scand. **16:**113, 1958.

Beust, T.: Morphology and biology of the enamel tufts with remarks on their relation to caries, J. Am. Dent. Assoc. **19:**488, 1932.

Bhussry, B.R., and Bibby, B.G.: Surface changes in enamel, J. Dent. Res. **36:**409, 1957.

Bibby, B.G., and Van Huysen, G.: Changes in the enamel surfaces; a possible defense against caries, J. Am. Dent. Assoc. **20:**828, 1933.

Bodecker, C.F.: Enamel of the teeth decalcified by the celloidin decalcifying method and examined by ultraviolet light, Dent. Rev. **20:**317, 1906.

Bodecker, C.F.: The color of the teeth as an index of their resistance to decay, Int. J. Orthod. **19:**386, 1933.

Brabant, H., and Klees, L.: Histological contribution to the study of lamellae in human dental enamel, Int. Dent. J. **8**:539, 1958.

Brudevold, F., and Söremark, R.: Chemistry of the mineral phase of enamel. In Miles, A.E.W., editor: Structural and chemical organization of teeth, vol. II, New York, 1967, Academic Press, Inc.

Burgess, R.C., Nikoforuk, G., and Maclaren, C.: Chromatographic studies of carbohydrate components in enamel, Arch. Oral Biol. **1**:8, 1960.

Chase, S.W.: The number of enamel prisms in human teeth, J. Am. Dent. Assoc. **14**:1921, 1927.

Crabb, H.S., and Darling. A.I.: The pattern of progressive mineralization in human dental enamel, Int. Ser. Monogr. Oral Biol. **2**:1, 1962.

Decker, J.D.: Fixation effects on the fine structure of enamel crystal-matrix relationships, J. Ultrastruct. Res. **44**:58, 1973.

Eastoe, J.E.: In Stack, M.V., and Fearnhead, R.W., editors: Tooth enamel, Bristol, 1965, John Wright & Sons, Ltd.

Eastoe, J.E.: Organic matrix of tooth enamel, Nature **187**:411, 1960.

Eggert, F. M., Allen G.A., and Burgess, R.C.: Amelogenins. Purification and partial characterization of proteins from developing bovine dental enamel. J. Biochem. **131**:471, 1973.

Engel, M.B.: Glycogen and carbohydrate-protein complex in developing teeth of the rat, J. Dent. Res. **27**:681, 1948.

Fincham, A.G., Burkland, G.A., and Shapiro, I.M.: Lipophilia of enamel matrix. A chemical investigation of the neutral lipids and lipophilic proteins of enamel, Calcif. Tissue Res. **9**:247, 1972.

Frank, R.M., and Brendel, A.: Ultrastructure of the approximal dental plaque and the underlying normal and carious enamel, Arch. Oral. Biol. **11**:883, 1966.

Frank, R.M., Sognnaes, R.F., and Kern, R.: In Sognnaes, R.F., editor: Calcification in biological systems, Washington, D.C., 1960, American Association for the Advancement of Science.

Frazier, P.D.: Adult human enamel: an electron microscopic study of crystallite size and morphology, J. Ultrastruct. Res. **22**:1, 1968.

Glas, J.E., and Omnell, K.A.: Studies on the ultrastructure of dental enamel, J. Ultrastruct. Res. **3**:334, 1960.

Glimcher, M.J., Bonar, L.C., and Daniel, E.J.: The molecular structure of the protein matrix of bovine dental enamel, J. Mol. Biol. **3**:541, 1961.

Gottlieb, B.: Dental caries, Philadelphia, 1947, Lea & Febiger.

Gray, J.A., Schweizer, H.C., Rosevear, F.B., and Broge, R.W.: Electron microscopic observations of the differences in the effects of stannous fluoride and sodium fluoride on dental enamel, J. Dent. Res. **37**:638, 1958.

Gustafson, A.-G.: A morphologic investigation of certain variations in the structure and mineralization of human dental enamel, Odont. Tidskr. **67**:361, 1959.

Gustafson, G.: The structure of human dental enamel, Odontol. Tidskr. **53**(suppl.), 1945.

Gustafson, G., and Gustafson, A.-G.: Human dental enamel in polarized light and contact microradiography, Acta Odontol. Scand. **19**:259, 1961.

Gustafson, G., and Gustafson, A.-G.: Micro-anatomy and histochemistry of enamel. In Miles, A.E.W., editor: Structural and chemical organization of teeth, vol. II, New York, 1967, Academic Press, Inc.

Gwinnett, A.J.: The ultrastructure of the "prismless" enamel of deciduous teeth, Arch. Oral. Biol. **11**:1109, 1966.

Gwinnett, A.J.: The ultrastructure of the "prismless" enamel of permanent human teeth, Arch. Oral Biol. **12**:381, 1967.

Gwinnett, A.J.: Human prisimless enamel and its influence on sealant penetration, Arch. Oral Biol. **18**:441, 1973.

Helmcke, J.-G.: Ultrastructure of enamel. In Miles, A.E.W., editor: Structural and chemical organization of teeth, vol. II, New York, 1967, Academic Press, Inc.

Hinrichsen, C.F.L., and Engel, M.B.: Fine structure of partially demineralized enamel, Arch. Oral Biol. **11**:65, 1966.

Hodson, J.J.: An investigation into the microscopic structure of the common forms of enamel lamellae with special reference to their origin and contents, Oral Surg. **6**:305, 1953.

Houver, G., and Frank R.M.: Ultrastructural significance of histochemical reactions on the enamel surface of erupted teeth, Arch. Oral Biol. **12**:1209, 1967.

Leach, S.A., and Saxton, C.A.: An electron microscopic study of the acquired pellicle and plaque formed on the enamel of human incisors, Arch. Oral Biol. **11**:1081, 1966.

Listgarten, M.A.: Phase-contrast and electron microscopic study of the junction between reduced enamel epithelium and enamel in unerupted human teeth, Arch. Oral Biol. **11**:999, 1966.

Listgarten, M.A.: Electron microscopic study of the gingivo-dental junction of man, Am. J. Anat. **119**:147, 1966.

Meckel, A.H.: The formation and properties of organic films on teeth, Arch. Oral Biol. **10**:585, 1965.

Meckel, A.H., Griebstein, W.J., and Neal, R.J.: Structure of mature human dental enamel as observed by electron microscopy, Arch. Oral Biol. **10**:775, 1965.

Muhler, J.C.: Present status of topical fluoride therapy, J. Dent. Child. **26**:173, 1959.

Muhler, J.C., and Radike, A.W.: Effect of a dentifrice containing stannous fluoride on adults. II. Results at the end of two years of unsupervised use, J. Am. Dent. Assoc. **55**:196, 1957.

Nikiforuk, G., and Sognnaes, R.F.: Dental enamel, Clin. Orthop. **47**:229, 1966.

Orban, B.: Histology of enamel lamellae and tufts, J. Am. Dent. Assoc. **15**:305, 1928.

Osborn, J.W.: Three-dimensional reconstructions of enamel prisms, J. Dent. Res. **46**:1412,1967.

Osborn, J.W.: Directions and interrelationship of prisms in cuspal and cervical enamel of human teeth, J. Dent. Res. **47**:395, 1968.

Osborn, J.W.: A relationship between the striae of Retzius and prism directions in the transverse plane of the human tooth, Arch. Oral Biol. **16**:1061, 1971.

Pautard, F.G.E.: An x-ray diffraction pattern from human enamel matrix, Arch. Oral Biol. **3**:217, 1961.

Piez, K.A.: The nature of the protein matrix of human enamel, J. Dent. Res. **39**:712, 1960.

Piez, K.A., and Likins, R.C.: The nature of collagen. II. Vertebrate collagens, In Sognnaes, R.F., editor: Calcification in biological systems, Washington, D.C., 1960, American Association for the Advancement of Science, p. 411.

Ripa, L.W., Gwinnett, A.J., and Buonocore, M.G.: The "prismless" outer layer of deciduous and permanent enamel, Arch. Oral Biol. **11**:41, 1966.

Robinson, C., Weatherell, J.A., and Hallsworth, S.A.: Variation in composition of dental enamel within thin ground tooth sections, Caries Res. **5**:44, 1971.

Rönnholm, E.: The amelogenesis of human teeth as revealed by electron microscopy. II. The development of the enamel crystallites, J. Ultrastruct. Res. **6**:249, 1962.

Rushton, M.A.: On the fine contour lines of the enamel of milk teeth, Dent. Rec. **53**:170, 1933.

Schmidt, W.J., and Keil, A.: Die gesunden und die erkrankten Zahngewebe des Menschen und der Wirbeltiere im Polarisationsmikroskop (Normal and pathological tooth structure of humans and vertebrates in the polarization microscope), Munich, West Germany, 1958, Carl Hanser Verlag.

Schour, I.: The neonatal line in the enamel and dentin of the human deciduous teeth and first permanent moler, J. Am. Dent. Assoc. **23**:1946, 1936.

Schour, I., and Hoffman, M.M.: Studies in tooth development. I. The 16 microns rhythm in the enamel and dentin from fish to man, J. Dent. Res. **18**:91, 1939.

Scott, D.B.: The electron microscopy of enamel and dentin. J. New York Acad. Sci. **60**:575, 1955.

Scott, D.B.: The crystalline component of dental enamel, Fourth International Conference on Electron Microscopy, Berlin, 1960, Springer Verlag.

Scott, D.B., Kaplan, H., and Wyckoff, R.W.G.: Replica studies of changes in tooth surfaces with age, J. Dent. Res. **28**:31, 1949.

Scott, D.B., Ussing, M.J., Sognnaes, R.F., and Wyckoff, R.W.G.: Electron microscopy of mature human enamel, J. Dent. Res. **31**:74, 1952.

Scott, D.B., and Wyckoff, R.W.G.: Typical structures on replicas of apparently intact tooth surfaces, Public Health Rep. **61**:1397, 1946.

Scott, D.B., and Wyckoff, R.W.G.: Studies of tooth surface structure by optical and electron microscopy, J. Am. Dent. Assoc. **39**:275, 1959.

Selvig, K.A.: The crystal structure of hydroxyapatite in dental enamel as seen with the electron microscope, J. Ultrastruct. Res. **41**:369, 1972.

Shaw, J.H.: Fluoridation as a public health measure, Washington, D.C., 1954, American Association for the Advancement of Science.

Skillen, W.C.: The permeability of enamel in relation to stain, J. Am. Dent. Assoc. **11**:402, 1924.

Sognnaes, R.F.: The organic elements of the enamel. III. The pattern of the organic framework in the region of the neonatal and other incremental lines of the enamel, J. Dent. Res. **28**:558, 1949.

Sognnaes, R.F.: The organic elements of the enamel. IV. The gross morphology and the histological relationship of the lamellae to the organic framework of the enamel, J. Dent. Res. **29**:260, 1950.

Sognnaes, R.F.: Microstructure and histochemical characteristics of the mineralized tisues, J. New York Acad. Sci. **60**:545, 1955.

Sognnaes, R.F., Shaw, J.H., and Bogoroch, R.: Radiotracer studies of bone, cementum, dentin and enamel of rhesus monkeys, Am. J. Physiol. **180**:408, 1955.

Spiers, R.L.: The nature of surface enamel, Br. Dent. J. **107**:209, 1959.

Stack, M.V.: Organic constituents of enamel, J. Am. Dent. Assoc. **48**:297, 1954.

Stack. M.V.: Chemical organization of the organic matrix of enamel, In Miles, A.E.W., editor: Structural and chemical organization of teeth, vol. II, New York, 1967, Academic Press, Inc.

Swancar, J.R., Scott, D.B., and Njemirovskij, Z.: Studies on the structure of human enamel by the replica method, J. Dent. Res. **49**:1025, 1970.

Wainwright, W.W., and Lemoine, F.A.: Rapid diffuse penetration of intact enamel and dentin by carbon[14]-labeled urea, J. Am. Dent. Assoc. **41**:135, 1950.

Warshawsky, H.: A light and electron microscopic study of the nearly mature enamel of rat incisors, Anat. Rec. **169**:559, 1971.

Watson, M.L.: The extracellular nature of enamel in the rat, J. Biophys. Biochem. Cytol. **7**:489, 1960.

Weber, D.F., and Glick, P.L.: Correlative microscopy of enamel prism orientation. Am. J. Anat. **144**:407, 1975.

Yoon, S.H., Brudwold, F., Gardner, D.E., and Smith, F.A.: Distribution of fluoride in teeth from areas with different levels of fluoride in the water supply, J. Dent. Res. **39**:845, 1960.

Development

Allan, J.H.: Investigations into the mineralization pattern of human dental enamel, J. Dent. Res. **38**:1096, 1959.

Allan, J.H.: Maturation of enamel, In Miles, A.E.W., editor: Structural and chemical organization of teeth, vol. I, New York, 1967, Academic Press, Inc.

Angmar-Måsson, B.: A quantitative microradiographic study on the organic matrix of developing humen enamel in relation to the mineral content, Arch. Oral Biol. **16**:135, 1971.

Bawden, J.W., and Wennberg, A.: In vitro study of cellular influence on ^{45}Ca uptake in developing rat enamel, J. Dent. Res. **56**:313, 1977.

Boyde, A.: The structure of developing mamalian dental enamel. In Stack, M.V., and Fearnhead, R.W., editors: Tooth enamel, Bristol, 1965, John Wright & Sons, Ltd.

Boyde, A., and Reith, E.J.: Scanning electron microscopy of the lateral cell surfaces of rat incisor ameloblasts, J. Anat. **122**:603, 1976.

Crabb, H.S.M.: The pattern of mineralization of human dental enamel, Proc. R. Soc. Med. **52**:118, 1959.

Crabb, H.S.M., and Darling, A.I.: The gradient of mineralization in developing enamel, Arch. Oral Biol. **2**:308, 1960.

Deakins, M.: Changes in the ash, water, and organic content of pig enamel during calcification, J. Dent. Res. **21**:429, 1942.

Deakins, M., and Burt, R.L.: The deposition of calcium, phosphorus, and carbon dioxide in calcifying dental enamel, J. Biol. Chem. **156**:77, 1944.

Dean, H.T.: Chronic endemic dental fluorosis, J.A.M.A. **107**:1269, 1936.

Decker, J.D.: The development of a vascular supply to the rat molar enamel organ, Arch. Oral Biol. **12**:453, 1967.

Engel, M.B.: Some changes in the connective tissue ground substance associated with the eruption of the teeth, J. Dent. Res. **30**:322, 1951.

Fearnhead, R.W.: Mineralization of rat enamel, Nature **189**:509, 1960.

Fosse, G.: A quantitative analysis of the numerical density and distributional pattern of prisms and ameloblasts in dental enamel and tooth germs. VII. The numbers of cross-sectioned ameloblasts and prisms per unit area in tooth germs. Acta Odontol. Scand. **26**:573, 1968.

Frank, R.M., and Nalbandian, J.: Ultrastructure of amelogenesis. In Miles, A.E.W., editor: Structural and chemical organization of teeth, vol. I, New York, 1967, Academic Press, Inc.

Garant, P.R., and Gillespie, R.: The presence of fenestrated capillaries in the papillary layer of the enamel organ, Anat. Rec. **163**:71, 1969.

Garant, P.R., and Nalbandian, J.: The fine structure of the papillary region of the mouse enamel organ, Arch. Oral Biol. **13**:1167, 1968.

Garant, P.R., and Nalbandian, J.: Observations on the ultrastructure of ameloblasts with special reference to the Golgi complex and related components, J. Ultrastruct. Res. **23**:427, 1968.

Glick, P.L., and Eisenmann, D.R.: Electron microscopic and microradiographic investigation of a morphologic basis for the mineralization pattern in rat incisor enamel, Ant. Rec. **176**:289, 1973.

Glimcher, M.J., Brickley-Parsons, D., and Levine, P.T.: Studies of enamel proteins during maturation, Calcif. Tissue Res. **24**:259, 1977.

Glimcher, M.J., Friberg, V.A., and Levine, P.T.: The isolation and amino acid composition of the enamel proteins of erupted bovine teeth, Biochem. J. **93**:202, 1964.

Gustafson, A.-G.: A morphologic investigation of certain variations in the structure and mineralization of human dental enamel, Odont. Tidskr. **67**:361, 1959.

Hals, E.: Fluorescence microscopy of developing and adult teeth, Oslo, 1953, Norwegian Academic Press.

Hammarlund-Essler, E.: A microradiographic, microphotometric and x-ray diffraction study of human developing enamel, Trans. R. Schools Dent., Stockholm and Umea **4**:15, 1958.

Irving, J.T.: The pattern of sudanophilia in developing rat molar enamel, Arch. Oral Biol. **18**:137, 1973.

Kallenbach, E.: Fine structure of rat incisor ameloblasts during enamel maturation, J. Ultrastruct. Res. **22**:90, 1968.

Kallenbach, E.: The fine structure of Tomes' process of rat incisor ameloblasts and its relationship to the elaboration of enamel, Tissue Cell **5**:501, 1973.

Kallenbach, E.: Fine structure of rat incisor ameloblasts in transition between enamel secretion and maturation stages, Tissue Cell **6**:173, 1974.

Kallenbach, E.: Fine structure of differentiating ameloblasts in the kitten, Am. J. Anat. **145**:283, 1976.

Kallenbach, E.: Fine structure of ameloblasts in the kitten, Am. J. Anat. **148**:479, 1977.

Kreshover, S.J., and Hancock, J.A., Jr.: The pathogenesis of abnormal enamel formation in rabbits inoculated with vaccinia, J. Dent. Res. **35**:685, 1936.

Listgarten, M.A.: Phase-contrast and electron microscopic study of the junction between reduced enamel epithelium and enamel in unerupted human teeth, Arch. Oral Biol. **11**:99, 1966.

Matthiessen, M.E., and Møllgard, K.: Cell junctions of the human enamel organ, Z. Zellforsch. Mikrosk. Anat. **146**:69, 1973.

Morningstar, C.H.: Effect of infection of the deciduous molar on the permanent tooth germ, J. Am. Dent. Assoc. **24**:786, 1937.

Nylen, M.U., and Scott, D.B.: An electron microscopic study of the early stages of dentinogenesis, Pub. 613, U.S. Public Health Serivce, Washington, D.C., 1958, U.S. Government Printing Office.

Nylen, M.U., and Scott, D.B.: Electron microscopic studies of odontogenesis, J. Indiana State Dent. Assoc. **39**:406, 1960.

Nylen, M.U., Eanes, E.D., and Omnell, K.-A.: Crystal growth in rat enamel, J. Cell. Biol. **18**:109, 1963.

Orban, B., Sicher, H., and Weinmann, J.P.: Amelogenesis (a critique and a new concept), J. Am. Coll. Dent. **10**:13, 1943.

Osborn, J.W.: The mechanism of ameloblast movement: a hypothesis, Calcif. Tissue Res. **5**:344, 1970.

Pannese, E.: Observations on the ultrastructure of the enamel organ. I. Stellate reticulum and stratum intermedium, J. Ultrastruct Res. **4**:372, 1960.

Pannese, E.: Observations on the ultrastructure of the enamel organ. II. Involution of the stellate reticulum, J. Ultrastruct. Res. **5**:328, 1961.

Pannese, E.: Observations on the ultrastructure of the enamel organ. III. Internal and external enamel epithelial, J. Ultrastruct. Res. **6**:186, 1962.

Reith, E.J.: The ultrastructure of ameloblasts during matrix formation and the maturation of enamel, J. Biophys. Biochem. Cytol. **9**:825, 1961.

Reith, E.J.: The ultrastructure of ameloblasts during early stages of maturation of enamel, J. Cell. Biol. **18**:691, 1963.

Reith, E.J., and Butcher, E.O.: Microanatomy and histochemistry of amelogenesis. In Miles, A.E.W., editor: Structural and chemical organization of teeth, vol. I, New York, 1967, Academic Press, Inc.

Reith, E.J., and Cotty, V.F.: The absorptive activity of ameloblasts during the maturation of enamel, Anat. Rec. **157**:577, 1967.

Reith, E.J., and Ross, M.H.: Morphological evidence for the presence of contractile elements in secretory ameloblasts of the rat, Arch. Oral Biol. **18**:445, 1973.

Rönnholm, E.: An electron microscopic study of the amelogenesis in human teeth. I. The fine structure of the ameloblasts, J. Ultrastruct. Res. **6**:229, 1962.

Rönnholm, E.: The amelogenesis of human teeth as revealed by electron microscopy. II. The development of the enamel crystallites, J. Ultrastruct. Res. **6**:249, 1962.

Rönnholm, E.: The amelogenesis of human teeth as revealed by electron microscopy. III. The structure of the organic stroma of human enamel during amelogenesis, J. Ultrastruct. Res. **6**:368, 1962.

Sarnat, B.G., and Schour, I.: Enamel hypoplasia (chronologic enamel aplasia) in relation to systemic disease, J. Am. Dent. Assoc. **28**:1989, 1941; **29**:67, 1942.

Scott, D.B., and Nylen, M.U.: Changing concepts in dental histology, Ann. New York Acad. Sci. **85**:133, 1960.

Scott, D.B., and Nylen, M.U.: Organic-incorganic interrelationships in enamel and dentin—a possible key to the mechanism of caries, Int. Dent. J. **12**:417, 1962.

Scott, D.B., Nylen, M.U., and Takuma, S.: Electron microscopy of developing and mature calcified tissues, Rev. Belg. Sci. Dent. **14**:329, 1959.

Slavkin, H.C., Mino, W., and Bringas, P., Jr.: The biosynthesis and secretion of precursor enamel protein by ameloblasts as visualized by autoradiography after tryptophan administration, Anat. Rec. **185**:289, 1976.

Suga, S.: Amelogenesis—some histological and histochemical observations, Int. Dent. J. **9**:394, 1959.

Travis, D.F., and Glimcher, M.J.: The structure and organization of and the relationship between the organic matrix and the inorganic crystals of embryonic bovine enamel, J. Cell Biol. **23**:447, 1964.

Ussing, M.J.: The development of the epithelial attachment, Acta Odontol, Scand. **13**:123, 1955; reprinted in J. West. Soc. Periodont. **3**:71, 1955.

Wasserman, F.: Analysis of the enamel formation in the continuously growing teeth of normal and vitamin C deficient guinea pigs, J. Dent. Res. **23**:463, 1944.

Watson, M.L.: The extracellular nature of enamel in the rat, J. Biophys. Biochem. Cytol. **7**:489, 1960.

Watson, M.L., and Avery, J.K.: The development of the hamster lower incisor as observed by electron microscopy, Am. J. Anat. **95**:109, 1954.

Weber, D.F., and Eisenmann, D.R.: Microscopy of the neonatal line in developing human enamel, Am. J. Anat. **132**:375, 1971.

Weinmann, J.P.: Developmental disturbances of the enamel, Bur. **43**:20, 1943.

Weinmann, J.P., Svoboda, J.F., and Woods, R.W.: Hereditary disturbances of enamel formation and calcification, J. Am. Dent. Assoc. **32**:397, 1945.

Weinmann, J.P., Wessinger, G.D., and Reed, G.: Correlation of chemical and histological investigations on developing enamel, J. Dent. Res. **21**:171, 1942.

Weinstock, A.: Matrix development in mineralizing tissues as shown by radioautography: formation of enamel and dentin. In Slavkin, H.C., and Bavetta, L.A., editors: Developmental aspects of oral biology, New York, 1972, Academic Press, Inc.

Weinstock, A., and Leblond, C.P.: Elaboration of the matrix glycoprotein of enamel by the secretory ameloblasts of the rat incisor as revealed by radioautography after galactose-^{3}H injection, J. Cell Biol. **51**:26, 1971.

4
DENTIN

The dentin provides the bulk and general form of the tooth and is characterized as a hard tissue with tubules throughout its thickness. Since it forms slightly before the enamel, it determines the shape of the crown, including the cusps and ridges, and the number and size of the roots. As a living tissue it contains within its tubules the processes of the specialized cells, the odontoblasts. Although the cell bodies of the odontoblast are arranged along the pulpal surface of the dentin, the cells are morphologically cells of the dentin, because the odontoblasts produce the dentin as well as the odontoblast processes existing within it. Physically and chemically the dentin closely resembles bone. The main morphologic difference between bone and dentin is that some of the osteoblasts that form bone become enclosed within its matrix substance as osteocytes, whereas the dentin contains only the processes of the cells that form it. Both are considered vital tissue because they contain living protoplasm.

PHYSICAL PROPERTIES

In the teeth of young individuals the dentin usually is light yellowish in color, becoming darker with age. Unlike enamel, which is very hard and brittle, dentin is elastic and subject to slight deformation. It is somewhat harder than bone but considerably softer than enamel. The lower content of mineral salts in dentin renders it more radiolucent than enamel.

CHEMICAL COMPOSITION

Dentin consists of 35% organic matter and water and 65% inorganic material. The organic substance consists of collagenous fibrils and a ground substance of mucopolysaccharides (proteoglycans and glycos aminoglycans). The inorganic component has been shown by X-ray diffraction to consist of hydroxyapatite, as in

101

bone, cementum, and enamel. Each hydroxyapatite crystal is composed of several thousand unit cells. The unit cells have a formula of $3Ca_3(PO_4)_2 \cdot Ca(OH)_2$. The crystals are plate shaped and much smaller than the hydroxyapatite crystals in enamel. Dentin also contains small amounts of phosphates, carbonates, and sulfates. Organic and inorganic substances can be separated by decalcification or incineration. In the process of decalcification the organic constituents can be retained and maintain the shape of the dentin. This is why decalcified teeth and bone can be sectioned and provide clear histologic visualization. The enamel, being over 90% mineral in composition, is lost after decalcification. Organic constituents may be removed from the mineral by incineration or organic chelation.

STRUCTURE

The dentinal matrix of collagen fibers is arranged in a random network. As dentin calcifies, the hydroxyapatite crystals mask the individual collagen fibers. Collagen fibers are only visible at the electron microscopic level.

As indicated earlier, the bodies of the odontoblasts are arranged in a layer on the pulpal surface of the dentin, and only their cytoplasmic processes are included in the tubules in the mineralized matrix. Each cell gives rise to one process, which traverses the predentin and calcified dentin within one tubule and terminates in a branching network at the junction with enamel or cementum. Tubules are found throughout normal dentin and are therefore characteristic of it.

Dentinal tubules. The course of the dentinal tubules follows a gentle curve in the crown, less so in the root, where it resembles an S in shape (Fig. 4-1). Starting at right angles from the pulpal surface, the first convexity of this doubly curved course is directed toward the apex of the tooth. These tubules end perpendicular to the dentinoenamel and dentinocementum junctions. Near the root tip and along

Fig. 4-1. Ground section of human incisor. Observe that course of dentinal tubules is S curved in crown but rather straight at incisal tip and in root.

the incisal edges and cusps the tubules are almost straight. Over their entire lengths the tubules exhibit minute, relatively regular secondary curvatures that are sinusoidal in shape.

The ratio between the outer and inner surfaces of dentin is about 5:1. Accordingly, the tubules are farther apart in the peripheral layers and are more closely packed near the pulp (Fig. 4-2). In addition, they are larger in diameter near the pulpal cavity (3 to 4 μm) and smaller at their outer ends (1 μm). The ratio

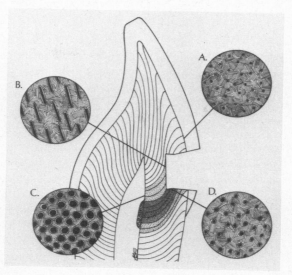

Fig. 4-2. Diagram illustrating the curvature, size, and distance between dentinal tubules in human outer, mid, and inner dentin. Bacterial penetration follows line of least resistance to reach the pulp. Appearance of cut tubules in the floor and walls of a cavity may be different; 1 mm^2 of cavity exposes 30,000 tubules.

Fig. 4-3. A, Secondary branches of human dentinal tubules in mid dentin. Note that branches join neighboring as well as distant tubules.

Continued.

Enamel

Enamel spindle

Terminal branches
of dentinal
tubules

Dentinoenamel
junction

B

Branching
of dentinal
tubules

Branching
of dentinal
tubules

Fig. 4-3, cont'd. B, Secondary branches of human dentinal tubules in outer dentin. Note relationships of these tubules to adjacent tubules and their occasional anastomoses. (Courtesy Dr. Gerrit Bevelander, Houston.)

between the numbers of tubules per unit area on the pulpal and outer surfaces of the dentin is about 4:1. Near the pulpal surface of the dentin the number per square millimeter varys between 50,000 and 90,000. There are more tubules per unit area in the crown than in the root. The dentinal tubules have lateral branches throughout dentin, which are termed canaliculi or microtubules. These canaliculi are 1 μm or less in diameter and originate more or less at right angles to the main tubule (Figs. 4-3, *A* and *B*). Some of them enter adjacent or distant tubules while others end in the intertubular dentin. A few dentinal tubules extend through the dentinoenamel junction into the enamel for several millimeters. These are termed *enamel spindles.*

Peritubular dentin. The dentin that immediately surrounds the dentinal tubules is termed *peritubular dentin.* This dentin forms the walls

of the tubules in all but the dentin near the pulp. It is more highly mineralized (about 9%) than intertubular dentin. It is twice as thick in outer dentin (approximately 0.75 μm) than in inner dentin (0.4 μm). By its growth, it constricts the dentinal tubules to a diameter of 1 μm near the dentinoenamel junction. Studies with soft roentgen rays and with electron microscopy show the increased mineral density in the intertubular dentin (Fig. 4-4). A very delicate organic matrix has been demonstrated in this dentin that along with the mineral is lost after decalcification (Fig. 4-4, *B*). After decalcification the odontoblast process appears to be surrounded by an empty space. In decalcified dentin visualized with a light microscope the tubule diameter will therefore appear similar in inner and outer dentin because of the loss of the peritubular dentin. When peritubular dentin is visualized ultrastructurally in a calcified

Peritubular
dentin
Odontoblastic
process

Peritubular
dentin

Odontoblastic
process

A

B

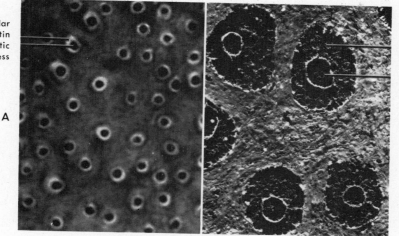

Fig. 4-4. Microscopic appearance of peritubular dentin. **A,** Undermineralized ground section of soft roentgen ray showing increased mineral density in peritubular zone. (×1000.) **B,** Electron micrograph of demineralized section of dentin showing loss of mineral in peritubular zone. Organic matrix in peritubular zone is sparse.

section of a tooth, the densely mineralized peritubular dentin appears structurally different than the intertubular dentin. The collagen fibers in the tubule wall are masked in peripheral dentin (Fig. 4-5, A). A comparison of the tubule wall in inner and outer dentin is shown in Figs. 4-5 and 4-6. Several investigators believe the calcified tubule wall has an inner organic lining termed the *lamina limitans*. This is described as a thin organic membrane, high in glucosamine glycan (GAG) and similar to the lining of lacunae in cartilage and bone. Other investigators believe this lining in the tubules is absent or limited and that instead only the cell wall of the odontoblast is present there.

Intertubular dentin. The main body of dentin is composed of intertubular dentin. It is located between the dental tubules or, more specifically, between the zones of peritubular dentin. Although it is highly mineralized, this matrix, like bone and cementum, is retained after decalcification, whereas peritubular dentin is not.

About one half of its volume is organic matrix, specifically collagen fibers, which are randomly oriented around the dentinal tubules (Figs. 4-5 and 4-6). The fibrils range from 0.5 to 0.2 μm in diameter and exhibit crossbanding at 64 μm (640 Å) intervals, which is typical for collagen (Fig. 4-6, A). Hydroxyapatite crystals, which average 0.1 μm in length, are formed along the fibers with their long axes oriented parallel to the collagen fibers.

Predentin. The predentin is located adjacent to the pulp tissue and is 2 to 6 μm wide, depending on the activity of the odontoblast. It is the first-formed dentin and is not mineralized (Figs. 4-7 and 4-2). As the collagen fibers undergo mineralization at the predentin-dentin front, the predentin then becomes dentin and a new layer of predentin forms circumpulpally.

Odontoblast process. The odontoblast processes are the cytoplasmic extensions of the odontoblasts. The odontoblasts reside in the peripheral pulp at the pulp-predentin border

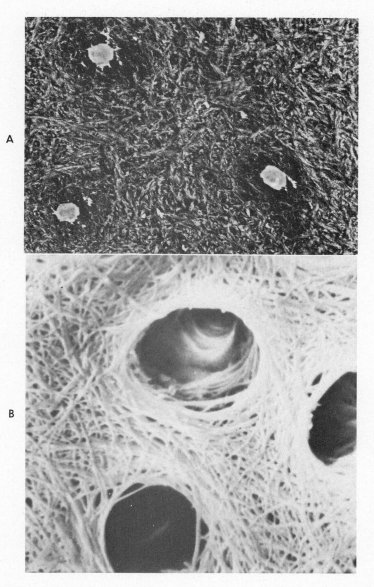

Fig. 4-5. A, Cross section of undecalcified peripheral human dentin, showing crisscross arrangement of collagen matrix fibers. Observe the more densely calcified peritubular dentin. **B,** Scanning electron microscope picture of pulpal surface of dentin illustrating random arrangement of calcifying collagen fibers of matrix surrounding the dentinal tubules. (×15,000.) (Courtesy A. Boyde, London.)

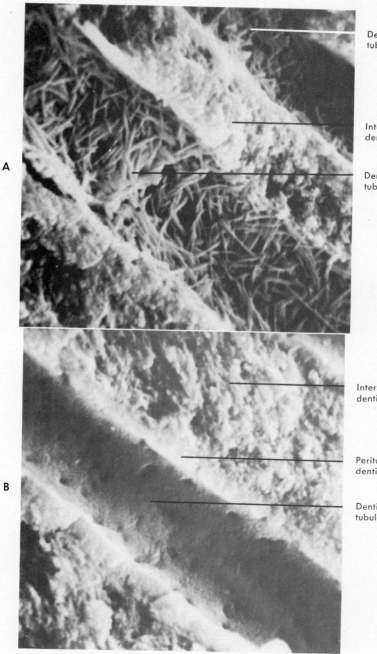

Dentinal
tubules

Intertubular
dentin

Dentinal
tubules

Intertubular
dentin

Peritubular
dentin

Dentinal
tubule

Fig. 4-6. A, Dentinal tubule representative of inner dentin near formative front as seen by scanning electron microscopy. Collagen fibers are evident, composing the walls of the dentinal tubules. (×18,000.) **B,** Same dentinal tubule as in **A,** further peripheral in calcified dentin viewed by scanning electron microscopy. Peritubular dentin masks collagen fibers in the tubule wall. Observe numerous side branches (canaliculi) of dentinal tubule. (×15,000.) (From Boyde, A.: Beitr Electronmikroskop Direktabb. Oberfl. **1**[S]:213-222, 1968.)

and their processes extend into the dentinal tubules (Fig. 4-7). The processes are largest in diameter near the pulp (3 to 4 μm) and taper to approximately 1 μm further into the dentin. The odontoblast cell bodies are approximately 7 μm in diameter and 40 μm in length. Consequently the processes narrow to about half the size of the cell as they enter the tubules (Fig. 4-7). There is disagreement among investigators whether the odontoblast processes extend through the thickness of mature human dentin. Good evidence is shown by transmission electron microscopy that dentinal tubules

200 to 300 μm from the pulp contain processes (Fig. 4-8, *A*). Other investigators, using scanning electron microscopy, have shown what appear to be processes at the dentinoenamel junctions (Fig. 4-8, *B*). Recently cryofractured human teeth revealed the odontoblast process to extend to the dentinoenamel junction (Fig. 4-9). The initial group of investigators believe the findings in Fig. 4-8 and 4-9 represent the organic lining membrane of the tubule (lamina limitans) and not the living process of the odontoblast. Further investigations using immunofluorescent techniques revealed tubulin (an in-

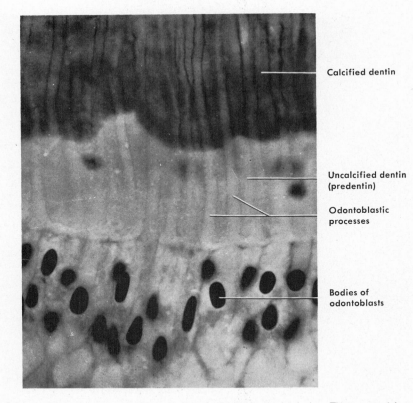

Calcified dentin

Uncalcified dentin
(predentin)

Odontoblastic
processes

Bodies of
odontoblasts

Fig. 4-7. Odontoblast processes (Tomes' fibers) within dentinal tubules. They extend from the cell body below at the pulp-predentin junction into the dentin above.

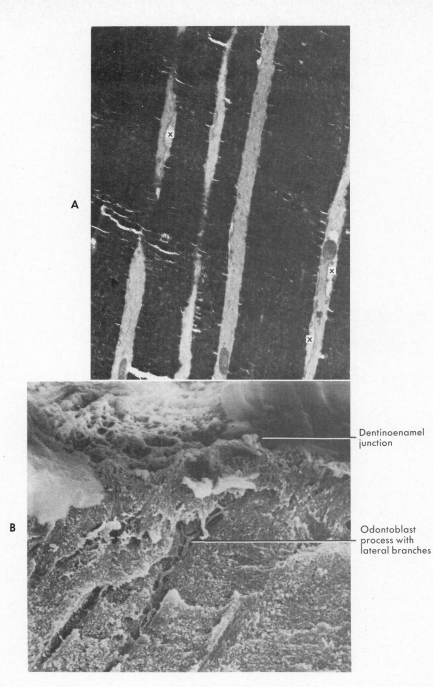

Fig. 4-8. A, Transmission electron micrograph of odontoblast processes in dentin tubules approximately 200 to 300 μm from the pulp. These processes contain microfilaments, a few vesicles, and an occasional mitochondrion enclosed in the plasma membrane of the process. (Calcified section; ×6000.) **B,** Scanning electron micrograph of presumed odontoblast process in dentinal tubule at the dentinoenamel junction. Side branches of the process are seen entering the canaliculi.

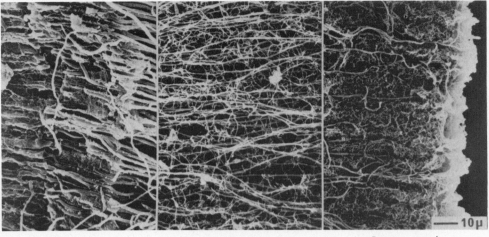

Near pulp   Mid-dentin   Dentinoenamel
                                                            junction

Fig. 4-9. Low magnification scanning electron micrograph of the human odontoblast processes in intact crown dentin. The odontoblasts and their processes are seen in the pulp predentin *(left)* extending through mid-dentin *(center),* and reaching the dentinoenamel junction *(right).* (× 1000.) (Courtesy Dr. Toshimoto Yamada.)

tracellular protein of microtubules) throughout the thickness of dentin (Fig. 4-10). It is appropriate to consider that some odontoblast processes traverse the thickness of dentin. In other areas a shortened process may be characteristic in tubules that are narrow or obliterated by mineral deposit.

The odontoblast process is composed of microtubules of 20 μm (200 to 250 Å) in diameter and small filaments 5 to 7.5 μm (50 to 75 Å) in diameter. Occasionally mitochondria, dense bodies resembling lysosomes, microvesicles, and coated vesicles that may open to the extracellular space are also seen (Fig. 4-8, *A*). The odontoblast processes divide near the dentinoenamel junction and may indeed extend into enamel in the *enamel spindles.* Periodically along the course of the processes side branches appear that extend laterally into adjacent tubules (Fig. 4-9).

PRIMARY DENTIN

Mantle dentin is the name of the first-formed dentin in the crown underlying the dentinoenamel junction. It is thus the outer or most peripheral part of the primary dentin and is about 20 μm thick. The fibrils formed in this zone are perpendicular to the dentinoenamel junction and the organic matrix is composed of collagen fibrils. It is thus the area of initial dentin matrix formation (see Fig. 4-16).

Circumpulpal dentin forms the remaining primary dentin or bulk of the tooth. It is the circumpulpal dentin that represents all of the dentin formed prior to root completion. The collagen fibrils in circumpulpal dentin are much smaller in diameter (0.05 μm) and are more closely packed together. The circumpulpal dentin may contain slightly more mineral than mantle dentin.

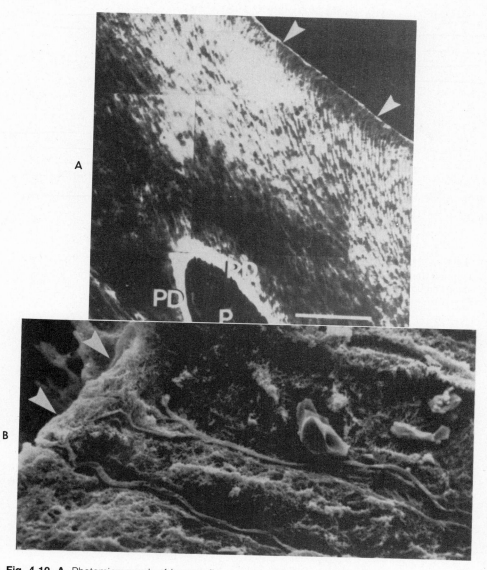

Fig. 4-10. A, Photomicrograph of immunofluorescence labeling of tubulin. This is a subunit protein of the microtubules present in the odontoblast process. *P,* pulp, *Pd,* predentin, *arrows,* dentinoenamel junction (DEJ). **B,** Scanning EM picture of odontoblast processes at DEJ *(arrowheads).* One of the processes displays terminal branching near the DEJ. (Courtesy Drs. J.E. Aubin, A.R. Ten Cate, and S. Pitaru, Medical Research Council Group in Periodontal Physiology and Faculty of Dentistry.)

SECONDARY DENTIN

Secondary dentin is a narrow band of dentin bordering the pulp and representing that dentin formed after root completion. This dentin contains fewer tubules than primary dentin. There is usually a bend in the tubules where primary and secondary dentin interface (Fig. 4-11).

INCREMENTAL LINES

The incremental lines (von Ebner), or imbrication lines, appear as fine lines or striations in dentin. They run at right angles to the dentinal tubules and correspond to the incremental lines in enamel or bone (Fig. 4-12). These lines reflect the daily rhythmic, recurrent deposition of dentin matrix as well as a hesitation in the daily formative process. The distance between lines varies from 4 to 8 μm in the crown to much less in the root. The daily increment decreases after a tooth reaches functional occlusion. The course of the lines indicates the growth pattern of the dentin.

Occasionally some of the incremental lines are accentuated because of disturbances in the matrix and mineralization process. Such lines are readily demonstrated in ground sections and are known as *contour lines* (Owen), (Fig.

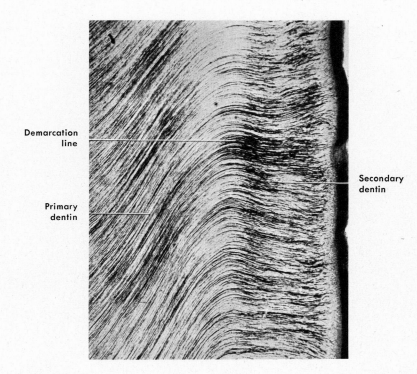

Demarcation line

Primary dentin

Secondary dentin

Fig. 4-11. Dentinal tubules bend sharply as they pass from primary into secondary dentin. The dentinal tubules are somewhat irregular in secondary dentin. Pulpal surface on the right. Ground section human dentin. (Courtesy Dr. Gerrit Bevelander, Houston.)

4-13). Analysis with soft x-ray has shown these lines to represent hypocalcified bands.

In the deciduous teeth and in the first permanent molars, where dentin is formed partly before and partly after birth, the prenatal and postnatal dentin are separated by an accentuated contour line. This is termed the *neonatal line* and is seen in enamel as well as dentin (Fig. 4-14). This line reflects the abrupt change in environment that occurs at birth. The dentin matrix formed prior to birth is usually of better quality than that formed after birth, and the neonatal line may be a zone of hypocalcification.

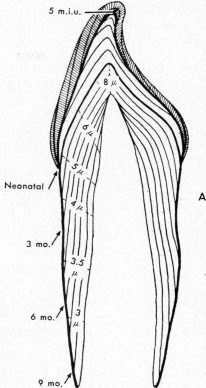

Fig. 4-12. A, Diagram of incremental appositional pattern in dentin in a human deciduous central incisor in a 5-month fetus. In the crown as much as 8 μm/day was deposited and in the root 3 to 4 μm. **B,** Incremental lines in dentin. Also known as imbrication lines or incremental lines of von Ebner. Ground section human tooth. (**A** from Schour, I., and Massler, M.: J. Am. Dent. Assoc. **23:**1946, 1936.)

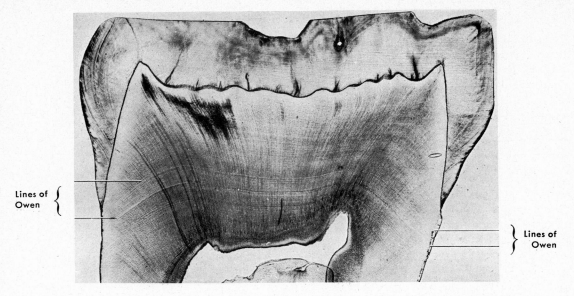

Lines of
Owen

Lines of
Owen

Fig. 4-13. Accentuated incremental lines are termed contour lines (of Owen). Ground section human tooth.

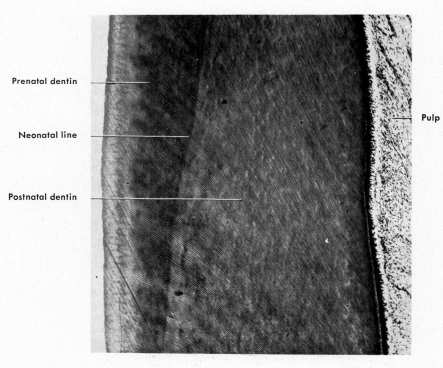

Prenatal dentin

Neonatal line

Postnatal dentin

Pulp

Fig. 4-14. Postnatal formed dentin is separated from prenatal formed dentin by an accentuated incremental line termed the neonatal line. (From Schour, I., Poncher, H.G.: Am. J. Dis. Child. **54:**757, 1937.)

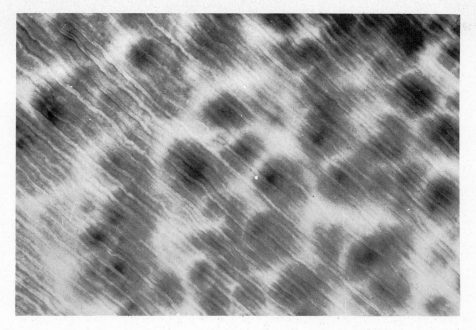

Fig. 4-15. Interglobular dentin as seen in decalcified section of dentin. Dentin tubules pass uninterrupted through uncalcified and hypocalcified areas.

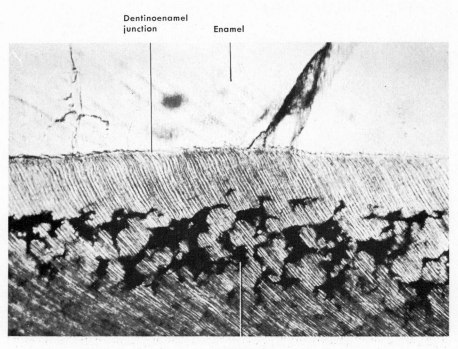

Dentinoenamel junction   Enamel

Interglobular areas

Fig. 4-16. Interglobular dentin appears in the crown dentin a short distance from the dentinoenamel junction. Interglobular spaces are seen in a dry ground section and appear as air-filled spaces that appear black in transmitted light. (Courtesy Dr. Gerrit Bevelander, Houston.)

INTERGLOBULAR DENTIN

Sometimes mineralization of dentin begins in small globular areas that fail to fuse into a homogenous mass. This results in zones of hypomineralization between the globules. These zones are known as *interglobular dentin*. Interglobular dentin forms in the crowns of teeth in the circumpulpal dentin just below the mantle dentin, and it follows the incremental pattern (Fig. 4-15). The dentinal tubules pass uninterruptedly through interglobular dentin, thus demonstrating a defect of mineralization and not of matrix formation (Fig. 4-15). In dry ground sections some of the interglobular dentin may be lost and a space results that appears black in transmitted light (Fig. 4-16). However, spaces in interglobular dentin are not believed to occur naturally.

GRANULAR LAYER

When dry ground sections of the root dentin are visualized in transmitted light, there is a

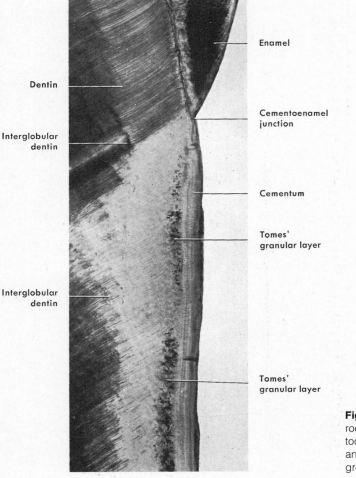

Dentin

Enamel

Interglobular dentin

Cementoenamel junction

Cementum

Tomes' granular layer

Interglobular dentin

Tomes' granular layer

Fig. 4-17. Granular layer (Tomes') appears in root dentin a short distance from the cementodentinal junction. The spaces are air filled and appear black in transmitted light in a ground section.

zone adjacent to the cementum that appears granular (Fig. 4-17). This is known as (Tomes') granular layer. This zone increases slightly in amount from the cementoenamel junction to the root apex and is believed to be caused by a coalescing and looping of the terminal portions of the dentinal tubules. The cause of development of this zone is probably similar to the branching and beveling of the tubules at the dentinoenamel junction. In any case the differentiating odontoblast initially interacts with ameloblasts or the root sheath cells through the basal lamina. In the crown extensive branching of the odontoblast process occurs, and in the root there is branching and coalescing of adjacent processes.

INNERVATION OF DENTIN

Intertubular nerves. Dentinal tubules contain numerous nerve endings in the predentin and inner dentin no farther than 100 to 150 μm from the pulp. Most of these small vesiculated endings are located in tubules in the coronal zone, specifically in the pulp horns. The nerves and their terminals are found in close association with the odontoblast process within the tubule. There may be single terminals (Fig. 4-18) or several dilated and constricted portions (Fig. 4-19). In either case, the nerve endings are packed with small vesicles, either electron dense or lucent, which probably depends on whether there has been discharge of their neurotransmitter substance. In any case, they interdigitate with the odontoblast process, indicating an intimate relationship to this cell. It is believed that most of these are terminal processes of the myelinated nerve fibers of the dental pulp. The primary afferent somatosensory nerves of the dentin and pulp project to the main sensory nucleus of the midbrain.

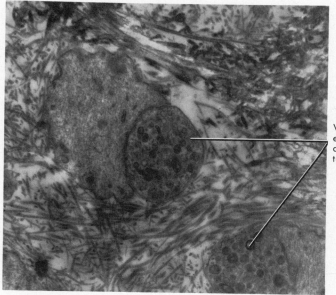

Vesiculated nerve endings in adjacent tubules

Fig. 4-18. Nerve endings in dentinal tubules in region of predentin. The vesiculated endings are seen in adjacent tubules lying in contact with the odontoblast processes.

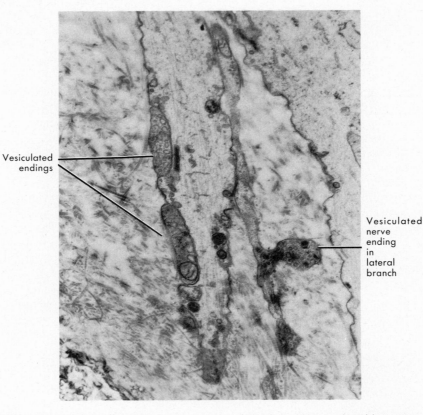

Vesiculated endings

Vesiculated nerve ending in lateral branch

Fig. 4-19. On the left two nerve endings in a dentinal tubule along with an odontoblast process. On the right a nerve ending extends into the side branch of a dentinal tubule in region of predentin. (Transmission electron micrograph.)

Theories of pain transmission through dentin. There are three basic theories of pain conduction through dentin. The first is that of *direct neural stimulation*, meaning that stimuli, in some manner as yet unknown, reach the nerve endings in the inner dentin. There is little scientific support of this theory. The second and most popular theory is the fluid or *hydrodynamic theory*. Various stimuli such as heat, cold, air blast desiccation, or mechanical pressure affect fluid movement in the dentinal tubules. This fluid movement, either inward or outward, stimulates the pain mechanism in the tubules by mechanical disturbance of the nerves closely associated with the odontoblast and its process. Thus these endings may act as mechanoreceptors as they are affected by mechanical displacement of the tubular fluid. The third theory is the *transduction theory*, which presumes that the odontoblast process is the primary structure excited by the stimulus and that the impulse is transmitted to the nerve endings in the inner dentin. This is not a popular theory since there are no neurotransmitter

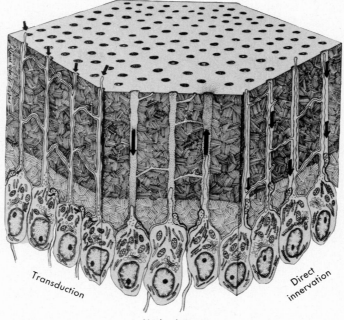

Fig. 4-20. A diagram of the three main explanations of pain transmission through dentin. On the left is shown the *transduction* theory in which the membrane of the odontoblast process conducts an impulse to the nerve endings in the predentin, odontoblast zone, and pulp. In the center is the *hydrodynamic* theory. Stimuli cause an inward or outward movement of fluid in the tubule, which in turn produces movement of the odontoblast and its process. This in turn stimulates the nerve endings. On the right is the *direct conduction* theory in which stimuli directly effect the nerve endings in the tubules.

vesicles in the odontoblast process to facilitate the synapse. The three theories are further explained in Fig. 4-20.

AGE AND FUNCTIONAL CHANGES

Vitality of dentin. Since the odontoblast and its process are an integral part of the dentin, there is no doubt that dentin is a vital tissue. Again, if vitality is understood to be the capacity of the tissue to react to physiologic and pathologic stimuli, dentin must be considered a vital tissue. Dentin is laid down throughout life, although after the teeth have erupted and have been functioning for a short time, dentin-

ogenesis slows, and further dentin formation is at a much slower rate. This is the secondary dentin described earlier in this chapter.

Pathologic effects of dental caries, abrasion, attrition, or the cutting of dentin of operative procedures cause changes in dentin. These are described as the development of *dead tracts*, *sclerosis*, and the addition of *reparative dentin*. The formation of reparative dentin pulpally underlying an area of injured odontoblast processes can be explained on the basis of increased dentinogenic activity of the odontoblasts. The mechanisms underlying the series of events that occur in the development of

Dentin

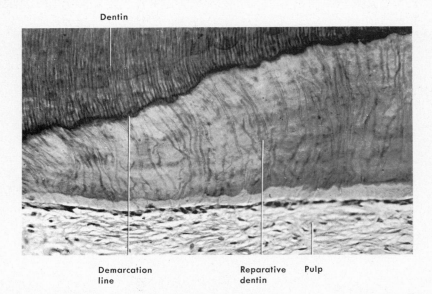

Demarcation   Reparative   Pulp
line   dentin

Fig. 4-21. Reparative dentin stimulated by penetration of caries into dentin. Dentinal tubules are irregular and less numerous than in regular dentin. Decalcified section.

reparative dentin, dead tracts, and sclerosis are not yet fully understood although the histology has been clearly described.

Reparative dentin. If by extensive abrasion, erosion, caries, or operative procedures the odontoblast processes are exposed or cut, the odontoblasts die or, if they live, deposit reparative dentin. The majority of odontoblasts in this situation degenerate, but a few may continue to form dentin. Some of the odontoblasts that are killed are replaced by the migration of undifferentiated cells arising in deeper regions of the pulp to the dentin interface. It is believed that the origin of the new odontoblast is from undifferentiated perivascular cells. Both the damaged and the newly differentiated odontoblasts then begin deposition of reparative dentin. This action to seal off the zone of injury occurs as a healing process initiated by the pulp, resulting in resolution of the inflammatory process and removal of dead cells. The

hard tissue thus formed is best termed reparative dentin although tertiary dentin or response dentin are also used. Reparative dentin is characterized as having fewer and more twisted tubules than normal dentin (Fig. 4-21). Dentin-forming cells are often included in the rapidly produced intercellular substance. In other instances a combination of osteodentin and tubular dentin are seen (Fig. 4-22). It is believed that bacteria, living or dead, or their toxic products, as well as chemical substances from restorative materials, migrate down the tubules to the pulp and stimulate pulpal response leading to reparative dentin formation. All of the events in this process are not yet known.

Dead tracts. In dried ground sections of normal dentin the odontoblast processes disintegrate, and the empty tubules are filled with air. They appear black in transmitted and white in reflected light (Fig. 4-23). Loss of odontoblast processes may also occur in teeth containing vi-

REPARATIVE DENTIN

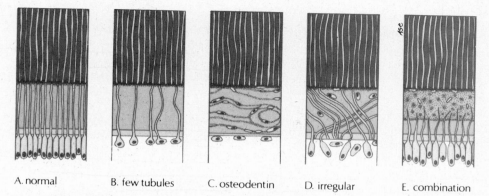

A. normal   B. few tubules   C. osteodentin   D. irregular   E. combination

Fig. 4-22. Diagrammatic illustration of normal *(A)* and reparative dentin *(B to E)*. Reparative dentin contains fewer than normal tubules *(B)*, or it includes cells within its matrix *(C)*, shows irregularly arranged tubules *(D)*, or is a combination of different types *(E)*.

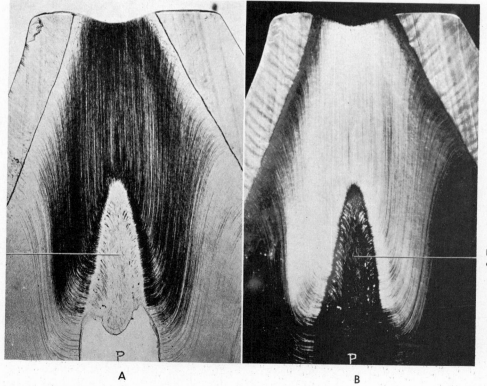

Reparative dentin

Reparative dentin

A   B

Fig. 4-23. Dead tracts in vital tooth caused by attrition and exposure of dentin. **A,** They appear black in transmitted light and, **B,** white in reflected light. Reparative dentin underlies exposed dentinal tubules and, because of the absence of tubules, appears light in transmitted and dark in reflected light. (Courtesy Dr. Gerrit Bevelander, Houston.)

tal pulp as a result of caries, attrition, abrasion, cavity preparation, or erosion (Figs. 4-23 and 4-24). Their degeneration is often observed in the area of narrow pulpal horns (Fig. 4-24) because of crowding of odontoblasts. Again, where reparative dentin seals dentinal tubules at their pulpal ends, dentinal tubules fill with fluid or gaseous substances. In ground sections such groups of tubules may entrap air and appear black in transmitted and white in reflected light. Dentin areas characterized by degenerated odontoblast processes give rise to dead tracts. These areas demonstrate decreased sensitivity and appear to a greater extent in older teeth. Dead tracts are probably the initial step in the formation of sclerotic dentin.

Sclerotic or transparent dentin. Stimuli may not only induce additional formation of reparative dentin but also lead to protective changes in the dentin itself. In cases of caries, attrition, abrasion, erosion, or cavity preparation, sufficient stimuli are generated to cause collagen fibers and apatite crystals to begin appearing in the dentinal tubules. This condition is preva-

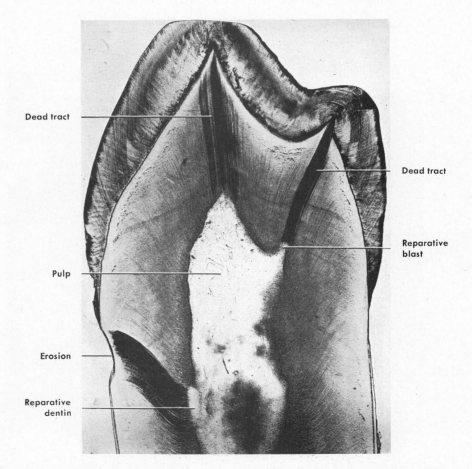

Fig. 4-24. Dead tracts in dentin of vital human tooth caused by crowding and degeneration of odontoblasts in narrow pulpal horns and by exposure of tubules to erosion.

Tubule

Fig. 4-25. Scanning electron micrograph of partially sclerosed dentin. Platelike crystals form a mesh-work occluding tubule lumen. (×9700). (From Lester, K.S., and Boyde, A.: Virchows Arch. [Zell-pathol.] **344:**196, 1968.)

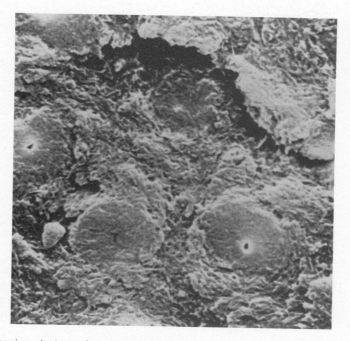

Fig. 4-26. Scanning electron micrograph of fractured cross section of dentin located between at-trited surface and the pulp. Various degrees of closure of the tubule lumen are seen. Complete obliteration *(T)* is seen as well as a minute lumen in other tubules. (×5800). (From Brannstrom, M.: Dentin and pulp in restorative dentistry, London, 1982, Wolfe Medical Publications, Ltd.)

lent in older individuals. In such cases blocking of the tubules may be considered a defensive reaction of the dentin. Apatite crystals are initially only sporadic in a dentinal tubule but gradually fill it with a fine meshwork of crystals (Fig. 4-25). Gradually, the tubule lumen is obliterated with mineral, which appears very much like the peritubular dentin (Fig. 4-26). The refractive indices of dentin in which the

Fig. 4-27. Sclerotic dentin in apical area of root dentin. Absence of the tubules causes transparent appearance of the dentin. (Courtesy Dr. A.E.W. Miles.)

tubules are occluded are equalized, and such areas become *transparent*. Transparent or sclerotic dentin can be observed in the teeth of elderly people, especially in the roots (Fig. 4-27). Sclerotic dentin may also be found under slowly progressing caries (Fig. 4-28). Mineral density is greater in this area of dentin, as shown both by radiography and permeability studies. It appears transparent or light in transmitted light and dark in reflected light (Fig. 4-28).

DEVELOPMENT

Dentinogenesis. Dentinogenesis begins at the cusp tips after the odontoblasts have differentiated and begin collagen production. As the odontoblasts differentiate they change from an ovoid to a columnar shape, and their nuclei become basally oriented at this early stage of development (Fig. 4-29). One or several processes arise from the apical end of the cell in contact with the basal lamina. The length of the odontoblast then increases to approximately 40 μm, although its width remains constant (7 μm). Proline appears in the rough surface endoplasmic reticulum and Golgi apparatus. The proline then migrates into the cell process in dense granules and is emptied into the extracellular collagenous matrix of the predentin. As the cell recedes it leaves behind a single extension, and the several initial processes join into one, which becomes enclosed in a tubule. As the matrix formation continues, the odontoblast process lengthens, as does the dentinal tubule. Initially daily increments of approximately 4

Fig. 4-28. Sclerotic dentin under carious area viewed by, **A,** transmitted light, **B,** reflected light and, **C,** radiograph (grenz ray). Dentinal tubules in dried ground section may be filled with air and appear black in transmitted light **(A),** and white in reflected light, **(B).** Sclerotic dentin with mineral-filled tubules will appear transparent in transmitted light, dark in reflected light, and white in radiographs. (Courtesy Dr. E. Applebaum, New York.)

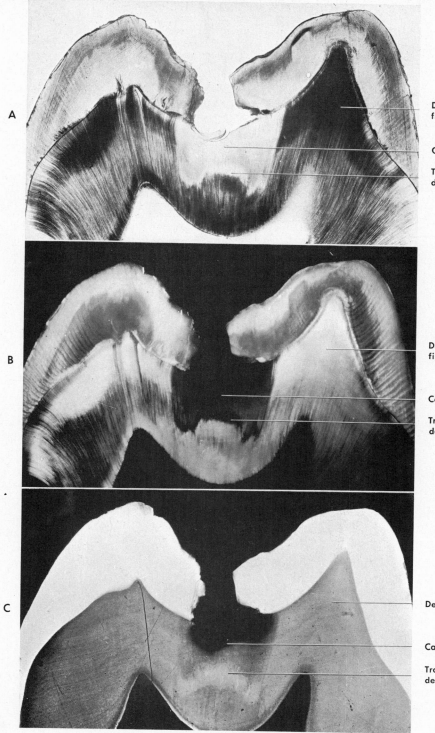

A — Dentinal tubules filled with air

Carious dentin

Transparent dentin

B — Dentinal tubules filled with air

Carious dentin

Transparent dentin

C — Dentin

Carious dentin

Transparent dentin

μm of dentin are formed. This continues until the crown is formed and the teeth erupt and move into occlusion. After this time dentin production slows to about 1 μm/day. After root development is complete, dentin formation may decrease further, although reparative dentin may form at a rate of 4 μm/day for several months after a tooth is restored. Dentinogenesis is a two-phase sequence in that collagen matrix is first formed and then calcified. As each increment of predentin is formed along the pulp border, it remains a day before it is calcified and the next increment of predentin forms (Fig. 4-30). Korff's fibers have been described as the initial dentin deposition along the cusp tips. Because of the argyrophilic reaction (stain black with silver) it was long believed that bundles of collagen formed among the odontoblasts (Fig. 4-31). Recently, ultrastructural studies revealed that the staining is of the ground substance among the cells and not collagen. Consequently, all predentin is formed in the apical end of the cell and along the forming tubule wall (Fig. 4-30). The finding of formation of collagen fibers in the immediate vicinity of the apical ends of the cells is in agreement with the general concept of collagen synthesis in connective tissue and bone. The odontoblasts secrete both collagen and the intercollagen substance proteoglycans.

Mineralization. The mineralization sequence in dentin appears to be as follows. The earliest crystal deposition is in the form of very fine plates of hydroxyapatite on the surfaces of the

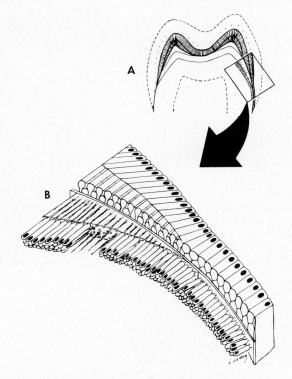

Fig. 4-29. A, Diagram of early dentinogenesis and amelogenisis. Lower right, on **B,** is site of differentiation of odontoblasts and ameloblasts. As the odontoblasts move away from the dentinoenamel junction, increments of dentin are formed.

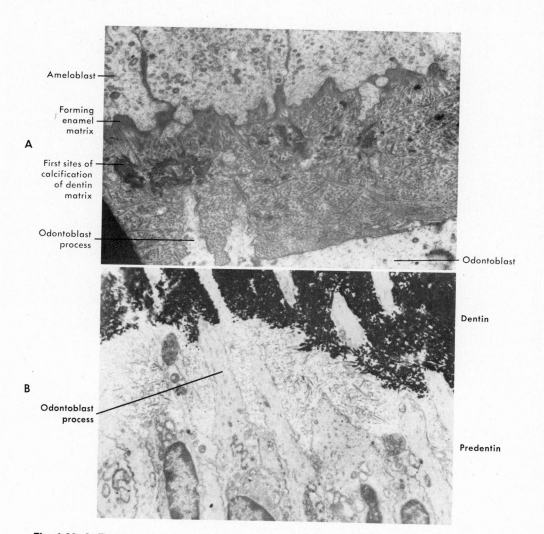

A — Ameloblast

Forming enamel matrix

First sites of calcification of dentin matrix

Odontoblast process

Odontoblast

Dentin

B — Odontoblast process

Predentin

Fig. 4-30. A, First-formed dentin, showing cytoplasm of apical zone of ameloblast, above, and first-formed enamel matrix at the dentinoenamel junction. Below the junction collagen fibers of dentin matrix are seen with calcification sites appearing near the first-formed enamel. Predentin zone is seen below these sites with the odontoblast process extending from the odontoblasts at bottom of field. **B,** Predentin and dentin as visualized in a later developing tooth. Observe calcified (black) dentin above, predentin composed of collagen fibers below, odontoblast processes, and the cell body. (Transmission electron micrographs.)

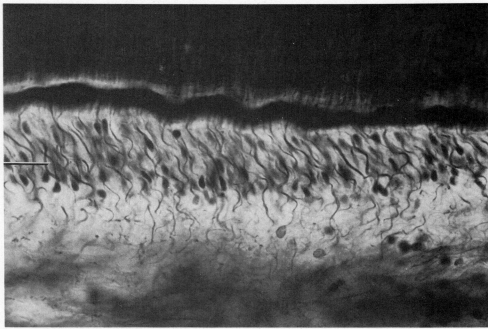

Argyrophilic
staining
substance

Fig. 4-31. Light micrograph of a silver-stained section of early forming dentin. The argyrophilic nature of the ground substances among the odontoblasts appears like bundles of collagen fibers. Fig. 4-27 illustrates that the collagen formed by the odontoblast is apical to the cell body in the area of the forming predentin.

collagen fibrils and in the ground substance (Fig. 4-30, *A*). Subsequently, crystals are laid down within the fibrils themselves. The crystals associated with the collagen fibrils are arranged in an orderly fashion, with their long axes paralleling the fibril axes, and in rows conforming to the 64 nm (640 Å) striation pattern. Within the globular islands of mineralization, crystal deposition appears to take place radially from common centers, in a so-called spherulite form (Fig. 4-30, *A*).

The general calcification process is gradual, but the peritubular region becomes highly mineralized at a very early stage. Although there is obviously some crystal growth as dentin matures, the ultimate crystal size remains very

small, about 3 nm (30 Å) in thickness and 100 nm (1000 Å) in length. The apatite crystals of dentin resemble those found in bone and cementum. They are 300 times smaller than those formed in enamel (Fig. 4-32). It is interesting that two cells so closely allied at the dentinoenamel junction produce crystals of such a size difference but at the same time produce chemically the same hydroxyapatite crystals. Calcospherite mineralization is seen occasionally along the pulp-predentin-forming front (Fig. 4-33).

CLINICAL CONSIDERATIONS

The cells of the exposed dentin should not be insulted by bacterial toxins, strong drugs,

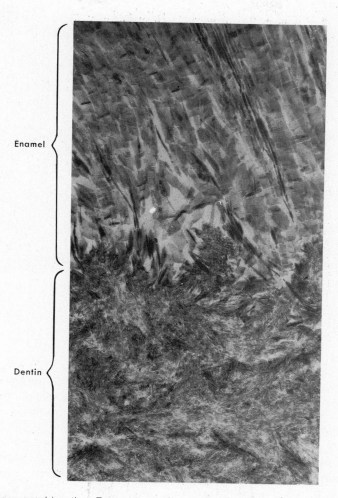

Enamel

Dentin

Fig. 4-32. Dentinoenamel junction. Enamel is above and dentin below. Note difference in size and orientation between crystallites of enamel and dentin. Whereas crystals of human enamel may be 90 nm (900 Å) in width and 0.5 to 1 μm in length, those of dentin are only 3nm (30 Å) in width and 100 nm (1000 Å) in length. Crystals of dentin are similar in size to bone. (Electron microphotograph; ×35,000.)

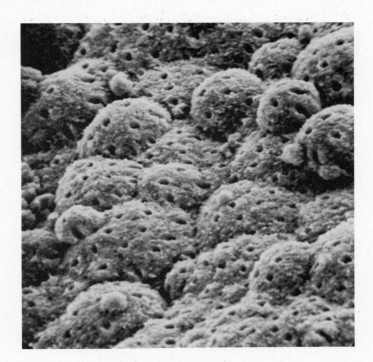

Fig. 4-33. Scanning electron micrograph of globular dentin (calcospherite mineralization) formation at predentin-forming front. Later-forming dentin may be linear, causing interglobular spaces to appear among earlier-formed globular dentin. Treated with ethylene diamine to remove organic material. (Courtesy A. Boyde, London.)

undue operative trauma, unnecessary thermal changes, or irritating restorative materials. One should bear in mind that when 1mm² of dentin is exposed, about 30,000 living cells are damaged. It is advisable to seal the exposed dentin surface with a nonirritating, insulating substance.

The rapid penetration and spread of caries in the dentin is the result of the tubule system in the dentin (Fig. 4-2). The enamel may be undermined at the dentinoenamel junction, even when caries in the enamel is confined to a small surface area. This is due in part to the spaces created at the dentinoenamel junction by enamel tufts, spindles, and open and branched dentinal tubules. The dentinal tu-

bules form a passage for invading bacteria that may thus reach the pulp through a thick dentinal layer.

Electron micrographs of carious dentin show regions of massive bacterial invasion of dentinal tubules (Fig. 4-34). The tubules are enlarged by the destructive action of the microorganisms. Dentin sensitivity of pain, unfortunately, may not be a symptom of caries until the pulp is infected and responds by the process of inflammation, leading to toothache. Thus patients are surprised at the extent of damage to their teeth with little or no warning from pain. Undue trauma from operative instruments also may damage the pulp. Air-driven cutting instruments cause dislodgement of the odonto-

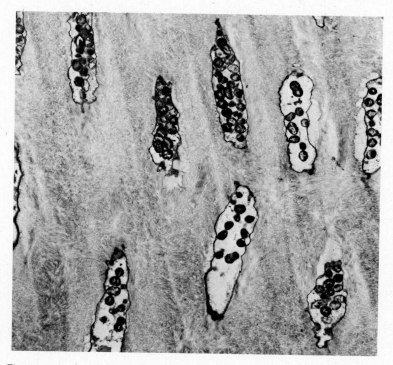

Fig. 4-34. Electron micrograph of dentin underlying carious lesion. Coccoid bacteria are present in the tubules. The peritubular dentin has been destroyed causing the enlargement of the tubules. (×10,000).

blasts from the periphery of the pulp and their "aspiration" within the dentinal tubule. This could be an important factor in survival of the pulp if the pulp is already inflamed. Repair requires the mobilization of the macrophage system as healing takes place; as this progresses there is the contribution of deeper pulpal cells, through cytodifferentiation into odontoblasts, which will be active in formation of reparative dentin.

The sensitivity of the dentin has been explained by the concept that alteration of the fluid and cellular contents of the dentinal tubules causes stimulation of the nerve endings in contact with these cells (Fig. 9-20). This theory explains pain throughout dentin since fluid movement will occur at the dentinoenamel junction as well as near the pulp.

Because we know that reparative dentin stimulates cavity lining materials and that dentin forms throughout the life of a tooth, it is now possible to save teeth that previously were lost by extraction or treated by endodontic therapy. Again, teeth with deep, penetrating, carious lesions can be treated by only partial removal of carious dentin and insertion of a "dressing" containing calcium hydroxide, for example, for a period of a few weeks or months. During this period the odontoblasts form new dentin along the pulpal surface underlying the carious lesion, and the dentist can then reopen the cavity and remove the remain-

Fig. 4-35. Dentinal surface of prepared cavity with smear layer*(S)* covering ends of tubules. Below surface are dentinal tubules *(T)* and in one tubule a debris plug *(P)*.

ing bacteria-laden decay without endangering the pulp. This treatment is termed indirect pulp capping. Today there is considerable research on the permeability of dentin. A number of factors have been noted to interfere with fluid flow in the tubules, such as when protein and apatite crystals are present in the tubules. Perhaps the most surprising is the effect of the

smear layer on the cavity floor created during cavity preparation (Fig. 4-35). Although it reduces permeability temporarily, it is a bacteria-laden mass and it is important to remove it because toxic products will migrate to the pulp. A cavity liner is then recommended to line the cavity.

A summary diagram illustrating the relationship of the odontoblast and its process to the dentin matrix is shown in Fig. 4-36.

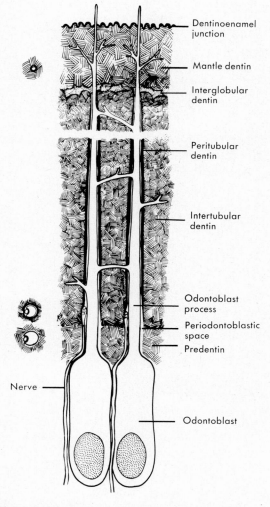

Dentinoenamel junction

Mantle dentin

Interglobular dentin

Peritubular dentin

Intertubular dentin

Odontoblast process

Periodontoblastic space

Predentin

Nerve

Odontoblast

Fig. 4-36. Diagram of the odontoblast and its process in the dental tubule. Note the relationship of the process to the periodontoblastic space and the pertitubular dentin.

REFERENCES

Anderson, D.J., and Ronning, G.A.: Osmotic excitants of pain in human dentine, Arch. Oral Biol. **7**:513, 1962.

Applebaum, E., Hollander, F., and Bodecker, C.F.: Normal and pathological variations in calcification of teeth as shown by the use of soft x-rays, Dent. Cosmos **75**:1097, 1933.

Arwill, T.: Innervation of the teeth, Stockholm, 1958, Ivar Haeggströms Boktryckeri AB.

Bergman, G., and Engfeldt, B.: Studies on mineralized dental tissues. II. Microradiography as a method for studying dental tissue and its application to the study of caries, Acta Odontol. Scand. **12**:99, 1954.

Bernick, S.: Innervation of the human tooth, Anat. Rec. **101**:81, 1948.

Bevelander, G.: The development and structure of the fiber system of dentin, Anat. Rec. **81**:79, 1941.

Bhaskar, S.N., and Lilly, G.E.: Intrapulpal temperature during cavity preparation, J. Dent. Res. **44**:644, 1965.

Boyde, A.: Scanning electron microscopy of collagen-free calcified connective tissues. Beitr elektronenikroskop, direktabb. Oberfl. 1[s]:213-222, Munster, 1968.

Boyde, A., and Lester, K.S.: An electron microscope study of fractured dentinal surfaces, Calcif. Tissue Res. **1**:122, 1967.

Bradford, E.W.: The maturation of the dentine, Br. Dent. J. **105**:212, 1958.

Bradford, E.W.: The dentine, a barrier to caries, Br. Dent. J. **109**:387, 1960.

Brannstrom, M., and Garberoglio, R.: Occlusion of dentinal tubules under superficial attrited dentine, Swed. Dent. J. **4**:87, 1980.

Brannstrom, M.: Dentin and pulp in restorative dentistry, London, 1982, Wolfe Medical Publications Ltd.

Cape, A.T., and Kitchin, P.C.: Histologic phenomena of tooth tissues as observed under polarized light, J. Am. Dent. Assoc. **17**:193, 1930.

Ebner, V. von: Ueber die Entwicklung der leimge-benden Fibrillen im Zahnbein (Development of collagenous fibrils in the dentin), Sitzungsber. Akad. Wissensch. Vienna **115**:281, 1906; Anat. Anz. **29**:137, 1906.

Fearnhead, R.W.: Histological evidence for the innervation of human dentine, J. Anat. **91**:267, 1957.

Frank, R.M.: Electron microscopy of undecalcified sections of human adult dentine, Arch. Oral Biol. **1**:29, 1959.

Harcourt, J.K.: Further observations on the peritubular translucent zone in human dentine, Aust. Dent. J. **9**:387, 1964.

Hess, W.C., Leo, D.Y., and Peckham, S.C.: The lipid content of enamel and dentin, J. Dent. Res. **35**:273, 1956.

Holland, G.R.: The dentinal tubule and odontoblast process in the cat, J. Anat. **12**:1169, 1975.

Korff, K. von: Die Entwicklung der Zahnbein Grundsubstanz der Säugetiere (The development of the dentin matrix in mammals), Arch. Mikrosk. Anat. **67**:1, 1905.

Korff, K. von: Wachstum der Dentingrundsubstanz verschiedener Wirbeltiere (Growth of the dentin matrix of different vertebrates), Z. Mikrosk. Anat. Forsch. **22**:445, 1930.

Kramer, I.R.H.: The distribution of collagen fibrils in the dentine matrix, Br. Dent. J. **91**:1, 1951.

Jessen, H.: The ultrastructure of odontoblasts in perfusion fixed, demineralized incisors of adult rats, Acta Odontol. Scand. **25**:491, 1967.

Lester, K.S., and Boyde, A.: Electron microscopy of predentinal surfaces, Calcif. Tissue Res. **1**:44, 1967.

Lester, K.S., and Boyde, A.: Some preliminary observations on caries ("remineralization") crystals in enamel and dentine by surface electron microscopy, Virchows Arch. [Pathol. Anat.] **344**:196-212, 1968.

Martens, P.J., Bradford, E.W., and Frank, R.M.: Tissue changes in dentine, Int. Dent. J. **9**:330, 1959.

Miller, J.: The micro-radiographic appearance of dentine, Br. Dent. J. **97**:72, 1954.

Nalbandian, J., Gonzales, F., and Sognnaes, R.F.: Sclerotic age changes in root dentin of human teeth as observed by optical, electron, and x-ray microscopy, J. Dent. Res. **39**:598, 1960.

Noble, H., Carmichael, A., and Rankine, D.: Electron microscopy of human developing dentine, Arch. Oral Biol. **7**:399, 1962.

Nylen, M.U., and Scott, D.B.: An electron microscopic study of the early stages of dentinogenesis, Pub. 613, U.S. Public Health Service, Washington, D.C., 1958, U.S. Government Printing Office.

Nylen, M.U., and Scott, D.B.: Basic studies in calcification, J. Dent. Med. **15**:80, 1960.

Orban, B.: The development of the dentin, J. Am. Dent. Assoc. **16**:1547, 1929.

Pashley, D.H., Kepler, E.E., Williams, E.C., and O'Meara, J.A.: The effect on dentine permeability of time following cavity preparation in dogs, Arch. Oral Biol. **29**:1, 65, 1984.

Piez, K.A., and Likens, R.C.: The nature of collagen. II. Vertebrate collagens. In Sognnaes, R.F., editor: Calcification in biological systems, Washington, D.C., 1960, American Association for the Advancement of Science.

Schour, I., and Hoffman, M.M.: The rate of apposition of enamel and dentin in man and other animals, J. Dent. Res. **18**:161, 1939.

Schour, I., and Massler, M.: The neonatal line in enamel and dentin of the human deciduous teeth and first permanent molar, J. Am. Dent. Assoc. **23**:1946, 1936.

Schour, I., and Massler, M.: Studies in tooth development: the growth pattern of the human teeth, J. Am. Dent. Assoc. **27**:1778, 1940.

Schour, I., and Poncher, H.G.: The rate of apposition of human enamel and dentin as measured by the effects of acute fluorosis, Am. J. Dis. Child **54**:757, 1937.

Scott, D.B., and Nylen, M.U.: Changing concepts in dental histology, Ann. New York Acad. Sci. **85**:133, 1960.

Selvig, K.A.: Ultrastructural changes in human dentine exposed to a weak acid, Arch. Oral Biol. **13**:719, 1968.

Shroff, F.R., Williamson, K.I., and Bertaud, W.S.: Electron microscope studies of dentin, Oral Surg. **7**:662, 1954.

Shroff, F.R.: Further electron microscope studies on dentin: the nature of the odontoblast process, Oral Surg. **9**:432, 1956.

Sicher, H.: The biology of dentin, Bur **46**:121, 1946.

Sognnaes, R.F.: Microstructure and histochemical characteristics of the mineralized tissues, J. New York Acad. Sci. **60**:545, 1955.

Takuma, S.: Electron microscopy of the structure around the dentinal tubule, J. Dent. Res. **39**:973, 1960.

Takuma, S., and Kurahashi, Y.: Electron microscopy of various zones in the carious lesion in human dentine, Arch. Oral Biol. **7**:439, 1962.

Ten Cate, A.R.: An analysis of Tomes' granular layer, Anat. Rec. **172**(2):137, 1972.

Ten Cate, A.R., Melcher, A.H., Pudy, G., and Wagner, D.: The non-fibrous nature of the von Korff fibers in developing dentine. A light and electron microscope study, Anat. Rec. **168**(4):491, 1970.

Watson, M.L., and Avery, J.K.: The development of the hamster lower incisor as observed by electron microscopy, Am. J. Anat. **95**:109, 1954.

Yamada, T., Nakamura, K., Iwaku, M., and Fusayama, T.: The extent of the odontoblast process in normal and carious human dentin, J. Dent. Res. **62**(7):798, 1983.

5
PULP

ANATOMY

General features. The dental pulp occupies the center of each tooth and consists of soft connective tissue. Every person normally has a total of 52 pulp organs, 32 in the permanent and 20 in the primary teeth. Each of these organs has a shape that conforms to that of the respective tooth. They have a number of morphologic characteristics that are similar. Each pulp organ resides in a pulp chamber surrounded by dentin containing the peripheral extensions of the cells that formed it. The total volumes of all the permanent teeth pulp organs is 0.38 cc, and the mean volume of a single adult human pulp is 0.02 cc. Molar pulps are three to four times larger than incisor pulps (Fig. 5-1). Table 2 gives the variation in the size of pulp organs in different permanent teeth.

The gross description of the pulps of the maxillary and the mandibular teeth is as follows.

MAXILLARY TEETH

Central incisor: It is somewhat shovel shaped coronally with three short horns on the coronal roof, tapering down to a triangle root in cross section, with the point of the triangle pointing lingually.

Lateral incisor: It has a small spoon shape coronally going to a round evenly tapering root to the apex.

Cuspid: It is the longest pulp with an elliptical cross section buccolingually and a distally inclined apex.

First premolar: It has a large occlusocervical pulp chamber with a mesial concavity from the root surface onto the cervical third of the chamber. The chamber divides into two smooth funnel-shaped roots.

Second premolar: It is similar coronally to the first premolar, except it has only one root, which begins to taper at about its midpoint.

Molars: The molars are generally all similar, having

135

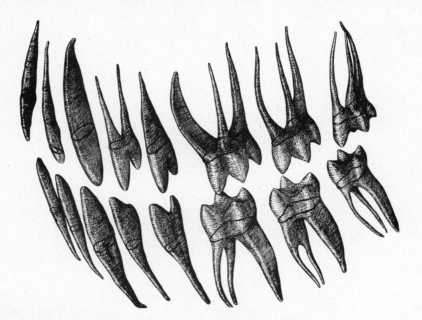

Fig. 5-1. Pulp organs of permanent human teeth. *Upper row,* Maxillary arch; left central incisor through third molar. *Lower row,* Mandibular arch; left central incisor through third molar.

Table 2. Pulp volumes for the permanent human teeth from a preliminary investigation of 160 teeth*

	Maxillary (cubic centimeters)	Mandibular (cubic centimeters)
Central incisor	0.012	0.006
Lateral incisor	0.011	0.007
Canine	0.015	0.014
First premolar	0.018	0.015
Second premolar	0.017	0.015
First molar	0.068	0.053
Second molar	0.044	0.032
Third molar	0.023	0.031

*Figures for volumes from Fanibunda, K.B.: Personal communication, University of Newcastle upon Tyne, Department of Oral Surgery, Newcastle upon Tyne, England.

a roughly rectangular cervical cross section with the greatest dimension buccolingually and also demonstrating mesiobuccal prominence; there are three roots; the lingual is longest and the distobuccal is shortest and straight, whereas the me-siobuccal is curved and flattened buccolingually with its convex surface mesially. From the first to third molars the crowns get smaller and the roots get closer together.

MANDIBULAR TEETH

Central incisor: It is one of the smallest pulps in the dentition and is long and narrow with a flattened elliptical shape in cross section buccolingually.

Lateral incisor: It is the same as the central incisor, only smaller in all dimensions.

Cuspid: It is similar to, but shorter than, the maxillary canine, and its root begins tapering at about its midpoint, ending in a distally inclined apex.

First premolar: It looks like a small mandibular canine with an insignificant or missing lingual pulp horn.

Second premolar: The lingual horn is much smaller than the buccal horn and is about the dimension of the mandibular canine. In cervical section it is often roundly triangular or sometimes rectangular.

Molars: The mandibular molars are all similar. The coronal cross section is usually rectangular with the mesiodistal dimension greatest, and it also dis-

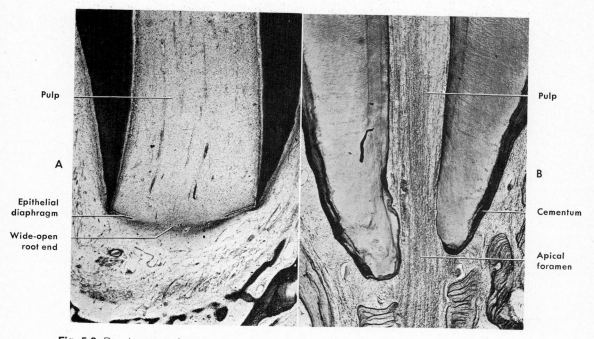

Fig. 5-2. Development of apical foramen. **A,** Undeveloped root end. Wide opening at end of root, partly limited by epithelial diaphragm. **B,** Apical foramen fully formed. Root canal straight. Apical foramen surrounded by cementum. (From Coolidge, E.D.: J. Am. Dent. Assoc. **16:**1456, 1929.)

plays a mesiobuccal prominence. The horn heights from highest to lowest are mesiobuccal, mesiolingual, distobuccal, distolingual. There are two roots, the distal being shorter and straighter and singular whereas the mesial is longer, curved, and often double. From first to third, the roots get smaller and closer together.

Coronal pulp. Each pulp organ is composed of a coronal pulp located centrally in the crowns of teeth and a root or radicular pulp. The coronal pulp in young individuals resembles the shape of the outer surface of the crown dentin. The coronal pulp has six surfaces: the occlusal, the mesial, the distal, the buccal, the lingual, and the floor. It has pulp horns, which are protrusions that extend into the cusps of each tooth. The number of these horns thus depends on the cuspal number. The cervical region of the pulp organs constricts as does the contour of the crown, and at this zone the cor-

onal pulp joins the radicular pulp (Fig. 5-1). Because of continuous deposition of dentin, the pulp becomes smaller with age. This is not uniform through the coronal pulp but progresses faster on the floor than on the roof or side walls.

Radicular pulp. The radicular pulp is that pulp extending from the cervical region of the crown to the root apex. In the anterior teeth the radicular pulps are single and in posterior ones multiple. They are not always straight and vary in size, shape, and number. The radicular portions of the pulp organs are continuous with the periapical connective tissues through the apical foramen or foramina. The dentinal walls taper, and the shape of the radicular pulp is tubular. During root formation the apical root end is a wide opening limited to an epithelial diaphragm (Fig. 5-2, *A*). As growth proceeds, more dentin is formed, so that when the root of the tooth has matured the radicular pulp is narrower. The apical pulp canal is made smaller also because of apical cementum deposition (Fig. 5-2, *B*).

Apical foramen. The average size of the apical foramen of the maxillary teeth in the adult is 0.4 mm. In the mandibular teeth it is slightly smaller, being 0.3 mm in diameter.

The location and shape of the apical foramen may undergo changes as a result of functional influences on the teeth. A tooth may be tipped from horizontal pressure, or it may migrate mesially, causing the apex to tilt in the opposite direction. Under these conditions the tissues entering the pulp through the apical foramen may exert pressure on one wall of the foramen, causing resorption. At the same time, cementum is laid down on the opposite side of the apical root canal, resulting in a relocation of the original foramen (Fig. 5-3, *A*).

Sometimes the apical opening is found on the lateral side of the apex (Fig. 5-3, *B*), although the root itself is not curved. Frequently, there are two or more foramina separated by a portion of dentin and cementum or by cementum only.

Accessory canals. Accessory canals leading from the radicular pulp laterally through the root dentin to the periodontal tissue may be seen anywhere along the root but are particularly numerous in the apical third of the root (Fig. 5-4, *A*). The mechanism by which they are formed is not known, but it is likely that they occur in areas where there is premature loss of root sheath cells because these cells induce the formation of the odontoblasts. Accessory canals may also occur where the developing root encounters a blood vessel. If the vessel is located in the area where the dentin is forming, the hard tissue may develop around it, making a lateral canal from the radicular pulp.

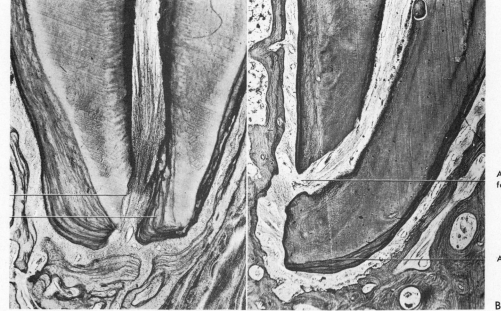

Resorption of dentin

Apposition of cementum

Apical foramen

Apex

Fig. 5-3. Variations of apical foramen. **A,** Shift of apical foramen by resorption of dentin and cementum on one surface and apposition of cementum on the other. **B,** Apical foramen on side of apex. (From Coolidge, E.D.: J. Am. Dent. Assoc. **16:**1456, 1929.)

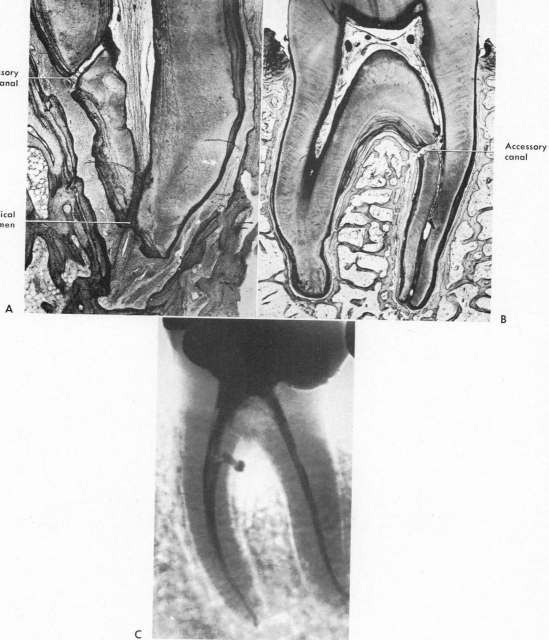

Accessory
canal

Apical
foramen

A

Accessory
canal

B

C

Fig. 5-4. A and **B,** Sections through teeth with accessory canals. **A,** Close to apex. **B,** Close to bifurcation. **C,** Roentgenogram of lower molar with accessory canal filled. (**C** from Johnston, H.B., and Orban, B.: J. Endodont. **3:**21, 1948.)

STRUCTURAL FEATURES

The central region of both the coronal and the radicular pulp contains large nerve trunks and blood vessels. Peripherally, the pulp is circumscribed by the specialized *odontogenic* region composed of (1) the odontoblasts (the dentin-forming cells), (2) the cell-free zone (Weil's zone), and (3) the cell-rich zone (Fig. 5-5). The cell-free zone is a space in which the odonto-

blast may move pulpward during tooth development and later to a limited extent in functioning teeth. This may be why the zone is inconspicuous during early stages of rapid dentinogenesis since odontoblast migration would be greatest at that time. The cell-rich layer is composed principally of fibroblasts and undifferentiated mesenchymal cells. The latter are distinctive because they lack a ribosome-

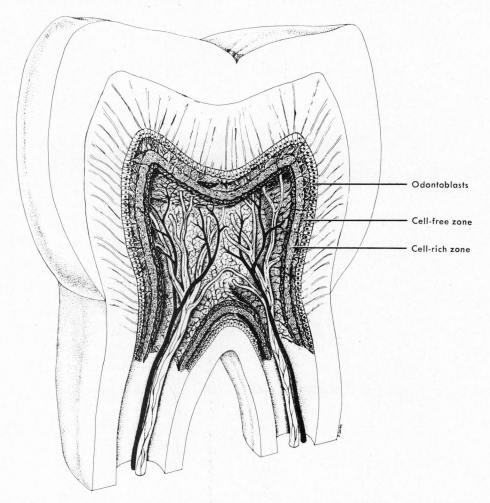

Fig. 5-5. Diagram of pulp organ, illustrating architecture of large central nerve trunks *(dark)* and vessels *(light)* and peripheral cell-rich, cell-free, and odontoblast rows. Observe small nerves on blood vessels.

Odontoblasts

Cell-free zone

Cell-rich zone

studded endoplasmic reticulum and have mitochrondria with readily discernible cristernae. During early dentinogenesis there are also many young collagen fibers in this zone.

Intercellular substance. The intercellular substance is dense and gellike in nature, varies in appearance from finely granular to fibrillar, and appears more dense in some areas, with clear spaces left between various aggregates. It is composed of both acid mucopolysaccharides and protein polysaccharide compounds (glycosaminoglycans and proteoglycans). During early

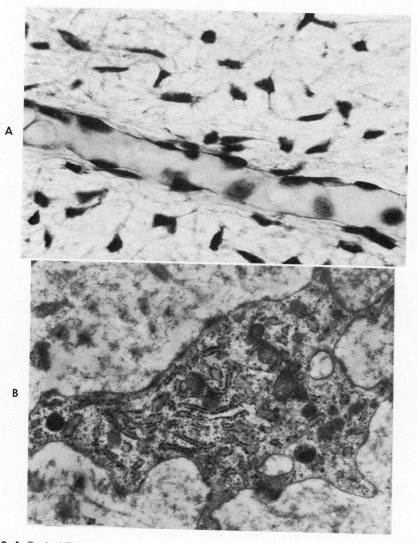

Fig. 5-6. A, Typical fibroblasts of pulp are stellate in shape with long processes. **B,** Electron micrograph of pulp fibroblast.

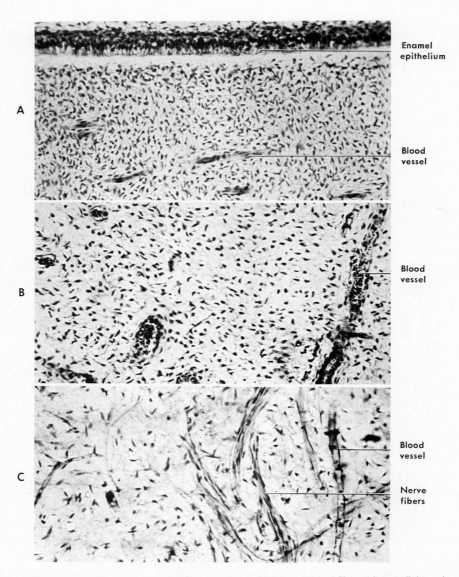

Fig. 5-7. Age changes of dental pulp. Cellular elements decrease and fibrous intercellular substance increases with advancing age. **A,** Newborn infant, **B,** Infant 9 months of age. **C,** Adult.

development, the presence of chondroitin A, chondroitin B, and hyaluronic acid has been demonstrated in abundance. Glycoproteins are also present in the ground substance. The aging pulp contains less of all of these substances. The ground substance lends support to the cells of the pulp while it also serves as a means for transport of nutrients from the blood vessels to the cells, as well as for transport of metabolites from cells to blood vessels.

func-
tion

Fibroblasts. The pulp organ is said to consist of specialized connective tissue because it lacks elastic fibers. Fibroblasts are the most numerous cell type in the pulp. As their name implies, they function in collagen fiber formation throughout the pulp during the life of the tooth. They have the typical stellate shape and extensive processes that contact and are joined by intercellular junctions to the processes of other fibroblasts (Fig. 5-6, A). Under the light microscope the fibroblast nuclei stain deeply with basic dyes, and their cytoplasm is lighter stained and appears homogeneous. Electron micrographs reveal abundant rough-surfaced endoplasmic reticulum, mitochondria, and other organelles in the fibroblast cytoplasm (Fig. 5-6, B). This indicates these cells are active in pulpal collagen production. There is some differ-

ence in appearance of these cells depending on the age of the pulp organ. In the young pulp the cells divide and are active in protein synthesis, but in the older pulp they appear rounded or spindle shaped with short processes and exhibit fewer intracellular organelles. They are then termed *fibrocytes*. In the course of development the relative number of cellular elements in the dental pulp decreases, whereas the fiber population increases (Fig. 5-7). In the embryonic and immature pulp the cellular elements predominate, while in the mature pulp the fibrous components predominate. The fibroblasts of the pulp, in addition to forming the pulp matrix, also have the capability of ingesting and degrading this same matrix. These cells thus have a dual function with pathways for both synthesis and degradation in the same cell.

Fibers. The collagen fibers in the pulp exhibit typical cross striations at 64 nm (640Å) and range in length from 10 to 100 nm or more (Fig. 5-8). Bundles of these fibers appear throughout the pulp. In very young pulp fine fibers ranging in diameter from 10 to 12 nm (100 to 120 Å) have been observed. Their significance is unknown. Pulp collagen fibers do not contribute to dentin matrix production,

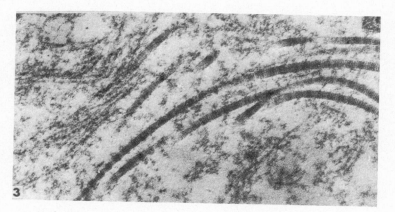

Fig. 5-8. Typical collagen fibers of the pulp with 640 Å banding.

which is the function of the odontoblast. After root completion the pulp matures and bundles of collagen fibers increase in number. They may appear scattered throughout the coronal or radicular pulp, or they may appear in bundles. These are termed *diffuse* or *bundle collagen* depending on their appearance, and their presence may relate to environmental trauma. Fiber bundles are most prevalent in the root canals, especially near the apical region.

Undifferentiated mesenchymal cells. Undifferentiated mesenchymal cells are the primary cells in the very young pulp, but a few are seen in the pulps after root completion. They appear larger than fibroblasts and are polyhedral in shape with peripheral processes and large oval staining nuclei. They are found along pulp vessels, in the cell-rich zone and scattered throughout the central pulp. Viewed from the side, they appear spindle shaped (Fig. 5-9). They are believed to be a totipotent cell and when need arises they may become odontoblasts, fibroblasts, or macrophages. They decrease in number in old age.

Odontoblasts. Odontoblasts, the second most prominent cell in the pulp, reside adjacent to the predentin with cell bodies in the pulp and cell processes in the dentinal tubules. They are approximately 5 to 7 μm in diameter and 25 to 40 μm in length. They have a constant location adjacent to the predentin, in what is termed the "odontogenic zone of the pulp" (Fig. 5-10). The cell bodies of the odontoblasts are columnar in appearance with large oval nuclei, which fill the basal part of the cell (Fig. 5-10). Immediately adjacent to the nucleus basally is rough-surfaced endoplasmic reticulum and the Golgi apparatus. The cells in the odontoblastic row lie very close to each other, and the plasma membranes of adjacent cells exhibit junctional complexes, (Fig. 5-11). Further toward the apex of the cell appears an abundance of rough-surfaced endoplasmic reticulum. Near the pupal-predentin junction the cell cytoplasm is devoid of organelles. The clear terminal part of the cell body and the adjacent intercellular junction is described by some as the terminal bar apparatus of the odontoblast. At this zone the cell constricts to a diameter of 3 to 4 μm, where the cell process enters the predentinal tubule (Fig. 5-10). The process of the cell contains no endoplasmic reticulum, but during the early period of active dentinogenesis it does contain occasional mitochondria and vesicles. During the later stages of dentinogenesis these are less frequently seen. There is also a striking difference in the cytoplasm of the young cell body, active in dentinogenesis, and the older cell. During this early active phase the Golgi apparatus is more prominent, the rough-surfaced endoplasmic reticulum is more abundant, and numerous mitochondria appear throughout the odontoblast. A great number of vesicles are seen along the periphery of the process where there is evidence of protein synthesis along the tubule wall. The cell actually increases in size as its process lengthens during dentin formation. When the cell process becomes 2 mm long, it is then many times greater in volume than the cell body. The form and arrangement of the bodies of the odontoblasts are not uniform throughout the pulp. They are more cylindrical and longer (tall columnar) in the crown (Fig. 5-12, *A*) and more cuboid in the middle of the root (Fig. 5-12, *B*). Close to the apex of an adult tooth the odontoblasts are ovoid and spindle shaped, appearing more like osteoblasts than odontoblasts, but they are recognized by their processes extending into the dentin. In areas close to the apical foramen the dentin is irregular in appearance (Fig. 5-12, *C*).

Defense cells. In addition to fibroblasts, odontoblasts, and the cells that are a part of the neural and vascular systems of the pulp, there are cells important to the defense of the pulp. These are histiocytes, or macrophages, mast cells, and plasma cells. In addition, there are the blood vascular elements such as the neutrophils (PMNs), eosinophils, basophils, lymphocytes, and monocytes. These latter cells emi-

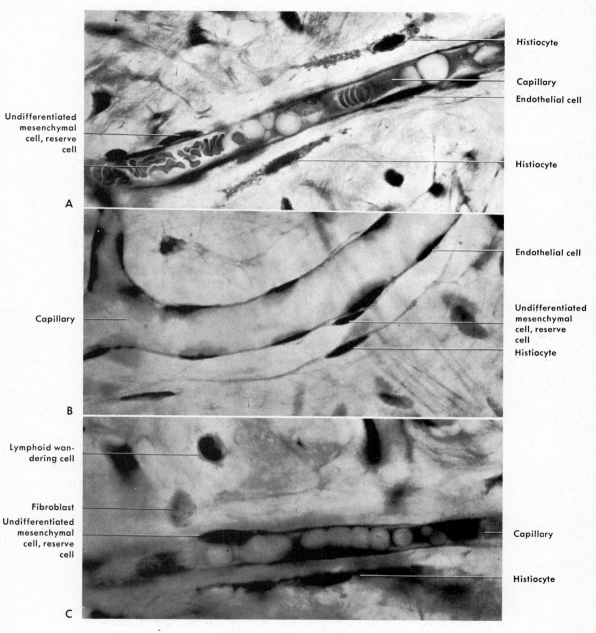

Histiocyte

Capillary

Endothelial cell

Histiocyte

Undifferentiated
mesenchymal
cell, reserve
cell

A

Endothelial cell

Capillary

Undifferentiated
mesenchymal
cell, reserve
cell

Histiocyte

B

Lymphoid wan-
dering cell

Fibroblast

Undifferentiated
mesenchymal
cell, reserve
cell

Capillary

Histiocyte

C

Fig. 5-9. Defense cells in pulp.

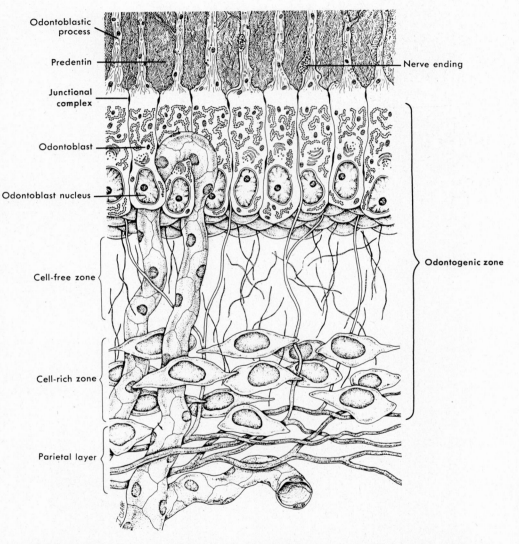

Fig. 5-10. Diagram of odontogenic zone illustrating odontoblast, cell-free, and cell-rich zones, with blood vessels and nonmyelinated nerves among odontoblasts.

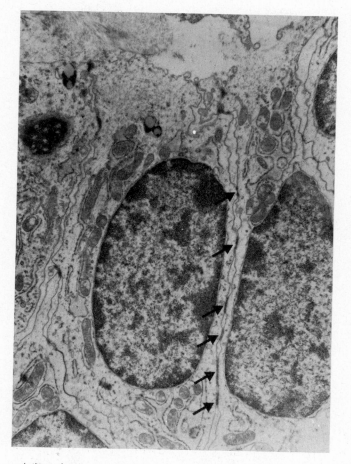

Fig. 5-11. Close relation of adjacent odontoblasts. Note junctional complexes between cells *(arrows).*

grate from the pulpal blood vessels and develop characteristics in response to inflammation.

The histiocyte, or macrophage, is an irregularly shaped cell with short blunt processes (Figs. 5-9 and 5-13). In the light microscope the nucleus is somewhat smaller, more rounded, and darker staining than that of fibroblasts, and it exhibits granular cytoplasm. When the macrophages are inactive and not in the process of ingesting foreign materials, one has difficulty distinguishing them from fibroblasts. In the case of a pulpal inflammation these cells exhibit granules and vacuoles in their cytoplasm, and their nuclei increase in size and exhibit a prominent nucleolus. Their presence is disclosed by intravital dyes such as trypan blue. These cells are usually associated with small blood vessels and capillaries. Ultrastructurally the macrophage exhibits a rounded outline with short, blunt processes (Fig. 5-13). Invaginations of the plasma membrane are noted, as are mitochon-

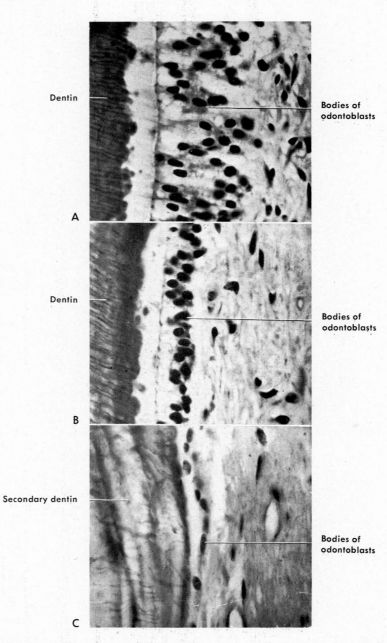

Dentin

Bodies of
odontoblasts

A

Dentin

Bodies of
odontoblasts

B

Secondary dentin

Bodies of
odontoblasts

C

Fig. 5-12. Variation of odontoblasts in different regions of one tooth. **A,** High columnar odontoblasts
in pulp chamber. **B,** Low columnar odontoblasts in root canal. **C,** Flat odontoblasts in apical region.

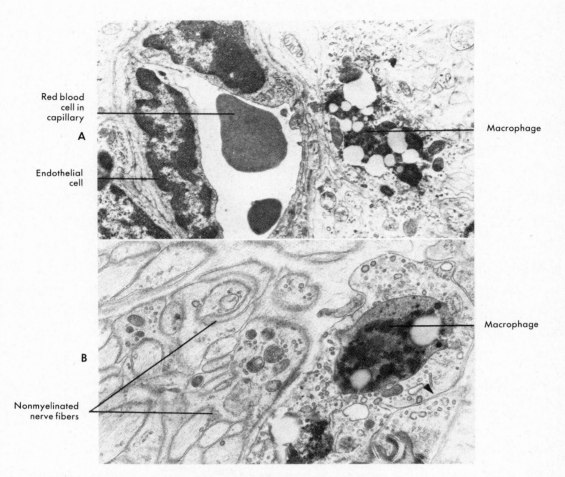

Red blood cell in capillary

A

Endothelial cell

Macrophage

B

Nonmyelinated nerve fibers

Macrophage

Fig. 5-13. A, This histiocyte or macrophage is located adjacent to capillary in peripheral pulp. Characteristic aggregation of vesicles, vacuoles, and phagocytized dense bodies is seen to right of capillary wall. **B,** Multivesiculated body characteristic of macrophage. Note typical invagination of cell plasma membrane *(arrow)*. This cell is located adjacent to group of nonmyelinated nerve fibers seen on left.

dria, rough-surfaced endoplasmic reticulum, free ribosomes, and also a moderately dense nucleus. The distinguishing feature of macrophages is aggregates of vesicles, or phagosomes, which contain phagocytized dense irregular bodies (Fig. 5-13).

Both lymphocytes and eosinophils are found extravascularly in the normal pulp (Fig. 5-14), but during inflammation they increase noticea-

bly in number. Mast cells are also seen along vessels in the inflamed pulp. They have a round nucleus and contain many dark-staining granules in the cytoplasm, and their number increases during inflammation.

The plasma cells are seen during inflammation of the pulp (Fig. 5-15). With the light microscope the plasma cell nucleus appears small and concentric in the cytoplasm. The chroma-

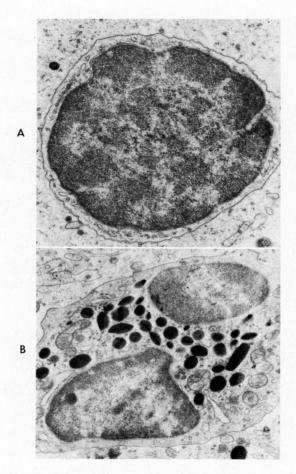

Fig. 5-14. A, Small lymphocyte located in pulp. Cytoplasm forms narrow rim around large oval-to-round nucleus. **B,** Eosinophil in extravascular location in pulp organ. Nucleus is polymorphic, and granules in cytoplasm are characteristically banded.

tin of the nucleus is adherent to the nuclear membrane and gives the cell a cartwheel appearance. The cytoplasm of this cell is basophilic with a light-stained Golgi zone adjacent to the nucleus. Under the electron microscope these cells have a densely packed, rough-surfaced endoplasmic reticulum. Both immature and mature cells may be found. The mature type exhibits a typical small eccentric nucleus and more abundant cytoplasm (Fig. 5-15). The plasma cells function in the production of antibodies.

Blood vessels. The pulp organ is extensively vascularized. It is known that the blood vessels of both the pulp and the periodontium arise from the inferior or superior alveolar artery and also drain by the same veins in both the man-

dibular and maxillary regions. The communication of the vessels of the pulp with the periodontium, in addition to the apical connections, is further enhanced by connections through the accessory canals. These relationships are of considerable clinical significance in the event of a potential pathologic condition in either the periodontium or the pulp, because it has a potential to spread through the accessory and apical canals. Although branches of the alveolar arteries supply both the tooth and its supporting tissues, those entering the pulp are different in structure from the branches to the periodontium. As the vessels enter the tooth their walls become considerably thinner than those surrounding the tooth.

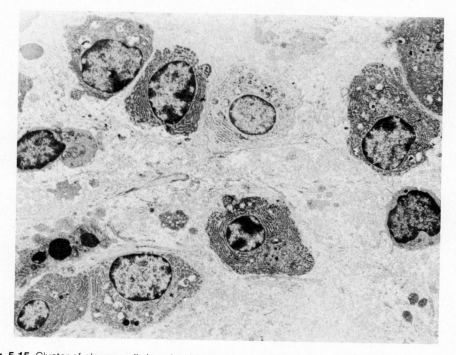

Fig. 5-15. Cluster of plasma cells in pulp with early caries pulpitis. Observe dense peripheral nuclear chromatin and cytoplasm with cisternae of rough endoplasmic reticulum. (Courtesy C. Torneck, University of Toronto Dental School.)

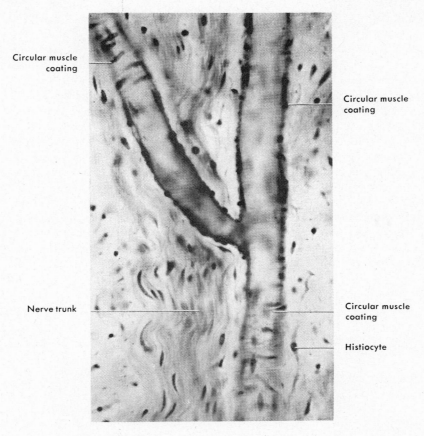

Circular muscle coating

Circular muscle coating

Nerve trunk

Circular muscle coating

Histiocyte

Fig. 5-16. Branching artery and nerve trunk in the pulp.

Small arteries and arterioles enter the apical canal and pursue a direct route to the coronal pulp (Fig. 5-16). Along their course they give off numerous branches in the radicular pulp that pass peripherally to form a plexus in the odontogenic region (Fig. 5-17). Pulpal blood flow is more rapid than in most areas of the body. This is perhaps attributable to the fact that the pulpal pressure is among the highest of body tissues. The flow of blood in arterioles is 0.3 to 1 mm per second, in venules approximately 0.15 mm per second, and in capillaries about 0.08 mm per second. The largest arteries in the human pulp are 50 to 100 μm in diame-

ter, thus equaling in size arterioles found in most areas of the body. These vessels possess three layers. The first, the tunica intima, consists of squamous or cuboid endothelial cells surrounded by a closely associated basal lamina. Where the endothelial cells contact, they appear overlapped to varying degrees. The second layer, the tunica media, is approximately 5 μm thick and consists of one to three layers of smooth muscle cells (Figs. 5-18 and 5-19). A basal lamina surrounds and passes between these muscle cells and separates the muscle cell layer from the intima. Occcasionally the endothelial cell wall is in contact with the mus-

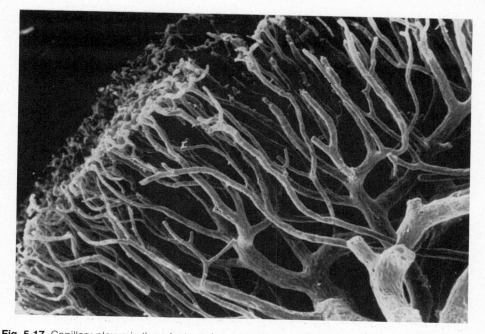

Fig. 5-17. Capillary plexus in the odontogenic region. These vessels penetrate the odontoblastic cell row and loop back to join venules of central pulp. (Courtesy K. Takahashi, Kanagawa Dental College, Kanagawa, Japan.)

cle cells. This is termed a myoendothelial junction. The third and outer layer, the tunica adventitia, is made up of a few collagen fibers forming a loose network around the larger arteries. This layer becomes more conspicuous in vessels in older pulps. Arterioles with diameters of 20 to 30 μm with one or occasionally two layers of smooth muscle cells are common throughout the coronal pulp (Fig. 5-18). The tunica adventita blends with the fibers of the surrounding intercellular tissue. Terminal arterioles with diameters of 10 to 15 μm appear peripherally in the pulp. The endothelial cells of these vessels contain numerous micropinocytotic vesicles, which function in transendothelial fluid movement. A single layer of smooth muscle cells surrounds these small vessels. Occasionally a fibroblast or pericyte lies on the surface of these vessels. Pericytes are capillary-associated fibroblasts, and their nuclei can be distiguished as round or slightly oval bodies closely associated with the outer surface of the terminal arterioles or precapillaries (Figs. 5-19 and 5-20). Some authors call the smaller diameter arterioles "precapillaries." They are slightly larger than the terminal capillaries and exhibit a complete or incomplete single layer of muscle cells surrounding the endothelial lining. These range in size from 8 to 12 μm

Veins and venules that are larger than the arteries also appear in the central region of the root pulp. They measure 100 to 150 μm in diameter, and their walls appear less regular than those of the arteries because of bends and irregularities along their course. The microscopic appearance of the veins is similar to that of the arteries except that they exhibit much thinner walls in relation to the size of the lumen. The

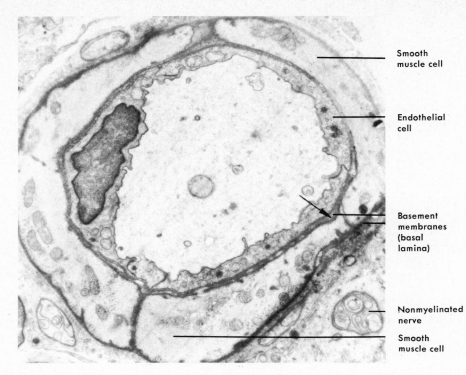

Smooth
muscle cell

Endothelial
cell

Basement
membranes
(basal
lamina)

Nonmyelinated
nerve

Smooth
muscle cell

Fig. 5-18. Small arteriole near central pulp exhibiting relatively thick layer of muscle cells. Dense basement membrane interspersed between endothelial and muscle cells *(arrow)*.

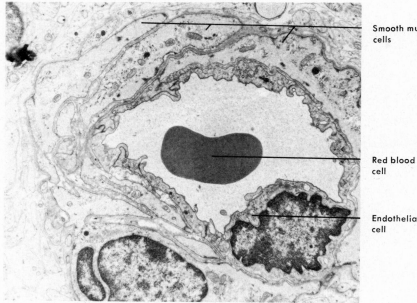

Smooth muscle
cells

Red blood
cell

Endothelial
cell

Fig. 5-19. Peripheral pulp and small arteriole or precapillary exhibiting two thin layers of smooth muscle cells surrounding the endothelial cell lining of vessel. Nucleus at bottom left of figure belongs to pericyte.

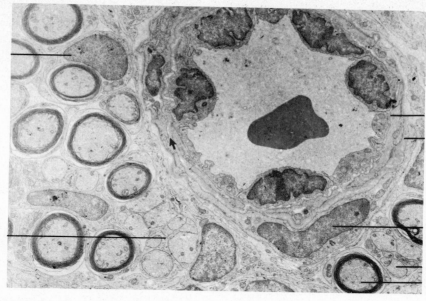

Schwann cell

Endothelial cell lining

Basement membrane

Nonmyelinated axons

Pericyte

Nonmyelinated axon

Myelinated axon

Fig. 5-20. Area near subodontoblastic plexus showing both myelinated and nonmyelinated axons adjacent to large capillary or precapillary. Endothelial cell lining is surrounded by basement membrane *(arrow)* and pericytes.

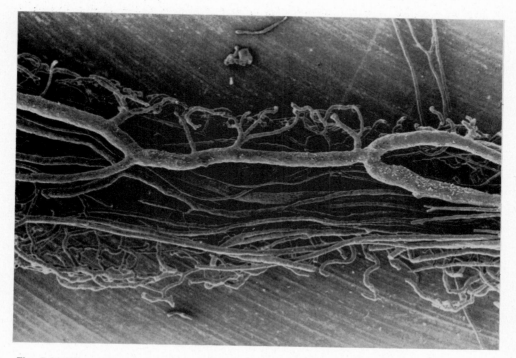

Fig. 5-21. Venous loops seen at the left and right of the field with connecting branch in center. Scanning electron micrograph. (Courtesy K. Takahashi, Kanagawa Dental College, Kanagawa, Japan.)

endothelial cells appear more flattened, and their cytoplasm does not project into the lumen. Fewer intracytoplasmic filaments appear in these cells than in the arterioles. The tunica media consists of a single layer or two of thin smooth muscle cells that wrap around the endothelial cells and appear discontinuous or absent in the smaller venules. The basement membranes of these vessels are thin and less distinct than those of arterioles. The adventitia is lacking or appears as fibroblasts and fibers continuous with the surrounding pulp tissue. Occasionally two venous loops will be seen connected by an anastomosing branch (Fig. 5-21). Both venous-venous anastomosis and arteriole-venous anastomosis occur in the pulp. The arteriole-venous shunts may have an important role in regulation of pulpal blood flow. Frequently arteriole or precapillary loops with capillaries are found underlying the odontogenic zone in the coronal pulp (Fig. 5-22).

Blood capillaries, which appear as endothelium-lined tubes, are 8 to 10 μm in diameter. The nuclei of these cells may be lobulated and have cytoplasmic projections into the luminal surface. The terminal network of capillaries in the coronal pulp appears nearly perpendicular to the main trunks. The vascular network passes among the odontoblasts and underlies them as well (Fig. 5-17). A few peripheral capillaries found among the odontoblasts have fenestrations in the endothelial cells. These pores are located in the thin part of the capillary wall and are spanned only by the thin diaphragm of contacting inner and outer plasma membranes of endothelial cells, (Fig. 5-23). These fenestrated capillaries are assumed to be involved in rapid transport of metabolites at a time when the odontoblasts are active in the process of dentinal matrix formation and its subsequent calcification. Both fenestrated and continuous terminal capillaries are found in the odontogenic region. During active dentinogenesis capillaries appear among the odontoblasts adjacent to the predentin (Fig. 5-17). Later, after

the teeth have reached occlusion and dentinogenesis slows down, these vessels usually retreat to a subodontoblastic position.

Lymph vessels. The presence of lymph vessels in the dental pulp is questioned by some and agreed upon by other investigators. Support for this system stems from investigators who use injection of fine particulate substances into the dentin or peripheral pulp, which are subsequently reported present in some of the thin-walled vessels that exit through the apical foramen. Lymph capillaries are described as endothelium-lined tubes that join thin-walled lymph venules or veins in the central pulp. The larger vessels have an irregular-shaped lumen composed of endothelial cells surrounded by an incomplete layer of pericytes or smooth muscle cells or both. They are further characterized by absence of red blood cells and presence of lymphocytes. Absence of basal lamina adjacent to the endothelium has also been reported (Fig. 5-30). Lymph vessels draining the pulp and periodontal ligament have a common outlet. Those draining the anterior teeth pass to the submental lymph nodes; those of the posterior teeth pass to the submandibular and deep cervical lymph nodes.

Nerves. The abundant nerve supply in the pulp follows the distribution of the blood vessels. The majority of the nerves that enter the pulp are nonmyelinated. Many of these gain a myelin sheath later in life. The nonmyelinated nerves are found in close association with the blood vessels of the pulp and many are sympathetic in nature. They have terminals on the muscle cells of the larger vessels and function in vasoconstriction (Fig. 5-18). Thick nerve bundles enter the apical foramen and proceed to the coronal area where their fibers separate and radiate peripherally to the odontogenic zone (Fig. 5-24). The number of fibers in these bundles varies greatly, from as few as 150 to more than 1200. The larger fibers range between 5 and 13 μm, although the majority are smaller than 4 μm. The large myelinated fibers

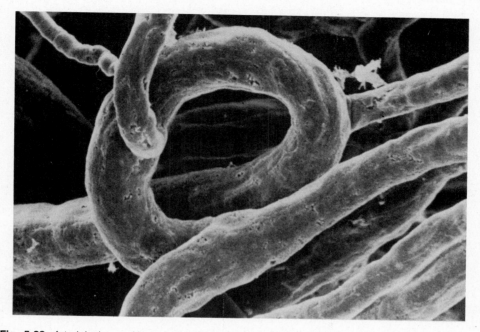

Fig. 5-22. Arteriole loop with associated capillaries located in subodontoblastic zone. Scanning electron micrograph. (Courtesy K. Takahashi, Kanagawa Dental College, Kanagawa, Japan.)

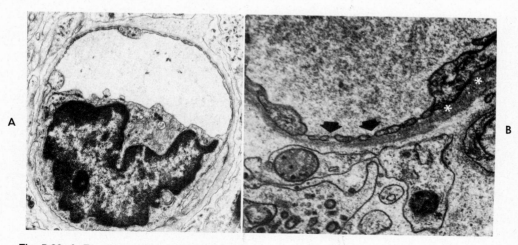

Fig. 5-23. A, Terminal capillary loops located among odontoblasts may be fenestrated. These capillaries have both thick and thin segments in their walls. **B,** Endothelial cell wall bridges pores *(arrows)* and is supported only by basement membrane (**).

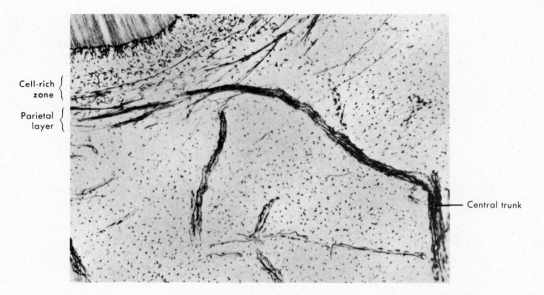

Cell-rich
zone

Parietal
layer

Central trunk

Fig. 5-24. Major nerve trunks branch in pulp and pass to parietal layer, which lies adjacent to cell-rich zone. Cell-rich zone curves upward to right.

Cell-rich
zone

Fig. 5-25. Parietal layer of nerves is composed of myelinated nerve fibers. Cell-rich zone curves upward to right.

mediate the sensation of pain that may be caused by external stimuli. The peripheral axons form a network of nerves located adjacent to the cell-rich zone. This is termed the *parietal layer of nerves,* also known as the plexus of Rashkow (Figs. 5-24 and 5-25). Both myelinated axons, ranging from 2 to 5 μm in diameter, and minute nonmyelinated fibers of approximately 200 to 1600 μm (2000 to 16,000 Å) in size make up this layer of nerves. The parietal layer develops gradually, becoming prominent when root formation is complete.

Nerve endings. Nerve axons from the parietal zone pass through the cell-rich and cell-free zones and either terminate among or pass between the odontoblasts to terminate adjacent to the odontoblast processes at the pulp-predentin border or in the dentinal tubules (Fig. 5-26). Nerve terminals consisting of round or oval enlargements of the terminal filaments contain microvesicles, small, dark, granular bodies, and mitochondria (Fig. 5-27). These terminals are very close to the odontoblast plasma membrane, separated only by a 20 μm (200 Å) cleft (Fig. 5-28). Many of these indent the odontoblast surface and exhibit a special relationship to these cells. Most of the nerve endings located among the odontoblasts are believed to be sensory receptors. Some sympathetic endings are found in this location as well. Whether they have some function relative to the capillaries or the odontoblast in dentinogenesis is not known. The nerve axons found among the odontoblasts and in the cell-free and cell-rich zones are nonmyelinated but are enclosed in a Schwann cell covering. It is presumed that these fibers lost their myelin sheath as they passed peripherally from the parietal zone. More nerve fibers and endings are found in the pulp horns than in other peripheral areas of the coronal pulp (Fig. 5-5).

Recently a great deal of information has been reported regarding the types of potential neurotransmitters that are present in the nerves of the dental pulp. Substances such as substance P, 5-hydroxytryptamine, vasoactive intestinal peptide, somatostatin, and prostaglandins, as well as acetylcholine and norepinephrine have been found throughout the pulp. The majority of these putative transmitters have been shown

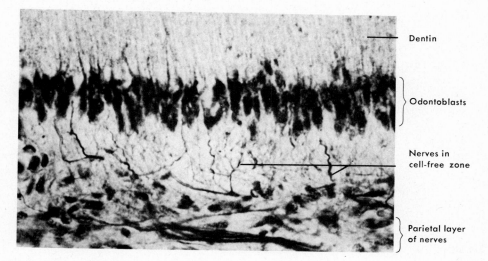

Fig. 5-26. Terminal nerve endings located among odontoblasts. These arise from subjacent parietal layer.

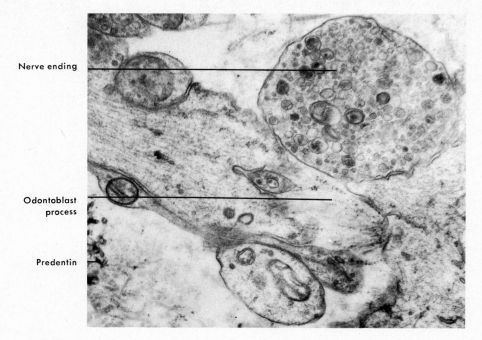

Nerve ending

Odontoblast process

Predentin

Fig. 5-27. Vesiculated nerve endings in predentin in zone adjacent to odontoblast process.

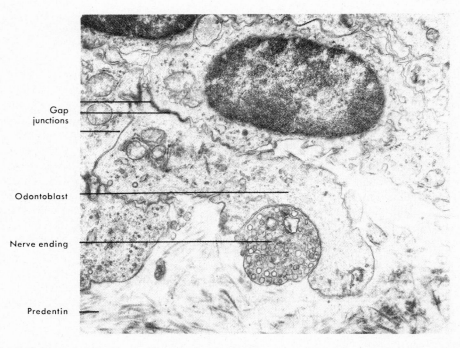

Gap junctions

Odontoblast

Nerve ending

Predentin

Fig. 5-28. Vesiculated nerve ending lying in apposition to odontoblast process adjacent to predentin. Gap junction appears between odontoblasts.

to affect vascular tone and subsequently modify the excitability of the nerve endings. Further, it has been suggested that these changes in vascular tone can also affect the incremental growth of dentin.

It is a feature unique to dentin receptors that environmental stimuli always elicit pain as a response. Sensory response in the pulp cannot differentiate between heat, touch, pressure, or chemicals. This is because the pulp organs lack those types of receptors that specifically distinguish these other stimuli.

FUNCTIONS

Inductive. The first role of the pulp anlage is to induce oral epithelial differentiation into dental lamina and enamel organ formation. The pulp anlage also induces the developing enamel organ to become a particular type of tooth.

Formative. The pulp organ cells produce the dentin that surrounds and protects the pulp. The pulpal odontoblasts develop the organic matrix and function in its calcification. Through the development of the odontoblast processes, dentin is formed along the tubule wall as well as at the pulp-predentin front.

Nutritive. The pulp nourishes the dentin through the odontoblasts and their processes and by means of the blood vascular system of the pulp.

Protective. The sensory nerves in the tooth respond with pain to all stimuli such as heat, cold, pressure, operative cutting procedures, and chemical agents. The nerves also initiate reflexes that control circulation in the pulp. This sympathetic function is a reflex, providing stimulation to visceral motor fibers terminating on the muscles of the blood vessels.

Defensive or reparative. The pulp is an organ with remarkable reparative abilities. It responds to irritation, whether mechanical, thermal, chemical, or bacterial, by producing reparative dentin and mineralizing any affected dentinal tubules. Both the reparative dentin

created in the pulp and the calcification of the tubules (sclerosis) are attempts to wall off the pulp from the source of irritation. Also, the pulp may become inflamed due to bacterial infection or by cutting action and placement of an irritating restorative material. The pulp has macrophages, lymphocytes, neutrophils, monocytes, and plasma and mast cells, all of which aid in the process of repair of the pulp. Although the rigid dentinal wall has to be considered as a protection of the pulp, it also endangers its existence under certain conditions. During inflammation of the pulp, hyperemia and exudate may lead to the accumulation of excess fluid outside the capillaries. An imbalance of this type, limited by the unyielding enclosure, can lead to partial or complete vascular collapse resulting in necrosis of the pulp. In most cases, if the inflammation is not too severe, however, the pulp will heal since it has excellent regenerative properties.

PRIMARY AND PERMANENT PULP ORGANS

Primary pulp organs. The primary pulp organs function for a shorter period of time than do the permanent pulps. The average length of time a primary pulp functions in the oral cavity is only about 8.3 years. This amount of time may be divided into three time periods—that of *pulp organ growth*, which takes place during the time the crown and roots are developing; that period of time after the root is completed until root resorption begins, which is termed the time of *pulp maturation;* and finally the period of *pulp regression*, which is the time from beginning root resorption until the time of exfoliation. Let us consider the average time of pulp life based on figures for the entire primary dentition. These three periods (growth, maturation, and regression) are not of equal lengths. Tooth eruption to root completion is about 1 year (11.85 months), and the time of root completion to beginning root loss (based on completion of the permanent crown) is 45.3

months, or 3 years, 9 months. Finally, the time of pulp regression based on the beginning of root resorption to exfoliation is 3 years, 6 months. The amount of time the primary pulp is undergoing changes relative to growth based on both the *prenatal* crown formation and the postnatal root completion is about 4 years, 2 months, 11 months of which are involved in crown completion from the time of beginning of crown calcification to its completion. The period of time the primary radicular pulp is regressing is based on the time from when the permanent crown is completed till the time of permanent tooth eruption. In some cases, root loss commences before the root is entirely complete. The maximum life of the primary pulp including both prenatal and postnatal times of development and the period of regression is approximately 9.6 years.

Permanent pulp organs. During crown formation the pulps of primary and permanent teeth are morphologically nearly identical. In the permanent teeth this is a process requiring about 5 years. During this time the organs are highly cellular, exhibiting a high mitotic rate especially in the cervical region. The young differentiating odontoblasts exhibit few organelles until dentin formation begins; then they rapidly change into protein-synthesizing cells. Both the primary and the permanent pulps are highly vascularized; however, the primary teeth never attain the extent of neural development that occurs in the permanent teeth. This is caused in part by the loss of neural elements during the root-resorption period. The greater the extent of root resorption, the greater the degenerative changes seen in the primary pulps. The architecture of the primary and permanent pulps is similar in appearance to the cell-free and cell-rich zones, parietal layer, and the large nerve trunks and vessels in the central pulp.

The periods of development for the pulps of the permanent teeth are, as might be expected, longer than those required for completion of the same processes in the primary teeth. As mentioned above, crown completion, based on the time during which the crown is completing formation and calcification, averages 5 years, 5 months. From the time of crown completion to eruption the time in both arches averages 3 years, 6 months. The time from eruption to root completion is 3 years, 11 months. Thus the pulp of the permanent teeth undergoes development for about 12 years, 4 months (based on the time from beginning prenatal crown calcification to root completion). This is in contrast to the 4 years, 2 months it takes in the primary teeth. Furthermore, the permanent roots take over twice as long to reach completion (7 years, 5 months) as do those of the primary pulps (average 3 years, 3 months).

The period of pulp aging is much accelerated in the primary teeth and occupies the time from root completion to exfoliation, or about 7 years, 5 months. Aging of the pulp in the permanent teeth, on the other hand, requires much of the adult life span.

Finally, one should note in passing that for both the primary and permanent teeth the maxillary arches require slightly longer to complete each process of development than do the mandibular arches.

REGRESSIVE CHANGES

Cell changes. In addition to the appearance of fewer cells in the aging pulp, the cells are characterized by a decrease in size and number of cytoplasmic organelles. The typical active pulpal fibrocyte or fibroblast has abundant rough-surfaced endoplasmic reticulum, notable Golgi complex, and numerous mitochondria with well-developed cristae. The fibroblasts in the aging pulp exhibit less perinuclear cytoplasm and possess long, thin cytoplasmic processes. The intracellular organelles are reduced in number and size; the mitochondria and endoplasmic reticulum are good examples of this.

Fibrosis. In the aging pulp accumulations of both diffuse fibrillar components as well as bundles of collagen fibers usually appear. Fiber

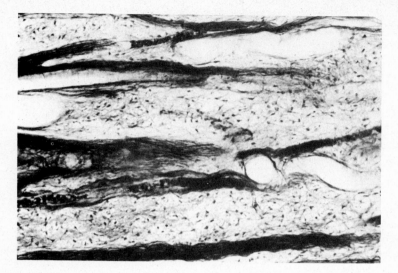

Fig. 5-29. Bundles of collagen fibers around and among blood vessels of pulp.

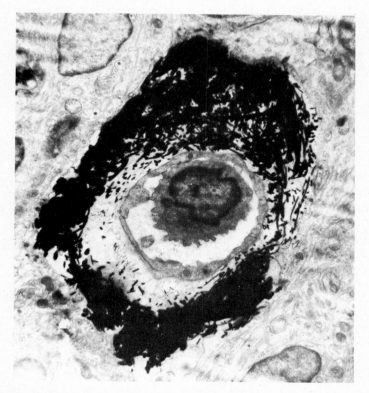

Fig. 5-30. Small vessel containing lymphocyte. Its wall exhibits no basement membrane around endothelial cells. It is probably a lymphatic capillary. Calcification appears around periphery of vessel.

bundles may appear arranged longitudinally in bundles in the radicular pulp and in a random, more diffuse arrangement in the coronal area. This condition is variable, with some older pulps showing surprisingly small amounts of collagen accumulation, others displaying considerable amounts (Fig. 5-29). The increase in fibers in the pulp organ is gradual and is generalized throughout the organ. Any external trauma such as dental caries or deep restorations usually causes a localized fibrosis or scarring effect. Collagen increase is noted in the medial and adventitial layers of blood vessels as well. The increase in collagen fibers may be more apparent than actual, being attributable to the decrease in the size of the pulp, which makes the fibers present occupy less space, and hence they become more concentrated without increasing in total volume.

Vascular changes occur in the aging pulp organ as they do in any organ. Plaques may appear in pulpal vessels. In other cases the outer diameter of vessel walls becomes greater as collagen fibers increase in the medial and adventitial layers. Also calcifications are found that surround vessels (Fig. 5-30). Calcification in the walls of blood vessels is found most often in the region near the apical foramen.

Pulp stones or denticles. Pulp stones, or denticles, are nodular, calcified masses appearing in either or both the coronal or root portions of the pulp organ. They often develop in teeth that appear to be quite normal in other respects. They have been seen in functional as well as embedded unerupted teeth.

Pulp stones are classified, according to their structure, as *true denticles, false denticles,* and *diffuse calcifications.* The structure of true denticles is similar to dentin in that they exhibit dental tubuli containing the processes of the odontoblasts that formed them and that exist on their surface (Fig. 5-31, *A*). True denticles are comparatively rare and are usually located close to the apical foramen. A theory has been advanced that the development of the

true denticle is caused by the inclusion of remnants of the epithelial root sheath within the pulp. These epithelial remnants induce the cells of the pulp to differentiate into odontoblasts, which then form the dentin masses called true pulp stones. (In tooth development the epithelial root sheath cells cause differentiation of odontoblasts along their pulpal boundary as they grow apically.)

False denticles do not exhibit dentinal tubules but appear instead as concentric layers of calcified tissue (Fig. 5-31, *B*). In some cases these calcification sites appear within a bundle of collagen fibers (Fig. 5-32). Other times they appear in a location in the pulp free of collagen accumulations (Fig. 5-31, *B*). Some false pulp stones undoubtedly arise around vessels as seen in Fig. 5-30. In the center of these concentric layers of calcified tissue there may be remnants of necrotic and calcified cells (Fig. 5-32). Calcification of thrombi in blood vessels, called phleboliths, may also serve as nidi for false denticles. All denticles begin as small nodules but increase in size by incremental growth on their surface. The surrounding pulp tissue may appear quite normal. Pulp stones may eventually fill substantial parts of the pulp chamber.

Diffuse calcifications. Diffuse calcifications appear as irregular calcific deposits in the pulp tissue, usually following collagenous fiber bundles or blood vessels (Fig. 5-31, *C*). Sometimes they develop into larger masses but usually persist as fine calcified spicules. The pulp organ may appear quite normal in its coronal portion without signs of inflammation or other pathologic changes but may exhibit these calcifications in the roots. Diffuse calcifications are usually found in the root canal and less often in the coronal area, whereas denticles are seen more frequently in the coronal pulp.

In addition to being classified according to their structure as true or false denticles and diffuse calcifications, pulp stones are also classi-

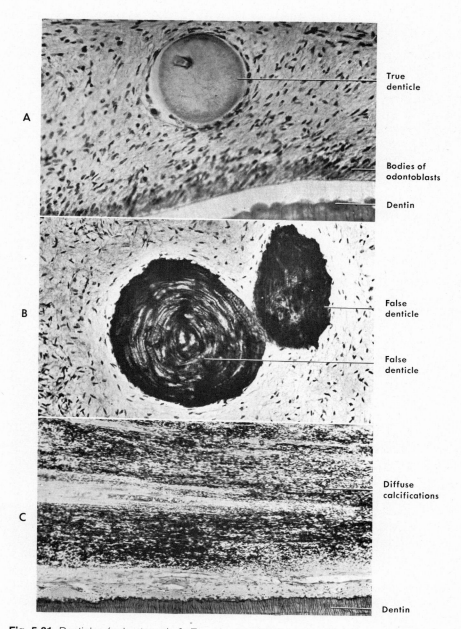

Fig. 5-31. Denticles (pulp stones). **A,** True denticle. **B,** False denticle. **C,** Diffuse calcifications.

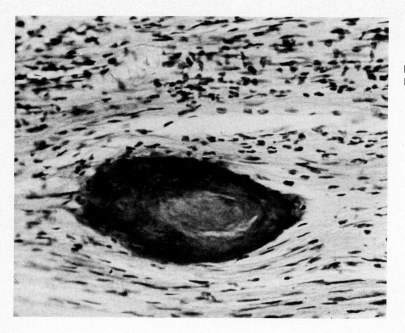

Fig. 5-32. Pulp stone within collagen bundle of coronal pulp.

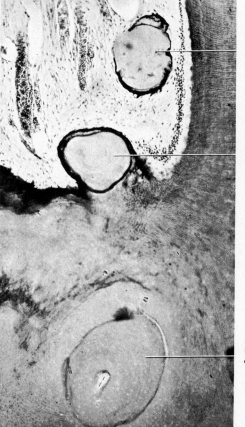

Free denticle

Attached denticle

Embedded denticle

Fig. 5-33. Free, attached, and embedded denticles.

fied according to their location in relation to the surrounding dentinal wall. Free, attached, and embedded denticles can be distinguished (Fig. 5-33). The free denticles are entirely surrounded by pulp tissue, attached denticles are partly fused with the dentin, and embedded denticles are entirely surrounded by dentin. All are believed to be formed free in the pulp and later to become attached or embedded as dentin formation progresses. Pulp stones may appear close to blood vessels and nerve trunks (Fig. 5-34). This is believed to be because they are large and grow so that they impinge on whatever structures are in their paths. The occurrence of pulp stones has been shown to be more prevalent through histologic study of human teeth than can be determined radiographically. It is believed that only a relatively small number of them are sufficiently large enough to be detected in roentgenograms. The inci-

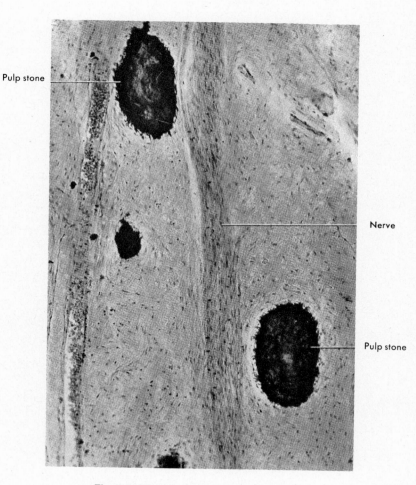

Fig. 5-34. Pulp stones in proximity to nerve.

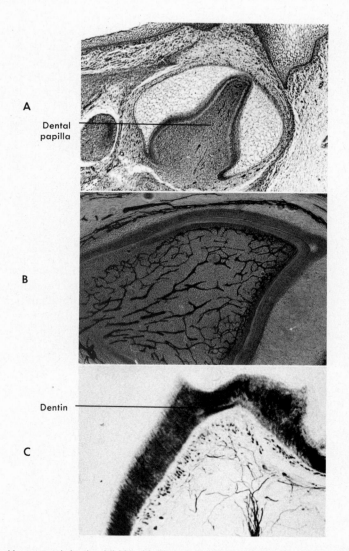

A

Dental
papilla

B

Dentin

C

Fig. 5-35. A, Young tooth bud exhibiting highly cellular dental papilla. Compare dense cell population to that of adjacent connective tissue. **B,** Young tooth with blood vessels injected with india ink
to demonstrate extent of vascularity of pulp. Large vessels located centrally and smaller ones peripherally among odontoblasts. Pulp surrounded by dentin and enamel. **C,** Young tooth stained with
silver to demonstrate neural elements. Myelinated nerves appear in pulp horn only after considerable amount of dentin has been laid down.

dence as well as the size of pulp stones increases with age. According to one estimate, 66% of teeth in persons 10 to 30 years of age, 80% in those between 30 and 50 years, and 90% in those over 50 years of age contain calcifications of some type.

DEVELOPMENT

The tooth pulp is initially called the dental papilla. This tissue is designated as "pulp" only after dentin forms around it. The dental papilla controls early tooth formation. In the earliest stages of tooth development it is the area of the proliferating future papilla that causes the oral epithelium to invaginate and form the enamel organs. These organs then enlarge to enclose the dental papillae in their central portions (Fig. 5-35, *A*). The dental papilla further controls whether the forming enamel organ is to be an incisor or a molar. At the location of the future incisor the development of the dental pulp begins at about the eighth week of embryonic life in the human. Soon thereafter, the more posterior tooth organs begin differentiating. The cell density of the dental papilla is great as a result of proliferation of the cells within it (Fig. 5-35, *A*). The young dental papilla is highly vascularized, and a well-organized network of vessels appears by the time dentin formation begins (Fig. 5-35, *B*). Capillaries crowd among the odontoblasts during this period of active dentinogenesis. The cells of the dental papilla appear as undifferentiated mesenchymal cells. Gradually these cells differentiate into stellate-shaped fibroblasts. After the inner and enamel organ cells differentiate into ameloblasts, the odontoblasts then differentiate from the peripheral cells of the dental papilla and dentin production begins. As this occurs, the tissue is no longer called dental papilla but is now designated the pulp organ. Few large myelinated nerves are found in the pulp until the dentin of the crown is well advanced. At that time nerves reach the odontogenic zone in the pulp horns. The sympathetic nerves, however, follow the blood vessels into the dental papilla as the pulp begins to organize.

CLINICAL CONSIDERATIONS

For all operative procedures the shape of the pulp chamber and its extensions into the cusps, the pulpal horns, is important to remember. The wide pulp chamber in the tooth of a young person will make a deep cavity preparation hazardous, and it should be avoided, if possible. In some instances of developmental disturbances the pulpal horns project high into the cusps, and the exposure of a pulp can occur when it is least anticipated. Sometimes a roentgenogram will help to determine the size of a pulp chamber and the extent of the pulpal horns.

If opening a pulp chamber for treatment becomes necessary, its size and variation in shape must be taken into consideration. With advancing age, the pulp chamber becomes smaller (Fig. 5-36), and because of excessive dentin formation at the roof and floor of the chamber, it is sometimes difficult to locate the root canals. In such cases it is advisable when one opens the pulp chamber to advance toward the distal root in the lower molar and toward the lingual root in the upper molar. In this region one is most likely to find the opening of the pulp canal without risk of perforating the floor of the pulp chamber. In the anterior teeth the coronal part of the pulp chamber may be filled with secondary dentin; thus locating the root canal is made difficult. Pulpstones lying at the opening of the root canal may cause considerable difficulty when an attempt is made to locate the canals.

The shape of the apical foramen and its location may play an important part in the treatment of root canals. When the apical foramen is narrowed by cementum, it is more readily located because further progress of the broach will be stopped at the foramen. If the apical opening is at the side of the apex, as shown in

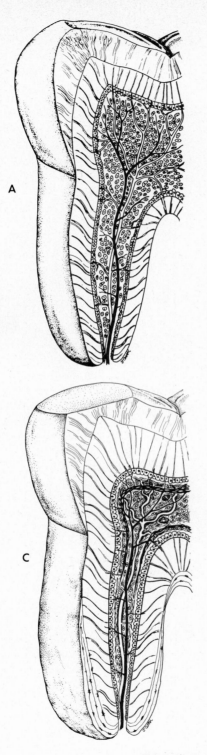
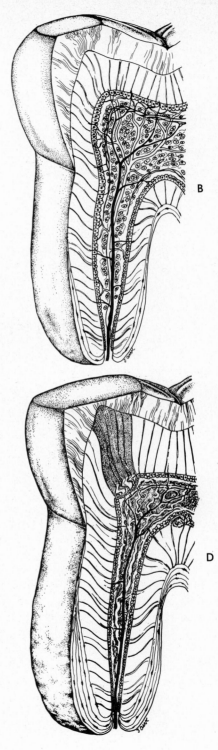

Fig. 5-36. For legend see opposite page.

Fig. 5-3, *B*, not even roentgenograms will reveal the true length of the root canal, and this may lead to misjudgment of the length of the canal and the root canal filling.

Since accessory canals are rarely seen in roentgenograms, they are not treated in root canal therapy. In any event it would be mechanically difficult or impossible to reach them. Fortunately, however, the majority of them do not affect the success of endodontic therapy.

When accessory canals are located near the coronal part of the root or in the bifurcation area (Fig. 5-4, *B*), a deep periodontal pocket may cause inflammation of the dental pulp. Thus periodontal disease can have a profound influence on pulp integrity. Conversely, a necrotic pulp can cause spread of disease to the periodontium through an accessory canal. It is recognized that pulpal and periodontal disease may spread by their common blood supply.

Fig. 5-37. Mild pulp response with loss of odontoblast identity and inflammatory cells obliterating cell-free zone.

Fig. 5-36. These four diagrams depict pulp organ throughout life. Observe first the decrease in size of pulp organ. **A** to **D,** Dentin is formed circumpulpally but especially in bifurcation zone. Note decrease in cells and increase in fibrous tissue. Blood vessels *(white)* organize early into odontoblastic plexus and later are more prominent in subodontoblastic zone, indicating decrease in active dentinogenesis. Observe sparse number of nerves in young pulp, organization of pariental layer of nerves. They are less prominent in aging pulp. Reparative dentin and pulpstones are apparent in oldest pulp, at lower right, **D.**

Until recently, some clinicians believed that an exposed pulp meant a lost pulp. This is no longer necessarily so. The fact that defense cells have been recognized in the pulp and that new odontoblasts can differentiate and form reparative dentin has changed this concept. Extensive experimental work has shown that exposed pulps can be preserved if proper pulp capping procedures are applied. This is especially true in noninfected or minimally infected, accidentally exposed pulps in individuals of any age. In these instances dentin is formed at the site of the exposure; thus a dentin barrier or bridge is developed and the pulp retains vitality. Pulp capping of primary teeth has been shown to be remarkably successful.

All operative procedures cause an initial response on the pulp, which is dependent on the severity of the insult. The pulp is highly responsive to stimuli. Even a slight stimulus will cause inflammatory cell infiltration (Fig. 5-37). A severe reaction is characterized as one with increased inflammatory cell infiltration adjacent to the cavity site, hyperemia, or localized abscesses. Hemorrhage may be present, and the odontoblast layer is either destroyed or greatly disrupted. It is of interest that most compounds containing calcium hydroxide readily induce reparative dentin underlying a cavity (Fig. 5-38). Most restorative materials also induce reparative dentin formations (Fig. 5-39). Usually the closer a restoration is to the pulp organ the greater will be the pulp response.

Since dehydration causes pulpal damage, operative procedures producing this condition should be avoided. When filling materials contain harmful chemicals (e.g., acid in silicate cements and monomer in the composites), an appropriate cavity liner should be used prior to the insertion of restorations. Most important is the effect of bacteria and bacterial toxins on the health of the pulp.

A vital pulp is essential to good dentition. Although modern endodontic procedures can

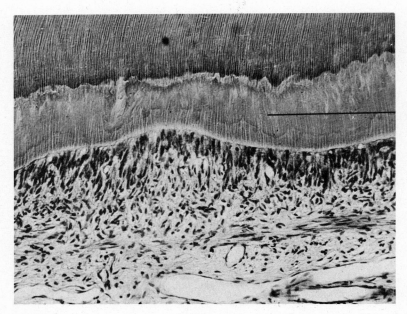

Reparative dentin

Fig. 5-38. Moderate cell response with formation of reparative dentin underlying cavity. Note viable odontoblasts have deposited tubular, reparative dentin.

prolong the usefulness of a tooth, a nonvital tooth becomes brittle and is subject to fractures. Therefore, every precaution should be taken to preserve the vitality of a pulp.

In clinical practice, instruments called vitalometers, which test the reaction of the pulp to electrical stimuli, or thermal stimuli (heat and cold) are often used to test the "vitality" of the pulp. These methods provide information about the status of the nerves supplying the pulpal tissue and therefore check the "sensitivity" of the pulp and not its "vitality." The vitality of the pulp depends on its blood supply, and one can have teeth with damaged nerve but normal blood supply (as in cases of traumatized teeth). Such pulps do not respond to electrical or thermal stimuli but are completely viable in every respect.

The preservation of the health of the pulp during operative procedures and its successful management in cases of disease is one of the most important challenges to the clinical dentist.

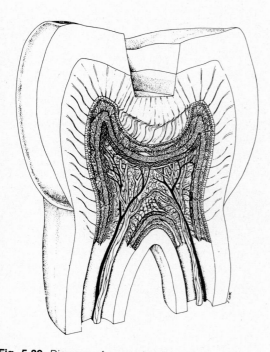

Fig. 5-39. Diagram of reparative function of pulp organ to cavity preparation and subsequent restoration. Reparative dentin is limited to zone of stimulation.

REFERENCES

Avery, J.K.: Structural elements of the young normal human pulp. In Siskin, M., editor: The biology of the human dental pulp, St. Louis, 1973. The C.V. Mosby Co. (Available only through American Association of Endodontists, Atlanta, Ga.)

Avery, J.K., and Han, S.S.: The formation of collagen fibrils in dental pulp, J. Dent. Res. **40**(6):1248, 1961.

Beveridge, E.E., and Brown, A.C.: The measurement of human dental intrapulpal pressure and its response to clinical variables, Oral Surg. **19**(5):655, 1965.

Bhussry, B.R.: Modification of the dental pulp organ during development and aging. In Finn, S.B., editor: Biology of the dental pulp organ: a symposium, University of Alabama, 1968, University of Alabama Press.

Corpron R.E., and Avery, J.K.: The ultrastructure of intradental nerves in developing mouse molars, Anat. Rec. **175**(3):585, 1973.

Corpron, R.E., Avery, J.K., and Lee, S.D.: Ultrastructure of terminal pulpal blood vessels in mouse molars, Anat. Rec. **179**(4):527, 1974.

Dahl, E., and Mjör, I.A.: The fine structure of the vessels in the human dental pulp, Acta Odontol. Scand. **31**(4):223, 1973.

Fanibunda, K.B.: Volume of the dental pulp cavity-method of measurement. British I.A.D.R. Abstr. No. 150, J. Dent. Res. **52**(suppl.):971, 1973.

Fanibunda, K.B.: A preliminary study of the volume of the pulp in the permanent human teeth. Unpublished. Personal communication, 1975.

Fearnhead, R.W.: The histological demonstration of nerve fibers in human dentin. In Anderson, D.J., editor: Sensory mechanisms in dentin, Oxford, England, 1963, Pergamon Press.

Finn, S.B.: Biology of the dental pulp organ: a symposium, University of Alabama, 1968, University of Alabama Press.

Graf, W., and Björlin, G.: Diameters of nerve fibers in human tooth pulps, J. Am. Dent. Assoc. **43**:186, 1951.

Green, D.A.: Stereoscopic study of the root apices of 400 maxillary and mandibular anterior teeth, Oral Surg. **9**:1224, 1956.

Griffin, C.J., and Harris, R.: The ultrastructure of the blood vessels of the human dental pulp following injury, Aust. Dent. J. **17**:303, 1972.

Griffin, C.J., and Harris, R.: The ultrastructure of the blood vessels of the human dental pulp following injury, Aust. Dent. J. **18**:88-96, 1973.

Han, S.S., and Avery, J.K.: The ultrastructure of capillaries and arterioles of the hamster dental pulp, Anat. Rec. **145**(4):549, 1963.

Han, S.S., Avery, J.K., and Hale, L.E.: The fine structure of differentiating fibroblasts in the incisor pulp of the guinea pig, Anat. Rec. **153**(2):187, 1965.

Han, S.S., and Avery, J.K.: The fine structure of intercellular substances and rounded cells in the incisor pulp of the guinea pig. Anat. Rec. **151**(1):41, 1965.

Harrop, T.J., and MacKay, B.: Electron microscopic observations of healing in dental pulp in the rat, Arch. Oral Biol. **13**(43):365, 1968.

Kim, S.: Regulation of blood flow of the dental pulp of dogs: macrocirculation and microcirculation studies, Thesis, 1981, Columbia University, New York.

Kollar, E.J., and Baird, G.R.: The influence of the dental papilla on the development of tooth shape in embryonic mouse tooth germs. J. Embryol. Exp. Morphol. **21**:131, 1969.

Kollar, E.J., and Baird, G.R.: Tissue interactions in embryonic mouse tooth germs. II. The indicative role of the dental papilla, J. Embryol. Exp. Morphol. **24**:173, 1970.

Kollar, E.J., and Baird, G.R.: Tissue interactions in embryonic mouse tooth germs. I. Reorganization of the dental epithelium during tooth-germ reconstruction, J. Embryol. Exp. Morphol. **24**:159, 1970.

Kovacs, I. A systematic description of dental roots. In Dahlberg, A.A., editor: Dental morphology and evaluation. Chicago, 1971. University of Chicago Press.

Kramer, I.R.H.: The vascular architecture of the human dental pulp, Arch. Oral Biol. **2**:177, 1960.

Langeland, K.: Tissue changes in the dental pulp, Odont. Tidskr. **65**(4):239, 1957.

Mjör, I.A., and Pindborg, J.J.: Histology of the human tooth, Copenhagen, 1973, Munksgaard, International Booksellers & Publishers, Ltd.

Nishijima, S., Imanishi, I., and Aka, M.: An experimental study on the lymph circulation in dental pulp, J. Osaka Dent. School **5**:45, 1965.

Nygaard-Ostby, B., and Hjortdal, O.: Tissue formation in the root canal following pulp removal, Scand. J. Dent. Res. **79**:333, 1971.

Ogilvie, A.L., and Ingle, J.E.: An atlas of pulpal and periapical biology. Philadelphia, 1965, Lea & Febiger.

Orban, B.J.: Contribution to the histology of the dental pulp and periodontal membrane, with special reference to the cells of "defense" of these tissues, J. Am. Dent. Assoc. **16**(6):965, 1929.

Rapp, R., Avery, J.K., and Rector, R.A.: A study of the distribution of nerves in human teeth. J. Can. Dent. Assoc. **23**:447, 1957.

Rapp, R., Avery, J.K., and Strachan, D.S.: The distribution of nerves in human primary teeth, Anat. Rec. **159**(1):89, 1967.

Ruben, M.P., Prieto-Hernandez, J.R., Gott, F.K., et al.: Visualization of lymphatic microcirculation of oral tissues. II. Vital retrograde lymphography. J. Periodontal. **42**:774, Dec. 1971.

Saunders, R.L. de C.H., and Röckert, H.Ö.E.: Vascular supply of dental tissues, including lymphatics. In Miles, A.E.W., editor: Structural and chemical organization of teeth, vol. 1, New York, 1967, Academic Press, Inc.

Schroff, F.R.: Physiologic path of changes in the dental pulp. Oral Surg. **6**:1455, 1953.

Seltzer, S., and Bender, I.B.: The dental pulp, Philadelphia, 1965, J.B. Lippincott Co.

Stanley, H.R., and Rainey, R.R.: Age changes in the human dental pulp, Oral Surg. **15**:1396, 1962.

Takahashi, K., Yoshiaki, K., and Kim, S.: A scanning electron microscope study of the blood vessels of dog pulp using corrosion resin casts. J. Endod. **8**(3):131, 1982.

Ten Cate, A.R.: Oral histology: development, structure, and function, ed. 2, St. Louis, 1985, The C.V. Mosby Company.

Torneck, C.D.: Changes in the fine structure of the dental pulp in human caries pulpitis. II. Inflammatory infiltration. J. Oral Pathol. **3**:83, 1974.

Torneck, C.D.: Changes in the fine structure of the dental pulp in human caries pulpitis. I. Nerves and blood vessels. J. Oral Pathol. **3**:71, 1974.

Weinstock, M., and Leblond, C.P.: Formation of collagen, Fed. Proc. **33**(5):1205, 1974.

Weinstock, M., and Leblond, C.P.: Synthesis migration and release of precursor collagen by odontoblasts as visualized by radioautography after [^{3}H] proline administration, J. Cell Biol. **60**:92, 1974.

Yankowitz, D.: An investigation of the existence and magnitude of intrapulpal pressure in dog teeth, Master's thesis, 1963, University of Washington, Seattle.

Zachrisson, B.V.: Mast cells in human dental pulp, Arch. Oral Biol. **16**:555, 1971.

Zerlotti, E.: Histochemical study of the connective tissue of the dental pulp, Arch. Oral Biol. **9**:149, 1964.

6
CEMENTUM

Cementum is the mineralized dental tissue covering the anatomic roots of human teeth. It was first demonstrated microscopically in 1835 by two pupils of Purkinje. It begins at the cervical portion of the tooth at the cementoenamel junction and continues to the apex. Cementum furnishes a medium for the attachment of collagen fibers that bind the tooth to surrounding structures. It is a specialized connective tissue that shares some physical, chemical, and structural characteristics with compact bone. Unlike bone, however, human cementum is avascular.

PHYSICAL CHARACTERISTICS

The hardness of fully mineralized cementum is less than that of dentin. Cementum is light yellow in color and can be distinguished from enamel by its lack of luster and its darker hue. Cementum is somewhat lighter in color than dentin. The difference in color, however, is slight, and under clinical conditions it is not possible to distinguish cementum from dentin based on color alone. Under some experimental conditions cementum has been shown to be permeable to a variety of materials.

CHEMICAL COMPOSITION

On a dry weight basis, cementum from fully formed permanent teeth contains about 45% to 50% inorganic substances and 50% to 55% organic material and water. The inorganic portion consists mainly of calcium and phosphate in the form of hydroxyapatite. Numerous trace elements are found in cementum in varying amounts. It is of interest that cementum has the highest fluoride content of all the mineralized tissues.

The organic portion of cementum consists primarily of type I collagen and protein polysaccharides (proteoglycans). Amino acid analyses of collagen obtained from the cementum of human teeth indicate close similarities to the collagens of dentin and alveolar bone. The chemical nature of the protein polysaccharides or ground substance of cementum is virtually unknown.

CEMENTOGENESIS

Cementum formation in the developing tooth is preceded by the deposition of dentin along the inner aspect of Hertwig's epithelial root sheath. Once dentin formation is under way, breaks occur in the epithelial root sheath allowing the newly formed dentin to come in direct contact with connective tissue of the dental follicle (Fig. 6-1). Cells derived from this connective tissue are responsible for cementum formation.

At the ultrastructural level, breakdown of Hertwig's epithelial root sheath involves degeneration or loss of its basal lamina on the cemental side. Loss of continuity of the basal lamina is soon followed by the appearance of collagen fibrils and cementoblasts between epithelial cells of the root sheath. Some sheath cells migrate away from the dentin toward the dental sac, whereas others remain near the developing tooth and ultimately are incorporated into the cementum. Sheath cells that migrate toward the dental sac become the epithelial rests of Malassez found in the periodontal ligament of fully developed teeth.

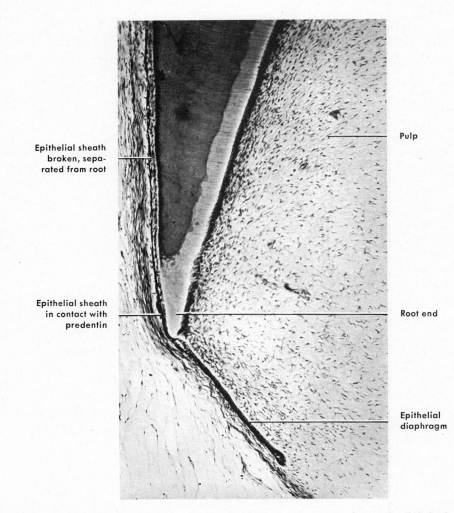

Fig. 6-1. Hertwig's epithelial root sheath at end of forming root. At side of root, sheath is broken up, and cementum formation begins. (From Gottlieb, B.: J. Periodontol. **13:**13, 1942.)

Cementoblasts. Soon after Hertwig's sheath breaks up, undifferentiated mesenchymal cells from adjacent connective tissue differentiate into cementoblasts (Fig. 6-2). Cementoblasts synthesize collagen and protein polysaccharides (proteoglycans), which make up the organic matrix of cementum. These cells have numerous mitochondria, a well-formed Golgi apparatus, and large amounts of granular endoplasmic reticulum (Fig. 6-3). These ultrastructural features are not unique to cementoblasts and can be observed in other cells actively producing proteins and polysaccharides.

After some cementum matrix has been laid down, its mineralization begins. The uncalcified matrix is called cementoid. Calcium and phosphate ions present in tissue fluids are deposited into the matrix and are arranged as unit cells of hydroxyapatite. Mineralization of cementoid is a highly ordered event and not the random precipitation of ions into an organic matrix.

Cementoid tissue. Under normal conditions growth of cementum is a rhythmic process, and as a new layer of cementoid is formed, the old one calcifies. A thin layer of cementoid can

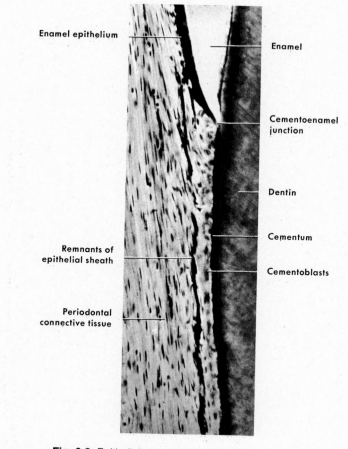

Enamel epithelium

Enamel

Cementoenamel junction

Dentin

Cementum

Cementoblasts

Remnants of epithelial sheath

Periodontal connective tissue

Fig. 6-2. Epithelial sheath is broken and separated from root surface by connective tissue.

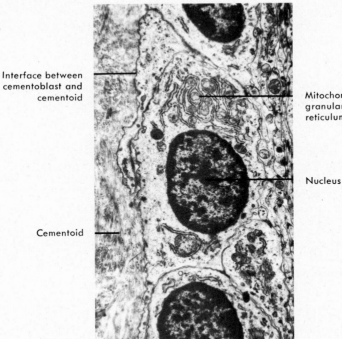

Interface between
cementoblast and
cementoid

Mitochondria and
granular endoplasmic
reticulum

Nucleus

Cementoid

Fig. 6-3. Cementoblasts on surface of cementoid. Mitochondria and granular endoplasmic reticulum are visible. (Electron micrograph; ×8000.) (Courtesy S.D. Lee, Ann Arbor, Mich.)

usually be observed on the cemental surface (Fig. 6-4). This cementoid tissue is lined by cementoblasts. Connective tissue fibers from the periodontal ligament pass between the cementoblasts into the cementum. These fibers are embedded in the cementum and serve to attach the tooth to surrounding bone. Their embedded portions are known as Sharpey's fibers (Fig. 6-5). Each Sharpey's fiber is composed of numerous collagen fibrils that pass well into the cementum (Fig. 6-6).

STRUCTURE

With the light microscope two kinds of cementum can be differentiated: acellular and cellular. The term "acellular cementum" is unfortunate. As a living tissue, cells are an integral part of cementum at all times. However,

some layers of cementum do not *incorporate* cells, the spiderlike cementocytes, whereas other layers do contain such cells in their lacunae. It is probably best to view cementum as a unit consisting of cementoblasts, cementoid, and fully mineralized tissue.

Acellular cementum may cover the root dentin from the cementoenamel junction to the apex, but is is often missing on the apical third of the root. Here the cementum may be entirely of the cellular type. Cementum is thinnest at the cementoenamel junction (20 to 50 μm) and thickest toward the apex (150 to 200 μm). The apical foramen is surrounded by cementum. Sometimes cementum extends to the inner wall of the dentin for a short distance, and so a lining of the root canal is formed.

In decalcified specimens of cementum, col-

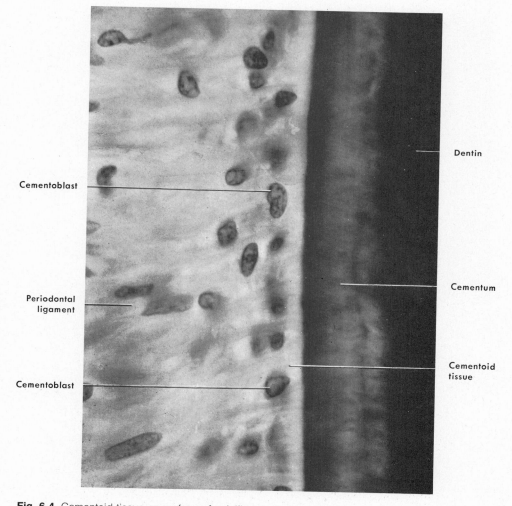

Cementoblast

Dentin

Periodontal
ligament

Cementum

Cementoblast

Cementoid
tissue

Fig. 6-4. Cementoid tissue on surface of calcified cementum. Cementoblasts between fibers.

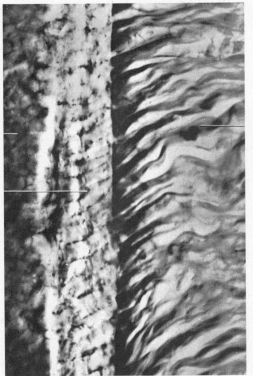

Dentin

Cementum

Fibers of
periodontal
ligament

Fig. 6-5. Fibers of periodontal ligament continue into surface layer of cementum as Sharpey's fibers.

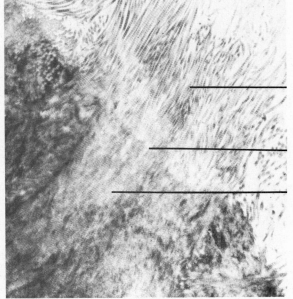

Collagen fibrils
of periodontal
ligament

Surface of
cementum

Collagen fibrils
embedded in
cementum

Fig. 6-6. Collagen fibrils from periodontal ligament continue into cementum. Numerous collagen fibrils embedded in cementum are collectively referred to as Sharpey's fibers. (Decalcified human molar; electron micrograph; ×17,000.)

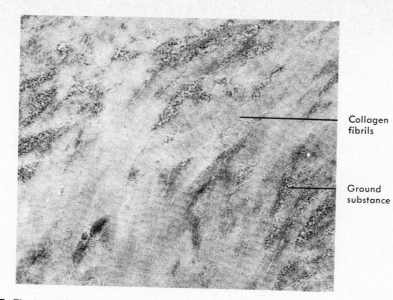

Fig. 6-7. Electron micrograph of human cementum showing ground substance interspersed between collagen fibrils. (Decalcified specimen; ×42,000.)

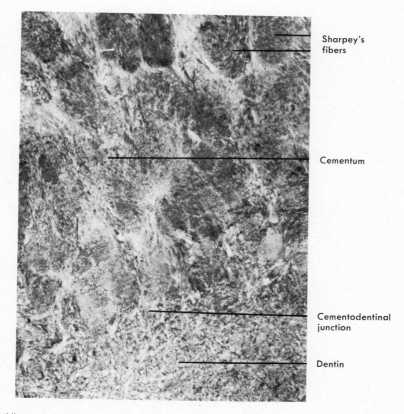

Fig. 6-8. Ultrastructural view of cementodentinal junction of human incisor. In this tangential section, Sharpey's fibers are visible as discrete bundles of collagen fibrils. (Decalcified specimen; electron micrograph; ×5000.)

lagen fibrils make up the bulk of the organic portion of the tissue. Interspersed between some collagen fibrils are electron-dense reticular areas, which probably represent protein polysaccharide materials of the ground substance (Fig. 6-7). Collagen fibrils of both acellular and cellular cementum are arranged in a very complex fashion with little discernible pattern. In some areas, however, relatively discrete bundles of collagen fibrils can be seen, particularly in tangential sections (Fig. 6-8). These bundles are Sharpey's fibers, which make up a substantial portion of the cementum.

In mineralized specimens it has been observed that cemental collagen is not totally mineralized. This is particularly true in a zone 10 to 50 μ wide near the cementodentinal junc-

tion, where unmineralized areas about 1 to 5 μm in diameter are seen. These areas probably represent poorly mineralized cores of Sharpey's fibers.

The cells incorporated into cellular cementum, cementocytes, are similar to osteocytes. They lie in spaces designated as lacunae. A typical cementocyte has numerous cell processes or canaliculi radiating from its cell body. These processes may branch, and they frequently anastomose with those of a neighboring cell. Most of the processes are directed toward the periodontal surface of the cementum. The full extent of these processes does not show up in routinely prepared histologic sections. They are best viewed in mineralized ground sections (Fig. 6-9). The cytoplasm of cementocytes in deeper layers of cementum contains few organ-

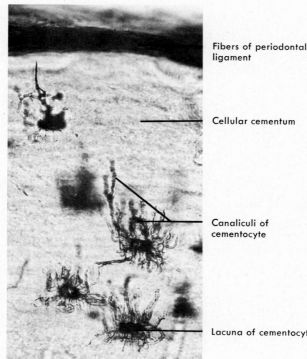

Fibers of periodontal ligament

Cellular cementum

Canaliculi of cementocyte

Lacuna of cementocyte

Fig. 6-9. Cellular cementum from human premolar. Note lacunae of spiderlike cementocytes with numerous canaliculi or cell processes. (Ground section; ×480.)

elles, the endoplasmic reticulum appears dilated, and mitochondria are sparse. These characteristics indicate that cementocytes are either degenerating or are marginally active cells. At a depth of 60 μm or more cementocytes show definite signs of degeneration, such as cytoplasmic clumping and vesiculation. At the light microscopic level, lacunae in the deeper layers of cementum appear to be empty, suggesting complete degeneration of cementocytes located in these areas (Fig. 6-10).

Both acellular and cellular cementum are separated by incremental lines into layers, which indicate periodic formation (Figs. 6-10 and 6-11). Incremental lines can be seen best in decalcified specimens prepared for light-microscopic observation. They are difficult to identify at the ultrastructural level. Histochemical studies indicate that incremental lines are highly mineralized areas with less collagen and more ground substance than other portions of the cementum.

When cementum remains relatively thin, Sharpey's fibers cross the entire thickness of the cementum. With further apposition of cementum, a larger part of the fibers is incorporated in the cementum. The attachment proper is confined to the most superficial or recently formed layer of cementum (Fig. 6-5). This would seem to indicate that the thickness of cementum does not enhance functional efficiency by increasing the strength of attachment of the individual fibers.

The location of acellular and cellular cementum is not definite. As a general rule, however, acellular cementum usually predominates on the coronal half of the root, whereas cellular cementum is more frequent on the apical half.

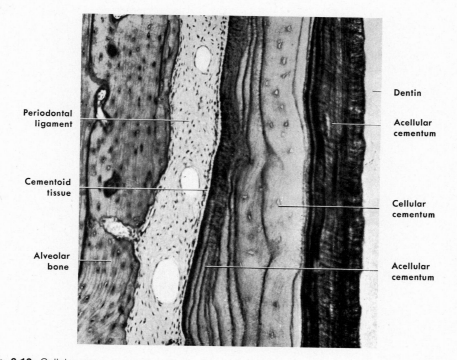

Fig. 6-10. Cellular cementum on surface of acellular cementum and again covered by acellular cementum (incremental lines). Lacunae of cellular cementum appear empty, indicating degeneration of cementocytes.

Layers of acellular and cellular cementum may alternate in almost any pattern. Acellular cementum can occasionally be found on the surface of cellular cementum (Fig. 6-10). Cellular cementum is frequently formed on the surface of acellular cementum (Fig. 6-10), but it may comprise the entire thickness of apical cementum (Fig. 6-12). It is always thickest around the apex and, by its growth, contributes to the length of the root (Fig. 6-13).

Extensive variations in the surface topography of cementum can be observed with the scanning electron microscope. Resting cemental surfaces, where mineralization is more or less complete, exhibit low, rounded projections corresponding to the centers of Sharpey's fibers (Fig. 6-14). Cemental surfaces with actively mineralizing fronts have numerous small openings that correspond to sites where individual Sharpey's fibers enter the tooth (Fig. 6-15). These openings represent unmineralized cores of the fibers. Numerous resorption bays and irregular

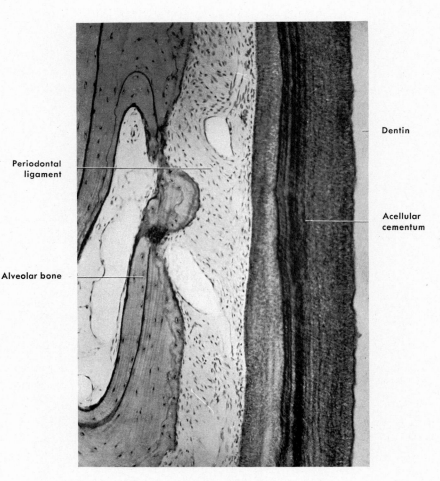

Periodontal ligament

Alveolar bone

Dentin

Acellular cementum

Fig. 6-11. Incremental lines in acellular cementum.

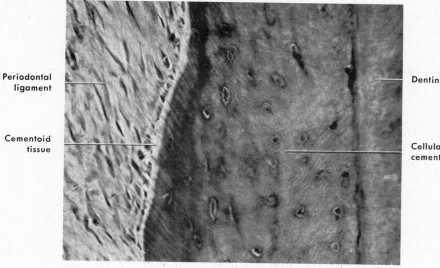

Periodontal ligament

Cementoid tissue

Dentin

Cellular cementum

Fig. 6-12. Cellular cementum forming entire thickness of apical cementum. (From Orban, B.: Dental histology and embryology, Philadelphia, 1929, P. Blakiston's Son & Co.)

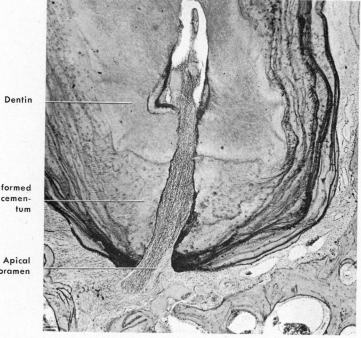

Dentin

Apex formed by cementum

Apical foramen

Fig. 6-13. Cementum thickest at apex, contributing to length of root.

Fig. 6-14. Scanning electron micrograph of resting cemental surface of human premolar. Rounded projections correspond to insertion sites of Sharpey's fibers. (Anorganic preparation; ×3400.) (Courtesy A. Boyde, London.)

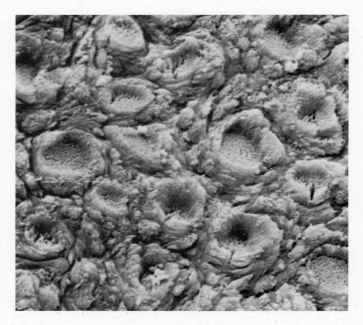

Fig. 6-15. Scanning electron micrograph of cemental surface of human molar with actively mineralizing front. Peripheral portions of Sharpey's fibers are more mineralized than their centers. (Anorganic preparation; approximately ×1500.) (From Jones, S.J., and Boyde, A.: Z. Zellforsch. **130:**318, 1972.)

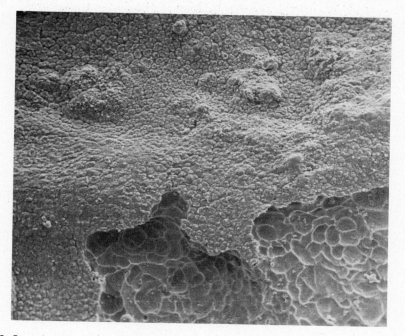

Fig. 6-16. Scanning electron micrograph of cemental surface of human molar showing numerous projections of Sharpey's fibers. Note large multiloculate resorption bay at bottom of field. (Anorganic preparation; ×250.) (From Jones, S.J., and Boyde, A.: Z. Zellforsch. **130:**318, 1972.)

ridges of cellular cementum are also frequently observed on root surfaces (Fig. 6-16).

CEMENTODENTINAL JUNCTION

The dentin surface upon which cementum is deposited is relatively smooth in permanent teeth. The cementodentinal junction in deciduous teeth, however, is sometimes scalloped. The attachment of cementum to dentin in either case is quite firm although the nature of this attachment is not fully understood.

The interface between cementum and dentin is clearly visible in decalcified and stained histologic sections using the light microscope (Figs. 6-10 and 6-11). In such preparations cementum usually stains more intensely than does dentin. When observed with the electron microscope, the cementodentinal junction is not as distinct as when observed with the light microscope. A narrow interface zone between the two tissues, however, can be detected with the electron microscope. In decalcified preparations, cementum is more electron dense than dentin and some of its collagen fibrils are arranged in relatively distinct bundles while those of dentin are arranged somewhat haphazardly (Fig. 6-8). Since collagen fibrils of cementum and dentin intertwine at their interface in a very complex fashion, it is not possible to precisely determine which fibrils are of dentinal and which are of cemental origin.

Sometimes dentin is separated from cementum by a zone known as the intermediate cementum layer, which does not exhibit charac-

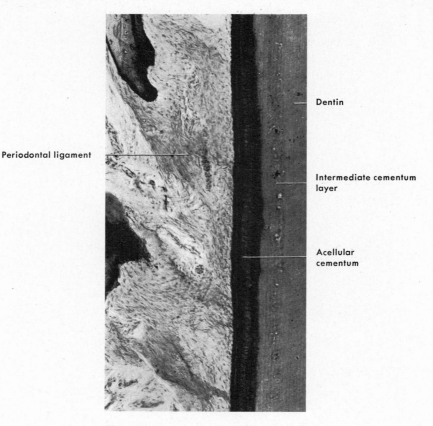

Periodontal ligament

Dentin

Intermediate cementum layer

Acellular cementum

Fig. 6-17. Intermediate layer of cementum.

teristic features of either dentin or cementum (Fig. 6-17). This layer is predominately seen in the apical two thirds of roots of molars and premolars and is only rarely observed in incisors or deciduous teeth. It is believed that this layer represents areas where cells of Hertwig's epithelial sheath become trapped in a rapidly deposited dentin or cementum matrix. Sometimes it is a continuous layer. Sometimes it is found only in isolated areas.

CEMENTOENAMEL JUNCTION

The relation between cementum and enamel at the cervical region of teeth is variable. In approximately 30% of all teeth, cementum meets the cervical end of enamel in a relatively sharp line (Fig. 6-18, *A*). In about 10% of the teeth, enamel and cementum do not meet. Presumably this occurs when enamel epithelium in the cervical portion of the root is delayed in its separation from dentin. In such cases there is no cementoenamel junction. Instead, a zone of the root is devoid of cementum and is, for a time, covered by reduced enamel epithelium.

In approximately 60% of the teeth, cementum overlaps the cervical end of enamel for a short distance (Fig. 6-18, *B*). This occurs when the enamel epithelium degenerates at its cervical termination, permitting connective tissue

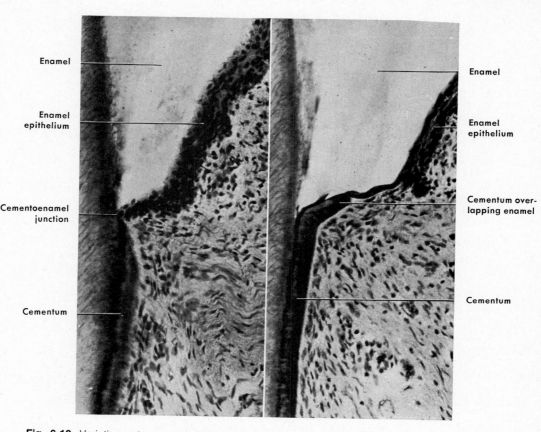

Fig. 6-18. Variations at cementoenamel junction. **A,** Cementum and enamel meet in sharp line. **B,** Cementum overlaps enamel.

to come in direct contact with the enamel surface. Electron microscopic evidence indicates that when connective tissue cells, probably cementoblasts, come in contact with enamel they produce a laminated, electron-dense, reticular material termed afibrillar cementum. Afibrillar cementum is so named because it does not possess collagen fibrils with a 64 nm (640 Å) periodicity. If such afibrillar cementum remains in contact with connective tissue cells for a long enough time, fibrillar cementum with characteristic collagen fibrils may subsequently be deposited on its surface; thus the thickness of cementum that overlies enamel increases.

FUNCTION

The primary function of cementum is to furnish a medium for the attachment of collagen fibers that bind the tooth to alveolar bone. Since collagen fibers of the periodontal ligament cannot be incorporated into dentin, a connective tissue attachment to the tooth is impossible without cementum. This is dramatically demonstrated in some cases of hypophosphatasia, a rare heredity disease in which loosening and premature loss of anterior deciduous teeth occurs. The exfoliated teeth are characterized by an almost total absence of cementum.

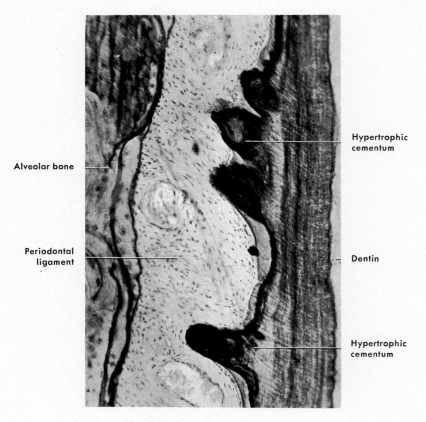

Alveolar bone

Periodontal ligament

Hypertrophic cementum

Dentin

Hypertrophic cementum

Fig. 6-19. Pronglike excementoses.

The continuous deposition of cementum is of considerable functional importance. In contrast to the alternating resorption and new formation of bone, cementum is not resorbed under normal conditions. As the most superficial layer of cementum ages, a new layer of cementum must be deposited to keep the attachment apparatus intact. The repeated apposition of cemental layers represents the aging of the tooth as an organ. In other words, a tooth is, functionally speaking, only as old as the last layer of cementum laid down on its root. The functional age of a tooth may be considerably less than its chronologic age.

Cementum serves as the major reparative tissue for root surfaces. Damage to roots such as fractures and resorptions can be repaired by the deposition of new cementum. Cementum can also be viewed as the tissue that makes functional adaptation of teeth possible. For example, deposition of cementum in an apical area can compensate for loss of tooth substance from occlusal wear.

HYPERCEMENTOSIS

Hypercementosis is an abnormal thickening of cementum. It may be diffuse or circumscribed. It may affect all teeth of the dentition, be confined to a single tooth, or even affect only parts of one tooth. If the overgrowth im-

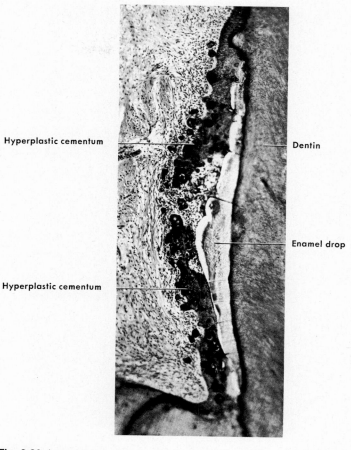

Hyperplastic cementum

Dentin

Enamel drop

Hyperplastic cementum

Fig. 6-20. Irregular hyperplasia of cementum on surface of enamel drop.

proves the functional qualities of the cementum, it is termed a cementum hypertrophy. If the overgrowth occurs in nonfunctional teeth or if it is not correlated with increased function, it is termed hyperplasia.

In localized hypertrophy a spur or pronglike extension of cementum may be formed (Fig. 6-19). This condition frequently is found in teeth that are exposed to great stress. The pronglike extensions of cementum provide a larger surface area for the attaching fibers; thus a firmer anchorage of the tooth to the surrounding alveolar bone is assured.

Localized hypercementosis may sometimes be observed in areas in which enamel drops have developed on the dentin. The hyperplastic cementum covering the enamel drops (Fig. 6-20) occasionally is irregular and sometimes contains round bodies that may be calcified epithelial rests. The same type of embedded calcified round bodies frequently are found in localized areas of hyperplastic cementum (Fig. 6-21). Such knoblike projections are designated as excementoses. They too develop around degenerated epithelial rests.

Extensive hyperplasia of cementum is occa-

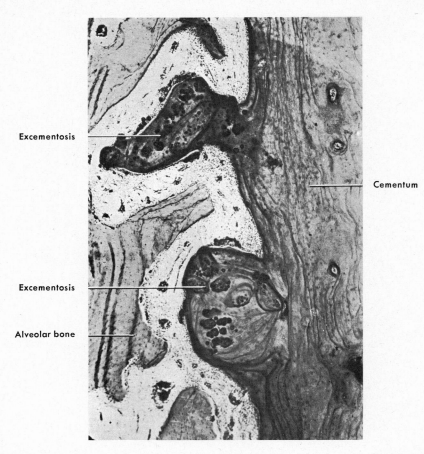

Excementosis

Cementum

Excementosis

Alveolar bone

Fig. 6-21. Excementoses in bifurcation of molar. (From Gottlieb, B.: Oesterr. Z. Stomatol. **19:**515, 1921.)

sionally associated with chronic periapical inflammation. Here the hyperplasia is circumscribed and surrounds the root like a cuff.

A thickening of cementum is often observed on teeth that are not in function. The hyperplasia may extend around the entire root of the nonfunctioning teeth or may be localized in small areas. Hyperplasia of cementum in nonfunctioning teeth is characterized by a reduction in the number of Sharpey's fibers embedded in the root.

The cementum is thicker around the apex of all teeth and in the furcation of multirooted teeth than it is on other areas of the root. This thickening can be observed in embedded as well as in newly erupted teeth.

In some cases an irregular overgrowth of cementum can be found, with spikelike extensions and calcification of Sharpey's fibers and accompanied by numerous cementicles. This type of cemental hyperplasia can occasionally be observed on many teeth of the same dentition and is, at least in some cases, the sequela of injuries to the cementum (Fig. 6-22).

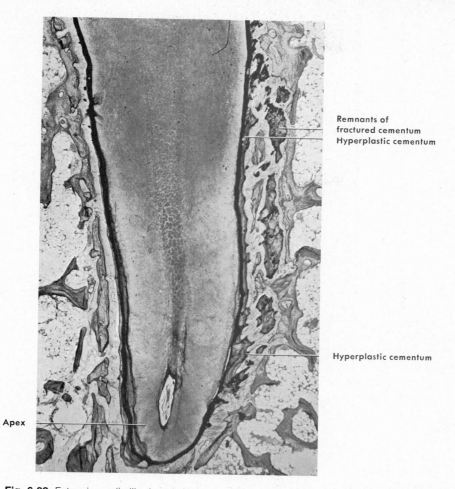

Remnants of
fractured cementum
Hyperplastic cementum

Hyperplastic cementum

Apex

Fig. 6-22. Extensive spikelike hyperplasia of cementum formed during healing of cemental tear.

CLINICAL CONSIDERATIONS

Cementum is more resistant to resorption than is bone, and it is for this reason that orthodontic tooth movement is made possible. When a tooth is moved by means of an orthodontic appliance, bone is resorbed on the side of the pressure, and new bone is formed on the side of tension. On the side toward which the tooth is moved, pressure is equal on the surfaces of bone and cementum. Resorption of bone as well as of cementum may be antici-

pated. However, in careful orthodontic treatment, cementum resorption is minimal or absent but bone resorption leads to tooth migration.

The difference in the resistance of bone and cementum to pressure may be caused by the fact that bone is richly vascularized, whereas cementum is avascular. Thus degenerative processes are much more easily effected by interference with circulation in bone, whereas cementum with its slow metabolism (as in other avascular tissues) is not damaged by a pressure equal to that exerted on bone.

Cementum resorption can occur after trauma or excessive occlusal forces. In severe cases cementum resorption may continue into the dentin. After resorption has ceased, the damage usually is repaired, either by formation of acellular (Fig. 6-23, *A*) or cellular (Fig. 6-23, *B*) cementum or by alternate formation of both (Fig. 6-23, *C*). In most cases of repair there is a tendency to reestablish the former outline of the root surface. This is called *anatomic repair.* However, if only a thin layer of cementum is deposited on the surface of a deep resorption, the root outline is not reconstructed, and a baylike recess remains. In such areas some-

times the periodontal space is restored to its normal width by formation of a bony projection, so that a proper functional relationship will result. The outline of the alveolar bone in these cases follows that of the root surface (Fig. 6-24). In contrast to anatomic repair, this change is called *functional repair.*

If teeth are subjected to a severe blow, fragments of cementum may be severed from the dentin. The tear occurs frequently at the cementodentinal junction, but it may also be in the cementum or dentin.

Transverse fractures of the root may occur after trauma, and these may heal by formation of new cementum.

Frequently, hyperplasia of cementum is sec-

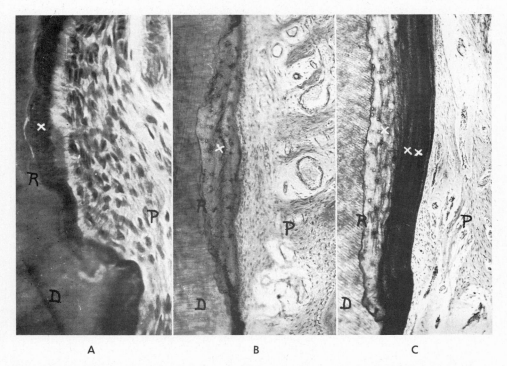

A   B   C

Fig. 6-23. Repair of resorbed cementum. **A,** Repair by acellular cementum, *x.* **B,** Repair by cellular cementum, *x.* **C,** Repair first by cellular, *x,* and later by acellular, *xx,* cementum, *D,* Dentin. *R,* Line of resorption. *P,* Periodontal ligament.

ondary

ondary to periapical inflammation or extensive occlusal stress. This is of practical significance because the extraction of such teeth may necessitate the removal of bone. This also applies to extensive excementoses, as shown in Fig. 6-21. These can anchor the tooth so tightly to the socket that the jaw or parts of it may be fractured in an attempt to extract the tooth. This possibility indicates the necessity for taking roentgenograms before any extraction. Small fragments of roots left in the jaw after extraction of teeth may be surrounded by cementum and remain in the jaw without causing any disturbance.

In periodontal pockets, plaque and its by-products can cause numerous alterations in the physical, chemical, and structural characteristics of cementum. The surface of pathologically exposed cementum becomes hypermineralized because of the incorporation of calcium, phosphorus, and fluoride from the oral environment. At the light-microscopic level no major structural changes occur in the surface of exposed cementum. However, at the ultrastruc-

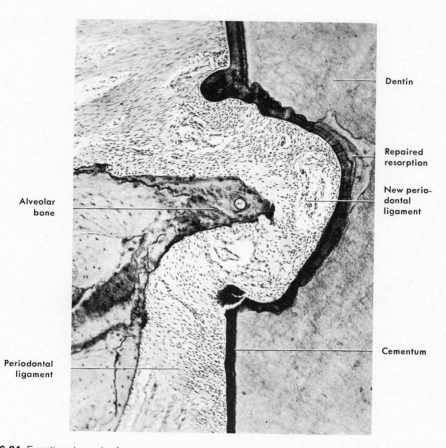

Fig. 6-24. Functional repair of cementum resorption by bone apposition. Normal width of periodontal ligament reestablished.

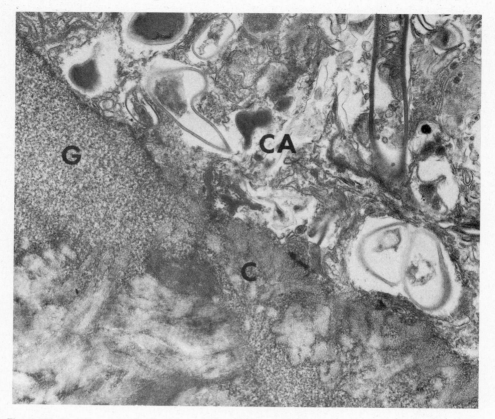

Fig. 6-25. Electron micrograph of exposed cemental surface from tooth with periodontal disease. Collagen fibrils at cemental surface *(C)* have lost their cross-striations or have been replaced by finely granular electron dense material *(G)*. Cell envelopes of bacteria can be observed in calculus *(CA)* on cemental surface. (Decalcified specimen; × 26,000.) (From Armitage, G.C.: Periodont. Abs. **25:**60, 1977.)

tural level there is a loss or decrease in the cross-striations of collagen near the surface (Fig. 6-25). Endotoxin originating from plaque can be recovered from exposed cementum, but it is not known if the distribution of the cementum-bound endotoxin is limited to the cemental surface (adsorbed) or if it penetrates into deeper portions of the root (absorbed). Alterations in exposed cementum are of particular interest to the periodontal therapist since it is believed that they may interfere with healing during periodontal therapy. Consequently, in periodontal therapy, various procedures (mechanical and chemical) have been proposed that are intended to remove this altered cemental surface.

REFERENCES

Armitage, G.C.: Alterations in exposed human cementum, Periodont. Abs. **25:**60, 1977.
Beumer, J., Trowbridge, H.O., Silverman, S., Jr., and Eisenberg, E.: Childhood hypophosphatasia and the premature loss of teeth. A clinical and laboratory study of seven cases, Oral Surg. **35:**631, 1973.

Blackwood, H.J.J.: Intermediate cementum, Br. Dent. J. 102:345, 1957.

Bruckner, R.J., Rickles, N.H., and Porter, D.R.: Hypophosphatasia with premature shedding of teeth and aplasia of cementum, Oral Surg. 15:1351, 1962.

Denton, G.B.: The discovery of cementum, J. Dent. Res. 18:239, 1939.

Eastoe, J.E.: Composition of the organic matrix of cementum, J. Dent. Res. 54 (special issue, abstr. L547):L137, 1975.

El Mostehy, M.R., and Stallard, R.E.: Intermediate cementum, J. Periodont. Res. 3:24, 1968.

Furseth, R.: A microradiographic and electron microscopic study of the cementum of human deciduous teeth, Acta Odontol. Scand. 25:613, 1967.

Furseth, R.: The fine structure of the cellular cementum of young human teeth, Arch. Oral Biol. 14:1147, 1969.

Gedalia, I., Nathan, H., Schapira, J., et al.: Fluoride concentration of surface enamel, cementum, lamina dura and subperiosteal bone from the mandibular angle of Hebrews, J. Dent. Res. 44:452, 1965.

Gottlieb, B.: Zementexostosen, Schmelztropfen und Epithelnester (Exostosis of cementum, enamal drops and epithelial rests), Osterr. Z. Stomatol. 19:515, 1921.

Gottlieb, B.: Biology of the cementum, J. Periodontal, 13:13, 1942.

Jones, S.J., and Boyde, A.: A study of human root cementum surfaces as prepared for and examined in the scanning electron microscope, Z. Zellforsch. 130:318, 1972.

Kronfeld, R.: The biology of cementum, J. Am. Dent. Assoc. 25:1451, 1938.

Kronfeld, R.: Coronal cementum and coronal resorption, J. Dent. Res. 17:151, 1938.

Lester, K.S.: The incorporation epithelial cells by cementum. J. Ultrastruct. Res. 27:63, 1969.

Linden, L-åA: Microscopic observations of fluid flow through cementum and dentine. An in vitro study on human teeth, Odontol. Revy 19:367, 1968.

Listgarten, M.A.: Phase-contrast and electron microscopic study of the junction between reduced enamel epithelium and enamel in unerupted human teeth, Arch. Oral Biol. 11:999, 1966.

Nihei, I.: A study on the hardness of human teeth, J. Osaka Univ. Dent. Soc. 4:1, 1959.

Olsen, T., and Johansen, E.: Inorganic composition of sound and carious human cementum. Preprinted abstracts. Fiftieth General Meeting of the International Association for Dental Research, Abstr. no. 174, 1972, p. 91.

Orban, B.: Dental histology and embryology, ed. 2, Philadelphia, 1929, P. Blakiston's Son & Co.

Paynter, K.J., and Pudy, G.: A study of the structure, chemical nature, and development of cementum in the rat, Anat. Rec. 131:233, 1958.

Rautiola, C.A., and Craig, R.G.: The microhardness of cementum and underlying dentin of normal teeth and teeth exposed to periodontal disease, J. Periodontal. 32:113, 1961.

Rodriguez, M.S., and Wilderman, M.N.: Amino acid composition of the cementum matrix from human molar teeth. J. Periodontal. 43:438, 1972.

Schroeder, H.E., and Listgarten, M.A.: Fine structure of the developing epithelial attachment of human teeth. In Wolsky, A., editor: Monographs in developmental biology, vol. 2, Basel, 1971, S. Karger, AG.

Selvig, K.A.: Electron microscopy of Hertwig's epithelial root sheath and of early dentin and cementum formation in the mouse incisor. Acta Odontol. Scand. 21:175, 1963.

Selvig, K.A.: An ultrastructural study of cementum formation, Acta Odontol. Scand. 22:105, 1964.

Selvig, K.A.: The fine structure of human cementum, Acta Odontol. Scand. 23:423, 1965.

Selvig, K.A., and Hals, E.: Periodontally diseased cementum studied by correlated microradiography, electron probe analysis and electron microscopy, J. Periodont. Res. 12:419, 1977.

Van Kirk, L.E.: Variations in structure of human enamel and dentin, J. Am. Dent. Assoc. 15:1270, 1928.

Zander, H.A., and Hürzeler, B.: Continuous cementum apposition, J. Dent. Res. 37:1035, 1958.

Zipkin, I.: The inorganic composition of bones and teeth. In Schraer, H., editor: Biological calcification, New York, 1970, Appleton-Century-Crofts.

7
PERIODONTAL LIGAMENT

The periodontium is a connective tissue organ, covered by epithelium, that attaches the teeth to the bones of the jaws and provides a continually adapting apparatus for support of the teeth during function. The periodontium comprises four connective tissues, two mineralized and two fibrous. The two mineralized connective tissues are cementum and alveolar bone (see Chapters 6 and 8) and the two fibrous connective tissues are the periodontal ligament and the lamina propria of the gingiva (see Chapter 9). The periodontium is attached to the dentin of the root of the tooth by cementum and to the bone of the jaws by alveolar bone. The periodontal ligament occupies the periodontal space, which is located between the cementum and the periodontal surface of the alveolar bone, and extends coronally to the most apical part of the lamina propria of the gingiva. By definition therefore the coronal part of the periodontal ligament is marked by the most superficial fibers appearing to extend from cementum to alveolar bone. Collagen fibers of the periodontal ligament are embedded in cementum and alveolar bone, so that the ligament provides soft-tissue continuity between the mineralized connective tissues of the periodontium.

The periodontal ligament is a fibrous connective tissue that is noticeably cellular (Fig. 7-1) and vascular. All connective tissues, the periodontal ligament included, comprise cells as well as extracellular substance consisting of fibers and ground substance. The majority of the fibers of the periodontal ligament are collagen, and the ground substance is composed of a variety of macromolecules, the basic constituents of which are proteins and polysaccharides. It is

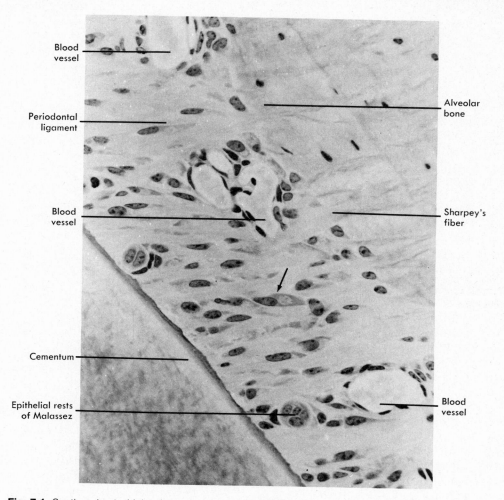

Blood vessel

Periodontal ligament

Blood vessel

Cementum

Epithelial rests of Malassez

Alveolar bone

Sharpey's fiber

Blood vessel

Fig. 7-1. Section, 1 μm thick, of mouse molar periodontal ligament. Tissue is cellular and vascular. The arrow indicates fibroblast that exhibits a negative image of the Golgi complex. (Hematoxylin and eosin; ×700.)

important to remember that the extracellular substance is produced and can be removed by the cells of the connective tissue.

The periodontal ligament has a number of functions, which include attachment and support, nutrition, synthesis and resorption, and proprioception. Over the years it has been described by a number of terms. Among these are desmodont, gomphosis, pericementum,

dental periosteum, alveolodental ligament, and periodontal membrane. "Periodontal membrane" and "periodontal ligament" are the terms that are now most commonly used. Neither term describes the structure and its functions adequately. It is neither a typical membrane nor a typical ligament. However, because it is a complex soft connective tissue providing continuity between two mineralized connective tissues, the term "periodontal ligament" appears to be the more appropriate.

EVOLUTION

There is a fundamental difference between the attachment of reptilian and mammamlian teeth. In the ancestral reptiles the teeth are ankylosed to the bone. In mammals they are suspended in their sockets by ligaments. The evolutionary step from reptile to mammal included a series of coordinated changes in the jaws. The central point of these changes is the radical "reconstruction" of the mandible. In reptiles the mandible consists of a series of bones united by sutures. Only the uppermost bone, the dentary, carries the ankylosed teeth. The mandibular articulation is formed by a separate bone of the mandible, the articulare, and a separate bone of the cranium, the quadratum. During the period of transition from advanced types of reptiles to the first mammals, the dentary attains larger proportions, whereas the other mandibular bones are reduced in size. Finally, only the dentary forms the mammalian mandible. The other bony components of the reptilian mandible are either lost or changed into two of the ossicles of the middle ear: the articulare survives as the malleus and the quadratum as the incus. Before this change could take place, the dentary, growing a condylar process, formed a "new" temporomandibular articulation that, for a time, functioned together with the old articulare-quadratum joint. Such "double-jointed" forms are now known.

The change from the many-boned reptilian to the single-boned mammalian mandible brings with it a radical change in the mode of growth. In the reptile of the growth of the mandible is "sutural," in the same manner as the growth of the cranium. In the mammal the newly acquired cartilage of the condyle takes over as the most important growth site of the mandible. In the reptile, growth of the mandibular body in height occurs in the mandibular sutures, whereas in the mammal is occurs by growth at the free margins of the alveolar process. In the reptile the mandibular (and maxillary) teeth "move" with the bones to which they are fused. In the mammal the teeth have to "move" as units independent of the bones, and this movement is made possible by the remodeling of the periodontium. The evolutionary change from the reptiles to mammals replaces the ankylosis of tooth and bone to a ligamentous suspension of the tooth.

DEVELOPMENT

The dental organ (enamel organ) and, later in tooth development, Hertwig's epithelial root sheath are surrounded by a condensation of cells, the dental sac (see p. 26). A thin layer of these cells that apparently is continuous with the cells of the dental papilla lies adjacent to the dental organ. It has been suggested that the term *dental follicle* be reserved for this layer of cells, and the term *perifollicular mesenchyme* for the cells that surround the dental follicle (Fig. 7-2). The cells of the dental follicle give origin to the cementoblasts that deposit cementum on the developing root, to the fibroblasts of the developing periodontal ligament, and possibly to the osteoblasts of the developing alveolar bone. The formation of the periodontal ligament occurs after the cells of Hertwig's epithelial root sheath (see p. 37) have separated, forming the strands known as the *epithelial rests of Malassez* (see p. 38). This separation permits the cells of the dental follicle to migrate to the external surface of the newly formed root dentin. These migrant follicle cells then differentiate into cementoblasts and deposit cementum on the surface of the dentin. Other cells of the dental follicle differentiate into fibroblasts, which synthesize the fibers and ground substance of the periodontal ligament. The fibers of the periodontal ligament become embedded in newly developed cementum and alveolar bone and, as the tooth erupts, are oriented in characteristic fashion (see Chapter 11).

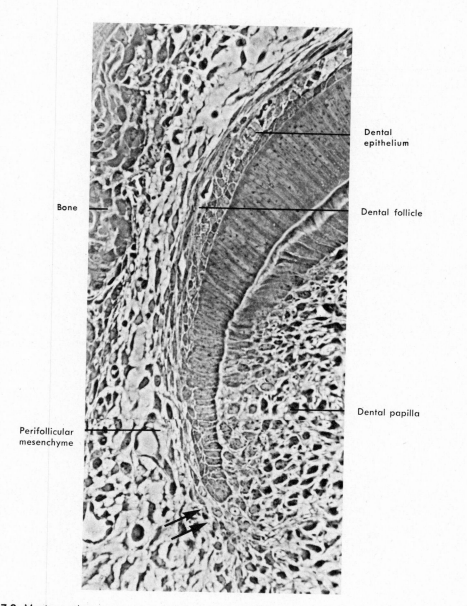

Dental
epithelium

Bone

Dental follicle

Dental papilla

Perifollicular
mesenchyme

Fig. 7-2. Montage phase-contrast photomicrograph of first molar tooth germ of 1-day-old mouse showing the dental follicle, which is continuous with the dental papilla around the cervical loop *(arrows).* (×450.) (From Freeman, E., and Ten Cate, A.R.: J. Periodontol. **42:**387, 1971.)

CELLS

The principal cells of the healthy, functioning periodontal ligament are the differentiated cells and their progenitors. The differentiated cells are concerned with the synthesis and resorption of alveolar bone and the fibrous connective tissue of the ligament and cementum. Consequently, the cells of the periodontal ligament may be divided into three main categories:

Synthetic cells
    Osteoblasts
    Fibroblasts
    Cementoblasts
Resorptive cells
    Osteoclasts
    Fibroblasts
    Cementoclasts
Progenitor cells

There are, in addition, epithelial cells present in the ligament:

Epithelial rests of Malassez

And there are other types of connective tissue cells:

Mast cells
Macrophages

Synthetic cells

There a certain general cytologic criteria that distinguish all cells that are synthesizing proteins for secretion (e.g., extracellular substance of connective tissue), and these criteria can be applied equally to osteoblasts, cementoblasts, and fibroblasts. For a cell to produce protein, it must, among other activities, transcribe ribonucleic acid (RNA) synthesize ribosomes in the nucleolus and transport them to the cytoplasm, and increase its complement of rough endoplasmic reticulum (RER) and Golgi membranes for translation and transport of the protein. It must also have the means to produce an adequate supply of energy. Each of these functional activities is reflected morphologically

when synthetically active tissues are viewed by the electron and light microscopes. Increased transcription of RNA and production of ribosomes is reflected by a large open-faced or vesicular nucleus containing prominent nucleoli. The development of large quantities of rough endoplasmic reticulum covered by ribosomes is readily recognized in the electron microscope and is reflected by hematoxyphilia of the cytoplasm when the cell is seen in the light microscope after staining by hematoxylin and eosin. The hematoxyphilia is the result of interaction of the RNA with the acid hematein in the stain. The Golgi saccules and vesicles are also readily seen in the electron microscope but are not stained by acid hematein and so, in the light microscope, they are seen in appropriate sections as a clear, unstained area in the otherwise hematoxyphilic cytoplasm. The increased requirement for energy is reflected in the electron microscope by the presence of relatively large numbers of mitochondria. Accommodation of all these organelles in the cell requires a large amount of cytoplasm. Thus a cell that is actively secreting extracellular substance will be seen in the light microscope to exhibit a large, open-faced or vesicular nucleus with prominent nucleoli and to have abundant cytoplasm that tends to be hematoxyphilic, with, if the plane of section is favorable, a clear area representing the Golgi membranes (Fig. 7-1).

Cells with the morphology described above, if found at the periodontal surface of the alveolar bone, are active osteoblasts; if lying in the body of the soft connective tissue, they are active fibroblasts; and, if found at cementum, they are active cementoblasts. These cells all have, in addition to the features described above, the particular characteristics of osteoblasts, fibroblasts, and cementoblasts. Detailed descriptions of the first two types of cells can be found in appropriate textbooks and of the third in the chapter on cementum (Chapter 6).

Synthetic cells in all stages of activity are

present in the periodontal ligament, and this is reflected directly by the degree to which the characteristics described above are developed in each cell. Cells having a paucity of cytoplasm (i.e., cytoplasm that virtually cannot be distinguished in the light microscope) and having very few organelles and a close-faced nucleus are also found in the ligament. Some cells that are not actively synthesizing extracellular substance may be progenitor cells and will be discussed below.

Osteoblasts. The osteoblasts covering the periodontal surface of the alveolar bone constitute a modified endosteum and not a perios-

teum. A periosteum can be recognized by the fact that it comprises at least two distinct layers, an inner cellular or cambium layer and an outer fibrous layer. A cellular layer, but not an outer fibrous layer, is present on the periodontal surface of the alveolar bone. The surface of the bone lining the dental socket must therefore be regarded as an interior surface of bone, akin to that lining medullary cavities, and not an external surface, which would be covered by periosteum. The surface of the bone is covered largely by osteoblasts in various stages of differentiation, (Figs. 7-1, 7-3, and 7-4), by progenitor cells, as well as by occasional osteoclasts.

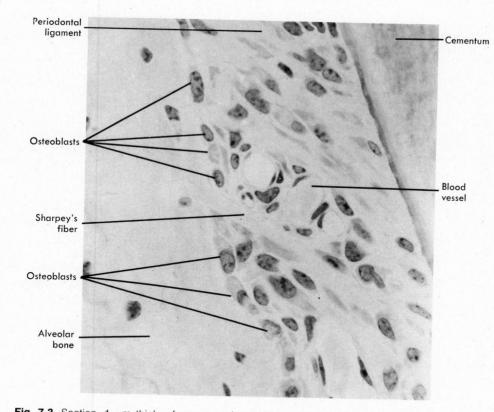

Fig. 7-3. Section, 1 μm thick, of mouse molar periodontal ligament. Note osteoblasts lining periodontal surface of alveolar bone, some of which exhibit a negative image of the Golgi complex. (Hematoxylin and eosin; ×1100.)

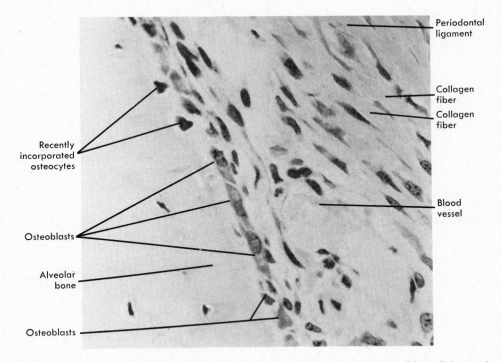

Fig. 7-4. Section, 1 μm thick, of mouse molar periodontal ligament. Note osteoblasts lining periodontal surface of alveolar bone, some of which exhibit a negative image of the Golgi complex. (Hematoxylin and eosin; ×1000.)

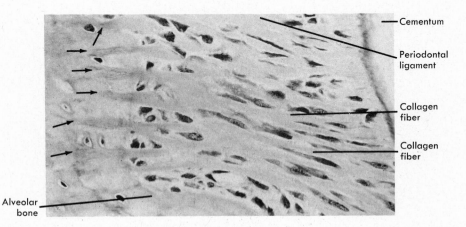

Fig. 7-5. Section, 1 μm thick, of mouse molar periodontal ligament. Collagen fibers from ligament that pass between osteoblasts to penetrate alveolar bone as Sharpey's fibers are shown by arrows. (Hematoxylin and eosin; ×1000.)

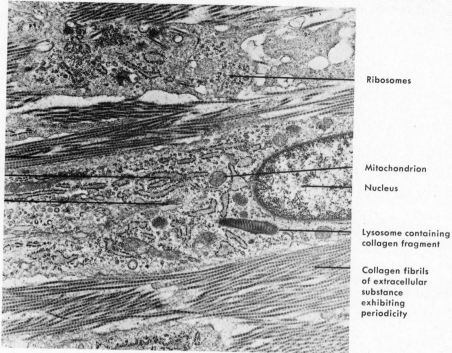

Ribosomes

Collagen fibrils
of extracellular
substance

Mitochondrion

Rough endoplasmic
reticulum

Nucleus

Golgi membranes

Lysosome containing
collagen fragment

Collagen fibrils
of extracellular
substance
exhibiting
periodicity

Fig. 7-6. Fibroblasts and collagen fibrils of mouse molar periodontal ligament. The lower cell exhibits numerous profiles of rough endoplasmic reticulum and Golgi membranes, and a collagen-containing lysosome. (×9250.) (Courtesy Dr. A.R. Ten Cate, Toronto.)

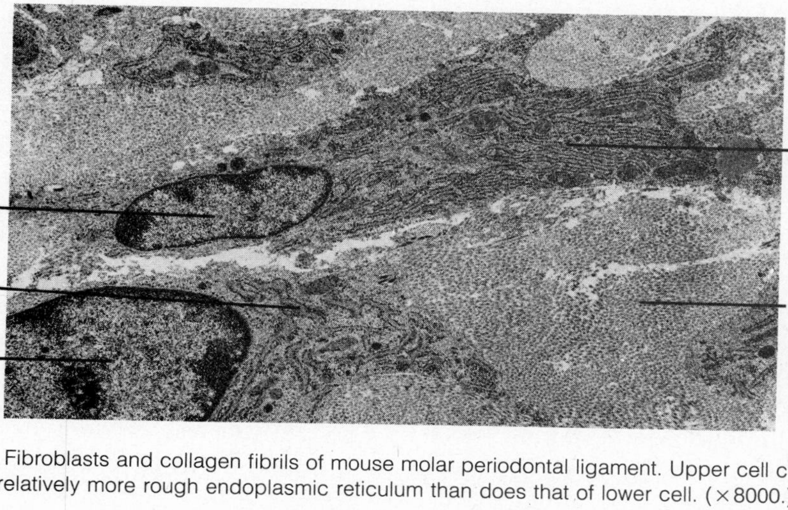

Rough
endoplasmic
reticulum

Nucleus

Rough
endoplasmic
reticulum

Collagen
fibrils
cut in
cross section

Nucleus

Fig. 7-7. Fibroblasts and collagen fibrils of mouse molar periodontal ligament. Upper cell cytoplasm exhibits relatively more rough endoplasmic reticulum than does that of lower cell. (×8000.)

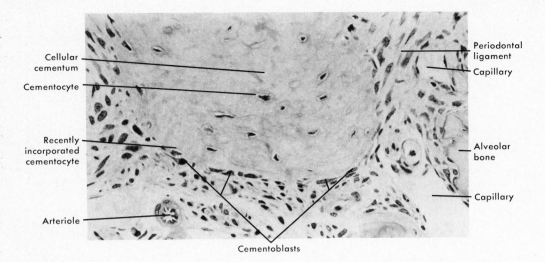

Cellular cementum

Cementocyte

Recently incorporated cementocyte

Arteriole

Periodontal ligament

Capillary

Alveolar bone

Capillary

Cementoblasts

Fig. 7-8. Section, 1 μm thick, of mouse molar periodontal ligament. Note cementoblasts on surface of cellular cementum at apex of root. (Hematoxylin and eosin; ×650.)

Collagen fibers of the ligament that penetrate the alveolar bone intervene between the cells (Figs. 7-1, 7-3, and 7-5).

Fibroblasts. Fibroblasts in various stages of differentiation, and their progenitors, are found in the periodontal ligament, where they are surrounded by fibers and ground substance (Figs. 7-6 and 7-7). In longitudinal sections viewed by light microscopy, the cells of the ligament frequently appear to be oriented parallel to the oriented bundles of collagen fibers (Figs. 7-4 and 7-5).

Cementoblasts. The distribution on the tooth surface of variously differentiated cementoblasts (Fig. 7-8) and their progenitors is similar to the distribution of osteoblasts on the bone surface.

Resorptive cells

Osteoclasts. Osteoclasts are cells that resorb bone and tend to be large and multinucleated (Fig. 7-9) but can be small and mononuclear. Multinucleated osteoclasts are formed by fusion of precursor cells. Circulating monocytes

have been identified as precursor cells. These characteristic multinucleated giant cells usually exhibit an eosinophilic cytoplasm and are easily recognizable. When viewed in the light microscope, the cells may sometimes appear to occupy bays in bone (Howship's lacunae) or surround the end of a bone spicule. In the electron microscope their cytoplasm is seen to exhibit numerous mitochondria and lysosomes, abundant Golgi saccules, and free ribosomes, but little RER. The part of the plasma membrane lying adjacent to bone that is being resorbed is raised in characteristic folds and is termed the *ruffled* or *striated border*. The ruffled border is separated from the rest of the plasma membrane by a zone of specialized membrane that is closely applied to the bone, the underlying cytoplasm of which tends to be devoid of organelles and has been called the *clear zone* (Fig. 7-10). The bone related to the ruffled border can be seen to be undergoing resorption. Resorption occurs in two stages: the mineral is first removed from a narrow zone at the bone margin, and this is followed by disin-

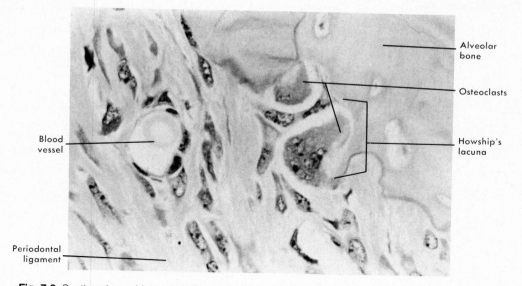

Fig. 7-9. Section, 1 μm thick, illustrating osteoclasts located on periodontal surface of alveolar bone of mouse molar. (Hematoxylin and eosin; ×1600.)

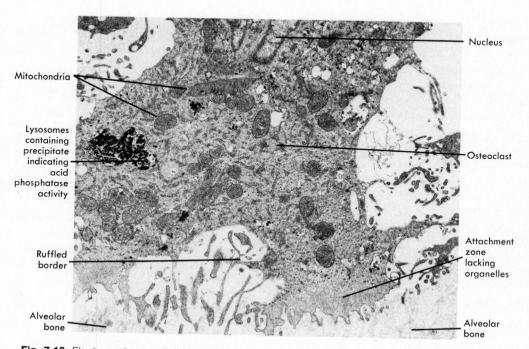

Fig. 7-10. Electron micrograph illustrating acid phosphatase activity in lysosomes of osteoclast located on periodontal surface of alveolar bone of mouse molar. (×14,000.)

tegration of the recognizable exposed organic matrix. The osteoclast appears to accomplish both demineralization and disaggregation of the organic matrix, the latter possibly being achieved by the secretion of appropriate enzymes. However, this question has not been settled. The ruffled border disappears in inactive osteoclasts. Light and electron microscopic histochemical tests can be used to show that osteoclasts are rich in acid phosphatase, which is contained in lysosomes (Fig. 7-10).

The presence of osteoclasts on the periodontal surface of the alveolar bone (Figs. 7-9 and 7-10) indicates that resorption was active or had recently ceased in that area at the time the tissue was removed. Osteoclasts are seen regularly in normal functioning periodontal ligament, in which the cells play a part in the removal and the deposition of bone that are responsible for its remodeling, a process that allows functional changes in the position of teeth that must be accommodated by the supporting tissues (see Chapter 11).

Fibroblasts. It has recently become evident that the collagen fibrils of mammalian periodontal ligament can be resorbed under physiologic conditions by mononuclear fibroblasts. These cells exhibit lysosomes that contain fragments of collagen that appear to be undergoing digestion (Fig. 7-6). The activity of these fibroclastic cells does not appear necessarily to be restricted to destruction of collagen, because large portions of their cytoplasm may be filled with the organelles normally associated with protein synthesis. It must be made clear that there does not appear to be a unique cell that resorbs the extracellular substance of soft connective tissue, but that the fibroblast may be capable of both synthesis and resorption. Collagen-resorbing fibroblasts are inhabitants of normal functioning periodontal ligament, and their presence, like that of osteoclasts in relation to bone, indicates resorption of fibers occurring during physiologic turnover or remodeling of periodontal ligament.

Cementoclasts. Cementoclasts resemble osteoclasts and are occasionally found in normal functioning periodontal ligament. This observation is consistent with the knowledge that cementum is not remodeled in the fashion of alveolar bone and periodontal ligament but that it undergoes continual deposition during life. However, resorption of cementum can occur under certain circumstances, and in these instances mononuclear cementoclasts or multinucleated giant cells, often located in Howship's lacunae, are found on the surface of the cementum. The origin of cementoclasts is unknown, but it is conceivable that they arise in the same manner as osteoclasts.

Progenitor cells

All connective tissues, including periodontal ligament, contain progenitor cells that have the capacity to undergo mitotic division. If they were not present, there would be no cells available to replace differentiated cells dying at the end of their life span or as a result of trauma. It is believed that generally, after division, one of the daughter cells differentiates into a functional type of connective tissue cell (i.e., any one of the cell types described above) while the other remains an undifferentiated progenitor cell retaining the capacity to divide when stimulated appropriately. Progenitor cells may have a small, close-faced nucleus and very little cytoplasm, or they may exhibit the characteristics of more differentiated cells.

Little is known about the progenitor cells of the ligament. For example, it is not known whether a single population of progenitor cells gives rise to all of the specialized synthetic cells in the ligament or if there are a number of populations, each of which gives rise to a different specialized cell. That progenitor cells are present is evident from the burst of mitoses that occur after application of pressure to a tooth as in orthodontic therapy or after wounding, maneuvers that stimulate differentiation of cells of periodontal ligament. The cells that divide in

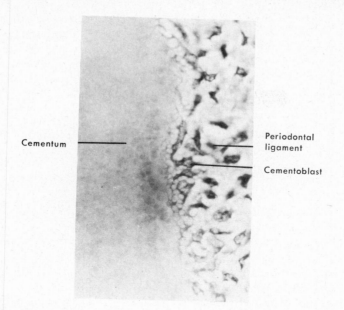

Fig. 7-11. Surface of monkey cementum illustrating cementoblast processes. (Hematoxylin and eosin; ×600.)

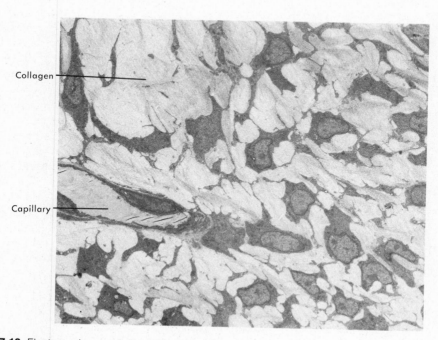

Fig. 7-12. Electron micrograph illustrating fibroblasts and mutual contacts made by their processes in periodontal ligament of mouse molar. (×3040.)

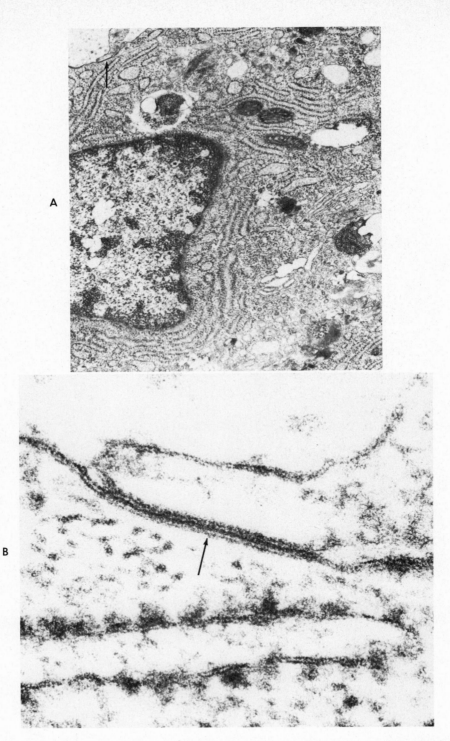

Fig. 7-13. Specialized structure, possibly gap junction *(arrow)*, marking contact between plasma membranes of adjacent fibroblasts in periodontal ligament of mouse molar. (**A,** ×32,000; **B,** ×450,000.)

response to normal biologic requirements and to wounding of the periodontal ligament appear to be located predominantly in the vicinity of blood vessels.

Relationship between cells

The cells of the periodontal ligament form a three-dimensional network, and, in appropriately oriented sections, their processes can be seen to surround the collagen fibers of the extracellular substance. Cells of periodontal ligament associated with bone, fibrous connective tissue, and cementum are not separated from one another, but adjacent cells generally are in contact with their neighbors, usually through their processes (Figs. 7-11 and 7-12). The site of some of the contacts between adjacent cells may be marked by modification of the structure of the contiguous plasma membranes (Fig. 7-13). The nature of these junctions has not yet been elucidated satisfactorily. Although many appear to be zonulae occludentes, it is conceivable that they are in fact gap junctions. Gap junctions in other tissues occur between cells that have been found to be in direct communication with one another. It is evident that some form of communication must exist between the cells of the periodontal ligament;

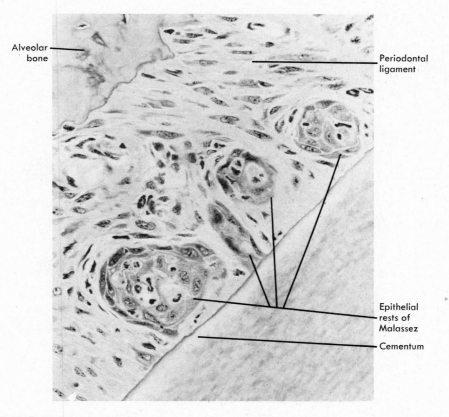

Fig. 7-14. Epithelial rests of Malassez in 1 μm thick section of periodontal ligament of mouse molar. (Hematoxylin and eosin; ×640.)

otherwise, it is difficult to see how the homeo-static mechanisms that are known to operate in the periodontal ligament could function.

Epithelial rests of Malassez

The periodontal ligament contains epithelial cells that are found close to the cementum (Fig. 7-14). These cells were first described by Malassez in 1884 and are the remnants of the epithelium of Hertwig's epithelial root sheath (see Chapter 2). At the time of cementum for-mation the continuous layer of epithelium that covers the surface of the newly formed dentin breaks into lacelike strands (Fig. 7-15). The ep-ithelial rests persist as a network, strands, is-lands, or tubulelike structures near and parallel to the surface of the root (Figs. 7-14 and 7-16).

Only in sections almost parallel to the root can the true arrangment of these epithelial strands be seen. When the tooth is sectioned longitudinally or transversely, the strands of the network are cut in cross section or obliquely and,

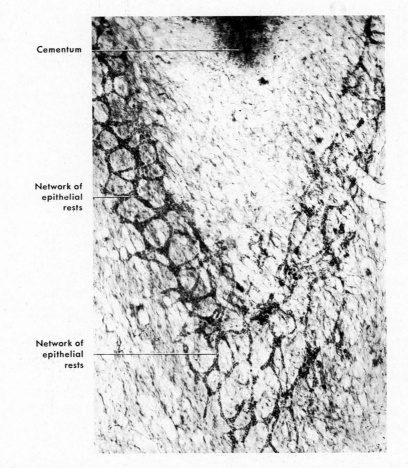

Cementum

Network of epithelial rests

Network of epithelial rests

Fig. 7-15. Network of epithelial rests in periodontal ligament. (Tangential section almost parallel to root surface.)

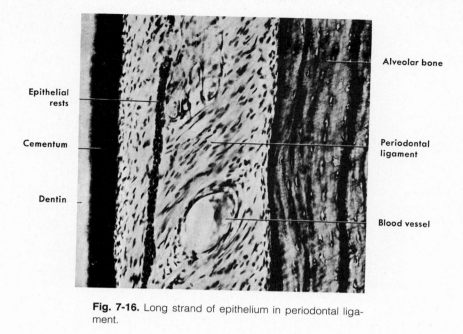

Fig. 7-16. Long strand of epithelium in periodontal ligament.

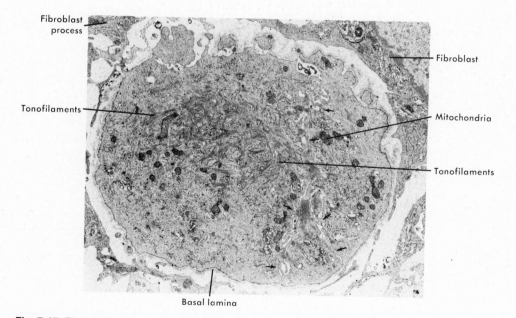

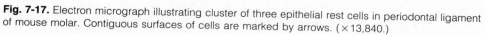

Fig. 7-17. Electron micrograph illustrating cluster of three epithelial rest cells in periodontal ligament of mouse molar. Contiguous surfaces of cells are marked by arrows. (×13,840.)

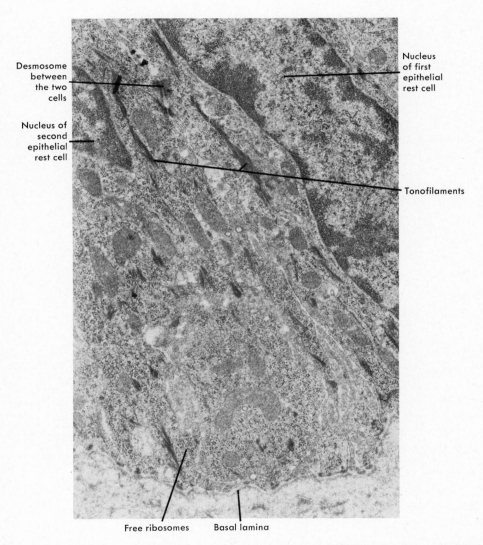

Desmosome between the two cells

Nucleus of second epithelial rest cell

Nucleus of first epithelial rest cell

Tonofilaments

Free ribosomes   Basal lamina

Fig. 7-18. Electron micrograph illustrating portions of two adjacent epithelial rest cells in periodontal ligament of mouse molar. (×20,790.)

as a result, appear as isolated islands when viewed in the light microscope. The cause of disintegration of the epithelium and any inductive influence that it may have on the cells of the dental follicle has not been elucidated.

In rat and mouse molars, most but not all of the epithelium of the developing root is incorporated into the cementum, and consequently the epithelial rests of Malassez are sparse.

Electron microscopic observations show that the epithelial rest cells exhibit tonofilaments (Figs. 7-17 and 7-18) and that they are attached to one another by desmosomes (Fig. 7-18). The epithelial cell clusters are isolated from the connective tissue cells by a basal lamina similar to that occurring at the junction of epithelium and connective tissue elsewhere in the body (Fig. 7-18). It is evident from their ultrastructure, response to histochemical tests, and behavior in cell, tissue, and organ culture that, although the epithelial cells appear in some mammals to decrease with age, they are not effete. However, their physiologic role, if any, in the functioning periodontal ligament is unknown. The fact that methods have recently been described for culturing the rest cells in vitro suggests that it may soon be possible to obtain new information about their functional capabilities. When certain pathologic conditions are present, cells of the epithelial rests can undergo rapid proliferation and can produce a variety of cysts and tumors that are unique to the jaws.

Mast cells

The mast cell is a relatively small, round or oval cell having a diameter of about 12 to 15 μm. The cells are characterized by numerous cytoplasmic granules, which frequently obscure the small, round nucleus. The granules stain with basic dyes but are most readily demonstrated by virtue of their capacity to stain metachromatically with metachromatic dyes such as azure A. They are also positively stained by the periodic acid–Schiff reaction. The granules have been shown to contain heparin and histamine and, in some animals, serotonin. In some preparations, mast cells may be seen to have degranulated so that many or all of the granules are located outside the cell.

Electron microscopy shows the mast cell cytoplasm to contain free ribosomes, short profiles of granular endoplasmic reticulum, few round mitochondria, and a prominent Golgi apparatus. The granules average about 0.5 to 1 μm in diameter and are membrane bound.

The physiologic role of heparin in mast cells does not appear to be clear. Mast cell histamine plays a role in the inflammatory reaction, and mast cells have been shown to degranulate in response to antigen-antibody formation on their surface. Occasional mast cells may be seen in the healthy periodontal ligament.

Macrophages

Macrophages may also be present in the ligament. However, it is important to understand that the only certain criterion by which macrophages can be distinguished from fibroblasts in the light microscope is by the presence of phagocytosed material in their cytoplasm and, further, that differentiated cells in the periodontal ligament are capable of phagocytosis. The wandering type of macrophage, probably derived from blood monocytes, has a characteristic ultrastructure that permits it to be readily distinguished from fibroblasts. It has a nucleus, generally of regular contour, which may be horseshoe or kidney shaped and which exhibits a dense uneven layer of peripheral chromatin. Nucleoli are rarely seen. The surface of the cell is generally raised in microvilli, and the cytoplasm contains numerous free ribosomes. The rough endoplasmic reticulum is relatively sparse and is adorned with widely spaced polysomes that are composed of only two to four ribosomes each. The Golgi apparatus is not well developed, but the cytoplasm contains numerous lysosomes in which identifiable material may be seen.

EXTRACELLULAR SUBSTANCE

The extracellular substance of the periodontal ligament comprises the following:

Fibers
    Collagen
    Oxytalan
Ground substance
    Proteoglycans
    Glycoproteins

Fibers

The fibers in human periodontal ligament are made up of collagen and oxytalan. Elastic fibers are restricted almost entirely to the walls of the blood vessels. The majority of fibers in the periodontal ligament are collagen.

Collagen. Collagen is a specific, high-molecular-weight protein to which are attached a small number of sugars. There are at least five different types of collagen, all basically similar in chemical structure, but each exhibiting certain specific and unique chemical characteristics. Periodontal ligament appears to be made up predominantly of type I collagen, but there is evidence that type III collagen is also present. Collagen *macromolecules* are rod-like, being very long in relation to their diameter, and are arranged to form *fibrils*. These fibrils show a highly ordered periodic banding pattern that is definitive for collagen when viewed in longitudinal section in the electron microscope (Fig. 7-6), but because of their small diameter, they cannot be resolved by light microscopy. However, the fibrils are packed side by side to form bundles or *fibers*, which, when of diameter greater than 0.2 μm, can be seen at the highest magnification of the light microscope. Fibers are the smallest order of collagen that can be resolved by light microscopy. Collagen fibers are further gathered together to form larger bundles, and these are readily resolved by light microscopy. The collagen fibrils of periodontal ligament, when examined by transmission electron microscopy, are seen to be gathered

together to form fibers. When examined in the light microscope, many of the collagen fibers are found to be gathered into bundles having clear orientation relative to the periodontal space, and these are termed *principal fibers.*

The principal fibers of the periodontal ligament (Figs. 7-19 to 7-21) are arranged in five particular groups, each group having a name, as follows:

1. *Alveolar crest group.* The fiber bundles of this group radiate from the crest of the alveolar process and attach themselves to the cervical part of the cementum.
2. *Horizontal group.* The bundles run at right angles to the long axis of the tooth, from the cementum to the bone.
3. *Oblique group.* The bundles run obliquely. They are attached in the cementum somewhat apically from their attachment to the bone. These fiber bundles are most numerous and constitute the main attachment of the tooth.
4. *Apical group.* The bundles are irregularly arranged and radiate from the apical region of the root to the surrounding bone.
5. *Interradicular group.* From the crest of the interradicular septum, bundles extend to the furcation of multirooted teeth.

There are also fiber bundles in the lamina propria of the gingiva that have specific orientation, and some of them lie immediately coronal to the periodontal ligament (see Chapter 9). The most superficial fibers of the alveolar crest group of principal fibers mark the coronal extremity of the periodontal ligament.

Collagen fibers are embedded into cementum on one side of the periodontal space and into alveolar bone on the other. The embedded fibers are termed *Sharpey's fibers* (Figs. 7-1, 7-3, and 7-5). There is some evidence from small rodents and monkeys that Sharpey's fibers may traverse the bone of the alveolar process, particularly in the crestal region, to continue interdentally as principal fibers in the adjacent

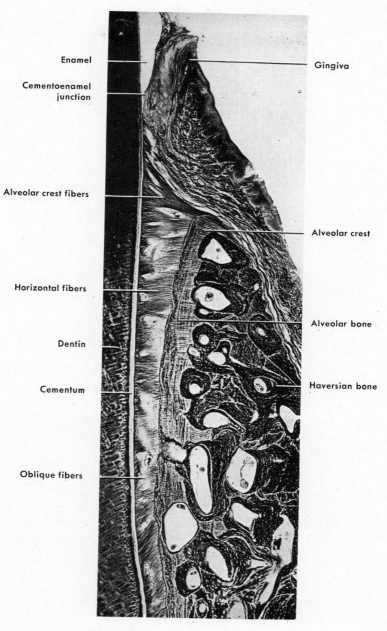

Enamel

Cementoenamel
junction

Alveolar crest fibers

Horizontal fibers

Dentin

Cementum

Oblique fibers

Gingiva

Alveolar crest

Alveolar bone

Haversian bone

Fig. 7-19. Fibers of periodontal ligament.

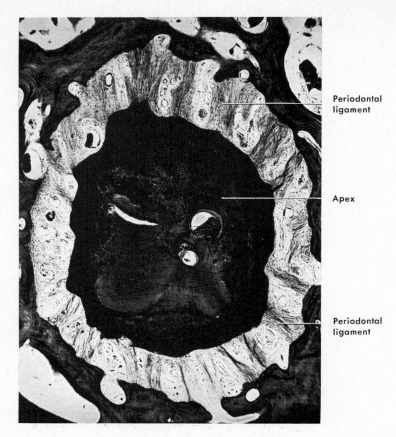

Periodontal
ligament

Apex

Periodontal
ligament

Fig. 7-20. Apical fibers of periodontal ligament. (From Orban, B.: Dental histology and embryology, Philadelphia, 1929, P. Blakiston's Son & Co.)

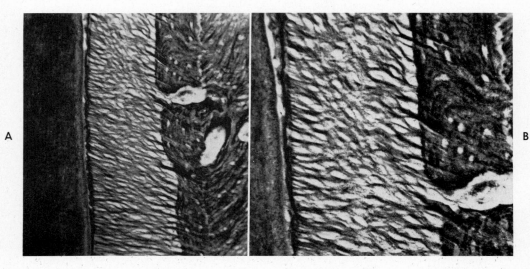

A

B

Fig. 7-21. A, Periodontal ligament of monkey premolar demonstrating principal fibers. **B,** Higher magnification of an area of **A.** Zone described as intermediate plexus is evident. (Silver impregnation, **A,** ×400; **B,** ×800.) (Courtesy Dr. I. Sciaky, Jerusalem.)

periodontal ligament or to mingle buccally and lingually with the fibers of the periosteum covering the alveolar process. However, this may not be a common finding, and, in any event, the arrangement would be unlikely to persist after bone of the alveolar process has been remodeled.

The principal fibers frequently run a wavy course from cementum to bone. It may appear in some sections examined in the light microscope as though fibers arising from cementum and bone are joined in the midregion of the periodontal space, giving rise to a zone of distinct appearance, the so-called *intermediate plexus* (Fig. 7-21, *B*) (see also Chapter 11). It used to be believed that the intermediate plexus provides a site where rapid remodeling of fibers occurs, allowing adjustments in the ligament to be made to accommodate small movements of the tooth. However, evidence derived from electron microscopy, radioautography, and surgical experiments on teeth of limited eruption provide no support for this belief. The so-called intermediate plexus is evidently an artifact arising out of the plane of section and may be attributable to the fact that the collagen fibers do not course only in one bundle but may move from one bundle to the other. This is most readily seen in horizontal sections of the ligament, where the fibers are found to be arranged in many small bundles on the tooth side but in a few large bundles on the bone side.

It no longer seems important to question whether fibers run continuously from tooth to bone. First, the length of a single fibril is not known; second, the fibers evidently run from one bundle to the other, probably to be spliced with, and incorporated into, their neighbors; and, third, the protein of the periodontal ligament is continually being remodeled, and so parts of fibrils must continually be removed and new pieces added, an alteration that does not necessarily reestablish continuity in the old orientation.

Oxytalan. Although elastic fibers are found in the periodontal ligaments of some animals, they are largely restricted to the walls of the blood vessels in humans. A fiber termed oxytalan, which may be an immature elastic fiber, is found in human periodontal ligament. Oxytalan fibers can be demonstrated in the light microscope in tissue stained by certain methods used to color elastic fibers, provided that the tissues are oxidized prior to staining. In the electron microscope, fibers believed to be oxytalan resemble developing elastic fibers.

The orientation of the oxytalan fibers is quite different from that of the collagen fibers. Instead of running from bone to tooth, they tend to run in an axial direction (Fig. 7-22), one end being embedded in cementum or possibly bone and the other often in the wall of a blood vessel. In the vicinity of the apex they form a complex network. The function of the oxytalan fibers is unknown, but it has been suggested that they may play a part in supporting the blood vessels of the periodontal ligament.

Ground substance

The space between cells, fibers, blood vessels, and nerves in the periodontal space is occupied by ground substance. Indeed, the ground substance is present in every nook and cranny, including the interstices between fibers and between fibrils. It is important to understand that all anabolites reaching the cells from the microcirculation in the ligament and all catabolites passing in the opposite direction must pass through the ground substance. Its integrity is essential if the cells of the ligament are to function properly. The importance of the ground substance is frequently overlooked, possibly because it is a difficult substance to investigate and also because it is not demonstrated and therefore not recognizable in tissue prepared by routine methods for light and electron microscopy.

In essence, the ground substance is made up of two major groups of substances, *proteogly-*

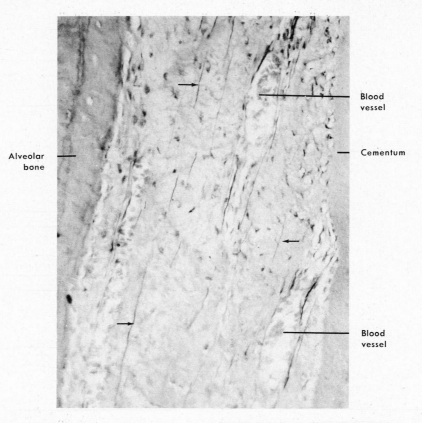

Alveolar bone

Blood vessel

Cementum

Blood vessel

Fig. 7-22. Oxytalan fibers *(arrows)* in monkey periodontal ligament. (×245.)

çans and *glycoproteins*. Both groups are composed of proteins and polysaccharides, but of different type and arrangement, and proteoglycans carry a much stronger negative charge than do the glycoproteins. The interested reader will find detailed descriptions of protein polysaccharides and glycoproteins in texts concerned with connective tissue biochemistry.

It was mentioned above that neither of these substances is demonstrated by routine histologic or electron-microscopic methods; they are demonstrated only by histochemical methods. A histochemical method is in essence a technique that attaches a material that can be rec-ognized microscopically to specific chemical groups in the substance to be demonstrated (see Chapter 15). For light microscopy, the specific substance is a dye that can be recognized by its color, and for electron microscopy, an electron-dense material. In the case of the proteoglycans, a number of methods that utilize their strong negative charges have been developed to demonstrate the location of these substances in both the light and electron microscopes. Examples of the chemicals used are Alcian blue 8GX and toluidine blue for light microscopy and ruthenium red for electron microscopy. Glycoproteins possess comparatively unique chemical groups (1, 2, glycols) that can

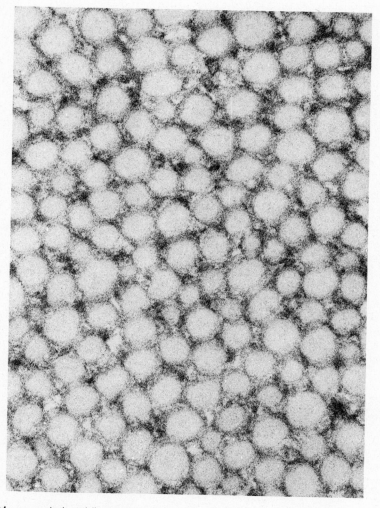

Fig. 7-23. Mouse periodontal ligament stained with ruthenium red to demonstrate proteoglycan of ground substance. Collagen are electron lucent, and ground substance is electron dense. (×162,000.)

be demonstrated in light microscopy by the periodic acid–Schiff method and in electron microscopy by the periodic acid–silver methanamine technique. These methods show quite clearly that ground substance is a significant constituent of the periodontal ligament (Fig. 7-23).

A particular glycoprotein, *fibronectin*, occurs in filamentous form in the periodontal ligament. It contains chemical groups that attach to the surface of the fibroblasts and to collagen, certain proteoglycans, and fibrin. Its orientation may be related to that of the microfila-

ments in the cytoplasm of contiguous fibroblasts.

Interstitial tissue

Some of the blood vessels, lymphatics, and nerves of the periodontal ligament are surrounded by loose connective tissue, and these areas can readily be recognized in the light microscope (Fig. 7-24). These areas have been termed *interstitial tissue*, but it is not known whether they have any particular biologic significance.

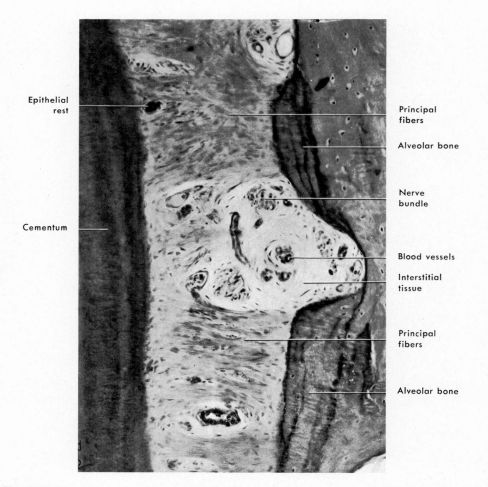

Fig. 7-24. Interstitial spaces in periodontal ligament contain loose connective tissue, vessels, and nerves. (From Orban, B.: J. Am. Dent. Assoc. **16:**405, 1929.)

STRUCTURES PRESENT IN CONNECTIVE TISSUE

The following discrete structures are present in the connective tissue of the periodontal ligament:

Blood vessels
Lymphatics
Nerves
Cementicles

Blood vessels. The arterial vessels of the periodontal ligament are derived from three sources:

Branches in the periodontal ligament from apical vessels that supply the dental pulp.

Branches from intra-alveolar vessels. These branches run horizontally, penetrating the alveolar bone to enter the periodontal ligament (Fig. 7-25) (see Chapter 8).

Branches from gingival vessels. These enter the periodontal ligament from the coronal direction.

The arterioles and capillaries of the microcirculation ramify in the periodontal ligament, forming a rich network of arcades that is more evident in the half of the periodontal space ad-

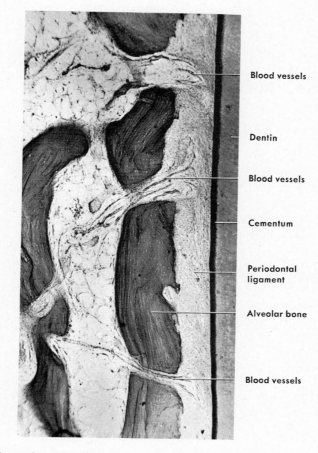

Fig. 7-25. Blood vessels enter periodontal ligament through openings in alveolar bone. (From Orban, B.: Dental histology and embryology, Philadelphia, 1929, P. Blakiston's Son & Co.)

jacent to bone than that adjacent to cementum. There is a particularly rich vascular plexus at the apex and in the cervical part of the ligament. The venous vessels tend to run axially to drain to the apex. There are numerous arteriovenous anastomoses between the two sides of the microcirculation, as well as glomerulus-like structures, and these are possibly involved in the role that the circulation plays in supporting the teeth during function.

Lymphatics. A network of lymphatic vessels, following the path of the blood vessels, provides the lymph drainage of the periodontal ligament. The flow is from the ligament toward and into the adjacent alveolar bone.

Nerves. Nerves, which usually are associated with blood vessels, pass through foramina in the alveolar bone, including the apical foramen, to enter the periodontal ligament. In the region of the apex, they run toward the cervix, whereas along the length of the root they branch and run both coronally and apically.

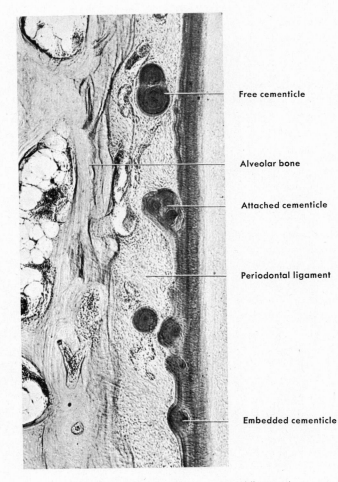

Free cementicle

Alveolar bone

Attached cementicle

Periodontal ligament

Embedded cementicle

Fig. 7-26. Cementicles in periodontal ligament.

The nerve fibers are either of large diameter and myelinated or of small diameter, in which case they may or may not be myelinated. The small fibers appear to end in fine branches throughout the ligament, and the large fibers in a variety of endings, for example, knoblike, spindlelike, and Meissner-like, but these seem to vary among the species. The large-diameter fibers appear to be concerned with discernment of touch and the small-diameter ones with pain. Some of the unmyelinated small-diameter fibers evidently are associated with blood vessels and presumably are autonomic.

Cementicles. Calcified bodies called cementicles are sometimes found in the periodontal ligament. These bodies are seen in older individuals, and they may remain free in the connective tissue, they may fuse into large calcified masses, or they may be joined with the cementum (Fig. 7-26). As the cementum thickens with advancing age, it may envelop these bodies. When they are adherent to the cementum, they form excementoses. The origin of these calcified bodies is not established. It is possible that degenerated epithelial cells form the nidus for their calcification.

FUNCTIONS

The periodontal ligament has the following functions:

Supportive
Sensory
Nutritive
Homeostatic

Supportive. When a tooth is moved in its socket as a result of forces acting on it during mastication or through application of an orthodontic force, part of the periodontal space will be narrowed and the periodontal ligament contained in these areas will be compressed. Other parts of the periodontal space will be widened. The compressed periodontal ligament provides support for the loaded tooth. The collagen fibers in the compressed ligament

act as a cushion for the displaced tooth. The ground substance, which, as a result of its chemical constitution, binds large quantities of water, does the same. The pressure of blood in the numerous vessels also provides a hydraulic mechanism for the support of the teeth. It has often been suggested that the collagen fibers in the widened parts of the periodontal space are extended to their limit when a force is applied to a tooth and, being nonelastic, prevent the tooth from being moved too far. However, evidence to support this contention appears to be lacking, and the role of the collagen fibers seems to be largely restricted to (1) attaching the cementum that is fused to the dentin of the root to alveolar bone and (2) acting as a cushion. The collagen fibers may be extended when a tooth is rotated excessively.

Sensory. The periodontal ligament, through its nerve supply, provides a most efficient proprioceptive mechanism, allowing the organism to detect the application of the most delicate forces to the teeth and very slight displacement of the teeth. Anyone who has bitten into soft food containing a small hard object such as stone or shot knows the importance of this mechanism in protecting both the supporting structures of the tooth and the substance of the crown from the effects of excessively vigorous masticatory movements.

Nutritive. The ligament transmits blood vessels, which provide anabolites and other substances required by the cells of the ligament, by the cementocytes, and presumably by the more superficial osteocytes of the alveolar bone. Extirpation of the ligament results in necrosis of underlying cementocytes. The blood vessels are also concerned with removal of catabolites. Occlusion of blood vessels leads to necrosis of cells in the affected part of the ligament; this occurs when too heavy a force is applied to a tooth in orthodontic therapy.

Homeostatic. It is evident that the cells of the periodontal ligament have the capacity to resorb and synthesize the extracellular substance of the connective tissue of the ligament, alveolar bone, and cementum. It is also evident that these processes are not activated sporadically or haphazardly but function continuously, with varying intensity, throughout the life of the tooth. Alveolar bone appears to be resorbed and replaced (i.e., remodeled) at a rate higher than other bone tissue in the jaws. Furthermore the collagen of the periodontal ligament is turned over at a rate that may be the fastest of all connective tissues in the body, and the cells in the bone half of the ligament may be more active than those on the cementum side. Visual evidence for the high turnover of *protein* in the periodontal ligament is provided by the numerous silver grains seen in radioautographs of the tissue removed from animals a few hours after they have received an injection of a radioactive precursor, for example, ³H-proline (Fig. 7-27). On the other hand, depo-

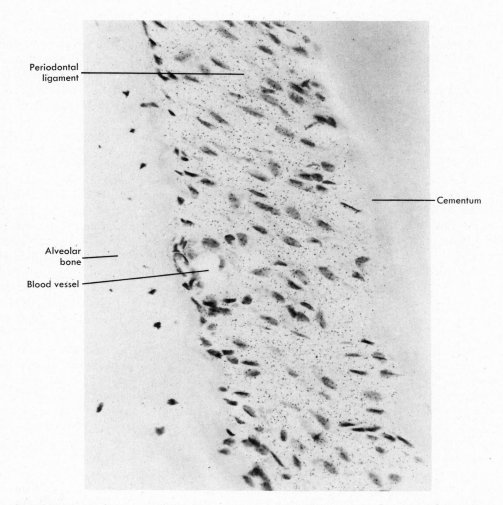

Fig. 7-27. Light microscope radioautograph of periodontal ligament of molar of mouse that had received an intraperitoneal injection of ³H-proline 24 hours before death. Black dots (silver grains) mark sites where isotope was incorporated into protein. (×650.)

sition of cementum by cementoblasts appears to be a slow, continuous process, and resorption is not a regular occurrence.

The mechanisms whereby the cells responsible for these processes of synthesis and resorption are controlled is unfortunately unknown. It is evident that the processes are exquisitely controlled as, under normal conditions of function, the various tissues of the periodontium maintain their integrity and relationship to one another. However, when these homeostatic mechanisms are upset, derangement of the periodontium occurs. If periodontal ligament, either in part or whole, is irreparably destroyed, bone will be deposited in the periodontal space, obliterating it, and this will result in ankylosis between bone and tooth. If the balance between synthesis and resorption is disturbed, the quality of the tissues will be changed. For example, if an experimental animal is deprived of substances essential for collagen synthesis, such as vitamin C or protein, resorption of collagen will continue unabated, but its synthesis and replacement will be markedly reduced. This will result in progressive destruction and loss of extracellular substance of periodontal ligament, more advanced on the bone side of the ligament than on the cementum side. This eventually will lead to loss of attachment between bone and tooth and finally to loss of the tooth, such as occurs in scurvy when vitamin C is absent from the diet.

From kinetic studies using labeled precursors of DNA and radioautography, it appears that the connective tissue cells of the periodontal ligament are also turned over. That is, there is apparently a continual slow death of cells, and these are replaced by new cells that are provided by cell division of progenitor cells in the ligament.

Another aspect of homeostasis relates to function. A periodontal ligament supporting a fully functional tooth exhibits all the structural features described above. However, with loss of function, much of the extracellular substance of the ligament is lost, possibly because of diminished synthesis of substances required to replace structural molecules resorbed during normal turnover, and the width of the periodontal space is subsequently decreased (Fig. 7-28). These changes are accompanied by increased deposition of cementum but by a decrease in the mass of alveolar bone tissue per unit volume. The process is reversible if the tooth is returned to function, but the precise nature of the stimuli that control the changed activity of the cells is unknown.

CLINICAL CONSIDERATIONS

The primary role of the periodontal ligament is to support the tooth in the bony socket. Its thickness varies in different individuals, in different teeth in the same person, and in different locations on the same tooth, as is illustrated in Tables 3 and 4.

The measurements shown in Tables 3 and 4 indicate that it is not feasible to refer to an average figure of normal width of the periodontal ligament. Measurements of a large numer of ligaments range from 0.15 to 0.38 mm. The fact that the periodontal ligament is thinnest in the middle region of the root seems to indicate that the fulcrum of physiologic movement is in this region. The thickness of the periodontal ligament seems to be maintained by the functional movements of the tooth. It is thin in functionless and embedded teeth and wide in teeth that are under excessive occlusal stresses (Fig. 7-28).

For the practice of restorative dentistry the importance of these changes in structure is obvious. The supporting tissues of a tooth long out of function are poorly adapted to carry the load suddenly placed on the tooth by a restoration. This applies to bridge abutments, teeth opposing bridges or dentures, and teeth used as anchorage for removable bridges. This may account for the inability of a patient to use a restoration immediately after its placement. Some time must elapse before the supporting tissues become adapted again to the new functional demands. An adjustment period, like-

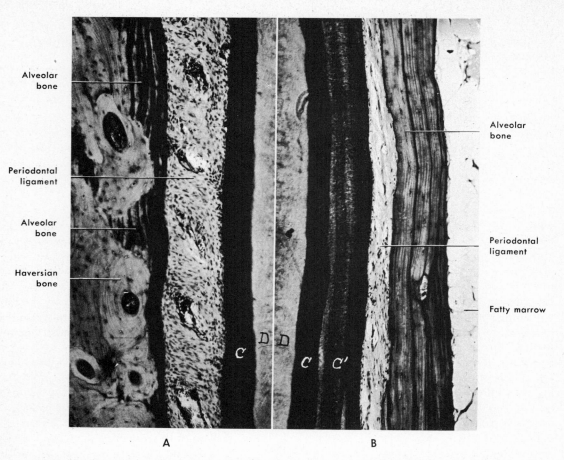

Fig. 7-28. Periodontal ligament of a functioning, **A,** and of a nonfunctioning, **B,** tooth. In the functioning tooth, the periodontal ligament is wide, and principal fibers are present. Cementum, *C,* is thin. In the nonfunctioning tooth, the periodontal space is narrow, and no principal fiber bundles are seen. Cementum is thick, *C* and *C'.* Alveolar bone is lamellated. *D,* Dentin.

Table 3. Thickness of periodontal ligament of 154 teeth from 14 human jaws*

	Average at alveolar crest (mm)	Average at midroot (mm)	Average at apex (mm)	Average for entire tooth (mm)
Ages 11-16 (83 teeth from 4 jaws)	0.23	0.17	0.24	0.21
Ages 32-50 (36 teeth from 5 jaws)	0.20	0.14	0.19	0.18
Ages 51-67 (35 teeth from 5 jaws)	0.17	0.12	0.16	0.15

From Coolidge, E.D.: J. Am. Dent. Assoc. **24:**1260, 1937.
*The table shows that the width of the periodontal ligament decreases with age and that it is wider at the crest and at the apex than at the midroot.

Table 4. Comparison of periodontal ligament in different locations around the same tooth (subject 11 years of age)*

	Mesial (mm)	Distal (mm)	Labial (mm)	Lingual (mm)
Upper right central incisor, mesial and labial drift	0.12	0.24	0.12	0.22
Upper left central incisor, no drift	0.21	0.19	0.24	0.24
Upper right lateral incisor, and labial drift	0.27	0.17	0.11	0.15

From Coolidge, E.D.: J. Am. Dent. Assoc. **24:**1260, 1937.
*The table shows the variation in width of the mesial, distal, labial, and lingual sides of the same tooth.

wise, must be permitted after orthodontic treatment.

Acute trauma to the periodontal ligament, accidental blows, condensing of foil, or rapid mechanical separation may produce pathologic changes such as fractures or resorption of the cementum, tears of fiber bundles, hemorrhage, and necrosis. The adjacent alveolar bone is resorbed, the periodontal ligament is widened, and the tooth becomes loose. When trauma is eliminated, repair usually take place. Occlusal trauma is always restricted to the intra-alveolar tissues and does not cause changes of the gingiva such as recession or pocket formation or gingivitis.

Orthodontic tooth movement depends on resorption and formation of both bone and periodontal ligament. These activities can be stimulated by properly regulated pressure and tension. The stimuli are transmitted through the medium of the periodontal ligament. If the movement of teeth is within physiologic limits (which may vary with the individual), the initial compression of the periodontal ligament on the pressure side is compensated for by bone resorption, whereas on the tension side bone apposition is seen. Application of large forces results in necrosis of periodontal ligament and alveolar bone on the pressure side, and movement of the tooth will occur only after the necrotic bone has been resorbed by osteoclasts located on its endosteal surface.

The periodontal ligament in the periapical area of the tooth is often the site of a pathologic lesion. Inflammatory diseases of the pulp progress to the apical periodontal ligament and replace its fiber bundles with granulation tissue. This lesion, called a dental granuloma, may contain epithelial cells that undergo proliferation and produce a cyst. The dental granuloma and the apical cyst are the most common pathologic lesions of the jaws.

Safeguarding the integrity of the periodontal ligament (and the alveolar bone) is one of the most important challenges for the clinician. Gingivitis, or inflammation of the gingiva, is the most common dental disease of humans. If not controlled or treated, it invariably extends to the periodontal ligament and bone and produces their slow but progressive destruction. Once destroyed by this slow inflammatory process, the periodontal ligament and the alveolar bone are very difficult to regenerate. Therefore the diseases of the periodontal ligament are often irreversible. Their control is the primary aim of good clinical practice and much of what is called preventive dentistry is directed toward these goals.

REFERENCES

Anderson, D.J., Hannam, A.G., and Mathews, B.: Sensory mechanisms in mammalian teeth and their supporting structures, Physiol. Rev. **50:**171, 1970.

Arnim, S.S., and Hagerman, D.A.: The connective tissue fibers of the marginal gingiva, J. Am. Dent. Assoc. **47:**271, 1953.

Bernick, S.: Innervation of teeth and periodontium after enzymatic removal of collagenous elements. Oral Surg. **10**:323, 1957.

Bernick, S., Levy, B.M., Dreizen, S., and Grant, D.A.: The intraosseous orientation of the alveolar component of Marmoset alveodental fibers, J. Dent. Res. **56**:1409, 1977.

Box, K.F.: Evidence of lymphatics in the periodontium, J. Can. Dent. Assoc. **15**:8, 1949.

Brunette, D.M., Kanoza, R.J., Marmary, Y., et al.: Interactions between epithelial and fibroblast-like cells in cultures derived from monkey periodontal ligament, J. Cell Sci. **27**:127-140, 1977.

Brunette, D.M., Melcher, A.H., and Moe, H.K.: Culture and origin of epithelium-like and fibroblast-like cells from porcine periodontal ligament explants and cell suspensions, Arch. Oral Biol. **21**:393, 1976.

Bruszt, P.: Ueber die netzartige Anordnung des paradentalen Epithels (The network arrangement of the epithelium in the periodontal membrane), Z. Stomatol. **30**:679, 1932.

Butler, W.T., Birkedal-Hansen, H., Beegle, W.F., et al.: Proteins of the periodontium. Identification of collagens with the $[\alpha 1(I)]_{2\alpha 2}$ and $[\alpha_1(III)]_3$ structures in bovine periodontal ligament, J. Biol. Chem. **250**:8907, 1975.

Carmichael, G.G., and Fullmer, H.M.: The fine structure of the oxytalan fiber. J. Cell Biol. **28**:33, 1966.

Cohn, S.A.: Disuse atrophy of the periodontium in mice, Arch. Oral Biol. **10**:909, 1965.

Cohn, S.A.: A re-examination of Sharpey's fibres in alveolar bone of the mouse, Arch. Oral Biol. **17**:255, 1972.

Cohn, S.A.: A re-examination of Sharpey's fibres in alveolar bone of the marmoset (*Saguinus fuscicollis*), Arch. Oral Biol. **17**:261, 1972.

Cohn, S.A.: Transalveolar fibres in the human periodontium. Arch. Oral Biol. **20**:257, 1975.

Connor, N.S., Aubin, J.E., and Melcher, A.H.: The distribution of fibronectin in rat tooth and periodontal tissue: an immunofluorescence study using a monoclonal antibody, J. Histochem. Cytochem. **32**:565, 1984.

Coolidge, E.D.: The thickness of the human periodontal membrane. J. Am. Dent. Assoc. **24**:1260, 1937.

Deporter, D.A., and Ten Cate, A.R.: Fine structural localization of acid and alkaline phosphatase in collagen-containing vesicles of fibroblasts, J. Anat. **114**:457, 1973.

Folke, L.E.A., and Stallard, R.E.: Periodontal microcirculation as revealed by plastic microspheres, J. Periodont. Res. **2**:53, 1967.

Freeman, E., and Ten Cate, A.R.: Development of the periodontium: an electron microscopic study, J. Periodontol. **42**:387, 1971.

Fullmer, H.M.: Connective tissue components of the periodontium. In Miles, A.E.W., editor: Structural and chemical organization of the teeth. vol. 2, New York, 1967, Academic Press, Inc.

Garant, P.R., Cho, M.I., and Cullen, M.R.: Attachment of periodontal ligament fibroblasts to the extracellular matrix in squirrel monkey, J. Periodont. Res. **17**:70, 1982.

Garfunkel, A., and Sciaky, I.: Vascularization of the periodontal tissues in the adult laboratory rat, J. Dent. Res. **50**:880, 1971.

Goldman, H.M.: The effects of dietary deprivation and of age on periodontal tissues of the rat and spider monkey, J. Periodontol. **25**:87, 1954.

Goldman, H.M., and Gianelly, A.A.: Histology of tooth movement, Dent. Clin. North Am. **16**:439, 1972.

Gould, T.R.L., Melcher, A.H., and Brunette, D.M.: Location of progenitor cells in periodontal ligament of mouse molar stimulated by wounding, Anat. Rec. **188**:133, 1977.

Griffin, C.J.: Unmyelinated nerve endings in the periodontal membrane of human teeth, Arch. Oral Biol. **13**:1207, 1968.

Ham, A.W.: Histology, ed. 7, Philadelphia, 1974, J.B. Lippincott Co.

Holtrop, M.E., Raisz, L.G., and Simmons, H.A.: The effects of parathormone, colchicine and calcitonin on the ultrastructure and the activity of osteoclasts in organ culture, J. Cell Biol. **60**:346, 1974.

Ishimitsu, K.: Beitrag zur Kenntnis der Morphologie and Entwicklungsgeschichte der Glomeruli periodontii (Contribution to the knowledge of morphology and development of the periodontal glomeruli), Yokohama Med. Bull. **11**:415, 1960.

Kindlova, M., and Matena, V.: Blood vessels of the rat molar, J. Dent. Res. **41**:650, 1962.

Kvam, E.: Cellular dynamics on the pressure side of the rat periodontium following experimental tooth movement, Scand. J. Dent. Res. **80**:369-383, 1972.

Leblond, C.P., Messier, B., and Kopriwa, B.: Thymidine-^{3}H as a tool for the investigation of the renewal of cell populations, Lab. Invest. **8**:296, 1959.

Leibovich, S.J., and Ross, R.: The role of the macrophage in wound repair. A study with hydrocortisone and anti-macrophage serum. Am. J. Pathol. **78**:71, 1975.

Malassez, M.L.: Sur l'existence de masses epitheliales dans le ligament alveolodentaire (On the existence of epithelial masses in the periodontal membrane), Comp. Rend. Soc. Biol. **36**:241, 1884.

Malkani, K., Luxembourger, M.-M., and Rebel, A.: Cytoplasmic modifications at the contact zone of osteoclasts and calcified tissue in the diaphyseal growing plate of foetal guinea-pig tibia. Calcif. Tissue Res. **11**:258, 1973.

McCulloch, C.A.G., and Melcher, A.H.: Cell density and cell generation in the periodontal ligament of mice, Am. J. Anat. **167**:43, 1983.

Melcher, A.H.: Repair of wounds in the periodontium of the rat. Influence of periodontal ligament on osteogenesis. Arch. Oral Biol. **15**:1183, 1970.

Melcher, A.H., and Correia, M.A.: Remodeling of periodontal ligament in erupting molars of mature rats, J. Periodont. Res. 6:118, 1971.

Picton, D.C.A.: The effects of external forces in the periodontium. In Melcher, A.H., and Bowen, W.H., editors: Biology of the periodontium, New York, 1969, Academic Press, Inc.

Revel, J.P., and Karnovsky, M.J.: Hexagonal array of subunits in intercellular junctions of the mouse heart and liver, J. Cell Biol. 33:C7, 1967.

Roberts, W.E., Chase, D.C., and Jee, W.S.S.: Counts of labelled mitoses in the orthodontically-stimulated periodontal ligament in the rat. Arch. Oral Biol. 19:665, 1974.

Rygh, P.: Ultrastructural cellular reactions in pressure zones of rat molar periodontium incident to orthodontic tooth movement, Acta Odontol. Scand. 30:575, 1972.

Sakamoto, S., Goldhaber, P., and Glimcher, M.J.: The further purification and characterization of mouse bone collagenase. Calcif. Tissue Res. 10:142, 1972.

Sodek, J.: A comparison of the rates of synthesis and turnover of collagen and non-collagen proteins in adult rate periodontal tissues and skin using a microassay, Arch. Oral Biol. 22:655, 1977.

Sodek, J.: A new approach to assessing collagen turnover by using a microassay. A highly efficient and rapid turnover of collagen in rat periodontal tissues. Biochem. J. 160:243, 1976.

Stallard, R.E.: The utilization of ^{3}H-proline by the connective tissue elements of the periodontium, Periodontics 1:185, 1963.

Svoboda, E.L.A., Brunette, D.M., and Melcher, A.H.: *In vitro* phagocytosis of exogenous collagen by fibroblasts from the periodontal ligament: an electronmicroscopic study, J. Anat. 128:301, 1979.

Ten Cate, A.R., and Mills, C.: The development of the periodontium: the origin of alveolar bone, Anat. Rec. 173:69, 1972.

Ten Cate, A.R., Mills, C., and Solomon, G.: The development of the periodontium. A transplantation and autoradiographic study, Anat. Rec. 170:365, 1971.

Vaes, G.: Excretion of acid and of lysosomal hydrolytic enzymes during bone resorption induced in tissue culture by parathyroid extract. Exp. Cell Res. 39:470, 1965.

Valderhaug, J.P., and Nylen, M.U.: Function of epithelial rests as suggested by their ultrastructure, J. Periodont. Res. 1:69, 1966.

Waerhaug, J.: Effect of C-avitaminosis on the supporting structures of the teeth. J. Periodontol. 29:87, 1958.

Walker, D.G.: Bone resorption restored in osteopetrotic mice by transplants of normal bone marrow and spleen cells, Science 190:784, 1975.

Yajima, T., and Rose, G.C.: Phagocytosis of collagen by human gingival fibroblasts in vitro, J. Dent. Res. 56:1271, 1977.

Zwarych, P.D., and Quigley, M.B.: The intermediate plexus of the periodontal ligament: history and further observations, J. Dent. Res. 44:383, 1965.

8
MAXILLA AND MANDIBLE (ALVEOLAR PROCESS)

DEVELOPMENT OF MAXILLA AND MANDIBLE
Maxilla
Mandible

DEVELOPMENT OF ALVEOLAR PROCESS
STRUCTURE OF ALVEOLAR PROCESS

PHYSIOLOGIC CHANGES IN ALVEOLAR PROCESS
INTERNAL RECONSTRUCTION OF BONE
CLINICAL CONSIDERATIONS

DEVELOPMENT OF MAXILLA AND MANDIBLE

In the beginning of the second month of fetal life the skull consists of three parts:
1. The chondrocranium, which is cartilaginous, is made up of the base of the skull with the otic and nasal capsules.
2. The desmocranium, which is membranous, forms the lateral walls and roof of the braincase.
3. The appendicular or visceral part of the skull, which is cartilaginous, consists of the skeletal rods of the branchial arches.

The bones of the skull develop either by endochondral ossification, replacing the cartilage, or by intramembranous ossification in the mesenchyme. Intramembranous bone may develop in proximity to cartilaginous parts of the skull or directly in the membranous capsule of the brain called desmocranium (see Plate 2).

The endochondral bones are the bones of the base of the skull: ethmoid bone; inferior concha (turbinate bone); body, lesser wings, basal part of the greater wings, and the lateral plate of the pterygoid process of the sphenoid bone; petrosal part of the temporal bone; and basilar,

lateral, and lower part of the squamous portion of the occipital bone. The following bones develop in the desmocranium: frontal bones; parietal bones; squamous and tympanic parts of the temporal bone; parts of the greater wings and the medial plate of the pterygoid process of the sphenoid bone; and the upper part of the squamous portion of the occipital bone. All the bones of the upper face develop by intramembranous ossification, most of them close to the cartilage of the nasal capsule. The mandible develops as intramembranous bone, lateral to the cartilage of the mandibular arch. This cartilage, Meckel's cartilage, is in its proximal parts the primordium for two of the auditory ossicles: the incus (anvil) and the malleus (hammer). The third auditory ossicle, the stapes (stirrup), develops from the proximal part of the skeleton in the second branchial arch, which also gives rise to the styloid process, the stylohyoid ligament, and part of the hyoid bone. The latter is completed by the derivatives of the third arch. The fourth and fifth arches form the skeleton of the larynx.

Maxilla. The human maxilla is homologous to two bones, the maxilla proper and the premax-

A

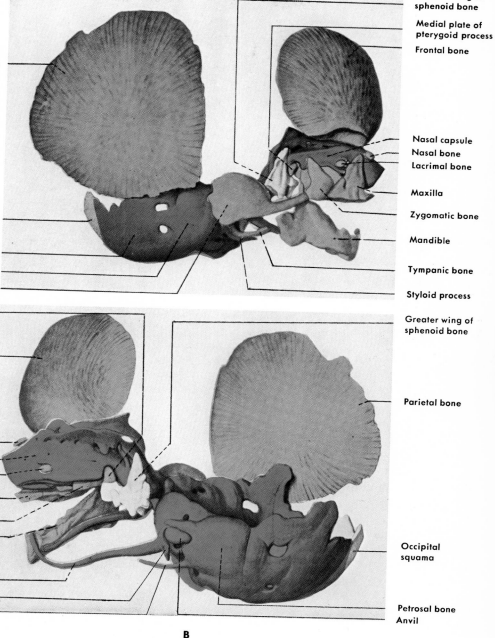

Greater wing of
sphenoid bone

Medial plate of
pterygoid process

Frontal bone

Parietal bone

Nasal capsule
Nasal bone
Lacrimal bone

Maxilla

Occipital
squama

Zygomatic bone

Lateral part of
occipital bone

Mandible

Petrosal bone

Squama of
temporal bone

Tympanic bone

Medial plate of
pterygoid
process

Styloid process

Greater wing of
sphenoid bone

Frontal bone

Parietal bone

Nasal bone
Nasal capsule
Lacrimal bone

Maxilla

Palatine bone

Mandible

Occipital
squama

Meckel's
cartilage

Hammer

Petrosal bone

Styloid process

Anvil

B

Plate 2. Reconstruction of skull of human embryo 80 mm in length. Cartilage, *green.* Intramembranous bones, *pink.* Endochondral bones, *white.* **A,** Right lateral view. **B,** Left lateral view after removal of left intramembranous bones. (From Sicher, H., and Tandler, J.: Anatomie Für Zahnärzte [Anatomy for dentists], Vienna, 1928, Julius Springer Verlag.)

illa. The latter, in most animals a separate bone, carries the incisors and forms the anterior part of the hard palate and the rim of the piriform aperture. The ossification centers of the premaxilla and maxilla may be separate for a very short time, or only one center of ossification, common to both the premaxilla and maxilla, appears. That humans therefore may not have an independent premaxilla, even in the first developmental stages, does not change the fact that they possess the homologue of a premaxilla. The composition of the human maxilla from premaxilla and maxilla is indicated by the incisive fissure, which is clearly visible in young skulls. It is seen on the palate, where it extends from the incisive foramen to the alveolus of the canine.

Mandible. The mandible makes its appearance as a bilateral structure in the sixth week of fetal life as a thin plate of bone lateral to, and at some distance from, Meckel's cartilage (Fig. 8-1). The latter is a cylindric rod of cartilage. Its proximal end (close to the base of the skull) gives rise to the malleus and the incus

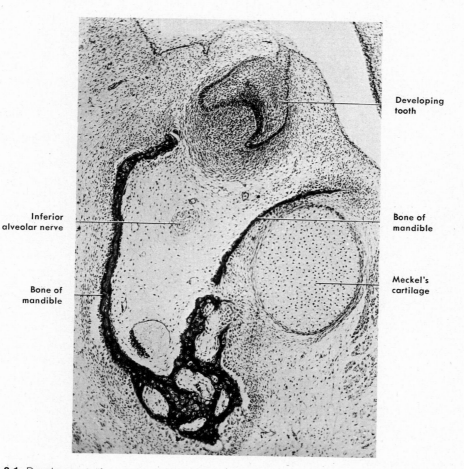

Fig. 8-1. Development of mandible as intramembranous bone lateral to Meckel's cartilage (human embryo 45 mm in length).

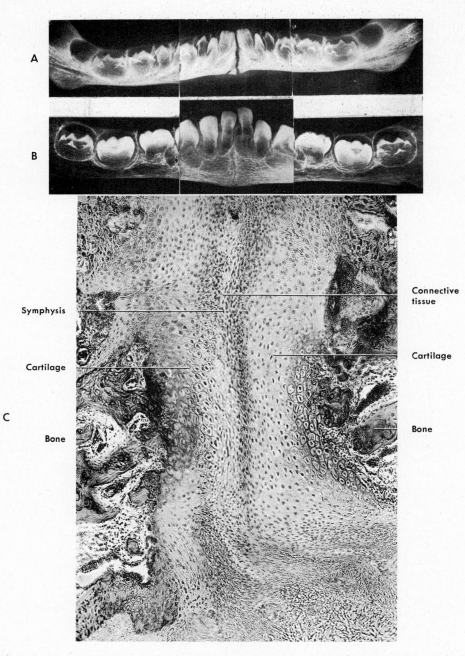

Fig. 8-2. Development of mandibular symphysis. **A,** Newborn infant. Symphysis wide open. Mental ossicle (roentgenogram). **B,** Child 9 months of age. Symphysis partly closed. Mental ossicles fused to mandible (roentgenogram). **C,** Frontal section through mandubular symphysis of newborn infant. Connective tissue in midline connects plates of cartilage on either side. Cartilage is later replaced by bone.

and therefore is continuous with and lies in contact with these bones, respectively. Its distal end at the midline is bent upward and is in contact with the cartilage of the other side (see Plate 2). The greater part of Meckel's cartilage disappears without contributing to the formation of the bone of the mandible. Only a small part of the cartilage, some distance from the midline, is the site of endochondral ossification. Here the cartilage calcifies and is destroyed by chrondoclasts, being replaced by connective tissue and then by bone. Throughout fetal life the mandible is a paired bone. Right and left mandibles are joined in the midline by fibrocartilage in the mandibular symphysis. The cartilage at the symphysis is not derived from Meckel's cartilage but differentiates from the connective tissue in the midline. In it, small irregular bones known as the mental ossicles develop and at the end of the first year fuse with the mandibular body. At the same time the two halves of the mandible unite by ossification of the symphyseal fibrocartilage. (Fig. 8-2).

DEVELOPMENT OF ALVEOLAR PROCESS

Near the end of the second month of fetal life the maxilla as well as the mandible forms a groove that is open toward the surface of the oral cavity (Fig. 8-1). The tooth germs are contained in this groove, which also includes the alveolar nerves and vessels. Gradually bony

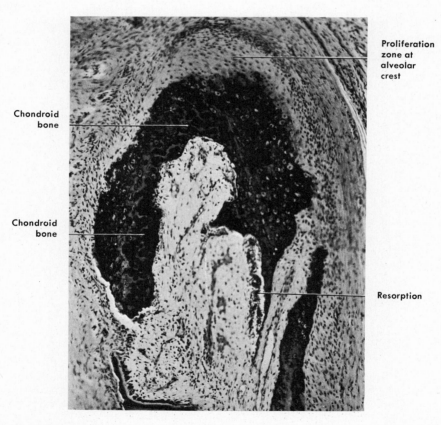

Proliferation zone at alveolar crest

Chondroid bone

Chondroid bone

Resorption

Fig. 8-3. Verticle growth of mandible at alveolar crest. Formation of chondroid bone that later is replaced by typical bone.

septa develop between the adjacent tooth germs, and much later the primitive mandibular canal is separated from the dental crypts by a horizontal plate of bone.

An alveolar process in the strict sense of the word develops only during the eruption of the teeth. It is important to realize that during growth part of the alveolar process is gradually incorporated into the maxillary or mandibular body, while it grows at a fairly rapid rate at its free borders. During the period of rapid growth a tissue may develop at the alveolar crest that combines characteristics of cartilage and bone. It is called chondroid bone (Fig. 8-3). The alveolar process forms with the devel-

opment and the eruption of teeth, and, conversely, it gradually diminishes in height after the loss of teeth.

STRUCTURE OF ALVEOLAR PROCESS

The alveolar process may be defined as that part of the maxilla and the mandible that forms and supports the sockets of the teeth (Fig. 8-4). Anatomically, no distinct boundary exists between the body of the maxilla or the mandible and their respective alveolar processes. In some places the alveolar process is fused with, and partly masked by, bone that is not functionally related to the teeth. In the anterior part of the maxilla the palatine process fuses

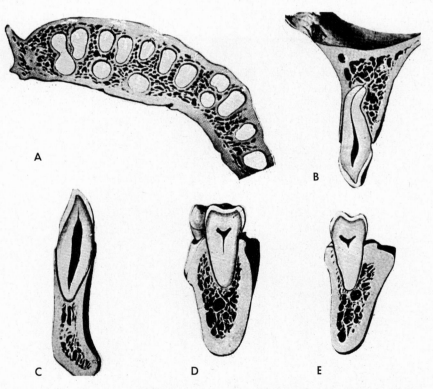

Fig. 8-4. Gross relations of alveolar processes. **A,** Horizontal section through upper alveolar process. **B,** Labiolingual section through upper lateral incisor. **C,** Labiolingual section through lower canine. **D,** Labiolingual section through lower second molar. **E,** Labiolingual section through lower third molar. (From Sicher, H., and Tandler, J.: Anatomie für Zahnärzte [Anatomy for dentists], Vienna, 1928, Julius Springer Verlag.)

with the oral plate of the alveolar process. In the posterior part of the mandible the oblique line is superimposed laterally on the bone of the alveolar process (Fig. 8-4, *D* and *E*).

As a result of its adaptation to function, two parts of the alveolar process can be distinguished. The first consists of a thin lamella of bone that surrounds the root of the tooth and gives attachment to principal fibers of the periodontal ligament. This is the *alveolar bone proper*. The second part is the bone that surrounds the alveolar bone proper and gives support to the socket. This has been called *supporting alveolar bone*. The latter, in turn, consists of two parts: (1) cortical plates, which

consist of compact bone and form the outer and inner plates of the alveolar processes, and (2) the spongy bone, which fills the area between these plates and the alveolar bone proper (Figs. 8-4 and 8-5).

The cortical plates, continuous with the compact layers of the maxillary and mandibular body, are generally much thinner in the maxilla than in the mandible. They are thickest in the premolar and molar regions of the lower jaw, especially on the buccal side. In the maxilla the outer cortical plate is perforated by many small openings through which blood and lymph vessels pass. In the lower jaw the cortical bone of the alveolar process is dense. In the

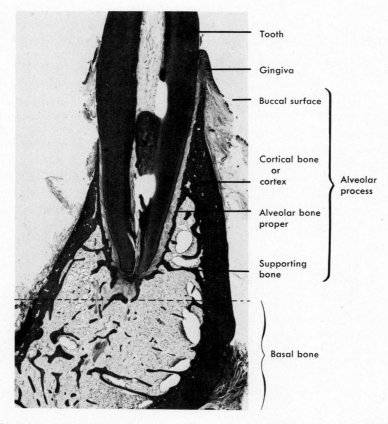

Fig. 8-5. Section through mandible showing relationship of tooth to alveolar process and basal bone. (From Bhaskar, S.N.: Synopsis of oral histology, St.Louis, 1962, The C.V. Mosby Co.)

region of the anterior teeth of both jaws the supporting bone usually is very thin. No spongy bone is found here, and the cortical plate is fused with the alveolar bone proper (Fig. 8-4, *B* and *C*). In such areas, notably in the premolar and molar regions of the maxilla, defects of the outer alveolar wall are fairly common. Such defects, where periodontal tissues and covering mucosa fuse, do not impair the firm attachment and function of the tooth.

The shape of the outlines of the crest of the alveolar septa in the roentgenogram is dependent on the position of the adjacent teeth. In a healthy mouth the distance between the cementoenamel junction and the free border of the alveolar bone proper is fairly constant. If the neighboring teeth are inclined, therefore, the alveolar crest is oblique. In the majority of individuals the inclination is most pronounced in the premolar and molar regions, with the teeth being tipped mesially. Then the cementoenamel junction of the mesial tooth is situated in a more occlusal plane than that of the distal tooth, and the alveolar crest therefore slopes distally (Fig. 8-6).

The interdental and interradicular septa contain the perforating canals of Zuckerkandl and Hirschfeld (nutrient canals), which house the interdental and interradicular arteries, veins, lymph vessels, and nerves (Fig. 8-7).

Histologically, the cortical plates consist of longitudinal lamellae and haversian systems (Fig. 8-8). In the lower jaw, circumferential or basic lamellae reach from the body of the mandible into the cortical plates.

The study of roentgenograms permits the classification of the spongiosa of the alveolar process into two main types. In type I the interdental and interradicular trabeculae are regular and horizontal in a ladderlike arrangement (Fig. 8-9, *A* to *C*). Type II shows irregularly arranged, numerous, delicate interdental and interradicular trabeculae (Fig. 8-9, *D*). Both types show a variation in thickness of trabeculae and size of marrow spaces. The architecture of type I is seen most often in the mandible and fits well into the general idea of a trajectory pattern of spongy bone. Type II, although evidently functionally satisfactory, lacks a distinct trajectory pattern, which seems to be compensated for by the greater number of trabeculae in any given area. This arrangement is more common in the maxilla. From the apical part of the socket of lower molars, trabeculae are sometimes seen radiating in a slightly distal direction. These trabeculae are less prominent in the upper jaw because of the proximity of the nasal cavity and the maxillary sinus. The marrow spaces in the alveolar process may contain hematopoietic marrow, but usually they contain fatty marrow. In the condylar process, in the angle of the mandible, in the maxillary

Fig. 8-6. Diagram of relation between cementoenamel junction of adjacent teeth and shape of crests of alveolar septa. (From Ritchey, B., and Orban, B.: J. Periodontol. **24:**75, 1953.)

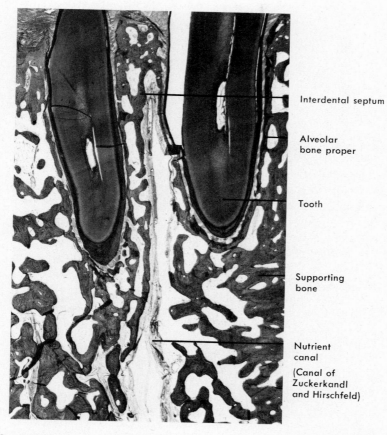

Interdental septum

Alveolar
bone proper

Tooth

Supporting
bone

Nutrient
canal

(Canal of
Zuckerkandl
and Hirschfeld)

Fig. 8-7. Section through jaw showing nutrient canal of Zuckerkandl and Hirschfeld in interdental bony septum. (From Bhaskar, S.N.: Synopsis of oral histology, St. Louis, 1962, The C.V. Mosby Co.)

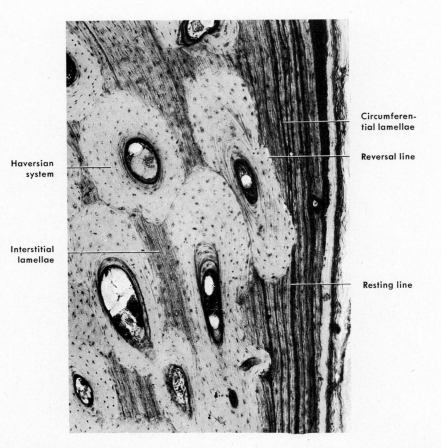

Fig. 8-8. Appositional growth of mandible by formation of circumferential lamellae. These are replaced by haversian bone; remnants of circumferential lamellae in the depth persisting as interstitial lamellae.

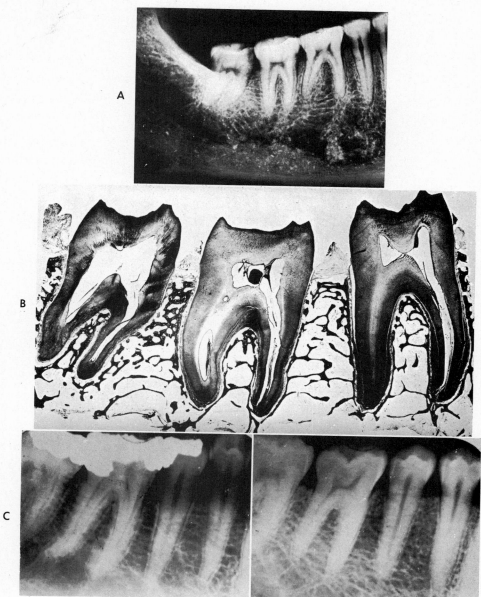

Fig. 8-9. Supporting trabeculae between alveoli. **A,** Roentgenogram of mandible. **B,** Mesiodistal section through mandibular molars showing alveolar bone proper and supporting bone. **C,** Type I alveolar spongiosa. Note regular horizontal trabeculae. **D,** Type II alveolar spongiosa. Note irregularly arranged trabeculae. (Courtesy Dr. N. Brescia, Chicago.)

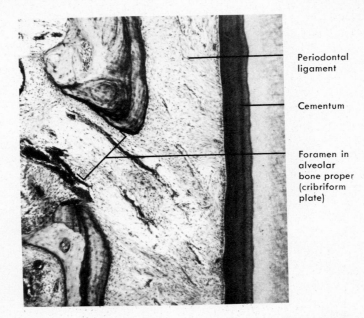

Periodontal ligament

Cementum

Foramen in alveolar bone proper (cribriform plate)

Fig. 8-10. Histologic section showing foramen in alveolar bone proper (cribriform plate). (From Bhaskar, S.N.: Synopsis of oral histology, St. Louis, 1962, The C.V. Mosby Co.)

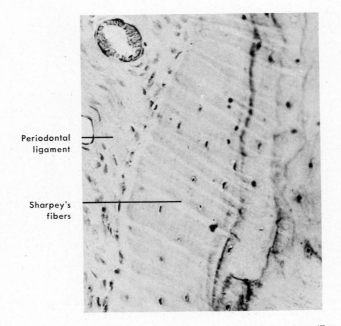

Periodontal ligament

Sharpey's fibers

Fig. 8-11. Histologic section showing Sharpey's fibers in alveolar bone proper. (From Bhaskar, S.N.: Synopsis of oral histology, St. Louis, 1962, The C.V. Mosby Co.)

tuberosity, and in other isolated foci, hemato-poietic cellular marrow is found.

The alveolar bone proper, which forms the inner wall of the socket (Fig. 8-10), is perforated by many openings that carry branches of the interalveolar nerves and blood vessels into the periodontal ligament (see Chapter 7), and it is therefore called the *cribriform plate*. The alveolar bone proper consists partly of *lamellated* and partly of *bundle bone*. Some lamellae of the lamellated bone are arranged roughly parallel to the surface of the adjacent marrow spaces, whereas others form haversian systems.

Bundle bone is that bone in which the principal fibers of the periodontal ligament are anchored. The term "bundle bone" was chosen because the bundles of the principal fibers continue into the bone as Sharpey's fibers (Fig. 8-11). The bundle bone is characterized by the scarcity of the fibrils in the intercellular substance. These fibrils, moreover, are all arranged at right angles to Sharpey's fibers. The bundle bone contains fewer fibrils than does lamellated bone, and therefore it appears dark in routine hematoxylin and eosin stained sections and much lighter in preparations stained with

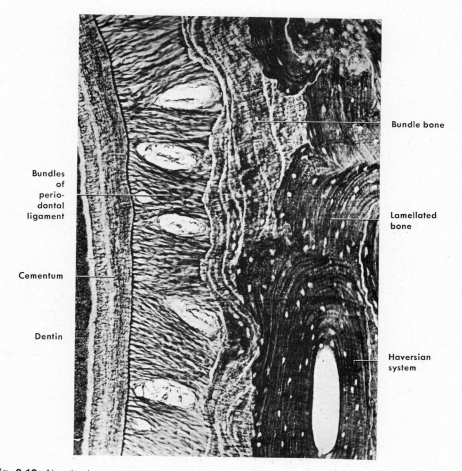

Fig. 8-12. Alveolar bone proper consisting of bundle bone and haversian bone on distal alveolar wall. A reversal line separates the two (silver impregnation).

silver than does lamellated bone (Fig. 8-12). In some areas the alveolar bone proper consists mainly of bundle bone. Since bundle bone contains more calcium salts per unit area than other types of bone tissue, such areas are seen in roentgenograms as dense radiopacities.

PHYSIOLOGIC CHANGES IN ALVEOLAR PROCESS

Bone consists of about 65% inorganic and 35% organic material. The inorganic material is hydroxyapatite while the organic material is primarily type I collagen, which lies in a ground sustance of glycoproteins and proteoglycans. The glycoproteins are proteins with a small amount of monosaccharide, disaccharide, polysaccharide, or oligosaccharide, while the proteoglycans are sulfated and nonsulfated glycosaminoglycans (high-molecular-weight carbohydrates), with a small amount of proteins. The approximate composition of bone is as follows:

Inorganic = 65%
Organic = 35%
  Collagen = 88%-89%
  Noncollagen = 11%-12%
    Glycoproteins = 6.5%-10%
    Proteoglycans = 0.8%
    Sialoproteins = 0.35%
    Lipids = 0.4%

The inorganic material almost exclusively consists of calcium and inorganic orthophosphate in the form of hydroxyapatite crystals. These crystals are deposited on and in between the molecules of collagen (which form the collagen fibrils) as well as in the noncollagen, organic material that makes up the ground substance of bone. The exact mechanism by which hydroxyapatite crystals are deposited in the bone matrix produced by osteoblasts is still unknown. However, certain enzymes have been shown to participate in this process. These are alkaline phosphatase, ATPase, and pyrophosphatase.

The internal structure of bone is adapted to mechanical stresses. It changes continuously during growth and alteration of functional stresses. In the jaws, structural changes are correlated to the growth, eruption, movements, wear, and loss of teeth. All these processes are made possible only by a coordination of destructive and formatie activities. Specialized cells, the osteoclasts, have the function of eliminating overage bony tissue or bone that is no longer adapted to mechanical forces, whereas osteoblasts produce new bone. Osteoblasts secrete the type I collagen as well as the noncollagenous matrix of bone. Their ultrastructure is characteristic of any actively secreting cell, that is, a prominent Golgi apparatus, rough endoplasmic reticulum, mitochondria, nucleoli, and many secretory vesicles and vacuoles. Osteoblasts differentiate from progenitor or precursor cells of the connective tissue at the site of bone formation. The mechanisms that determine bone formation at any given site are unknown. They must, however, be varied and in part must be determined on a genetic and functional basis. Although there are many theories and opinions in this regard, specific details have not yet been elucidated. As the osteoblasts secrete the organic matrix of bone, it is at first devoid of mineral salts, and at this stage it stains pink in routine hematoxylin and eosin stains and is called *osteoid tissue*. As this material is produced, some of the osteoblasts become embedded in it and form the *osteocytes*. In areas of bone formation, mineralization always lags behind the production of bone matrix, and therefore in such areas a superficial layer of osteoid tissue is always seen. In routine sections mineralized bone is basophilic and can be easily distinguished from the osteoid tissue.

Osteoclasts are, as a rule, multinucleated giant cells (Fig. 8-13, *left*). The number of nuclei in one cell may rise to a dozen or more. However, occasionally uninucleated osteoclasts are found. The cell body is irregularly oval or club shaped and may show many branching processes. In general, osteoclasts are found in baylike depressions in the bone called *Howship's lacunae*. Osteoclasts have prominent mi-

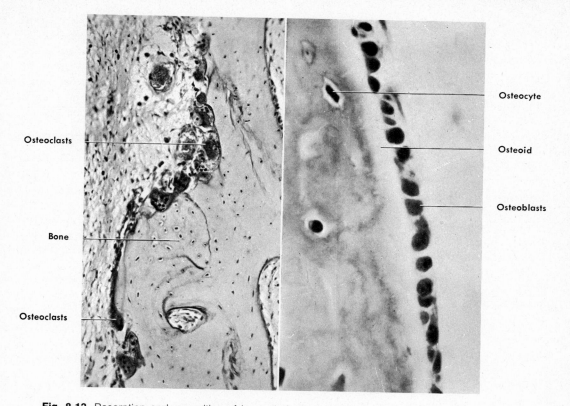

Osteoclasts

Bone

Osteoclasts

Osteocyte

Osteoid

Osteoblasts

Fig. 8-13. Resorption and apposition of bone. *Left,* Osteoclasts in Howship's lacunae. *Right,* Osteoblasts along bone trabecula. Layer of osteoid tissue is a sign of bone formation.

tochondria, lysosomes, vacuoles, and little rough endoplasmic reticulum. Their many nuclei have condensed chromatin and a single nucleolus. The part of the cell in contact with the bone shows a convoluted surface, the *ruffled border,* which is the site of great activity. Here, pieces of bone are broken off and released into the extracellular spaces. The ruffled border is surrounded by a *clear zone* that has no organelles but only fine granular cytoplasm with microfilaments. Osteoclasts are probably derived from the circulating blood cells (monocytes), but they may differentiate from the mesenchymal cells in situ. Whereas the major position of bone resorption occurs through the mediation of osteoclasts, on rare occasions bone

resorption by osteocytes has been described (osteocytic osteolysis). This, however, is only of academic interest.

It has been suggested that bone resorption at any site is a chemotactic phenomenon; that is, it is initiated by the release of some soluble factor that attracts monocytes to the target site. In the bone resorption phenomenon, however, the role of genetic and functional influences cannot be underestimated. It might be that aging osteocytes, in their degeneration and death, liberate the substances that cause the differentiation of the osteoclasts.

During bone resorption, three processes occur in more or less rapid succession: (1) decalcification, (2) degradation of matrix, and (3) transport of soluble products to the extracellular fluid or the blood vascular system. Since calcified matrix is resistant to proteases of all kinds, bone must first be decalcified; this is achieved at the ruffled border of the osteoclasts by secretion of organic acids (citric and lactic acid), which chelate bone, and by H^+, which increases the solubility of hydroxyapatite. After this decalcification process, pieces of matrix are released by the activity of cathepsin B-1 (lysosomal acid protease) and collagenase enzymes. (Collagenase is secreted as a proenzyme that is activated by specific neutral proteases.) Collagenolytic activity takes place outside the osteoclast and occurs at a specific site on the tropocollagen (collagen) molecule. This site is one third the distance from the caboxyl end of the molecule. The broken fragments of collagen are further decalcified, and breakdown of collagen by proteases other than collagenase continues.

Collagenolysis occurs outside the osteoclast, and only calcium phosphate can be identified within these cells. After the degradation of the matrix, the breakdown products of bone must be transported to the extracellular fluids and to the blood vascular system, but the details of this mechanism are yet unknown.

INTERNAL RECONSTRUCTION OF BONE

The bone in the alveolar process is identical to bone elsewhere in the body and is in a constant state of flux. During the growth of the maxilla and the mandible, bone is deposited on the outer surfaces of the cortical plates. In the mandible, with its thick, compact cortical plates, bone is deposited in the shape of basic or circumferential lamellae (Fig. 8-8). When the lamellae reach a certain thickness, they are replaced from the inside by haversian bone. This reconstruction is correlated to the func-

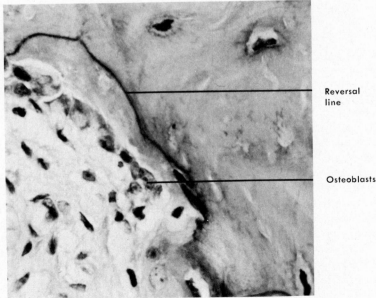

Reversal line

Osteoblasts

Fig. 8-14. Reversal line in bone. (From Bhaskar, S.N.: Synopsis of oral histology, St. Louis, 1962, The C.V. Mosby Co.)

tional and nutritional demands of the bone. In the haversian canals, closest to the surface, osteoclasts differentiate and resorb the haversian lamellae and part of the circumferential lamellae. The resorbed bone is replaced by proliferating loose connective tissue. This area of resorption is sometimes called the *cutting cone* or the *resorption tunnel*. After a time the resorption ceases and new bone is apposed onto the old. The scalloped outline of Howship's lacunae that turn their convexity toward the old bone remains visible as a darkly stained cementing line, a *reversal line* (Fig. 8-14). This is in contrast to those cementing lines that correspond to a rest period in an otherwise continuous process of bone apposition. They are called *resting lines* (Fig. 8-8). Resting and reversal lines are found between layers of bone of varying age.

Wherever a muscle, tendon, or ligament is attached to the surface of bone, Sharpey's fibers can be seen penetrating the basic lamellae. During replacement of the latter by haversian systems, fragments of bone containing Sharpey's fibers remain in the deeper layers. Thus the presence of these lamellae containing Sharpey's fibers indicates the former level of the surface.

Alterations in the structure of the alveolar bone are of great importance in connection with the physiologic eruptive movements of the teeth. These movements are directed mesio-occlusally. At the alveolar fundus the continual apposition of bone can be recognized by resting lines separating parallel layers of bundle bone. When the bundle bone has reached a certain thickness, it is resorbed partly from the marrow spaces and then replaced by lamellated bone or spongy trabeculae. The presence of bundle bone indicates the level at which the alveolar fundus was situated previously. During the mesial drift of a tooth, bone is apposed on the distal and resorbed on the mesial alveolar wall (Fig. 8-15). The distal wall is made up almost entirely of bundle bone. However, the

osteoclasts in the adjacent marrow spaces remove part of the bundle bone when it reaches a certain thickness. In its place, lamellated bone is deposited (Fig. 8-12).

On the mesial alveolar wall of a drifting tooth, the sign of active resorption is the presence of Howship's lacunae containing osteoclasts (Fig. 8-15). Bundle bone, however, on this side is always present in some areas but forms merely a thin layer (Fig. 8-16). This is because the mesial drift of a tooth does not occur simply as a bodily movement. Thus resorption does not involve the entire mesial surface of the alveolus at one and the same time. Moreover, periods of resorption alternate with periods of rest and repair. It is during these periods of repair that bundle bone is formed, and detached periodontal fibers are again secured. Islands of bundle bone are separated from the lamellated bone by reversal lines that turn their convexities toward the lamellated bone (Fig. 8-16).

During these changes, compact bone may be replaced by spongy bone or spongy bone may change into compact bone. This type of internal reconstruction of bone can be observed in physiologic mesial drift or in orthodontic mesial or distal movement of teeth. In these movements an interdental septum shows apposition on one surface and resorption on the other. If the alveolar bone proper is thickened by apposition of bundle bone, the interdental marrow spaces widen and advance in the direction of apposition. Conversely, if the plate of the alveolar bone proper is thinned by resorption, apposition of bone occurs on those surfaces that face the narrow spaces. The result is a reconstructive shift of the interdental septum.

CLINICAL CONSIDERATIONS

Bone, although one of the hardest tissues of the human body, is biologically a highly plastic tissue. Where bone is covered by a vascularized connective tissue, it is exceedingly sensi-

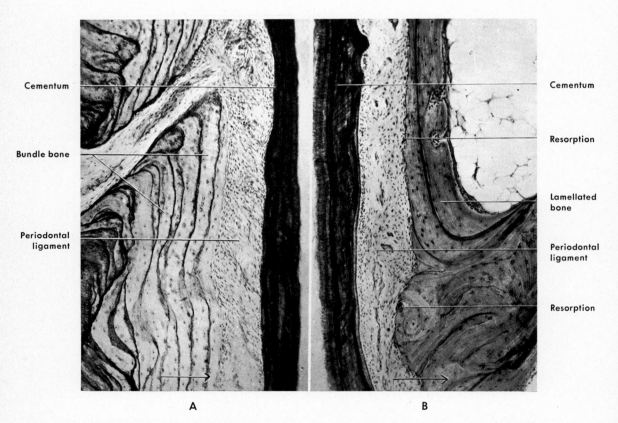

Fig. 8-15. Mesial drift indicated by arrow. **A,** Apposition of bundle bone on distal alveolar wall. **B,** Resorption of bone on mesial alveolar wall. (From Weinmann, J.P.: Angle Orthod. **11:**83, 1941.)

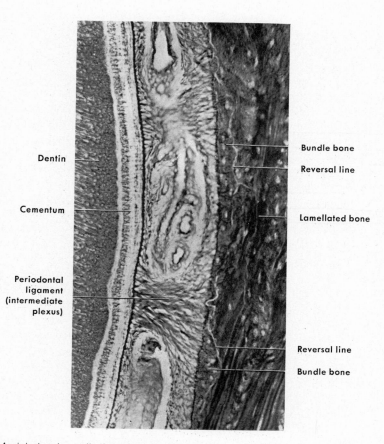

Fig. 8-16. Mesial alveolar wall where alveolar bone proper consists mostly of lamellated bone and islands of bundle bone, which anchor principal fibers of periodontal ligament.

tive to pressure, whereas tension acts generally as a stimulus to the production of new bone. It is this biologic plasticity that enables the orthodontist to move teeth without disrupting their relations to the alveolar bone. Bone is resorbed on the side of pressure and apposed on the side of tension; thus the entire alveolus is allowed to shift with the tooth. It has been shown that on the pressure side there is an increase in the level of cyclic adenosine monophosphate (cAMP) in cells. This may play some role in bone resorption.

The adaptation of bone to function is quantitative as well as qualitative. Whereas increase in functional forces leads to formation of new bone, decreased function leads to a decrease in the volume of bone. This can be observed in the supporting bone of teeth that have lost their antagonists. Here the spongy bone around the alveolus shows pronounced rarefaction: the bone trabeculae are less numerous and very thin (Fig. 8-17). The alveolar bone proper, however, is generally well preserved because it continues to receive some stimuli from the tension of the periodontal tissues.

During healing of fractures or extraction wounds an embryonic type of bone is formed, which only later is replaced by mature bone. The embryonic bone, also called immature or coarse fibrillar bone, is characterized, among other aspects, by the greater number, size, and irregular arrangement of the osteocytes than are found in mature bone (Fig. 8-18). The greater number of cells and the reduced volume of calcified intercellular substance render this immature bone more radiolucent than mature bone. This explains why bony callus cannot be seen in roentgenograms at a time when histologic examination of a fracture reveals a well-developed union between the fragments and why a socket after an extraction wound appears to be empty at a time when it is almost filled with immature bone. The visibilty in radiographs lags 2 or 3 weeks behind actual formation of new bone.

The most frequent and harmful change in the alveolar process is that which is associated with periodontal disease. The bone resorption is almost universal, occurs more frequently in posterior teeth, is usually symmetrical, occurs in episodic spurts, is both of the horizontal and vertical type (i.e., occurs from the gingival and tooth side, respectively), and is intimately related to bacterial plaque and pocket formation. Recent studies have revealed some of the mechanisms that contribute to this process. It has been shown, for example, the endotoxins produced by the gram-negative bacteria of the plaque lead to an increase in cAMP, which increases the osteoclastic activity. Furthermore, a peptide called *osteoclast activating factor* (OAF) has been demonstrated in the lymphocytes near the periodontal pocket. This substance is capable of increasing cAMP and osteoclastic activity and reducing osteoblastic activity at the target site. The insidious and progressive loss of alveolar bone in periodontal disease is difficult to control, and once lost, this bone is even more difficult to repair or regenerate. The therapeutic replacement or regeneration of just a few millimeters of bone tissue lost from the alveolar process in periodontal disease is the greatest challenge to the speciality of periodontics.

In the last few years synthetic materials have been introduced that are intended to replace bone tissue lost through disease or injury. These materials are of two types: the nonresorbable hydroxyapatite and the resorbable tricalcium phosphate. These synthetic inorganic products are currently being used for the augmentation of the alveolar ridges and for filling bone defects produced by periodontal disease. Although these materials are safe to use, it has not yet been determined if they can produce a functional periodontal support for the subject teeth. Results of their use in augmenting edentulous ridges have been more promising.

Since the alveolar process of the maxilla and mandible develops and is maintained for the support of the teeth, when the teeth are lost it undergoes gradual atrophy. In some instances

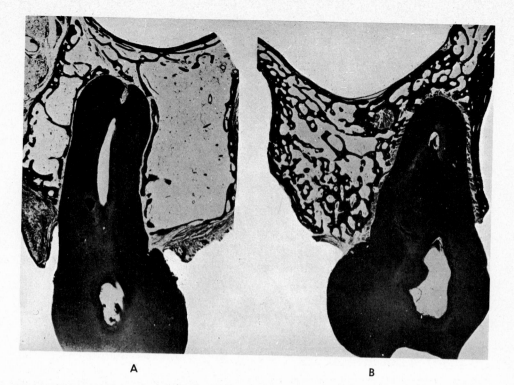

A   B

Fig. 8-17. Osteoporosis of alveolar process aused by inactivity of tooth that has no antagonist:
Labiolingual sections through upper molars of same individual. **A,** Disappearance of bony trabeculae after loss of function. Plane of mesiobuccal root. Alveolar bone proper remains intact. **B,** Normal
spongy bone in plane of mesiobuccal root of functioning tooth. (From Kellner, E.: Z. Stomatol. **18:**59,
1920.)

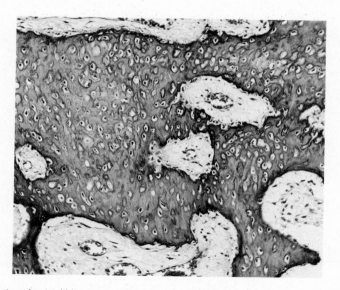

Fig. 8-18. Immature bone. Note many osteocytes and absence of lamellae or resting lines. (From
Bhaskar, S.N.: Synopsis of oral histology, St. Louis, 1962, The C.V. Mosby Co.)

the resorption may be so severe that the fabrication of a functional prosthesis may become difficult or even impossible. Studies in both humans and animals have shown that if during the extraction of teeth the root portion is retained in the alveolar process, this structure does not undergo a noticeable reduction in its height. Since leaving the roots of the teeth within the jaws requires time-consuming endodontic therapy, an alternative experimental procedure has been suggested that consists of implanting artificial root-shaped replicas made of hydroxyapatite into the empty sockets. If proven effective in long-term studies, this procedure may be very useful in maintaining the height of the alveolar process in edentulous patients.

REFERENCES

Bhaskar, S.N.: Radiographic interpretation for the dentist, ed. 3, St. Louis, 1979, The C.V. Mosby Co.

Bhaskar, S.N., Mohammed, C., and Weinmann, J.: A morphological and histochemical study of osteoclasts, J. Bone Joint Surg. [Am.] **38:**1335, 1956.

Brodie, A.G.: Some recent obsevations on the growth of the mandible, Angle Orthod. **10:**63, 1940.

Brodie, A.G.: On the growth pattern of the human head from the third month to the eighth year of life, Am. J. Anat. **68:**209, 1941.

Council on Dental Materials, Instruments and Equipment, American Dental Association: Hydroxyapatite, beta tricalcium phosphate and autogenous and allogeneic bone for filling periodontal defects, alveolar ridge augmentation, and pulp capping, J. Am. Dent. Assoc. **108:**822, 1984.

Glimcher, M.J.: Composition, structure and organization of bone and other mineralized tissues and the mechanism of calcification. In Greep, R.O., and Aptwood, E.B., editors: The handbook of physiology. Section 7, vol. 7, Washington, D.C., 1976, American Physiological Society, pp. 25-116.

Hall, D.A.: Glycoproteins and proteoglycans. In The aging of connective tissue, New York, 1976, Academic Press, Inc.

Ham, A., and Leeson, T.: Histology, ed. 4, Philadelphia, 1961, J.B. Lippincott Co.

Kaban, L.B., and Glowacki, J.: Augmentation of rat mandibular ridge with demineralized bone implants, J. Dent. Res. **63:**998, 1984.

Marks, S.C., and Schneider, G.: Evidence for a relationship between lymphoid cells and osteoclasts, Am. J. Anat. **152:**331, 1978.

Orban, B.: A contribution to the knowledge of the physiologic changes in the periodontal membrane, J. Am. Dent. Assoc. **16:**405, 1929.

Ritchey, B., and Orban, B.: The crests of the interdental alveolar septa, J. Periodontol. **24:**75, 1953.

Schaffer, J.: Die Verknöcherung des Unterkiefers (Ossification of the mandible), Arch. Mikrosk. Anat. **32:**266, 1888.

Sicher, H., and DuBrul, E.L.: Oral anatomy, ed. 6, St. Louis, 1975, The C.V. Mosby Co.

Urist, M.R.: Biochemistry of calcification. In Bourne, G.H., editor: The biochemistry and physiology of bone, vol. 4, ed. 2, New York, 1976, Academic Press, Inc.

Weinmann, J.P.: Das Knochenbild bei Störungen per physiologischen Wanderung der Zähne (Bone in disturbances of the physiologic mesial drift), Z. Stomatol. **24:**397, 1926.

Weinmann, J.P.: Bone changes related to eruption of the teeth. Angle Orthod. **11:**83, 1941.

Weinmann, J.P., and Sicher, H.: Bone and bones; fundamentals of bone biology, ed. 2, St. Louis, 1955, The C.V. Mosby Co.

9
ORAL MUCOUS MEMBRANE

The oral cavity is unique in structure. It contains the teeth. The salivary glands discharge their secretions into it. It contains the taste buds and can be used to perceive and sense in other ways. Thus it serves a variety of functions.

Food first enters the digestive tract through the oral cavity. Here the food is tasted, masticated, and mixed with saliva. Hard inedible particles are sensed and expectorated. Saliva secreted into the oral cavity lubricates the food and facilitates swallowing. Enzymes in the saliva initiate digestion.

Body cavities that communicate with the external surface are lined by mucous membranes, which are coated by serous and mucous secre-

tions. The surface of the oral cavity is a mucous membrane. Its structure varies in an apparent adaptation to function in different regions of the oral cavity. Areas involved in the mastication of food, such as the gingiva and the hard palate, have a much different structure than does the floor of the mouth or the mucosa of the cheek.

Basing classification on these functional criteria, the oral mucosa may be divided into three major types:

1. Masticatory mucosa (gingiva and hard palate)
2. Lining or reflecting mucosa (lip, cheek, vestibular fornix, alveolar mucosa, floor of mouth and soft palate)

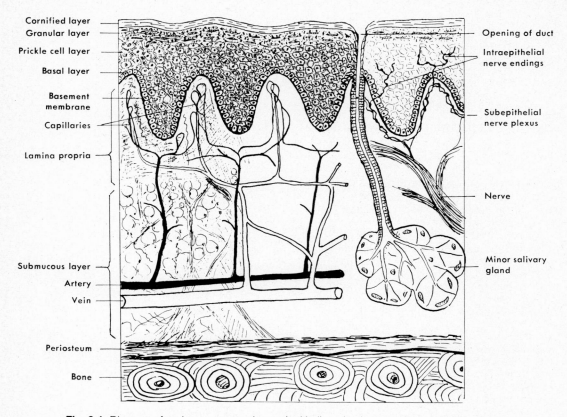

Cornified layer
Granular layer
Prickle cell layer
Basal layer
Basement membrane
Capillaries
Lamina propria
Submucous layer
Artery
Vein
Periosteum
Bone

Opening of duct
Intraepithelial nerve endings
Subepithelial nerve plexus
Nerve
Minor salivary gland

Fig. 9-1. Diagram of oral mucous membrane (epithelium, lamina propria, and submucosa).

3. Specialized mucosa (dorsum of the tongue and taste buds)

The masticatory mucosa tends to be bound to bone and does not stretch. It bears forces generated when food is chewed. The lining mucosa is not equally exposed to such forces. However, it covers the musculature and is distensible, adapting itself to the contraction and relaxation of cheeks, lips, and tongue and to movements of the mandible produced by the muscles of mastication. It makes up all the surfaces of the mouth except for the dorsum of the tongue and the masticatory mucosa. The specialized (sensory) mucosa is so-called because it bears the taste buds, which have a sensory function. These will be discussed below as will two areas with a slightly different structure—the dentogingival junction (the attachment of the gingiva to the tooth) and the red zone or vermilion border of the lips.

DEFINITIONS AND GENERAL CONSIDERATIONS

The structure of the oral mucous membrane resembles the skin in many ways. It is composed of two layers, epithelium and connective tissue (Fig. 9-1). The connective tissue component of oral mucosa is termed the *lamina propria*. The comparable part of skin is known as dermis or corium.

The two layers form an interface that is folded into corrugations. Papillae of connective tissue protrude toward the epithelium (Fig. 9-2) carrying blood vessels and nerves. Although

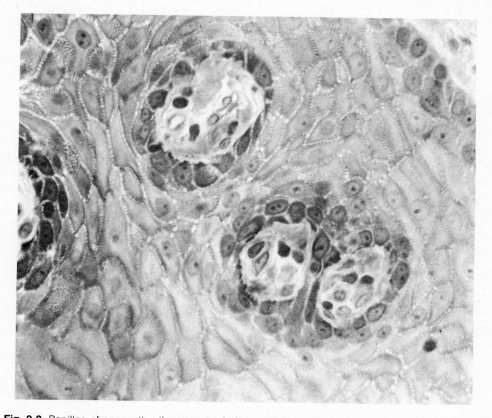

Fig. 9-2. Papillae of connective tissue protrude into epithelium. Blood vessels, fibroblasts, and collagen fibers are seen within them. Cells surrounding papillae are basal cells. The other cells are mainly spinous cells.

some of the nerves actually pass into it, the epithelium does not contain blood vessels. The epithelium in turn is formed into ridges that protrude toward the lamina propria. These ridges interdigitate with the papillae and are called epithelial ridges. When the tissue is sectioned for microscopy, these ridges look like pegs as they alternate with the papillae, forming a serpentine interface. At one time, the epithelial ridges were mistakenly called epithelial pegs.

Although the two tissues are intimately connected, they are separate. At their junction there are two different structures with very similar names, the basal lamina and the basement membrane. The basal lamina is evident at the electron microscopic level and is epithelial in origin (Fig. 9-3). The basement membrane is evident at the light microscopic level. It is found at the interface of epithelial and connective tissue, within the connective tissue. It is a zone that is 1 to 4 μm wide and is relatively cell free. This zone stains positively with the periodic acid–Schiff method, indicating that it contains neutral mucopolysaccharides (glycosaminoglycans) (Fig. 9-4). It also contains fine argyrophilic reticulin fibers (Fig. 9-5), as well as special anchoring fibrils (Fig. 9-6). In the skin the basement membrane zone has been shown to contain fibronectin and laminin (glycoproteins), heparin sulfate, proteoglycans, type IV collagen, as well as some special antigens (p. 262).

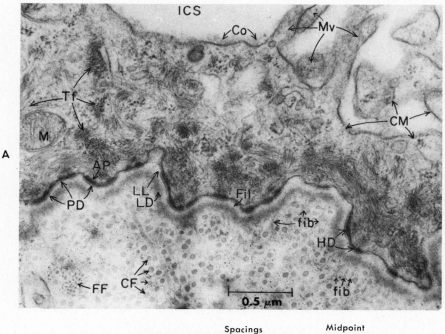

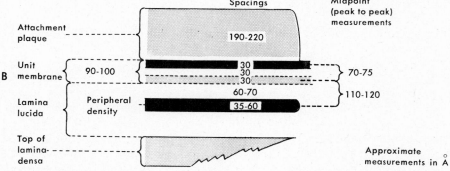

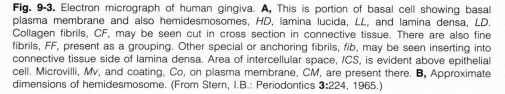

Fig. 9-3. Electron micrograph of human gingiva. **A,** This is portion of basal cell showing basal plasma membrane and also hemidesmosomes, *HD,* lamina lucida, *LL,* and lamina densa, *LD.* Collagen fibrils, *CF,* may be seen cut in cross section in connective tissue. There are also fine fibrils, *FF,* present as a grouping. Other special or anchoring fibrils, *fib,* may be seen inserting into connective tissue side of lamina densa. Area of intercellular space, *ICS,* is evident above epithelial cell. Microvilli, *Mv,* and coating, *Co,* on plasma membrane, *CM,* are present there. **B,** Approximate dimensions of hemidesmosome. (From Stern, I.B.: Periodontics **3:**224, 1965.)

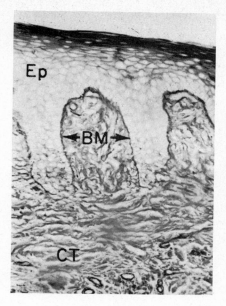

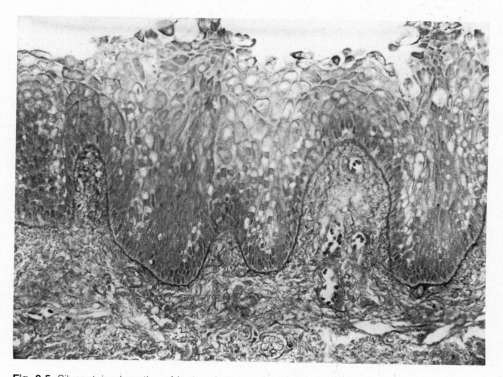

Fig. 9-4. Photomicrograph of human gingiva (PAS stain). PAS-positive basement membrane appears as dense line at epithelium–connective tissue junction. Note that blood vessels in lamina propria also have PAS-positive basement membrane. *BM,* Basement membrane; *CT,* connective tissue; *Ep,* epithelial cells. (× 160.) (From Stern, I.B.: Periodontics **3:**224, 1965.)

Fig. 9-5. Silver-stained section of human fetal tongue showing basement membrane as dense line separating epithelium above from connective tissue below. Extending for variable distance into connective tissue are dark-stained reticular fibers, which are found in greatest number immediately below basement membrane. This zone, known as reticular zone, is found whether or not papillae are present. Papillae delineate extent of papillary zone.

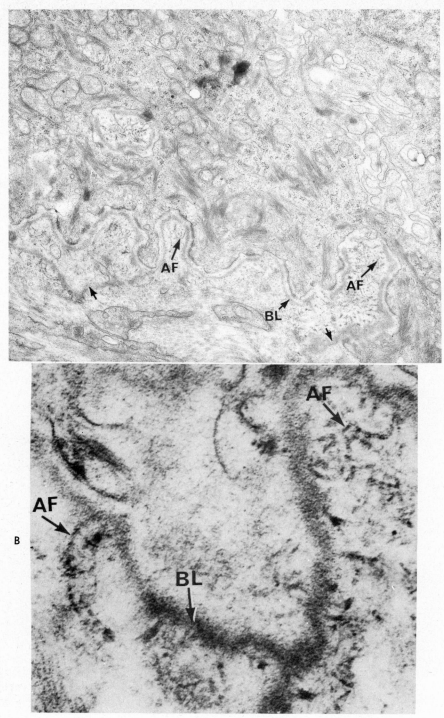

Fig. 9-6. A, Anchoring fibrils, *AF,* and basal lamina, *BL,* which are cut tangentially at places *(arrows).* **B,** These fibrils branch, loop, and exhibit banding.

Lamina propria. The lamina propria may be described as a connective tissue of variable thickness that supports the epithelium. It is divided for descriptive reasons into two parts—papillary and reticular. The papillary portion is named for the papillae, the reticular portion for the reticular fibers. Since there is considerable variation in length and width of the papillae in different areas, the papillary portion is also of variable depth. A portion of the lamina propria subjacent to the basement membrane can be distinguished from the connective tissue because it has the property of taking up silver stain more strongly (argyrophilia) (Fig. 9-5). Fine immature collagen fibers that are argyrophilic and have a trellislike or latticelike arrangement are termed reticulin. This portion as well as the papillary portion contains reticular fibers. The two portions are not separate. They are a continuum but the two terms are used to describe this region in different ways. The reticular zone is always present. The papillary zone may be absent in some areas such as the alveolar mucosa when the papillae are either very short or lacking.

The interlocking arrangement of the connective tissue papillae and the epithelial ridges and the even finer undulations and projections found at the base of each epithelial cell increases the area of contact between the lamina propria and epithelium (Fig. 9-7). This additional area facilitates exchange of material between the epithelium and the blood vessels in the connective tissue. In addition, cells with heavily arranged pedicles (serrations) may serve to strengthen the attachment to the connective tissue. Cells with flatter basal surfaces may be preparing to undergo cell division and will either remain in the proliferative pool in the basal layer or will become determined as keratinocytes, cells destined to migrate to the tissue surface to become part of the stratum corneum.

The lamina propria may attach to the periosteum of the alveolar bone, or it may overlay the submucosa, which varies in different regions of the mouth such as the soft palate and floor of the mouth.

Submucosa. The submucosa consists of connective tissue of varying thickness and density. It attaches the mucous membrane to the underlying structures. Whether this attachment is loose or firm depends on the character of the submucosa. Glands, blood vessels, nerves, and also adipose tissue are present in this layer. It is in the submucosa that the larger arteries divide into smaller branches, which then enter the lamina propria. Here they again divide to form a subepithelial capillary network in the papillae. The veins originating from the capillary network course back along the path taken by the arteries. The blood vessels are accompanied by a rich network of lymph vessels. The sensory nerves of the mucous membrane tend to be more concentrated toward the anterior part of the mouth (rugae, tip of tongue, etc.). The nerve fibers are myelinated as they traverse the submucosa but lose their myelin sheath before splitting into their end arborzations. Sensory nerve endings of various types are found in the papillae (Fig. 9-8, *A*). Some of the fibers enter the epithelium, where they terminate between the epithelial cells as free nerve endings (Fig. 9-8, *B*). The blood vessels are accompanied by nonmyelinated visceral nerve fibers that supply their smooth muscles. Other visceral fibers supply the glands.

In studying any mucous membrane, the following features should be considered: (1) type of covering epithelium, (2) structure of the lamina propria, its density and thickness, and the presence or lack of elasticity, (3) the form of junction between the epithelium and lamina propria, and (4) the membrane's fixation to the underlying structures, that is, the submucous layer. Considered as a separate and well-defined layer, submucosa may be present or absent. Looseness or density of its texture determines whether the mucous membrane is

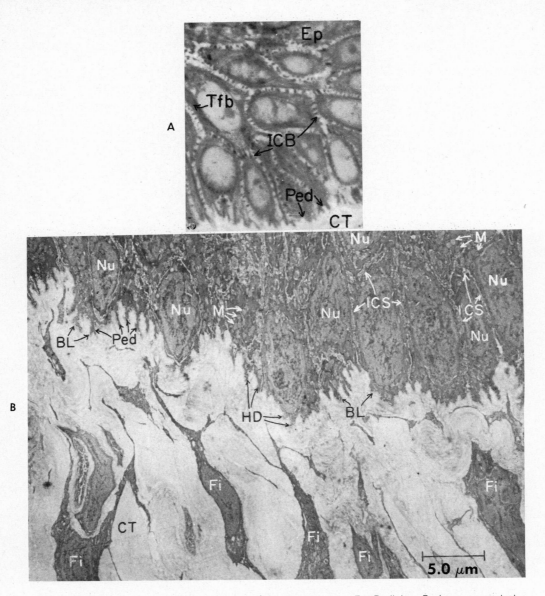

Fig. 9-7. A, Photomicrograph of human gingival epithelial cells, *Ep.* Pedicles, *Ped,* are present at base of basal cells and extend toward connective tissue, *CT.* Tonofibrils, *Tfb,* are evident both in cells and apparently coursing across intercellular bridges, *ICB.* **B,** Electron micrograph of rat gingiva. Several basal cells with apparent pedicles, *Ped,* extending toward connective tissue, *CT,* but separated from it by basal lamina, *BL,* which is barely visible. Fibroblasts, *Fi,* may be noted within connective tissue. Epithelial cells contain prominent nucleus, *Nu,* and are demarcated from adjacent cells by lighter appearance of intercellular spaces, *ICS.* Small, round, light areas in epithelial cells are mitochondria, *M.* Pedicles, *Ped,* in this electron micrograph are of a much smaller dimension than larger undulations of basal cell surface outlined by arrows at *HD.* These, in turn, are smaller than ridges shown in Fig. 9-7, *A.* (**A,** × 1400.) (From Stern, I.B.: Periodontics **3:**224, 1965.)

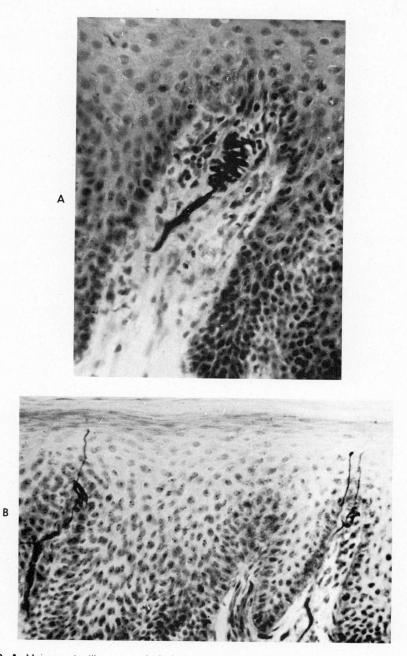

Fig. 9-8. A, Meissner tactile corpuscle in human gingiva (silver impregnation after Bielschowsky-Gros). **B,** Intraepithelial "ultraterminal" extensions and nerve endings in human gingiva (silver impregnation after Bielschowsky-Gros). (From Gairns, F.W., and Aitchison, J.A.: Dent. Rec. **70:**180, 1950.)

movably or immovably attached to the deeper layers. Presence or absence and location of adipose tissue or glands should also be noted.

Epithelium. The epithelium of the oral mucous membrane is of the stratified squamous variety. It may be keratinized, parakeratinized, or nonkeratinized, depending on location. In humans the epithelial tissues of the gingiva and the hard palate (masticatory mucosa) are keratinized (Fig. 9-9, *A*), although in many individuals the gingival epithelium is parakeratinized (Fig. 9-9, *C*). The cheek, faucial, and sublingual tissues are normally nonkeratinized (Fig. 9-9, *B*).

A common feature of all epithelial cells is that they contain keratin intermediate filaments as a component of their cytoskeleton. This is one of the distinguishing features of an epithelial cell, regardless of its function. The analogous components of connective tissue cells are called *vimentin;* in muscle cells they are called *desmin,* and in nerve cells *neural filaments.* These words are used with a biochemical orientation. All intermediate filaments resemble tonofilaments, are 7 to 11 nm in width, and can be reconstituted in vitro from the isolated filaments.

Keratinizing oral epithelium has four cell layers: basal, spinous, granular, and cornified. These are also referred to in Latin as *stratum basale, stratum spinosum, stratum granulosum,* and *stratum corneum.* These layers take their names from their morphologic appearance. A single cell is, at different times, a part of each layer. After mitosis, it may remain in the basal layer and divide again or it may become determined, during which time it migrates and is pushed upward. During its migration, as a keratinocyte it becomes committed to biochemical and morphologic changes (differentiation), which culminate in the formation of a keratinized squama, a dead cell filled with densely packed protein contained within a toughened cell membrane. After reaching the surface it desquamates. This whole process from the onset of determination is called keratinization. A determined keratinocyte can no longer divide. In order for the tissue to remain in a steady state, undifferentiated cells must remain in the basal layer and form one differentiated cell for each cell that desquamates.

The basal layer is made up of cells that synthesize DNA and undergo mitosis, thus providing new cells (Fig. 9-10). Most new cells are generated in the basal layer. However, some mitotic figures may be seen in spinous cells just beyond the basal layer. Therefore the basal cells and the parabasal spinous cells are referred to as the *stratum germinativum.*

It has been proposed that the basal cells are made up of two populations. One population is serrated and heavily packed with tonofilaments, which are adaptations for attachment, and the other is nonserrated and is composed of slowly cycling stem cells. The stem cells give rise to a population of cells amplified for cell division, the proliferative compartment.

The serrated basal cells are a single layer of cuboid or high cuboid cells that have protoplasmic processes (pedicles) projecting from their basal surfaces toward the connective tissue (Fig. 9-7). Specialized structures called *hemidesmosomes,* which abut on the basal lamina (Fig. 9-3), are found on the basal surface. They consist of a single attachment plaque, the adjacent plasma membrane, and an associated extracellular structure that appears to attach the epithelium to the connective tissue. The basal lamina is made up of a clear zone *(lamina lucida)* just below the epithelial cells and a dark zone *(lamina densa)* beyond the lamina lucida and adjacent to the connective tissue (Fig. 9-3; also see p. 317). The lamina lucida thus far has been shown to contain laminin and bullous pemphigoid antigen. The lamina densa contains type IV collagen and an antigen bound by the antibody KF-1. Below the lamina densa is a fibrillar zone (sublamina densa fibrils) that is not of epithelial origin.

The lateral borders of adjacent basal cells are closely apposed and connected by desmosomes

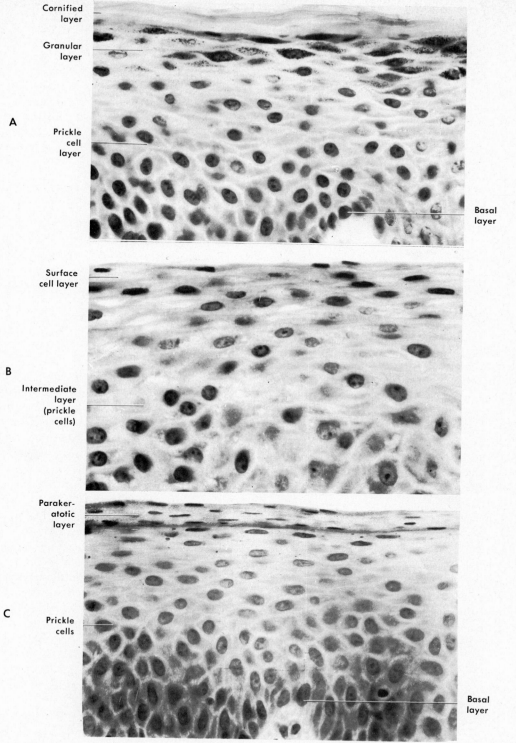

Fig. 9-9. Variations of gingival epithelium. **A,** Keratinized. **B,** Nonkeratinized. **C,** Parakeratinized.

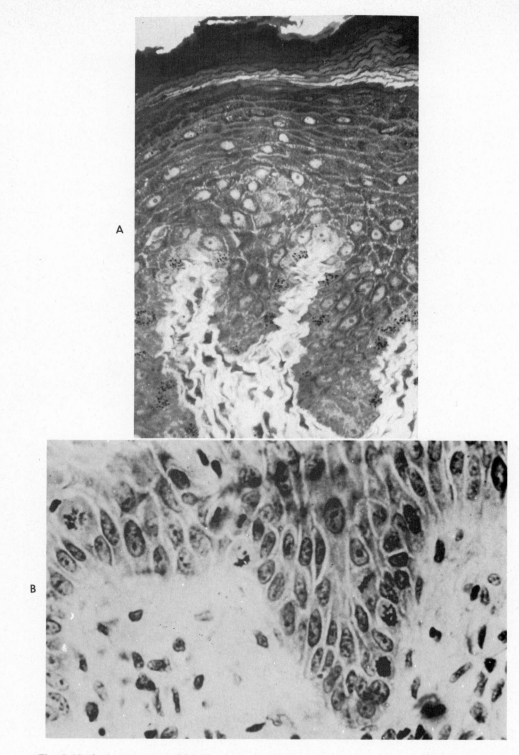

Fig. 9-10. A, Arrangement of labeling in oral epithelium 30 minutes after administration of tritiated thymidine. Grains are localized over nuclei in stratum basale. **B,** Oral epithelium showing many mitotic figures. (**A** from Anderson, G.S., and Stern, I.B.: Periodontics **4:**115, 1966.)

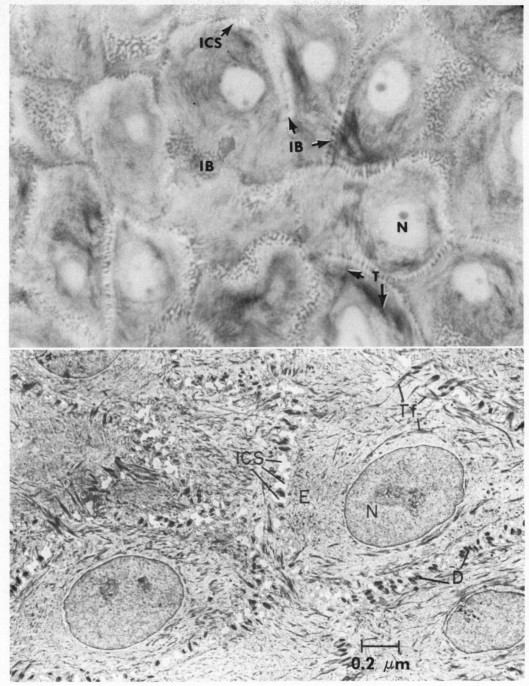

Fig. 9-11. A, High magnification light micrograph showing epithelial cells with nuclei, *N;* intercellular spaces, *ICS;* tonofibrils, *T;* and intercellular bridges, *IB.* Speckled areas are intercellular bridges (desmosomes) cut tangentially or "en face." **B,** Electron micrograph of prickle cells of human gingiva. Portions of epithelial cells, *E,* are evident, separated by intercellular space, *ICS.* Several nuclei, *N,* are evident. Tonofilaments, *Tf,* are present in cytoplasm and extend toward desmosomes, *D,* located at periphery of cells. (**B** from Grant, D.A., Stern, I.B., and Everett, F.G.: Periodontics in the tradition of Orban and Gottlieb, ed. 5, St. Louis, 1979. The C.V. Mosby Co.)

(Fig. 9-11, *B*). These are specializations of the cell surface, consisting of adjacent cell membranes and a pair of denser regions (attachment plaques) as well as intervening extracellular structures (Fig. 9-12). The basal cells contain tonofilaments, which course toward, and in some way are attached to, the attachment plaques. There are other types of cell junctions such as tight, close, and gap junctions present. There are also ribosomes and elements of rough-surfaced endoplasmic reticulum, indicative of protein-synthesizing activity. Basal cells synthesize some of the proteins of the basal lamina. They also synthesize proteins, which form the intermediate filaments of the basal cells.

The *spinous cells* (stratum spinosum) are irregularly polyhedral and larger than the basal cells. On the basis of light microscopy, it appears that the cells are joined by "intercellular bridges" (Fig. 9-11, *A*). Tonofibrils seem to course from cell to cell across these bridges. Electron microscopic studies have shown that the "intercellular bridges" are desmosomes and the tonofibrils are bundles of tonofilaments (Fig. 9-13). The tonofilaments turn or loop adjacent to the attachment plaques and do not cross over into adjacent cells. It is suspected that an agglutinating material joins them to the attachment plaques. The desmosome attachment plaques contain the polypeptides desmoplakin I and II. Monoclonal antibodies to these polypeptides can be used to detect carcinomas (an epithelial tumor) by immunofluorescent microscopy. The intercellular spaces contain glycoprotein, glycosaminoglycans, and fibronectin.

The tonofilament network and the desmosomes appear to make up a tensile supporting system for the epithelium. The percentage of cell membrane occupied by hemidesmosomes is higher in basal cells of gingiva and palate than in alveolar mucosa, buccal mucosa, and tongue. The intercellular spaces of the spinous cells in keratinizing epithelia are large or distended; thus the desmosomes are made more prominent and these cells are given a prickly

appearance. The spinous (prickle) cells resemble a cockleburr or sticker that has each spine ending at a desmosome. Of the four layers, the spinous cells are the most active in protein synthesis. These cells synthesize additional proteins that differ from those made in the basal cells. This change indicates their biochemical commitment to keratinization. In terms of number and length the desmosomes of the spinous layer occupy more of the membrane in the tongue, gingiva, and palate than in either alveolar or buccal mucosa.

The next layer *(stratum granulosum)* contains flatter and wider cells. These cells are larger than the spinous cells. This layer is named for the basophilic keratohyalin granules (blue staining with hematoxylin and eosin) (Fig. 9-14, *A* to *C*) that it contains. The nuclei show signs of degeneration and pyknosis. This layer still synthesizes protein, but reports of synthesis rates at this level differ. However, as the cell approaches the stratum corneum, the rate diminishes. Tonofilaments are more dense in quantity and are often seen associated with keratohyalin granules (Fig. 9-14, *D* to *F*). Sometimes dense networks of tonofilaments and keratohyalin granules are evident. Keratohyalin granules contain at least two types of proteins, one that (solubilized by solutions containing high concentrations of salt) is rich in the amino acid histidine. It functions in the stratum corneum as a keratin filament matrix protein (see below).

In the stratum granulosum the cell surfaces become more regular and more closely applied to adjacent cell surfaces. At the same time the lamellar granule, a small organelle (also known as keratinosome, Odland body, or membrane-coating granule) forms in the upper spinous and granular cell layers. It has an internal lamellated structure (Fig. 9-15, *A* and *B*). Lamellar granules discharge their contents into the intercellular space forming an intercellular lamellar material, which contributes to the permeability barrier (Fig. 9-15, *C*). This barrier forms at the junction of granular and cornified

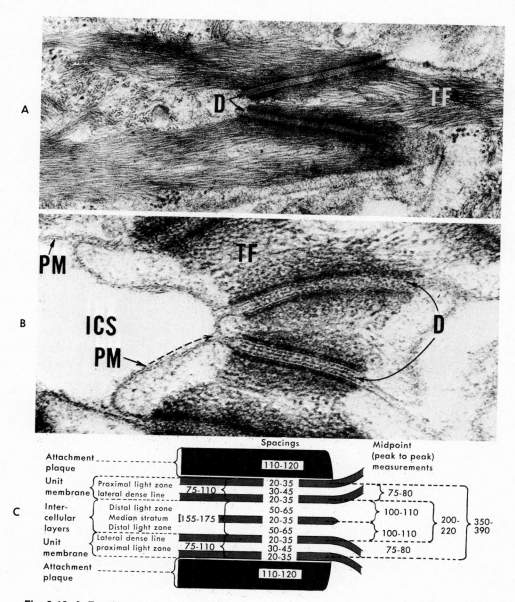

Fig. 9-12. A, Tonofilaments, *Tf,* extending to series of desmosomes, *D.* Tonofilaments are sectioned in long axis (human gingiva). **B,** Higher magnification of two desmosomes, *D,* showing substructure. Tonofilaments are cross sectioned. Intercellular space, *ICS,* is bounded by adjacent cell membranes, *PM,* whose unit membrane is clearly evident *(dashed arrow).* Unit membranes form part of substructure of desmosome. **C,** Diagrammatic cross-sectioned representation of desmosome and dimensions (in Angstrom units) of various components. (From Stern, I.B.: Periodontics **3:**224, 1965.)

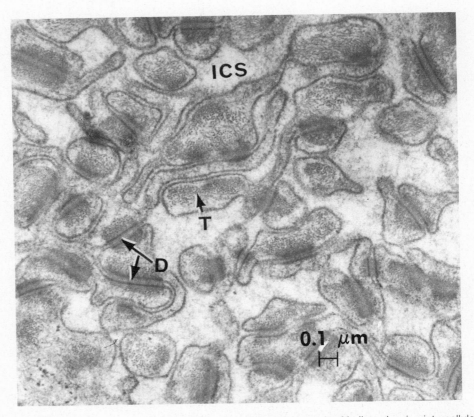

Fig. 9-13. Electron micrograph of prickle cell layer of human gingival epithelium showing intercellular bridges and tonofibrils. Here desmosomes are cut tangentially or "en face" as shown in light micrograph (Fig. 9-11, *A*). Note relatively close adaptation of cell processes ending in desmosomes, *D*. These processes contain tonofilaments, *T*, cut on end, which appear as fine dots. Relatively large intercellular space, *ICS*, contains cell-coating material.

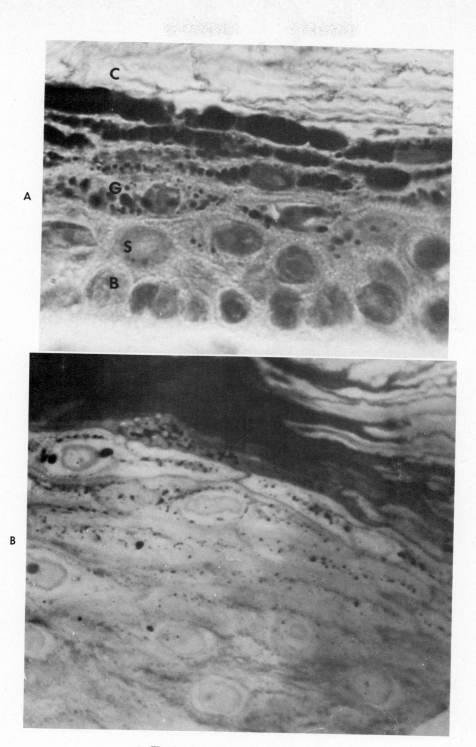

Fig. 9-14. For legend see page 270.

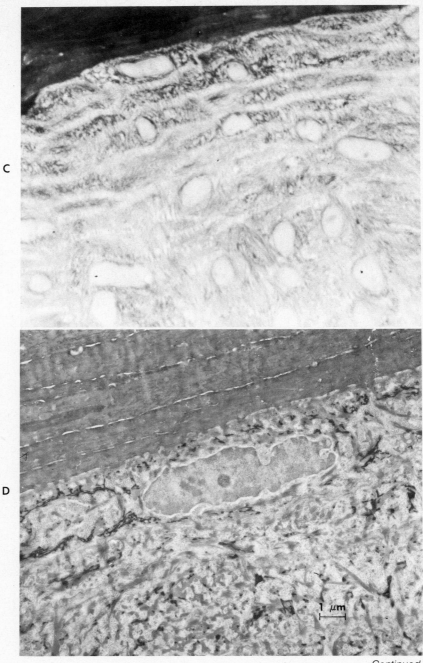

C

D

1 µm

Continued.

Fig. 9-14. A, Light micrograph of newborn rat skin showing basal cells, *B,* spinous cells, *S,* granular cells with numerous dense granules, *G,* and cornified (keratinized) components. **B** and **C,** Keratohyalin is formed as discrete spherical granules in some tissues or is formed as angular amorphous material in other tissues. **D** and **E,** Angular form is associated with tonofilaments primarily *(arrow);* **F,** whereas spherical form is surrounded by ribosomes *(arrow)* and may contain more than one material *(small arrows).*

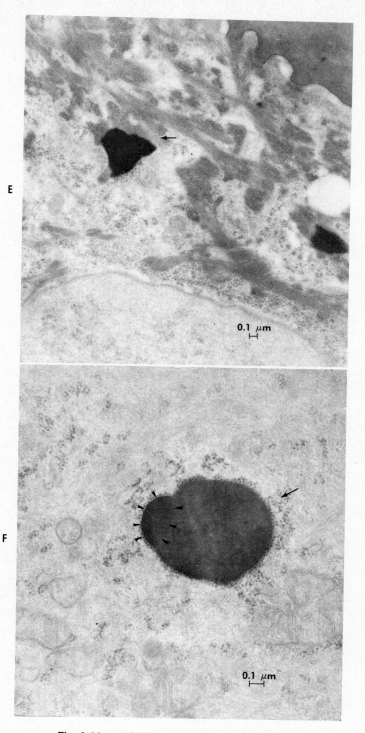

Fig. 9-14, cont'd. For legend see opposite page.

cell layers. The intercellular space of this region has a lamellar structure similar to that of the lamellar granule (Fig. 9-15, *C*) and has been shown to contain glycolipid. In nonkeratinizing oral epithelium a similar small organelle forms. It differs in that its contents are granular rather than lamellar. It may serve a similar function. At approximately the same time during differentiation, the inner unit of the cell membrane thickens forming the "cornified cell envelope." Several proteins contribute to this structure, among which is involucrin (keratolinin), which is present at the upper half of the stratum spinosum. Thereafter the thickened membrane contains sulfur-rich proteins stabilized by covalent crosslinks. It forms a highly resistant structure.

The *stratum corneum* is made up of keratinized squamae, which are larger and flatter than the granular cells. Here all of the nuclei and other organelles such as ribosomes and mitochondria have disappeared (Fig. 9-14, *D*).* The layer is acidophilic (red staining with hematoxylin and eosin) and is histologically amorphous. The keratohyalin granules have disappeared. Ultrastructurally the cells of the cornified layer are composed of densely packed filaments developed from the tonofilaments, altered, and coated by the basic protein of the keratohyalin granule, filaggrin.

The cells of the stratum corneum are densely packed with filaments in this nonfibrous interfilamentous matrix protein, filaggrin (named for its function in filament aggregation). When the purified solubilized matrix protein obtained from the epithelium is combined with solubi-

---

*In states such as dandruff the rates of cell division and desquamation increase and nuclei may persist in the desquamating cells.

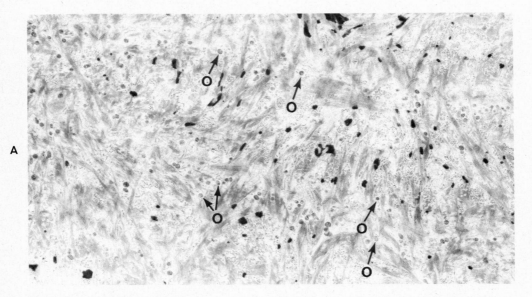

Fig. 9-15. A, Lamellar granules, *O,* are found close to cell membrane *(arrows)* and desmosomes in granular cells. **B,** Lamellar granules, *O,* lying close to plasma membrane *(arrows)* and in cells containing ribosome-associated keratohyalin granules. Note that some of keratohyalin granules have two densities and perhaps two components. Lamellar granules contain an internal lamellar structure. **C,** Lamellar structure in intercellular space *(arrows)*. It is presumed that these lamellae are derived from lamellar granules that are no longer present.

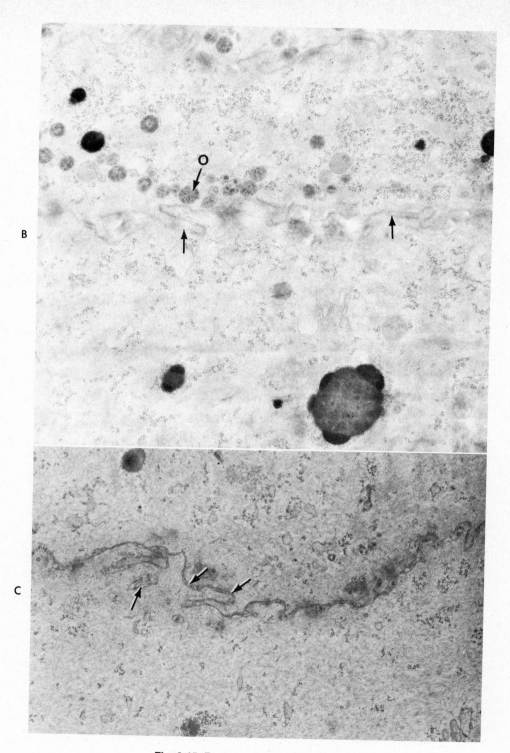

Fig. 9-15. For legend see opposite page.

lized keratin filaments in vitro, aggregates of matrix and highly oriented filaments form instantaneously (Fig. 9-16). Their ultrastructural appearance is similar to that of the contents of the stratum corneum. The active matrix protein, filaggrin, is derived from a precursor in the keratohyalin granules. Studies of the interaction of matrix and filaments have been performed with filaggrin and keratin filaments obtained from epidermis; however, the same proteins can also be demonstrated in keratinizing oral epithelium. The keratinized cell becomes compact and dehydrated and covers a greater surface area than does the basal cell from which it developed. It does not synthesize protein. It is closely applied to adjacent squamae. The cell surface and desmosomes are altered, and the plasma membrane is denser and thicker than in the cells of deeper layers.

Epithelial cells that ultimately keratinize are called *keratocytes* or *keratinocytes*. Keratinocytes increase in volume in each successive layer from basal to granular. The cornified cells, however, are smaller in volume than the granular cells. The cells of each successive layer cover a larger area than do the cells of the layers immediately below.

Nonkeratinizing epithelia differ from keratinizing epithelia primarily because they do not produce a cornified surface layer, but there are other differences as well. The layers in nonkeratinizing epithelium are referred to as basal, intermediate, and superficial (*stratum basale, stratum intermedium, stratum superficiale*) (Fig. 9-9, *B*). The basal cells of both types are

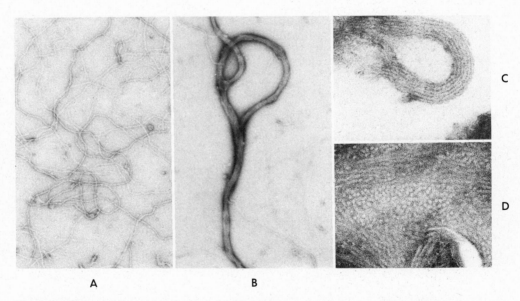

A   B   C

D

Fig. 9-16. A, Rat epidermal keratin subunits formed after solubilization in 8M urea. **B,** Aggregates of epidermal keratin filaments and filaggrin after mixing the proteins in a 0.25:1 ratio and allowing mixture to stand for 5 minutes. **C,** Aggregates formed as in **B,** fixed and embedded for transmission electron microscopy. Filaments 7 to 11 nm wide appear in longitudinal section separated at a uniform distance by darker staining material. **D,** Cross section of inner portion of rat stratum corneum showing keratin pattern of electron-lucent filament 7 to 11 nm wide embedded in an osmiophilic interfilamentous matrix. (**A** and **B** negatively stained with 1% uranyl acetate; ×62, 700. **C** and **D,** ×105,000.) (**A** to **C** from Dale, B.A., Holbrook, K.A., and Steinert, P.M.: Nature **276:**129, 1978. **D** courtesy B.A. Dale, K.A. Holbrook, and P.M. Steinert.)

similar. The cells of the stratum intermedium are larger than cells of the stratum spinosum. The intercellular space is not obvious or distended and hence the cells do not have a prickly appearance. Nevertheless, the cells of the stratum intermedium often are referred to as spinous or prickle cells, even though morphologically they are not spinous and biochemically they do not keratinize. These cells do contain some intermediate keratin filaments, but they differ biochemically from those in keratinizing epithelia and are sparsely distributed within the cells. The cells of the stratum intermedium are attached by desmosomes and other junctions, and their cell surfaces are more closely applied than are spinous cells. There is no stratum granulosum (although incomplete or vestigial granules may form), nor is there a stratum corneum. Nucleated cells exist at the surface (Fig. 9-17). These cells ultimately desquamate, as do the cornified squamae.

In *orthokeratinization*, keratinized squamae form as has been described. In *parakeratinization*, the cells retain pyknotic and condensed nuclei and other partially lysed cell organelles until they desquamate. There are signs of condensation of the superficial cells, which appear almost as if they were keratinizing. Tissues that are not keratinized at one stage of development may keratinize at another (Fig. 9-15). Similarly, tissues may be modulated from keratinized-parakeratinized and nonkeratinized variants in pathologic states. While the terms "keratinized" and "parakeratinized" may be used interchangeably with the terms "parakeratosis" and "keratosis," the former terms refer to physiologic and the latter terms refer to pathologic stages. When keratinization occurs in a normally nonkeratinized tissue, it is referred to as keratosis. When normally keratinizing tissue such as the epidermis becomes parakeratinized, it is referred to as parakeratosis.

The oral epithelium in addition contains melanocytes, Langerhans' cells, Merkel's cells, and various white blood cells (see p. 287).

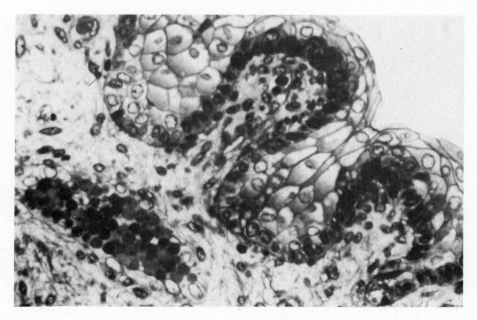

Fig. 9-17. Section of human fetal tongue showing three cell strata of nonkeratinized epithelium.

SUBDIVISIONS OF ORAL MUCOSA

For descriptive purposes the oral mucosa may be divided into the following areas:

Keratinized areas
   Masticatory mucosa
   Vermilion border of lip
Nonkeratinized areas
   Lining mucosa
   Specialized mucosa

Keratinized areas

Masticatory mucosa (gingiva and hard palate)

The masticatory mucosa is keratinized and is made up of the gingiva and the hard palate. They have similarities in thickness and keratinization of epithelium; in thickness, density, and firmness of lamina propria; and in being immovably attached. However, there are differences in their submucosa.

Hard palate. The mucous membrane of the hard palate is tightly fixed to the underlying periosteum and therefore immovable. Like the gingiva it is pink. The epithelium is uniform in form with a rather well-keratinized surface. The cells of the stratum corneum exhibit stacking, and in the rat there are complementary grooves and ridges between the apposing surfaces of the cells. The pedicles (see p. 259), the increase in number and length of desmosomes, the density of the tonofilaments, and the complementary grooves and ridges all appear to be adaptations of keratinizing epithelium to resist forces and to bind the epithelium to the connective tissue. The lamina propria, a layer of dense connective tissue, is thicker in the anterior than in the posterior parts of the palate and has numerous long papillae. Various regions in the hard palate differ because of the varying structure of the submucous layer. The following zones can be distinguished (Fig. 9-18):

1. Gingival region, adjacent to the teeth
2. Palatine raphe, also known as the median area, extending from the incisive or palatine papilla posteriorly

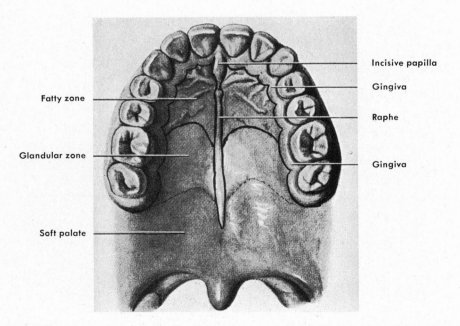

Fatty zone

Glandular zone

Soft palate

Incisive papilla

Gingiva

Raphe

Gingiva

Fig. 9-18. Surface view of hard and soft palates. The different zones of palatine mucosa are outlined.

3. Anterolateral area or fatty zone between the raphe and gingiva
4. Posterolateral area or glandular zone between the raphe and gingiva

Except for narrow and specific zones, the palate has a distinct submucous layer. The zones that do not have a submucous layer occur peripherally where the palatine tissue is identical with the gingiva and along the midline for the entire length of the hard palate (the palatine raphe) (Fig. 9-18). The marginal area shows the same structure as the other regions of the gingiva. Only the lamina propria and periosteum are present below the epithelium (Fig. 9-19). Similarly, a submucosa is not found below the palatine raphe, or median area (Fig. 9-20). The lamina propria blends with the periosteum. If a palatine torus is present, the mucous membrane is thinner. The otherwise narrow raphe is widened and spreads over the entire torus.

The submucous layer occurs in wide regions extending between the palatine gingiva and palatine raphe. Despite this extensive submucosa, the mucous membrane is immovably attached to the periosteum of the maxillary and palatine bones. This attachment is formed by dense bands and trabeculae of fibrous connective tissue that join the lamina propria of the mucosa membrane to the periosteum. The submucous space is thus subdivided into irregular intercommunicating compartments of various sizes. These are filled with adipose tissue in the anterior part and with glands in the posterior part of the hard palate. The presence of fat or glands in the submucous layer acting as a cushion is comparable to the subcutaneous tissue of the palm of the hand and the sole of the foot.

When the submucosa of hard palate and that of gingiva are compared, there are pronounced differences. The dense connective tissue that makes up the lamina propria of gingiva is bound to the periosteum of the alveolar process or to the cervical region of the tooth. A sub-

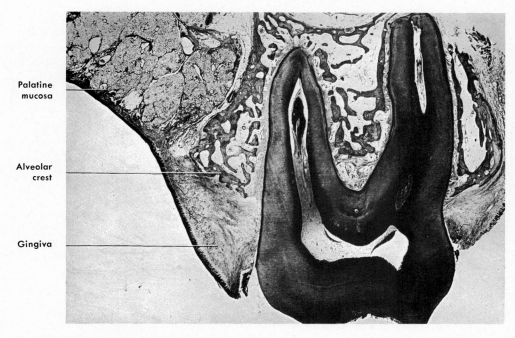

Palatine mucosa

Alveolar crest

Gingiva

Fig. 9-19. Structural differences between gingiva and palatine mucosa. Region of first molar.

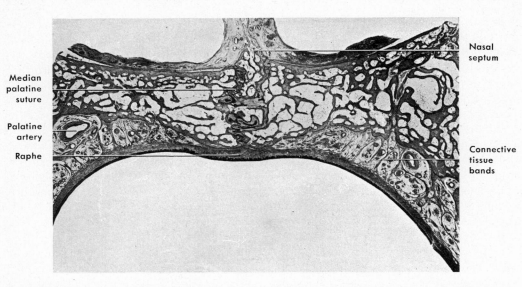

Median
palatine
suture

Palatine
artery

Raphe

Nasal
septum

Connective
tissue
bands

Fig. 9-20. Transverse section through hard palate. Palatine raphe. Fibrous bands connecting mucosa and periosteum in lateral areas. Palatine vessels. (From Pendleton, E.C.: J. Am. Dent. Assoc. **21:**488, 1934.)

mucous layer, as such, cannot generally be recognized. In the lateral areas of the hard palate (Fig. 9-20), in both fatty and glandular zones, the lamina propria is fixed to the periosteum by bands of dense fibrous connective tissue. These bands are arranged at right angles to the surface and divide the submucous layer into irregularly shaped spaces. The distance between lamina propria and periosteum is smaller in the anterior than in the posterior parts. In the anterior zone the connective tissue contains fat (Fig. 9-21), whereas in the posterior part it contains mucous glands (Fig. 9-21). The glandular layers of the hard palate and of the soft palate are continuous.

At the junction of the alveolar process and the horizontal plate of the hard palate the anterior palatine vessels and nerves course, surrounded by loose connective tissue. This wedge-shaped area (Fig. 9-22) is large in the posterior part of the palate and smaller in the anterior part. It is important for oral surgeons

and periodontists to know the distribution of these vessels.

Incisive papilla. The oral incisive (palatine) papilla is formed of dense connective tissue. It contains the oral parts of the vestigial nasopalatine ducts. They are blind ducts of varying length lined by simple or pseudostratified columnar epithelium, rich in goblet cells. Small mucous glands open into the lumen of the ducts. These ducts sometimes become cystic in humans. Frequently the ducts are surrounded by small, irregular islands of hyaline cartilage, which are the vestigial extensions of the paraseptal cartilages. In most mammals the nasopalatine ducts are patent and, together with Jacobson's organ, are considered as auxiliary olfactory sense organs. Jacobson's organ (the vomeronasal organ) is a small ellipsoid (cigar-shaped) structure lined with olfactory epithelium that extends from the nose to the oral cavity. In humans, Jacobson's organ is apparent in the twelfth to fifteenth fetal week, after which

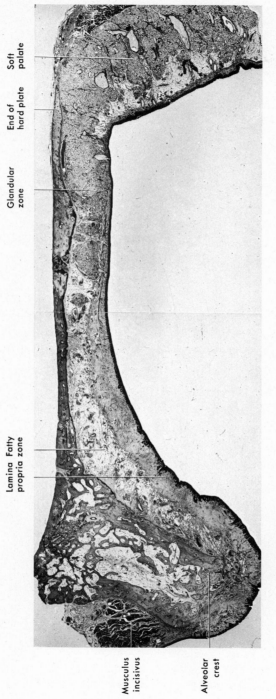

Soft palate

End of hard plate

Glandular zone

Lamina propria

Fatty zone

Musculus incisivus

Alveolar crest

Fig. 9-21. Longitudinal section through hard and soft palates lateral to midline. Fatty and glandular zones of hard palate.

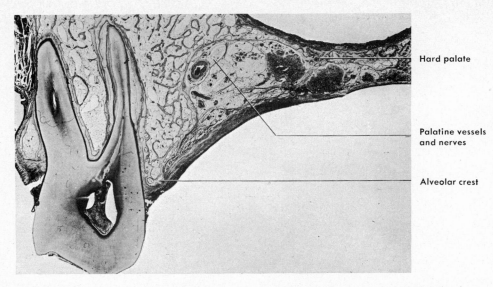

Hard palate

Palatine vessels
and nerves

Alveolar crest

Fig. 9-22. Transverse section through posterior part of hard palate, region of second molar. Loose connective tissue in groove between alveolar process and hard palate around palatine vessels and nerves.

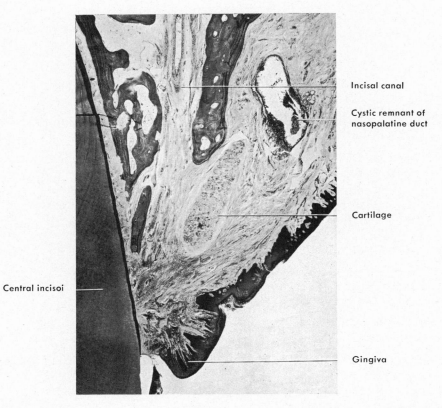

Incisal canal

Cystic remnant of
nasopalatine duct

Cartilage

Central incisoi

Gingiva

Fig. 9-23. Sagittal section through palatine papilla and anterior palatine canal. Note cartilage in papilla.

it undergoes involution. In humans cartilage is sometimes found in the anterior parts of the papilla. In this location it bears no relation to the nasopalatine ducts (Fig. 9-23).

Palatine rugae (transverse palatine ridges). The palatine rugae, irregular and often asymmetric in humans, are ridges of mucous membrane extending laterally from the incisive papilla and the anterior part of the raphe. Their core is made of a dense connective tissue layer with fine interwoven fibers.

Epithelial pearls. In the midline, especially in the region of the incisive papilla, epithelial pearls may be found in the lamina propria. They consist of concentrically arranged epithelial cells that are frequently keratinized. They are remnants of the epithelium formed in the line of fusion between the palatine processes (see Chapter 1).

Gingiva. The gingiva extends from the dentogingival junction to the alveolar mucosa and is subject to the friction and pressure of mastication. The morphology of both epithelium and connective tissues indicates its adaptation to these forces. It is made up of stratified squamous epithelium, which may be keratinized or nonkeratinized but most often is parakeratinized. The epithelium covers a dense lamina propria. The

collagen fibers of the lamina propria may either insert into the alveolar bone and the cementum or blend with the periosteum.

The gingiva is limited on the outer surface of both jaws by the mucogingival junction, which separates it from the alveolar mucosa (Fig. 9-24). The alveolar mucosa is red and contains numerous small vessels close to the surface. On the inner surface of the lower jaw a line of demarcation is found between the gingiva and the mucosa on the floor of the mouth. On the palate the distinction between the gingiva and the peripheral palatal mucosa is not so sharp.*

---

*These surfaces are frequently referred to as buccal or labial, lingual or palatal. The oral cavity can be divided into two parts: the vestibulum oris (vestibule) and the cavum oris proprium (oral cavity proper). The term "vestibular" is used to describe those surfaces that face the vestibule; thus the need of differentiating between buccal and labial is eliminated. This tends to simplify descriptions and coincides with proper anatomic usage. The vestibular cavity is bounded anterolaterally by the mucous membranes of the lips and cheeks and internally by the teeth and gingiva. Vestibular would therefore apply to any tooth surface facing the vestibular cavity. Similarly the term "oral" describes the palatal and lingual. The oral cavity proper is bounded anterolaterally by the teeth and gingiva, superiorly by the soft and hard palate, inferiorly by the tongue and mucous membranes of the floor of the mouth, and posteriorly by the pillars of the fauces, the opening into the oral pharynx.

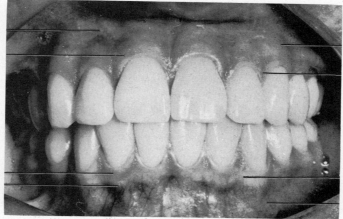

Fig. 9-24. Vestibular surface of gingiva of yound adult. (Courtesy Dr. A. Ogilvie, Vancouver, British Columbia.)

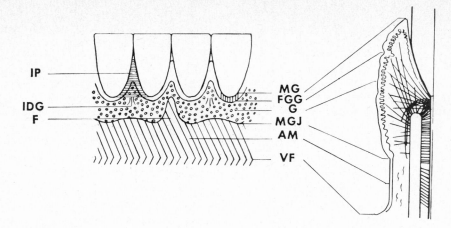

Fig. 9-25. Diagrammatic illustration of surface characteristics of the clinically normal gingiva. *IP,* Interdental papilla; *IDG,* interdental groove; *F,* frenum; *MG,* marginal gingiva; *FGG,* free gingival groove; *G,* gingiva; *MGJ,* mucogingival junction; *AM,* alveolar mucosa; *VF,* vestibular fornix. (From Grant, D.A., Stern, I.B., and Everett, F.G.: Periodontics in the tradition of Orban and Gottlieb, ed. 5, St. Louis, 1979, The C.V. Mosby Co.)

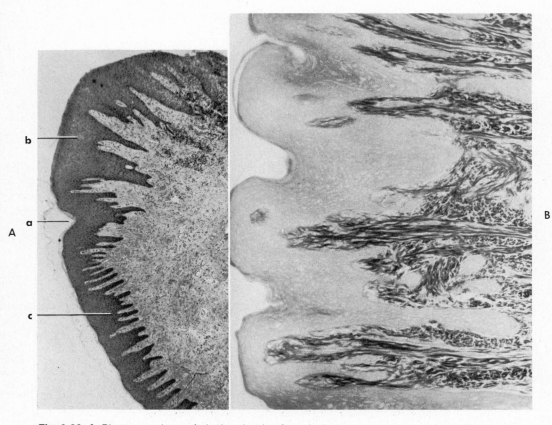

Fig. 9-26. A, Biopsy specimen of gingiva showing free gingival groove, *a,* and corresponding heavy epithelial ridge; *b,* free gingiva; *c,* gingiva. **B,** Gingival specimen showing stippling. Note relation of connective tissue fiber bundles to stippled surface (Mallory stain). (From Grant, D.A., Stern, I.B., and Everett, F.G.: Periodontics in the tradition of Orban and Gottlieb, ed. 5, St. Louis, 1979, The C.V. Mosby Co.)

The gingiva can be divided into the *free gingiva*, the "attached" *gingiva** (Fig. 9-25), and the *interdental papilla*. The dividing line between the free gingiva and the gingiva is the *free gingival groove*, which runs parallel to the margin of the gingiva at a distance of 0.5 to 1.5 mm. The free gingival groove, not always visible microscopically, appears in histologic sections (Fig. 9-26, *A*) as a shallow V-shaped notch at a heavy epithelial ridge. The free gingival groove develops at the level of, or somewhat apical to, the bottom of the gingival sulcus. In some cases the free gingival groove is not so well defined as in others, and then the division between the free gingiva and the gingiva is not clear. The free gingival groove and the epithelial ridge are probably caused by functional impacts on the free gingiva. In the absence of a sulcus there is no free gingiva.

The gingiva is characterized by a surface that appears stippled (Fig. 9-26, *B*). Portions at the epithelium appear to be elevated, and between the elevations there are shallow depressions, the net result of which is stippling. The depressions correspond to the center of heavier epithelial ridges. There may be protuberances of the epithelium as well as stippling. They probably are functional adaptations to mechanical impacts. The disappearance of stippling is an indication of edema, an expression of an involvement of the gingiva in a progressing gingivitis.

Although the degree of stippling (Fig. 9-24) and the texture of the collagenous fibers vary with different individuals, there are also differences according to age and sex. In younger females the connective tissue is more finely textured than in the male. However, with increasing age the collagenous fiber bundles become more coarse in both sexes. Males tend to have more heavily stippled gingivae than do females. Like the human epidermis, the cells

---

*At the International Conference on Research in the Biology of Periodontal Disease, Chicago, June 12-15, 1977, it was voted to drop the use of "attached" and simply refer to gingiva.

of the oral epithelium show another sex difference. In females the majority of the nuclei contain a large chromatin particle adjacent to the nuclear membrane.

The gingiva appears slightly depressed between adjacent teeth, corresponding to the depression on the alveolar process between eminences of the sockets. In these depressions the gingiva sometimes forms slight vertical folds called interdental grooves.

The interdental papilla is that part of the gingiva that fills the space between two adjacent teeth. When viewed from the oral or vestibular aspect, the surface of the interdental papilla is triangular. In a three-dimensional view the interdental papilla of the posterior teeth is tent shaped, whereas it is pyramidal between the anterior teeth. When the interdental papilla is tent shaped, the oral and the vestibular corners are high, whereas the central part is like a valley. The central concave area fits below the contact point, and this depressed part of the interdental papilla is called the *col*. The col is covered by thin nonkeratinized epithelium, and it has been suggested that the col (the nonkeratinized epithelium) is more vulnerable to periodontal disease.

The lamina propria of the gingiva consists of a dense connective tissue that does not contain large vessels. Small numbers of lymphocytes, plasma cells, and macrophages are present in the connective tissue of normal gingiva (Fig. 9-27) subjacent to the sulcus and are involved in

Fig. 9-27. Macrophages in normal gingiva. (Rio Hortega stain; × 1000.) (From Aprile, E.C. de: Arch. Hist. Normal Pat. **3:**473, 1947.)

defense and repair. The papillae of the connective tissue are characteristically long, slender, and numerous. The presence of these high papillae makes for ease in the histologic differentiation of gingiva and alveolar mucosa, in which the papillae are quite low (Fig. 9-28). The tissue of the lamina propria contains only few elastic fibers and for the most part they are confined to the walls of the blood vessels. Other elastic fibers known as oxytalan fibers (because of special staining qualities) are also present. On the other hand, the alveolar mucosa and the submucosa contain numerous elastic fibers. These fibers are thickest in the submucosa.

The gingival fibers of the periodontal ligament enter into the lamina propria, attaching the gingiva firmly to the teeth (see Chapter 7). The gingiva is also immovably and firmly attached to the periosteum of the alveolar bone. Because of this arrangement it is often referred to as mucoperiosteum. Here a dense connective tissue, consisting of coarse collagen bundles (Fig. 9-29, *A*), extends from the bone to the lamina propria. In contrast, the submucosa underlying the alveolar mucous membrane is loosely textured (Fig. 9-29, *B*). The fiber bundles of the lamina propria of the alveolar mucosa are thin and regularly interwoven.

The gingiva contains dense fibers of collagen, sometimes referred to as the gingival ligament, which are divided into the following major groups.

1. *Dentogingival.* Extends from the cervical cementum into the lamina propria of the gingiva. The fibers of the gingival ligament constitute the most numerous group of gingival fibers.

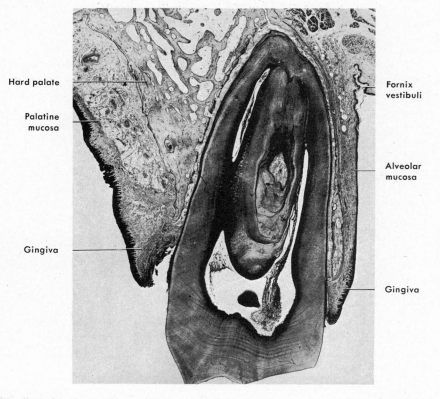

Fig. 9-28. Structural differences between gingiva and alveolar mucosa. Upper premolar.

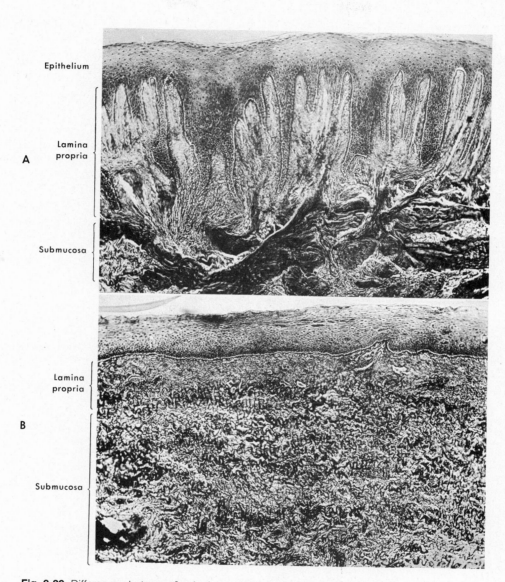

Epithelium

Lamina
propria

A

Submucosa

Lamina
propria

B

Submucosa

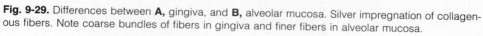

Fig. 9-29. Differences between **A,** gingiva, and **B,** alveolar mucosa. Silver impregnation of collagen-ous fibers. Note coarse bundles of fibers in gingiva and finer fibers in alveolar mucosa.

2. *Alveologingival.* The fibers arise from the alveolar crest and extend into the lamina propria.
3. *Circular.* A small group of fibers that circle the tooth and interlace with the other fibers.
4. *Dentoperiosteal.* These fibers can be followed from the cementum into the periosteum of the alveolar crest and of the vestibular and oral surfaces of the alveolar bone.

There are also accessory fibers that extend interproximally between adjacent teeth and are also referred to as transseptal fibers. These fibers make up the *interdental ligament.*

The gingiva is normally pink but may sometimes have a grayish tint. The color depends in part on the surface (keratinized or not) and thickness and in part on pigmentation. The surface may be translucent or transparent, permitting the color of the underlying tissues to be seen. The reddish or pinkish tint is attributable to the color given the underlying tissue by the blood vessels and the circulating blood.

The presence of melanin pigment in the epithelium may give it a brown to black coloration. Pigmentation is most abundant at the base of the interdental papilla. It may increase considerably in a number of pathologic stages. Melanin is stored by the basal cells in the form of melanosomes, but these cells do not produce the pigment (Fig. 9-30). Melanin is elaborated by specific cells, *melanocytes,* residing in the basal layer and is transferred to the basal cells. The melanocytes are derived from the embryologic neural crest and migrate into the ep-

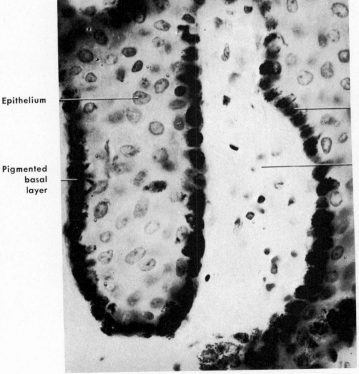

Epithelium

Pigmented basal layer

Pigmented basal layer

Connective tissue

Fig. 9-30. Basal cells of gingiva showing pigmentation.

ithelium (Fig. 9-31). Oral pigmentation can be studied by use of either the dopa reaction or silver-staining techniques. In the dopa reaction the cells containing tyrosinase enzyme appear dark. Therefore the melanin-producing cells, which contain tyrosinase (dopa oxidase), are demonstrated. Silver stains also dye the melanin pigment. The dopa reaction is likewise found in certain connective tissue cells of the lamina propria that contain melanin (melanophages). These cells obtain the pigment from the melanocytes. Melanocytes appear as clear cells in hematoxylin sections. Silver stains reveal a spiderlike (dendritic) appearance. Thus melanocytes are referred to as clear cells or dendritic cells. The number of melanocytes per square millimeter is quite constant for any particular region, and no difference in their numbers is found in the mucosa of blacks and whites.

The following are three types of epithelial surface layers that result from differences in differentiation (Fig. 9-9).

1. *Keratinization,* in which the superficial cells form scales of keratin and lose their nuclei. A stratum granulosum is present.
2. *Parakeratinization,* in which the superficial cells retain pyknotic nuclei and show some signs of being keratinized; however, the stratum granulosum is generally absent.
3. *Nonkeratinization,* in which the surface cells are nucleated and show no signs of keratinization.

The gingiva is parakeratinized 75%, keratinized 15%, and nonkeratinized 10% of the time. It has been suggested that inflammation, which is seen in almost all gingival specimens, interferes with keratinization. The more highly keratinized the tissue, the whiter and less translucent is the tissue.

The Langerhans cell is another clear cell or dendritic cell found in the upper layers of the skin and the mucosal epithelium restricted to zones of orthokeratinization. There is a correlation in the occurrence of a stratum granu-losum and Langerhans' cells. This cell is free of melanin and does not give a dopa reaction. It stains with gold chloride, ATPase, and immunofluorescent markers. Neither the Langerhans cell nor the melanocyte forms desmosomal attachments to the epithelial cells. The Langerhans cell is a cell of hematopoietic origin that contains antigen markers in common with killer lymphocytes. Langerhans cells are involved in the immune response. They contain Ia antigens, which they present to primed T cells (thymocytes). They may function, as do macrophages, by picking up antigen and presenting it to lymphocytes, either locally or at lymph nodes. They have vimentin-type intermediate filaments (see p. 262).

There is still another cell found among the basal cells. This cell, called the *Merkel cell,* has nerve tissue immediately subjacent and is presumed to be a specialized neural pressure-sensitive receptor cell. It is believed to be slow acting, to have neurosecretory activity, and to be a migrant from the neural crest. Thus the oral epithelium not only contains the normal population of keratinocytes arranged in strata according to degree of differentiation, it also contains as residents three types of cells, the *melanocytes* (derived from the neural crest), the *Langerhans* cells (originating in the bone marrow), and the *Merkel cells* (neurally interrelated, though lacking neural filaments).

Other cells, such as lymphocytes and polymorphonuclear leukocytes, are also found at various levels of the epithelium. These cells are transients and can pass through the epithelium to the surface.

Blood and nerve supply. The blood supply of the gingiva is derived chiefly from the branches of the alveolar arteries that pass upward through the interdental septa. The interdental alveolar arteries perforate the alveolar crest in the interdental space and end in the interdental papilla, supplying it and the adjacent areas of the buccal and lingual gingiva. In the gingiva these branches anastomose with superficial branches of arteries that supply the oral and

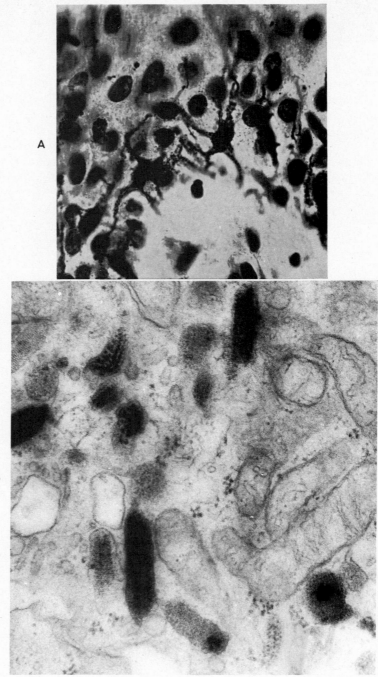

Fig. 9-31. A, Dendritic cells (melanocytes) in basal layer of epithelium. Biopsy of normal gingiva.
B, Ultrastructural photographs of melanosomes from human gingiva showing substructural form. It
is believed that tyrosinase associated with these structures is responsible for melanization and when
melanosomes become fully melanized they are transferred to keratinocytes. (×1000.) (**A** from Aprile,
E.C. de: Arch. Hist. Normal Pat. **3:**473, 1947.)

vestibular mucosa and marginal gingiva, for instance, with branches of the lingual, buccal, mental, and palatine arteries. The numerous lymph vessels of the gingiva lead to submental and submandibular lymph nodes.

The gingiva is well innervated. Different types of nerve endings can be observed, such as the Meissner or Krause corpuscles, end bulbs, loops, or fine fibers that enter the epithelium as "ultraterminal" fibers (Fig. 9-8).

Vermilion border of lip

The transitional zone between the skin of the lip and the mucous membrane of the lip is the red zone, or the vermilion border. It is found only in humans (Fig. 9-32). The skin on the outer surface of the lip is covered by a moderately thick, keratinized epithelium with a rather thick stratum corneum. The papillae of the connective tissue are few and short. Many sebaceous glands are found in connection with the hair follicles. Sweat glands occur between them.

The boundary between the red zone and the mucous membrane of the inner surface of the lip occurs where the keratinization of the transitional zone ends. The epithelium of the mucous membrane of the lip is not keratinized.

The transitional region is characterized by numerous, densely arranged, long papillae of the lamina propria, reaching deep into the epithelium and carrying large capillary loops close to the surface. Thus blood is visible through the thin parts of the translucent epithelium and

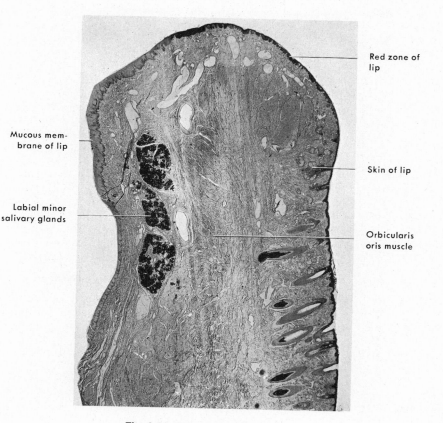

Fig. 9-32. Section through the lip.

gives the red color to the lips. Because this transitional zone contains only occasional sebaceous glands, it is subject to drying and therefore requires moistening by the tongue.

Nonkeratinized areas
Lining mucosa

Lining mucosa is found on the lip, cheek, vestibular fornix, and alveolar mucosa. All the zones of the lining mucosa are characterized by a relatively thick nonkeratinized epithelium and a thin lamina propria. Different zones of lining mucosa vary from one another in the

structure of their submucosa. Where the lining mucosa reflects from the movable lips, cheeks, and tongue to the alveolar bone, the submucosa is loosely textured. The reflectory mucosa found in the fornix vestibuli and in the sublingual sulcus at the floor of the oral cavity has a submucosa that is loose and of considerable volume. The mucous membrane is movably attached to the deep structures and does not restrict the movement of lips and cheeks and the tongue.

Where lining mucosa covers muscle, as on the lips, cheeks, and underside of the tongue,

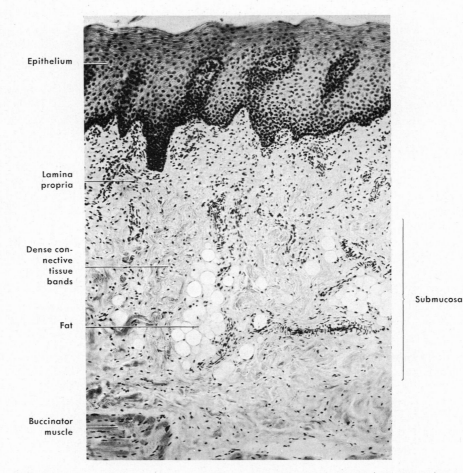

Fig. 9-33. Section through mucous membrane of cheek. Note bands of dense connective tissue attaching lamina propria to fascia of buccinator muscle.

the mucosa is fixed to the epimysium or fascia. In these regions the mucosa is also highly elastic. These two characteristics permit the mucosa to maintain a relatively smooth surface during muscular movement. Thus heavy folding, which could lead to injury during chewing if such folds were caught between the teeth, does not occur.

The mucosa of the soft palate is intermediate between this type of lining mucosa and the reflecting mucosa.

Lip and cheek. The epithelium of the mucosa of the lips (Fig. 9-32) and of the cheek (Fig. 9-33) is stratified squamous nonkeratinized epithelium. The lamina propria of the labial and buccal mucosa consists of dense connective tissue and has short, irregular papillae.

The submucous layer connects the lamina propria to the thin fascia of the muscles and consists of strands of densely grouped collagen fibers. There is loose connective tissue containing fat and small mixed glands between these strands. The strands of dense connective tissue limit the mobility of the mucous membrane, holding it to the musculature and preventing its elevation into folds. This prevents the mucous membrane of the lips and cheeks from lodging between the biting surfaces of the teeth during mastication. The mixed minor salivary glands of the lips are situated in the submucosa, whereas in the cheek the glands are

larger and are usually found between the bundles of the buccinator muscle and sometimes on its outer surface. The cheek, lateral to the corner of the mouth, may contain isolated sebaceous glands called Fordyce's spots (Fig. 9-34). These may occur lateral to the corner of the mouth and are often seen opposite the molars.

A comparison of masticatory and buccal mucosa shows that in the keratinized tissue the epithelium is thinner. It has a granular layer, the basal cells are larger, but the average cell size is smaller, and the cells have an angular shape. Furthermore, it is characterized by having many tonofibrils, wider intercellular spaces, and "prickles" that form "intercellular bridges." The cells of both tissues are joined by desmosomes. The appearance of the two differs by the heightened prominence of the "prickles" in the keratinized tissues, brought about by the increased width of the intercellular space and the greater density of the tonofibrils. Even the lamina propria of the two differ. In masticatory mucosa the basement membrane contains more reticular fibers, and its papillae are high and more closely spaced.

Vestibular fornix and alveolar mucosa. The mucosa of the lips and cheeks reflects from the vestibular fornix to the alveolar mucosa covering the bone. The mucous membrane of the cheeks and lips is attached firmly to the buccinator muscle in the cheeks and orbicularis oris

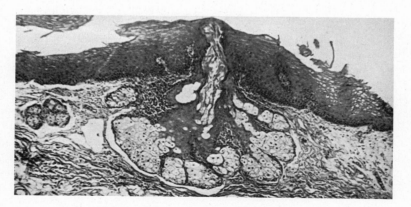

Fig. 9-34. Sebaceous gland in cheek (Fordyce spot).

muscle in the lips. In the fornix the mucosa is loosely connected to the underlying structures, and so the necessary movements of the lips and cheeks are permitted. The mucous membrane covering the outer surface of the alveolar process (alveolar mucosa) is attached loosely to the periosteum. It is continuous with, but different from, the gingiva, which is firmly attached to the periosteum of the alveolar crest and to the teeth.

The median and lateral labial frenula are folds of the mucous membrane containing loose connective tissue. No muscle fibers are found in these folds.

Gingiva and alveolar mucosa are separated by the mucogingival junction. The gingiva is stippled, firm, and thick, lacks a separate submucous layer, is immovably attached to bone and teeth by coarse collagen fibers, and has no glands. The gingival epithelium is thick and mostly parakeratinized or keratinized. The epithelial ridges and the papillae of the lamina propria are high. The alveolar mucosa is thin and loosely attached to the periosteum by a well-defined submucous layer of loose connective tissue (Fig. 9-29, *B*), and it may contain small mixed glands. The epithelium is thin and nonkeratinized, and the epithelial ridges and papillae are low and often entirely missing. These differences cause the variation in color between the pale pink gingiva and the red lining mucosa.

Epithelium

Lamina propria

Submucosa

Minor sublingual gland

Fig. 9-35. Mucous membrane from floor of mouth.

Inferior surface of tongue; floor of oral cavity. The mucous membrane on the floor of the oral cavity is thin and loosely attached to the underlying structures to allow for the free mobility of the tongue. The epithelium is nonkeratinized, and the papillae of the lamina propria are short (Fig. 9-35). The submucosa contains adipose tissue. The sublingual glands lie close to the covering mucosa in the sublingual fold. The sublingual mucosa and the lingual gingiva have a junction corresponding to the mucogingival junction on the vestibular surface. The sublingual mucosa reflects onto the lower surface of the tongue and continues as the ventrolingual mucosa.

The mucosa membrane of the inferior surface of the tongue is smooth and relatively thin (Fig. 9-36). The epithelium is nonkeratinized. The papillae of the connective tissue are numerous but short. Here the submucosa cannot be identified as a separate layer. It binds the mucous membrane tightly to the connective tissue surrounding the bundles of the muscles of the tongue.

Soft palate. The mucous membrane on the oral surface of the soft palate is highly vascular-

ized and reddish in color, noticeably differing from the pale color of the hard palate. The papillae of the connective tissue are few and short. The stratified squamous epithelium is nonkeratinized (Fig. 9-37). The lamina propria shows a distinct layer of elastic fibers separating it from the submucosa. The latter is relatively loose and contains an almost continuous layer of mucous glands. It also contains taste buds. Typical oral mucosa continues around the free border of the soft palate for a variable distance and is then replaced by nasal mucosa with its pseudostratified, ciliated columnar epithelium.

Specialized mucosa

Dorsal lingual mucosa. The superior surface of the tongue is rough and irregular (Fig. 9-38). A V-shaped line divides it into an anterior part, or body, and a posterior part, or base. The former comprises about two thirds of the length of the organ, and the latter forms the posterior one third. The fact that these two parts develop embryologically from different visceral arches (see Chapter 1) accounts for the different source of nerves of the general senses: the an-

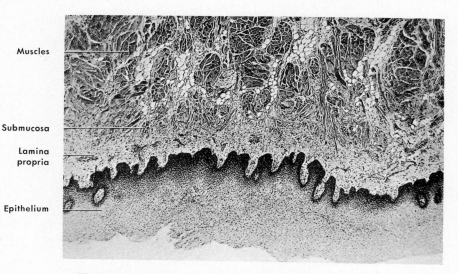

Muscles

Submucosa

Lamina propria

Epithelium

Fig. 9-36. Mucous membrane on inferior surface of tongue.

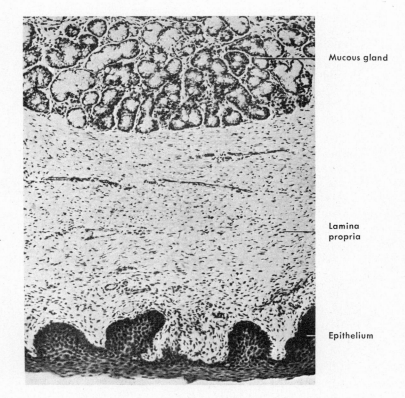

Mucous gland

Lamina propria

Epithelium

Fig. 9-37. Mucous membrane from oral surface of soft palate.

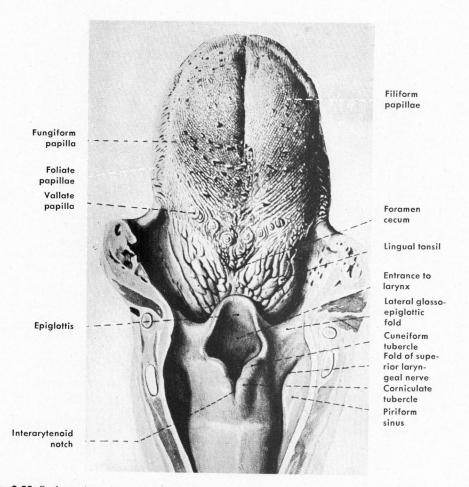

Filiform
papillae

Fungiform
papilla

Foliate
papillae

Vallate
papilla

Foramen
cecum

Lingual tonsil

Entrance to
larynx

Lateral glosso-
epiglottic
fold

Cuneiform
tubercle

Fold of supe-
rior laryn-
geal nerve

Corniculate
tubercle

Piriform
sinus

Epiglottis

Interarytenoid
notch

Fig. 9-38. Surface view of human tongue. (From Sicher, H., and Tandler, J.: Anatomie für Zahnärzte, Vienna, 1928, Julius Springer Verlag.)

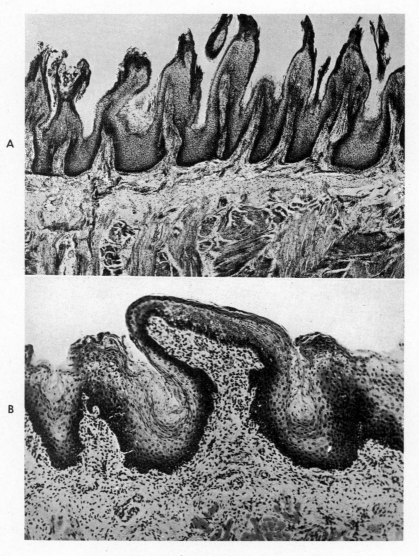

Fig. 9-39. A, Filiform, and **B,** fungiform papillae.

terior two thirds are supplied by the trigeminal nerve through its lingual branch and the posterior one third by the glossopharyngeal nerve.

The body and the base of the tongue differ widely in the structure of the mucous membrane. The anterior part can be termed the "papillary" and the posterior part the "lymphatic" portion of the dorsolingual mucosa. On the anterior part are found numerous fine-pointed, cone-shaped papillae that give it a velvetlike appearance. These projections, the filiform (thread-shaped) papillae, are epithelial structures containing a core of connective tissue from which secondary papillae protude toward the epithelium (Fig. 9-39, *A*). The covering epithelium is keratinized and forms tufts at the apex of the dermal papilla. The filiform papillae do not contain taste buds.

Interspersed between the filiform papillae are the isolated fungiform (mushroom-shaped) papillae (Fig. 9-39, *B*), which are round, reddish prominences. Their color is derived from a rich capillary network visible through the relatively thin epithelium. Fungiform papillae contain a few (one to three) taste buds found only on their dorsal surface.

In front of the dividing V-shaped terminal sulcus, between the body and the base of the tongue, are eight to ten vallate (walled) papillae (Fig. 9-40). They do not protrude above the surface of the tongue but are bounded by a deep circular furrow so that their only connection to the substance of the tongue is at their narrow base. Their free surface shows numerous secondary papillae that are covered by a thin, smooth epithelium. On the lateral surface

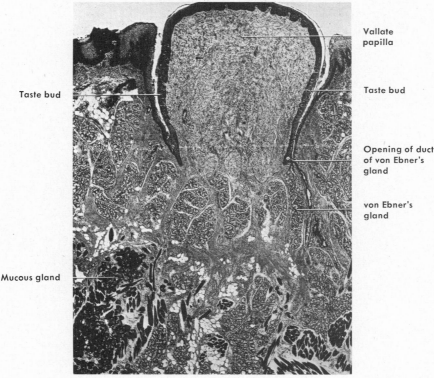

Fig. 9-40. Vallate (or circumvallate) papilla.

of the vallate papillae, the epithelium contains numerous taste buds. The ducts of small serous glands called von Ebner's glands open into the trough. They may serve to wash out the soluble elements of food and are the main source of salivary lipase.

On the lateral border of the posterior parts of the tongue, sharp parallel clefts of varying length can often be observed. They bound narrow folds of the mucous membrane and are the vestige of the large foliate papillae found in many mammals. They contain taste buds.

Taste buds. Taste buds are small ovoid or barrel-shaped intraepithelial organs about 80 μm high and 40 μm thick (Fig. 9-41). They extend from the basal lamina to the surface of the epithelium. Their outer surface is almost cov-

ered by a few flat epithelial cells, which surround a small opening, the *taste pore* (a taste bud may have more than one taste pore). It leads into a narrow space lined by the supporting cells of the taste bud. The outer supporting cells are arranged like the staves of a barrel. The inner and shorter ones are spindle shaped. Between the latter are arranged 10 to 12 neuroepithelial cells, the receptors of taste stimuli. They are slender, dark-staining cells that carry fingerlike processes at their superficial end. The fingerlike processes are visible at the ultrastructural level and resemble hairs at the light microscope level. The hairs reach into the space beneath the taste pore.

A rich plexus of nerves is found below the taste buds. Some fibers enter the epithelium

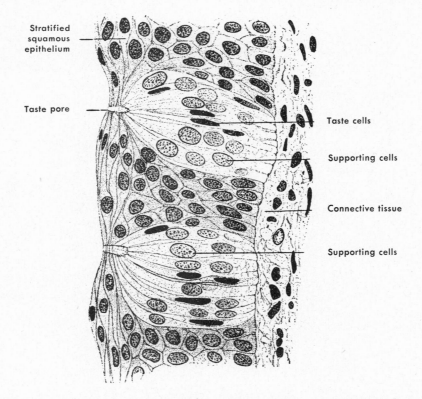

Fig. 9-41. Taste buds from slope of vallate papilla. (From Schaffer, J.: Lehrbuch der histologie und histogenese, ed. 2, Leipzig, 1922, Wilhelm Engelmann.)

and end in contact with the sensory cells of the taste bud.

Taste buds are numerous on the inner wall of the trough surrounding the vallate papillae, in the folds of the foliate papillae, on the posterior surface of the epiglottis, and on some of the fungiform papillae at the tip and the lateral borders of the tongue (Fig. 9-42).

The classic view maintains that the primary taste sensations, that is sweet, salty, bitter, and sour, are perceived in different regions of the tongue and palate (sweet at the tip, salty at the lateral border of the tongue, bitter and sour on the palate and also in the posterior part of the tongue—bitter in the middle and sour in the lateral areas of the tongue). The classic view also diagrammatically and arbitrarily correlates the distribution of the receptors for primary taste qualities with the different types of papillae (vallate papillae with bitter, foliate papillae with sour, taste buds of the fungiform papillae at the tip of the tongue with sweet and at the borders with salty taste). Bitter and sour taste sensations are mediated by the glossopharyngeal nerve, and sweet and salty taste are mediated by the intermediofacial nerve by the chorda tympani.

On the other hand, many authorities believe that taste cannot be broken down into these four primaries, sweet, sour, salty, and bitter, but that it consists of a range of stimuli that form a spectrum of sensations making up all taste senses. Taste occurs when a chemical substance contacts a receptor cell in the taste bud. Each taste bud is innervated by many fibers. The reception of a chemical substance fires the nerve fiber. Thus taste may be a continuum or a composite of the firing of many fibers.

At the angle of the V-shaped terminal groove on the tongue is located the foramen cecum, which represents the remnant of the thyroglossal duct (see Chapter 1). Posterior to the terminal sulcus, the surface of the tongue is irregularly studded with round or oval prominences, the lingual follicles. Each of these

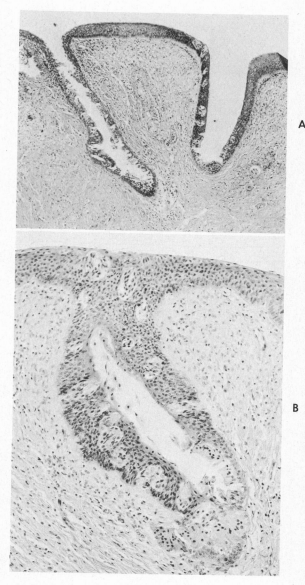

A

B

Fig. 9-42. A, Circumvallate papilla showing trough and numerous taste buds *(light areas)* **B,** Higher magnification of trough and taste buds.

shows one or more lymph nodules, sometimes containing a germinal center (Fig. 9-43). Most of these prominences have a small pit at the center, the lingual crypt, which is lined with stratified squamous epithelium. Innumerable lymphocytes migrate into the crypts through the epithelium. Ducts of the small posterior lingual mucous glands open into the crypts. Together the lingual follicles form the lingual tonsil.

GINGIVAL SULCUS AND DENTOGINGIVAL JUNCTION
Gingival sulcus

The gingival sulcus or crevice is the name given to the invagination made by the gingiva as it joins with the tooth surface. The gingiva does not join the tooth at the gingival margin. It forms a small infolding known as the *sulcus*. The sulcus extends from the free gingival margin to the dentogingival junction. In health its depth is at the approximate level of the free gingival groove on the outer surface of the gingiva. The sulcus may be responsible for the formation of the groove since it leaves the gingival

margin without firm support. The groove is believed to be formed by the functional folding of the free gingival margin during mastication. The sulcular (crevicular) epithelium is nonkeratinized in humans. It lacks epithelial ridges and so forms a smooth interface with the lamina propria. It is thinner than the epithelium of the gingiva. The sulcular epithelium is continuous with the gingival epithelium and the attachment epithelium. These three epithelia have a continuous and coextensive basal lamina.

One view holds that the gingival sulcus is universally and normally present and as it deepens a pathologic entity, the pocket, is formed. Another view holds that initially the gingiva attaches to the tooth completely, and its gradual detachment to form the sulcus is the result of pathologic phenomena.

Dentogingival junction

The junction of the gingiva and the tooth is of great physiologic and clinical importance. This union is unique in many ways and may be a point of lessened resistance to mechanical

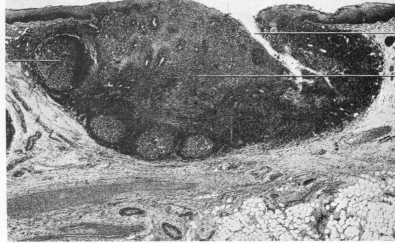

Lingual lymph follicle with germinal center

Follicular crypt

Lingual follicle

Fig. 9-43. Lingual lymph follicle.

forces and bacterial attack. The gingiva consists of two tissues maintaining the junction intact. Their biology differs. The dense, resilient lamina propria takes up impacts produced during mastication. In a similar sense so does the keratinized or parakeratinized surface of the gingiva. When the epithelium is injured, the turnover of cells and their ability to migrate repair the wound. When the connective tissue is injured, ribosomes within the fibroblasts form molecules of the precursor protein of collagen (procollagen) and ground substances as well, contributing to repair.

Defense against bacterial injury is a function of the defense mechanism of the body. Macrophages, lymphocytes, plasma cells, and white blood cells protect against invasion and form antibodies against bacterial antigens.

Both epithelium and connective tissue are attached to the tooth, and in health each contributes to the integrity of the dentogingival junction. Again, the firmness of this junction is maintained by the gingival division of the periodontal ligament. It is weakened by any situation that causes the collagen to break down (collagenolysis). The adherence of epithelium to the tooth is a function of the attachment (junctional) epithelium. It is weakened by any cause that injures the epithelium.

Development of junctional (attachment) epithelium. When the ameloblasts finish formation of the enamel matrix, they leave a thin membrane on the surface of the enamel, the *primary enamel cuticle*. This cuticle may be connected with the interprismatic enamel substance and the ameloblasts. The ameloblasts shorten after the primary enamel cuticle has been formed, and the epithelial enamel organ is reduced to a few layers of flat cuboid cells, which are then called *reduced enamel epithelium*. Under normal conditions it covers the entire enamel surface, extending to the cementoenamel junction (Fig. 9-44), and remains attached to the primary enamel cuticle. During eruption, the tip of the tooth approaches the

oral mucosa, and the reduced enamel epithelium and the oral epithelium meet and fuse (Fig. 9-45). The remnant of the primary enamel cuticle after eruption is referred to as Nasmyth's membrane.

The epithelium that covers the tip of the crown degenerates in its center, and the crown emerges through this perforation into the oral cavity (Fig. 9-46). The reduced enamel epithelium remains organically attached to the part of the enamel that has not yet erupted. Once the tip of the crown has emerged, the reduced enamel epithelium is termed the *primary attachment epithelium.** At the margin of the gingiva the attachment epithelium is continuous with the oral epithelium (Fig. 9-47). As the tooth erupts, the reduced enamel epithelium grows gradually shorter. A shallow groove, the *gingival sulcus* (Fig. 9-47), may develop between the gingiva and the surface of the tooth and extend around its circumference. It is bounded by the attachment epithelium at its base and by the gingival margin laterally. The gingiva encompassing the sulcus is the free, or marginal, gingiva.

Although the firmness and mechanical strength of the dentogingival junction is mainly attributable to the connective tissue attachment, the attachment of the epithelium to the enamel is by no means loose or weak. This can be demonstrated with ground histologic sections of frozen specimens where enamel and soft tissues are retained in their normal rela-

---

*Some confusion may result if the student refers to the older literature in which the attachment epithelium is referred to as the epithelial attachment. It was first named the epithelial attachment (*Epithelansatz*) by Gottlieb, but after it was examined electron microscopically it was renamed the junctional, or attachment, epithelium by Stern. This epithelium synthesizes the material that attaches it to the tooth. This material, its morphology, mode, and mechanism of function, is what is now called the epithelial attachment. Thus the cellular structure is referred to as junctional or attachment epithelium and its extracellular tooth-attaching substance is referred to as the epithelial attachment.

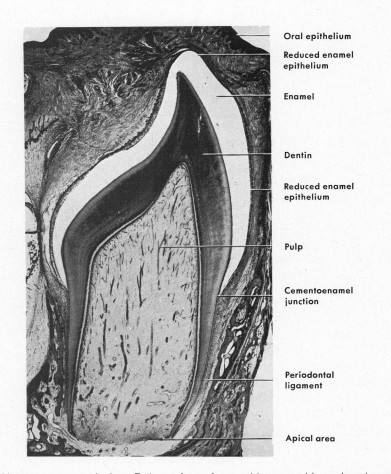

Oral epithelium

Reduced enamel epithelium

Enamel

Dentin

Reduced enamel epithelium

Pulp

Cementoenamel junction

Periodontal ligament

Apical area

Fig. 9-44. Human permanent incisor. Entire surface of enamel is covered by reduced enamel epithelium. Mature enamel is lost by decalcification. (From Gottlieb, B., and Orban, B.: Biology of the investing structures of the teeth. In Gordon, S.M., editor: Dental science and dental art, Philadelphia, 1938, Lea & Febiger.)

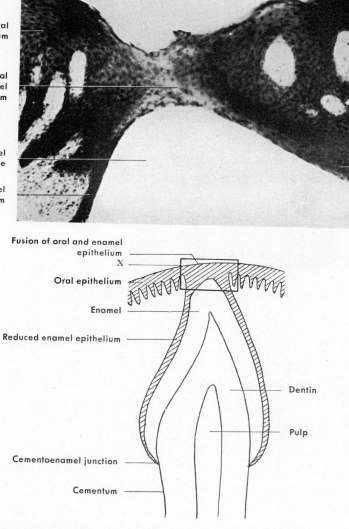

Oral
epithelium

Fusion of oral
and enamel
epithelium

Enamel
space

Reduced enamel
epithelium

Oral
epithelium

Reduced enamel
epithelium

Fusion of oral and enamel
epithelium

X

Oral epithelium

Enamel

Reduced enamel epithelium

Dentin

Pulp

Cementoenamel junction

Cementum

Fig. 9-45. Reduced enamel epithelium fuses with oral epithelium. X in diagram indicates area from which photomicrograph was taken.

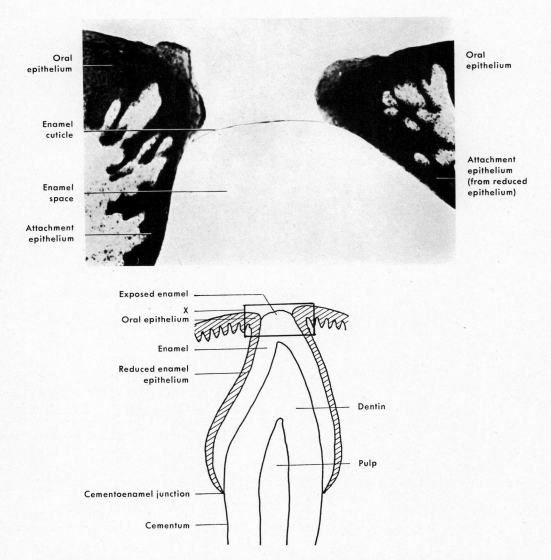

Fig. 9-46. Tooth emerges through perforation in fused epithelia. X in diagram indicates area from which photomicrograph was taken.

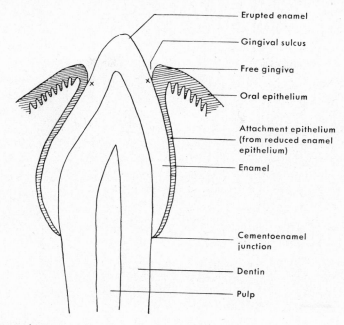

Erupted enamel

Gingival sulcus

Free gingiva

Oral epithelium

Attachment epithelium
(from reduced enamel
epithelium)

Enamel

Cementoenamel
junction

Dentin

Pulp

Fig. 9-47. Diagram of attached epithelial cuff and gingival sulcus at an early stage of tooth eruption. Bottom of sulcus at X.

tion. When an attempt is made to detach the gingiva from the tooth in these preparations, the epithelium tears but does not peel off from the enamel surface (Fig. 9-48). Similar results are obtained surgically when gingival flaps are pulled away from teeth.

Shift of dentogingival junction. The position of the gingiva on the surface of the tooth changes with time. When the tip of the enamel first emerges through the mucous membrane of the oral cavity, the epithelium covers almost the entire enamel (Fig. 9-49). The tooth erupts until it reaches the plane of occlusion (see Chapter 11). The attachment epithelium separates from the enamel surface gradually while the crown emerges into the oral cavity. When the tooth first reaches the plane of occlusion, one third to one fourth of the enamel still remains covered by the gingiva (Fig. 9-50). A gradual exposure of the crown follows. The ac-

tual movement of the teeth toward the occlusal plane is termed active eruption. This applies to the preclinical phase of eruption also. The separation of the primary attachment epithelium from the enamel is termed passive eruption. Further recession exposing the cementum may ultimately occur. At that stage the reduced enamel epithelium has disappeared and the primary attachment epithelium is replaced by a *secondary attachment epithelium* derived from the gingival epithelium.

There is a conceptual construct, called passive eruption, that may be useful in describing the various levels of attachment that may occur as the gingiva recedes onto the cementum. Some persons believe passive eruption to be a normal occurrence with aging. The belief that this is a "normal" occurrence is probably incorrect. Crown exposure involving passive eruption and further recession has been described in four stages.

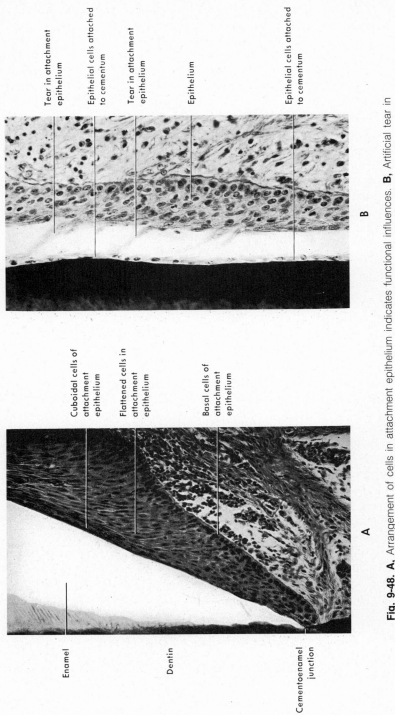

Fig. 9-48. A, Arrangement of cells in attachment epithelium indicates functional influences. **B,** Artificial tear in attachment epithelium. Some cells remain attached to cementum, while others bridge tear. (**A** from Orban, B.: Z. Stomatol. **22:**353, 1924; **B** from Orban, B., and Mueller, E.: J. Am. Dent. Assoc. **16:**1206, 1929.)

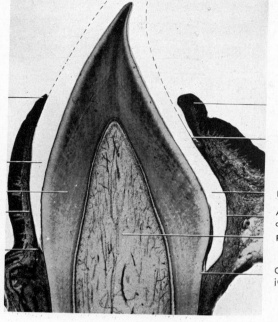

Free gingiva
(gingival
sulcus)

Enamel
space

Dentin
Junctional or
attachment
epithelium

Cementoenamel
junction

Free gingiva

Gingival sulcus

Enamel space

Attachment epithelium
cuff

Pulp

Cementoenamel
junction

Fig. 9-49. Attachment epithelium and gingival sulcus in erupting tooth. *Dotted line,* Erupted part of enamel. Enamel is lost in decalcification. (From Kronfeld, R.: J. Am. Dent. Assoc. **18:**382, 1936.)

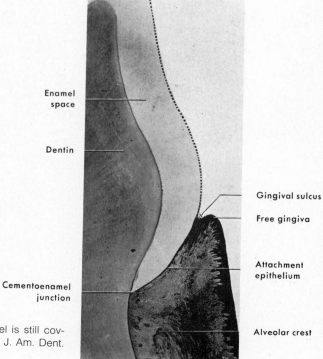

Enamel
space

Dentin

Cementoenamel
junction

Gingival sulcus

Free gingiva

Attachment
epithelium

Alveolar crest

Fig. 9-50. Tooth in occlusion. One fourth of enamel is still covered by attachment epithelium. (From Kronfeld, R.: J. Am. Dent. Assoc. **18:**382, 1936.)

The first two may be physiologic. Many conceive of the last two as normal also, but there is a strong possibility that they are pathologic.

First stage. The bottom of the gingival sulcus remains in the region of the enamel-covered crown for some time, and the apical end of the attachment epithelium (reduced enamel epithelium) stays at the cementoenamel junction (Fig. 9-51). This relation persists in primary teeth almost up to 1 year of age before shedding and, in permanent teeth, usually to the age of 20 or 30 years. However, this relation is subject to a wide range of variation (Fig. 9-52).

Second stage. The bottom of the gingival sulcus is still on the enamel, and the apical end of the attachment epithelium has shifted to the surface of the cementum (Fig. 9-53).

The downgrowth of the attachment epithelium along the cementum is but one facet of the shift of the dentogingival junction. This entails dissolution of fiber bundles that were anchored in the cervical parts of the cementum, now covered by the epithelium, and an apical shift of the gingival and transseptal fibers. The destruction of the fibers may be caused by en-

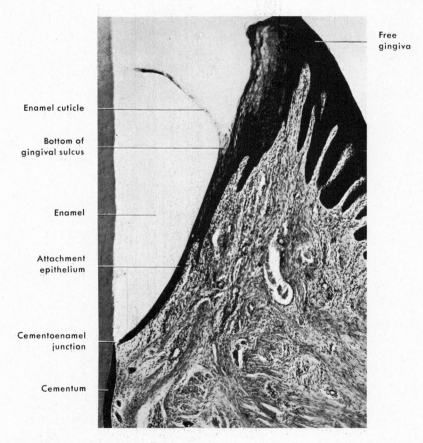

Enamel cuticle

Bottom of gingival sulcus

Enamel

Attachment epithelium

Cementoenamel junction

Cementum

Free gingiva

Fig. 9-51. Attachment epithelium on enamel. First stage of crown exposure. (From Gottlieb, B., and Orban, B.: Biology and pathology of the tooth [translated by M. Diamond], New York, 1938, The Macmillan Co.)

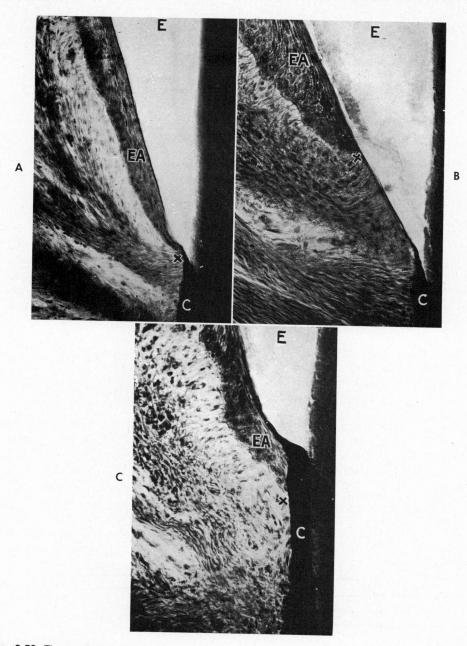

Fig. 9-52. Three sections of same tooth showing different relations of tissues at cementoenamel junction. **A,** Attachment epithelium reaching to cementoenamel junction. **B,** Attachment epithelium ends coronally to cementoenamel junction. **C,** Attachment epithelium covers part of cementum. Cementum overlaps edge of enamel. *C,* Cementum; *E,* enamel (lost in decalcification); *EA,* attachment epithelium; *X,* end of attachment epithelium. (From Orban, B.: J. Am. Dent. Assoc. **17:**1977, 1930.)

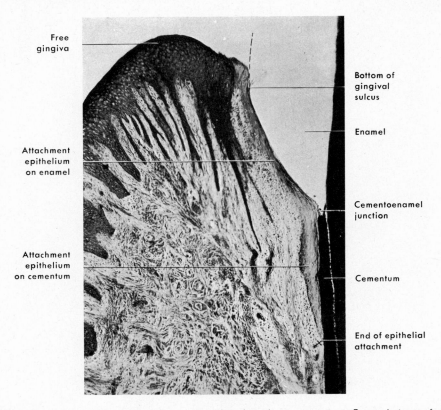

Free gingiva

Attachment epithelium on enamel

Attachment epithelium on cementum

Bottom of gingival sulcus

Enamel

Cementoenamel junction

Cementum

End of epithelial attachment

Fig. 9-53. Attachment epithelium partly on enamel and partly on cementum. Second stage of passive tooth exposure. (From Gottlieb, B., and Orban, B.: Biology and pathology of the tooth [translated by M. Diamond], New York, 1938, The Macmillan Co.)

zymes formed by the epithelial cells, by plaque metabolites or enzymes, or by immunologic reactions as manifestations of periodontal disease. This stage of tooth exposure may persist to the age of 40 years or later.

Third stage. When the bottom of the gingival sulcus is at the cementoenamel junction, the epithelium attachment is entirely on the cementum, and the enamel-covered crown is fully exposed (Fig. 9-54). This stage in the exposure of a tooth no longer is a passive manifestation. The epithelium shifts gradually along the surface of the tooth and does not remain at the cementoenamel junction. This more or less continuous but slow process is regarded as the

body's attempt to maintain an intact dentogingival junction in the face of factors that cause its deterioration.

Fourth stage. The fourth stage represents recession of the gingiva. When the entire attachment is on cementum, the gingiva may appear normal, but the process is regarded as pathologic (Figs. 9-55 and 9-56). It may occur without perceptible evidence of inflammatory periodontal disease.

The rates of crown exposure and recession vary in different persons. In some cases the fourth stage is observed in persons during their twenties. In others, even at 50 years of age or older the teeth are still in the first or second

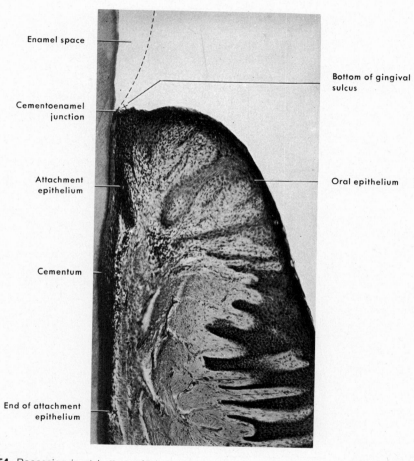

Enamel space

Bottom of gingival
sulcus

Cementoenamel
junction

Attachment
epithelium

Oral epithelium

Cementum

End of attachment
epithelium

Fig. 9-54. Recession is at bottom of gingival sulcus at cementoenamel junction, and attachment epithelium is on cementum. (From Gottlieb, B.: J. Am. Dent. Assoc. **14:**2178, 1927.)

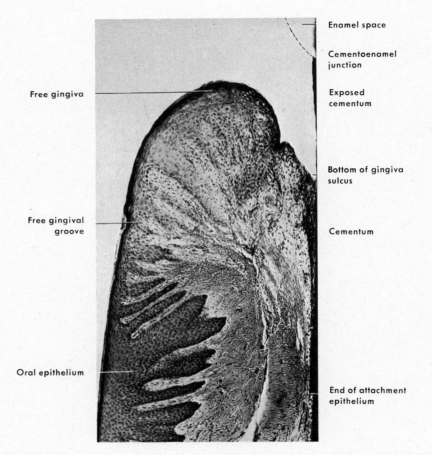

Free gingiva

Free gingival
groove

Oral epithelium

Enamel space

Cementoenamel
junction

Exposed
cementum

Bottom of gingiva
sulcus

Cementum

End of attachment
epithelium

Fig. 9-55. Recession. Bottom of gingival sulcus and attachment epithelium both on cementum. Continued recession may reduce the width of gingiva. (From Gottlieb, B.: J. Am. Dent. Assoc. **14:**2178, 1927.)

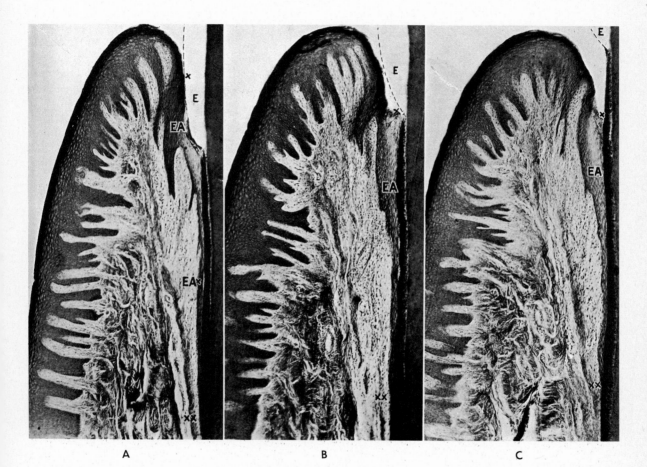

Fig. 9-56. Three sections of same tooth showing different relation of soft to hard tissues. **A,** Bottom of sulcus on enamel (second stage). **B,** Bottom of sulcus at cementoenamel junction (third stage). **C,** Bottom of sulcus on cementum (fourth stage). *E,* Enamel lost in decalcification *(dotted line); EA,* attachment epithelium; X, bottom of gingival sulcus; XX, end of attachment epithelium.

stage. The rate varies also in different teeth of the same jaw and on different surfaces of the same tooth. One side may be in the first stage and the other in the second or even the fourth stage (Fig. 9-56).

Gradual exposure of the tooth makes it necessary to distinguish between the anatomic and the clinical crowns of the tooth (Fig. 9-57). That part of the tooth covered by enamel is the anatomic crown. The clinical crown is the part of the tooth exposed in the oral cavity. In the first and second stages the clinical crown is smaller than the anatomic crown. With recession (third stage) the entire enamel-covered part of the tooth is exposed, and the clinical crown is equal to the anatomic crown. Later the clinical crown is larger than the anatomic crown because parts of the root have been exposed (fourth stage). This type of crown exposure is to be differentiated from crown exposure that is produced by pocket formation.

Sulcus and cuticles. For a long time the epithelium was believed only to contact but not attach to the enamel. The contact was supposed to be maintained by the turgor of the connective tissue elements of the gingiva. Thus a capillary space was supposed to exist between the gingiva and enamel to the cementoenamel junction. Gottlieb and Orban demonstrated the presence of an organic attachment, which they termed the epithelial attachment. The mode or mechanism of the epithelial attachment is very important. The classic view proposed by Gottlieb and Orban involves the primary cuticle mediating an organic union between ameloblasts and the enamel. When the ameloblasts are replaced by the oral epithelium, a secondary cuticle is formed. When the epithelium proliferates beyond the cementoenamel junction, the cuticle extends along the cementum (Figs. 9-58 and 9-59). Secondary enamel cuticle and the cemental cuticle are referred to as den-

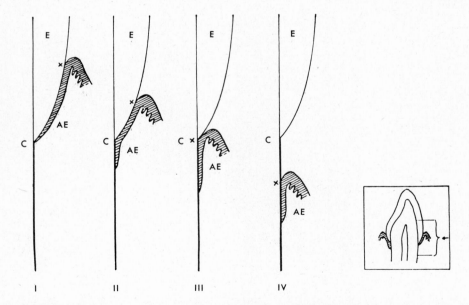

Fig. 9-57. Diagram of four stages in eruption. In stages I and II (passive eruption), anatomic crown in larger than clinical crown. III and IV represent recession. In stage III, anatomic and clinical crowns are equal. In stage IV, clinical crown is larger than anatomic crown. Arrow in small diagram indicates area from which drawings were made. *C,* Cementoenamel junction; *E,* enamel; *AE,* attachment epithelium; *X,* bottom of gingival sulcus.

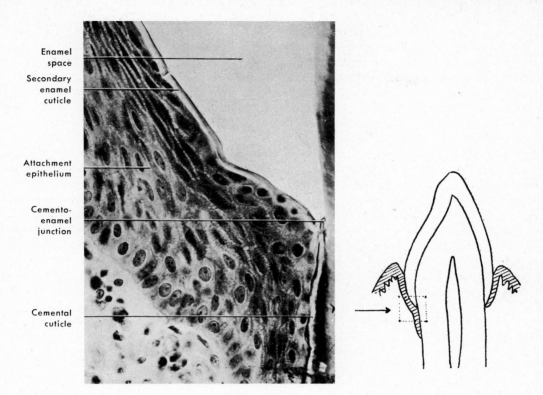

Enamel
space

Secondary
enamel
cuticle

Attachment
epithelium

Cemento-
enamel
junction

Cemental
cuticle

Fig. 9-58. "Secondary enamel cuticle" follows attachment epithelium to cementum forming the dental cuticle. Arrow in diagram indicates area from which photomicrograph was taken.

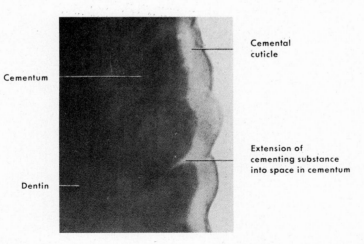

Cemental
cuticle

Cementum

Extension of
cementing substance
into space in cementum

Dentin

Fig. 9-59. Cemental cuticle extending into cementum. (From Gottlieb, B., and Orban, B.: Biology and pathology of the tooth [translated by M. Diamond], New York, 1938, The Macmillan Co.)

tal cuticle. These cuticles are microscopically evident as an amorphous material between the attachment epithelium and the tooth.

Deepening of sulcus (pocket formation). The gingival sulcus forms when the tip of the crown emerges through the oral mucosa. It deepens as a result of separation of the reduced dental epithelium from the actively erupting tooth. At first after the tip of the crown has appeared in the oral cavity the epithelium separates rapidly from the surface of the tooth. Later, when the tooth comes to occlude with its antagonist, the separation of the attachment from the surface of the tooth slows down.

The formation and relative depth of the gingival sulcus at different ages is a subject of considerable interest. At one time, it was believed that from the time the tip of the crown had pierced the oral mucosa the gingival sulcus extended to the cementoenamel junction (Fig. 9-60, I). It was assumed that the attachment of the gingival epithelium to the tooth occurred only at the cementoenamel junction. The con-

cept of the epithelial attachment introduced by Gottlieb and Orban showed that no cleft existed between epithelium and enamel and that these tissues were organically connected. The gingival sulcus was shown to be a shallow groove, the bottom of which is at the point of separation of attached epithelium from the tooth (Fig. 9-60, II).

Some investigators contended that the deepening of the gingival sulcus was caused by a tear in the attached epithelium (Fig. 9-60, III). Others believed, however, that deepening occurred as a result of the downgrowth of the oral epithelium alongside the reduced enamel epithelium (primary attachment epithelium), as shown in Fig. 9-60, IV.

What the depth of a normal gingival sulcus should be has been a frequent source of argument. Under normal conditions the depth of the sulcus is variable; 45% of all measured sulci are below 0.5 mm. The average sulcus is 1.8 mm. The more shallow a sulcus, the more likely that the gingival margin is not inflamed.

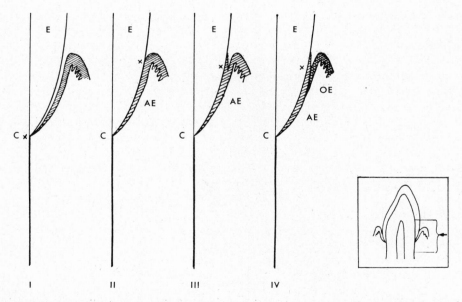

Fig. 9-60. Diagram of different views on formation of gingival sulcus as discussed in text. Arrow in small diagram indicates area from which drawings were made.

Lymphocytes and plasma cells are routinely seen in the connective tissue at the bottom of the gingival sulcus and below the attachment epithelium. This is a defense reaction to the bacteria in the gingival sulcus and constitutes a barrier against the invasion of bacteria and the penetration of toxins. The bacterial products may act directly or indirectly via the immune responses.

Epithelial attachment. In the past decade the ultrastructural attachment of the ameloblasts (primary attachment epithelium) to the tooth was first shown by Stern and confirmed by Listgarten and Schroeder, among others, to be basal lamina to which hemidesmosomes are attached (Fig. 9-6, A). This mode of attachment is referred to as the *epithelial attachment.* The secondary attachment epithelium composed of cells derived from the oral epithelium forms an epithelial attachment identical with that of the primary attachment epithelium, that is, a basal lamina and hemidesmosomes. Both reduced ameloblasts and gingival epithelial cells have been shown to form an electron microscopic basal lamina on enamel and cementum. Hemidesmosomes of these cells attach to the basal lamina in the same manner as all basal cells. Thus there is an epithelial attachment. It is submicroscopic, approximately 40 nm (400 Å) wide, and formed by the attachment epithelium. Its exact biochemical nature is unknown, but some of its constituents have been grossly identified. Apparently these constituents are produced by the epithelium. The adhesive forces in this zone are molecular in nature and act across a distance smaller than 40 nm (400 Å).

The epithelial attachment resembles an electron microscopic basal lamina. The cells of the attachment epithelium are held to this structure by hemidesmosomes (Fig. 9-61).

Migration of attachment epithelium. Mitotic figures have been observed in cells adjacent to the tooth. When tritiated thymidine is administered to experimental animals, cells about to undergo DNA synthesis pick up radioactive thymidine. The radioactivity can be detected in histologic sections by the use of photographic emulsion. After the administration of the tritiated thymidine, labeled cells are found in the attachment epithelium. Can it be that the cells of the attachment epithelium adjacent to the tooth are basal cells?

When cells leave the stratum germinativum, they become specialized. For instance, in oral epithelium cells specialize and undergo keratinization. In attachment epithelium the cells specialize and synthesize a basal lamina (the epithelial attachment). They then migrate over it, with their attachment being maintained by the hemidesmosomes. In general a cell once specialized neither synthesizes DNA nor divides.

The time it takes for labeled attachment epithelial cells to migrate and desquamate is called transit time. It is less than 144 hours for the continuously growing incisor of rodents (Fig. 9-62), about 72 to 120 hours for primates, and presumably much the same for humans.

How can the cells be attached to the tooth if they are actively migrating? The same mechanism is present at the epidermis–connective tissue junction. The epithelial cells are affixed to the connective tissue through the basal lamina, yet they can detach from it and migrate toward the surface. Similarly in healing wounds the epithelial cells form a basal lamina on the connective tissue and migrate over it to epithelialize the wound. At no time is the epithelium loose from the connective tissue. The two tissues are in intimate connection. Picture the epithelial attachment as the basal lamina of the attachment (junctional) epithelium. It turns about the most apical cell and extends up along the tooth surface. The cells can then migrate along this basal lamina (Fig. 9-61). The hemidesmosomes hold the cells to this structure so that the strength of the attachment is not diminished despite the migration. The physical integrity of the attachment is maintained during the four stages of tooth exposure by this same biologic mechanism.

The reduced ameloblasts do not divide; on the other hand, basal cells adjacent to the tooth do divide and then migrate up and along the tooth, desquamating in 4 to 6 days. They seem to migrate from a mitotically DNA-synthetic active area, a locus of proliferation, in the basal layer at the junction of the oral and the attachment epithelia.

While the reduced ameloblasts are still present, the cells of the oral epithelium join them by forming desmosomes. Gradually the re-duced enamel epithelium is lost, and the cells of the oral epithelium contact the tooth surface, there forming hemidesmosomes and a lamina lucida, by means of which the cells attach themselves to the tooth. The apical migration of the sulcus is the result of a detachment of basal cells and a reestablishment of their epithelial attachment at a more apical level. It is not the result of degeneration and peeling off of the most coronal cells of the attachment epithelium. The mechanism of sulcus deepening by the deepening of splits and so forth (Fig. 9-62) is not accurate, since the attachment is

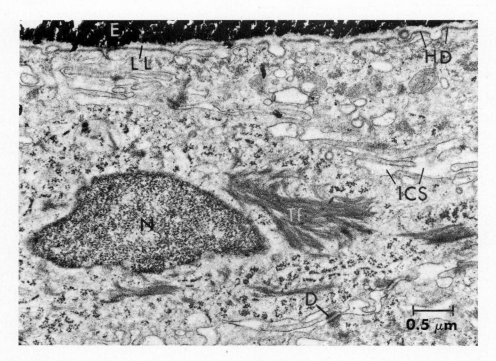

Fig. 9-61. Electron micrograph of cells of attachment epithelium of rat incisor adjacent to enamel, *E*. Hemidesmosomes, *HD*, abut on and attach to lamina lucida, *LL*. Lamina densa is fully calcified and cannot be demonstrated in this calcified specimen. Lamina lucida is approximately 40 nm (400 Å) wide. Note that intercellular space, *ICS,* is wider than lamina lucida. Cells are attached to each other by desmosomes, *D. N*, Nucleus; *Tf,* a bundle of tonofilaments. (From Grant, D.A., Stern, I.B., and Everett, F.G.: Periodontics in the tradition of Orban and Gottlieb, ed. 5, St. Louis, 1979, The C.V. Mosby Co.)

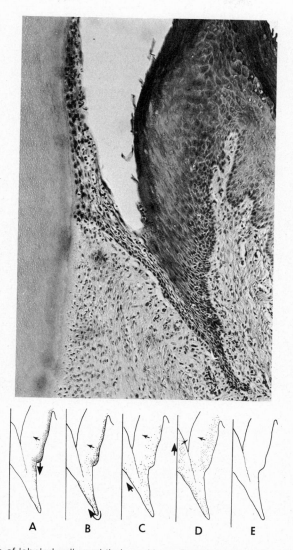

Fig. 9-62. Composite of labeled cells and their positions: **A,** ½ hour; **B,** 6 hours; **C,** 24 hours; **D,** 72 hours; and **E,** 144 hours after administration of tritiated thymidine to rats. Diagram of morphology of attachment epithelium and adjacent tissues is representative of gingiva on cemental (oral) surface of continuously growing rat incisor. *Large arrows,* Migration of attachment (junctional) epithelium toward and along tooth surface. *Small arrows,* Migration of cells toward sulcus. (From Anderson, G.C., and Stern, I.B.: Periodontics **4:**115, 1966.)

formed by the more deeply located basal cells (Fig. 9-63). Perhaps toxic or inflammatory influences diminish the ability of the basal cells to synthesize DNA or otherwise interfere with the physiology of these cells. Perhaps collagenolysis destroys the subjacent collagen fibers, permitting the epithelium to migrate apically. Perhaps immunologically competent cells or antibody complexes produce tissue damage and permit the epithelium to migrate apically. In any event the junctional epithelium moves apically, replicates a new basal lamina, and reestablishes the epithelial attachment. If this results in a deepening of the sulcus, as gauged by a difference in the position of the top of the epithelial attachment relative to the marginal gingiva, a pocket will have formed.

CLINICAL CONSIDERATIONS

It is essential to be thoroughly familiar with the structure and biologic interrelations of the various periodontal tissues in order to understand the pathogenesis of periodontal disease. Periodontal disturbances produce a deepened gingival sulcus as a response to plaque toxins and the subsequent immunologic response. Reduction in pocket depth is the primary objective of treatment. Treatment methods should be judged by their ability to reduce the depth of pockets and to prevent their recurrence.

The level of the gingival attachment to the tooth plays an important role in restorative dentistry. In young persons the clinical crown is smaller than the anatomic crown. It is therefore very difficult to prepare a tooth properly for an abutment or crown in young individuals. Moreover, when recession occurs at a later time, the restoration may require replacement.

When the root is exposed by recession and a restoration is to be placed, the preparation need not extend to the gingiva. The first requirement is that the restoration be adapted to mechanical needs. In extension of the gingival margin of any restoration the following rules should be observed. If the gingiva is still on the enamel and the gingival papilla fills the entire interdental space, the gingival margin of a cavity should be placed at the sulcus. Special care should be taken to avoid injury to the gingiva and the dentogingival junction and to prevent premature recession of the gingiva. When periodontal disease is present, treatment should precede the placing of a restoration. If the gingiva has receded to the cementum and the gingival papilla does not fill the interdental space, the margin of a cavity need not necessarily be carried to the gingiva.

With gingival recession and exposure of the cervical part of the anatomic root, cemental caries or abrasion may occur. Improperly constructed clasps, overzealous scaling, and strongly abrasive dentrifrices may result in pronounced abrasion. After loss of the cementum, the dentin may be extremely sensitive to thermal or chemical stimuli. Desensitizing drugs, judiciously applied, may be used to accelerate sclerosis of the tubules and reparative dentin formation.

The difference in the structure of the submucosa in various regions of the oral cavity is of great clinical importance. Whenever the submucosa consists of a layer of loose connective tissue, edema or hemorrhage can cause much swelling, and infection can spread speedily and extensively.

Injections should be made into loose submucous connective tissue (i.e., the fornix and the alveolar mucosa). The only place in the palate where larger amounts of fluid can be injected without damaging the tissues is the loose connective tissue in the furrow between the palatal and the alveolar processes (Fig. 9-21).

The gingiva is exposed to heavy mechanical stresses during mastication. Moreover, the epithelial attachment to the tooth is relatively weak, and injuries or infections can cause permanent damage. Keratinization of the gingiva may afford relative protection. Therefore steps taken to increase keratinization can be considered preventive measures. One of the methods

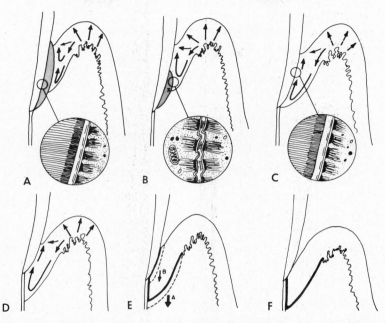

Fig. 9-63. Current concept of dynamics of migration of junctional epithelium. **A,** Primary junctional epithelium consists of reduced ameloblasts attached by hemidesmosomes to lamina lucida, an electron microscopic structure. Gingival epithelial cells migrate either to gingival surface and keratinize or *(arrows)* toward reduced enamel epithelium, to which they attach. **B,** Primary junctional epithelium gradually degenerates and is replaced by secondary junctional epithelium. Cells of the two tissues are joined by desmosomes and by tight junctions. At point of fusion of two tissues, X, mitotic activity is increased. Here cells of outer enamel epithelium and possibly stratum intermedium join cells of gingival epithelium in forming a locus of proliferation. **C,** With complete degeneration of primary attachment epithelium, secondary attachment epithelium contacts enamel and attaches by same mechanism as shown in **A. D,** In time, secondary junctional epithelium may be found attaching to both enamel and cementum. How did this apical migration occur? **E,** Secondary junctional epithelium renews itself in a matter of days, as does gingival epithelium. Cells migrate in pathways denoted by arrows in **D.** Cells of junctional epithelium travel along electron microscopic basal lamina lucida-lamina densa complex, which forms epithelial attachment *(heavy line).* When basal cells at *A* migrate apically, they form a new attachment. Junctional cells at *B* may also migrate apically, with either resultant deepening of sulcus or progressive exposure of clinical crown (recession). **F,** When junctional epithelium has completely migrated into cementum, attachment is mediated by lamina lucida and by hemidesmosomes, as it was on enamel. (From Grant, D.A., Stern, I.B., and Everett, F.G.: Periodontics in the tradition of Orban and Gottlieb, ed. 5, St. Louis, 1979, The C.V. Mosby Co.)

of including keratinization is by massage or brushing, which acts directly by stimulation and by minimizing plaque accumulation.

Unfavorable mechanical irritation of the gingiva may ensue from sharp edges of carious cavities, overhanging fillings or crowns, and accumulation of plaque. These may cause chronic inflammation of the gingival tissue.

Many systemic diseases cause characteristic changes in the oral mucosa. For instance, metal poisoning (lead, bismuth) causes characteristic discoloration of the gingival margin. Leukemia, pernicious anemia, and other blood dyscrasias can be diagnosed by characteristic infiltrations of the oral mucosa. In the first stages of measles, small red spots with bluish white centers can be seen in the mucous membrane of the cheeks, even before the skin rash appears. They are known as Koplik's spots. Endocrine disturbances, including those of the sex hormones and of the pancreas, may be reflected in the oral mucosa.

Changes of the tongue are sometimes diagnostically significant. In scarlet fever the atrophy of the lingual mucosa causes the peculiar redness of the strawberry tongue. Systemic diseases such as pernicious anemia and vitamin deficiencies, especially vitamin B—complex deficiency, lead to characteristic changes such as magenta tongue and beefy red tongue.

In denture construction it is important to observe the firmness or looseness of the mucous membrane. In denture-bearing areas the mucosa should be firm.

In old age the mucous membrane of the mouth may atrophy. It is then thin and parchmentlike. The atrophy of the lingual papillae leaves the upper surface of the tongue smooth, shiny and varnished in appearance. Atrophy of the major and minor salivary glands may lead to xerostomia (dry mouth) and sometimes to a secondary atrophy of the mucous membrane. In a large percentage of individuals the sebaceous glands of the cheek are visible as fairly large, yellowish patches called Fordyce's spots. They do not represent a pathologic change (Fig. 9-33).

REFERENCES

Adams, D.: Surface coatings of cells in the oral epithelium of the human fetus, J. Anat. **118:**61, 1974.

Ainamo, J., and Löe, H.: Anatomical characteristics of gingiva. A clinical and microscopic study of the free and attached gingiva, J. Periodontol. **37:**5, 1966.

Anderson, G.S., and Stern, I.B.: The proliferation and migration of the attachment epithelium on the cemental surface of the rat incisor, Periodontics **4:**115, 1966.

Arvidson, D., and Friberg, U.: Human taste: response and tastebud number in fungiform papillae, Science **209:**807, 1980.

Barker, D.S.: The dendritic cell system in human gingival epithelium, Arch. Oral Biol. **12:**203, 1967.

Barnett, M.L.: Mast cells in the epithelial layer of human gingiva, J. Ultrastruct. Res. **43:**247, 1973.

Barnett, M.L., and Szabó, G.: Gap junctions in human gingival keratinized epithelium, J. Periodont. Res. **8:**117, 1973.

Baume, L.J.: The structure of the epithelial attachment revealed by phase contrast microscopy, J. Periodontol. **24:**99, 1953.

Beagrie, G.S., and Skougaard, M.R.: Observations in the life cycle of the gingival epithelial cells of mice as revealed by autoradiography, Acta Odontol. Scand. **20:**15, 1962.

Beidler, L.N., and Smallman, R.L.S.: Renewal of cells within taste buds, J. Cell Biol. **27:**263, 1965.

Bergstrasser, P.R., Tigelaar, R.E., and Streilein, J.W.: Thy-1 antigen-bearing dendritic cells in murine epidermis are derived from bone marrow precursors, J. Invest. Dermatol. **83:**83, 1984.

Bjercke, S., Elgo, J., Braathen, L., and Thorsby, E.: Enriched epidermal cells are potent antigen-presenting cells for T cells, J. Invest. Dermatol. **83:**286, 1984.

Bolden, T.E.: Histology of oral pigmentation, J. Periodontol. **31:**361, 1960.

Bradley, R.M., and Mistretta, C.M.: The morphological and functional development of fetal gustatory receptors. In Emmelin, N., and Zotterman, Y., editors: Oral physiology, Oxford, England, 1972, Pergamon Press, Ltd.

Bradley, R.M., and Stern, I.B.: The development of the human taste bud during the foetal period, J. Anat. **101:**743, 1967.

Breathnach, S.M., Fox, P.A., Neises, G.R., Stanley, J.R., and Katz, S.J.: A unique epithelial basement membrane antigen defined by a monoclonal antibody (KF-1), J. Invest. Dermatol. **80:**392, 1983.

Buck, D.: The uptake of H[3] proline in the guinea pig gingiva and palate, J. Periodontol **4:**94, 1969.

Chen, S.-Y., Gerson, S., and Meyer, J.: The fusion of Merkel cell granules with a synapse-like structure, J. Invest. Dermatol. **61:**290, 1973.

Clark, R.A.F.: Fibronectin in the skin, J. Invest. Dermatol. **81:**475, 1983.

Cleaton-Jones, P., and Fleisch, L.: A comparative study of the surface of keratinized and non-keratinized oral epithelia, J. Periodont. Res. 8:366, 1973.

Dale, B.A.: Purification and characterization of a basic protein from the stratum corneum of mammalian epidermis, Biochem. Biophys. Acta 491:193, 1977.

Dale, B.A., Holbrook, K.A., and Steinert, P.M.: Assembly of stratum corneum basic protein and keratin filaments in macrofibrils, Nature 276:729, 1978.

Dale, B.A., and Ling, S.-Y.: Evidence of a precursor form of stratum corneum basic protein in rat epidermis, Biochemistry 18:35, 1979.

Dale, B.A., and Ling, S.-Y.: Immunologic cross-reaction of stratum corneum basic protein and a keratohyalin granule protein, J. Invest. Dermatol. 72:257, 1979.

Dale, B.A., Lonsdale-Eccles, J.D., and Lynley, A.M.: Two dimensional analysis of rat oral epithelium and epidermis, Arch. Oral Biol. 27:529, 1982.

Dale, B.A., Smith S., Clausen H., Vedtofte, P., and Dabelsteen, E.: Use of antibodies to epithelial keratins and keratohyalin, IADR Abst. 63:167, (special issue), 1984. (Abstract).

Dale, B.A., and Stern, I.B.: Keratohyalin granule proteins, J. Dent. Res. 53:143, 1975.

Dale, B.A., and Stern, I.B.: SDS polyacrylamide electrophoresis of proteins of newborn rat skin. I. Cell strata and nuclear proteins. J. Invest. Dermatol. 65:220, 1975.

Dale, B.A., and Stern, I.B.: SDS polyacrylamide electrophoresis of proteins of newborn rat skin. II. Keratohyalin and stratum corneum proteins, J. Invest. Dermatol. 65:223, 1975.

Dale, B.A., Stern, I.B., and Clagett, J.: Initial characterization of the proteins of keratinized epithelium of rat oral mucosa. Arch. Oral Biol. 22:75, 1977.

Dale, B.A., Stern, I.B., Rabin, M., and Huang, L.-Y.: The identification of fibrous proteins in fetal rat epidermis by electrophoretic and immunologic techniques, J. Invest. Dermatol. 66:230, 1976.

Dale, B.A., and Thompson, W.L.: Stratum corneum basic protein of keratinized rat oral epithelia, J. Dent. Res. 57:222, 1978.

Dale, B.A., Thompson, W.B., and Stern, I.B.: Distribution of histidine-rich basic protein, a possible keratin matrix protein, in the oral epithelium, Arch. Oral Biol. 27:535, 1982.

Daniels, T.E.: Human mucosal Langerhans cells: postmortem identification of regional variations in oral mucosa, J. Invest. Dermatol. 82:21, 1984.

DeHan, R., and Graziadei, P.P.C.: Functional anatomy of frog's taste organs, Experientia 27:823, 1971.

DeWaal, R.M.W., Semeijn, J.T., Cornelissen, I.M.H., and Rameaeker, F.C.S.: Epidermal Langerhans cells contain intermediate sized filaments of the Vimentin type: an immunocytologic study, J. Invest. Dermatol. 82:602, 1984.

Egelberg, J.: The blood vessels of the dentogingival junction, J. Periodont. Res. 1:163, 1966.

El-Labban, N.G., and Kramer, I.R.H.: On the so-called microgranules in the non-keratinized buccal epithelium, J. Ultrastruct. Res. 48:377, 1974.

Emslie, R.D., and Weinmann, J.P.: The architectural pattern of the boundary between epithelium and connective tissue of the gingiva in the rhesus monkey, Anat. Rec. 105:35, 1949.

Farbman, A.I.: Electron microscope study of a small cytoplasmic structure in rat oral epithelium, J. Cell Biol. 21:491, 1964.

Farbman, A.I.: Electron microscope study of the developing taste bud in rat fungiform papillae, Dev. Biol. 11:110, 1965.

Farbman, A.I.: Plasma membrane changes during keratinization, Anat. Rec. 156:269, 1966.

Farbman, A.I.: Structure of chemoreceptors. In Symposium on foods. Chemistry and physiology of flavors, Westport, Conn., 1967, Avi Publishing Co.

Flotte, T.J., Murphy, G.F., and Bhan, A.K.: Demonstration of T-200 on human Langerhans cell surface membranes, J. Invest. Dermatol. 82:535, 1984.

Fortman, G.J., and Winkelmann, R.K.: The Merkel cell in oral human mucosa, J. Dent. Res. 56:1303, 1977.

Frank, R.M., and Cimasoni, G.: Electron microscopic study of the human epithelial attachment, J. Dent. Res. 49:691, 1970.

Frank, R.M., and Cimasoni, G.: Ultrastructure de l'épithélium cliniquement normal du sillon et de la jonction gingivo-dentaires, Z. Zellforsch. Mikrosk. Anat. 109:356, 1970.

Franke, W.W., Moll, R., Mueller, H., Schmid, E., Kuhn, C., Krepler, R., Artlieb, U., and Denk, H.: Immunocytochemical identification of epithelium-derived human tumors with antibodies to desmosomal plaque proteins, Proc. Natl. Acad. Sci. USA 80:543, 1983.

Frithiof, L.: Ultrastructural changes in the plasma membrane in human oral epithelium, J. Ultrastruct. Res. 32:1, 1970.

Frithiof, L., and Wersall, J.: A highly ordered structure in keratinizing human oral epithelium, J. Ultrastruct. Res. 12:371, 1965.

Gavin, J.B.: The ultrastructure of the crevicular epithelium of cat gingiva, Am. J. Anat. 123:283, 1968.

Geisenheimer, J., and Han, S.S.: A quantitative electron microscopic study of desmosomes and hemidesmosomes in human crevicular epithelium, J. Peridontol. 42:396, 1971.

Gorbsky, G., and Steinberg, M.S.: Isolation of intercellular glycoproteins of desmosomes, J. Cell Biol. 90:243, 1981.

Gottlieb, B.: Der Epithelansatz am Zähne, Dtsch. Monatsschr, Zahnheilkd. 39:142, 1921.

Gottlieb, B.: Zur Biologie des Epithelansatzes und des Alveolarrandes, Dtsch. Zahnaerztl. Wochenschr. **25:**434, 1922.

Gottlieb, B., and Orban, B.: Biology and pathology of the tooth (translated by M. Diamond), New York, 1938, The MacMillan Co.

Grant, D.A., and Orban, B.: Leukocytes in the epithelial attachment, J. Periodontol. **31:**87, 1960.

Grant, D.A., Stern, I.B., and Everett, F.G.: Periodontics in the tradition of Orban and Gottlieb, ed. 5, St. Louis, 1979, The C.V. Mosby Co.

Grossman, E.S., and Austin, J.C.: The ultrastructural response to loading of the oral mucosa of the vervet monkey J. Periodont. Res. **18:**474, 1983.

Grossman, E.S., and Austin, J.C.: A quantitative electron microscopic study of desmosomes and hemidesmosomes in vervet monkey oral mucosa. J. Periodont. Res. **18:**580, 1983.

Hamilton, A.I., and Blackwood, H.J.J.: Cell renewal of oral mucosal epithelium of the rat, J. Anat. **117:**313, 1974.

Hansen, E.R.: Mitotic activity of the gingival epithelium in colchicinized rats, Odont. Tidskr. **74:**229, 1966.

Hashimoto, K.: The fine structure of the Merkel cell in human oral mucosa, J. Invest. Dermatol. **58:**381, 1972.

Hashimoto, K., Dibella, R.J., and Shklar, G.: Electron microscopic studies of the normal human buccal mucosa, J. Invest. Dermatol. **47:**512, 1966.

Hayward, A.F., and Hackemann, M.M.: Electron microscopy of membrane-coating granules and a cell surface coat in keratinized and nonkeratinized human oral epithelium, J. Ultrastruct. Res. **43:**205, 1973.

Hayward, A.F., Hamilton, A.I., and Hackemann, M.M.: Histological and ultrastructural observations on the keratinizing epithelia of the palate of the rat, Arch. Oral Biol. **18:**1041, 1973.

Huang, L.Y., Stern, I.B., Clagett, J.A., and Chi, E.Y.: Two polypeptide chain constituents of the major protein of the cornified layer of newborn rat epidermis, Biochemistry **14:**3573, 1975.

Hutchens, L.M., Sagebiel, R.W., and Clark, M.A.: Oral epithelial cells of the Rhesus monkey—histologic demonstration, fine structure and quantitative distribution. J. Invest. Dermatol. **56:**325, 1971.

Ito, H., Enomoto, S., and Kobayashi, K.: Electron microscopic study of the human epithelial attachment, Bull. Tokyo Med. Dent. Univ. **14:**267, 1967.

Karring, T., and Löe, H.: The three dimensional concept of the epithelium-connective tissue boundary of gingiva, Acta Odontol. Scand. **28:**917, 1970.

Katz, S.I., Tamaki, K., and Sachs, D.I.: Epidermal Langerhans cells are derived from cells originating in bone marrow, Nature **282:**324, 1979.

Klavan, B., Genco, R., Löe, H., et al.: Proceedings of the International Conference on Research in the Biology of Periodontal Disease, Chicago, 1977, University of Illinois College of Dentistry.

Kobayashi, K., Rose, G.G., and Mahan, C.J.: Ultrastructure of the dento-epithelial junction, J. Periodont. Res. **11:**313, 1976.

Kobayashi, K., Rose, G.G., and Mahan, C.J.: Ultrastructural histochemistry of the dento-epithelial junction. I.J. Periodont. Res. **12:**351, 1977.

Korman, M., Rubinstein, A., and Gargiulo, A.: Preservation of palatal mucosa. I. Ultrastructural changes and freezing technique, J. Periodontol. **44:**464, 1973.

Kubo, M., Norris, D.A., Howell, S.E., Ryan, S.R., and Clark, R.A.F.: Human keratinocytes synthesize, secrete, and deposit fibronectin in the pericellular matrix, J. Invest. Dermatol. **82:**580, 1984.

Kurahashi, Y., and Takuma, S.: Electron microscopy of human gingival epithelium, Bull. Tokyo Dent. Col. **3:**29, 1962.

Landay, M.A., and Schroeder, H.E.: Quantitative electron microscopic analysis of the stratified epithelium of normal human buccal mucosa, Cell Tissue Res. **177:**383, 1977.

Lange, D., and Schroeder, H.E.: Cytochemistry and ultrastructure of gingival sulcus cells, Helv. Odontol. Acta **15:**65, 1971.

Lavker, R.M.: Membrane coating granules: the fate of the discharged lamellae, J. Ultrastruc. Res. **55:**79, 1976.

Lavker, R.M., and Sun, T.T.: Heterogeneity in epidermal basal keratinocytes: morphological and functional correlations, Science **215:**1239, 1982.

Lavker, R.H., and Sun, T.T.: Epidermal stem cells, J. Invest. Dermatol. **81**(suppl.):121s, 1983.

Listgarten, M.A.: The ultrastructure of human gingival epithelium, Am. J. Anat. **114:**49, 1964.

Listgarten, M.A.: Electron microscopic study of the gingivo-dental junction of man, Am. J. Anat. **119:**147, 1966.

Listgarten, M.A.: Phase contrast and electron microscopic study of the junction between reduced enamel epithelium and enamel in unerupted human teeth, Arch. Oral Biol. **11:**999, 1966.

Listgarten, M.A.: Changing concepts about the dento-epithelial junction, J. Can. Dent. Assoc. **36:**70, 1970.

Löe, H., Karring, T., and Hara, K.: The site of mitotic activity in rat and human oral epithelium, Scand. J. Dent. Res. **80:**111, 1972.

Loening, T., Caselitz, J., Seifert, G., Weber, K., and Osborn, M.: Identification of Langerhans cells: simultaneous use of sera to intermediate filaments, T6 and HLA-DR antigens on oral mucosa, human epidermis and their tumors. Virchows Arch. [Pathol. Anat.] **398:**119, 1982.

Luzardo-Baptista, M.: Intraepithelial nerve fibers in the human oral mucosa, Oral Surg. 35:372, 1973.

Mahrle, G., and Orfanos, C.E.: Merkel cells as human cutaneous neuroceptor cells. Their presence in dermal neural corpuscles and in the external hair root sheath of human adult skin, Arch. Dermatol. Forsch. 251:19, 1974.

Mattern, C.F.T., Daniel, W.A., and Henkin, R.I.: The ultrastructure of the human circumvallate papilla. I. Cilia of the papillary crypt, Anat. Rec. 167:175, 1970.

Matusim, D.F., Takahashi, Y., Labib, R.S., Anhalt, G.J., Patel, H.P., and Diaz, L.A.: A pool of bullous pemphigoid antigen(s) is intracellular and associated with the basal cell cytoskeleton–hemidesmosome complex, J. Invest. Dermatol. 84:47, 1985.

McDougall, W.A.: pathways of penetration and effects of horseradish peroxidase in rat molar gingiva, Arch. Oral Biol. 15:621, 1970.

McHugh, W.D.: Keratinization of gingival epithelium in laboratory animals, J. Periodontol. 35:338, 1964.

McMillan, M.D.: A scanning electron study of keratinized epithelium of the hard palate of the rat, Arch. Oral Biol. 19:225, 1974.

McMillan, M.D.: The complementary structure of the superficial and deep surfaces of the cells of the stratum corneum of the hard palate of the rat. A scanning and transmission electron microscope study, J. Periodont. Res. 14:492, 1979.

Melcher, A.H., and Bowen, W.H.: Biology of the periodontium, London, 1969, Academic Press, Inc.

Meyer, J., and Gerson, S.J.: A comparison of human palatal and buccal mucosa, Periodontics 2:284, 1964.

Meyer, M., and Schroeder, H.E.: A quantitative electron microscopic analysis of the keratinizing epithelium of normal human hard palate, Cell Tissue Res. 158:177, 1975.

Mignon, M.L.: Ultrastructure of the gingival epithelium in the newborn cat—some characteristics of the intercellular junctions, J. Dent. Res. 53:1484, 1974.

Mihara, M., Hashimoto, K., Ueda, K., and Kumakiri, M.: The specialized junctions between Merkel cell and neurite: an electron microscopic study, J. Invest. Dermatol. 73:325, 1979.

Mueller, H., and Franke, W.: Biochemical and immunological characterization of desmoplakin I and II, the major polypeptides of the desmosomal plaque. J. Mol. Biol. 163:647, 1983.

Munger, B.: Neural-epithelial interactions in sensory receptors, J. Invest. Dermatol. 69:27, 1977.

Murphy, G.F.: Cytokeratin typing of cutaneous tumors: Q new immunocytochemical probe for cellular differentiation and malignant transformation, J. Invest. Dermatol. 84:1, 1985.

Murray, R.G., Murray, A., and Fujimoto, S.: Fine structure of gustatory cells in rabbit taste buds, J. Ultrastruct. Res. 27:444, 1969.

Negus, V.E.: The comparative anatomy of the nose and paranasal sinuses, London, 1958, E. and S. Livingstone.

Newcomb, G.M., Seymour, G.J., and Powell, R.N.: Association between plaque accumulation and Langerhans cell numbers in the oral epithelium of attached gingiva, J. Clin. Periodontol. 9:197, 1982.

Nuki, K., and Hock, J.: The organization of the gingival vasculature, J. Periodont. Res. 9:305, 1974.

Orban, B.: Zahnfleischtasche und Epithelansatz, Z. Stomatol. 22:353, 1924.

Orban, B.: Hornification of the gums, J. Am. Dent. Assoc. 17:1977, 1930.

Orban, B.: Clinical and histologic study of the surface characteristics of the gingiva, Oral Surg. 1:827, 1948.

Orban, B., Bhattia, H., et al.: Epithelial attachment (the attached gingival cuff). J. Periodontol. 27:167, 1956.

Orban, B., and Mueller, E.: The gingival crevice, J. Am. Dent. Assoc. 16:1206, 1929.

Orban, B., and Sicher, H.: The oral mucosa, J. Dent. Educ. 10:94, 1946.

Osborn, M.: Components of the cellular cytoskeleton: a new generation of markers of histogenetic origin? J. Invest. Dermatol. 82:443, 1984.

Palade, G.E., and Farquhar, M.G.: A special fibril of the dermis, J. Cell Biol. 27:215, 1965.

Peterson, L.I., Zettergren, J.G., and Wuepper, K.D.: Biochemistry of transglutaminases and cross-linking in the skin, J. Invest. Dermatol. 81(suppl.):95s, 1983.

Petitet, N.F., and Stern, I.B.: Ultrastructure de l'épithélium gingival humain. In Favard, P., editor: Microscopie électronique, vol. 3, Paris, 1970, Société Française de Microscopie Électronique.

Pruniéras, M., Régnier, M., Fougère, S., and Woodley, D.: Keratinocytes synthesize basal-lamina proteins in culture, J. Invest. Dermatol. 81(suppl.):28s, 1983.

Sage, H.: Collagens of basement membranes, J. Invest. Dermatol. 79:515, 1982.

Saito, I., Watanabe, O., Kawahara, H., Igarashi, Y., Yamamura, T., and Shimono, M.: Intercellular junctions and the permeability barrier in the junctional epithelium. A study with freeze-fracture and thin sectioning, J. Periodont. Res. 16:467, 1981.

Sauder, D.N., Carter, C.S., Katz, S.J., and Oppenheim, J.J.: Epidermal cell production of thymocyte activating factor (ETAF). J. Invest. Dermatol. 79:34, 1982.

Sauder, D.N., Dinerello, C.A., and Morhenn, V.B.: Langerhans cell production of interleukin-1, J. Invest. Dermatol. 82:605, 1984.

Schroeder, H.E.: Differentiation of human oral stratified epithelia, Basel, Switzerland, 1981, Karger.

Schroeder, H.E.: Melanin-containing organelles in cells of the human gingiva. I. Epithelial melanocytes, J. Periodont. Res. 4:1, 1969.

Schroeder, H.E., and Listgarten, M.A.: Fine structure of the developing epithelial attachment of human teeth (Monographs in developmental biology, vol. 2), ed. 2, Basel, Switzerland, 1977, Karger.

Schroeder, H.E., and Munzel-Pedrazzoli, S.: Correlated morphometric and biochemical analysis of gingival tissue, J. Microscopy 99:301, 1973.

Schroeder, H.E., and Theilade, J.: Electron microscopy of normal human gingival epithelium, J. Periodont. Res. 1:95, 1966.

Schuler, G.: The dendritic Thy-1-positive cell of murine epidermis: a new epidermal cell type of bone marrow origin, J. Invest. Dermatol. 83:81, 1984.

Schweizer, J., and Marks, F.: A developmental study of the distribution and frequency of Langerhans cells in relation to formation and patterning in mouse tail epidermis, J. Invest. Dermatol. 69:198, 1977.

Silberberg-Sinakin, I., Thorbecke, G.J., Baer, R.L., Rosenthal, S.A., and Berezowsky, V.: Antigen-bearing Langerhans cells in the skin, dermis and in lymph developments. Cell. Immunol. 25:137, 1976.

Skillen, W.G.: the morphology of the gingiva of the rat molar, J. Am. Dent. Assoc. 17:645, 1930.

Skougaard, M.R.: Cell renewal, with special reference to the gingival epithelium, Adv. Oral Biol. 4:261, 1970.

Smith, C.J.: Gingival epithelium. In Melcher, A.H., and Bowen, W.H., editors: Biology of the periodontium, New York, 1969, Academic Press, Inc.

Squier, C.A.: The permeability of keratinized and nonkeratinized oral epithelium to horseradish peroxidase, J. Ultrastruct. Res. 43:160, 1973.

Squier, C.A., and Meyer, J.: Current concepts of the histology of oral mucosa, Springfield, Ill., 1971, Charles C Thomas, Publisher.

Squier, C.A., and Waterhouse, L.P.: The ultrastructure of the melanocyte in human gingival epithelium, J. Dent. Res. 46:112, 1967.

Stanley, J.R., Hawley-Nelson, P., Yaar, M., Martin, G.R., and Katz, S.I.: Laminin and bullous pemphigoid antigen are distinct basement membrane proteins synthesized by epidermal cells, J. Invest. Dermatol. 78:457, 1982.

Stern, I.B.: Electron microscopic observations of oral epithelium. I. Basal cells and the basement membrane, Periodontics 3:224, 1965.

Stern, I.B.: The fine structure of the ameloblast-enamel junction in rat incisors, epithelial attachment and cuticular membrane, vol. B, Fifth International congress for Electron Microscopy, New York, 1966, Academic Press, Inc.

Stern, I.B.: Further electron microscopic observations of the epithelial attachment, Int. Assoc. Dent. Res. Abstr. no. 325, 45th general meeting, 1967, p. 118.

Stern, I.B.: Current concepts of the dentogingival junction: the epithelial and connective tissue attachments to the tooth, J. Periodontol. 52:465, 1981.

Stern, I.B., Dayton, L., and Duecy, J.: The uptake of tritiated thymidine by the dorsal epidermis of the fetal and newborn rat, Anat. Rec. 170:225, 1971.

Stern, I.B., and Sekeri-Pataryas, K.H.: The uptake of ^{14}C-leucine and ^{14}C-histidine by cell suspension of isolated strata of neonatal rat epidermis, J. Invest. Dermatol. 59:251, 1972.

Streilein, J.W.: Skin-associated lymphoid tissue (SALT): origins and functions, J. Invest. Dermatol. 80(suppl.):12s, 1983.

Saurat, J-H., Merot, Y., Didierjean, L., and Dahl, D.: Normal rabbit Merkel cells do not express neurofilament proteins, J. Invest. Dermatol. 82:641, 1984.

Saurat, J-H., Didierjean, L., Skalli, O., Siegenthaler, G., and Gabbian, G.: The intermediate filaments of rabbit normal epidermal Merkel cells and cytokeratins, J. Invest. Dermatol. 83:431, 1984.

Susi, F.R.: Histochemical, autoradiographic and electron microscopic studies of keratinization in oral mucosa, Ph.D. thesis, Tufts University, October 1967.

Susi, F.R.: Studies of cellular renewal and protein synthesis in mouse oral mucosa utilizing H^3-thymidine and H^3-cystine, J. Invest. Dermatol. 51:403, 1968.

Susi, F.R.: Anchoring fibrils in the attachment of epithelium to connective tissue in oral mucous membranes, J. Dent. Res. 48:144, 1969.

Susi, F.R., Belt, W.D., and Kelly, J.W.: Fine structure of fibrillar complexes associated with the basement membrane in human oral mucosa, J. Cell Biol. 34:686, 1967.

Svejda, J., and Janota, M.: Scanning electron microscopy of the papillae foliatae of the human tongue, Oral Surg. 37:208, 1974.

Thilander, H., and Bloom, G.D.: Cell contacts in oral epithelia, J. Periodont. Res. 3:96, 1968.

Toto, P.D., and Grundel, E.R.: Acid mucopolysaccharides in the oral epithelium, J. Dent. Res. 45:211, 1966.

Toto, P.D., and Sicher, H.: The epithelial attachment, Periodontics 2:154, 1964.

Vidic, B., et al.: The structure and prenatal morphology of the nasal septum in the rat, J. Morphol. 137:131, 1972.

Weinmann, J.P.: The keratinization of the human oral mucosa, J. Dent. Res. 19:57, 1940.

Weinmann, J.P., and Meyer, J.: Types of keratinization in the human gingiva, J. Invest. Dermatol. 32:9, 1959.

Weinstock, M., and Wilgram, G.F.: Fine structural observations on the formation and enzymatic activity of keratinosomes in mouse tongue filiform papilla, J. Ultrastruct. Res. **30:**262, 1970.

Wentz, F.M., Maier, A.W., and Orban, B.: Age changes and sex differences in the clinically "normal" gingiva, J. Periodontol. **23:**13, 1952.

Wertz, P.W., and Downing, D.T.: Glycolipids in mammalian epidermis: structure and function in the water barrier, Science **217:**1261, 1982.

Wiebkin, O.W., and Thonard, J.C.: Mucopolysaccharide localization in gingival epithelium. I. An autoradiographic demonstration, J. Periodont. Res. **16:**600, 1981.

Winkelmann, R.K.: The Merkel cell system and a comparison between it and the neurosecretory or APUD cell system, J. Invest. Dermatol. **69:**41, 1977.

Wolff, K., and Stingl, G.: The Langerhans cell, J. Invest. Dermatol. **80:**(suppl):17s, 1983.

10
SALIVARY GLANDS

The salivary glands are exocrine glands whose secretions flow into the oral cavity. There are three pairs of large glands, located extraorally, known as the major salivary glands (Plate 3), and numerous small glands widely distributed in the mucosa and submucosa of the oral cavity, known as the minor salivary glands. Both the major and minor glands are composed of parenchymal elements invested in and supported by connective tissue. The parenchymal elements are derived from the oral epithelium and consist of terminal secretory units leading into ducts that eventually open into the oral cavity. The connective tissue forms a capsule around the gland and extends into it, dividing groups of secretory units and ducts into lobes and lobules. The blood and lymph vessels and nerves that supply the gland are contained within the connective tissue. The production of saliva is the most important function of the salivary glands. Saliva contains various organic and inorganic substances, provides the primary natural protection for the teeth and soft tissues of the oral cavity, and assists in the mastication, deglutition, and digestion of food.

STRUCTURE AND FUNCTION OF SALIVARY GLAND CELLS

The terminal secretory units are composed of serous, mucous, and myoepithelial cells arranged into acini or secretory tubules (Fig. 10-1). The secretions of these units are collected by the intercalated ducts, which empty into the striated ducts. The structure and function of each of these components will be considered in detail, followed by a description of the connective tissue elements and nerves.

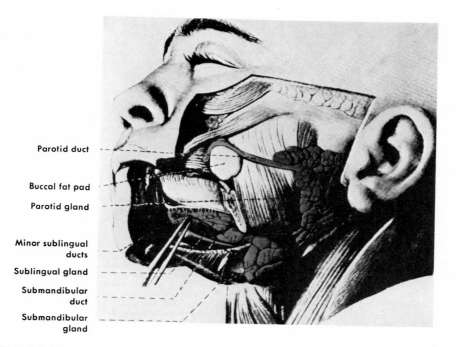

Parotid duct

Buccal fat pad

Parotid gland

Minor sublingual ducts

Sublingual gland

Submandibular duct

Submandibular gland

Plate 3. Salivary glands of major secretion. Part of mandible and mylohyoid muscle removed. (From Sicher, H., and Tandler, J.: Anatomie für Zahnärtze [Anatomy for dentists], Berlin, 1928, Julius Springer Verlag.)

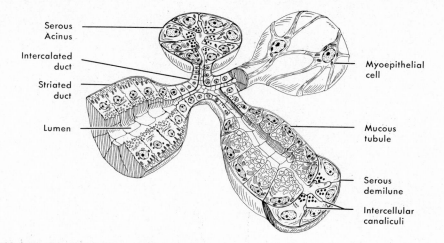

Fig. 10-1. Main features of parenchymal cells of salivary glands and their arrangement to form ducts and terminal secretory units.

Labels on figure: Serous Acinus, Intercalated duct, Striated duct, Lumen, Myoepithelial cell, Mucous tubule, Serous demilune, Intercellular canaliculi

Serous cells

Serous cells are specialized for the synthesis, storage, and secretion of proteins. The typical serous cell is pyramidal in shape, with its broad base resting on a thin basal lamina and its narrow apex bordering on the lumen (Figs. 10-1 and 10-2). The spherical nucleus is located in the basal region of the cell; occasionally binucleated cells are observed.

The most prominent feature of the serous cell is the accumulation of secretory granules in the apical cytoplasm (Fig. 10-2). These granules are about 1 μm in diameter and by electron microscopy are observed to have a distinct limiting membrane and a dense content (Fig. 10-3). In some salivary glands, including those of humans, the serous granules may contain a dense core or a twisted skeinlike structure within a lighter matrix. The granules may be very closely apposed to one another, the plasma membrane, or other organelles, but in unstimulated cells they retain their individual nature and do not fuse with other structures. In routine histologic preparations, the serous granules are usually not well resolved because

of section thickness and the conditions of fixation, and the apical portion of the cell may appear as an acidophilic mass. However, in semithin (1 μm) sections of plastic-embedded tissue, stained with toluidine blue or by specific cytochemical techniques (Fig. 10-2), the secretory granules are clearly seen.

The basal portion of the cytoplasm is filled with ribosome-studded (rough) endoplasmic reticulum (RER), a closed system of membranous sacs or cisternae (Figs. 10-3 and 10-4, A). The ribosomes, consisting of ribonucleic acid (RNA) and proteins, are the basic units of protein synthesis. Acting under the direction of messenger RNA from the nucleus, the ribosomes translate the encoded message, adding the appropriate amino acids in their proper sequence in the protein being synthesized. Proteins destined for secretion are synthesized as *pre*proteins, with an NH_2-terminal extension of about 16 to 30 amino acids, called a signal sequence. As the signal sequence emerges from the ribosome it is recognized by specific proteins in the RER membrane, which direct attachment of the ribosome to the membrane. The

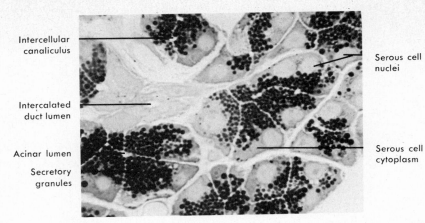

Intercellular
canaliculus

Serous cell
nuclei

Intercalated
duct lumen

Acinar lumen

Secretory
granules

Serous cell
cytoplasm

Fig. 10-2. Light micrograph of rat parotid gland illustrating general arrangement and cytologic features of serous cells. Gland was incubated in cytochemical medium to demonstrate the secretory enzyme peroxidase, resulting in unstained nuclei, lightly stained cytoplasm, and heavily stained secretory granules. Cells of intercalated duct are unreactive. (1 μm; × 990.)

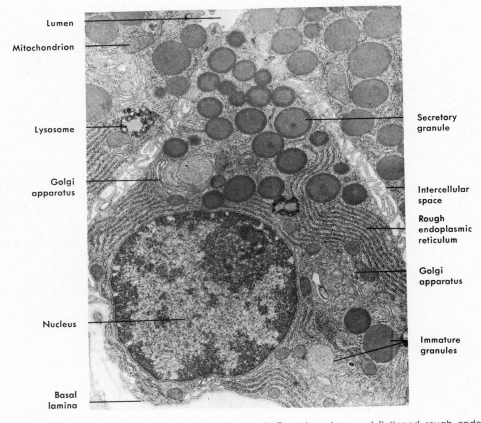

Lumen

Mitochondrion

Lysosome

Golgi
apparatus

Nucleus

Basal
lamina

Secretory
granule

Intercellular
space

Rough
endoplasmic
reticulum

Golgi
apparatus

Immature
granules

Fig. 10-3. Electron micrograph of typical serous cell. Round nucleus and flattened rough endoplasmic reticulum cisternae are located in basal half of cell. Golgi apparatus and immature granules are located apical and lateral to nucleus, and dense secretory granules are located in cell apex. Folds of cell membranes interdigitate in intercellular spaces. Note faint basal lamina. (Rat parotid gland; ×10,500.) (From Hand, A.R.: Am. J. Anat. **135**:71, 1972; reprinted by permission of the Wistar Institute Press.)

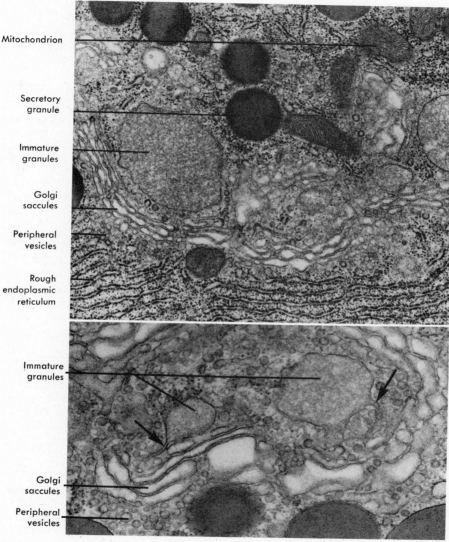

Fig. 10-4. Electron mictrographs of Golgi apparatus of serous cells. Saccules at *cis* face of Golgi apparatus are wider than saccules at *trans* face. Immature granules, located at *trans* face, are frequently larger than the mature granules and are often connected to membranes in Golgi region *(arrows).* Numerous vesicles are present between rough endoplasmic reticulum and *cis* saccules and at the *trans* face. (Rat parotid gland; **A,** ×25,400; **B,** ×46,400.)

signal sequence also assists in transfer of the growing polypeptide chain across the RER membrane. When the newly synthesized protein reaches the cisternal space of the RER, the signal sequence is removed by a proteolytic enzyme called signal peptidase, and the protein assumes its characteristic three-dimensional structure. In cells that produce large amounts of protein for secretion, the RER is well developed and arranged in parallel stacks, usually basal and lateral to the nucleus.

A second system of membranous cisternae, the Golgi apparatus, is located apical or lateral to the nucleus (Figs. 10-3 and 10-4). The Golgi apparatus consists of several stacks of four to six smooth-surfaced saccules that are slightly curved or cup shaped, with the concave or *trans* face usually oriented toward the secretory surface of the cell. The Golgi apparatus is functionally interconnected with the RER through vesicles budding from the ends of the RER cisternae that approach the periphery (convex or *cis* face) of the Golgi apparatus (Fig. 10-4, *A*). The newly synthesized secretory proteins within the RER are transported to the Golgi apparatus via these small vesicles. Although the exact route followed by the secretory proteins has not been firmly established, it is believed that in most cells the RER-derived vesicles fuse with the *cis* Golgi saccule, contributing their content to it. The proteins apparently then move through the Golgi saccules toward the *trans* face of the Golgi apparatus, where they are packaged into vacuoles of variable size and density (Figs. 10-3 and 10-4). These vacuoles are the forming secretory granules and are called *immature granules, prosecretory granules,* or *condensing vacuoles.* The immature granules have direct connections with smooth membranes at the *trans* face of the Golgi apparatus (Fig. 10-4, *B*); additionally, their limiting membrane is frequently irregular, suggesting that some secretory material may reach the immature granules through fu-

sion with small vesicles. The smaller immature granules have a light flocculent content; as they increase in size, their content increases in density until it is near that of the mature granules. The increase in density of the secretory material suggests that it is being concentrated as it is being transported and packaged for storage in granules.

Following their synthesis, many secretory proteins undergo one or more covalent structural modifications prior to their secretion. The most common modification of salivary proteins is glycosylation (i.e., the addition of carbohydrate side chains to the amino acids asparagine, serine, and threonine in the protein). The carbohydrates of secretory glycoproteins include galactose, mannose, fucose, glucosamine, galactosamine, and sialic acid. Glycosylation is a multistep process that begins in the RER and is completed in the Golgi apparatus. Other modifications of the secretory proteins may include the addition of phosphate or sulfate groups and specific proteolytic cleavage to produce the final secretory product.

The completed secretory proteins are stored in the secretory granules in the cell apex. Secretion or discharge of the granule content occurs by a process called exocytosis. This involves fusion of the granule membrane with the plasma membrane at the lumen or intercellular canaliculus, followed by the opening of the fused portion (Fig. 10-5). In this manner, the granule membrane becomes continuous with the plasma membrane, and the granule content is exteriorized without loss of cytoplasm. During rapid secretion, such as that occurring after stimulation by various pharmacologic agents, a second granule may fuse with the membrane of a previously discharged granule; continuation of this process, termed compound exocytosis, can lead to a long string of interconnected granule profiles extending into the cytoplasm. The addition of granule membranes results in a great enlargement of the plasma membrane at the secretory surface; during recovery, this excess membrane is retrieved by the cell as small vesicles. A portion of these vesicles fuses with lysosomes (see below) where the membrane may be degraded, while some may return to the Golgi apparatus where the membrane may be reutilized during the formation of new secretory granules. The proportion of the membrane undergoing degradation or reutilization has not been established.

In summary, then, secretory proteins are synthesized by membrane-bound ribosomes, transferred to the cisternal space of the RER, and migrate to the Golgi apparatus, where carbohydrate addition and other posttranslational modifications are completed, and they are packaged into secretory granules. After a variable period of storage in the cell apex, they are discharged by exocytosis at the secretory surface of the cell. The incorporation of amino acids into the secretory proteins and the movement of the proteins through the various compartments of the cell has been studied by electron microscope radioautography (Fig. 10-6, A). Counts of the radioautographic grains at various times after administration of radioactive amino acids reveal the sequential flow of proteins from the RER to the Golgi apparatus, their accumulation in immature granules, and finally storage in the mature secretory granules (Fig. 10-6, B). Enzyme and immunocytochemical studies have confirmed the presence of specific secretory proteins in these various intracellular compartments.

The serous cell contains several other cytoplasmic organelles, which are also found in most other salivary gland cells. Free or unattached ribosomes are located in the cytoplasm throughout the cell; they are concerned with the synthesis of nonsecretory cellular proteins. Mitochondria are also found throughout the cell, most frequently between the RER cisternae, around the Golgi apparatus, and along the lateral and basal plasma membranes. The mitochondria contain the enzymes of the citric acid cycle, electron transport, and oxidative phosphorylation; hence they are the major

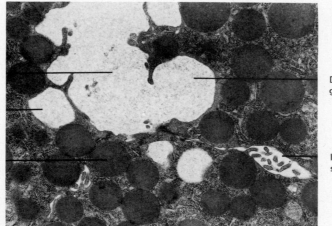

Lumen

Discharged granule

Secretory granule

Discharged granules

Intercellular space

Fig. 10-5. Electron micrograph of three serous cells illustrating exocytosis of secretory granules induced by isoproterenol. Flask-shaped invaginations of lumen into cell apices mark sites of granule discharge. (Rat lingual serous gland; ×12,800.)

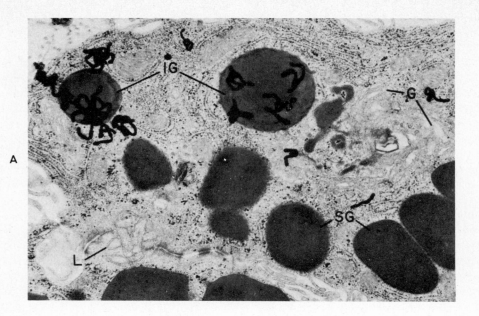

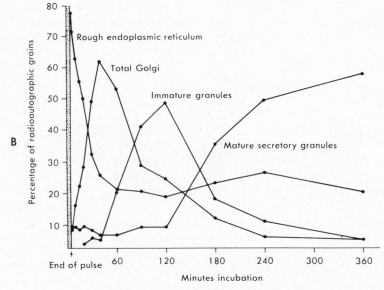

Fig. 10-6. **A,** Electron microscope radioautograph of serous cell of rabbit parotid gland, pulse labeled with ^{3}H-leucine for 4 minutes and incubated in vitro for 116 minutes. Radioautographic grains, indicating presence of radioactive leucine incorporated into newly synthesized proteins, are concentrated over immature granules, *IG*. A few grains are present over rough endoplasmic reticulum (RER) and Golgi apparatus, *G,* but none are localized over mature secretory granules, *SG*. Lumen, *L*. **B,** Radioautographic grain counts of rabbit parotid serous cells after pulse-labeling with ^{3}H-leucine and in vitro incubation. Newly synthesized proteins move in wavelike fashion from RER, through Golgi apparatus, and into immature and finally mature secretory granules. About 20% of label remains with RER as nonsecretory proteins. (**A,** ×17,100.) (**A** and **B** from Castle, J.D., Jamieson, J.D., and Palade, G.E.: J. Cell Biol. **53:**290, 1972; reprinted by permission of the Rockefeller University Press.)

source of high-energy compounds necessary for the numerous synthetic and transport processes that occur in the cell. Lysosomes, organelles that contain potent hydrolytic enzymes, are also occasionally seen. They function to destroy foreign materials taken up by the cells, as well as portions of the cells themselves, such as worn-out mitochondria or other membranous organelles. Their typical heterogeneous content of granular and membranous debris and lipidlike droplets probably reflects the role of lysosomes in this latter process. A few peroxisomes (microbodies), small organelles containing the enzyme catalase and other oxidative enzymes, can be demonstrated in the serous cells by cytochemical techniques; their exact functions are unknown, but they probably participate in certain aspects of lipid metabolism. Bundles of tonofilaments, associated with desmosomes, and microfilaments may be seen in the cytoplasm, as well as an occasional microtubule.

Mucous cells

The mucous cell, like the serous cell, is specialized for the synthesis, storage, and secretion of a secretory product. However, its structure differs from that of the serous cell. In routine histologic preparations, the apex of the cell appears empty except for thin strands of cytoplasm forming a trabecular network (Fig. 10-7, A). The nucleus and a thin rim of cytoplasm are compressed against the base of the cell.

In the electron microscope the mucous cell is seen to be filled with pale, electron-lucent secretory droplets containing scattered flocculent material (Fig. 10-8). These droplets are usually larger than serous granules and may be irregular or compressed in shape. Adjacent mucous droplets are separated by thin strands of cytoplasm, or they may be so closely apposed that their membranes fuse. During fixation and processing of the tissue for micros-

copy, these tenuous partitions are often disrupted; several droplets may thus form a larger mass. The secretory products of most mucous cells differ from those of serous cells in two important respects: (1) they have little or no enzymatic activity and probably serve mainly for lubrication and protection of the oral tissues, and (2) the ratio of carbohydrate to protein is greater, and larger amounts of sialic acid and occasionally sulfated sugar residues are present. The differences in the carbohydrate content of the secretory material of mucous and serous cells can be demonstrated by histochemical staining techniques (Fig. 10-7, B).

The nucleus of the mucous cell is oval or flattened in shape and located just above the basal plasma membrane (Fig. 10-8). The RER is limited to a narrow band of cytoplasm along the base and lateral borders of the cell and to an occasional patch of cytoplasm between the mucous droplets. The mitochondria and other organelles are also primarily limited to this band of basal and lateral cytoplasm. The Golgi apparatus is large, consisting of several stacks of 10 to 12 saccules sandwiched between the basal RER and mucous droplets forming from the *trans* face. The Golgi apparatus plays an important role in these cells because of the large amount of carbohydrate that it adds to the secretory products.

The secretion of mucous droplets occurs by a somewhat different mechanism than the exocytotic process seen in the serous cells. When a single droplet is discharged, its limiting membrane fuses with the apical plasma membrane, resulting in a single membrane separating the droplet from the lumen. This separating membrane may then fragment, being lost with the discharge of mucus, or the droplet may be discharged with the membrane intact, surrounding it. During rapid droplet discharge, the apical cytoplasm may not seal itself off, and the entire mass of mucus may be spilled into the lumen.

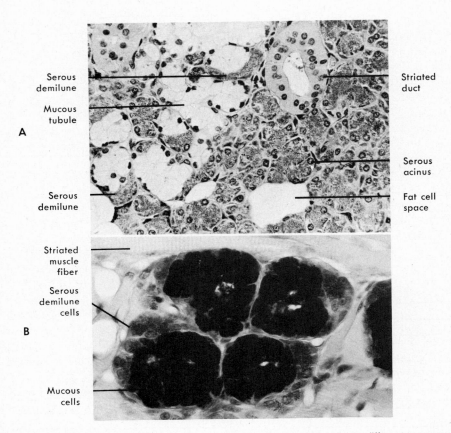

Fig. 10-7. A, Light micrograph of human submandibular gland illustrating different appearance of mucous and serous cells. Mucous tubules are capped by serous demilunes. Two striated ducts are cut in cross section. **B,** Light micrograph of posterior lingual mucous gland of rat, stained with alcian blue and periodic acid–Schiff (PAS). Mucous secretory glycoprotein stains with both alcian blue and PAS, indicating acidic carbohydrate residues. Granules of serous demilune cells stain only with PAS, indicating neutral glycoproteins. (**A,** ×265; **B,** ×420.)

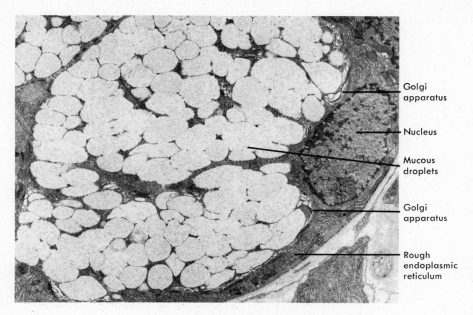

Golgi apparatus

Nucleus

Mucous droplets

Golgi apparatus

Rough endoplasmic reticulum

Fig. 10-8. Electron micrograph of mucous cell. Pale mucous droplets have flocculent content and tend to coalesce into larger masses. Golgi apparatus is well developed; rough endoplasmic reticulum and nucleus are compressed against base of cell. (×7000.)

Myoepithelial cells

Myoepithelial cells are closely related to the secretory and intercalated duct cells, lying between the basal lamina and the basal membranes of the parenchymal cells (Figs. 10-1 and 10-9). The body of the cell is small, filled mostly with a flattened nucleus, and numerous branching cytoplasmic processes radiate out to embrance the parenchymal cells. Myoepithelial cells are difficult to identify in routine histologic preparations, but their typical stellate shape can be observed in sections stained by special histochemical or immunofluorescent techniques (Fig. 10-9, *A*). Their appearance is reminiscent of a basket cradling the secretory unit; hence the name "basket cell" in the older literature.

The usual appearance of myoepithelial cells in electron micrographs is a section through one of their processes lying in a groove on the surface of a secretory or duct cell (Fig. 10-11, *B*). The processes are filled with longitudinally oriented fine filaments about 6 nm (60 Å) thick (Fig. 10-9, *C*). Small dense bodies are frequently present between the thin filaments; these are also present in smooth muscle cells, where they appear to form a cytoskeletal network in association with 10 nm (100 Å) diameter filaments. The usual cytoplasmic organelles are largely restricted to the perinuclear cytoplasm. The body of the cell, containing the nucleus, often lies in the space where the basal regions of two or three parenchymal cells come together (Fig. 10-9, *B*). The plasma membrane of the myoepithelial cell closely parallels the basal membrane of the parenchymal cell, and the two are joined by occasional desmosomes. Numerous micropinocytotic vesicles or caveolae are located on the plasma membranes of the myoepithelial cells.

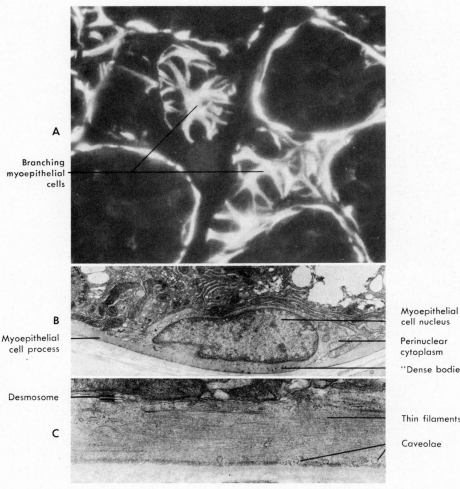

Fig. 10-9. A, Fluorescent micrograph of rat sublingual gland treated with antibody to smooth muscle myosin, to localize myosin present in myoepithelial cells. Tangential sections of acini reveal branching nature of myoepithelial cells; myoepithelial cell processes cut in cross and longitudinal section surround adjacent acini. **B,** Electron micrograph of myoepithelial cell body showing concentration of organelles in perinuclear cytoplasm and processes filled with fine filaments. The "dense bodies" are characteristic of myoepithelium and smooth muscle. Rat sublingual gland. **C,** Higher magnification of myoepithelial cell process filled with longitudinally arranged thin filaments. Several caveolae are located along basal surface of cell, and a desmosome attaches the process to mucous cell. (Rat sublingual gland; **A,** ×700; **B,** ×6000; **C,** ×19,100.) (**A** courtesy D. Drenckhahn, Kiel, Federal Republic of Germany.)

Myoepithelial cells are considered to have a contractile function, helping to expel secretions from the lumina of the secretory units and ducts. Although direct evidence is lacking for the salivary glands, the following observations suggest that this may be the case: (1) the structure of myoepithelium is similar to that of smooth muscle; (2) immunofluorescent studies indicate the presence of actin, myosin, and related proteins in myoepithelial cells (Fig. 10-9, A); (3) measurements of ductal pressure after appropriate stimulation suggest a contractile process; and (4) cinemicrography of individual secretory units stimulated to secrete in vitro reveals a regular pulsatile movement of the entire unit. Studies of sweat and mammary glands, where myoepithelial cells are abundant, also support a contractile function.

Arrangement of cells in the terminal secretory units

The structure of the terminal secretory units is different for different glands. When the gland consists entirely of serous secretory units, such as the human parotid, the serous cells are clustered in a roughly spherical fash-

ion around a central lumen, forming an acinus (Fig. 10-1). At the apical ends of adjoining cells, the lumen is sealed off from the lateral intercellular spaces by junctional complexes, consisting of a tight junction (zonula occludens), an intermediate junction (zonula adherens), and one or more desmosomes (maculae adherens). These junctions serve to hold the cells together as well as prevent leakage of the luminal contents into the intercellular spaces. Experimental studies have demonstrated, however, that following secretory stimulation the tight junctions may become more permeable to macromolecules and other organic substances. Fingerlike branches of the lumen, called intercellular canaliculi, extend between adjacent cells almost to their base (Fig. 10-10); they increase the area of the secretory surface and are sealed by junctional complexes along their length. A fine microfilament network, containing actin and myosin (Fig. 10-10), is located in the apical cytoplasm adjacent to the secretory surface. The remainder of the apposed lateral surfaces are joined by frequent desmosomes and an occasional gap junction.

In glands composed entirely of mucous se-

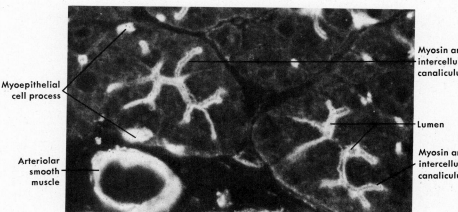

Fig. 10-10. Fluorescent micrograph of rat parotid gland treated with antibody to smooth muscle myosin. Myosin is located in myoepithelial cell processes, cut in cross section, and in apical cytoplasm of acinar cells, outlining lumen and intercellular canaliculi. (×900.) (From Drenckhahn, D., Gröschel-Stewart, U., and Unsicker, K.: Cell Tissue Res. **183:**273, 1977; reprinted by permission of Springer-Verlag.)

cretory units the arrangement of the secretory cells is similar. Rather than a spherical acinus, however, a tubular secretory end piece may be formed (Fig. 10-1). The central lumen is usually larger than in serous acini, and intercellular canaliculi are not usually present, although they have been observed between the mucous cells of the human labial glands.

In mixed glands the proportion of serous and mucous cells may vary from predominantly serous, as in the human submandibular gland, to predominantly mucous, as in the human sublingual gland. Separate serous and mucous units may exist, in addition to secretory units composed of both cell types. In the latter arrangement, the mucous cells form a typical tubular portion that is capped at the blind end by crescents of several serous cells, known as demilunes (Figs. 10-1 and 10-7). The secretion of the serous demilune cells reaches the lumen through the intercellular canaliculi.

The disposition of the myoepithelium in re-

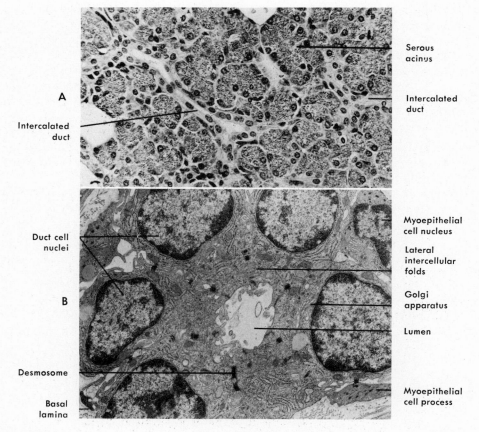

Fig. 10-11. A, Light micrograph of human parotid gland showing long branching intercalated ducts between serous acini. **B,** Electron micrograph of intercalated duct cut in cross section. Duct cells contain a moderate amount of rough endoplasmic reticulum and a prominent Golgi apparatus but few or no secretory granules. Prominent desmosomes and interlocking folds are present between adjacent cells. Myoepithelial cell processes extend longitudinally along duct, inside basal lamina. (Rat parotid gland. **A,** ×265; **B,** ×9600.)

lation to the parenchymal cells has already been described. In some glands the cell bodies may be restricted to the intercalated ducts with only the branching processes reaching the acini. Myoepithelial cells are not usually present along the striated ducts.

Ducts

The duct system of the salivary glands is formed by the confluence of small ducts into ones of progressively larger caliber. Within a lobule, the smallest ducts are the intercalated ducts (Fig. 10-1); they are thin branching tubes of variable length that connect the terminal secretory units to the next larger ducts, the striated ducts. In the interlobular connective tissue the ducts continue to join one another, increasing in size until the main excretory duct is formed.

Intercalated ducts. The intercalated ducts (Figs. 10-1, 10-2, and 10-11, *A*) are lined by a single layer of low cuboid cells with relatively empty-appearing cytoplasm. They are often difficult to identify in the light microscope because they are compressed between secretory units. In electron micrographs the intercalated duct cells share several characteristics of serous

cells (Fig. 10-11, *B*). A small amount of RER is located in the basal cytoplasm, and a Golgi apparatus of moderate size is found apically. In proximally located cells (near the secretory units) a few small secretory granules may be found. The lateral membranes of adjacent cells are joined apically by junctional complexes and several desmosomes. One or two areas of prominent interlocking folds of the lateral surface are located further basally. At the periphery of the duct, processes and cell bodies of myoepithelial cells may be found, attached by desmosomes to the duct cells.

Striated ducts. The striated ducts are lined by a layer of tall columnar epithelial cells with large, spherical, centrally placed nuclei (Figs. 10-1, 10-7, *A* and 10-12). The cytoplasm is abundant and eosinophilic and shows prominent striations at the basal ends of the cells, perpendicular to the basal surface. An occasional basally located cell can be identified by the position of its nucleus, below the level of those of the other cells (Fig. 10-7, *A*).

In electron micrographs the basal cytoplasm of the striated duct cells is partitioned by deep infoldings of the plasma membrane, producing numerous sheetlike folds that extend beyond

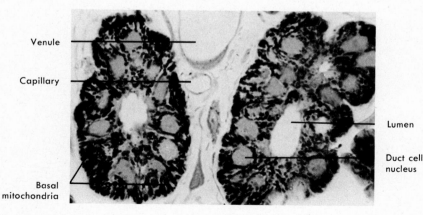

Fig. 10-12. Light micrograph of two striated ducts cut in cross section. Large, primarily radially oriented mitochondria, stained for cytochrome oxidase activity, fill basal regions of duct cells. Unstained nuclei are centrally located, and small mitochondria are found in apical cytoplasm. (Rat parotid gland; ×990.)

the lateral boundaries of the cell and interdigitate with similar folds of adjacent cells (Figs. 10-13 and 10-14, *B*). Abundant large mitochondria, usually radially oriented, are located in portions of the cytoplasm between the membrane infoldings (Figs. 10-12 to 10-14). The combination of infoldings and mitochondria accounts for the striations seen in the light microscope. A few short RER cisternae and a small Golgi apparatus are found in the perinuclear cytoplasm. Apically the cytoplasm may contain a variable amount of branching, tubular, smooth endoplasmic reticulum, small secretory granules of moderate density (Fig. 10-14, *A*) or small, empty-appearing vesicles. Several lysosomes, numerous small peroxisomes, bundles of cytoplasmic filaments, free ribosomes, and a

moderate amount of glycogen are also usually present. Numerous short microvilli on the apical surfaces project into the lumen, and adjacent cells are joined by apical junctional complexes and several desmosomes along their lateral surfaces.

In the interlobular (excretory) ducts the epithelium becomes pseudostratified, with increasing numbers of smaller basal cells between the tall columnar cells. The characteristics of the striated cells are maintained to a variable degree, becoming less pronounced as the duct increases in size. In the largest ducts occasional mucous goblet cells and ciliated cells may be found, and the epithelium of the main duct gradually becomes stratified as it merges with the epithelium of the oral cavity.

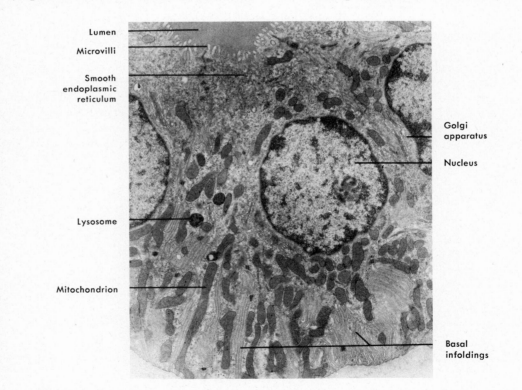

Fig. 10-13. Electron micrograph of striated duct cells of rat parotid gland. Numerous mitochondria are located between infoldings of basal plasma membrane. A few lysosomes and the Golgi apparatus are located in perinuclear region, and smooth endoplasmic reticulum is found in cell apices. Short microvilli project into lumen. (×7400.)

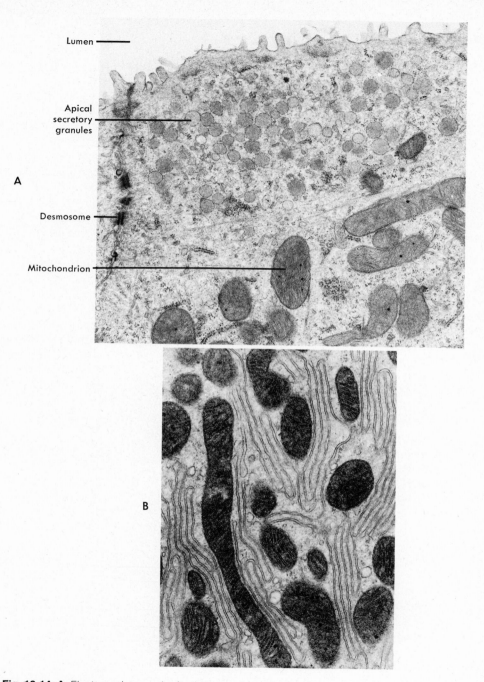

Lumen

Apical
secretory
granules

A

Desmosome

Mitochondrion

B

Fig. 10-14. A, Electron micrograph of apical cytoplasm of striated duct cell. Small secretory granules are located near lumen. Mitochondria and smooth and rough endoplasmic reticulum are also present. **B,** Basal region of striated duct cell, showing mitochondria and infolded plasma membranes. (Rat sublingual gland; **A,** ×22,200; **B,** ×23,900.)

Functions of salivary ducts. The main function of the salivary gland ducts is to convey the primary saliva secreted by the terminal secretory units to the oral cavity. The ducts are not just passive conduits, however; they actively modify the primary saliva by secretion and reabsorption of electrolytes and secretion of proteins. The intercalated duct cells often contain secretory granules in their apical cytoplasm, and two of the antibacterial proteins present in saliva, lysozyme and lactoferrin, have been localized to these ducts by immunofluorescent procedures. The striated duct cells contain kallikrein, an enzyme found in saliva, and synthesize secretory glycoproteins, which are stored in the apical granules. Recent studies have shown that both intercalated and striated duct cells are also capable of reabsorbing proteins from the lumen by endocytic mechanisms.

The structure of the striated duct cells (i.e., basal infoldings and numerous mitochondria) is typical of tissues involved in water and electrolyte transport, such as the kidney tubules and the choroid plexus. The role of the salivary ducts in electrolyte secretion and reabsorption has been extensively studied in experimental animals under various conditions of secretion. Analysis of the primary secretion, obtained by micropuncture techniques from the lumen of an intercalated duct draining several secretory units, reveals that it is isotonic or slightly hypertonic to plasma, with Na^+ and Cl^- concentrations approximately equal to those in plasma. K^+ concentration is low compared to that of Na^+, but it is significantly higher than the K^+ concentration of plasma. HCO_3^- concentration is variable depending upon the specific gland. Analysis of fluid collected from the excretory ducts reveals that it is hypotonic, with low Na^+ and Cl^- and high K^+ concentrations. Furthermore, the concentration of these electrolytes varies with the flow rate of the saliva: with increasing flow, Na^+ and Cl^- increase, as does HCO_3^-, while K^+ decreases. It is believed that the striated ducts actively reabsorb Na^+ from the primary secretion and secrete K^+ and HCO_3^-; Cl^- tends to follow the electrochemical gradient established by Na^+ reabsorption. At increased flow rates Na^+ reabsorption becomes less efficient and the secretion is in contact with the ductal epithelium for a shorter time; hence Na^+ concentrations of the saliva tend to increase. Supporting this postulated role in Na^+ and Cl^- reabsorption, the basal regions of the striated duct cells contain high concentrations of the transport enzyme $(Na^+ + K^+)$–activated adenosine triphosphatase, as shown by binding of the specific inhibitor ^3H-ouabain. Microperfusion studies of the main excretory duct have shown that it too is able to reabsorb Na^+ and secrete K^+ and HCO_3^-.

Essentially all of the water enters saliva at the level of the terminal secretory units; the striated and excretory ducts appear to be relatively impermeable to water. The ductal reabsorption of Na^+ and Cl^- exceeds the secretion of K^+ and HCO_3^-, leaving a hypotonic luminal fluid. Since active transport of water does not occur, the ducts cannot secrete water against the osmotic gradient to produce the final hypotonic saliva.

Connective tissue elements

The cells found in the connective tissue of the salivary glands are the same as those in other connective tissues of the body and include fibroblasts, macrophages, mast cells, occasional leukocytes, fat cells, and plasma cells. The cells, along with collagen and reticular fibers, are embedded in a ground substance composed of proteoglycans and glycoproteins. The vascular supply to the glands is also embedded within the connective tissue, entering the glands along the excretory ducts and branching to follow them into the individual lobules. The ducts, to the level of the intralobular striated ducts, are supplied with a dense capillary network; the capillary loops to the intercalated ducts and terminal secretory units are less extensive. A system of arteriovenous anastomoses around the larger interlobular ducts has also been described.

Nerves. The main branches of the nerves supplying the glands follow the course of the vessels, breaking up into terminal plexuses in the connective tissue adjacent to the terminal portions of the parenchyma. Nerve bundles, consisting of unmyelinated axons surrounded by cytoplasmic processes of Schwann cells, are distributed to the smooth muscle of the arterioles, the secretory cells and myoepithelium, and possibly the intercalated and striated ducts.

The secretory cells receive their innervation by one of two patterns. In the intraepithelial type, the axons split off from the nerve bundle and penetrate the basal lamina, lying adjacent to or between the secretory cells (Fig. 10-15, A). As the axons pass through the basal lamina, the Schwann cell covering is usually lost; occasionally it may be continued into the parenchyma and lie between the axons and the secretory cell. The site of innervation (neuroeffector site) is considered to be at varicosities of the axon, which contain small vesicles and mitochondria. The vesicles are believed to contain the chemical neurotransmitters norepinephrine and acetylcholine and presumably release them by an exocytosis-like process. The membranes of the axon and secretory cell are separated by a space of only 10 to 20 nm (100 to 200 Å), but no specializations of the plasma membranes have been detected at these sites. A single axon may have several varicosities along its length, making contact with the same cell or with two or more cells.

The second type of innervation is subepithelial. Instead of penetrating the basal lamina, the axons remain associated with the nerve bundle in the connective tissue (Fig. 10-15, B). Where the nerve bundles approach the secretory cells, some of the axonal varicosities, which contain the small neurotransmitter vesicles, lose their covering of Schwann cell cytoplasm. Presumably, these bared axonal varicosities are the sites of transmitter release. The axons remain separated from the secretory cells by 100 to 200 nm (1000 to 2000 Å), and the transmitters must diffuse across this space, which includes the basal laminae of the secretory cells and the nerve bundle.

The pattern of innervation varies between glands in the same animal and between the same gland in different species. The parotid serous cells and the sublingual mucous cells of the rat receive an intraepithelial type of innervation, as do the mucous cells of human labial glands. In contrast, the innervation of the secretory cells of the rat submandibular and the serous cells of the human parotid and submandibular glands is of the subepithelial type.

Both divisions of the autonomic nervous system may participate in the innervation of the secretory cells. In some glands both sympathetic (adrenergic) (Fig. 10-15, C) and parasympathetic (cholinergic) terminals (distinguished by special fixation and cytochemical techniques) have been observed in proximity to the secretory cells. Similarly, physiologic studies indicate that the cells of some glands respond to both sympathetic and parasympathetic stimulation by changes in their membrane potential. However, the extent of participation by each division varies between glands and animals, and the composition of the saliva secreted in response to stimulation of each division is distinctly different. In general, a copious flow of watery saliva is secreted in response to parasympathetic stimulation, whereas that produced by sympathetic stimulation is thicker, higher in organic content, and comparatively less in quantity.

The innervation of duct cells is not clear. Intraepithelial terminals in ducts have been observed only rarely, but histochemical studies suggest that cholinergic and adrenergic nerves are found in the connective tissue around the ducts. Physiologic studies indicate that the ductal system is responsive to autonomic stimulation or administration of autonomic drugs: membrane potential changes in duct cells have been recorded, as well as changes in the transductal ion flux.

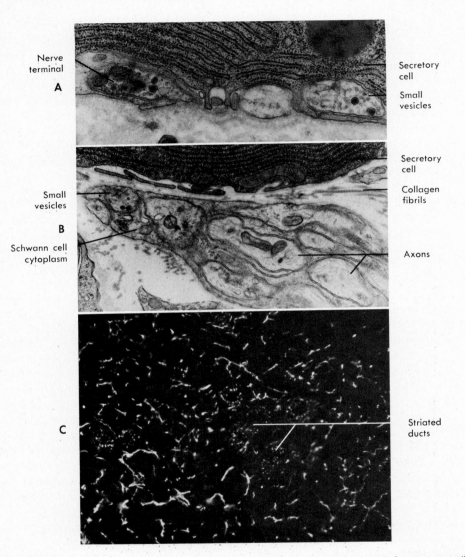

Fig. 10-15. A, Electron micrograph of three nerves of intraepithelial type at base of secretory cell. Axons are on epithelial side of basal lamina, in close contact with secretory cell. Nerve terminals contain mitochondria, small vesicles, a few larger dense-cored vesicles, and microtubules. Rat parotid gland. **B,** Electron micrograph of nerve bundle of subepithelial type in rat submandibular gland. Several axons are enclosed by a Schwann cell; innervation of secretory cell presumably occurs where axonal varicosities are bared of covering Schwann cell cytoplasm. **C,** Light micrograph of human parotid gland treated with formaldehyde vapor, which causes fluorescence of adrenergic nerves. Fluorescent structures seen here are nerve bundles, such as that in Fig. 10-15, *B,* located in connective tissue around parenchymal cells. The extensive nature of the sympathetic innervation of human parotid is evident. Fluorescence of striated ducts is caused by lysosomes. (**A,** ×20,800; **B,** ×15,200; **C,** courtesy J.R. Garrett, London.)

CLASSIFICATION AND STRUCTURE OF HUMAN SALIVARY GLANDS

The salivary glands have been classified in a variety of ways by different histologists; the two most commonly used groupings are based on (1) the size and location and (2) the histochemical nature of the secretory products. In this chapter the former classification will be used, although the latter is not without considerable merit. To a large extent, the nature of the secretion produced by a gland depends on its cellular makeup in terms of serous and mucous cells. However, all serous cells are not alike; they may differ considerably in the type and amount of enzymes and other proteins they produce and in the amount and nature of the carbohydrates attached to the secretory proteins. Mucous cells show a similar variability in the nature of their carbohydrate component. Furthermore, in salivary glands of some animals, the secretory cells may have a structure that cannot be readily classified as serous or mucous; histochemical characterization of their secretory products is useful for comparisons with other glands.

Major salivary glands

The largest of the glands are the three bilaterally paired major salivary glands (Plate 3). They are all located extraorally and their secretions reach the mouth by variably long ducts.

Parotid gland. The parotid gland is enclosed within a well-formed connective tissue capsule, with its superficial portion lying in front of the external ear and its deeper part filling the retromandibular fossa. The main excretory duct (Stensen's duct) opens into the oral cavity on the buccal mucosa opposite the maxillary second molar. The opening is usually marked by a small papilla.

The parotid gland is a pure serous gland (Fig. 10-16, *A*); all the acinar cells are similar in structure to the serous cells described earlier. In the infant, however, a few mucous secretory units may be found. Electron micro-scopic studies indicate that the serous granules may have a dense central core. The intercalated ducts of the parotid are long and branching (Fig. 10-11, *A*), and the pale-staining striated ducts are numerous and stand out conspicuously against the more densely stained acini. The connective tissue septa in the parotid contain numerous fat cells, which increase in number with age and leave an empty space in histologic sections.

Submandibular gland. The submandibular gland is also enveloped by a well-defined capsule; it is located in the submandibular triangle behind and below the free border of the mylohyoid muscle, with a small extension lying above the mylohyoid. The main excretory duct (Wharton's duct) opens at the *caruncula sublingualis*, a small papilla at the side of the lingual frenum on the floor of the mouth. Some isolated smooth muscle cells have been reported around the duct.

The submandibular gland is a mixed gland, with both serous and mucous secretory units (Figs. 10-7, *A* and 10-16, *B*). The serous units predominate, but the proportions may vary from one lobule to the next. The mucous terminal portions are capped by demilunes of serous cells. Although they appear similar by light microscopy, notable differences between submandibular and parotid serous cells are observed in the electron microscope (Fig. 10-17). The basal and lateral plasma membranes are thrown into numerous folds, interdigitating with similar processes from adjacent cells. The serous granules exhibit a variable substructure, from a granular matrix with a dense core or crescent, to an irregular skein of dense material dispersed in the matrix. The intercalated ducts tend to be somewhat shorter than those of the parotid, whereas the striated ducts are usually longer.

Sublingual gland. The sublingual gland lies between the floor of the mouth and the mylohyoid muscle; it is composed of one main gland and several smaller glands. The main duct

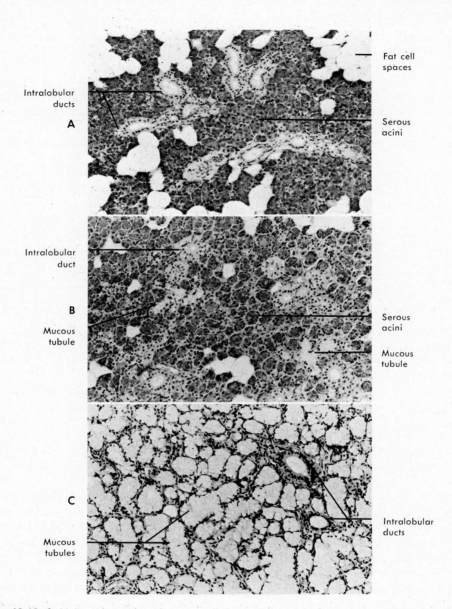

Fat cell spaces

Intralobular ducts

A

Serous acini

Intralobular duct

B

Serous acini

Mucous tubule

Mucous tubule

C

Intralobular ducts

Mucous tubules

Fig. 10-16. A, Light micrograph of human parotid gland, showing serous acini, several intralobular striated ducts, and numerous fat cell spaces. **B,** Light micrograph of human submandibular gland. Serous acini predominate, but a few mucous secretory units are present. Several intralobular striated ducts are cut in cross section. **C,** Light micrograph of human sublingual gland showing large mucous secretory units with typical tubular structure. Serous demilunes are difficult to distinguish at low magnification. Intralobular ducts are poorly developed. (**A** to **C,** ×90.)

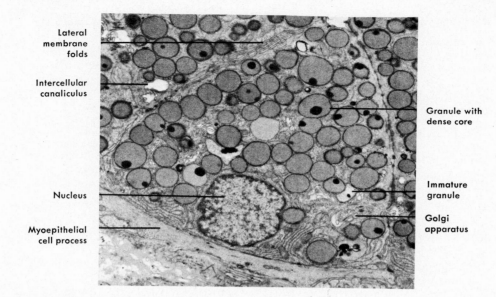

Lateral membrane folds

Intercellular canaliculus

Granule with dense core

Nucleus

Immature granule

Golgi apparatus

Myoepithelial cell process

Fig. 10-17. Electron micrograph of serous cell of human submandibular gland, showing secretory granules with dense core. Immature granules with similar cores are seen in Golgi regions. Several intercellular canaliculi are cut in cross section, and extensive folding of lateral cell membranes occurs between adjacent cells. Myoepithelial cell process is present at base of cell. (×6600.) (From Tandler, B., and Erlandson, R.A.: Am. J. Anat. **135:**419, 1972; reprinted by permission of the Wistar Institute Press.)

(Bartholin's duct) opens with or near the submandibular duct, and several smaller ducts open independently along the sublingual fold. The capsule is poorly developed, but the connective tissue septa are particularly prominent within the gland.

The sublingual is also a mixed gland, but the mucous secretory units greatly outnumber the serous units (Fig. 10-16, *C*). The mucous cells are usually arranged in a tubular pattern; serous demilunes may be present at the blind ends of the tubules. Pure serous acini are rare or absent. The intercalated and striated ducts are poorly developed; mucous tubules may open directly into ducts lined with cuboid or columnar cells without typical basal striations.

Minor salivary glands

The minor salivary glands are located beneath the epithelium in almost all parts of the oral cavity. These glands usually consist of several small groups of secretory units opening via short ducts directly into the mouth. They lack a distinct capsule, instead mixing with the connective tissue of the submucosa or muscle fibers of the tongue or cheek.

Labial and buccal glands. The glands of the lips and cheeks classically have been described as mixed, consisting of mucous tubules with serous demilunes. However, ultrastructural studies of the labial glands have revealed the presence of mucous cells only. Intercellular canaliculi have also been observed between the mucous cells. The intercalated ducts are variable in length, and the intralobular ducts possess only a few cells with basal striations. Although the buccal glands have not been examined by electron microscopy, they are usually described as a continuation of the labial glands with a similar structure.

Glossopalatine glands. The glossopalatine glands are pure mucous glands. They are principally localized to the region of the isthmus in the glossopalatine fold but may extend from the posterior extension of the sublingual gland to the glands of the soft palate.

Palatine glands. The palatine glands are also of the pure mucous variety. They consist of several hundred glandular aggregates in the lamina propria of the posterolateral region of the hard palate and in the submucosa of the soft palate and uvula. The excretory ducts may have an irregular contour with large distensions as they course through the lamina propria. The openings of the ducts on the palatal mucosa are often large and easily recognizable.

Lingual glands. The glands of the tongue can be divided into several groups. The anterior lingual glands (glands of Blandin and Nuhn) are located near the apex of the tongue. The anterior regions of the glands are chiefly mucous in

character, while the posterior portions are mixed. The ducts open on the ventral surface of the tongue near the lingual frenum. The posterior lingual mucous glands (Fig. 10-18) are located lateral and posterior to the vallate papillae and in association with the lingual tonsil. They are purely mucous in character, and their ducts open onto the dorsal surface of the tongue. The posterior lingual serous glands (von Ebner's glands) are an extensive group of purely serous glands located between the muscle fibers of the tongue, below the vallate papillae (Fig. 10-18). Their ducts open into the trough of the vallate papillae, and at the rudimentary foliate papillae on the sides of the tongue.

Of all of the minor salivary glands, the posterior lingual serous glands are among the most interesting. Classically, their secretions have been described as serving to wash out the trough of the papillae and ready the taste re-

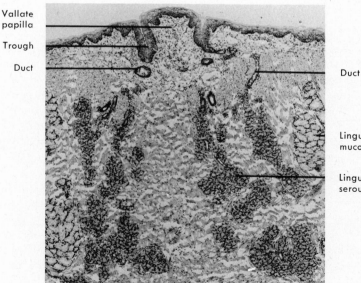

Fig. 10-18. Light micrograph of minor salivary glands of rat tongue. Lingual serous (von Ebner's) gland is located between muscle fibers of tongue below vallate papilla. Its ducts empty into trough around papilla. Posterior lingual mucous glands are located lateral to serous glands; their ducts open onto surface of tongue. (×36.) (From Hand, A.R.: J. Cell Biol. **44:**340, 1970; reprinted by permission of the Rockefeller University Press.)

ceptors (located in the epithelium of the trough) for a new stimulus. Although this may be a part of their function, recent studies suggest that these glands have significant protective and digestive functions. Histochemical studies have localized the antibacterial enzymes peroxidase and lysozyme to these glands in humans. Biochemical studies of the lingual serous glands have demonstrated the presence of a secretory enzyme with lipolytic activity; similar lipolytic activity has been detected in aspirates from the esophagus and stomach. This lingual lipase has an acid pH optimum, so that it is capable of hydrolyzing triglycerides in the stomach. The fatty acids, monoglycerides and diglycerides produced by lingual lipase, help to emulsify the remaining fat and increase the efficiency of pancreatic lipase in the intestine. In the newborn, when fat intake is high and levels of pancreatic lipase are low, lingual lipase probably plays a significant role in lipid digestion. Amylase activity has also been detected in the lingual serous glands of some species.

SPECIES VARIATION

From the preceding sections it is obvious that a number of differences exist between the individual salivary glands of humans. Numerous differences in the structure and biochemical makeup of the salivary glands also exist between various mammalian species. Although many of them are relatively minor, such as variations in the proportion of serous to mucous cells or the extent of innervation by each division of the autonomic nervous system, others are important because they represent variations in the structure and biochemistry of the parenchymal cells, they may be a reflection of the particular diet of the animal, or they occur in animals that are widely used for research purposes.

The parotid gland of ruminants is specialized for the production of large amounts of fluid, up to 60 liters per day. The structure of the secretory cells reflects this function; they have little RER and few secretory granules, but extensive

lateral membrane folds and numerous apical microvilli are present. The submandibular gland of rodents is one of the most interesting and widely studied of the salivary glands. In the rat and mouse the acinar cells are intermediate in structure and carbohydrate content between serous and mucous cells; they are usually termed "seromucous." The rodent submandibular gland, at the time of sexual maturity, also undergoes a specialization of the proximal portion of the intralobular striated ducts (adjacent to the intercalated ducts) to form a granular tubule segment. The cells of the granular tubule are large, with basally situated nuclei and remnants of basal infoldings and a large number of electron-dense granules of various size in the apical cytoplasm. The granular tubules are very sensitive to hormonal influences. They are generally smaller in the female than in the male, but administration of testosterone to females results in development of a malelike structure; conversely, castration of males results in female-like glands. Interrelationships between the pituitary, thyroid, and submandibular gland have also been shown. The male mouse shows the greatest development of the granular tubules, and several proteins with unique biologic activities are found in the cells of the granular tubules. Many of these proteins exhibit trypsinlike esteroprotease activity or occur as complexes with protein components that have protease activity; however, their effects on tissues and animals are quite different. Nerve growth factor (NGF) stimulates the growth of neurites from embryonic dorsal root ganglia and sympathetic ganglia in culture; injection of antibodies to NGF into mice destroys their sympathetic neurons. Epidermal growth factor (EGF) is a potent mitogen for a variety of cell types and when injected into newborn mice causes increased keratinization and premature opening of the eyelids and eruption of the incisors.

Other growth-stimulating properties attributed to extracts of the mouse submandibular gland include mesenchyme stimulating factor, muscle differentiating factor, endothelial growth—stimulating factor, and thymotropic factor. Significant quantities of a glucagon-like substance, capable of producing hyperglycemia after intravenous injection, have been demonstrated in the rodent submandibular gland. Synthesis of an insulin-like substance has also been shown to occur in rodent salivary glands. Renin and kallikrein, proteolytic enzymes that act on plasma proteins to liberate vasoactive peptides, have been localized to the granular tubule cells by immunocytochemical techniques; kallikrein has also been found in the striated duct cells of other salivary glands and in other species. Most of these substances are found in high concentration in mouse submandibular saliva, indicating that they are exocrine secretory products of the granular tubule cells. The presence of many of these biologically active substances in blood plasma suggests that these cells may also have an endocrine function. However, several other tissues also produce these substances, and studies of the effects of gland extirpation on plasma concentrations have been contradictory. Thus their significance in the salivary glands remains obscure.

DEVELOPMENT AND GROWTH

During fetal life each salivary gland is formed at a specific location in the oral cavity through the growth of a bud of oral epithelium into the underlying mesenchyme. The primordia of the parotid and submandibular glands of humans appear during the sixth week, whereas the primordium of the sublingual gland appears after 7 to 8 weeks of fetal life. The minor salivary glands begin their development during the third month. The epithelial bud grows into an extensively branched system of cords of cells that are first solid but gradually develop a lumen and become ducts. The secretory portions develop later than the duct system and form by repeated branching and budding of the finer cell cords and ducts.

Studies of embryonic salivary glands in vitro have provided considerable information on the mechanism of glandular morphogenesis. The mesenchyme into which the glandular rudiment grows produces a factor or factors that stimulate the growth of the gland. If the mesenchyme and epithelium are separated and cultured on opposite sides of a filter, the growth of the epithelium proceeds normally; in the absence of the mesenchyme, the epithelium fails to grow. In the mouse the submandibular gland exhibits a specific requirement for submandibular mesenchyme; however, in the rat, parotid mesenchyme, and to a lesser extent lung mesenchyme, can support morphogenesis of the submandibular epithelium. The rat parotid and sublingual rudiments appear to be somewhat less specific in their mesenchymal requirements. The process of branching morphogenesis, that is, the formation of hollow, tubular glands from an initially flat epithelial surface, appears to be related to the presence of microfilaments in the epithelial cells. Microfilaments, about 5 to 7 nm (50 to 70 Å) thick, form a network beneath the cell membrane of almost all cells; they consist of the contractile protein actin. In developing salivary epithelium they are particularly prominent at the apical and basal ends of the cells; differential contraction could cause a group of cells to pucker outward, or clefts to form in a solid cord or sheet of cells, similar to the effect of pulling a purse string. Addition of the drug cytochalasin B, which disrupts the structure and function of microfilaments, to salivary gland rudiments growing in vitro prevents branching and cleft formation and causes newly formed clefts to disappear. Older clefts are unaffected, probably because they have been stabilized by the presence of mesenchymal cells and extracellular materials.

The presence of a functional innervation is also essential to proper growth and maintenance of salivary gland structure. Parasympathetic denervation of adult animals results in a

30% loss in glandular weight within 2 to 3 weeks. Sympathetic denervation causes variable responses, from atrophy of some glands to hypertrophy of others. Parasympathectomy of the developing rat parotid prevents attainment of adult gland size, cell number and size, and DNA and RNA content; sympathectomy has a moderate effect on cell and gland size only. Normal physiologic activity is also important for the proper growth of developing glands, as well as maintenance of adult structure and enzyme content. Feeding of a liquid diet to rats greatly diminishes the reflexly mediated secretory activity; the parotid rapidly decreases in weight and amylase content, and the normal diurnal pattern of synthesis and secretion is eliminated.

Conversely, chronically increased stimulation can cause an increase in glandular size. For example, increasing the bulk content of the food, which necessitates increased masticatory activity, results in hypertrophy of the rat parotid. Repeated amputation of the incisors, apparently acting reflexly through the superior cervical ganglion, also causes enlargement of the salivary glands. Treatment of mice and rats with isoproterenol, a β-adrenergic drug, causes several interesting changes in the salivary glands. A single injection results in the rapid and complete discharge of the stored secretory products and stimulation of protein synthesis; 20 to 30 hours after injection an increase in DNA synthesis and a wave of mitoses occur. Daily injections of isoproterenol cause cellular hypertrophy and hyperplasia, resulting in glandular enlargement up to five times that of untreated animals. The synthesis of certain proteins is enhanced, while that of others is reduced; additionally, several new proteins are synthesized by the enlarged glands. The effects on the salivary glands of isoproterenol and related adrenergic drugs have found wide application in experimental studies of cellular secretion, protein and nucleic acid synthesis, and regulation of gene expression.

CONTROL OF SECRETION

The physiologic control of salivary gland secretion is mediated through the activity of the autonomic nervous system. The release of neurotransmitters from the vesicles in the nerve terminals adjacent to the parenchymal cells stimulates them to discharge their secretory granules and secrete water and electrolytes. The molecular events that occur during this process, called stimulus-secretion coupling, have been extensively studied in the last few years. The parotid gland of the rat has served as an important experimental tissue in these studies.

The neurotransmitters interact with specific receptors located on the plasma membrane of the acinar cell. Norepinephrine, the sympathetic transmitter, interacts with both α- and β-adrenergic receptors, while acetylcholine interacts with the cholinergic receptor. Protein secretion is mediated primarily through the β-adrenergic receptor; stimulation of the α-adrenergic and cholinergic receptors also causes low levels of protein secretion, but these two receptors appear to be mainly involved in the secretion of water and electrolytes. Receptors for the peptide transmitter substance P are also present on salivary gland cells; substance P stimulates secretion similar to that caused by α-adrenergic and cholinergic agonists. Vasoactive intestinal polypeptide (VIP) is present in nerve endings in the salivary glands and has been shown to induce secretion by some glands.

Receptor stimulation results in increases in the intracellular concentration of "second messengers," which trigger additional events leading to the cellular response. In the case of α-adrenergic, cholinergic, and substance P receptors, the membrane permeability to Ca^{++} is increased and a marked influx of Ca^{++} into the cell occurs. Recent experiments have linked the activation of these receptors to rapid changes in membrane phospholipid metabolism and release of Ca^{++} from intracellular stores such as the endoplasmic reticulum or the plasma membrane. The increased cytoplasmic Ca^{++} concentration causes K^+ efflux, water and electrolyte secretion, and a low level of exocytosis. Stimulation of the β-adrenergic receptor activates the plasma membrane enzyme adenylate cyclase, which catalyzes the formation of $3',5'$-cyclic adenosine monophosphate (cyclic AMP) from adenosine triphosphate. The increased intracellular concentration of cyclic AMP activates cyclic AMP–dependent protein kinase, an enzyme that phosphorylates other proteins, which in turn may be involved in the process of exocytosis. Cyclic AMP may also stimulate release of Ca^{++} from intracellular stores, thereby increasing its cytoplasmic concentration. Thus Ca^{++} may be the common intracellular mediator for all of the receptors; the different cellular responses may reflect the different sources of Ca^{++} or differing local concentration or both. Ca^{++} may have additional effects, including stimulation of guanylate cyclase activity and an increase in the concentration of $3',5'$-cyclic guanosine monophosphate (cyclic GMP). However, the role of cyclic GMP in the secretory process has not yet been determined.

Adjacent secretory cells are joined to one another by specialized intercellular junctions called *gap junctions*. These junctions are permeable to ions and small molecules; thus changes in the intracellular concentration of these substances in one cell are reflected by parallel changes in the adjacent cells. Therefore physiologic stimulation probably results in a response by secretory units (acini) rather than individual cells.

The discharge of secretory granules involves translocation to the luminal cell surface and fusion of the granule membrane with the plasma membrane. Considerable evidence suggests that microtubules and microfilaments may act as a cytoskeletal framework or contractile mechanism for granule movement. Colchicine and vinblastine, drugs that disrupt microtubules, and cytochalasin B inhibit amylase release from rat parotid gland in vitro; similar effects are observed in several other tissues. However, contradictory results have been reported for some secretory cells, and definite associations of microtubules and microfilaments with secretory granules have been observed only rarely. Conclusions regarding the involvement of these organelles in salivary secretion cannot be made at present. The molecular mechanisms responsible for granule–plasma membrane fusion are poorly understood. Evidence for the presence of a Ca^{++}-dependent protein called synexin, which is involved in the initial fusion of granules with the plasma membrane, has been obtained for certain secretory cells. Other studies have indicated structural and compositional differences between the luminal membrane and the remainder of the plasma membrane, which may account for the specificity of the exocytotic process, as well as structural modifications of the membrane that may precede or occur concomitantly with exocytosis. These and other processes involved in exocrine secretion are currently under active investigation in a number of laboratories.

SALIVA: COMPOSITION AND FUNCTIONS

The most important function of the salivary glands is the production and secretion of saliva. It is important to make a distinction between pure glandular secretions, collected by special devices from the ducts, and whole saliva obtained from the mouth, usually by expectoration. In addition to the components contributed by the glands, whole saliva contains desquamated oral epithelial cells, leukocytes, microorganisms and their products, fluid from the gingival sulcus, and food remnants. The total volume of saliva secreted daily by humans is approximately 750 ml, of which about 60% is produced by the submandibular glands, 30% by the parotids, 5% or less from the sublinguals, and about 7% from the minor salivary glands. These proportions may change considerably with stimulation of various intensities, however. Water accounts for 99% or more of the saliva; inorganic ions, secretory proteins and glycoproteins, certain serum constituents, and other substances make up the remaining 1% or less. The major inorganic ions of saliva are Na^+, K^+, Cl^-, and HCO_3^-; the levels of these ions are variable, depending on the type of stimulation and rate of salivary flow. Other ions found in smaller amounts include Ca^{++}, Mg^{++}, HPO_4^{--}, I^-, SCN^-, and F^-. The pH of whole saliva varies from 6.7 to about 7.4, whereas parotid saliva may vary over a greater range, from pH 6.0 to 7.8. The primary buffering system of saliva is formed by HCO_3^-, but certain salivary proteins may also provide some buffer capacity.

Secretory proteins represent the main category of organic substances in the saliva. These include various enzymes, large carbohydrate-rich glycoproteins or mucins, antibacterial substances, and a group of proteins involved principally in enamel pellicle formation and calcium phosphate homeostasis in the saliva. Certain serum constituents such as albumin, blood clotting factors, β_2-microglobulin, and immunoglobulins are also found in saliva. Other organic molecules present in saliva include cyclic AMP and cyclic AMP–binding proteins, amino acids, urea, uric acid, various lipids, and corticosteroids.

Saliva participates in digestion by providing a fluid environment for solubilization of food and taste substances and through the action of its digestive enzymes, principally amylase. Several isoenzymes of amylase have been identified in humans; two of these, representing 25% to 30% of the total amylase protein, have small amounts of bound carbohydrate. The action of amylase on ingested carbohydrates to produce glucose and maltose begins in the mouth and may continue for up to 30 minutes in the stomach before the amylase is inactivated by the acid pH and proteolysis. Lingual lipase, produced by the lingual serous glands, initiates the digestion of dietary lipids, hydrolyzing triglycerides to monoglycerides and diglycerides and fatty acids. Other hydrolytic enzymes have been detected in saliva, but their significance in food digestion has not been established.

Saliva has several protective functions. It keeps the oral tissues moist and facilitates swallowing and speaking. The mucous glycoproteins, which may have up to 800 oligosaccharide groups attached to the protein core, provide lubrication for the movement of the oral tissues against each other, as well as protection from chemical and thermal insults. Saliva also helps to protect the teeth from dental caries by means of both the cleansing and buffering action of saliva and the control of calcium and phosphate concentrations in the saliva and around the teeth. A large group of salivary proteins, called proline-rich proteins because of their high content of the amino acid proline, and statherin, a small tyrosine-rich protein, inhibit the precipitation of calcium phosphate from the saliva. Along with other salivary glycoproteins, statherin and certain of the proline-rich proteins bind to the tooth surface, forming the acquired enamel pellicle. The resulting localized supersaturation of calcium and phosphate reduces dissolution and promotes remineralization of the tooth enamel.

Several substances that are capable of inhibiting the growth of microorganisms and possibly preventing infection are found in saliva. Certain of the high-molecular-weight salivary glycoproteins aggregate specific strains of oral microorganisms or prevent their adherence to oral tissues, facilitating clearance from the mouth by swallowing. The secretion of peroxidase by the acinar cells and thiocyanate by the duct system establishes a bactericidal system in saliva. In the presence of hydrogen peroxide, peroxidase catalyzes the formation of hypothiocyanite ($OSCN^-$), which is inhibitory to bacteria. Another antibacterial protein present in saliva is lysozyme, an enzyme that hydrolyzes the polysaccharide of bacterial cell walls, resulting in cell lysis. An important group of defensive substances in saliva are the immunoglobulins. The predominant salivary immunoglobulin is

IgA. Salivary or secretory IgA differs from serum IgA in that it is produced locally by plasma cells in the connective tissue stroma of the glands and consists of a dimer of two IgA molecules and a protein called J chain. There is an additional glycoprotein produced by the parenchymal cells, called secretory component, that is also a part of the secretory IgA molecule. Secretory component, acting as a specific receptor in the parenchymal cell membrane for dimeric IgA, facilitates the transfer of the IgA to the lumen, either by translation in the cell membrane or by endocytosis and secretion along with the secretory products of the parenchymal cells. Secretory component may also increase the resistance of the IgA molecule to denaturation or proteolysis in the oral cavity. Small amounts of IgG and IgM have also been detected in saliva, and occasional plasma cells in the glandular stroma can be stained by fluorescent antibodies specific for these immunoglobulins. Serum immunoglobulins may also enter the saliva through the gingival crevice. Salivary immunoglobulins may act primarily through their ability to inhibit the adherence of microorganisms to oral tissues. Another antibacterial substance found in saliva is lactoferrin, an iron-binding protein. In the presence of specific antibody, lactoferrin that is not saturated with iron enhances the inhibitory effect of the antibody on the microorganisms.

The salivary glands of animals other than humans have additional specialized functions, such as thermoregulation in mammals lacking sweat glands. In some reptiles and amphibians the homologous venom glands produce a variety of toxic substances. The salivary glands, as are many other tissues, are affected by secretions of the endocrine glands. The pronounced sexual dimorphism of the rodent submandibular gland has already been discussed. Thyroid and pituitary hormones have also been implicated in structural and functional changes of the salivary glands. The sodium and potassium content of saliva can be influenced by the administration of adrenocorticotropic hormone or mineralocorticoids, and alterations of salivary $Na^+:K^+$ ratios are observed in patients with Addison's disease and Cushing's syndrome. The possibility that the salivary glands may have an endocrine function in addition to their role in saliva production was discussed in relation to the variety of biologically active substances found in the rodent submandibular gland. There is some evidence, although not yet thoroughly accepted, that the human parotid gland produces a hormone called parotin. Parotin is said to promote the growth of mesenchymal tissues; it also lowers serum calcium levels in rabbits, stimulates calcification of rat incisor dentin, and increases bone marrow temperature with an accompanying increase in circulating leukocytes.

CLINICAL CONSIDERATIONS

An understanding of the anatomy, histology, and physiology of the salivary glands is essential for good dental practice. There is hardly any aspect of clinical practice in which salivary glands and saliva do not play an obvious or hidden role.

With the exception of a portion of the anterior part of the hard palate, salivary glands are seen everywhere in the oral cavity. They may, by developmental coincidence, even be included within the jaws. In the mandible this occurs in an area just posterior to the third molar teeth. In the maxilla salivary glands may be present in the nasopalatine canal. Because of these features, lesions of salivary glands, including tumors, can occur almost anywhere within the mouth. In a differential diagnosis of oral lesions therefore a salivary gland origin must always be kept in mind.

The salivary glands are subject to a number of pathologic conditions. These include inflammatory diseases, such as viral, bacterial, or allergic sialadenitis, a variety of benign and malignant tumors, autoimmune diseases, such as Sjögren's syndrome, and genetic diseases, such as cystic fibrosis. One of the most common surface lesions of the oral mucosa is a vesicular elevation called mucocele. This is produced from the severance of the duct of a minor salivary gland and pooling of the saliva in the tissues. A blockage of a salivary gland duct may occur after formation of a mucous or calcified plug within the duct. If this occurs in a minor salivary gland, it usually causes no symptoms, but in major glands such obstruction can be very painful and may require surgical treatment.

The salivary glands may also be affected by a variety of systemic and metabolic diseases. The major glands, especially the parotid, may become enlarged during starvation, protein deficiency, alcoholism, pregnancy, diabetes mellitus, and liver disease. The association of the major salivary glands with the cervical lymph nodes, brought about by a common area of development, necessitates the differentiation of pathologic conditions of these lymph nodes from salivary gland diseases.

Alteration of salivary gland function during disease states may have profound influences on the oral tissues. The quality and quantity of saliva has a relationship to the incidence of dental caries. In conditions associated with the reduction or absence of salivary flow (xerostomia) the incidence of decay increases. Reduction in salivary flow is most often seen in patients with sicca syndrome or Sjögren's syndrome, after irradiation of the head and neck region, or because of the use of various therapeutic pharmacologic agents. Inflammation and ulceration of the oral mucosa and frequent oral infections may also occur when salivary flow is reduced. Age changes in the salivary glands, particularly prominent in the parotid, consist of a gradual replacement of parenchyma with fatty tissue. Since the parotid is the major source of serous saliva, with advancing age, patients often complain of dryness and an increase in the viscosity of saliva. Recent studies have shown that in the aged the flow of saliva is reduced during resting conditions, but in quantity and composition stimulated saliva in healthy, aged individuals is similar to that of young adults.

Determination of the quantity and composition of the saliva, sialochemistry, is often of value in the diagnosis of glandular or systemic disease. Experimentally, sialochemistry has also been used to determine ovulation time. Saliva is frequently used to monitor plasma concentrations of certain drugs that exhibit consistent saliva: plasma ratios.

REFERENCES

Amsterdam, A., Ohad, I., and Schramm, M.: Dynamic changes in the ultrastructure of the acinar cell of the rat parotid gland during the secretory cycle, J. Cell Biol. 41:753, 1969.

Archer, F.L., and Kao, V.C.Y.: Immunohistochemical identification of actomyosin in myoepithelium of human tissues, Lab. Invest. 18:669, 1968.

Aub, D.L., McKinney, J.S., and Putney, J.W., Jr.: Nature of the receptor-regulated calcium pool in the rat parotid gland, J. Physiol. 331:557, 1982.

Ball, W.D.: Development of the rat salivary glands. III. Mesenchymal specificity in the morphogenesis of the embryonic submaxillary and sublingual glands of the rat, J. Exp. Zool. 188:277, 1974.

Barka, T.: Biologically active polypeptides in submandibular glands, J. Histochem. Cytochem. 28:836, 1980.

Batzri, S., Selinger, Z., Schramm, M., and Robinovitch, M.R.: Potassium release mediated by the epinephrine α-receptor in rat parotid slices. Properties and relation to enzyme secretion, J. Biol. Chem. 248:361, 1973.

Bdolah, A., and Schramm, M.: The function of 3'5'-cyclic AMP in enzyme secretion, Biochem. Biophys. Res. Commun. 18:452, 1965.

Bennick, A.: Salivary proline-rich proteins, Mol. Cell. Biochem. 45:83, 1982.

Bhaskar, S.N.: Synopsis of oral pathology, ed. 5, St. Louis, 1977, The C.V. Mosby Co.

Bhaskar, S.N.: Radiographic interpretation for the dentist, ed. 3, St. Louis, 1979, The C.V. Mosby Co.

Bienenstock, J., Tourville, D., and Tomasi, T.B., Jr.: The secretion of immunoglobulins by the human salivary glands. In Botelho, S.Y., Brooks, F.P., and Shelley, W.B., editors: The exocrine glands, Philadelphia, 1969, University of Pennsylvania Press.

Blobel, G.: Synthesis and segregation of secretory proteins: the signal hypothesis. In Brinkley, B.R., and Porter, K.R., editors: International cell biology, 1976-1977, New York, 1977, Rockefeller University Press.

Brandtzaeg, P.: Mucosal and glandular distribution of immunoglobulin components: differential localization of free and bound SC in secretory epithelial cells, J. Immunol. 112:1553, 1974.

Bullen, J.J., Rogers, H.J., and Griffiths, E.: Iron binding proteins and infection, Br. J. Haematol. 23:389, 1972.

Bundgaard, M., Møller, M., and Poulsen, J.H.: Localization of sodium pump sites in cat salivary glands, J. Physiol. **273:**339, 1977.

Case, R.M.: Synthesis, intracellular transport and discharge of exportable proteins in the pancreatic acinar cell and other cells, Biol. Rev. **53:**211, 1978.

Castle, J.D., Jamieson, J.D., and Palade, G.E.: Radioautographic analysis of the secretory process in the parotid acinar cell of the rabbit, J. Cell Biol. **53:**290, 1972.

Clamp, J.R., Allen, A., Gibbons, R.A., and Roberts, G.P.: Chemical aspects of mucus, Br. Med. Bull. **34:**25, 1978.

Code, C.F., editor: Handbook of Physiology, section 6, vol. 2, Washington, D.C., 1967, American Physiological Society.

Creutz, C.E., Pazoles, C.J., and Pollard, H.B.: Identification and purification of an adrenal medullary protein (synexin) that causes calcium-dependent aggregation of isolated chromaffin granules, J. Biol. Chem. **253:**2858, 1978.

De Camilli, P., Peluchetti, D., and Meldolesi, J.: Dynamic changes of the luminal plasmalemma in stimulated parotid acinar cells. A freeze-fracture study, J. Cell Biol. **70:**59, 1976.

Drenckhahn, D., Gröschel-Stewart, U., and Unsicker, K.: Immunofluorescence-microscopic demonstration of myosin and actin in salivary glands and exocrine pancreas of the rat, Cell Tissue Res. **183:**273, 1977.

Ekfors, T.O., and Hopsu-Havu, V.K.: Immunofluorescent localization of trypsin-like esteropeptidases in the mouse submandibular gland, Histochem. J. **3:**415, 1971.

Farquhar, M.G., and Palade, G.E.: The Golgi apparatus (complex)—(1954-1981)—from artifact to center stage, J. Cell Biol. **91:**77s, 1981.

Garrett, J.R.: The innervation of normal human submandibular and parotid salivary glands. Demonstrated by cholinesterase histochemistry, catecholamine fluorescence and electron microscopy, Arch. Oral Biol. **12:**1417, 1967.

Garrett, J.R.: Neuro-effector sites in salivary glands. In Emmelin, N., and Zotterman, Y., editors: Oral physiology, Oxford, England, 1972, Pergamon Press.

Gill, G.: Metabolic and endocrine influences on the salivary glands, Otolaryngol. Clin. North Am. **10:**363, 1977.

Gresik, E., Michelakis, A., Barka, T., and Ross, T.: Immunocytochemical localization of renin in the submandibular gland of the mouse, J. Histochem. Cytochem. **26:**855, 1978.

Grobstein, C.: Epithelio-mesenchymal specificity in the morphogenesis of mouse submandibular rudiments in vitro, J. Exp. Zool. **124:**383, 1953.

Hall, H.D., and Schneyer, C.A.: Salivary gland atrophy in rat induced by liquid diet, Proc. Soc. Exp. Biol. Med. **117:**789, 1964.

Hammer, M.G., and Sheridan, J.D.: Electrical coupling and dye transfer between acinar cells in rat salivary glands, J. Physiol. **275:**495, 1978.

Hamosh, M.: The role of lingual lipase in neonatal fat digestion. In Harries, J.T., editor: Pre- and post-natal development of mammalian absorptive processes. Ciba Found. Symp. **70:**69, 1979.

Hamosh, M., and Scow, R.O.: Lingual lipase and its role in the digestion of dietary fat, J. Clin. Invest. **52:**88, 1973.

Hand, A.R.: The fine structure of von Ebner's gland of the rat, J. Cell Biol. **44:**340, 1970.

Hand, A.R.: Morphology and cytochemistry of the Golgi apparatus of rat salivary gland acinar cells, Am. J. Anat. **130:**141, 1971.

Hand, A.R.: Synthesis of secretory and plasma membrane glycoproteins by striated duct cells of rat salivary glands as visualized by radioautography after ^{3}H-fucose injection, Anat. Rec. **195:**317, 1979.

Hand, A.R., and Oliver, C.: Cytochemical studies of GERL and its role in secretory granule formation in exocrine cells, Histochem. J. **9:**375, 1977.

Hand, A.R., and Oliver, C., editors: Basic mechanisms of cellular secretion: Methods in cell biology, vol. 23, New York, 1981, Academic Press, Inc.

Ito, Y.: Parotin: a salivary gland hormone, Ann. N.Y. Acad. Sci. **85:**228, 1960.

Jamieson, J.D., and Palade, G.E.: Intracellular transport of secretory proteins in the pancreatic exocrine cell. I. Role of the peripheral elements of the Golgi complex, J. Cell Biol. **34:**577, 1967.

Jamieson, J.D., and Palade, G.E.: Intracellular transport of secretory proteins in the pancreatic exocrine cell. II. Transport to condensing vacuoles and zymogen granules, J. Cell Biol. **34:**597, 1967.

Jamieson, J.D., and Palade, G.E.: Production of secretory proteins in animal cells. In Brinkley, B.R., and Porter, K.R., editors: International cell biology, 1976-1977, New York, 1977, Rockefeller University Press.

Johnson, D.A., and Sreebny, L.M.: Effect of food consistency and starvation on the diurnal cycle of the rat parotid gland, Arch. Oral Biol. **16:**177, 1971.

Johnson, D.A., and Sreebny, L.M.: Effect of increased mastication on the secretory process of the rat parotid gland, Arch. Oral Biol. **18:**1555, 1973.

Kauffman, D.L., Zager, N.I., Cohen, E., and Keller, P.J.: The isoenzymes of human parotid amylase, Arch. Biochem. Biophys. **137:**325, 1970.

Kim, S.K., Nasjleti, C.E., and Han, S.S.: The secretion processes in mucous and serous secretory cells of the rat's sublingual gland, J. Ultrastruct. Res. **38:**371, 1972.

Klebanoff, S.J., and Luebke, R.G.: The antilactobacillus system of saliva. Role of salivary peroxidase, Proc. Soc. Exp. Biol. Med. **118:**483, 1965.

Kleinberg, I., Ellison, S.A., and Mandel, I.D., editors: Saliva and dental caries, (special suppl.) Microbiology Abstracts, New York, 1979, Information Retrieval Inc.

Korsrud, F.R., and Brandtzaeg, P.: Characterization of epithelial elements in human major salivary glands by functional markers: Localization of amylase, lactoferrin, lysozyme, secretory component, and secretory immunoglobulins by paired immunofluorescence staining, J. Histochem. Cytochem. **30**:657, 1982.

Lawrence, A.M., Tan, S., Hojvat, S., and Kirsteins, L.: Salivary gland hyperglycemic factor: an extrapancreatic source of glucagon-like material, Science **195**:70, 1977.

Lawson, D., Raff, M.C., Gomperts, B., et al.: Molecular events during membrane fusion. A study of exocytosis in rat peritoneal mast cells, J. Cell Biol. **72**:242, 1977.

Lawson, K.A.: The role of mesenchyme in the morphogenesis and functional differentiation of rat salivary epithelium, J. Embryol. Exp. Morphol. **27**:497, 1972.

Leblond, C.P., and Bennett, G.: Role of the Golgi apparatus in terminal glycosylation. In Brinkley, B.R., and Porter, K.R., editors: International cell biology, 1976-1977, New York, 1977, Rockefeller University Press.

Leslie, B.A., Putney, J.W., Jr., and Sherman J.M.: α-Adrenergic, β-adrenergic and cholinergic mechanisms for amylase secretion by rat parotid gland *in vitro*, J. Physiol. **260**:351, 1976.

Levi-Montalcini, R., and Angeletti, P.U.: Nerve growth factor, Physiol. Rev. **48**:534, 1968.

Liang, T., and Cascieri, M.A.: Substance P receptor on parotid cell membranes, J. Neurosci. **1**:1133, 1981.

Mandel, I.D.: Human submaxillary, sublingual, and parotid glycoproteins and enamel pellicle. In Horowitz, M.I., and Pigman, W., editors: The glycoconjugates. Vol. 1. Mammalian glycoproteins and glycolipids, New York, 1977, Academic Press Inc.

Mandel, I.D.: Sialochemistry in diseases and clinical situations affecting salivary glands, CRC Crit. Rev. Clin. Lab. Sci. **12**:321, 1980.

Mason, D.K., and Chisholm, D.M.: Salivary glands in health and disease, London, 1975, W.B. Saunders Co.

Masson, P.L., Heremans, J.L., and Dive, C.: An iron-binding protein common to many external secretions, Clin. Chim. Acta **14**:735, 1966.

Mayo, J.W., and Carlson, D.M.: Protein composition of human submandibular secretions, Arch. Biochem. Biophys. **161**:134, 1974.

Mayo, J.W., and Carlson, D.M.: Isolation and properties of four α-amylase isozymes from human submandibular saliva, Arch. Biochem. Biophys. **163**:498, 1974.

Mazariegos, M.R., and Hand, A.R.: Regulation of tight junctional permeability in the rat parotid gland by autonomic agonists, J. Dent. Res. **63**:1102, 1984.

Mednieks, M.I., and Hand, A.R.: Cyclic AMP-dependent protein kinase in stimulated rat parotid gland cells: compartmental shifts after in vitro treatment with isoproterenol, Eur. J. Cell. Biol. **28**:264, 1982.

Mednieks, M.I., and Hand, A.R.: Cyclic AMP binding proteins in saliva, Experientia. **40**:945, 1984.

Mestecky, J., and Lawton, A.R., editors: The immunoglobulin A system, New York, 1974, Plenum Press.

Murakami, K., Tanaguchi, H., and Baba, S.: Presence of insulin-like immunoreactivity and its biosynthesis in rat and human parotid gland, Diabetologia **22**:358, 1982.

Murphy, R.A., Saide, J.D., Blanchard, M.H., and Young, M.: Molecular properties of the nerve growth factor secreted in mouse saliva, Proc. Natl. Acad. Sci. U.S.A. **74**:2672, 1977.

Myant, N.B.: Iodine metabolism of salivary glands, Ann. N.Y. Acad. Sci. **85**:208, 1960.

Nakamura, T., Nagura, H., Watanabe, K., Komatsu, N., Uchikoshi, S., Kobayashi, K., and Nabeshima, J.: Immunocytochemical localization of secretory immunoglobulins in human parotid and submandibular glands, J. Electron Microsc. **31**:151, 1982.

Neutra, M., and Leblond, C.P.: Synthesis of the carbohydrate of mucus in the Golgi complex shown by electron microscope radioautography of goblet cells from rats injected with glucose-H^3, J. Cell. Biol. **30**:119, 1966.

Nustad, K., Ørstavik, T.B., Gautvik, K.M., and Pierce, J.V.: Glandular kallikreins, Gen. Pharmacol. **9**:1, 1978.

Oliver, C., and Hand, A.R.: Uptake and fate of luminally administered horseradish peroxidase in resting and isoproterenol stimulated rat parotid acinar cells, J. Cell Biol. **76**:207, 1978.

Ørstavik, T.B., Brandtzaeg, P., Nustad, K., and Halvorsen, K.M.: Cellular localization of kallikreins in rat submandibular and sublingual salivary glands, Acta Histochem. **54**:183, 1975.

Palade, G.: Intracellular aspects of the process of protein secretion, Science **189**:347, 1975.

Parks, H.F.: On the fine structure of the parotid gland of mouse and rat, Am. J. Anat. **108**:303, 1961.

Petersen, O.H.: The electrophysiology of gland cells. London, 1980, Academic Press, Inc.

Poggioli, J., and Putney, J.W., Jr.: Net calcium fluxes in rat parotid acinar cells. Evidence for a hormone-sensitive calcium pool in or near the plasma membrane, Pflügers Arch. **392**:239, 1982.

Putney, J.W., Jr.: Inositol lipids and cell stimulation in mammalian salivary gland, Cell Calcium **3**:369, 1982.

Putney, J.W., Jr., Weiss, S.J., Leslie, B.A., and Marier, S.H.: Is calcium the final mediator of exocytosis in the rat parotid gland? J. Pharmacol. Exp. Ther. **203**:144, 1977.

Rasmussen, H.: Cell communication, calcium ion, and cyclic adenosine monophosphate, Science **170**:404, 1970.

Riva, A., and Riva-Testa, F.: Fine structure of acinar cells of human parotid gland, Anat. Rec. **176**:149, 1973.

Riva, A., Testa-Riva, F., Del Fiacco, M., and Lantini, M.S.: Fine structure and cytochemistry of the intralobular ducts of the human parotid gland, J. Anat. **122**:627, 1976.

Schneyer, C.A., and Hall, H.D.: Autonomic regulation of postnatal changes in cell number and size of rat parotid gland, Am. J. Physiol. **219**:1268, 1970.

Schneyer, L.H., Young, J.A., and Schneyer, C.A.: Salivary secretion of electrolytes, Physiol. Rev. **52**:720, 1972.

Schramm, M., and Selinger, Z.: The functions of cyclic AMP and calcium as alternative second messengers in parotid gland and pancreas, J. Cyclic Nucleotide Res. **1**:181, 1975.

Scott, B.L., and Pease, D.C.: Electron microscopy of the salivary and lacrimal glands of the rat, Am. J. Anat. **104**:115, 1959.

Selye, H., Veilleux, R., and Cantin, M.: Excessive stimulation of salivary gland growth by isoproterenol, Science **133**:44, 1961.

Shackleford, J.M., and Klapper, C.E.: Structure and carbohydrate histochemistry of mammalian salivary glands, Am. J. Anat. **111**:25, 1962.

Simson, J.A.V., Hazen, D., Spicer, S.S., et al.: Secretagogue-mediated discharge of nerve growth factor from granular tubules of male mouse submandibular glands: an immunocytochemical study, Anat. Rec. **192**:375, 1978.

Smith, P.H., and Patel, D.G.: Immunochemical studies of the insulin-like material in the parotid gland of rats, Diabetes **33**:661, 1984.

Spooner, B.S., and Wessells, N.K.: An analysis of salivary gland morphogenesis: role of cytoplasmic microfilaments and microtubules, Dev. Biol. **27**:38, 1972.

Sreebny, L.M., Johnson, D.A., and Robinovitch, M.R.: Functional regulation of protein synthesis in the rat parotid gland, J. Biol. Chem. **246**:3879, 1971.

Suddick, R.P., and Dowd, F.J.: The microvascular architecture of the rat submaxillary gland: possible relationship to secretory mechanisms, Arch. Oral Biol. **14**:567, 1969.

Tabak, L.A., Levine, M.J., Mandel, I.D., and Ellison, S.A.: Role of salivary mucins in the protection of the oral cavity, J. Oral Pathol. **11**:1, 1982.

Tamarin, A., and Sreebny, L.M.: The rat submaxillary salivary gland. A correlative study by light and electron microscopy, J. Morphol. **117**:295, 1965.

Tandler, B.: Ultrastructure of the human submaxillary gland. I. Architecture and histological relationships of the secretory cells, Am. J. Anat. **111**:287, 1962.

Tandler, B.: Ultrastructure of the human submaxillary gland. III. Myoepithelium, Z. Zellforsch. **68**:852, 1965.

Tandler, B., Denning, C.R., Mandel, I.D., and Kutscher, A.H.: Ultrastructure of human labial salivary glands. I. Acinar secretory cells, J. Morphol. **127**:383, 1969.

Tandler, B., and Erlandson, R.A.: Ultrastructure of the human submaxillary gland. IV. Serous granules, Am. J. Anat. **135**:419, 1972.

Taubman, M.A., and Smith, D.J.: Secretory immunoglobulins and dental disease. In Han, S.S., Sreebny, L., and Suddick, R., editors: Symposium on the mechanism of exocrine secretion, Ann Arbor, 1973, University of Michigan Press.

Taylor, T., and Erlandsen, S.L.: Peroxidase localization in von Ebner's gland of man, J. Dent. Res. **52**:635, 1973.

Testa-Riva, F.: Ultrastructure of human submandibular gland, J. Submicrosc. Cytol. **9**:251, 1977.

Testa-Riva, F., Puxeddu, P., Riva, A., and Diaz, G.: The epithelium of the excretory duct of the human submandibular gland: a transmission and scanning electron microscope study, Am. J. Anat. **160**:381, 1981.

Tomasi, T.B., Jr., Tan, E.M., Solomon, A., and Prendergast, R.A.: Characteristics of an immune system common to certain external secretions, J. Exp. Med. **121**:101, 1965.

Wells, H.: Functional and pharmacological studies on the regulation of salivary gland growth. In Schneyer, L.H., and Schneyer, C.A., editors: Secretory mechanisms of salivary glands, New York, 1967, Academic Press, Inc.

Young, J.A., and Schneyer, C.A.: Composition of saliva in mammalia, Aust. J. Exp. Biol. Med. Sci. **59**:1, 1981.

Young, J.A., and van Lennep, E.W.: The morphology of salivary glands, London, 1978, Academic Press.

Young, J.A., and van Lennep E.W.: Transport in salivary and salt glands. In Giebisch, G., Tosteson, D.C., and Ussing, H.H., editors: Membrane transport in biology, vol. 4B, Berlin, 1979, Springer, Verlag.

11
TOOTH ERUPTION

Although the word "eruption" properly refers to the cutting of the tooth through the gum (from the Latin *erumpere,* meaning "to break out") it is generally understood to mean the axial or occlusal movement of the tooth from its developmental position within the jaw to its functional position in the occlusal plane. However, eruption is only part of the total pattern of physiologic tooth movement, because teeth also undergo complex movements related to maintaining their position in the growing jaws and compensating for masticatory wear. Physiologic tooth movement is described as consisting of the following:

1. Preeruptive tooth movement
2. Eruptive tooth movement
3. Posteruptive tooth movement

Superimposed on these movements is the replacement of the entire deciduous dentition by the permanent dentition.

PATTERN OF TOOTH MOVEMENT

Preeruptive tooth movement. When deciduous tooth germs first differentiate, there is a good deal of space between them. However, because of their rapid growth, this available space is utilized and the developing teeth become crowded together, especially in the incisor and canine region. This crowding is relieved by growth in the length of the infant jaws, which provides room for the second deciduous molars to drift backward and the anterior teeth to drift forward. At the same time the tooth germs also move outward as the jaws increase in width, and upward (downward in the upper jaw) as the jaws increase in height.

Permanent teeth with deciduous predecessors also undergo complex movements before they reach the position from which they will erupt (see Chapter 12). For example, the permanent incisors and canines first develop lingual to the deciduous tooth germs at the level of their occlusal surfaces and in the same bony crypt. As their deciduous predecessors erupt, they move to a more apical position and occupy their own bony crypts (Fig. 11-1). Permanent premolars begin their development lingual to their predecessors at the level of their occlusal surfaces and in the same bony crypt. They also shift so that they are eventually situated in their own crypts beneath the divergent roots of the deciduous molars (Fig. 11-2).

The permanent molars, which have no deciduous predecessors, also move considerably from the site of their initial differentiation. For example, the upper permanent molars, which develop in the tuberosity of the maxilla, at first have their occlusal surfaces facing distally (Fig. 11-3) and swing around only when the maxilla has grown sufficiently to provide the necessary space. Similarly, mandibular molars develop with their occlusal surfaces inclined mesially and only become upright as room becomes available. All these movements are linked to jaw growth and may be considered as movements positioning the tooth and its crypt within the jaws preparatory to tooth eruption.

Eruptive tooth movement. During the phase of eruptive tooth movement the tooth moves from its position within the bone of the jaw to its functional position in occlusion, and the principal direction of movement is occlusal or axial. However, it is important to recognize that jaw growth is normally occurring while most teeth are erupting, so that movement in planes other than axial is superimposed on eruptive movement.

Posteruptive tooth movement. Posteruptive tooth movements are those that (1) maintain the position of the erupted tooth while the jaw continues to grow and (2) compensate for occlusal and proximal wear. The former movement,

Fig. 11-1. Buccolingual sections through central incisor region of mandible at representative stages of development from birth to 9 years of age. At birth both deciduous and permanent tooth germs occupy same bony crypt. Notice how, by eccentric growth and eruption of deciduous tooth, permanent tooth germ comes to occupy its own bony crypt apical to erupted incisor. At 4½ years, resorption of deciduous incisor has begun. At 6 years, deciduous incisor has been shed and its successor is erupting. Notice active deposition of new bone at base of socket at this time.

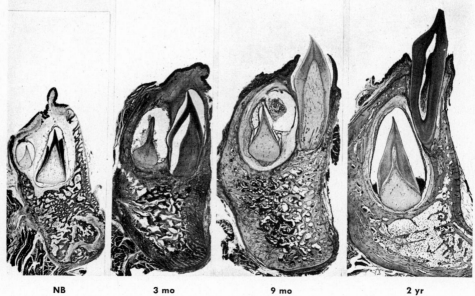

NB   3 mo   9 mo   2 yr

like eruptive movement, occurs principally in an axial direction to keep pace with the increase in height of the jaws. It involves both the tooth and its socket and ceases when jaw growth is completed. The movements compensating for occlusal and proximal wear continue throughout life and consist of axial and mesial migration, respectively.

HISTOLOGY OF TOOTH MOVEMENT

Preeruptive phase. During the preeruptive phase, positioning of the developing tooth within the growing jaws is achieved in two ways. First, there is a total bodily movement of the germ, and second, there is its excentric growth. Excentric growth means that one part of the developing tooth germ remains stationary while the remainder continues to grow, leading to a shift in its center. This type of growth explains, for example, how the deciduous incisors maintain their superficial position as the jaws grow in height (Fig. 11-1). Histologically, preeruptive tooth movement is reflected by bone remodeling at the periphery of the dental follicle, bringing about bone remodeling of the crypt wall. Thus during bodily movement of the tooth, osteoclastic bone resorption occurs on the surface of the crypt wall in advance of the moving tooth while bone deposition occurs on the crypt wall behind it. During

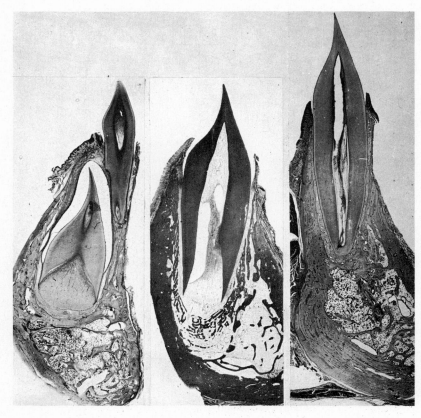

4½ yr   6 yr   9 yr

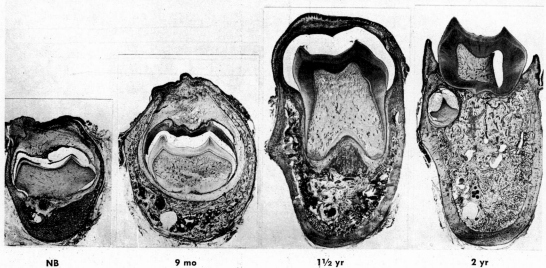

NB   9 mo   1½ yr   2 yr

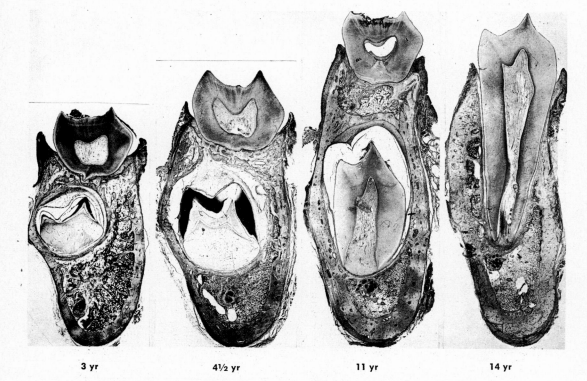

3 yr   4½ yr   11 yr   14 yr

Fig. 11-2. Buccolingual sections through deciduous first molar and first permanent premolar of mandible at representative stages of development from birth to 14 years. Notice how permanent tooth germ shifts its position. In section of 4½-year mandible, gubernacular canal is clearly visible. Lack of roots in the 2-, 3-, 4½-, and 11-year sections is not the result of resorption but of the section's being cut in midline of tooth with widely divergent roots.

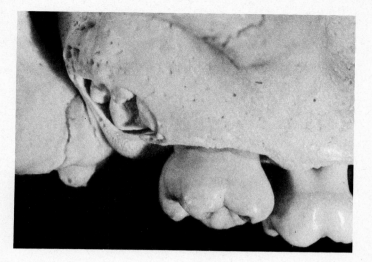

Fig. 11-3. Region of maxillary tuberosity of dried skull of 4-year-old child. At this stage of development, first permanent molar is still within its bony crypt. Notice how occlusal surface faces backward. With further growth of maxilla, molar swings down so that it eventually erupts into occlusal plane.

excentric movement bone resorption is seen on the surface of the crypt that faces the growing tooth germ.

Eruptive phase. During the eruptive phase of physiologic tooth movement, significant developmental changes occur, including the formation of the roots, periodontal ligament, and dentogingival junction of the tooth.

Root formation is initiated by proliferation of Hertwig's epithelial root sheath (see Chapter 2). The forming root first grows toward the floor of the bony crypt and, as a result, there is resorption of bone in this location to provide room for the advancing root tip. However, with the onset of eruptive tooth movement (probably coincident with periodontal ligament formation) space is created for the forming root, and resorption no longer occurs on the floor of the crypt. Indeed, in some instances the distance moved by the tooth outstrips the rate of root formation and bone deposition occurs on the crypt floor (Fig. 11-4).

As the roots of the tooth form, important changes associated with the development of the supporting apparatus of the tooth occur in the dental follicle. There is bone deposition on the crypt wall, cement deposition on the newly formed root surface, and organization of a periodontal ligament from the dental follicle (see Chapter 7). These changes lag behind root formation.

There are a number of important histologic features in the periodontal ligament that are important in explaining eruptive tooth movement. First is the occurrence of cell-to-cell contacts of the adherens type between periodontal ligament fibroblasts. Second is the demonstrated presence of contractile elements in ligament fibroblasts. Third is the occurrence of a structure called the fibronexus. This describes a morphologic relationship between intracellular microfilaments in the fibroblast, a corresponding increased density of fibroblast cell membrane, extracellular filaments, and fibronectin. Fibronectin is a sticky glycoprotein that sticks to a number of extracellular components, including collagen. Fourth is the active ingestion and degradation of old collagen fibrils

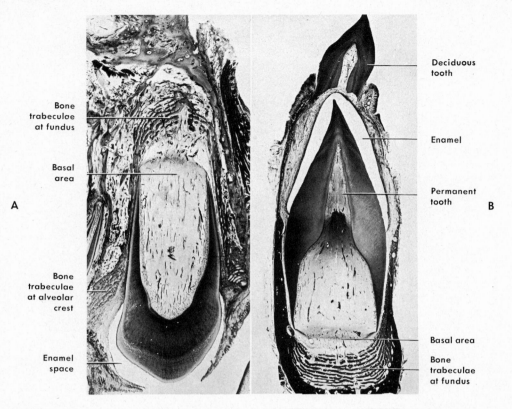

Fig. 11-4. Erupting upper deciduous canine, **A,** and lower permanent canine, **B.** Note formation of numerous parallel bone trabeculae at alveolar fundus. Formation of bone trabeculae at alveolar crest of deciduous canine, **A,** is a sign of rapid growth of maxilla in height. (From Kronfeld, R.: Dent. Cosmos **74:**103, 1932.)

by many of the fibroblasts of the ligament and the concurrent formation of new collagen fibrils (Fig. 11-5). Thus the continual degradation and synthesis of collagen by fibroblasts permits remodeling of the principal fiber bundles of the periodontal ligament.

Significant changes occur within the tissues that cover the erupting tooth. There is a loss of the intervening connective tissue between the reduced enamel epithelium covering the crown of the tooth and the overlying oral epithelium. Because of this loss the two epithelia proliferate and form a solid plug of cells in advance of the erupting tooth. The central cells of this epithelial mass degenerate and form an epithe-

lium-lined canal through which the tooth erupts without any hemorrhage. This epithelial cell mass is also involved in the formation of the dentogingival junction (see Chapter 9).

Once the tooth has broken through the oral mucosa, it continues to erupt at the same rate until it reaches the occlusal plane and meets its antagonist. Rapid eruptive movement then ceases. Root formation, however, is not yet complete, and because further occlusal movement is restricted, additional root growth is accommodated by removal of bone on the socket floor.

The above description generally applies to all teeth. Successional teeth, however, possess an

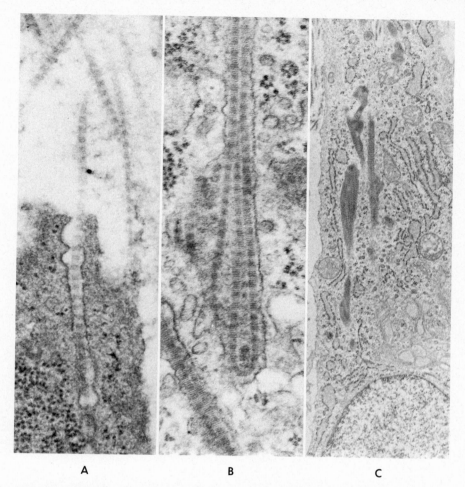

A   B   C

Fig. 11-5. Three electron micrographs illustrating role of fibroblast in periodontal ligament remodeling and turnover. **A,** Phagocytosis (ingestion) of collagen fibril. **B,** Once within fibroblast, lysosomes containing catabolic enzymes fuse with vesicle containing collagen, and **C,** degradation continues in phagolysosomes. (From Ten Cate, A.R.: Anat. Rec. **182:**1, 1975.)

additional anatomic feature, the *gubernacular canal* and its contents, the *gubernacular cord,* which may have an influence on eruptive tooth movement. When the successional tooth germ first develops within the same crypt as its deciduous predecessor, bone surrounds both tooth germs but does not completely close over them. As the deciduous tooth erupts, the permanent tooth germ becomes situated apically and entirely enclosed by bone (Figs. 11-1 and

11-2) except for a small canal that is filled with connective tissue and often contains epithelial remnants of the dental lamina. This connective tissue mass is termed the "gubernacular cord" (Figs. 11-6 and 11-7), and it may have a function in guiding the permanent tooth as it erupts.

Posteruptive phase. In the posteruptive phase the tooth makes movements primarily to accommodate the growth of the jaws. The prin-

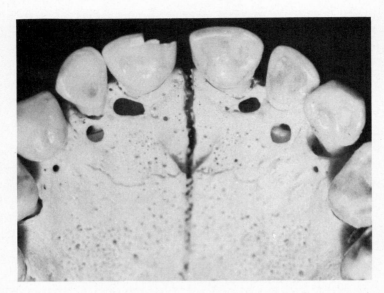

Fig. 11-6. Incisor region of dried maxilla of 4-year-old child. Notice foramina lingual to deciduous teeth. These are gubernacular canals.

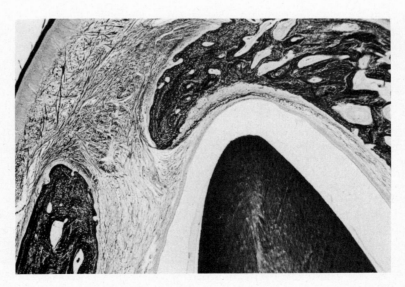

Fig. 11-7. Gubernacular cord consists mainly of connective tissue and often contains a central strand of epithelium surrounded by connective tissue.

cipal movement is in an axial direction. It occurs most actively between the ages of 14 and 18 and is associated with condylar growth, which separates the jaws and teeth. Although bone deposition occurs at the alveolar crest and on the socket floor, this is not responsible for tooth movement. The same forces responsible for eruptive tooth movement achieve axial posteruptive movement with bone deposition occurring later.

There are also movements made to compensate for occlusal and proximal wear of the tooth. It is generally assumed that the continuous deposition of cement around the apices of the roots of teeth is sufficient to compensate for occlusal wear. However, there is no evidence that this deposition of cement actually moves the tooth. It is more likely that the forces causing tooth eruption are still available to bring about sufficient axial movement of the tooth to compensate for occlusal wear. The cement deposition that occurs is probably an infilling phenomenon.

Wear also takes place at the contact points between teeth, and to maintain tooth contact mesial or proximal drift takes place. Histologically this drift is seen as a selective deposition and resorption of bone on the socket walls by osteoblasts and osteoclasts respectively, and with the electron microscope collagen remodeling in both the periodontal and transseptal ligaments is seen.

MECHANISM OF TOOTH MOVEMENT

The mechanism that brings about tooth movement is still debatable and is likely to be a combination of a number of factors. Although many possible causes have been proposed, only four merit serious consideration: (1) bone remodeling, (2) root growth, (3) vascular pressure, and (4) ligament traction. Briefly stated, the bone remodeling theory supposes that selective deposition and resorption of bone brings about eruption. The root growth theory supposes that the proliferating root impinges on a fixed case, thus converting an apically di-

rected force into occlusal movement. The vascular pressure theory supposes that a local increase in tissue fluid pressure in the periapical region is sufficient to move the tooth. The ligament traction theory proposes that the cells and fibers of the ligament pull the tooth into occlusion.

Bone remodeling. Bone remodeling of the crypt wall clearly is important to achieve tooth eruption, and in experiments where tooth germ is removed but the follicle is left in position, the eruptive pathway still forms in bone. Such experiments properly indicate dental follicle, not bone, as the major determinant in tooth eruption. Most bone remodeling is in response to other factors.

Root formation. Root formation is also unlikely to be the cause of tooth eruption, although at first glance this may seem an obvious mechanism. It has long been recognized that some teeth move a greater distance than the length of their fully formed roots, and others still erupt after root formation has been completed. Removal of roots does not prevent eruption. If root formation is responsible for eruption, it would be expected that the onset of root formation and eruptive movement would coincide, but, as has already been stated, the onset of root formation is not synchronous with the onset of axial tooth movement. Indeed, initial root formation results in bone resorption at the base of the socket. This is an important observation, for it illustrates a fundamental point of bone biology, which is that when pressure is applied to bone, the bone is usually removed by osteoclastic action. Thus for root formation to result in an eruptive force, the apical growth of the root needs to be translated into occlusal movement and requires the presence of a fixed base. No such fixed base exists. The bone at the base of the socket cannot act as a fixed base because pressure on bone results in its resorption. Advocates of the root growth theory of tooth eruption postulated the existence of a ligament, the cushion-hammock ligament, straddling the base of the socket from one bony wall to the other like a sling. Its function was to provide a fixed base for the growing root to react against. But the structure described as the cushion-hammock ligament is the pulp-delineating membrane that runs across the apex of the tooth and has no bony insertion. It cannot act as a fixed base.

Vascular pressure. It is known that teeth move in synchrony with the arterial pulse, so local volume changes can produce limited tooth movement. Ground substance can swell by up to 50% with the addition of water, and a differential pressure sufficient to cause tooth move-

ment between the tissues below and above an erupting tooth has been reported in the dog. Again, whether such pressures are the prime movers of teeth is debatable because surgical excision of the root, and therefore the local vasculature, does not prevent tooth eruption.

Periodontal ligament traction. There is a good deal of evidence that the eruptive force resides in the dental follicle–periodontal ligament complex. Experiments on the continuously erupting rodent incisor (Fig. 11-8), designed to eliminate the effects of root growth and vascular supply, also show that, so long as periodontal tissue is available, tooth movement

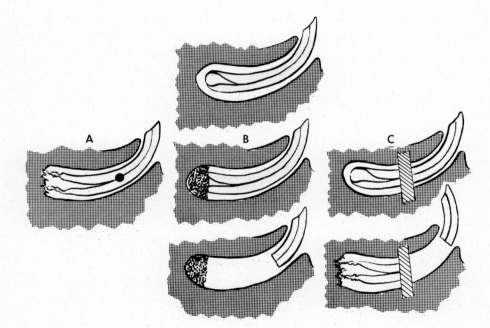

Fig. 11-8. Diagram summarizing some of the experimental work undertaken on continuously erupting rodent incisor. As this tooth erupts continuously, dental tissues are also continually being generated at base of tooth. If tooth is pinned, **A,** eruptive movement is prevented and continual generation of new dental tissue results in buckling of root and resorption of bone at base of socket. If, on the other hand, basal generative tissues are removed surgically, eliminating root growth and vascular pressure, **B,** distal fragment of tooth continues to erupt and undergoes eventual exfoliation. **C,** Combination of both experiments. Here, incisor is cut in half and an artificial rigid barrier is inserted between the two halves. Distal fragment continues to erupt and is eventually shed. The only tissue present here is a periodontal ligament. Proximal fragment can no longer erupt, and as a result the growing end buckles and also retreats.

occurs. Drugs that interrupt the proper formation of collagen in the ligament also interfere with eruption.

Tissue culture experiments have shown that ligament fibroblasts are able to contract a collagen-gel, which in turn brings about movement of a disc of root tissue attached to that gel. Thus there is no doubt that periodontal ligament fibroblasts have the ability to contract and transmit a contractile force to the extracellular environment and in particular to the collagen fiber bundles in vitro. All the morphologic features exist in vivo to permit similar movement. Thus the fibroblasts possess contractile filaments, are in contact with one another to permit summation of contractile forces, and exhibit fibronexuses by which such forces can be transmitted to the collagen fiber bundles. These not only remodel but are also inclined at the correct angle to bring about eruptive movement. This angulation of the ligament fiber bundles is a prerequisite for tooth movement, and the orientation is believed to be established by the developing root.

The follicle, before it becomes periodontal ligament, also plays a role in tooth eruption, even though it may not provide the actual eruptive force. It has already been mentioned that if tooth germs are removed and the follicle is left intact, the eruptive pathway still forms in bone. Equally, if a tooth is enucleated and substituted with a silicone replica within the follicle, the replica erupts, which again establishes the absolute requirement for a follicle-ligament complex to achieve tooth movement.

In summary, eruptive movement is brought about by a combination of events involving a force initiated by the fibroblast. This force is transmitted to the extracellular compartment via fibronexuses and to collagen fiber bundles, which, aligned in an appropriate inclination brought about by root formation, bring about tooth movement. These fiber bundles must have the ability to remodel for eruption to continue, and interference with this ability affects the process. The removal of bone to create the eruptive pathway is also dictated by the tissues surrounding the tooth.

In posteruptive tooth movement the mechanisms for moving the tooth axially during eruption are most likely also used to compensate for occlusal wear. It is worth noting that axial movement at this time does not require bone remodeling. Mesial, or proximal, drift involves a combination of two separate forces resulting from occlusal contact of teeth and contraction of the transseptal ligaments between teeth. When the jaws are clenched, bringing teeth into contact, force is generated in a mesial direction because of the summation of cuspal planes and because many teeth have a mesial inclination. This can be demonstrated in a number of ways. When opposing teeth are removed, the rate of mesial drift is slowed but not eliminated. Selective grinding of cuspal slopes can either enhance or counter the effect of occlusal force, and when this is done the rate of mesial drift is respectively enhanced or decreased but again not eliminated. These observations indicate that although an anterior component of occlusal force is responsible for mesial drift, it is not solely responsible.

Running between teeth across the alveolar process is the transseptal ligament, and there is evidence that this ligament has a key role in maintaining tooth position. For example, if a tooth is bisected, the two halves move away from each other, but if the transseptal ligament is previously removed, this separation does not occur. A simple but elegant experiment demonstrates that mesial drift is indeed multifactorial. By disking away the approximal contacts room is made to permit mesial drift, and the teeth begin to move to reestablish contact. If teeth are ground out of occlusal contact, however, the rate of drift is slowed. The conclusion must be that mesial drift is achieved by contraction of transseptal fibers and enhanced by occlusal forces.

CLINICAL CONSIDERATIONS

From all that has been written so far in this chapter it should be evident that the principal supporting tissues of the tooth, the periodontal ligament and the bone of the jaw, possess a remarkable "plasticity" that enables the tooth to react either favorably or unfavorably to its immediate environment. This plasticity of the supporting tissues is used by the orthodontist to achieve a favorable clinical response. By applying forces to the tooth and by relying on the biologic responses of bone and periodontal ligament, malalignment of teeth can often be corrected.

Table 5 gives the time of tooth emergence (in whites); one should note that there is considerable variation in these times. However, only teeth emerging significantly outside these

Table 5. Chronology of human dentition*

Tooth	Formation of enamel matrix and dentin begins	Amount of enamel matrix formed at birth	Enamel completed	Emergence into oral cavity	Root completed
Primary dentition					
Maxillary					
Central incisor	4 mo. in utero	Five sixths	1½ mo.	7½ mo.	1½ yr.
Lateral incisor	4½ mo. in utero	Two thirds	2½ mo.	9 mo.	2 yr.
Canine	5 mo. in utero	One third	9 mo.	18 mo.	3¼ yr.
First molar	5 mo. in utero	Cusps united	6 mo.	14 mo.	2½ yr.
Second molar	6 mo. in utero	Cusp tips still isolated	11mo.	24 mo.	3 yr.
Mandibular					
Central incisor	4½ mo. in utero	Three fifths	2½ mo.	6 mo.	1½ yr.
Lateral incisor	4½ mo. in utero	Three fifths	3 mo.	7 mo.	1½ yr.
Canine	5 mo. in utero	One third	9 mo.	16 mo.	3¼ yr.
First molar	5 mo. in utero	Cusps united	5½ mo.	12 mo.	2¼ yr.
Second molar	6 mo. in utero	Cusp tips still isolated	10 mo.	20 mo.	3 yr.
Permanent dentition					
Maxillary					
Central incisor	3-4 mo.			7-8 yr.	10 yr.
Lateral incisor	10-12 mo.			8-9 yr.	11 yr.
Canine	4-5 mo.			11-12 yr.	13-15 yr.
First premolar	1½-1¾ yr.			10-12 yr.	12-13 yr.
Second premolar	2-2¼ yr.			10-12 yr.	12-14 yr.
First molar	At birth	Sometimes a trace		6-7 yr.	9-10 yr.
Second molar	2½-3 yr.			12-13 yr.	14-16 yr.
Third molar	7-9 yr.			17-21 yr.	18-25 yr.
Mandibular					
Central incisor	3-4 mo.		4-5 yr.	6-7 yr.	9 yr.
Lateral incisor	3-4 mo.		4-5 yr.	7-8 yr.	10 yr.
Canine	4-5 mo.		6-7 yr.	9-10 yr.	12-14 yr.
First premolar	1¾-2 yr.		5-6 yr.	10-12 yr.	12-13 yr.
Second premolar	2¼-2½ yr.		6-7 yr.	11-12 yr.	13-14 yr.
First molar	At birth	Sometimes a trace	2½-3 yr.	6-7 yr.	9-10 yr.
Second molar	2½-3 yr.		7-8 yr.	11-13 yr.	14-15 yr.
Third molar	8-10 yr.		12-16 yr.	17-21 yr.	18-25 yr.

*From Logan, W.H.G., and Kronfeld, R.: J. Am. Dent. Assoc. **20**:379, 1933, slightly modified by McCall and Schour.

ranges should be considered as abnormal and indicative of some fault in eruptive movement. By far the greatest number of aberrations in eruption times are delayed eruptive movements. Premature eruption of teeth occurs infrequently. Sometimes infants are born with "erupted" lower central incisors, but this is an example of gross maldevelopment. Such teeth need to be extracted as soon as possible because they prevent suckling. Premature loss of a deciduous tooth without closure of the gap may lead to early eruption of its successor. Far more common, however, is the occurrence of delayed or retarded eruption. This may be caused by either local or systemic factors. Systemic factors include nutritional, genetic, and endocrine deficiencies. Local factors include such situations as loss of a deciduous tooth and drifting of opposing teeth to block the eruptive pathway. Severe trauma may eliminate the dental follicle, and hence periodontal ligament formation is prevented. When this happens, the bone of the jaw fuses with tooth, a condition known as ankylosis, and eruption is not possible.

Whites exhibit an evolutionary trend to a diminution in the size of the jaws. This trend has not been accompanied by a corresponding decrease in the size of the teeth, and as a result crowding is a common occurrence. The third molars are the last teeth to erupt, and frequently all the available space has been used. As a result these teeth become impacted. Canines are also often impacted because of their late eruption time. Finally, it has been shown that the moment a tooth breaks through the oral epithelium an acute inflammatory response occurs in the connective tissue adjacent to the tooth. This is seen even in the germ-free animals and is seen in varying degrees around all teeth throughout life. Clinically, as teeth break through the oral mucosa, there is often some pain, slight fever, and general malaise, all signs of an inflammatory process. In infants these symptoms are popularly called "teething."

REFERENCES

Beertsen, W., Everts, V., and van den Hoof, A.: Fine structure of fibroblasts in the periodontal ligament of the rat incisor and their possible role in tooth eruption, Arch. Oral Biol. **19**:1087, 1974.

Bellows, C.F., Melcher, A.H., and Aubin, J.E.: Contraction and organization of collagen gels by cells cultured from periodontal ligament, gingiva and bone suggest functional differences between cell types, J. Cell. Sci. **50**:299, 1981.

Bellows, C.F., Melcher, A.H., and Aubin, J.E.: An in-vitro model for tooth eruption utilizing periodontal ligament fibroblasts and collagen lattices, Arch. Oral Biol. **28**:715, 1983.

Berkovitz, B.K.B.: The effect of root transection and partial root resection on the unimpeded eruption rate of the rat incisor. Arch. Oral Biol. **16**:1033, 1971.

Berkovitz, B.K.B.: The healing process in the incisor tooth socket of the rat following root resection and exfoliation. Arch. Oral Biol. **16**:1045, 1971.

Berkovitz, B.K.B.: The effect of preventing eruption on the proliferative basal tissues of the rat lower incisor. Arch. Oral Biol. **17**:1279, 1972.

Berkovitz, B.K.B.: Mechanisms of tooth eruption. In Lavelle, C.L.B., editor: Applied physiology of the mouth, Bristol, England, 1975. John Wright and Sons Ltd.

Berkovitz, B.K.B., and Thomas, N.R.: Unimpeded eruption in the root resected lower incisor of the rat with a preliminary note on root transection. Arch. Oral Biol. **14**:771, 1969.

Brash, J.C.: The growth of the alveolar bone and its relation to the movements of the teeth, including eruption. Int. J. Orthodont. **14**:196, 283, 398, 487, 494, 1928.

Brodie, A.G.: The growth of alveolar bone and the eruption of the teeth. Oral Surg. **1**:342, 1948.

Bryer, L.W.: An experimental evaluation of physiology of tooth eruption. Int. Dent. J. **7**:432, 1957.

Cahill, D.R.: Eruption pathway formation in the presence of experimental tooth impaction in puppies, Anat. Rec. **164**:67, 1969.

Cahill, D.R.: The histology and rate of tooth eruption with and without temporary impaction in the dog. Anat. Rec. **166**:225, 1970.

Cahill, D.R.: Histological changes in the bony crypt and gubernacular canal of erupting permanent premolars during deciduous premolar exfoliation in beagles. J. Dent. Res. **53**:786, 1974.

Cahill, D.R., and Marks, S.C., Jr.: Tooth eruption: evidence for the central role of the dental follicle, J. Oral Pathol. **9**:189, 1980.

Cahill, D.R., and Marks, S.C., Jr.: Chronology and histology of exfoliation and eruption of mandibular premolars in dogs, J. Morphol. **171**:213, 1982.

Carollo, D.A., Hoffman, R.L., and Brodie, A.G.: Histology and function of the dental gubernacular cord. Angle Orthod. **41**:300, 1971.

Gowgiel, J.M.: Eruption of irradiation-produced rootless teeth in monkeys, J. Dent. Res. **40**:538, 1961.

Herzberg, F., and Schour, I.: Effects of the removal of pulp and Hertwig's sheath on the eruption of incisors in the albino rat. J. Dent. Res. **20**:264, 1941.

Jenkins, G.N.: The physiology of the mouth, ed. 3, Oxford, 1966, Blackwell Scientific Publications Ltd.

Logan, W.H.G., and Kronfeld, R.: Development of the human jaws and surrounding structures from birth to the age of fifteen years. J. Am. Dent. Assoc. **20**:379, 1933.

Magnusson, B.: Tissue changes during molar tooth eruption, Trans. R. Sch. Dent. Stockholm **13**:1, 1968.

Main, J.H.P.: A histological survey of the hammock ligament, Arch. Oral Biol. **10**:343, 1965.

Main, J.H.P., and Adams, D.: Experiments on the rat incisor into the cellular proliferation and blood pressure theories of tooth eruption. Arch. Oral Biol. **11**:163, 1966.

Manson, J.D.: Bone changes associated with tooth eruption. In The mechanisms of tooth support, a symposium, Oxford, 6-8 July, 1965, Bristol, England, 1967, John Wright & Sons Ltd.

Marks, S.C., Jr., Cahill, D.R., and Wise, G.E.: The cytology of the dental follical and adjacent alveolar bone during tooth eruption, Am. J. Anat. **168**:277, 1983.

Moss, J.P., and Picton, D.C.A.: Mesial drift of teeth in adult monkeys *(Macaca irus)* when forces from the cheeks and tongue had been eliminated, Arch. Oral Biol. **15**:979, 1970.

Moxham, B.J., and Berkovitz, B.K.B.: The periodontal ligament and physiological tooth movements. In Berkovitz, B.K.B., Moxham, B.J., and Newman, H.N., editors: The periodontal ligament in health and disease, Elmsford, N.Y., 1982, Pergamon Press, Inc.

Orban, B.: Growth and movement of the tooth germs and teeth. J. Am. Dent. Assoc. **15**:1004, 1928.

Sicher, H.: Tooth eruption: the axial movement of continuously growing teeth. J. Dent. Res. **21**:201, 1942.

Sicher, H.: Tooth eruption: the axial movement of teeth with limited growth. J. Dent. Res. **21**:395, 1942.

Sicher, H., and Weinmann, J.P.: Bone growth and physiological tooth movement, Am. J. Orthod. **30**:109, 1944.

Taylor, A.C., and Butcher, E.O.: The regulation of eruption rate in the incisor teeth of the white rat. J. Exp. Zool. **117**:165, 1951.

Ten Cate, A.R.: The mechanism of tooth eruption. In Melcher, A.H., and Bowen, W.H., editors: The biology of the periodontium, New York, 1969, Academic Press, Inc.

Ten Cate, A.R.: Physiological resorption of connective tissue associated with tooth eruption. An electron microscope study, J. Periodont. Res. **6**:168, 1971.

Ten Cate, A.R.: Morphological studies of fibrocytes in connective tissue undergoing rapid remodelling, J. Anat. **112**:401, 1972.

Thomas, N.R.: The properties of collagen in the periodontium of an erupting tooth. In The mechanisms of tooth support, a symposium, Oxford, 6-8 July, 1965, Bristol, England, 1967. John Wright & Sons Ltd.

Thomas, N.R.: The effect of inhibition of collagen maturation on eruption in rats. J. Dent. Res. **44**:1159, 1969.

Weinmann, J.P.: Bone changes related to eruption of the teeth, Angle Orthod. **11**:83, 1941.

12

SHEDDING OF DECIDUOUS TEETH

DEFINITION

PATTERN OF SHEDDING

HISTOLOGY OF SHEDDING

MECHANISM OF RESORPTION AND SHEDDING

CLINICAL CONSIDERATIONS

Remnants of deciduous teeth

Retained deciduous teeth

Submerged deciduous teeth

DEFINITION

The human dentition, like those of most mammals, consists of two generations. The first generation is known as the deciduous (primary) dentition and the second as the permanent (secondary) dentition. The necessity for two dentitions exists because infant jaws are small and the size and number of teeth they can support is limited. Since teeth, once formed, cannot increase in size, a second dentition, consisting of larger and more teeth, is required for the larger jaws of the adult. The physiologic process resulting in the elimination of the deciduous dentition is called *shedding* or *exfoliation*.

PATTERN OF SHEDDING

The shedding of deciduous teeth is the result of progressive resorption of the roots of teeth and their supporting tissue, the periodontal ligament. Most attention has been paid to the removal of the dental hard tissues, which is accomplished by easily identified multinuclear odontoblasts, cells in every way similar to osteoclasts (Fig. 12-1). In general, the pressure generated by the growing and erupting permanent tooth dictates the pattern of deciduous tooth resorption. At first this pressure is directed against the root surface of the deciduous

tooth itself (Fig. 12-2). Because of the developmental position of the permanent incisor and canine tooth germs and their subsequent physiologic movement in an occlusal and vestibular direction, resorption of the roots of the deciduous incisors and canines begins on their lingual surfaces (Fig. 12-3). Later, these developing tooth germs occupy a position directly apical to the deciduous tooth, which permits them to erupt in the position formerly occupied by the deciduous tooth (Fig. 12-4). Frequently, however, and especially in the case of the permanent mandibular incisors, this apical positioning of the tooth germs does not occur and the permanent tooth erupts lingual to the still functioning deciduous tooth (Fig. 12-5).

Resorption of the roots of deciduous molars often first begins on their inner surfaces because the early developing bicuspids are found between them (Fig. 12-6). This resorption occurs long before the deciduous molars are shed and reflects the expansion of their growing permanent successors. However, as a result of the continued growth of the jaws and occlusal movement of the deciduous molars, the successional tooth germs come to lie apical to the deciduous molars (Fig. 12-7). This change in position provides the growing bicuspids with adequate space for their continued develop-

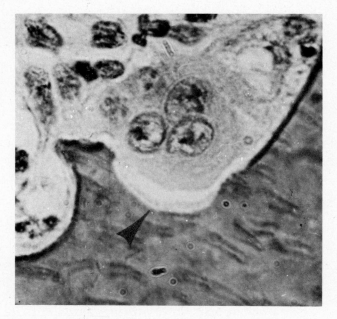

Fig. 12-1. Photomicrograph of odontoclast resorbing dentin. Note brush border *(arrow)* where odontoclast is in contact with dentin. (From Furseth, R.: Arch. Oral Biol. **13:**417, 1968.)

ment and also relieves the pressure on the roots of the overlying deciduous molars. The areas of early resorption are repaired by the deposition of a cementum-like tissue. When the bicuspids begin to erupt, resorption of the deciduous molars is again initiated and this time continues until the roots are completely lost and the tooth is shed (Fig. 12-8). The bicuspids thus erupt in the position of deciduous molars.

HISTOLOGY OF SHEDDING

The cells responsible for the removal of dental hard tissue are identical to osteoclasts, the highly specialized cells responsible for the removal of bone, and are called *odontoclasts.*

Odontoclasts are readily identifiable in the light microscope as large, multinucleated cells occupying resorption bays on the surface of a dental hard tissue. Their cytoplasm is vacuo-lated, and the surface of the cell adjacent to the resorbing hard tissue forms a "brush" border (Fig. 12-1). Histochemically, a characteristic feature of the odontoclast is a high level of activity of the enzyme acid phosphatase. These light-microscope observations have been confirmed and extended with the electron microscope (Fig. 12-9). The brush border is resolved as a ruffled border produced by extensive folding of the cell membrane into a series of invaginations 2 to 3 μm deep, with mineral crystallites within the depths of the invaginations. The cytoplasm of the odontoclast is characterized by an exceptionally high content of mitochondria and many vacuoles, which are especially concentrated adjacent to the ruffled border. Acid phosphatase activity occurs within these vacuoles (Fig. 12-10).

Odontoclasts are able to resorb all the dental hard tissues including, on occasions, enamel.

Text continued on page 385.

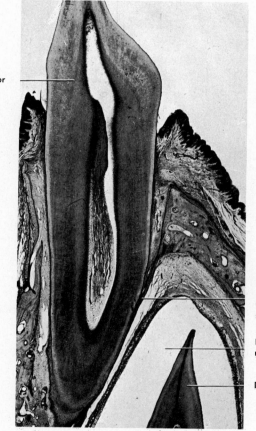

Deciduous incisor

Bone between
deciduous
tooth and
successor

Enamel of perma-
nent incisor

Dentin

Fig. 12-2. Thin lamella of bone separates permanent tooth germ from its predecessor.

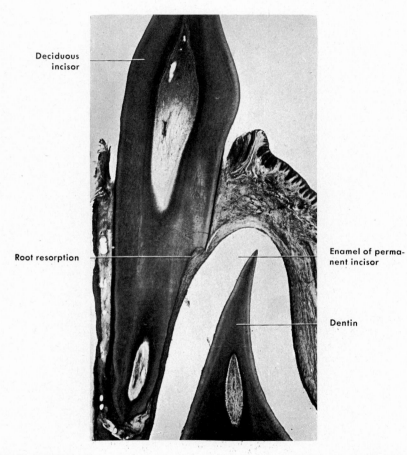

Deciduous
incisor

Root resorption

Enamel of perma-
nent incisor

Dentin

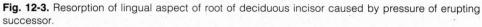

Fig. 12-3. Resorption of lingual aspect of root of deciduous incisor caused by pressure of erupting successor.

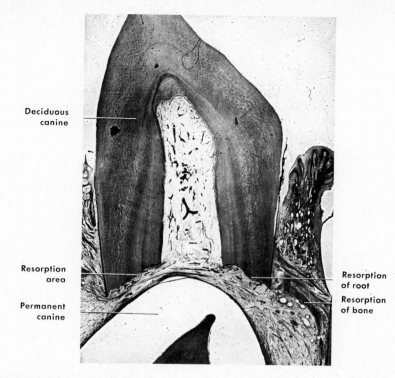

Deciduous
canine

Resorption
area

Permanent
canine

Resorption
of root

Resorption
of bone

Fig. 12-4. Resorption of root of deciduous canine. Note apical position of permanent successor. (From Kronfeld, R.: Dent. Cosmos **74:**103, 1932.)

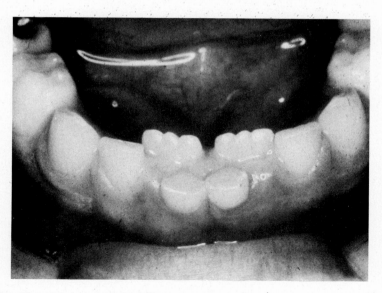

Fig. 12-5. Dentition of 6-year-old child showing how permanent incisors frequently erupt lingually to deciduous incisors before latter teeth are shed.

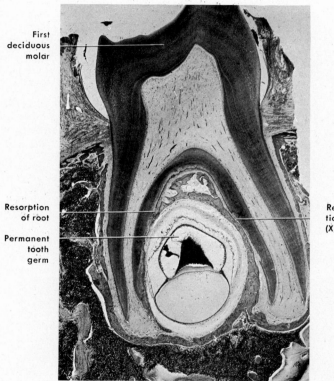

First
deciduous
molar

Resorption
of root

Permanent
tooth
germ

Repaired resorp-
tion of dentin
(X)

Fig. 12-6. Germ of lower first permanent premolar between roots of first deciduous molar. Repair of previously resorbed dentin has occurred at X. (See also Figs. 12-16 and 12-17.)

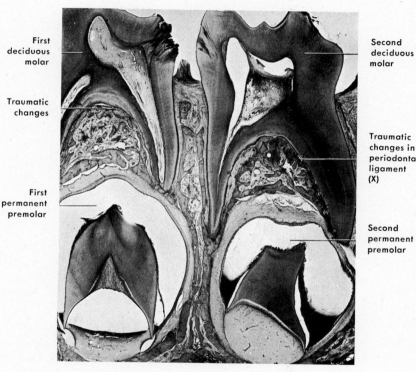

First
deciduous
molar

Traumatic
changes

First
permanent
premolar

Second
deciduous
molar

Traumatic
changes in
periodontal
ligament
(X)

Second
permanent
premolar

Fig. 12-7. Germs of permanent premolars below roots of deciduous molars.

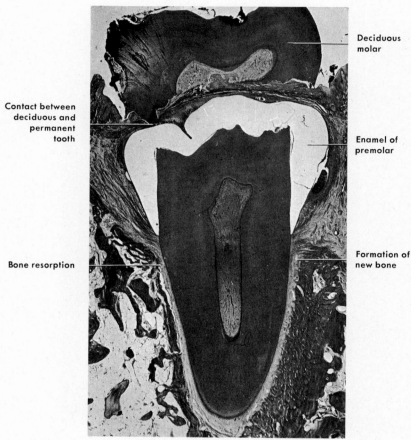

Contact between
deciduous and
permanent
tooth

Deciduous
molar

Enamel of
premolar

Bone resorption

Formation of
new bone

Fig. 12-8. Roots of primary molar completely resorbed. Dentin of primary tooth in contact with enamel of premolar. Resorption of bone on one side and formation of new bone on opposite side of premolar caused by transmitted eccentric pressure to premolar. (From Grimmer, E.A.: J. Dent. Res. **18:**267, 1939.)

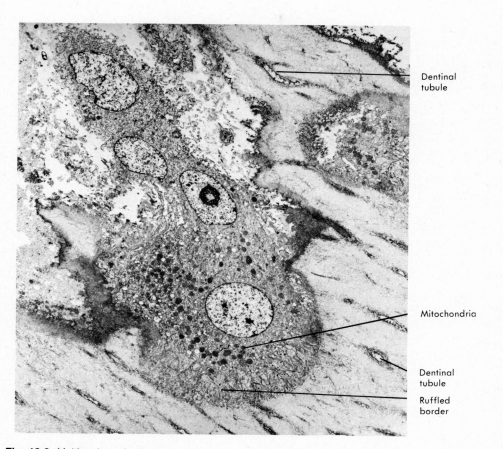

Dentinal
tubule

Mitochondria

Dentinal
tubule

Ruffled
border

Fig. 12-9. Multinucleated odontoclast displays ruffled border that is well adapted to resorption la-cuna in root dentin. Dense mitochondria are aggregated toward "resorptive" (lower) pole of cell, and most of the cytoplasm is highly vacuolated. Dentinal tubules are visible in oblique section (×3000.) (From Freilich, L.S.: J. Dent. Res. **50:**1047, 1971.)

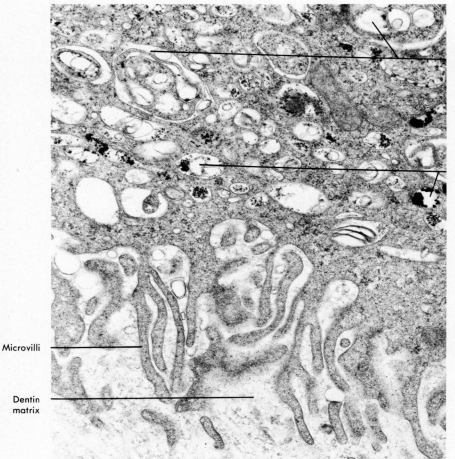

Autophagocytosed cellular material

Reaction product

Microvilli

Dentin matrix

Fig. 12-10. Interface between odontoclast ruffled border region (indicated by irregular microvilli) and disintegrated dentin matrix of root surface undergoing resorption. Numerous membrane-bound vacuoles in odontoclast cytoplasm show varied contents, including autophagocytosed cellular material and dense patches of reaction product, indicating acid phosphatase activity. (×40,000.) (From Freilich, L.S.: J. Dent. Res. **50:**1047, 1971.)

When dentin is being resorbed, the presence of the tubules provides a pathway for the easy extension of odontoclast processes (Fig. 12-11). Odontoclasts probably have the same origin as osteoclasts. The monocyte, circulating in the blood, originally gives rise to all the different tissue macrophages, including the osteoclast, but what is not certain is whether osteoclasts are further formed from mainly resident tissue macrophages or continuously from circulating monocytes.

Debate also exists concerning the distribution of odontoclasts during tooth resorption. Odontoclasts are most commonly found on surfaces of the roots in relation to the advancing permanent tooth. However, they have also been described in the root canals and pulp chambers of resorbing teeth lying against the predentin surface. Although their location in the pulp chamber has been disputed, the most likely reason is that different patterns of resorption exist for different teeth. For example,

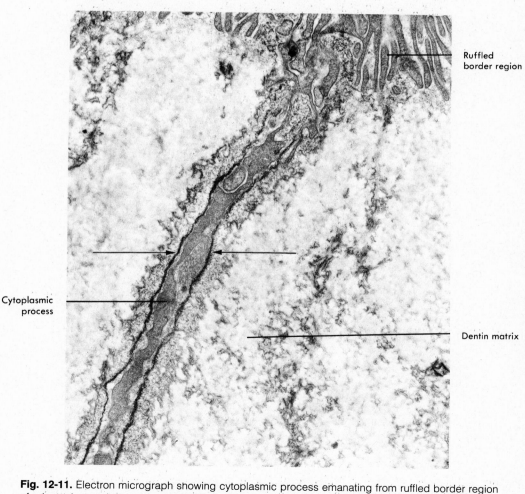

Fig. 12-11. Electron micrograph showing cytoplasmic process emanating from ruffled border region of odontoclast and occupying dentinal tubule. Dentin matrix occupies bulk of field. (×25,000.) (From Freilich, L.S.: J. Dent. Res. **50:**1047, 1971.)

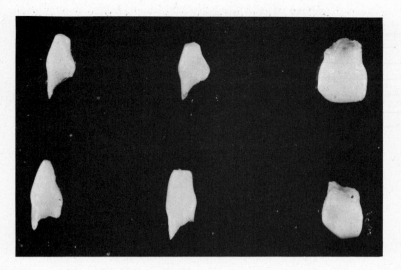

Fig. 12-12. Random selection of exfoliated deciduous incisor and canine teeth showing that considerable amount of root dentin remains at time of exfoliation.

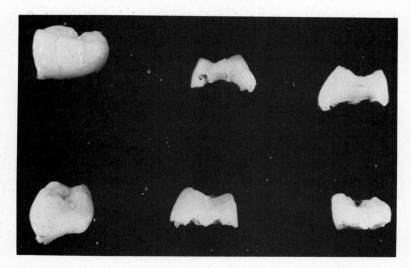

Fig. 12-13. Random selection of exfoliated deciduous molars showing that total loss of roots usually occurs before these teeth are shed. This photograph also shows occurrence of enamel resorption.

Fig. 12-14. Osteoclastic resorption in surface of coronal dentin of deciduous first molar. Odontoblast layer is absent and numerous odontoblasts can be seen lining pulp chamber. (From Weatherell, J.A., and Hargreaves, J.A.: Arch. Oral Biol. **11:**749, 1966.)

single-rooted teeth are usually shed before root resorption is complete (Fig. 12-12); therefore odontoclasts are not found within the pulp chambers of these teeth and the odontoblast layer remains intact. In molars, however, the roots are usually completely resorbed and the crown is also partially resorbed before exfoliation. When this happens (Fig. 12-13), the odontoblast layer is replaced by odontoclasts (Fig. 12-14), which resorb both primary and secondary dentin (Fig. 12-15). Sometimes all the dentin is removed, and the vascular connective tissue is visible beneath the translucent cap of enamel.

The process of tooth resorption is not continuous, since there are periods of rest and repair; however, in the long term, resorption predominates over repair. Repair is achieved by cells resembling cementoblasts that lay down a dense collagenous matrix in which spotty mineralization occurs. The final repair tissue resembles cellular cementum but is less mineralized (Figs. 12-16 and 12-17).

MECHANISM OF RESORPTION AND SHEDDING

The mechanisms involved in bringing about tooth resorption and exfoliation are not yet fully understood. It seems clear that pressure from the erupting successional tooth plays a key role because the odontoclasts differentiate at predicted sites of pressure.

How the odontoclast actually resorbs dental hard tissue is not known. The finding of mineral crystallites in the depths of the ruffled border and the fact that scanning electron microscopy indicates that the collagenous matrix of the dentin becomes exposed during resorption

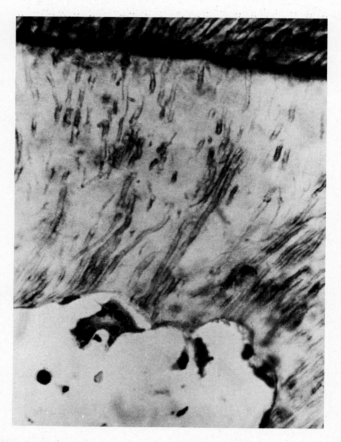

Fig. 12-15. Odontoclasts resorbing secondary dentin. (From Weatherell, J.A., and Hargreaves, J.A.: Arch. Oral Biol. **11:**749, 1966.)

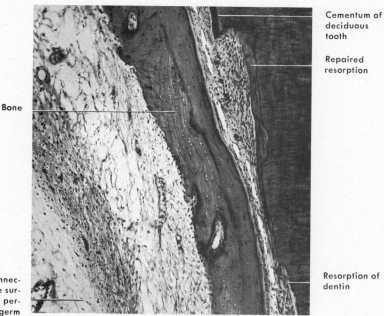

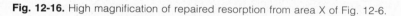

Fig. 12-16. High magnification of repaired resorption from area X of Fig. 12-6.

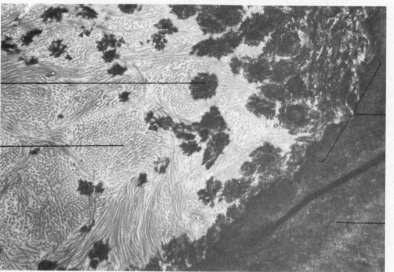

Fig. 12-17. Electron micrograph of resorption lacunae where repair of cementum is taking place. Newly deposited repair tissue is not as electron dense as underlying cementum. Note electron-dense reversal line and calcific globules in precementum. (From Furseth, R.: Arch. Oral Biol. **13:**417, 1968.)

suggest than mineral is removed first. How the dissolution of the crystallites is achieved is not known. Nor is it known how the organic matrix, that is, the collagen and associated ground substance, is dispersed. The acid phosphatase content of the vesicles close to the ruffled border suggests that these structures are phagosomes in which breakdown of ingested material is taking place. The most likely sequence of events in resorption of dental hard tissue by the odontoclast is an initial removal of mineral followed by extracellular dissolution of the organic matrix (mainly collagen) to smaller molecules, which are then taken up by the odontoclast and degraded further.

Although pressure obviously has a key role in initiating tooth resorption, other factors must also be involved. It is a common clinical observation that when a successional tooth germ is missing, shedding of the deciduous tooth is delayed. Also, experimental removal of a permanent tooth germ delays, but does not prevent, shedding of its deciduous predecessor. It is more than likely that the forces of mastication applied to the deciduous tooth are also capable of initiating the resorption. As an individual grows, the muscles of mastication increase in size and exert forces on the deciduous tooth greater than its periodontal ligament can withstand. This leads to trauma to the ligament and the initiation of resorption. That this is so has been established experimentally by placing a splint bridge into the mouth of an experimental animal in such a way as to protect the deciduous tooth from occlusal stress. When this is done, resorption of the deciduous tooth is halted and repair takes place.

In practice a combination of both factors likely determines the rate and pattern of resorption. As resorption of the roots initiated by pressure of the underlying tooth occurs, there is a progressive loss of surface area for attachment of the periodontal ligament fiber bundles. This weakening of tooth support occurs because it has to withstand increasingly greater occlusal forces generated by the growing muscles of mastication.

Finally, although the resorption of the dental hard tissues has been studied extensively, little, if anything, is known about how the dental soft tissues are removed, especially the periodontal ligament. Examination of histologic sections shows that the loss of the soft connective tissues is abrupt, and there is no evidence for the accumulation of macrophages or polymorphonuclear cells, which remove soft connective tissue at inflammatory sites. In the case of the periodontal ligament it may be speculated that, because this tissue has a high rate of turnover, a change in either the normal degradative or synthetic process might achieve removal of tissue in the absence of any inflammatory change.

CLINICAL CONSIDERATIONS

Remnants of deciduous teeth. Sometimes parts of the roots of deciduous teeth are not in the path of erupting permanent teeth and may escape resorption. Such remnants, consisting of dentin and cementum, may remain embedded in the jaw for a considerable time. They are most frequently found in association with the permanent premolars, especially in the region of the lower second premolars (Fig. 12-18). The reason is that the roots of the lower second deciduous molar are strongly curved or divergent. The mesiodistal diameter of the second premolars is much smaller than the greatest distance between the roots of the deciduous molar. Root remnants may later be found deep in the bone, completely surrounded by and ankylosed to the bone (Fig. 12-19). Frequently they are cased in heavy layers of cellular cementum. When they are close to the surface of the jaw (Fig. 12-20) they may ultimately be exfoliated. Progressive resorption of the root remnants and replacement by bone may cause the disappearance of these remnants.

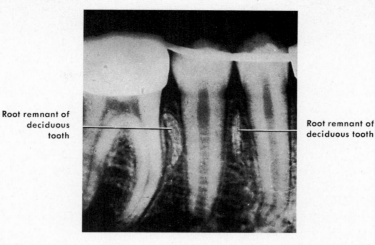

Root remnant of deciduous tooth   Root remnant of deciduous tooth

Fig. 12-18. Remnants of roots of deciduous molar embedded in interdental septa. (Courtesy Dr. G.M. Fitzgerald, University of California.)

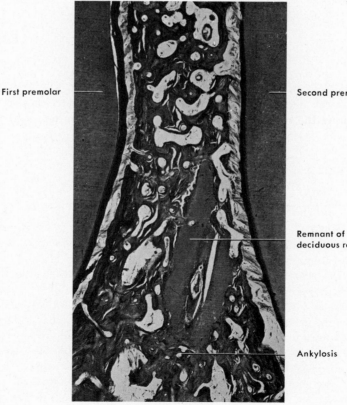

First premolar   Second premolar

Remnant of deciduous root

Ankylosis

Fig. 12-19. Remnant of deciduous tooth embedded in and ankylosed to the bone. (From Schoenbauer, F.: Z. Stomatol. **29:**892, 1931.)

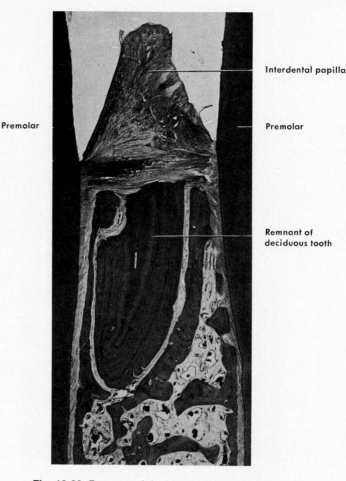

Interdental papilla

Premolar

Premolar

Remnant of
deciduous tooth

Fig. 12-20. Remnant of deciduous tooth at alveolar crest.

Retained deciduous teeth. Deciduous teeth may be retained for a long time beyond their usual shedding schedule. Such teeth are usually without permanent successors, or their successors are impacted. They are invariably out of function. Retained deciduous teeth are most often the upper lateral incisor (Fig. 12-21, *A*), less frequently the second permanent premolar, especially in the mandible (Fig. 12-21, *B*), and rarely the lower central incisor (Fig. 12-21, *C*). If a permanent tooth is ankylosed or

impacted, its deciduous predecessor may also be retained (Fig. 12-21, *D*). This is most frequently seen with the deciduous and permanent canine teeth.

If the permanent lateral incisor is missing, the deciduous tooth is often resorbed under the pressure of the erupting permanent canine. This resorption may be simultaneous with that of the deciduous canine (Fig. 12-22). Sometimes the permanent canine causes resorption of the deciduous lateral incisor only and erupts

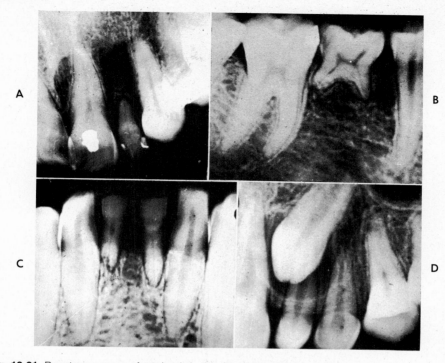

Fig. 12-21. Roentgenograms of retained deciduous teeth. **A,** Upper permanent lateral incisor missing, and deciduous tooth retained (age 56 years). **B,** Lower second premolar missing and deciduous molar retained. Roots partly resorbed. **C,** Permanent lower central incisors missing and deciduous teeth retained. **D,** Upper permanent canine embedded and deciduous canine retained. (**A** and **B** courtesy Dr. M.K. Hine, Indiana University. **C** and **D** courtesy Dr. Rowe Smith, Texarkana, Tex.)

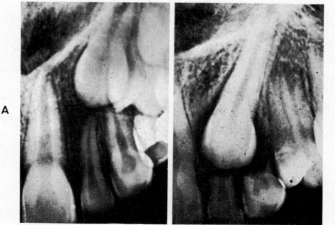

Fig. 12-22. Upper permanent lateral incisor missing. Deciduous lateral incisor and deciduous canine are resorbed because of pressure of erupting permanent canine. **A,** At 11 years of age. **B,** At 13 years of age.

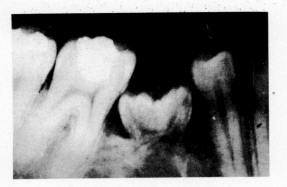

Fig. 12-23. Submerging deciduous lower second molar. Second premolar missing. (Courtesy Dr. M.K. Hine, Indiana University.)

in its place. In such cases the deciduous canine may be retained distally to the permanent canine. A supernumerary tooth or an adontogenic tumor may occasionally prevent the eruption of one or more of the permanent teeth. In such cases ankylosis of the deciduous tooth may occur.

Submerged deciduous teeth. Trauma may result in damage to either the dental follicle or the developing periodontal ligament. If this happens, the eruption of the tooth ceases and it becomes ankylosed to the bone of the jaw. Because of continued eruption of neighboring teeth and increased height of the alveolar bone, the ankylosed tooth may be either "shortened" (Fig. 12-23) or submerged in the alveolar bone. Submerged deciduous teeth prevent the eruption of their permanent successors or force them from their position. Submerged deciduous teeth should therefore be removed as soon as possible.

REFERENCES

Boyde, A., and Lester, K.S.: Electron microscopy of resorbing surfaces of dental hard tissues, Z. Zellforsch. **83:**538, 1967.

Freilich, L.S.: Ultrastructure and acid phosphatase cytochemistry of odontoclasts: effect of parathyroid extract. J. Dent. Res. **50:**1047, 1971.

Furseth, R.: The resorption processes of human deciduous teeth studied by light microscopy, microradiography and electron microscopy, Arch. Oral Biol. **13:**417, 1968.

Kronfeld, R.: The resorption of the roots of deciduous teeth, Dent. Cosmos **74:**103, 1932.

Morita, H., Yamashiya, H., Shimizu, M., and Sasaki, S.: The collagenolytic activity during root resorption of bovine deciduous tooth, Arch. Oral Biol. **15:**503, 1970.

Owen, M.: Histogenesis of bone cells, Calcif. Tissue Res. **25:**205, 1978.

Weatherell, J.A., and Hargreaves, J.A.: Effect of resorption on the fluoride content of human deciduous dentine, Arch. Oral Biol. **11:**749, 1966.

Westin, G.: Über Zahndurchbruch and Zahnwechsel, Z. Minkrosk. Anat. Forsch. **51:**393, 1942.

Yaeger, J.A., and Kraucunas, E.: Fine structure of the resorptive cells in the teeth of frogs, Anat. Rec. **164:**1, 1969.

13

TEMPOROMANDIBULAR JOINT

ANATOMIC REMARKS

DEVELOPMENT OF THE JOINT

HISTOLOGY
Bony structures
Articular fibrous covering
Articular disc
Articular capsule
Innervation and blood supply

CLINICAL CONSIDERATIONS

ANATOMIC REMARKS

The temporomandibular joint is formed by the articulation between the articular tubercle and the anterior part of the mandibular fossa of the temporal bone above and the condylar head of the mandible below.

The articular disc (meniscus) is interposed between the articular surfaces of the two bones. It is an oval, fibrous plate that fuses at its anterior margin with the fibrous capsule. Its posterior border is connected to the capsule by loose connective tissue, which allows its anterior movement (Fig. 13-1). Its medial and lateral corners are directly attached to the poles of the condyle. The articular space is divided into two compartments: a lower, between the condyle and the disc (condylodiscal), and an upper, between the disc and temporal bone (temporodiscal). The disc is biconcave in sagittal section, with a thin intermediate zone, a thick anterior band, and a thick posterior band. The latter is continuous with a loose fibroelastic portion called the bilaminar zone, which is highly vascular and highly innervated. The superior stratum or lamina of the bilaminar zone attaches to the posterior wall of the mandibular or glenoid fossa and the squamotympanic suture, while the inferior stratum attaches to the back of the mandibular condyle (Fig. 13-2).

Some fibers of the lateral pterygoid muscle attach to the anterior border of the disc. The disc produces a movable articulation for the condyle. In the inferior portion of the joint, rotational movement about an axis through the heads of the condyles permits opening of the jaws. This is designated as a hinge movement. The superior portion of the joint permits a translatory movement as the discs and the condyles traverse anteriorly along the inclines of the articular tubercles to provide an anterior and inferior movement of the mandibular head.

The articular capsule is a fibrous sac that attaches anteriorly to the articular tubercle and posteriorly to the lips of the squamotympanic fissure and between the two attachments to the margins of the mandibular fossa and to the neck of the mandible below. It is strengthened laterally by the temporomandibular ligament (lateral ligament). The inner aspect of the capsule is lined by a synovial membrane, which is

A   B

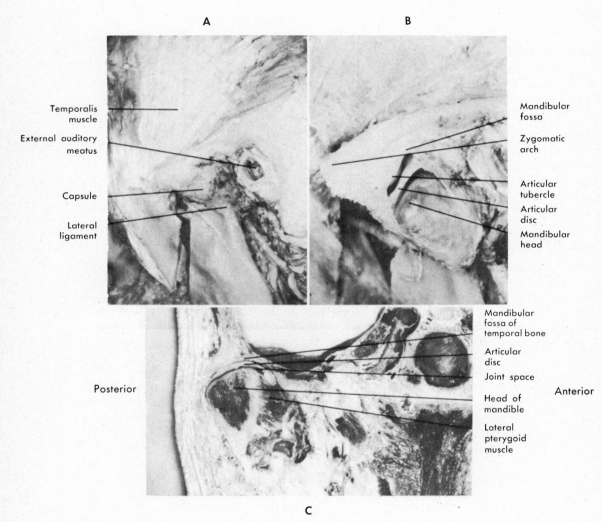

Temporalis
muscle

External auditory
meatus

Capsule

Lateral
ligament

Mandibular
fossa

Zygomatic
arch

Articular
tubercle

Articular
disc

Mandibular
head

Posterior

Mandibular
fossa of
temporal bone

Articular
disc

Joint space

Head of
mandible

Lateral
pterygoid
muscle

Anterior

C

Fig. 13-1. A, Lateral view of temporomandibular joint with capsule and lateral ligament in situ. **B,** Sagittal section through temporomandibular joint. **C,** Frontal section of head through condyle of mandible. (Courtesy Dr. F.R. Suarez, Georgetown University, Washington, D.C.)

A

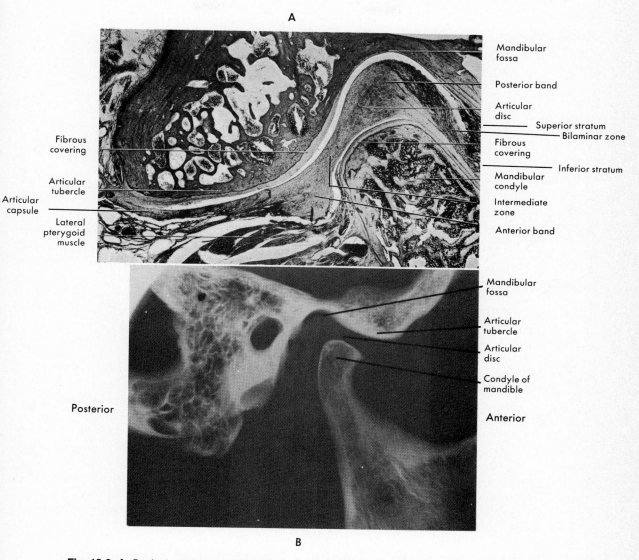

Mandibular
fossa

Posterior band

Articular
disc

Superior stratum

Bilaminar zone

Fibrous
covering

Inferior stratum

Mandibular
condyle

Intermediate
zone

Anterior band

Fibrous
covering

Articular
tubercle

Articular
capsule

Lateral
pterygoid
muscle

Mandibular
fossa

Articular
tubercle

Articular
disc

Condyle of
mandible

Posterior

Anterior

B

Fig. 13-2. A, Sagittal section through temporomandibular joint. **B,** Temporomandibular joint sagittal radiograph showing condyle of mandible, articular tubercle, mandibular fossa, and articular disc.

especially well developed behind the disc. It lines the capsule in each of the two cavities but does not extend over the surfaces of the discs, the articular tubercle, or the condyle.

DEVELOPMENT OF THE JOINT

At approximately 10 weeks, the components of the future joint show the first indication in the mesenchyme between the condylar cartilage of the mandible and the developing temporal bone. Two slitlike joint cavities and an intervening disc make their appearance in this region at 12 weeks. The mesenchyme around the joint begins to form the fibrous joint capsule. Very little is known about the significance of newly forming muscles in joint formation.

HISTOLOGY

Bony structures. The condyle of the mandible is composed of cancellous bone covered by a thin layer of compact bone (Fig. 13-2, A). The trabeculae are grouped in such a way that they radiate from the neck of the mandible and reach the cortex at right angles, thus giving maximal strength to the condyle. The large marrow spaces decrease in size with progressing age by a noticeable thickening of the trabeculae. The red marrow in the condyle is of the myeloid or cellular type. In older individuals it is sometimes replaced by fatty marrow.

During the period of growth a layer of hyaline catilage lies underneath the fibrous covering of the condyle. This cartilaginous plate grows by apposition from the deepest layers of the covering connective tissue. At the same time its deep surface is replaced by bone (Fig. 13-3). Remnants of this cartilage may persist into old age (Fig. 13-4).

The roof of the mandibular fossa (Fig. 13-2) consists of a thin, compact layer of bone. The articular tubercle is composed of spongy bone covered with a thin layer of compact bone. In rare cases islands of hyaline cartilage are found in the articular tubercle.

Articular fibrous covering. The condyle as well as the articular tubercle is covered by a rather thick layer of fibrous tissue containing a variable number of chondrocytes. The fibrous covering of the mandibular condyle is of fairly even thickness (Fig. 13-4). Its superficial layers consist of a network of strong collagenous fibers. Chondrocytes may be present, and they have a tendency to increase in number with age. They can be recognized by their thin capsule, which stains heavily with basic dyes. The deepest layer of the fibrocartilage is rich in chondroid cells as long as growing hyaline cartilage is present in the condyle. It contains only a few thin collagenous fibers. In this zone the appositional growth of the hyaline cartilage of the condyle takes place.

The fibrous layer covering the articulating surface of the temporal bone (Fig. 13-5) is thin in the articular fossa and thickens rapidly on the posterior slope of the articular tubercle (Fig. 13-2, A). In this region the fibrous tissue shows a definite arrangement in two layers, with a small transitional zone between them. The two layers are characterized by the different course of the constituent fibrous bundles. In the inner zone the fibers are at right angles to the bony surface. In the outer zone they run parallel to that surface. As in the fibrous covering of the mandibular condyle, a variable number of chondrocytes are found in the tissue on the temporal surface. In adults the deepest layer shows a thin zone of calcification.

There is no continuous cellular lining on the free surface of the fibrocartilage. Only isolated fibroblasts are situated on the surface itself. They are characterized by the formation of long, flat cytoplastmic processes.

Articular disc. In young individuals the articular disc is composed of dense fibrous tissue. The interlacing fibers are straight and tightly packed (Fig. 13-6). Elastic fibers are found only in relatively small numbers. The fibroblasts in the disc are elongated and send flat cytoplasmic wiglike processes into the interstices between the adjacent bundles.

With advancing age, some of the fibroblasts develop into chondroid cells, which later may

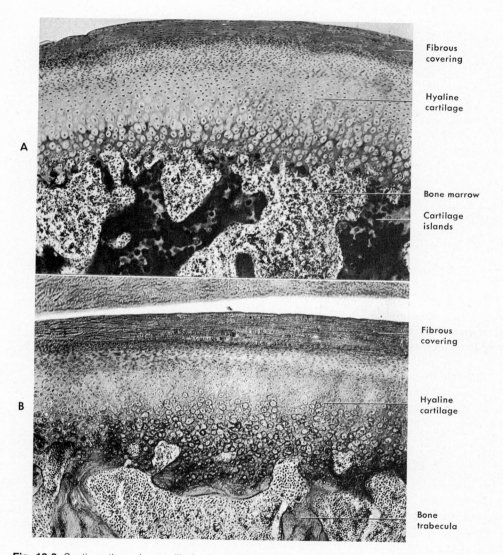

Fig. 13-3. Sections through mandibular head. **A,** Newborn infant. **B,** Young adult. Note transitional zone between fibrous covering and hyaline cartilage, characteristic for appositional growth of cartilage.

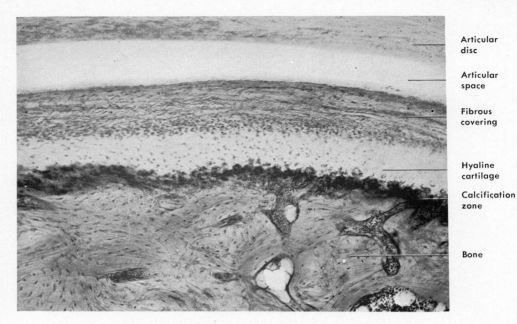

Articular disc

Articular space

Fibrous covering

Hyaline cartilage

Calcification zone

Bone

Fig. 13-4. Higher magnification of part of mandibular condyle shown in Fig. 13-2, *A*.

Bone

Calcification zone

Inner fibrous layer

Outer fibrous layer

Articular space

Articular disc

Fig. 13-5. Higher magnification of articular tubercle shown in Fig. 13-2, *A*.

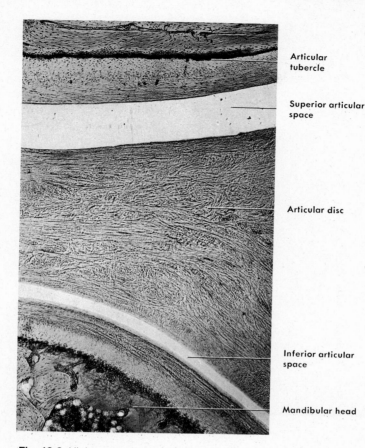

Articular
tubercle

Superior articular
space

Articular disc

Inferior articular
space

Mandibular head

Fig. 13-6. Higher magnification of articular disc shown in Fig. 13-2, *A*.

differentiate into true chondrocytes. Even
small islands of hyaline cartilage may be found
in the discs of older persons. Chondroid cells,
true cartilage cells, and hyaline ground substance develop in situ by differentiation of the
fibroblasts. In the discs, as well as in the fibrous tissue covering the articular surfaces, this
cellular change seems to be dependent on mechanical influences. The presence of chondrocytes may increase the resistance and resilience
of the fibrous tissue.

The fibrous tissue covering the articular eminence and mandibular condyle, as well as the
large central area of the disc, is devoid of blood
vessels and nerves and has limited reparative
ability.

Articular capsule. As in all other joints, the
articular capsule consists of an outer fibrous
layer that is strengthened on the lateral surface
to form the temporomandibular ligament. The
articular capsule is lined with synovial membrane, which folds synovial villi. Synovial villi
project into the joint spaces (Fig. 13-7). The synovial membrane consists of internal cells,
which do not form a continuous layer but show
gaps between the cells, and the subintimal connective tissue layer, rich in blood capillaries.
The internal cells are of three types. The first

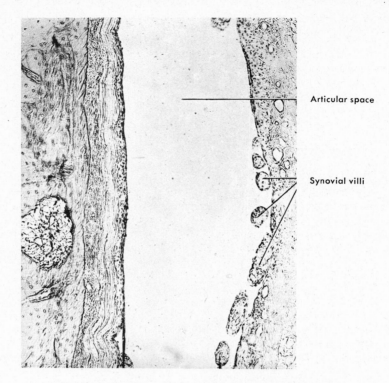

Fig. 13-7. Villi on synovial capsule of temporomandibular joint.

is rich in RER and is called the fibroblast-like or B cell. The second type is rich in Golgi complex, contains little or no RER, and is called the macrophage-like or A cell. The third type has a cellular morphology between cell types A and B.

A small amount of a clear, straw-colored viscous fluid, synovial fluid, is found in the articular spaces. It is a lubricant and also a nutrient fluid for the avascular tissues covering the condyle and the articular tubercle and for the disc. It is elaborated by diffusion from the rich capillary network of the synovial membrane, augmented by mucin possibly secreted by the synovial cells.

Innervation and blood supply. Sensations from the joint structures have usually been considered proprioceptive in nature. The auriculotemporal and masseteric nerves from the mandibular branch of the trigeminal nerve supply the joint. It has been suggested that there exist a number of free, complex, and encapsulated receptors among the synovial villi of the joint capsule. The primary source of blood supply comes from the superficial temporal and the maxillary arteries of the external carotid.

CLINICAL CONSIDERATIONS

The thinness of the bone in the articular fossa is responsible for fractures if the mandibular head is driven into the fossa by a heavy blow. In such cases injuries of the dura mater and the brain have been reported.

The finer structure of the bone and its fibrocartilaginous covering depends on mechanical influences. A change in force or direction of stress, occurring especially after loss of posterior teeth, may cause structural changes. These are characterized by degeneration of the fibrous covering of the articulating surfaces and of the disc. Abnormal functional activity produces injury to the fibrous covering and the articular bones. Compensation and partial repair may be accomplished by the development of hyaline cartilage on the condylar surface and in the disc. In severe trauma the articular bone is destroyed, and cartilage and new bone develop in the marrow spaces and at the periphery of the condyle. Then the function of the joint is severely impaired.

Normally, in the open position of the mandible, the interincisal distance is about 48 mm in males and 45.5 mm in females. In about 18% of the population the mandible deviates on opening, and in almost 86% of this group deviation is to the left. In about 35% of the population the temporomandibular joint produces sounds in opening movements. These joints display palpable irregularities and produce popping and clicking noises. However, use of a stethoscope reveals that about 65% of temporomandibular joints produce these sounds. This feature therefore, by itself, is not a sign of disease and does not need treatment.

The term "myofacial pain dysfunction syndrome" is used to indicate a dysfunction of the temporomandibular joint. It is characterized by (1) masticatory muscle tenderness (most frequently the lateral pterygoid and then, in order, the temporalis, medial pterygoid, and masseter), (2) limited opening of the mandible (less than 37 mm), and (3) joint sounds. This symptom complex is seen more often in females than in males. Its cause is usually the spasm of the masticatory muscles. It may be related to stress, and treatment should be as conservative as possible. Invasive therapy, splinting, occlusal grinding, occlusal reconstruction, and other similar measures are frequently implemented and usually represent overtreatment.

Dislocation of the temporomandibular joint may take place without the impact of an external force. The dislocation of the jaw is usually bilateral and the displacement is anterior. When the mouth is opened unusually wide during yawning, the head of the mandible may slip forward into the infratemporal fossa, causing articular dislocation of the joint.

REFERENCES

Bauer, W.: Anatomische and mikroskopische Untersuchungen über das Kiefergelenk (Anatomical and microscopic investigations on the temporomandibular joint), Z. Stomatol. **30**:1136, 1932.

Bauer, W.H.: Osteo-arthritis deformans of the temporomandibular joint, Am. J. Pathol. **17**:129, 1941.

Bernick, S.: The vascular and nerve supply to the temporomandibular joint of the rat, Oral Surg. **15**:488, 1962.

Breitner, C.: Bone changes resulting from experimental orthodontic treatment. Am. J. Orthod. **26**:521, 1940.

Cabrini, R., and Erausquin, J.: La articulación temporomaxilar de la rata (Temporomandibular joint of the rat), Rev. Odont. Buenos Aires, **29**:385, 1941.

Choukas, N.C., and Sicher, H.: The structure of the temporo-mandibular joint. Oral Surg. **13**:1263, 1960.

Cohen, D.W.: The vascularity of the articular disc of the temporo-mandibular joint, Alpha Omegan, Sept. 1955.

Cowdry, E.V.: Special cytology, ed. 2, New York, 1932, Paul B. Hoeber, Med. Book Div. of Harper & Brothers.

Gross, A., and Gale, E.N.: A prevalence study of the clinical signs associated with mandibular dysfunction, JADA **107**:932, 1983.

Kawamura, Y.: Recent concepts of physiology of mastication, Adv. Oral Biol. **1**:102, 1964.

Kreutziger, K.L., and Mahan, P.E.: Temporomandibular degenerative joint disease. Part I. Anatomy, pathophysiology and clinical description, Oral Surg. **40**:165, 1975.

Kreutziger, K.L. and Mahan, P.E.: Temporomandibular degenerative joint disease. Part II. Diagnostic procedure and comprehensive management, Oral Surg. **40**:297, 1975.

Lipke, D.P., et al.: An Electromyographic study of the human lateral pterygoid muscle, J. Dent. Res. **56** (Special issue B: B230), 1977, (Abstract no. 713).

McLeran, J.H., et al.: A Cinefluorographic analysis of the temporomandibular joint, J. Am. Dent. Assoc. **75**:1394, 1967.

McNamara, J.A.: The independent functions of the two heads of the lateral pterygoid muscle, Am. J. Anat. **138**:197, 1973.

Mathews, M.P., and Moffett, B.C.: Histologic maturation and initial aging of the human temporomandibular joint, J. Dent. Res. **53** (Special issue: 246), 1974, (Abstract no. 765).

Moffett, B.: The morphogenesis of the temporomandibular joint, Am. J. Ortho. **52**401, 1966.

Payne, G.S.: The effect of intermaxillary elastic force on the temporomandibular articulation in the growing macaque monkey, Am. J. Ortho. **60:**491, 1971.

Radin, E.L., et al.: Response of joints to impact loading III. Relationships between trabecular microfractures and cartilage degeneration, J. Biomech. **6:**51, 1973.

Ramfjord, S.P., and Ash, M.M.: Occlusion, Philadelphia, 1966, W.B. Saunders Co.

Rees, L.A.: Structure and function of the mandibular joint, Br. Dent. J. **96:**6, 1954.

Sarnat, B.G.: The temporomandibular joint, ed. 2, Springfield, Ill., 1964, Charles C Thomas, Publisher.

Schaffer, J.: Die Stützgewebe (Supporting tissues). In von Möllendorff, W., editor: Handbunch der mikroskopischen Anatomie des Menschen, Berlin, 1930, Julius Springer Verlag, vol. 2, pt. 2.

Shapiro, H.H., and Truex, R.C.: The temporomandibular joint and the auditory function, J. Am. Dent. Assoc. **30:**1147, 1943.

Sicher, H.: Some aspects of the anatomy and pathology of the temporomandibular articulation, N.Y. State Dent. J. **14:**451, 1948.

Sicher, H.: Temporomandibular articulation in mandibular overclosure, J. Am. Dent. Assoc. **36:**131, 1948.

Sicher, H.: Positions and movements of the mandible, J. Am. Dent. Assoc. **48:**620, 1954.

Sicher, H.: Structural and functional basis for disorders of the temporomandibular articulation, J. Oral Surg. **13:**275, 1955.

Steinhardt, G.: Die Beansprunchung der Gelenkflächen bei verschiedenen Bissarten (Investigations on the stresses in the mandibular articulation and their structural consequences), Deutsch. Zahnheilk. Vortr. **91:**1, 1934.

Strauss, F., et al.: The Architecture of the disk of the human temporomandibular joint, Helv. Odont. Acta. **4:**1, 1960.

Thilander, B.: Innervation of the temporomandibular joint capsule in man, tr. Roy Schools Den. Stockholm Umea **7:**1,1961.

Toller, P.A.: Osteoarthrosis of the mandibular condyle, Brit. Dent. J. **134:**223, 1973.

Toller, P.A.: Opaque arthrography of the temporomandibular joint, Int. J. Oral Surg. **3:**17, 1974.

Yavelow, I., and Arnold, G.S.: Temporomandibular joint clicking, Oral Surg. **32:**708, 1971.

14
MAXILLARY SINUS

DEFINITION

HISTORICAL REVIEW

DEVELOPMENTAL ASPECTS

DEVELOPMENTAL ANOMALIES

STRUCTURE AND VARIATIONS

MICROSCOPIC FEATURES

FUNCTIONAL IMPORTANCE

CLINICAL CONSIDERATIONS

DEFINITION

The maxillary sinus is the pneumatic space that is lodged inside the body of the maxilla and that communicates with the environment by way of the middle nasal meatus and the nasal vestibule.

HISTORICAL REVIEW

A recent publication entitled *Eighteen Hundred Years of Controversy: The Paranasal Sinuses* (Blanton and Biggs, 1969) reflects quite accurately the present confused state of knowledge about the pneumatic cavities. The maxillary sinus, more than any other of these cavities, has been subjected to peculiar interpretations throughout history. As early as the second century, Galen (AD 130-201) made the first known descriptive remarks about the adult maxillary sinus. In the following centuries many prominent scientists (Leonardo da Vinci, 1452-1519; Berengar, 1507-1527; Massa, 1542; Vesalius, 1542; Fallopius, 1600; Veslingius, 1637; Spigelius, 1645; Highmore, 1651; Schneider, 1655; Bartholinus, 1658; Morgagni, 1723; Boerhaave, 1735; and Haller, 1763—cited by Blanton and Biggs) contributed to the ever-increasing knowledge of the structure and function of the paranasal cavities.

Despite historical uncertainty about the specific contribution of each of these researchers, it is widely accepted that Highmore was the first to describe in detail the morphology of the maxillary sinus and to advance the idea of pneumatization by the sinuses. In later centuries the interest of investigators focused on the mechanism of pneumatizing processes and the functional significance of the paranasal sinuses as a whole, in addition to the structural, dimensional, sexual, racial, environmental, and developmental diversity among the sinuses.

DEVELOPMENTAL ASPECTS

The initial development of the maxillary sinus follows a number of morphogenic events in the differentiation of the nasal cavity in early gestation (about 32 mm crown-rump length [CRL] in an embryo). First, the horizontal shift of palatal shelves and subsequent fusion of the shelves with one another and with the nasal septum separate the secondary oral cavity from two secondary nasal chambers (see Chapter 1). This modification presumably influences further expansion of the lateral nasal wall in that the wall begins to fold; thus three nasal conchae and three subjacent meatuses arise. The inferior and superior meatuses remain as shallow depressions along the lateral nasal wall for approximately the first half of the intrauterine life; the middle meatus expands immediately into the lateral nasal wall. Because the cartilaginous skeleton of the lateral nasal capsule is already established, expansion of the middle meatus proceeds primarily in an inferior direction, occupying progressively more of the future maxillary body (Fig. 14-1).

The maxillary sinus thus established in the embryo of about 32 mm CRL expands vertically into the primordium of the maxillary body and reaches a diameter of 1 mm in the 50 mm CRL fetus (at this time the first glandular primordia from the maxillary sinus epithelium are apparent), 3.5 mm in the 160 mm CRL fetus, and 7.5 mm in the 250 mm CRL fetus (Vidić). In the perinatal period the human maxillary sinus measures about 7 to 16 mm (standard deviation [SD] 2.64) in the anteroposterior direction, 2 to 13 mm (SD 1.52) in the superoinferior direction, and 1 to 7 mm (SD 1.18) in the mediolateral direction (Cullen and Vidić). According to Shaeffer these diameters increase to 15, 6, and 5.5 mm, respectively, at the age of 1 year, to 31.5, 19, and 19.5 mm at the age of 15 years, and to 34, 33, and 23 mm in the adult. Although the exact time at which the hu-

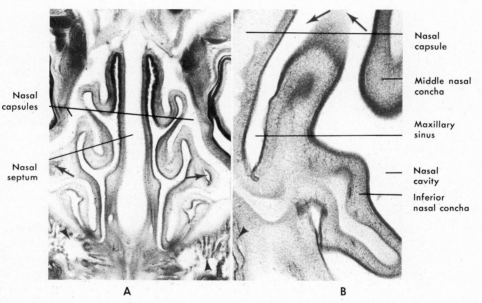

Fig. 14-1. A, This coronal section of a human fetal head (60 mm CRL) demonstrates both nasal cavities bordered by nasal septum medially, three conchae and subjacent meatuses laterally, and palate, which already shows extensive centers of ossification inferiorly *(arrowheads at bottom)*. Lateral to conchae and continuous with their cartilaginous skeletons is nasal capsule. Developing maxillary sinuses on both sides of midline are indicated by arrows. (Hematoxylin and eosin stain; ×27.) B, This coronal section demonstrates nasal cavity, maxillary sinus, and inferior and middle nasal conchae in a 69 mm CRL fetus. Sinus grows into maxilla in an inferior direction parallel to plane of cartilaginous nasal capsule. Skeleton of both conchae is cartilaginous, while in maxilla several centers of ossification *(arrowhead)* are present. Communication between middle nasal meatus and maxillary sinus is indicated by arrows. (Hematoxylin and eosin stain; ×67.5.) (**A,** No. 1183; **B,** No. 4291; courtesy Dr. Ronan O'Rahilly, Carnegie Laboratories of Embryology, Davis, Calif.)

man maxillary sinus attains its definite size is not known, the sinus appears to expand and modify in form until the time of eruption of all permanent teeth.

DEVELOPMENTAL ANOMALIES

Agenesis (complete absence), aplasia, and hypoplasia (altered development or underdevelopment) of the maxillary sinus occurs either alone or in association with other anomalies, for example, choanal atresia, cleft palate, high palate, septal deformity, absence of a concha, mandibulofacial dysostosis, malformation of the external nose, and the pathologic conditions of the nasal cavity as a whole (Gouzy et al., Rosenberger, Schürch, Blair et al., Mocellin, Fatin, Eckel and Beisser, and Blumenstein). The supernumerary maxillary sinus, on the other hand, is the occurrence of two completely separated sinuses on the same side. This condition is most likely initiated by outpocketing of the nasal mucosa into the primordium of the maxillary body from two points either in the middle nasal meatus or in the middle and superior or middle and inferior nasal meatuses, respectively. Consequently, the result is two permanently separated ostia of the sinus.

STRUCTURE AND VARIATIONS

The maxillary sinus is subject to a great extent of variation in shape, size and mode of developmental pattern. It is inconceivable therefore to propose any structural description that would satisy the majority of human maxillary sinuses. Usually, however, the sinus is described as a four-sided pyramid, the base of which is facing medially toward the nasal cavity and the apex of which is pointed laterally toward the body of the zygomatic bone (Fig. 14-2). The four sides are related to the surface of the maxilla in the following manner: (1) anterior, to the facial surface of the body; (2) inferior, to the alveolar and zygomatic processes; (3) superior, to the orbital surface; and (4) posterior, to the infratemporal surface. The four sides of the sinus, which are usually distant from one another medially, converge laterally and meet at an obtuse angle. The identity of each of the four sides is somewhat difficult to discern and the transition of the surface from one side to the other is usually poorly defined. Thus it is apparent that the comparison of the sinus space to a geometrically well-defined body is of pedagogic value only.

The base of the sinus, which is the thinnest of all the walls, presents a perforation, the ostium, at the level of the middle nasal meatus (Fig. 14-3). In some individuals, in addition to the main ostium, two or many more accessory ostia connect the sinus with the middle nasal meatus. In 5.5% of instances the main ostium is located within the anterior third of the hiatus semilunaris, in 11% within the middle third, and 71.7% within the posterior third, and in 11.3% the ostium is found outside and in a posterior position to the hiatus semilunaris. The accessory ostia are found in 23% of these instances in the middle nasal meatus (Van Alyea) and occur rarely in the inferior nasal meatus (Delaney and Morse).

In the course of development the maxillary sinus often pneumatizes the maxilla beyond the boundaries of the maxillary body. Some of the processes of the maxilla consequently become invaded by the air space. These expansions, referred to as the *recesses*, are found in the alveolar process (50% of all instances), zygomatic process (41.5% of all instances), frontal process (40.5% of all instances), and palatine process (1.75% of all instances) of the maxilla (Hajniš et al). The occurrence of the zygomatic recess usually brings the superior alveolar neurovascular bundles into proximity with the space of the sinus. The frontal recess invades and sometimes surrounds the content of the infraorbital canal, whereas the alveolopalatine recesses reduce the amount of the bone between the dental apices and the sinus space. The latter de-

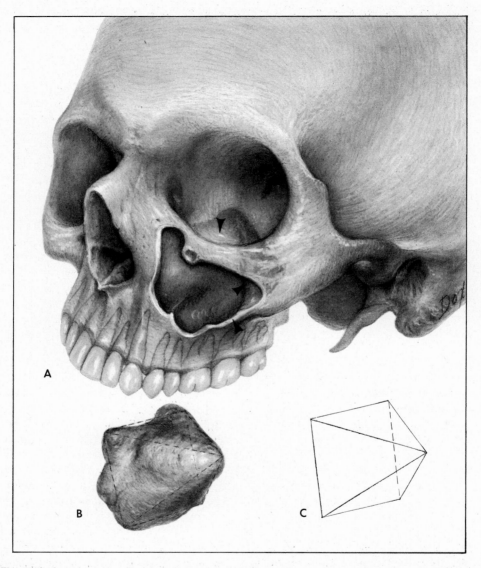

Fig. 14-2. A, Left semiprofile of skull showing maxillary sinus opened through anterior wall. Outline of superior, posterior, and inferior walls of sinus are respectively indicated in relation to floor of orbit *(upper arrowhead),* infratemporal surface of maxilla *(middle arrowhead),* and alveolar and zygomatic processes of maxilla *(lower arrowhead).* Basal wall separates space of sinus from nasal cavity. **B** and **C,** Polysulfate rubber cast of maxillary sinus about 10 ml in volume and an idealized geometric form of sinus, respectively. For convenience of visualizing the presumed pyramidal form of the sinus the orientation is the same for the sinus in situ, **A;** the cast of the sinus, **B;** and for closest geometric form to the sinus, **C.** (Courtesy Mr. and Mrs. B.F. Melloni, Department of Medical-Dental Communications, Georgetown University, Washington, D.C.)

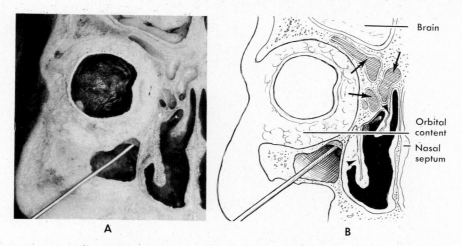

A   B

Fig. 14-3. A, Coronal section of adult female face was made approximately 7 mm anterior to ostium of maxillary sinus. **B,** Drawing made from **A** (proportion, 1:1). Probe indicates ostium, or communication between upper part of sinus lumen and middle nasal meatus. Several ethmoidal air cells *(arrows),* middle and inferior conchae *(two arrowheads),* orbital content, nasal septum, and frontal lobe of brain are also indicated. (**A,** Frontal section courtesy Dr. F.R. Suarez, Georgetown University, Washington, D.C. **B,** courtesy Mr. and Mrs. B.F. Melloni, Department of Medical-Dental Communications, Georgetown University, Washington, D.C.)

velopment most often pneumatizes the floor of the sinus adjacent to the roots of the first molar (Fig. 14-4) and less often to the roots of the second premolar, first premolar, and second molar, in that order of frequency (Osmont et al.). The fully developed alveolar recess is characterized by three depressions separated by two incomplete bony septa. The anterior depression, or fossa, corresponds to the original site of premolar buds, the middle to the molar buds, and the posterior to the third molar bud (Perović).

MICROSCOPIC FEATURES

Three microscopically distinct layers surround the space of the maxillary sinus: the epithelial layer, the basal lamina, and the subepithelial layer, including the periostium (Figs. 14-5 and 14-6). The epithelium, which is pseudostratified, columnar, and ciliated, is derived from the olfactory epithelium of the middle nasal meatus and therefore undergoes the same

pattern of differentiation as does the respiratory segment of the nasal epithelium proper. The most numerous cellular type in the maxillary sinus epithelium is the columnar ciliated cell. In addition, there are basal cells, columnar nonciliated cells, and mucus-producing, secretory goblet cells (Figs. 14-6 and 14-7). A ciliated cell encloses the nucleus and an electron-lucent cytoplasm with numerous mitochondria and enzyme-containing organelles. The basal bodies, which serve as the attachment of the ciliary microtubules to the cell, are characteristic of the apical segment of the cell. The cilia are typically composed of 9 + 1 pairs of microtubules, and they provide the motile apparatus to the sinus epithelium (Satir). By way of ciliary beating, the mucous blanket lining the epithelial surface moves generally from the sinus interior toward the nasal cavity.

The goblet cell displays all of the characteristic features of a secretory cell. In its basal segment the cell is occupied by, in addition to the

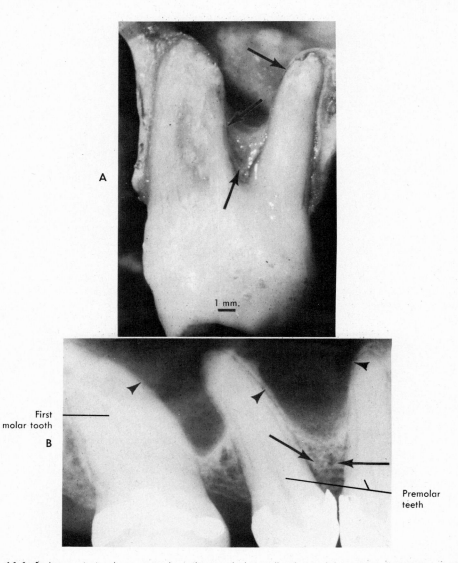

First
molar tooth

Premolar
teeth

Fig. 14-4. A demonstrates in a coronal section made immediately mesial to upper first molar tooth the gross relationships of the vestibular (left) and palatine (right) roots with floor of maxillary sinus. Amount of bone *(arrows)* interposed between roots and the space of the sinus is reduced at certain levels to a thin lamina. **B,** Lateral radiograph of both right upper premolar and first molar teeth. Maxillary sinus expands, in this instance, deep into alveolar process *(arrows)* between each two of the indicated teeth. Bony lamina separating dental roots from sinus *(arrowheads)* is extremely thin at certain levels. (**B** courtesy Dr. Donald Reynolds, Georgetown University, Washington, D.C.)

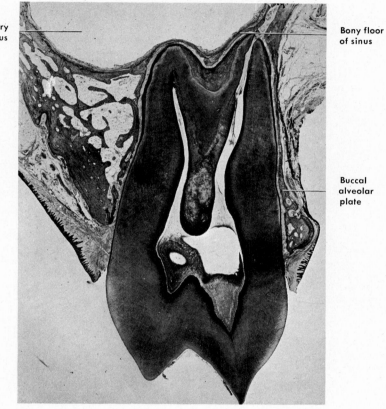

Maxillary
sinus

Bony floor
of sinus

Buccal
alveolar
plate

Fig. 14-5. Buccolingual section through first upper premolar. Apex is separated from sinus by thin
plate of bone.

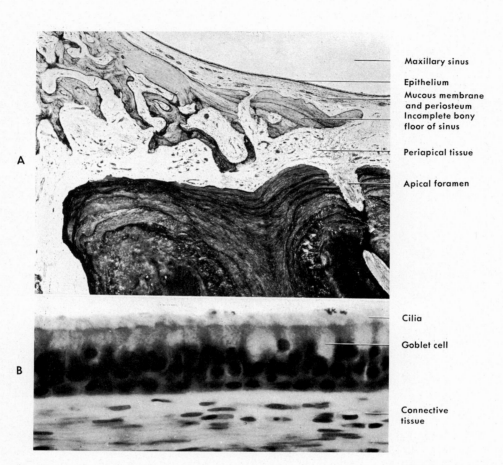

Fig. 14-6. Mucous membrane and epithelium of maxillary sinus. **A,** Apical region of second premolar. Lining of sinus is continuous with periapical tissue through openings in bony floor of sinus. **B,** High magnification of epithelium of maxillary sinus. (From Bauer, W.H.: Am. J. Orthodont. **29:**133, 1943.)

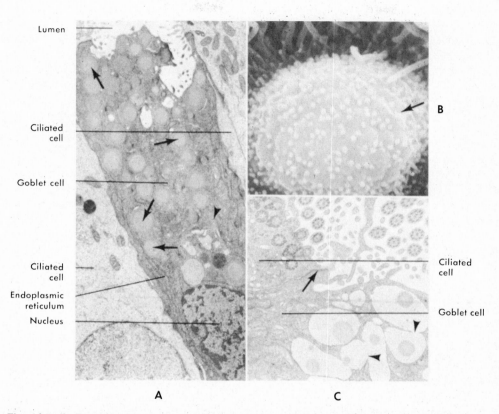

Lumen

Ciliated
cell

Goblet cell

Ciliated
cell

Endoplasmic
reticulum

Nucleus

A

B

Ciliated
cell

Goblet cell

C

Fig. 14-7. A, Electron micrograph of thin section (about 35 nm [350 Å]) taken from rat trachea. Goblet cell is surrounded by two ciliated cells. From nucleus toward lumen, goblet cell is occupied by endoplasmic reticulum, Golgi apparatus *(arrowhead),* and numerous secretory granules *(arrows).* Luminal surface of goblet cell is covered by short microvilli. **B,** Scanning electron micrograph taken from rat trachea demonstrates surface view of goblet cell *(arrow)* bordered above by numerous cilia from neighboring cells. In addition to microvilli, surface of goblet cell appears rough because of projection of apically situated secretory granules. **C,** Electron micrograph of thin section taken from human maxillary sinus demonstrates apical portions and surfaces of ciliated cell and goblet cell. Several secretory granules in goblet cell are demonstrated as either individual organelles or coalescing with one another *(arrowheads).* A junctional complex between the two cells is indicated by arrow. **(A,** Uranyl acetate and lead citrate stain; ×10,400. **B,** Fixed in aldehyde, dried by critical-point technique, and coated with a layer of gold-palladium about 20 nm [200 Å] thick; ×14,000. **C,** Uranyl acetate and lead citrate stain, ×22,400.)

nucleus, the cytocavitary network consisting of the rough and smooth endoplasmic reticulum and the Golgi apparatus, all of which are involved in the synthesis of the secretory mucosubstances. From the Golgi apparatus the zymogenic granules transport the mucopolysaccharides toward the cellular apex and finally release this material onto the epithelial surface by exocytosis (Fig. 14-8). In addition to the epithelial secretion, the surface of the sinus is provided with a mixed secretory product (se-

rous secretion consisting primarily of water with small amounts of neutral nonspecific lipids, proteins, and carbohydrates, and mucous secretion consisting of compound glycoproteins or mucopolysaccharides or both) from the subepithelial glands (Fig. 14-9). These are located in the subepithelial layer of the sinus and reach the sinus lumen by way of excretory ducts (Fig. 14-10), after the ducts have pierced the basal lamina.

On the basis of histochemical differentiation

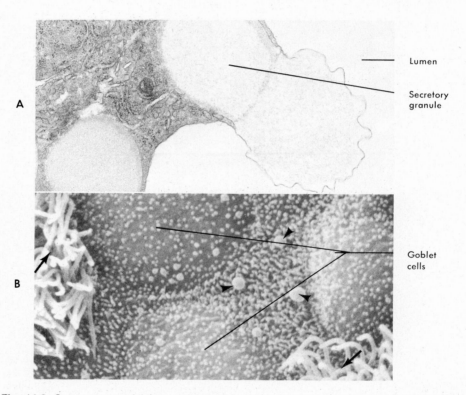

Lumen

Secretory granule

Goblet cells

Fig. 14-8. Secretory material from goblet cell is released into lumen by exocytosis. **A,** Electron micrograph of thin section taken from rat trachea shows a secretory granule in process of extrusion from cell into lumen. **B,** Scanning electron micrograph taken from rat trachea shows several goblet cells and parts of two ciliated cells *(arrows)*. Arrowheads indicate surface projection of secretory granules in process of extrusion from cell into lumen. (**A,** Uranyl acetate and lead citrate stain; ×48,000. **B,** Fixed in aldehydes, dried by critical-point technique, and coated with a layer of gold-palladium about 20 nm [200 Å] thick; ×11,600.)

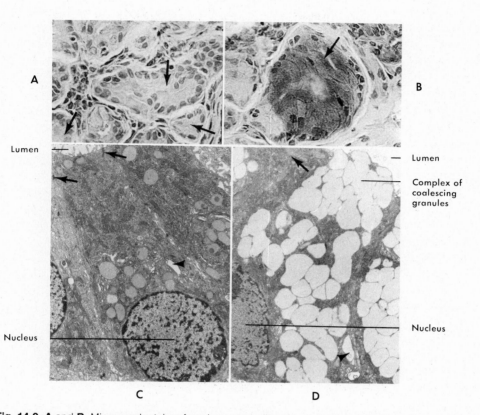

Fig. 14-9. A and **B,** Micrographs taken from human maxillary sinus demonstrate several serous acini (see arrows in **A**) and mucous acinus (see arrow in **B**). Note positive reaction of secretory material with alcian blue in mucous acinus and no reaction in serous gland. **C** and **D,** Electron micrographs illustrate respectively a thin section of several serous and mucous secretory cells taken from human submucosal maxillary gland. In both representative cells, from nucleus toward acinar lumen, cytoplasm is occupied by endoplasmic reticulum, mitochondria, secretory granules, and Golgi apparatus *(arrowheads)*. Note difference in electron opacity between the two types of secretory granules. Serous granules are separated from one another by respective membranes, while mucous granules frequently coalesce among them. (Note the complex of coalescing granules.) Junctional complexes between cells in both illustrations are indicated by arrows. (**A,** Alcian blue and fast red procedure; ×1900 for serous acini and ×2000 for mucous acinus. **B,** Uranyl acetate and lead citrate stain; ×7200 for serous gland and ×6750 for mucous gland.)

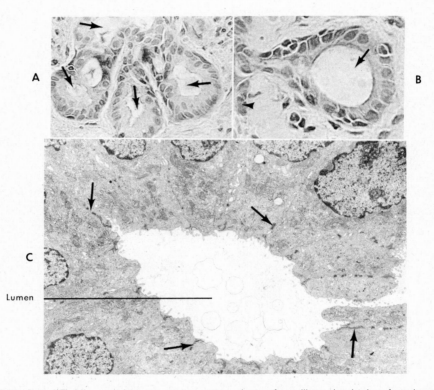

Fig. 14-10, A and **B,** Micrographs represent excretory ducts of maxillary gland taken from human maxillary sinus. Ductal cells, from cuboid to columnar in shape, surround lumen *(arrows),* which measures in these instances up to 12.5 μm in radius. **B,** Duct is demonstrated in a close apposition to epithelium of sinus *(arrowhead).* **C,** Thin section of several cells lining lumen of excretory duct from human maxillary gland. In addition to nucleus, these cells contain endoplasmic reticulum, Golgi apparatus, numerous mitochondria, lipid droplets, and occasional lysosomes. *Arrows,* Many junctional complexes between ductal cells. (**A** and **B,** Alcian blue and fast red procedure; **C,** uranyl acetate and lead citrate stain; **A** to **C,** ×1600, ×2400, ×6900.)

and fine structural characteristics (Vidić and Tandler) it is evident that the acini of subepithelial glands contain in varying proportions two types of secretory cells, serous and mucous. The serous cell is stained with ninhydrin-Schiff and sudan black B procedures and encloses an electron-dense, homogeneous secretory material. The mucous cell reacts positively with the alcian blue 8GX procedure for acid sialomucin or sulfomucin or both and produces an electron-lucent, heterogeneous secretory material. The myoepithelial cells (Fig. 14-11) surround the acini composed of either both secretory cells or a pure population of cells of either secretory type.

The secretion from these glands, like that of the other exocrine glands, is controlled by both divisions of the autonomic nervous system (Fig. 14-11). The autonomic axons, together with general sensory components, are supplied to

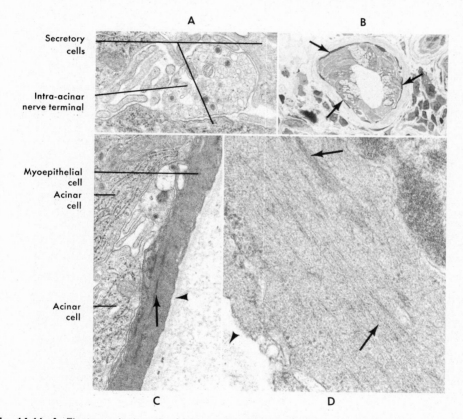

Fig. 14-11. A, Electron micrograph of intra-acinar nerve terminal in juxtaposition to two secretory cells taken from human maxillary gland. Note, in addition to mitochondria, two populations of small vesicles, dense and translucent, inside nerve terminal. **B,** Thin section (0.5 μm) of several mucous and ductal cells from human maxillary gland. Periphery of this acinus is surrounded by dark-appearing myoepithelial cells *(arrows).* **C** and **D** illustrate relationship between acinar cells and myoepithelial cell and numerous bundles of filaments *(arrows)* that occupy most of the cytoplasm of the myoepithelial cells, respectively. In both instances basement lamina adjacent to myoepithelial cell is indicated by arrowhead. **(A, C,** and **D,** Uranyl acetate and lead citrate stain; **B,** toluidine blue stain; **A** to **D,** ×37,500, ×1600, ×27,000, ×96,000.)

the maxillary sinus from the maxillary nerve complex. Numerous nonmyelinated and fewer myelinated axons are readily observable in the subepithelial layer of the sinus (Fig. 14-12). They are related here to the blood capillaries, fibroblasts, fibrocytes, collagen bundles, and other connective tissue elements.

FUNCTIONAL IMPORTANCE

Very little is known about the participation of the paranasal sinuses in the functioning of either the nasal cavity or the respiratory system as a whole. This is partially because of the relative inaccessibility of the sinuses to the sys-

temic functional studies, as well as because of the great variation in size of sinuses and their relationship to and communication with the nasal cavity. It is not surprising then that the theories of the functional importance of the sinuses range from no importance on the one hand to a multitude of involvements on the other hand.

The sinus is regarded by some as an accessory space to the nasal cavity, occurring only as a result of an inadequate process of ossification (Negus). In contrast, others report the functional contributions of the maxillary sinus in many aspects of olfactory and respiratory phys-

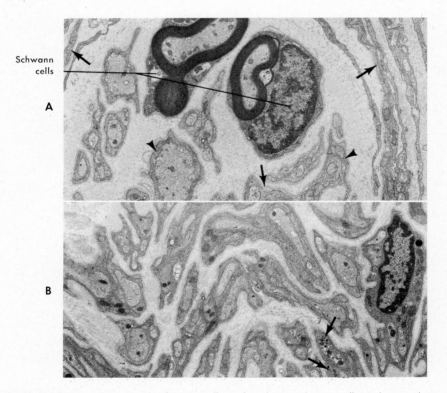

Fig. 14-12. A, Electron micrograph of two myelinated and several nonmyelinated axons in mucoperiosteal layer of human maxillary sinus. Schwann cells *(labeled and at arrowheads)* are intimately related to axons, which contain individual mitochondria and microtubules or microfilaments cut in different planes. *Arrows,* Connective tissue elements surrounding either individual axon or entire nerve. **B,** Nonmyelinated axons isolated from human mucoperiosteal layer of maxillary sinus. Most of them contain the same organelles as in **A.** However, some *(arrows)* are occupied by dense or translucent vesicles. (**A** and **B,** Uranyl acetate and lead citrate stain; **A,** ×11,200; **B,** ×19,200.)

iology. In individuals in whom the maxillary ostium is large enough and conveniently situated in the hiatus semilunaris the air pressure in the sinus fluctuates from ± 0.7 to ± 4 mm of water between the nasal expiration and inspiration (Lamm and Schaffrath). This dependence of the pressure in the sinus on the wave of respiration is, however, less probable in instances of either the small maxillary ostium or the ostium hidden in the depth of the hiatus semilunaris. On the basis of the same two conditions related to the structure and topography of the ostium some suggested functions attributed to the sinus by Koertvelyessy, Allen, Döderlein, Latkowski, and Doiteau (humidification and warming of inspired air and contribution to the olfaction, for instance) are subject to controversy. However, it is possible that if air is arrested in the sinus for a certain time, it quickly reaches body temperature and thus protects the internal structures, particularly the brain, against exposure to cold air (Koertvelyessy, Allen, Latkowski, and Mauer). The other contributions by paranasal cavities to the resonance of voice, lightening of the skull weight (Merideth, Nemours), enhancement of faciocranial resistance to mechanical shock, and the production of bactericidal lysozyme to the nasal cavity are reviewed in detail by Latkowski and by Blanton and Biggs.

CLINICAL CONSIDERATIONS

The section on developmental anomalies discusses several modifications of genetic and other origins in the developmental pathways of the maxillary sinus (agenesia, aplasia, hypoplasia, and supernumerary sinus). Some other criteria that correlate the extent of pneumatization by sinuses with the general dysfunctions of the endocrine system are by now developed. In the case of pituitary giantism, for example, all sinuses assume a much larger volume than in healthy individuals of the same geographic environment (Püschel and Schlosshauer). It is also known that in some congenital infections, such as by spirochetes in congenital syphilis,

the pneumatic processes are greatly suppressed, resulting in small sinuses (Richter).

In most respects the pathogenic relationship of the maxillary sinus to the orodental complexes is the result of topographic arrangement and of the functional and systemic association between the two territories. The transfer of a pathologic condition from the sinus to the orodental apparatus, or vice versa, is achieved either by mechanical connections or by way of the blood or lymphatic pathways. Since the upper first molar tooth is most often closest to the floor of the maxillary sinus, surgical manipulation on this tooth is most likely to break through the partitioning bony lamina and thus to establish an oroantral fistula (2.19% of all such fistulas are caused by first molars, 2.01% by second molars). If untreated, the lumen of such fistulas might epithelialize and permanently connect the maxillary space with the oral cavity. A similar condition might arise as a result of either a molar or a premolar radicular cyst, granuloma, or abscess. Hypercementosis of root apices and subsequent extraction of the affected tooth may also lead to a perforation. It is necessary therefore to consider on a radiograph the relationship between any such premolar or molar tooth with the floor of the maxillary sinus prior to surgical intervention.

The chronic infections of the mucoperiosteal layer of the sinus, on the other hand, might involve superior alveolar nerves, if these nerves are closely related to the sinus, and cause the neuralgia that mimics possible dental origin (Osmont et al.). In this instance the diagnosis must be based on a careful inspection of all the upper teeth as well as of the maxillary sinus to differentiate cause and eventual result of this condition. The neuralgia of the maxillary nerve (tic douloureux) could also have an etiologic origin in the superior dental apparatus or the mucoperiosteal layer of the sinus or both. For the diagnosis and treatment of this condition, it is most important to determine precisely the causal focus. Because of overlap of innervated territories and close topographic relationships between the teeth and the sinus, however, the causal focus is often difficult to assess.

The pathogenic association of the sinus with the orodental system, or vice versa, is based, in addition to a close topographic relationship, on an extensive vascular connection between these two regions by the superior alveolar vessels. As a consequence of this vascular arrangement, nonspecific bacterial sinusitis may be followed by some oral manifestations. Also the infections caused by the streptococci, staphylococci, pneumococci, or the virus of the common cold are likely to spread from either of the two regions to involve the other one. Finally, malignant lesions (adenocarcinoma, squamous cell carcinoma, osteosarcoma, fibrosarcoma, lymphosarcoma, etc.) of the maxillary sinus may produce their primary manifestation in the maxillary teeth. This may consist of pain, loosening supraeruption, or bleeding in their gingival tissue.

REFERENCES

Allen, B.C.: Applied anatomy of paranasal sinuses, J. Am. Osteopath. Assoc. **60**:978, 1961.

Ardouin, P.: Étude embryologique du développement du sinus maxillaire, Rev. Laryngol. Otol. Rhinol. (Bord.) **79**:834, 1958.

Blair, V.P., Brown, J.B., and Byars, L.T.: Observations on sinus abnormalities in congenital total and hemiabsence of the nose, Ann. Otol. Rhinol. Laryngol. **46**:592, 1937.

Blanton, P.L., and Biggs, N.L.: Eighteen hundred years of controversy: the paranasal sinuses, Am. J. Anat. **124**:135, 1969.

Blumenstein, G.: Die Entwicklung der Kieferhöhlen bei Rachenmandelhyperplasie, Nasenrachenfibrom, Choanalatresia und Dysostosis mandibulofacialis im Vergleich zur normalen Entwicklung, Hals-Nasen-Ohrenklinik der Westfälischen Wilhelms Universität (Thesis), Münster, Germany, 1963.

Cheraskin, E.: Diagnostic stomatology: a clinical pathologic approach, New York, 1961, McGraw-Hill Book Co.

Colby, R.A., Kerr, D.A., and Robinson, H.B.G.: Color atlas of oral pathology, Philadelphia, 1961, J.B. Lippincott Co.

Cullen, R.L., and Vidić, B.: The dimensions and shape of the human maxillary sinus in the perinatal period, Acta Anat. **83**:411, 1972.

Delaney, A.J., and Morse, H.R.: Inferior meatal accessory ostia: report of a case, Ann. Otol. Rhinol. Laryngol. **60**:635, 1951.

Döderlein, W.: Experimentelle Untersuchungen zur Physiologie der Nasen und Mundatmung und über die physiologische Bedeutung der Nasennebenhöhlen, Z. Hals-Nasen-u. Ohrenheilk. **30**:459, 1932.

Doiteau, R.: Contribution à l'étude de la physiologie des sinus de la face: renouvellement de l'air intrasinusien échanges gazeux permuqueux, Rev. Laryngol. Otol. Rhinol. (Bord.) **77**:900, 1956.

Eckel, W., and Beisser, D.: Untersuchungen zur Frage eines Einflusses der Gaumenspaltbildung auf die Kieferhöhlengrösse, Z. Laryngol. Rhinol. Otol. **40**:23, 1961.

Fatin, M.: A rare case of congenital malformation: total absence of half the nose, probably supporting the theory of bilateral nasal origin, J. Egypt. Med. Assoc. **38**(8):470, 1955.

Gouzy, J., Voilgue, G., and Jakubowicz, B.: A propos de deux cas d'agénésie du sinus maxillaire, J. Fr. Otorhinolaryngol. **17**:579, 1968.

Hajiniš, K., Kustra, T., Farkaš, L.G., and Feiglová, B.: Sinus maxillaris, Z. Morph. Anthropol. **59**:185, 1967.

Koertvelyessy, T.: Relationships between the frontal sinus and climatic conditions: a skeletal approach to cold adaptation, Am. J. Phys. Anthropol. **37**:161, 1972.

Lamm, H., and Schaffrath, H.: Druckmessungen im gesunden Sinus maxillaris bei verschiedenen Atmungstypen, Z. Laryngol. Rhinol. Otol. **46**:172, 1967.

Latkowski, B.: Poglady na znaczenie zatok bocznych nosa, Pol. Tyg. Lek. **19**:1206, 1964.

Maurer, R.: Zur Physiologie der Schädelpneumatisation, Arch. Ohren-Nasen-u. Kelhkopfheilk. **163**:471, 1953.

Merideth, H.W.: The paranasal sinuses, Rocky Mt. Med. J. **49**:343, 1952.

Mocellin, L.: Um caso de pan-agenesia dos seios paranasais, Rev. Bras. Cirurg. **48**(4):283, 1964.

Negus, V.: The function of the paranasal sinuses, A.M.A. Arch Otolaryngol. **66**:430, 1957.

Nemours, P.R.: A comparison of the accessory nasal sinuses of man with those of the lower vertebrates, Trans. Am. Laryngol. Otol. Rhinol. Soc. **37**:195, 1931.

Osmont, J., Jars, G., and Ged, S.: Anatomie chirurgicale du sinus maxillaire, Rev. Odontostomatol. Midi Fr. **25**:50, 1967.

Perović, D.: Medicinska Enciklopedija, vol. 6, Zagreb, 1962, Naklada Leksikografskog Zavoda F.N.R.J.

Püschel, L., and Schlosshauer, B.: Ueber den Einfluss des somatotropen und androgenen Hormons auf die Pneumatisation, Arch. Ohren-Nasen-u. Kehlkopfheilk. **167**:595, 1955.

Richter, H.: Ueber exogene Einflüsse auf die Entwicklung der Nasennebenhöhlen, Arch. Ohren-Naseu-U. Kehlkopfheilk. **143:**251, 1937.

Rosenberger, H.C.: Does sinus infection affect sinus growth? Laryngoscope **55:**62, 1945.

Satir, P.: How cilia move, Sci. Am. **231:**45, 1974.

Schaeffer, J.P.: The nose, paranasal sinuses, nasolacrimal passageways, and olfactory organ in man, Philadelphia, 1920, P. Blakiston's Son & Co.

Schaeffer, J.P.: The anatomy of the paranasal sinuses in children, Arch. Otolaryngol. **15:**657, 1932.

Schaeffer, J.P.: The clinical anatomy and development of the paranasal sinuses, Penn. Med. J. **65:**395, 1935.

Schürch, O.: Ueber die Beziehungen der Grössenvariationen der Highmorshöhle zum individuellen Schädelbau und deren praktische Bedeutung für die Therapie der Kieferhöhleneiterungen, Arch. Laryngol. Rhinol. **18:**229, 1906.

Scopp, I.W.: Oral medicine: a clinical approach with basic science correlation, ed. 2, St. Louis, 1973, The C.V. Mosby Co.

Terracol, J., and Ardouin, P.: Anatomie des fosses nasales et des cavités annexes, Paris, 1965, Librairie Maloine S.A.

Van Alyea, O.E.: The ostium maxillare, Arch Otolaryngol. **24:**553, 1936.

Vidić, B.: The morphogenesis of the lateral nasal wall in the early prenatal life of man, Am. J. Anat. **130:**121, 1971.

Vidić, B., and Tandler, B.: Ultrastructure of the secretory cells of the submucosal glands in the human maxillary sinus, J. Morphol. **150:**167, 1976.

15

HISTOCHEMISTRY OF ORAL TISSUES

Histochemistry is generally considered to be an empirical extension of routine histologic staining methods. However, this is not necessarily so since most histochemical techniques have had their origin with a chemist or an immunochemist in collaboration with a cell biologist. Histochemical techniques generally arise from precise chemical rationales for their ability to stain different biochemical substances. These techniques necessitate as many as, if not more stringent precautions to preserve the chemical integrity of the tissues than in a biochemical or an immunochemical assay. In fact, the histochemical techniques provide more exact in situ information on the chemical compo-

sition of a cell or groups of cells than do most biochemical methods.

Recent application of immunobiologic principles in the immunohistochemical localization of specific proteins, glycoproteins, and proteoglycans have opened new dimensions in the detection of a host of biologic molecules that play important roles in normal tissues during development and in different pathologic conditions. A case in point is the recent use of plant lectins as biochemical probes in the characterization of sugar moeities within glycoprotein molecules, be they of the secretory type or attached to cell surfaces. Use of these probes at the histochemical level has generated much more specific and meaningful in situ information than could be obtained from biochemical assays of total tissue homogenate pools. As a consequence, histochemistry has been advantageously used with increasing frequency in the diagnosis of disease and in studying changes in metabolic pathways of tissues under normal and altered physiologic environments.

Most histochemical techniques have generally been used for qualitative analysis of chemical substances in cells and tissues. However, many sophisticated techniques have been devised recently for quantitative analysis of histochemical reactions. These include use of the original microphotocell counter, double-beam recording microdensitometry, and more recently the scanning and integrating microdensitometry. This latter method has been used successfully by Chayen in measuring lysosomal membrane permeability and by Stuart and Simpson in measuring the activity of dehydrogenase enzymes in single cells from bone marrow biopsies of normal and leukemic patients. Phillips and co-workers have done a quantitative analysis of total mineral content in bone by combining microradiography and microdensitometry using a scanning autodensidater attachment.

Many new techniques, not precisely histochemical, are frequently utilized by histochemists in making qualitative as well as quantitative analysis of tissue substances, particularly mineral elements. These include x-ray and interference microscopy for measuring the dry mass of a biological substance or a reaction product, x-ray diffraction, x-ray spectrophotometry, and electron probe microanalysis. Recently, scanning electron microscopy (SEM) has come to occupy an important place in dental research. SEM has been used in the study of bone morphology and in the analysis of changes in bone architecture induced by the presence of surgically inserted metallic implants (Fig. 15-1) Neiders and co-workers have made conjunctive use of SEM with an electron probe attachment for obtaining a visual surface texture image of tooth cementum analyzed for its mineral content. Reith and Boyde have also used electron probe with SEM in a study of calcium transport across ameloblasts in the enamel organ (Fig. 15-2). Techniques of polarized light and x-ray analysis have been used for the study of enamel. Phosphorescence emitted as a result of tetracycline binding to mineralized tissues has been demonstrated in bone, dentin, and enamel at liquid nitrogen temperatures. Laser spectroscopy has also been used for qualitative and quantitative microanalysis of inorganic components of calcified tissues. Immunohistochemical techniques using fluorescein tags have been applied to the study of oral tissues (Fig. 15-3).

In the last 10 years, light microscopic histochemical techniques have been increasingly adapted for use in electron microscopic histochemistry. The visualization of carbohydrates, specific proteins, and phosphatases are some examples of such adaptive use (Fig. 15-4).

Radioautographic techniques play a vital role in histochemistry because of their ability to elucidate the uptake of chemical substances by the metabolic pathways of different tissues and by different regions of the cytoplasm (Figs. 15-5 and 15-6). Tissue sections taken from animals injected with a radioisotope are covered with a photographic film or emulsion and left in the dark. Radio waves emitted by the isotope hit the silver halides of the film, and these tracks are later developed by processing of the slide or the metal grid like a photographic film. The radioisotope appears as dark granules in the light microscope and as linear tracks of the radio waves in the electron microscopic autoradiographs.

STRUCTURE AND CHEMICAL COMPOSITION OF ORAL TISSUES

Oral structures are primarily composed of connective tissue and epithelial linings and associated glands. An understanding of these structures and their chemical composition is important in the consideration of biologic problems related to oral health. Significant chemical constituents of these tissues are proteoglycans, glycoproteins, mucins, and enzymes.

Connective tissue

Connective tissue is derived from the mesenchyme and consists of various types of cells and fibers that are embedded in an amorphous, semigel, colloidal ground substance. The connective tissue ground substance is primarily composed of proteoglycans and glycoproteins. Proteoglycans are large molecules formed of a protein core to which a large number of glycosaminoglycan (GAG) chains composed of repeating disaccharide units are attached. The GAGs may be unsulfated (hyaluronic acid) or sulfated (chondroitin sulfates, keratan sulfates,

and heparan sulfates). Heparin, a highly sulfated GAG, is secreted by connective tissue mast cells. Hyaluronic acid is synthesized as a very large, free, nonsulfated GAG that does not require a protein core and differs from chondroitin sulfates in having acetylglucosamines instead of acetylgalactosamines as its constituents. Hyaluronic acid binds proteoglycan molecules along its length to form large proteoglycan polymers. The numbers and types of GAGs attached to the protein core in a proteoglycan molecule, as well as the polymeric state of proteoglycans, determine the viscosity of the amorphous ground substance. Hyaluronic acid predominates in the loose connective tissues and, because of its high capacity to bind water, is primarily responsible for transport and diffusion of metabolic substances across tissues. Bacterial infections may occur as

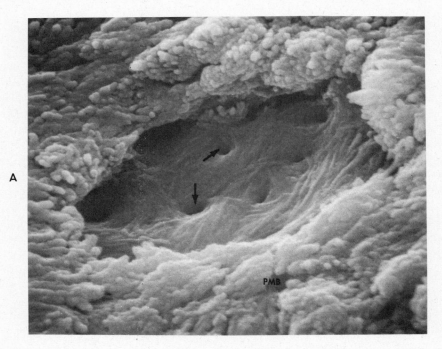

Fig. 15-1. Scanning electron micrographs of cancellous bone in deorganified dog mandible. A metallic implant was surgically inserted subperiosteally in previously edentulatized mandible and was kept in place for 24 months prior to removal of bone sample for microscopic analysis. **A,** A forming osteocytic lacuna with openings of canaliculi *(arrows)* visible on its inside while the newly deposited perilacunar mineral matrix *(PMB)* is seen roofing over and around it. (×7000.)

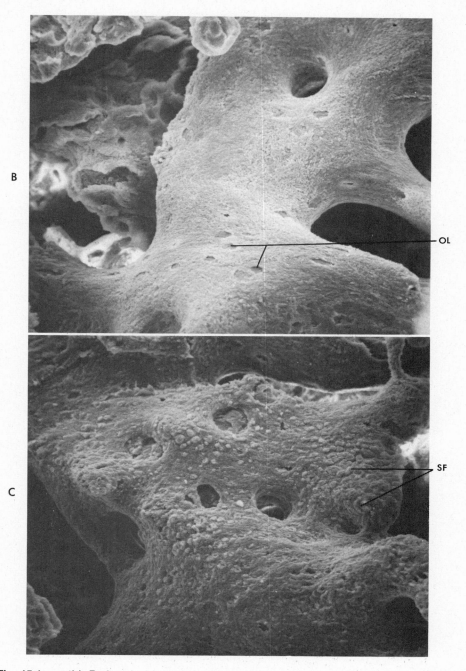

Fig. 15-1, cont'd. B, A typical bone trabeculum revealing several developing osteocytic lacunae *(OL)* on its forming surface. (×280.) **C,** An atypical bone trabeculum, which reveals a rough surface studded with fully or partially mineralized Sharpey fibers *(SF)*. (×280.) (From Russell, T.E., and Kapur, S.P.: J. Oral Implantol. **7:**415, 1977.)

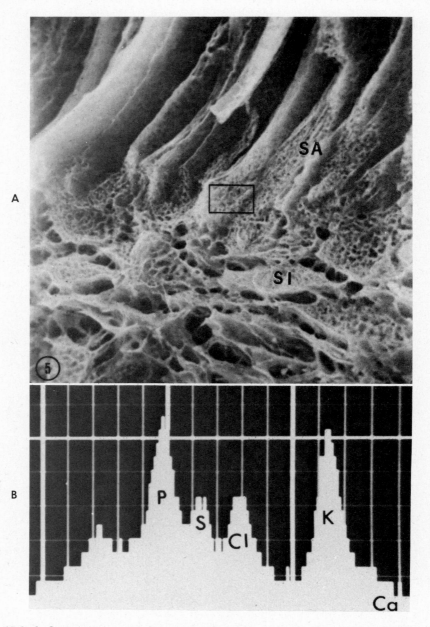

Fig. 15-2. A, Scanning electron microscopic view of freeze-fractured surface of secretory ameloblasts *(SA)* and stratum intermedium *(SI)* from developing rat molar tooth. (×2700.) Rectangular area marks proximal region of an ameloblast that was bombarded with electrons in order to obtain an electron probe spectrum, **B,** for an analysis of its mineral content. **B,** X-ray spectrum for this site reveals notable peaks for phosphorus *(P),* potassium *(K),* sulfur *(S),* and chloride *(Cl)* in order of descending heights. No significant calcium *(Ca)* peak is evident. (From Reith, E.J., and Boyde, A.: Histochemistry **55:**17, 1978.)

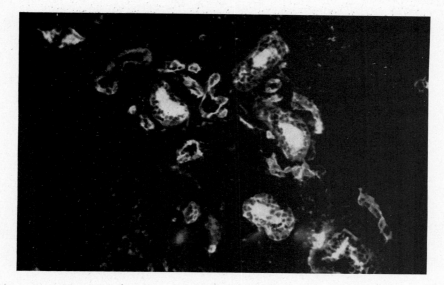

Fig. 15-3. Submaxillary gland secreting mucus in which the A antigen is demonstrated by immuno-fluorescence using rabbit anti-A serum and goat antirabbit serum: fluorescein. Human fetus, 8 cm crown-rump length. This secretion and mucus-borne antigen persist throughout life. (Courtesy Dr. A.E. Szulman, Pittsburgh.)

a result of the hydrolytic action of the bacterial enzyme hyaluronidase on the polymeric integrity of hyaluronic acid. Chondroitin sulfates predominate in the cartilage proteoglycans and are primarily responsible for the supportive and somewhat plastic texture of this tissue. Chondroitin sulfates constitute 1% of the total bone tissue, whereas only 0.5% is present in dentin. Other organic components of bone are approximately 93% type I collagen and 5% noncollagenous proteins including phosphoproteins, glycoproteins, osteocalcin (Gla protein), and osteonectin.

Glycoproteins present in the connective tissue ground substance are protein macromolecules, which contain fewer associated carbohydrate moieties than are present in the proteoglycans. Also, carbohydrates are not present in the form of regular repeating units. Several glycoproteins secreted into the connective tissue by epithelial cells and fibroblasts have been identified in the last few years. Of these, the most well known and characterized

are (1) *fibronectin* (secreted by fibroblasts, smooth muscle cells, and various other cell types); (2) *laminin* (secreted by epithelial cells and present in all basement membranes); (3) *chondronectin* (secreted by chondrocytes); and (4) *osteonectin* (secreted by osteoblasts). All of these glycoproteins promote attachment of cells to their extracellular collagen matrices and therefore not only maintain normal cell morphology but also control cell function.

Both proteoglycans and glycoproteins in connective tissues undergo alterations in various pathologic states. During inflammation or in early stages of wound healing there is a histochemically detectable increase in both glycoproteins and proteoglycans. However, as wound healing progresses there is a gradual decline in the levels of both substances until normal levels are restored. Levels of fibronectin and its cell membrane receptors are known to undergo a decline in certain forms of cancer. These reduced levels are correlated with the altered or transformed behavior of the cancer

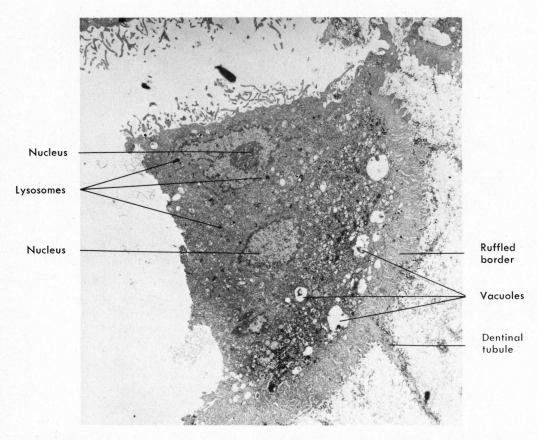

Fig. 15-4. Electron micrograph of odontoclast from mongrel-puppy primary tooth undergoing resorption. Notice acid phosphatase reaction product in form of a black precipitate along dentinal tubule, ruffled border, vacuoles, and in lysosomes. (Glutaraldehyde fixation, Gomori's metal substitution method; ×6250.) (From Freilich, L.S.: A morphological and histological study of the cells associated with physiological root resorption in human and canine primary teeth. Ph.D. thesis, 1972, Georgetown University, Washington, D.C.)

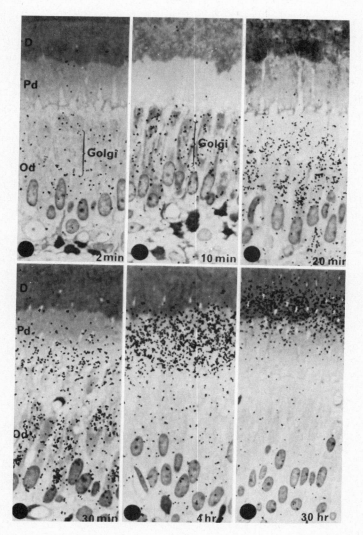

Fig. 15-5. Light microscopic radioautographs illustrate path of ^{3}H-proline (injected into a young rat) over odontoblasts, *OD;* predentin, *PD;* and dentin, *D,* at growing end of incisor tooth. Notice that silver grains representing path of ^{3}H-proline appear first in granular endoplasmic reticulum at 2 minutes and subsequently at 10 and 20 minutes in Golgi region of odontoblasts. Thirty minutes after injection silver grains start appearing in odontoblastic processes and predentin, whereas at 4 hours entire radioactivity is located in predentin. Thirty hours after injection, dentin is completely labeled with ^{3}H-proline, now incorporated into collagen fibrils of dental matrix. ($\times$1000.) (From Weinstock, M., and Leblond, C.P.: J. Cell Biol. **60:**92, 1974.)

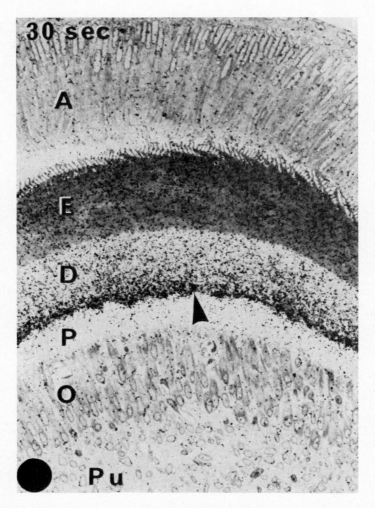

Fig. 15-6. Radioautograph of incisor tooth (undecalcified cross section) at its growing end in young rat killed 30 seconds after intravenous injection of ^{45}Ca. *A,* Ameloblasts; *E,* enamel; *D,* dentin; *P,* predentin; *O,* odontoblasts; *Pu,* pulp. Notice that ^{45}Ca is immediately incorporated into dentin over predentin-dentin junction at arrow. Some ^{45}Ca activity in form of few grains is seen in odontoblasts and predentin. (×250.) (From Munhoz, C.O.G., and Leblond, C.P.: Calcif. Tissue Res. **15:**221, 1974.)

cells. Some investigators have suggested that metastatic cancer cells preferentially bind to type IV collagen via laminin, both being components of the basement membranes. In contrast, it has been suggested that nonmalignant tumor cells do not use laminin for attachment.

Fibroblasts are the most common cell type in the connective tissues. They are responsible for the elaboration of glycoproteins, like fibronectin, and proteoglycans that form the amorphous ground substance. They also elaborate the fibrous components of the ground substance, including different types of collagen (especially types I and III), reticular fibers, and elastic fibers.

Current biochemical, histochemical, and ultrastructural evidence suggests that collagen is initially synthesized as much larger preprocollagen polypeptide chains. The prepeptide component is removed during or shortly after translocation in the rough endoplasmic reticulum. Posttranslational changes include hydroxylation of proline and lysine residues, glycosilation of hydroxylysine residues, formation of disulfide bonds between adjacent chains, and the formation of the characteristic triple helix. Procollagen is secreted on the cell surface where the propeptide sequence is deleted by a specific protease. This is immediately followed by the formation of the collagen microfibrils. The microfibrils serve as a template for initiation and extension of the polymerization and the accretion of more newly secreted monomeric tropocollagen into collagen fibrils. Hydroxylation steps are facilitated by vitamin C and are essential for providing conformation and stability to the triple helix. Deficiency of vitamin C results in the loss of molecular stability, resulting in the formation of abnormal, immature collagen and consequent collagen diseases.

The newly elaborated collagen fibrils, formed during development or in wound healing, are equivalent to reticular fibers in their electron microscopic structure. Both of these fibers stain positively for glycoproteins with silver stains and the periodic acid–Schiff (PAS) method. These reactions indicate the presence of a considerable packing of glycoprotein between aligned microfibrils of tropocollagen macromolecules.

Elastic fibers are elaborated by fibroblasts and also possibly by smooth muscle cells in the walls of blood vessels. They are composed of a protein component characterized by the presence of the amino acids desmocine and isodesmocine and glycosaminoglycans. Unlike collagen and reticular fibers, elastic fibers are not considered to be important constituents of the fully repaired tissues. Elastic fibers are stained by aldehyde fuchsin, resorcin fuchsin, and specifically by the dye orcein in histologic preparations. A fluorescent staining method, using tetraphenylporphine sulfonate in combination with silver or gold, has been developed by Albert and Fleischer for electron microscopic visualization of elastic fibers.

Besides fibroblasts, other cellular elements of connective tissue are macrophages, which scavenge on tissue debris; mast cells, which are rich in the highly sulfated proteoglycan heparin (an anticoagulant) and histamine (a vasodilator); and plasma cells, which elaborate immunoglobulins.

Epithelial tissues and derivatives

Salivary glands elaborate the so-called mucins or mucoids. The definition of these substances is exclusively chemical from the biochemical standpoint, but from a histochemical point of view this definition is in part based on color reactions. Histochemical detection of mucins is generally based on their glycosaminoglycan content, which affects certain staining reactions. The acidic nature is attributable to the presence of glucuronic acid, sulfate, or sialic acids. Histochemical observations show that a number of acid mucins lack sulfate esters. Histochemical characteristics of the oral epithelium, the epithelial components of the tooth germ, and the salivary glands will be considered in another section of this chapter.

Several histochemical studies have been made on the structural proteins of the salivary gland leukocytes or the so-called salivary corpuscles. Histochemical techniques are also being utilized in oral exfoliative cytology for the detection of oral cancer. Identification of lung carcinoma by analysis of normal and abnormal cells present in sputum is used clinically.

Enzymes

Histochemistry has enabled histologists to demonstrate the actual sites of cellular enzymatic activity. The topographic distribution of enzymes may be ascertained by the quantitative microchemical techniques developed by the Linderstrøm-Lang group or by techniques that result in the formation of visible reaction products in tissue sections. The latter approach is widely used in histochemical demonstrations of enzymes.

The most frequently studied enzymes in oral tissues are those related to the transfer of phosphate esters (specific and nonspecific phosphatases) in the organic matrix of bone, dentin, and enamel (alkaline phosphatase) and to resorption of bone and of dentin (acid phosphatase). Oxidases and dehydrogenases, reflecting the metabolic activity of different tissues in oral structures, have also been studied extensively. Esterases, generally associated with the hydrolysis of carboxylic acid esters of alcohol, have been studied in salivary glands and in the taste buds. More recently, studies on lysosomal sulfatase and on adenyl cyclase involved in the formation of cyclic adenosine monophosphate (cAMP) have been reported.

HISTOCHEMICAL TECHNIQUES
Fixation procedures

For histochemical study, a tissue block must be preserved in such a way that it causes minimal changes in the reactivity of the cytoplasmic and extracellular macromolecules, for example, enzymes, structural proteins, protein-carbohydrate complexes, lipids, and nucleic acids. This is accomplished by using optimum osmotic conditions, cold temperatures, controlled pH of the fixing solutions, and minimum possible exposure to the fixative.

Formaldehyde is considered to be one of the ideal fixatives, especially for enzymes and other proteins. This is because of its ability to react with major reactive groups of proteins to form polymeric or macromolecular networks, without affecting their native reactivity to histochemical procedures. Formaldehyde has a preservative effect on lipids by altering their relationship with the proteins. Use of electrolytes such as calcium or cadmium in formaldehyde or chromation of tissue blocks subsequent to fixation prevents dissolution of phospholipids. Formaldehyde is generally used as a 10% solution buffered to pH 7 at cold temperatures in the range of 0° to 4° C.

Acrolein and glutaraldehyde are other frequently used aldehydes, with the latter being routinely used for electron microscopy. Conjunctive use of colloids such as sucrose, ficoll, polyvinylpyrrolidone, and dextrans in the fixing solutions is often made to prevent osmotic rupture of cell organelles. This helps to improve the in situ localization of the histochemical reactions.

Other fixatives used for the study of glycogen, glycoproteins, proteoglycans, and nucleic acids are frequently mixtures of many chemical ingredients. *Rossman's fluid*, used for visualization of glycogen, glycoproteins, and proteoglycans, contains formaldehyde, alcohol, picric acid, and acetic acid. *Carnoy's mixture*, used for histochemical staining of nucleic acids, is composed of ethyl alcohol, acetic acid, and chloroform. Alcohol denatures proteins without causing irreversible chemical changes in the active groups but, being a poor fixative, is used in combination with acetic acid and chloroform. *Feulgen's reaction*, used for visualizing deoxyribonucleic acids (DNA), requires acid hydrolysis of the DNA polymers to expose the deoxyribose sugar residues of DNA molecules. The aldehyde groups thus exposed (on the deoxyribose sugar residues) are then chemically reacted with leucofuchsin (Schiff's reagent) to form a reddish purple reaction product.

Some enzyme systems, such as cytochrome oxidases, are highly labile and therefore cannot be preserved by chemical fixation. Visualization of such enzymes is performed on fresh frozen (cryostat) sections. However, to prevent diffusion and to preserve the in vivo status of the tissue macromolecules, one must fix the tissue blocks by a freeze-drying procedure. Tissues are frozen rapidly at very low temperatures, usually in liquid nitrogen, and then placed in a refrigerated vacuum chamber where ice, formed in the tissues, is removed by sublimation, that is, by direct transformation into vapor without going through a liquid phase. After dehydration in vacuum, tissue blocks are embedded in paraffin and sectioned routinely with a microtome. Freeze-dried tissues exhibit optimal enzyme activity, show excellent histologic characteristics, and do not show any shrinkage artifacts that are seen with routine fixation. Besides oxidative enzymes, freeze-drying is used for visualization of other enzyme systems, for example, phosphatases and dehydrogenases, and also for the precise localization of otherwise diffusible inorganic ions.

Techniques of freeze-fracture and freeze-etching have now been devised for use in electron microscopy to avoid use of chemicals in tissue preparation. This technique has enabled biologists to obtain excellent three-dimensional images of the surfaces of various cell membranes not previously observed.

Histochemical study of teeth and bone requires careful fixation and controlled decalcification procedures. Simultaneous fixation and decalcification with formaldehyde or glutaraldehyde and Versene (ethylenediaminetetraacetate [EDTA]) have been successfully employed in the study of teeth and bone for light and electron microscopic histochemistry. Decalcified ground sections have also been employed in histochemical studies of teeth and bone.

Techniques have been developed for sectioning freeze-dried, undecalcified tissues. Gray and Opdyke have described a saw for the preparation of 10 to 50 μm sections of undecalcified tissues. Such sectioning has been employed for histochemical studies of dental decay. Study of bone has been considerably enhanced by the process of deorganification (deproteinization). Swedlow and colleagues, and Kapur and Russell have studied bone architecture with SEM to great advantage after deorganification with concentrated hydrazine.

Specific histochemical methods

Histochemical techniques primarily used in the study of oral tissues may be categorized as (1) glycogen, glycoprotein, and proteoglycan methods; (2) protein and lipid methods; and (3) enzyme methods. They are all characterized by a direct staining reaction or by the formation of an insoluble dye or precipitate at the reactive sites.

Glycogen, glycoproteins, and proteoglycans. The best-known and most frequently used technique for detection of carbohydrate groupings is the periodic acid–Schiff (PAS) technique. The chemical basis of this method lies in the fact that periodic acid oxidizes the glycol groups to aldehydes and these in turn are revealed as a reddish purple dye product on treatment with leucofuchsin (Schiff reagent). Use of fluorescent reagent anthracene-9-carboxyaldehyde carbohydrazone as a substitute for Schiff reagent has been made by Cotelli and Livingston. Treatment of tissue sections with amylase prior to oxidation removes glycogen from the tissues, and this is reflected in a reduced Schiff reaction product. A comparison of the amylase digested and undigested sections is used in estimating the amounts of glycogen or other carbohydrate-protein molecules. Electron microscopic visualization of carbohydrates has been achieved, among other techniques, by use of phosphotungstic acid and lead citrate after oxidation with periodic acid. The periodate–thiocarbohydrazide–silver proteinate method of Theiry shows high specificity for glycol containing glycoconjugates at the electron microscope level.

Proteoglycans are well demonstrated by thiazine dyes such as toluidine blue, azure A, and Alcian blue. Toluidine blue produces a metachromatic reaction ranging from a purple to a red reaction product. This change of color (metachromasia) from the original (orthochromatic) blue color of the monomeric form of toluidine blue reflects the extent of polymerization of the dye molecules as they tag onto the anionic residues on the glycosaminoglycans molecule. Thus heparin present in the mast cell granules and chondroitin sulfates present in the intercellular ground substance of the cartilage or developing bone give an intense red metachromasia demonstrating the highly acidic or sulfated nature of these proteoglycans. Alcian blue staining has been used to considerable advantage in characterizing the specific types of acid radicals present within proteoglycans. When used at pH 2 to 2.8, Alcian blue stains weakly acid sulfated proteoglycans. However, when it is used at pH 1 to 1.2, Alcian blue binds to highly sulfated proteoglycans. By incubating tissue sections in the enzyme sialidase prior to staining with Alcian blue, distinction can be made between sialidase-resistant and sialidase nonresistant molecules.

Several techniques are available for the localization of proteoglycans at the electron microscopic level. The high iron diamine thiocarbohydrazide–silver proteinate method of Spicer provides high specificity for sulfated glycoconjugates. This method excludes reaction with carboxyl and phosphate groups. Several cationic dyes, including ruthenium red, silver tetraphenylporphine sulfonate, Alcian blue, bismuth nitrate, and cuprolinic blue, have been used successfully by several investigators in localizing proteoglycans in oral tissue.

Use of specific plant lectins in the identification or characterization of specific carbohydrate moeities within a glycoprotein molecule has recently been made in the study of carbohydrate histochemistry. Fluorescein dyes or horseradish peroxidase techniques are used as tags for the visualization of lectin binding sites on carbohydrate moeities within a glycoprotein molecule. Electron microscopic localization of sugar moeities in the glycoprotein molecules has been made by using ferritin and horseradish peroxidase as lectin tags.

Proteins and lipids. Histochemistry of proteins is based on classic reactions of protein chemistry involving various amino acid groups, that is, amino, imino, carboxyl, disulfide, and sulfhydryl groups. Reagents such as dinitrofluorbenzene, ninhydrin, or ferric ferricyanide are utilized to give insoluble colored reaction products.

Histochemical study of lipids frequently implies use of frozen or freeze-dried sections. Total lipids are studied by using fat colorant dyes such as Sudan dyes. Chromation of formol-calcium–fixed tissues and their subsequent staining with Sudan black has been employed for the identification of phospholipids. Extraction procedures with various lipid solvents are considered essential to accompany most histochemical staining procedures for lipids.

Enzymes. The enzyme techniques utilize many different principles. Some of the criteria used in deciding the application of a technique are related to avoidance of inhibition by the substrate and insolubility of the primary reaction product and its immediate coupling to the capture reagent to prevent diffusion and false localization of enzyme activity.

The Gomori method for phosphatases uses phosphoric esters of glycerol, glucose, or adenosine. The enzymatically liberated phosphate ion is converted into an insoluble salt, which can be visualized by polarized light or phase contrast, or the salt can be transformed into a cobalt or lead compound, which is black. Riboflavin 5′-phosphate has been used as substrate, which at the site of phosphatase activity results in the formation of a fluorescent precipitate. Electron microscopic demonstration of phosphatase is also based, with some modifica-

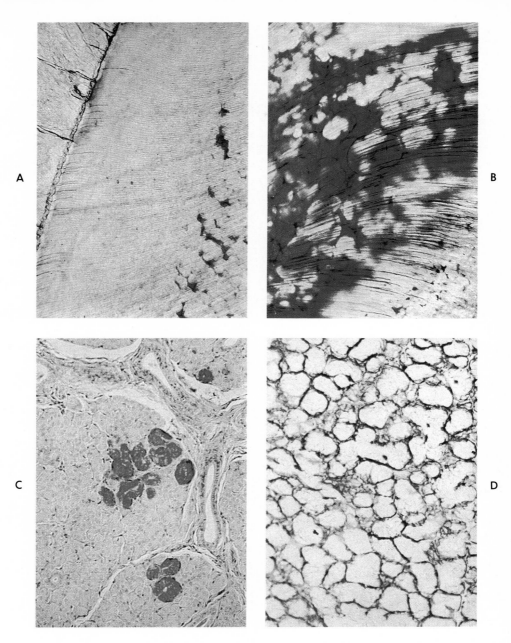

Plate 4. Photographs of several histochemical reactions. **A,** Ground section of human dentin. Dinitrofluorobenzene (DNFB) technique reveals reactive protein groups of interglobular areas. ($\times 143$.) **B,** Ground section of human odontoma showing PAS reaction of poorly calcified dentin. ($\times 85$.) **C,** Alcian blue staining of isolated group of mucous cells in dog parotid gland. Mucicarmine counterstain. ($\times 143$.) **D,** Alkaline phosphatase activity of basement membranes of rat sublingual gland. ($\times 143$.)

tion, on Gomori's original method of metal substitution. A technique employing ruthenium red has been used for demonstration of acid phosphatase in electron microscopic studies.

Another procedure employed for demonstration of phosphatases is the simultaneously coupling azo dye technique. This employs a naphthol phosphate or other type of ester. The enzymatically released naphthol is coupled in situ with a diazonium salt to form an insoluble colored reaction product. With regard to the original Gomori glycerophosphate technique, it has been shown that the calcium phosphate formed may diffuse and give false localization. Because of this and other considerations in calcified tissues, the azo dye techniques are better suited for the study of phosphatases in teeth and bones. Sophisticated new substrates for use with azo dye techniques have been developed in recent years. They facilitate precise microscopic localization of alkaline and acid phosphatases as well as esterases. Aminopeptidases can also be dectected by an azo dye method.

Immunohistochemistry. Precise localization of specific biologic molecules in different intracellular compartments, on cell surfaces, or in extracellular matrices is now made possible by the application of some basic principles of immunochemistry. The immunohistochemical techniques are based on the premise that protein-based antigens or immunogens bind avidly to their specific antibodies. Antibodies to specific antigens can be prepared by injecting the known antigen into an animal in order to provoke an immune response. This response results in the production of antibody immunoglobins, and these can be isolated from the serum of the injected animal.

When a solution containing an antibody or an antiserum is directly applied to a tissue section containing the antigen, the antibody binds specifically to that antigen. This antigen-antibody complex is subsequently attached to a second antibody, which is conjugated either to a fluorescent dye, like rhodamine or FITC (fluorescien isothiocynate), or to an enzyme conjugated to its antibody, like peroxidase-antiperoxidase (PAP). The antigen-antibody complexes bound to a fluorescent dye are examined in a fluorescence microscope. The antigenic sites fluoresce against a dark background and are immediately photographed on high speed film. The enzyme-bound antigen-antibody complexes are further developed histochemically by exposure to an enzyme substrate. This results in the development of a dark brown to a black color, which allows examination of the antigenic sites by light or electron microscopy. Ferritin and colloidal gold are also frequently used as heavy metal markers for the antigen-antibody complexes because of their electron density and also because of their specific particle size.

HISTOCHEMISTRY OF ORAL HARD TISSUES
Carbohydrates and protein

The PAS method is employed more than any other in studying the ground substance of teeth and bones. Under specific conditions, this method is believed to demonstrate the carbohydrate moiety as well as the glycoprotein complexes. The ground substance of normal mature bone and dentin exhibits little or no reactivity with the PAS technique (Fig. 15-7). However, developing or resorbing bone and dentin stain intensely with PAS. Newly formed bone and dentin are also rich in PAS-reactive carbohydrates. In addition to glycogen and glycoproteins, the mineralizing zone of developing bone and dentin matrix is also rich in chondroitin sulfate. Sulfated glycoconjugates accumulate in predentin and are either removed or masked to staining in the dentin.

Interglobular, less-calcified dentin exhibits a distinct PAS reaction (Fig. 15-7) as does abnormally and poorly calcified dentin matrix in dentinogenesis imperfecta and in odontomas (Plate 4, *B*).

Enamel matrix is essentially nonreactive with the PAS method. However, enamel

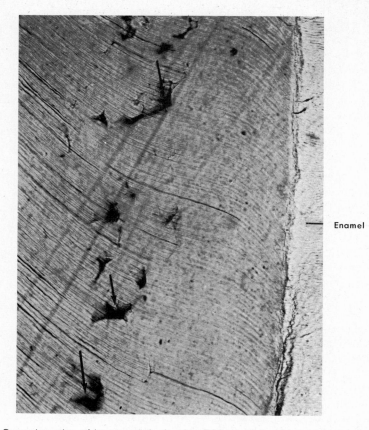

Enamel

Fig. 15-7. Ground section of human tooth showing PAS reactivity of interglobular dentin *(arrows).* (×143.)

lamellae are intensely stained in ground sections (Fig. 15-8). In some areas the rod interprismatic substance exhibits some reactivity.

Specific protein methods identify certain amino acids or their groupings, that is, amino, carboxyl, or sulfhydryl. Only a few of these techniques have been applied in the study of teeth and bone. Of interest are the dinitrofluorobenzene (DNFB) and ninhydrin-Schiff methods. The DNFB reagent combines with α-amino groups of proteins in tissue sections to form a pale yellow complex. An intense reddish color is subsequently revealed by a reduction

and diazotization technique, which results in the formation of an azo dye (Plate 4, A). The pattern of staining is essentially the same as seen with the PAS method in both normal and abnormal dentin. A modification of DNFB method wherein the final reaction product is a mercaptide of lead or silver has been used for electron microscopic histochemistry of amino groups. The ninhydrin Schiff method is dependent on the formation of imino groups that decompose to a keto acid to form aldehyde groups, and these are reacted with leucofuchsin (Schiff reagent) to form a final red-colored reaction product. Some of these techniques

Fig. 15-8. Ground section of human enamel. Lamella stains with PAS method. (×143.)

have been applied to the study of dental caries in bone and in dentin resorption.

Histochemical reactions imply a need for some specific protein groups to initiate the mineralization of predentin and osteoid. Everett and Miller have noted the absence of carboxyl and amino complexes in predentin and osteoid in contradistinction to the presence of these complexes in dentin and bone. Sulfhydryl groups are present optimally at the mineralizing front in predentin and osteoid while being present minimally in the mineralized regions of these tissues.

Immature, newly formed enamel in rats

shows histochemical staining for sulfhydryls and tyrosine residues characteristic of keratin. However, these protein residues are not demonstrable in the mature enamel.

Several immunohistochemical studies mapping distribution of basement membrane components such as type IV collagen, laminin, proteoglycans, and fibronectin during tooth development have shown close association of these proteins with different stages of cell differentiation and matrix secretion (Figs. 15-9 and 15-10). Disappearance of type III collagen and confinement of fibronectin surrounding preodontoblasts to the epitheliomesenchymal junction after the polarization of ondontoblasts during tooth development is another interesting observation emphasizing the significance of these proteins in the regulation of the complex developmental changes occurring in the tooth anlage. Significance of such observations is further highlighted in certain disease states. The presence of a bluish brown opalescence and a diminished pulpal chamber in the teeth of patients with *dentinogenesis imperfecta type II* or in *osteogenesis imperfecta* is associated with the localization of type III collagen in dentin. Type III collagen is absent in normal adult dentin.

Lipids

Biochemical studies indicate a rather low lipid content in the organic matrix of dentin. Lipids have been demonstrated by the sudanophilic reaction in the odontoblast processes

and enamel rod sheaths. Sudanophilia is based on the solubility of Sudan dyes with lipids of varied description. Sudanophilia is widespread in the developing tooth, being present in the zone of mineralization and predentin, and in the basal zone of the ameloblasts. These reactive zones of the predentin and ameloblasts imply a role of phospholipids in the process of mineralization of dentin and enamel matrices.

Enzyme histochemistry of hard tissue

Histochemical techniques are extremely useful in demonstrating specific enzymes in specific cellular and intercellular locations in bone and teeth.

Alkaline phosphatase. Alkaline phosphatase is capable of hydrolyzing phosphoric acid esters. In hard tissues, alkaline phosphatase has been implicated in the process of mineralization. However, some recent studies have

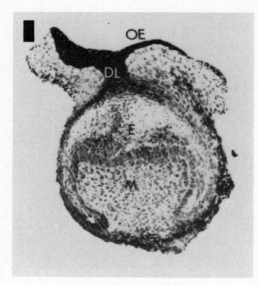

Fig. 15-9. Light micrograph of mandibular first molar of day 16 mouse embryo stained with periodic acid–Schiff *(PAS)*. This section corresponds to sections shown in Fig. 5-10. Notice epithelial enamel organ *(E)* surrounds a condensation of dental papilla mesenchyme *(M)*. Dental lamina *(DL)* connects tooth germ to oral epithelium *(OE)*. (From Thesleff, I., Barrach, H.J., Foidart, J.M., Vaheri, A., Pratt, R.M., and Martin, G.R.: Dev. Biol. **81:**182, 1981.)

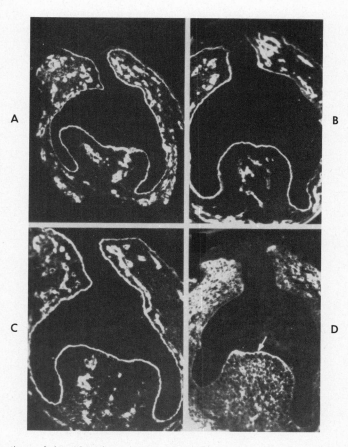

Fig. 15-10. Sections of day 16 embryonic molars stained immunohistochemically with, **A,** antibody to type IV collagen; **B,** antibody to laminin; **C,** antibody to basement membrane proteoglycan; and **D,** with antiserum to fibronectin. (see Fig. 15-9 for reference.) Linear deposits are seen in oral and dental basement membranes and in walls of blood vessels. Distributions of type IV collagen, laminin, and basement membrane proteoglycan appear identical. Immunofluorescence of fibronectin is particularly intense in dental basement membrane *(arrow)* and is prominent in mesenchyme of dental papilla. No stain is observed in enamel organ epithelial cells. (From Thesleff, I., Barrach, H.J., Foidart, J.M., Vaheri, A., Pratt, R.M., and Martin, G.R.: Dev. Biol. **81:**182, 1981.)

raised doubts concerning this assumption and instead suggest that the enzyme is involved in the synthesis of organic matrix only.

Alkaline phosphatase is observed to be associated with osteogenesis and dentinogenesis (Figs. 15-11 and 15-12). The osteoblasts and odontoblasts give an intense staining reaction for the enzyme (Figs. 15-12 and 15-13). No enzyme activity is found in bone or dentin matrices per se, except in close association with the matrix-synthesizing cells. At sites of intramembranous bone development, alkaline phosphatase activity is observed in the endosteum, periosteum, and osteocytes (Table 6). Wergedal and Baylink report no enzyme activity at the actual calcification sites. However, conflict-

ing views are reported in endochondral bone formation where the enzyme is localized in matrices and cells of the hypertrophic and provisional zones of calcification. The view that alkaline phosphatase is involved in actual calcification is also strengthened by observations in vitamin D treatment of human rickets wherein the increase in the calcification zone parallels an increase in serum alkaline phosphatase.

In the developing molar and incisor teeth, alkaline phosphatase is present in the stratum intermedium, the odontoblasts (Fig. 15-12) and subjacent Korff's fibers, and the ground substance. No activity is observed in the ameoloblasts (Fig. 15-12). However, in the incisors of

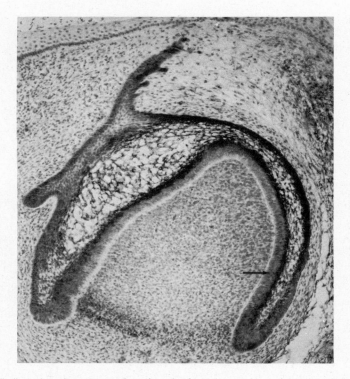

Fig. 15-11. Alkaline phosphatase reaction of tooth of monkey embryo. Ameloblastic layer *(arrow)* is nonreactive. (×87.)

Table 6. Enzyme activity of cells associated with bones and teeth*

	Alkaline phosphatase†	Acid phosphatase†	Amino-peptidase‡	Cytochrome oxidase‡	Succinic dehydrogenase‡
Bone					
Osteoblasts	+ +	0	+	+	+
Osteocytes	+ +	0	+	+	+
Osteoclasts	0	+ +	?	+ +	+ +
Cartilage					
Active chondrocyte	+ +	0	+ +	+ +	+ +
Resting chondrocyte	0	0	+	+	+
Hypertrophic chondrocyte	+ +	0	+	+	+
Tooth					
Stellate reticulum	+ +	0	+	+	+
Stratum intermedium	+ +	0	+	+	+
Ameloblasts (molar)	0	0	0	0 or +	0
Odontoblasts	+ or + +	0	+	+	+

*0, No staining; +, less active; + +, more active.
†Freeze-dried paraffin-embedded tissues.
‡Fresh-frozen tissues.

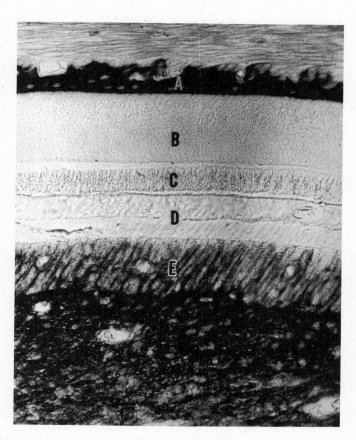

Fig. 15-12. Freeze-dried undecalcified incisor of hamster. *A*, Stratum intermedium; *B*, ameloblasts; *C*, enamel matrix; *D*, dentin matrix; *E*, odontoblasts. Note alkaline phosphatase reactivity of Korff's fibers and subjacent pulp. (×87.)

rodents enzyme activity is present in the ameloblasts and the reduced enamel organ at the growing end of the tooth (Table 6), with the remaining ameloblasts of the incisor being unreactive, as in the molars. The existence of alkaline phosphatase in the dentin proper has been reported.

Acid phosphatase. Acid phosphatase is less widely distributed than its alkaline counterpart (Table 6). Histochemical localization of intracellular acid phosphatase is generally more discrete than that of alkaline phosphatase because it is localized mainly in specific membrane-bound organelles, the lysosomes.

Osteoclasts in bone and odontoclasts in resorbing dentin exhibit an intense acid phospha-

tase activity (Figs. 15-13 and 15-14). The enzyme is localized in the part of the cytoplasm that lies apposed to the resorbing surface of bone and dentin (Figs. 15-14 and 15-15). Electron microscopic studies reveal that the enzyme is localized in the lysosomes, although activity is also seen extracellularly between the microvillus-like projections of the ruffled border (Figs. 15-4). Uptake of resorbed mineral, hydrolyzed collagen fibrils and injected radioactive substances at the site of resorption has been observed.

Several recent studies imply that acid phosphatase may (in addition to its function in bone and dentin resorption) confer "calcifiability" to the organic matrix by its hydrolytic action on

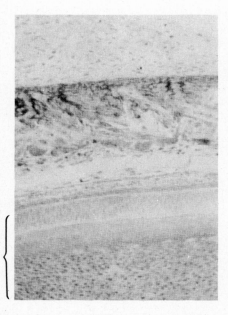

Fig. 15-13. Alkaline phosphatase *(dark areas)* in osteoblasts and acid phosphatase *(dark areas)* reaction in osteoclasts in resorbing bone adjacent to incisor tooth *(brace)* in 3-day-old hamster. (MX naphthol phosphate and red-violet LB salt incubation for alkaline phosphatase; GR naphthol phosphate and blue BBN salt incubation for acid phosphatase, ×143.) (From Burstone, M.S.: In Sognnaes, R.F., editor: Calcification in biological systems, Washington, D.C., 1960, American Association for the Advancement of Science Publications, p. 64.)

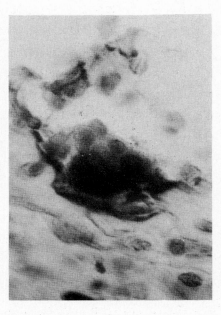

Fig. 15-14. Hamster osteoclast showing cytoplasmic acid phosphatase reaction *(dark),* with some enzyme activity also present in resorbing bone matrix. (AS-BI naphthol phosphate and red-violet LB salt incubation. Nuclear counterstain is hematoxylin; ×750.) (From Burstone, M.S.: In Sognnaes, R.F., editor: Calcification in biological systems, Washington, D.C., 1960, American Association for the Advancement of Science Publications, p. 64.)

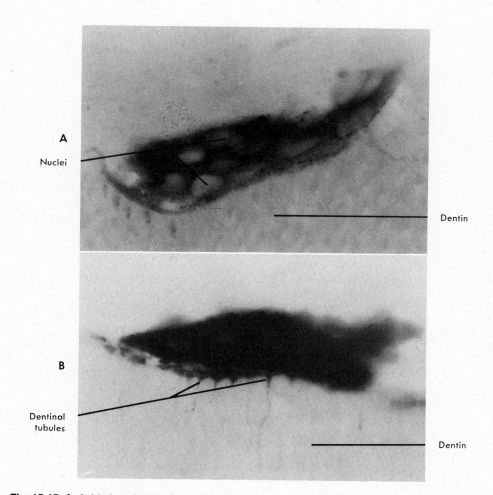

Fig. 15-15. A, Acid phosphatase activity in odontoclasts of mongrel-puppy primary tooth undergoing resorption. Notice reaction product is localized in discrete granules in cytoplasm. No reaction is seen in nuclei. **B,** Acid phosphatase activity in odontoclast and dentinal tubules of mongrel-puppy primary tooth undergoing resorption. (Alpha-naphthyl acid phosphatase and fast garnet GBC salt incubation; ×1060.) (From Freilich, L.S.: A morphological and histological study of the cells associated with physiological root resorption in human and canine primary teeth, Ph.D. thesis, Georgetown University, Washington, D.C., 1972.)

the protein-polysaccharide granules present in the zone of mineralization.

Esterase. According to histochemical definition, esterases hydrolyze simpler fatty acid esters in comparison to lipases, which hydrolyze complex fatty acid esters.

Most histochemical techniques for esterases do not reveal any activity in bone or dentin. However, with use of specific naphthol esters such as naphthol AS-D acetate, an intense staining reaction is observed in the calcifying matrices of bone and dentin. This reactive zone, situated in the tooth between predentin and dentin, is also sudanophilic, indicating the presence of phospholipids.

Considerable esterase activity has also been found in the cells and microorganisms associated with the formation of calculus deposits on teeth.

Aminopeptidase. Aminopeptidases are proteolytic enzymes that hydrolyze certain terminal peptide bonds. Azo dye techniques using L-leucyl-β-naphthylamide or DL-alanyl-β-naphthylamide have been developed for histochemical demonstration of this enzyme. Human osteoclasts give a strong reaction. Although no staining reaction occurs in the osteoclasts of rodents, the enzyme is demonstrated in the stratum intermedium and odontoblasts during dentinogenesis. Some staining reaction is also noticed in the periosteum, perichondrium, and chondrocytes (Table 6). It is significant that aminopeptidase has also been localized in the macrophages and certain sites associated with the breakdown of connective tissues.

Cytochrome oxidase. Cytochrome oxidase is an iron-porphyrin protein that enables cells to utilize molecular oxygen. Its histochemical localization therefore reflects the oxygen requirements of the cells and tissues and the levels of their metabolic and physiologic activity.

The original histochemical reaction, the "nadi" reaction, employing α-naphthol and N,N-dimethyl-p-phenylenediamine, is now considered inadequate because of the instability of the substrate solution, lipid solubility, crystallization, and fading of the reaction product—indophenol blue. New techniques using p-aminodiphenylamine in conjunction with p-methoxy-p-aminodiphenylamine or 8-aminotetrahydroquinoline have overcome these technical problems so that the reaction is discretely localized in the mitochondria.

Both osteoclasts and osteoblasts show oxidase activity, with the reaction being more predominant in the former. Stratum intermedium of both molars and incisors also exhibits oxidase activity (Table 6).

Succinic dehydrogenase. Succinic dehydrogenase is closely associated with cytochrome oxidase in the mitochondria. It is one of a series of citric acid cycle enzymes that catalyzes the removal of hydrogen, which in turn is removed by a hydrogen acceptor or carrier. This serves as the basis of the histochemical reaction used in demonstrating this enzyme. The enzyme present in the tissues acts on the substrate (usually sodium succinate), causing the removal of hydrogen, which is picked up by a synthetic acceptor (a tetrazolium compound) present in the incubating medium. The reduced acceptor substance, called formazan, appears as a colored reaction product.

The distribution of succinic dehydrogenase in oral hard tissues is essentially similar to that of cytochrome oxidase. The dehydrogenase activity is higher in osteoclasts than in osteoblasts. The stratum intermedium and the odontoblasts in developing teeth also reveal a positive reaction (Table 6).

Citric acid cycle in osteoblasts and osteoclasts. Besides observations on succinic dehydrogenase, studies on isocitric dehydrogenase, α-ketoglutaric dehydrogenase, and DPN-TPNH-diaphorases have also been reported. It is indicated that osteoclasts maintain a high rate of citrate and lactate production at the expense of glutamate and thereby actively promote decalcification of bone matrix and calcified cartilage.

Summary. A survey of the distribution of various enzymes associated with bone and teeth is given in Table 6. It is interesting to note that although acid phosphatase activity is associated with the osteoclasts only, distribution of other enzymes is widespread in hard oral tissues.

HISTOCHEMISTRY OF ORAL SOFT TISSUES
Polysaccharides, proteins, and mucins

Polysaccharides. The dye carmine is often used to demonstrate glycogen, but it is not as specific as the PAS method. Epithelial glycogen is known to increase during inflammation and repair. Attached human gingiva shows variation in the extent of its keratinization, and this variability is reflected in the glycogen content of the tissue. On the other hand, the non-keratinized alveolar mucosa virtually always shows constant levels of glycogen. Animal experiments involving benign and malignant epithelial proliferations demonstrate an increase in glycogen.

Proteoglycans with chondroitin sulfate and hyaluronic acid form a major intercellular component of human gingival epithelium. The molecular conformation and relatively rapid rate of synthesis and secretion of these macromolecules in the gingiva may explain the lack of susceptibility of this material to the degradative action of specific enzymes. When oral soft tissues are stained with the metachromatic dye toluidine blue, mast cells become visible in varying numbers in the loose connective tissue, particularly along the blood vessels. The metachromatic reaction given by the cytoplasmic granules of these cells is caused by the presence of heparin—a sulfated proteoglycan. The cytoplasmic granules also contain histamine—a vasodilator that can be demonstrated by fluorescence microscopy. Mast cells are present in particularly large numbers in the tongue and in the gingiva. The lack of these cells in acute necrotizing gingivitis is significant.

Proteins and protein groups. Keratinization is one of the important characteristics of the epidermis. Although under normal circumstances it occurs only in some areas of the oral epithelium, in pathologic conditions it occurs anywhere in the mouth. The mechanism by which the cells of the malpighian layer are altered to form keratin has been only partly elucidated. The disulfide bridges present in keratin are believed to result from the oxidation of sulfhydryl groups of cysteine. Sulfhydryl groups are demonstrated histochemically by the ferric ferricyanide method in which this compound is reduced to a Prussian blue color by these protein groups. Thus the extent of the blue reaction product reflects the degree of keratinization. Attempts to demonstrate sulfhydryl groups in electron microscopy have not been completely successful. However, electron microscopic demonstration of disulfide groups, using alkaline methenamine silver, has been made.

Mucins. Salivary mucins form semiviscous protective coatings over oral mucous membranes. They are composed of high-molecular-weight carbohydrate-protein complexes. Two types of mucins are recognized by the predominant carbohydrate component in their molecules—fucomucins, rich in L-fucose, and sialomucin, rich in sialic acid. The latter is believed to confer acidity on certain types of mucins. Both of these mucins are present together in saliva, with one predominating over the other.

Histochemical techniques have been very useful in our understanding of the salivary mucins present in various salivary glands and their chemical composition. The dyes mucicarmine and mucihematin are frequently used for nonspecific staining of mucins. PAS technique is used to identify neutral mucins (Fig. 15-16). Alcian blue, toluidine blue, colloidal iron, and aldehyde fuchsin methods are used to localize the acid mucins (Plate 4, *C*). These techniques reveal species differences in the mucins of different salivary glands.

Enzyme histochemistry

Alkaline phosphatase. Alkaline phosphatase activity in human gingiva is specifically demon-

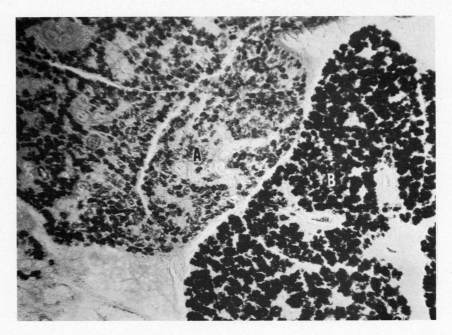

Fig. 15-16. Freeze-dried mouse submandibular gland, *A,* and sublingual gland, *B,* showing PAS reactivity of mucins. (× 140.)

strable in the capillary endothelium of the lamina propria (Fig. 15-17). The reaction product, observed in the gingival epithelium and in the collagen fibers, seems to be a diffusion artifact.

Oral epithelium of the rat exhibits an increased alkaline phosphatase activity during the estrous cycle, correlated to phosphatase changes in the vaginal epithelium. Alkaline phosphatase is implicated in the mechanism of keratinization, although its precise role in this process is still uncertain.

The basement membranes associated with salivary gland acini exhibit high alkaline phosphatase activity (Plate 4, *D*). Similar activity in taste buds of several species of animals has also been reported.

Acid phosphatase. Acid phosphatase activity in human gingiva seems related to the degree of keratinization, being very high in the zone of keratinization and low in nonkeratinized re-

gions. This pattern corresponds with that observed in the skin epidermis. Cells of the functional epithelium in the gingival sulcus have been reported by Lange and Schroeder to be rich in lysosomal enzymes in the normal healthy tissue.

Esterase. Little information is available on the esterase activity of human gingiva. Superficial layers including keratinizing zone show the presence of some esterase activity.

High esterase activity is demonstrable in the salivary gland ducts and also in the serous demilunes of the sublingual gland (Figs. 15-18 and 15-19). Similar activity is observed in the taste buds of several animal species, and this has been implicated in gustatory discrimination. Mast cells in oral tissues also have esterase activity (Fig. 15-20).

Aminopeptidase. The activity of this enzyme in human gingiva is low and is localized pri-

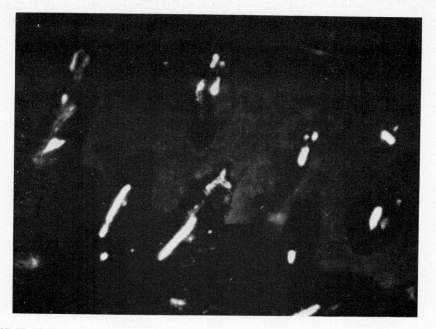

Fig. 15-17. Alkaline phosphatase activity of capillaries of lamina propria of human gingiva revealed by ultraviolet fluorescence. ($\times$80.)

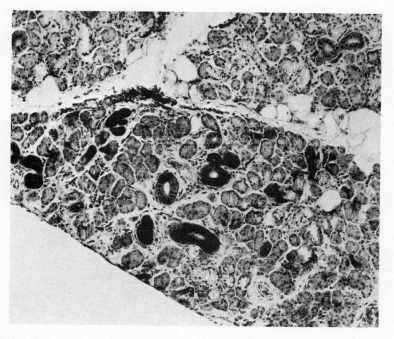

Fig. 15-18. Esterase activity of ducts of freeze-dried human parotid gland. (Nuclear counterstain; $\times$110.)

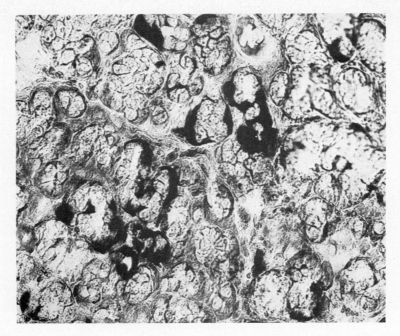

Fig. 15-19. Esterase activity of demilune cells of freeze-dried human sublingual gland. (×210.)

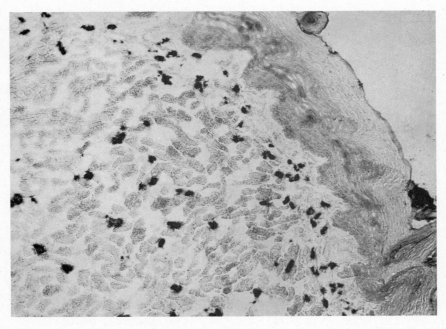

Fig. 15-20. Mast cells of rat tongue incubated with substrate solution containing naphthol AS-D chloroacetate. (×250.)

marily in the basal cell layers of the epithelium and in the underlying connective tissue. An increase in aminopeptidase activity during inflammation and in hyperplasia caused by the drug phenytoin has been reported.

Aminopeptidase is also observed in the salivary gland ducts.

β-Glucuronidase. β-Glucuronidases hydrolyze the β-glycoside linkage of glucuronides, are involved in conjugation of steroid hormones and in hydrolysis of conjugated glucuronides, and play a role in cell proliferation. The enzyme has been localized in the basal cell layers of the oral epithelium in humans and rats.

Cytochrome oxidase. Histochemical techniques demonstrate low levels of cytochrome oxidase activity in human gingiva. Specifically, this cytochrome oxidase activity is localized in the basal layers of the free and attached gingiva, crevicular epithelium, and epithelial attachment (Fig. 15-21). In chronic gingivitis a striking increase in cytochrome oxidase activity is observed in the epithelium from the free gingival groove through to the epithelial attachment. In chronic gingivitis, the underlying connective tissue also shows a variable increase in oxidase activity.

Cytochrome oxidase activity is also demon-

Fig. 15-21. Human attached gingiva showing cytochrome oxidase activity of basal cell layer and in connective tissue of lamina propria.

strated in the salivary glands, especially in the duct system (Fig. 15-22).

Succinic dehydrogenase and glucose 6-phosphate dehydrogenase. The distribution pattern of succinic dehydrogenase is similar to that of cytochrome oxidase. This dehydrogenase is observed primarily in the basal cell layers of the gingival epithelium and in the ducts of the salivary glands. Glucose 6-phosphate dehydrogenase is present in significant quantities in the human oral mucosal epithelium. The levels of this enzyme become highly elevated in malignant dysplastic lesions of the oral mucosa. Elevation of this enzyme is considered to assist in the diagnosis of oral cancers.

CLINICAL CONSIDERATIONS

Histochemical techniques are not only an important tool in dental research but are also frequently used in histopathologic diagnosis.

Although the tissue biopsy materials are usually stained with hematoxylin and eosin, there are numerous occasions when this type of staining technique does not permit a definitive diagnosis. In a differential diagnosis of an epithelial tumor in or around the oral cavity a histochemical stain for mucin may assist the oral pathologist in distinguishing a tumor of salivary gland origin from an odontogenic tumor or a tumor arising from nonglandular epithelium. Because these tumors often require different types of treatment, this distinction is of great practical importance.

A variety of fungi that infect humans contain mucopolysaccharides. The hyphae and spores of these fungi are present in the infected tissues, and their correct diagnosis can often be made only after special histochemical stains for mucin have been done. In the human oral cavity diagnosis of histoplasmosis, actino mycosis,

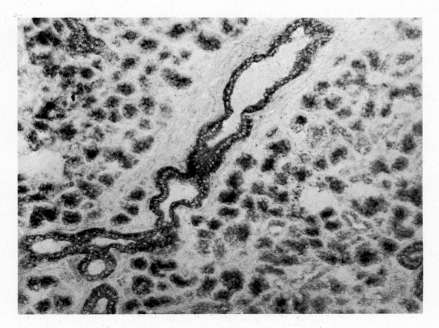

Fig. 15-22. Cytochrome oxidase activity of human parotid gland. (×143.)

blastomycosis, and coccidioidomycosis can often be made only after special histochemical stains.

Histochemical stains that reveal lipids are of value in correctly diagnosing tumors that arise from the fat cells (lipoma and liposarcoma). They are also an important aid in establishing the identity of vesicles that may appear in tumor cells of various benign and malignant lesions. Since cytoplasmic vesicles may represent lipid, mucin, glycogen, or intracellular edema, their true identity is sometimes important for correct diagnosis and therapy.

REFERENCES

Albert, E.N., and Fleischer, E.: A new electron dense stain for elastic tissue, J. Histochem. Cytochem. **18**:697, 1970.

Argyris, T.S.: Glycogen in the epidermis of mice painted with methylchol-anthrene, J. Natl. Cancer Inst. **12**:1159, 1952.

Baer, P.N., and Burstone, M.S.: Esterase activity associated with formation of deposits on teeth, Oral Surg. **12**:1147, 1959.

Balough, K.: Decalcification with versene for histochemical study of oxidative enzyme systems, J. Histochem. Cytochem. **10**:232, 1962.

Balough, K.: Histochemical study of oxidative enzyme systems in teeth and peridental tissues, J. Dent. Res. **42**:1457, 1963.

Baradi, A.F., and Bourne, G.H.: Gustatory and olfactory epithelia. In Bourne, G.H., and Danielli, J.F., editors: International review of cytology, vol. 2, New York, 1953, Academic Press, Inc.

Baradi, A.F., and Bourne, G.H.: Histochemical localization of cholinesterase in gustatory epithelia, J. Histochem. Cytochem. **7**:2, 1959.

Baradi, A.F., and Bourne, G.H.: New observations on alkaline glycophosphatase reaction in the papilla foliata, J. Biophys. Biochem. Cytol. **5**:173, 1959.

Bernhard, W., and Avrameas, S.: Ultrastructural visualization of cellular carbohydrate components by means of concanavalin A, Exp. Cell Res. **64**:232, 1971.

Birkedal-Hansen, H.: Effect of fixation on detection of carbohydrates in demineralized paraffin sections of rat jaw, Scand. J. Dent. Res. **82**:99, 1974.

Bourne, G.H., editor: The biochemistry and physiology of bone, New York, 1956, Academic Press, Inc.

Boyde, A., and Reith, E.J.: Electron probe analysis of maturation ameloblasts of the rat incisor and calf molar, Histochemistry **55**:41, 1978.

Bradfield, J.R.G.: Glycogen of the vertebrate epidermis, Nature **167**:40, 1951.

Burstone, M.S.: A cytologic study of salivary glands of the mouse tongue, J. Dent. Res. **32**:126, 1953.

Burstone, M.S.: The ground substance of abnormal dentin, secondary dentin, and pulp calcification, J. Dent. Res. **32**:269, 1953.

Burstone, M.S.: Esterase of the salivary glands, J. Histochem. Cytochem. **4**:130, 1956.

Burstone, M.S.: Histochemical observations on enzymatic processes in bones and teeth, Ann. N.Y. Acad. Sci. **85**:431, 1960.

Burstone, M.S.: Histochemical study of cytochrome oxidase in normal and inflamed gingiva, Oral Surg. **13**:1501, 1960.

Burstone, M.S.: Hydrolytic enzymes in dentinogenesis and osteogenesis. In Sognnaes, R.F., editor: Calcification in biological systems, Washington, D.C., 1960, American Association for the Advancement of Science.

Burstone, M.S.: Postcoupling, noncoupling and fluorescence techniques for the demonstration of alkaline phosphatase, J. Natl. Cancer Inst. **24**:1199, 1960.

Burstone, M.S.: Enzyme histochemistry and its application in the study of neoplasms, New York, 1962, Academic Press, Inc.

Burstone, M.S.: Enzyme histochemistry and cytochemistry. In Bourne, G.H., editor: Cytology and cell physiology, New York, 1964, Academic Press, Inc.

Burstone, M.S., and Folk, J.E.: Histochemical demonstration of aminopeptidase, J. Histochem. Cytochem. **4**:217, 1956.

Cabrini, R.L., and Carranza, F.A.: Histochemical distribution of acid phosphatase in human gingiva, J. Periodontol. **29**:34, 1958.

Cabrini, R.L., and Carranza, F.A.: Histochemical localization of β-glucuronidase in stratified squamous epithelium, Naturwissenschaften **22**:553, 1958.

Cabrini, R.L., and Carranza, F.A.: Histochemical distribution of beta-glucuronidase in gingival tissues, Arch. Oral Biol. **2**:28, 1960.

Carranza, F., and Cabrini, R.L.: Mast cells in human gingiva, Oral Surg. **8**:1093, 1955.

Chapman, J.A.: Fibroblasts and collagen, Br. Med. Bull. **18**:233, 1962.

Chayen, J., and Bitensky, L.: Lysosomal enzymes and inflammation with particular references to rheumatoid diseases, Ann. Rheum. Dis. **30**:522, 1971.

Chayen, J. Bitensky, L., and Butcher, R.G.: Practical histochemistry, New york, 1973, John Wiley & Sons.

Comar, C.L., and Bronner, F., editors: Mineral metabolism, vol. 1, New York, 1960, Academic Press, Inc.

Cotelli, D.C., and Livingston, D.C.: Fluorescent reagent for the periodic acid–Schiff and Feulgen reactions for cytochemical studies, J. Histochem. Cytochem. **24:**956, 1976.

Dorfman, A.: The biochemistry of connective tissues, J. Chronic Dis. **10:**403, 1959.

Eichel, B.: Oxidative enzymes of gingiva, Ann. N.Y. Acad. Sci. **85:**479, 1960.

Essner, E., Schrieber, J., and Griewski, R.A.: Localization of carbohydrate components in rat colon with fluoresceinated lectins, J. Histochem. Cytochem. Cytochem. **26:**452, 1978.

Etzler, M.E., and Branstrator, M.L.: Differential localization of cell surface and secretory components in rat intestinal epithelium by use of lectins. J. Cell Biol. **62:**329, 1974.

Evans, A.W., Johnson, N.W., and Butcher, R.G.: A quantitative histochemical study of glucose-6-phosphate dehydrogenase activity in premalignant and malignant lesions of human oral mucosa, Histochem. J. **15:**483, 1983.

Eveland, W.C.: Fluorescent antibody technique in medical diagnosis, Curr. Med. Dig. **31:**351, 1964.

Everett, M.M., and Miller, W.A.: Histochemical studies on calcified tissues. I. Amino acid histochemistry of fetal calf and human enamel matrix, Calcif. Tissue Res. **14:**229, 1972.

Everett, M.M., and Miller, W.A.: Histochemical studies on calcified tissues. II. Amino acid histochemistry of developing dentin and bone, Calcif. Tissue Res. **16:**73, 1974.

Felton, J.H., Person, P., and Stahl, S.S.: Biochemical and histochemical studies of aerobic oxidative metabolism of oral tissues. II. Enzymatic dissection of gingival and tongue epithelia from connective tissues, J. Dent. Res. **44:**392, 1965.

Fisher, E.R.: Tissue mast cells, J.A.M.A. **173:**171, 1960.

Freilich, L.S.: Ultrastructure and acid phosphatase cytochemistry of odontoclasts: effects of parathyroid extract, J. Dent. Res. **50:**1047, 1971.

Freilich, L.S.: A morphological and histological study of the cells associated with physiological root resorption in human and canine primary teeth, Ph.D. thesis, 1972, Department of Anatomy, Georgetown University, Washington, D.C.

Gersh, I., and Gatchpole, J.: The organization of ground substance and basement membrane and its significance in tissue injury, disease, and growth, Am. J. Anat. **85:**457, 1949.

Gerson, S.: Activity of glucose-6-phosphate dehydrogenase and acid phosphatase in nonkeratinized and keratinized oral epithelia and epidermis in rabbit, J. Periodont. Res. **8:**151, 1973.

Goldberg, M., and Septier, D.: Electron microscopic visualization of proteoglycans in rat incisor predentin and dentin with cuprolinic blue, Arch. Oral Biol. **28:**79, 1983.

Goldman, H.M., Ruben, M.P., and Sherman, D.: The application of laser spectroscopy for the qualitative and quantitative analysis of the inorganic components of calcified tissues, Oral Surg. **17:**102, 1964.

Gray, J.A., and Opdyke, D.L.: A device for thin sectioning of hard tissues, J. Dent. Res. **41:**172, 1962.

Gregg, J.M.: Analysis of tooth eruption and alveolar bone growth utilizing tetracycline fluorescence, J. Dent. Res. 43 (suppl.):887, 1964.

Greep, R.O., Fischer, C.J., and Morse, A.: Alkaline phosphatase in odontogenesis and osteogenesis and its histochemical demonstration after demineralization, J. Am. Dent. Assoc. **36:**427, 1948.

Gros, D., Obrenovitch, A., Challice, C.E., et al.: Ultrastructural visualization of cellular carbohydrate components by means of lectins on ultrathin glycol methacrylate sections, J. Histochem. Cytochem. **25:**104, 1977.

Gustafson, G.: The histopathology of caries of human dental enamel, Acta Odontol. Scand. **15:**13, 1957.

Hancox, N.M., and Boothroyd, B.: Structure-function relationship in the osteoclast. In Sognnaes, R.F., editor: Mechanism of hard tissue destruction, Washington, D.C., 1963, American Association for the Advancement of Science.

Hess, W.C., Lee, C.Y., and Peckham, S.C.: The lipid content of enamel and dentin, J. Dent. Res. **35:**273, 1956.

Hewitt, A.T., Klienman, H.K., Pennypacker, J.P., and Martin, G.R.: Identification of an attachment factor for chondrocytes, Proc. Natl. Acad. Sci. U.S.A. **77:**385, 1980.

Hoerman, K.C., and Mancewicz, S.A.: Phosphorescence of calcified tissues, J. Dent. Res. 43(suppl.):775, 1964.

Holliday, T.D.: Diagnostic exfoliative cytology, its value as an everyday hospital investigation, Lancet **1:**488, 1963.

Jackson, D.S.: Some biochemical aspects of fibrogenesis and wound healing, N. Engl. J. Med. **259:**814, 1958.

Kapur, S.P., and Russell, T.E.: Sharpey fiber bone development in surgically implanted dog mandible: a scanning electron microscopic study, Acta Anat. (Basel) **102:**260, 1978.

Kleinman, H.K.: Role of cell attachment proteins in defining cell matrix interactions. In Liotta, L.A., and Hart, I.R., editors: Tumor invasion and metatasis, Boston, 1982, Martinus Nijhoff Publishers.

Lange, D.E., and Schroeder, H.E.: Structural localization of lysosomal enzymes in gingival sulcus cells, J. Dent. Res. **51:**272, 1972.

Larmas, L.A., Makinen, K.K., and Paunio, K.U.: A histochemical study of arylaminopeptidases in hydantoin induced hyperplastic, healthy and inflamed human gingiva, J. Periodont. Res. 8:21, 1973.

Laurie, G.W., Leblond, C.P., and Martin, G.R.: Light microscopic immunolocalization of type IV collagen, laminin, heparan sulfate proteoglycan, and fibronectin in the basement membrane of a variety of rat organs, Am. J. Anat. 167:71, 1983.

Lesot, H., Osman, M., and Ruch, J.V.: Immunofluorescent localization of collagens, fibronectin, and laminin during terminal differentiation of odontoblasts, Dev. Biol. 82:371, 1981.

Lev, R., and Spicer, S.S.: Specific staining of sulphate groups with alcian blue at low pH, J. Histochem. Cytochem. 12:39, 1964.

Linde, A., Johansson, S., Jonsson, R., and Jontell, M.: Localization of fibronectin during dentinogenesis in rat incisor, Arch. Oral Biol. 27:1069, 1982.

Luft, J.H.: Ruthenium red and violet. I. Chemistry, purification, methods of use for electron microscopy and mechanism of action, Anat. Rec. 171:347, 1971.

Matsuzawa, T., and Anderson, H.C.: Phosphatases of epiphyseal cartilage studied by electron microscopic cytochemical methods, J. Histochem. Cytochem. 19:801, 1971.

Matukas, V.J., and Krikos, G.A.: Evidence for changes in protein-polysaccharide association with the onset of calcification in cartilage, J. Cell. Biol. 39:43, 1968.

Millard, H.D.: Oral exfoliative cytology as an aid to diagnosis, J. Am. Dent. Assoc. 69:547, 1964.

Mörnstad, H., and Sundström, B.: Cytochemical demonstration of adenyl cyclase in rat incisor enamel organ, Scand. J. Dent. Res. 82:146, 1974.

Munhoz, C.O., Cassio, O.G., and Leblond, C.P.: Deposition of calcium phosphate into dentin and enamel as shown by radioautography of sections of incisor teeth following injection of ^{45}Ca into rats, Calcif. Tissue Res. 14:221, 1974.

Narayanan, A.S., and Page, R.C.: Connective tissue of the periodontium: a summary of current work, Coll. Relat. Res. 3:33, 1983.

Neiders, M.E., Eick, J.D., Miller, W.A., and Leitner, J.W.: Electron probe microanalysis of cementum and underlying dentin in young permanent tooth, J. Dent. Res. 51:122, 1972.

Nicolson, G.L., and Singer, S.J.: Ferritin conjugated plant agglutinins as specific saccharide stains for electron microscopy: application to saccharides bound to cell membranes, Proc. Natl. Acad. Sci. U.S.A. 68:942, 1971.

Opdyke, D.L.: The histochemistry of dental decay, Arch. Oral Biol. 7:207, 1962.

Pearse, A.G.E.: Histochemistry, theoretical and applied, vol. 2, Baltimore, 1972, The Williams & Wilkins Co.

Perry, M.M.: Identification of glycogen in thin sections of amphibian embryos, J. Cell. Sci. 2:257, 1967.

Person, P., and Burnett, G.W.: Dynamic equilibria of oral tissues. II. Cytochrome oxidase and succinoxidase activity of oral tissues, J. Periodontol. 26:99, 1955.

Philips, F.R.: A short manual of respiratory cytology: a guide to the identification of carcinoma cells in the sputum, Springfield, Ill., 1964, Charles C Thomas, Publisher.

Phillips, H.B., Owen-Jones, S., and Chandler, B.: Quantitative histology of bone: a computerized method for measuring the total mineral content of bone, Calcif. Tissue Res. 26:85, 1978.

Piez, K.A., and Reddi, A.H., editors: Extracellular matrix biochemistry, New York, 1984, Elsevier Science Publishers.

Polak, J.M., and Noorden, S., editors: Immunocytochemistry: practical applications in pathology and biology, Boston, 1983, PSG/Wright Publishing Co., Inc.

Porter, K.R., and Pappas, G.D.: Collagen formation by fibroblasts of the chick embryo dermis, J. Biophys. Biochem. Cytol. 5:153, 1959.

Rabinowitz, J.L., Ruthberg, M., Cohen, D.W., and Marsh, J.B.: Human gingival lipids, J. Periodont. Res. 8:381, 1973.

Rasmussen, H., and Bordier, P.: They physiological and cellular basis of metabolic bone disease, Baltimore, 1974, The Williams & Wilkins Co.

Reith, E.J., and Boyde, A.: Histochemistry and electron probe analysis of secretory ameloblasts of developing molar teeth, Histochemistry 55:17, 1978.

Rovalstad, G.H., and Calandra, J.C.: Enzyme studies of salivary corpuscles, Dent. Progr. 2:21, 1961.

Russell, T.E., and Kapur, S.P.: Bone surfaces adjacent to a sub-periosteal implant: an SEM study, Oral Implant. 7:415, 1977.

Sandritter, W., and Schreiber, M.: Histochemie von Sputumzellen. I. Qualitative histochemische Untersuchungen, Frankfurt, Z. Pathol. 68:693, 1958.

Saulk, J.J., Gay, R., Miller, E.J., and Gay, S.: Immunohistochemical localization of type III collagen in the dentin of patients with osteogenesis imperfecta and hereditary opalescent dentin, J. Oral Pathol. 2:210, 1980.

Schajowicz, F., and Cabrini, R.L.: Histochemical studies on glycogen in normal ossification and calcification, J. Bone Joint Surg. 40:1081, 1958.

Shackleford, J.M., and Klapper, C.E.: Structure and carbohydrate histochemistry of mammalian salivary glands, Am. J. Anat. 111:825, 1962.

Shimizu, M., Glimcher, M.J., Travis, D., and Goldhaber, P.: Mouse bone collagenase: isolation, partial purification, and mechanism of action, Proc. Soc. Exp. Biol. Med. 130:1175, 1969.

Sognnaes, R.F.: Mechanism of hard tissue destruction, Washington, D.C., 1963, The American Association for the Advancement of Science.

Soyenkoff, R., Friedman, B.K., and Newton, M.: The lipids of dental tissues: a preliminary study, J. Dent. Res. **30**:599, 1951.

Spicer, S.S.: A correlative study of the histochemical properties of rodent acid mucopolysaccharides, J. Histochem. Cytochem. **8**:18, 1960.

Spicer, S.S.: Histochemical differentiation of mammalian mucopolysaccharides, Ann. N.Y. Acad. Sci. **106**:379, 1963.

Spicer, S.S., and Warren, L.: The histochemistry of sialic acid containing mucoproteins, J. Histochem. Cytochem. **8**:135, 1960.

Steinman, R.R., Hewes, C.G., and Woods, R.W.: Histochemical analysis of lesions in incipient dental caries, J. Dent. Res. **38**:592, 1959.

Stoward, P.J.: Fixation in histochemistry, London, 1973, Chapman & Hall Ltd.

Stuart, J., and Simpson, J.S.: Dehydrogenase enzyme cytochemistry of unfixed leucocytes, J. Clin. Pathol. **23**:517, 1970.

Swedlow, D.B., Harper, R.A., and Katz, J.L.: Evolution of a new preparative technique for bone examination in the SEM. Scanning electon microscopy (part II). Proceedings of the Workshop on Biological Specimen Preparation for SEM, Chicago, 1972, IIT Research Institute.

Symons, N.B.B.: Alkaline phosphatase activity in the developing tooth of the rat, J. Anat. **89**:238, 1955.

Symons, N.B.B.: Lipid distribution in the developing teeth of the rat, Br. Dent. J. **105**:27, 1958.

Takagi, M., Parmley, T.R., and Denys, F.R.: Ultrastructural localization of complex carbohydrates in odontoblasts, predentin and dentin, J. Histochem. Cytochem. **29**:747, 1981.

Takagi, M., Parmley, T.R., Spicer, S.S., Denys, F.R., and Setser, M.E.: Ultrastructural localization of acid glycoconjugates with the low iron diamine method, J. Histochem. **30**:471, 1982.

Terranova, V.P., Liotta, L.A., Russo, R.G., and Martin, G.R.: Role of laminin in the attachment and metastasis of murine tumor cells, Cancer Res. **42**:2265, 1982.

Thesleff, I., Barrach, H.J., Foidart, J.M., Vaheri, A., Pratt, R.M., and Martin, G.R.: Changes in the distribution of type IV collagen laminin, proteoglycan, and fibronectin during mouse tooth development, Dev. Biol. **81**:182, 1981.

Turesky, S., Crowly, J., and Glickman, I.: A histochemical study of protein-bound sulfhydryl and disulfide groups in normal and inflamed human gingiva, J. Dent. Res. **36**:255, 1957.

Turesky, S., Glickman, I., and Litwin, T.: A histochemical evaluation of normal and inflamed human gingiva, J. Dent. Res. **30**:792, 1951.

Vallotton, C.: Etude bio-histologique de la phosphatase dans la geneive humaine normale et dans les gingivites, Schweiz. Monatsschr. Zahnheilkd. **52**:512, 1942.

Van Scott, E.J., and Flesh, P.: Sulfhydryl groups and disulfide linkages in normal and pathological keratinization, Arch. Derm. Syph. **70**:141, 1954.

Veterans Administration Cooperative Study: Oral exfoliative cytology, Washington, D.C., 1962, U.S. Government Printing Office.

Walker, D.G.: Citric acid cycle in osteoblasts and osteoclasts, Bull. Johns Hopkins Hosp. **108**:80, 1961.

Weinmann, J.P., Meyer, J., and Mardfin, D.: Occurrence and role of glycogen in the epithelium of the alveolar mucosa and of the attached gingiva, Am. J. Anat. **104**:381, 1959.

Weinstock, M., and Leblond, C.P.: Synthesis, migration, and release of precursor collagen by odontoblasts as visualized by radioautography after [^{3}H] proline administration, J. Cell. Biol. **60**:92, 1974.

Weinstock, A., Weinstock, M., and Leblond, C.P.: Autoradiographic detection of ^{3}H-glucose incorporation into glycoprotein by odontoblasts and its deposition at the site of the calcification front in dentin, Calcif. Tissue Res. **8**:181, 1972.

Wergedal, J.E., and Baylink, D.J.: Distribution of acid and alkaline phosphatase activity in undemineralized sections of the rat tibial diaphysis J. Histochem. Cytochem. **17**:799, 1969.

Wiebkin, O.W., and Thonard, J.C.: Mucopolysaccharide localization in gingival epithelium. I. An autoradiographic demonstration, J. Periodont. Res. **16**:600, 1981.

Wiebkin, O.W., and Thonard, J.C.: Mucopolysaccharide localization in gingival epithelium: factors affecting biosynthesis of sulfated proteoglycans in organ cultures of gingival epithelium, J. Periodont. Res. **17**:629, 1982.

Wied, G.L., editor: Introduction to quantitative cytochemistry, New York, 1965, Academic Press, Inc.

Wisotzky, J.: Effects of neo-tetrazolium chloride on the phosphorescence of teeth, J. Dent. Res. **43**:659, 1964.

Wisotzky, J.: Effect of tetracycline on the phosphorescence of teeth, J. Dent. Res. **51**:7, 1972.

Yamada, K., and Shimizu, S.: The histochemistry of galactose residues of complex carbohydrates as studied by peroxidase labelled *Ricinus communis* agglutinin, Histochemistry **53**:143, 1977.

Yoshiki, S., and Kurahashi, Y.: A light and electron microscopic study of alkaline phosphatase activity in the early stage of dentinogenesis in the young rat, Arch. Oral Biol. **16**:1143, 1971.

Zander, H.A.L.: Distribution of phosphatase in gingival tissue, J. Dent. Res. **20**:347, 1941.

PREPARATION OF SPECIMENS FOR HISTOLOGIC STUDY

PREPARATION OF SECTIONS OF PARAFFIN-EMBEDDED SPECIMENS

Obtaining the specimen
Fixation of the specimen
Dehydration of the specimen
Infiltration of the specimen with paraffin
Embedding the specimen
Cutting the sections of the specimen
Mounting the cut sections on slides
Staining the sections

PREPARATION OF SECTIONS OF PARLODION-EMBEDDED SPECIMENS

Obtaining the specimen
Fixation of the specimen
Decalcification of the specimen
Washing the specimen
Dehydration of the specimen
Infiltration of the specimen with parlodion
Embedding the specimen in parlodion
Cutting the sections of the specimen
Staining the sections

PREPARATION OF GROUND SECTIONS OF TEETH OR BONE

PREPARATION OF FROZEN SECTIONS

TYPES OF MICROSCOPY

The morphologic study of oral tissues involves the preparation of tissue sections for microscopic examination. Knowledge of various types of microscopes and related histologic techniques will assist the student in interpretation of the structure and function of oral tissues.

The fundamental methods of tissue preparation for various types of microscopy, although basically similar to those for light microscopy, show differences in specific procedures. For example, differences in the tissue preparation for electron microscopy are necessitated by the lower penetrating power of electrons compared with the light and the greater resolving power of the electron microscope. Tissues for light microscopic study must be sufficiently thin to transmit light, and its components must have sufficient contrast for the parts to be distinguishable from each other. Routine histologic techniques involve the fixation of tissues in protoplasmic coagulating solution, dehydration in organic solvents, embedding in paraffin or plastics, and cutting of thin sections on a microtome. The sections are mounted on an appropriate supporting structure, stained, and examined under a microscope. The basic procedures are modified depending on the nature of the specimen and the type of microscope to be used for examination of structures of particular interest.

Four methods of preparation of oral tissues for microscopic examination are commonly used:

1. *Specimens may be embedded in paraffin and sectioned.* The most commonly used method of preparing soft tissues for study with an ordinary light microscope is that

455

of embedding the specimen in paraffin and then cutting sections 4 to 10 μm thick. The sections are mounted on microscope slides, passed through a selected series of stains, and covered with a cover glass.

2. *Specimens may be embedded in parlodion and sectioned.* Specimens containing bone or teeth require different preparation. Such specimens must be decalcified (the mineral substance removed) and usually embedded in parlodion rather than in paraffin prior to being sectioned on a microtome.

3. *Specimens of calcified tissue may be ground into thin sections.* Sections of undeclacified tooth or bone may be obtained by preparing a *ground section.* This is done by slicing the undeclacified specimen, which is ground down to a section of about 50 μm on a revolving stone or disk.

4. *Specimens of soft tissue may be frozen and sectioned.* When it is important that pathologic tissue specimens be examined immediately, or if the reagents used for paraffin or parlodion embedding would destroy the tissue characteristics that are to be studied, the fresh, unfixed, or fixed soft tissue may be frozen and sectioned without being embedded. Such tissue sections are usually referred to as *frozen sections.*

These four methods of specimen preparation will now be described in more detail.

PREPARATION OF SECTIONS OF PARAFFIN-EMBEDDED SPECIMENS

The method of preparing a specimen for sectioning by embedding it in paraffin is suitable for oral specimens that contain no calcified tissue, such as specimens of gingiva, cheek, and tongue.

Obtaining the specimen. Specimens taken from humans or experimental animals must be removed carefully, without crushing, either while the animal is alive or immediately after it has been killed.

Fixation of the specimen. Immediately after removal of the specimen it must be placed in a *fixing solution.* Specimens that have not been placed in such a solution are seldom any good. There are many good fixing solutions available. Sometimes the kinds of stains subsequently to be used determine the kind of solution to be chosen. One of the most commonly used fixatives for dental tissues is 10% neutral formalin.

The purposes of fixation are to coagulate the protein, thus reducing alteration by subsequent treatment, and to make the tissues more readily permeable to the subsequent applications of reagents. The fixation period varies from several hours to several days, depending on the size and density of the specimens and on the type of fixing solution used.

After fixation in formalin, the specimen is washed overnight in running water.

Dehydration of the specimen. Since it is necessary that the specimen be completely infiltrated with the paraffin in which it is to be embedded, it must first be infiltrated with some substance that is miscible with paraffin. Paraffin and water do not mix. Therefore after being washed in running water to remove the formalin, the specimen is gradually dehydrated by being passed through a series of increasing percentages of alcohol (40%, 60%, 80%, 95%, and absolute alcohol), remaining in each dish for several hours. (The time required for each step of the process depends on the size and density of the specimen.) To ensure that the water is replaced by alcohol, two or three changes of absolute alcohol are used. Then, since paraffin and alcohol are not miscible, the specimen is passed from alcohol through two changes of xylene, which is miscible with both alcohol and paraffin.

Infiltration of the specimen with paraffin. When xylene has completely replaced the alcohol in the tissue, the specimen is ready to be infiltrated with paraffin. It is removed from the xylene and placed in a dish of melted embedding paraffin, and the dish is put into a constant-temperature oven regulated to about 60 C. (The exact temperature depends on the melting point of the paraffin used.) During the course of several hours the specimen is changed to two or three successive dishes of paraffin so that all of the xylene in the tissue is replaced by paraffin. The time in the oven depends on the size and density of the specimen: a specimen the size of a 2 or 3 mm cube may need to remain in the oven only a couple of hours, whereas a larger, firmer specimen may require 12 to 24 hours to ensure complete paraffin infiltration.

Embedding the specimen. When the specimen is completely infiltrated with paraffin, it is embedded in the center of a block of paraffin. A small paper box, perhaps a 19 mm cube for a small specimen, is filled with melted paraffin, and with warm forceps the specimen is removed from the dish of melted paraffin and placed in the center of the box of paraffin. Attention must be given here to the orientation of the specimen so that it will be cut in the plane desired for examination. A good plan is to place the surface to be cut first toward the bottom of the box. The paper box containing the paraffin and the specimen is then immersed in cool water to harden the paraffin. The hardened paraffin block is removed from the paper box and is mounted on a paraffin-coated wooden cube (about a 19mm cube). The mounted paraffin block is trimmed with a razor blade so that there is about 3mm of paraffin surrounding the specimen on all four sides such that the edges are parallel. The specimen is now ready to be sectioned on a microtome.

Cutting the sections of the specimen. The wooden cube to which the paraffin block is attached is clamped on a precision rotary microtome, the microtome is adjusted to cut sections of the desired thickness (usually 4 to 10 μm), and the perfectly sharpened microtome knife is clamped into place for sectioning.

Mounting the cut sections on slides. Suitable lengths of the paraffin ribbon are then mounted on prepared microscope slides. The preparation of the slides is done by the coating of clean slides with a thin film of Meyer's albumin adhesive (egg albumin and glycerin). A short length of paraffin ribbon is floated in a pan of warm water (about 45 C). A prepared slide is slipped under the ribbon and then is lifted from the water with the ribbon, which of course contains the tissue sections, arranged on its upper surface. The slide is placed on a constant-temperature drying table, which is regulated to about 42 C, so that the sections will adhere to the slide. The slide is then allowed to dry on this table.

Staining the sections. There are innumerable tissue stains, methods of using stains, and methods of preparing tissues to receive stains. Some of the many factors that influence the choice of stains are the kind or kinds of tissue to be studied and the particular characteristics of immediate interest.

One combination of stains often used for routine microscopic study is hematoxylin and eosin, commonly known as H & E. A usual procedure for staining sections with hematoxylin and eosin follows.

The dried slides are placed vertically in glass staining trays; the trays are then passed through a series of staining dished that contain the various reagents (Table 7).

The slides are removed one at a time from the xylene, the sections are covered with a mounting medium, and a cover glass if affixed. When the mounting medium has hardened, the slides are ready for examination.

PREPARATION OF SECTIONS OF PARLODION-EMBEDDED SPECIMENS

Specimens that contain bone and teeth cannot be cut with a microtome knife unless the calcified tissues are first made soft by decalcification. Furthermore, if a specimen contains any appreciable amount of bone or teeth, the decalcified specimen is better embedded in parlodion (celloidin, pyroxylin) than in paraffin. It is extremely difficult, if not impossible, to

Table 7. Staining of sections

1. Xylene	2 min.	To remove paraffin from sections
2. Xylene	2 min.	To remove paraffin from sections
3. Absolute alcohol	2 min.	To remove xylene
4. 95% alcohol	1 min.	Approach to water
5. 80% alcohol	1 min.	Approach to water
6. 60% alcohol	1 min.	Approach to water
7. Distilled water	1 min.	Water precedes stains dissolved in water
8. Hematoxylin (Harris's)	3-10 min.	To stain nuclei
9. Distilled water	Rinse	To rinse off excess stain
10. Ammonium alum (saturated solution)	2-10 min.	To differentiate; nuclei will retain stain
11. Sodium bicarbonate (saturated solution)	1-2 min.	Makes stain blue
12. Distilled water	1 min.	Removes $NaHCO_3$
13. 80% alcohol	1 min.	Partially dehydrates
14. 95% alcohol	1 min.	Alcohol precedes stains dissolved in alcohol
15. Eosin (alcohol soluble)	1-2 min.	To stain cytoplasm and intercellular substance
16. 95% alcohol	Rinse, or longer	Alcohol destains eosin and should be used as long as needed
17. 95% alcohol	Rinse, or longer	To remove excess eosin
18. Absolute alcohol	1 min.	To dehydrate
19. Absolute alcohol	2 min.	To dehydrate
20. Xylene	2 min.	To remove alcohol and clear
21. Xylene	2 min.	To clear

get good sections of a large mandible containing teeth in situ if the specimen is embedded in paraffin.

Let us suppose that we are to section a specimen of dog mandible bearing two premolar teeth. One method is as follows.

Obtaining the specimen. The portion of the mandible containing the two premolar teeth is separated as carefully as possible from the rest of the mandible by means of a sharp scalpel and a bone saw. Unwanted soft tissue is removed. If the area of the specimen next to the line of sawing will be seriously damaged by the saw, the specimen should be cut a little larger than needed and then trimmed to the desired size after partial decalcification. It is better to have the mandible cut into several pieces before placing it in the fixative because a smaller specimen allows quicker penetration before placing it in the fixative because a smaller specimen allows quicker penetration of the fixing

solution to its center. If the tooth pulp is of interest, a bur should be used to open the root apex of the teeth to permit entrance of the fixing solution into the pulp chamber. This operation must be done with care so that too much heat does not burn the pulp tissue.

Fixation of the specimen. The specimen so cut and prepared is quickly rinsed in running water and for fixation is placed immediately in about 400 ml of 10% neutral formalin. It should remain in the formalin not less than a week and preferably longer. It may be stored in formalin for a long period.

Decalcification of the specimen. When fixation is complete, the specimen is then decalcified. Decalcification may be accomplished in several ways. One way is to suspend the specimen in about 400 ml of 5% nitric acid. The acid is changed daily for 8 to 10 days, and then the specimen is tested for complete decalcification.

One way to test for complete decalcification is to pierce the hard tissue with a needle. When the needle enters the bone and tooth easily, the tissue is probably ready for further treatment.

Another way to test for complete decalcification is to determine by a precipitation test whether there is calcium present in the nitric acid in which the specimen is immersed. This is done by placing in a test tube 5 or 6 ml of the acid in which the specimen has been standing and then adding 1 ml of concentrated ammonium hydroxide and several drops of a saturated aqueous solution of ammonium oxalate. A precipitate will form if any appreciable amount of calcium is present. If a precipitate forms, the acid covering the specimen should be changed and a couple of days later the test for complete decalcification should be repeated. If no precipitate is detected after the test tube has stood for an hour and after several additions of ammonium oxalate, it may be assumed that the specimen is almost completely decalcified. The specimen should be allowed to remain in the same acid for 48 hours longer and the test repeated.

The end point of decalcification is sometimes difficult to determine, but it is important. Specimens left in the acid too short a time are not completely decalcified and cannot be cut successfully, and specimens left in the acid too long a time do not stain well. Because of the adverse effect of prolonged exposure to acid on the staining quality of tissues, specimens should be reduced to their minimum size before decalcification is begun in order to keep the time necessary for acid treatment as short as possible.

Washing the specimen. When decalcification is complete, the specimen must be washed in running water for at least 24 hours to remove all of the acid.

Dehydration of the specimen. After washing, dehydration is accomplished by the placement of the specimen successively in increasing percentages of alcohol (40%, 60%, 80%, 95%, and absolute alcohol). The specimen should remain in each of the alcohols up to and including 95% for 24 to 48 hours, and it should then be placed in several changes of absolute alcohol over a period of 48 to 72 hours. It is necessary to remove, as much as possible, all of the water from the tissues in order to have good infiltration of parlodion.

From absolute alcohol the specimen is transferred to ether-alcohol (1 part anhydrous ether, 1 part absolute alcohol), because parlodion is dissolved in ether-alcohol. There should be several changes of ether-alcohol over a period of 48 to 72 hours.

Infiltration of the specimen with parlodion. Parlodion is purified nitrocellulose dissolved in ether-alcohol. From the ether-alcohol in which it has been standing, the specimen is transferred to 2% parlodion, covered tightly to prevent evaporation, and allowed to stand for a period of from 2 weeks to a month.

From 2% parlodion the specimen is transferred to increasing percentages of parlodion (4%, 6%, 10%, and 12%). The estimation of the time required for the infiltration of a specimen is a matter of experience, with the determining factors being the size of the specimen and the amount of bone and tooth material present. For the specimen of mandible being described here, the time required for complete parlodion infiltration might vary from several weeks to several months.

Embedding the specimen in parlodion. When infiltration with parlodion is complete, the specimen is embedded in the center of a block of parlodion. A glass dish with straight sidewalls and a lid is a good container to use for embedding. Some 12% parlodion is poured into the dish, and the specimen is placed in the parlodion. Then more parlodion is added so that there is about 13mm of parlodion above the specimen, the additional amount being necessary to allow for shrinkage during hardening.

Orientation of the specimen at this point to ensure the proper plane of cutting is important. If this piece of dog mandible is to be sectioned in such a way that the premolar teeth are cut in a mesiodistal plane and the first sections are cut from the buccal surface, then the buccal surface of the mandible should be placed toward the bottom of the dish when the specimen is embedded.

The dish is now covered with a lid that fits loosely enough to permit very slow evaporation of the ether-alcohol in which the parlodion is dissolved. As the ether-alcohol evaporates, the parlodion will become solidified and will eventually acquire a consistency *somewhat* like that of hard rubber.

This process of hardening the parlodion may require 2 or 3 weeks. When the block is very firm, it is removed from the dish and placed in chloroform until it sinks. It is then transferred to several changes of 70% alcohol to remove the chloroform.

Blocks of parlodion-embedded material must never be allowed to dry out. The blocks should be stored in 70% alcohol to allow the parlodion to harden further. Blocks that are to be stored for many months or years should eventually be transferred to a mixture of 70% alcohol and glycerin for storage.

Cutting the sections of the specimen. The hardened block of the parlodion-embedded specimen is fastened with liquid parlodion to a fiber block or to a metal object holder so that it can be clamped onto the precision sliding microtome. (This is a different instrument from the rotary microtome used for cutting paraffin.) Sections are cut with a sharp microtome knife. For the specimen of dog mandible being described here, the sections may have to be cut at a thickness of as much as 15 μm. Unlike paraffin sections, these parlodion sections must be handled one at a time. As each section is cut, it is straightened out with a camel's hair brush on the top surface of the horizontally placed microtome knife and is then removed from the knife and placed flat in a dish of 70% alcohol. It must not be allowed to become dry. If it is important that the sections be kept in serial order, a square of paper should be inserted after every fourth or fifth section as they are stored in the dish of alcohol.

Staining the sections. Ordinarily the parlodion is not removed from the sections, and the sections are not mounted on slides until after staining, dehydrating, and clearing are completed. The sections are passed through the series of reagents separately or in groups of three or four, using a perforated section lifter to make the transfer.

From the 70% alcohol in which they are stored when cut, the sections may be stained with hematoxylin and eosin as follows.

Referring to Table 7, omit steps 1 to 3 and start with step 4; that is, transfer the parlodion sections from 70% to 95% alcohol. Follow each step down through step 17, which is 95% alcohol. At this point, for the absolute alcohol specified in steps 18 and 19, substitute carbolxylene (75 ml xylene plus 25 ml melted carbolic acid crystals). This substitution is made because the parlodion is slightly soluble in absolute alcohol. From carbolxylene the sections are transferred to xylene (steps 20 and 21).

The sections should not be allowed to become folded or rolled up during the staining process. When they are put into the carbolxylene, they must be flattened out carefully, because the xylene that follows will slightly harden the parlodion sections so that they cannot easily be flattened.

To mount the stained section on a slide, slip the clean side (no adhesive is used) into the dish of xylene beneath the section, and lift the section onto the slide from the liquid, straightening it carefully, and quickly and firmly press it with a small piece of filter paper. The slide bearing the section is then quickly dipped back into the xylene and drained, and mounting medium is flowed over the section and a cover glass is dropped into place.

A modification of this embedding method, using acid celloidin instead of parlodion, will preserve much of the organic matrix of tooth enamel during the process of decalcification.

For variations in the hematoxylin and eosin stain and for information on the many other kinds of stains useful for both paraffin-embedded and parlodion-embedded specimens, the histology student must refer to books on microtechnique.

PREPARATION OF GROUND SECTIONS OF TEETH OR BONE

Decalcification of bone and teeth often obscures the structures. Teeth in particular are damaged because tooth enamel, being about 96% mineral substance, is usually completely destroyed by ordinary methods of decalcification. Undecalcified teeth and undecalcified bone may be studied by making thin ground sections of the specimens.

The equipment used for making ground sections includes a laboratory lathe, a coarse and a fine abrasive lathe wheel, a stream of water directed onto the rotating wheel and a pan beneath to catch the water, a wooden block (about a 25 mm cube), some 13 mm adhesive tape, a camel's hair brush, ether, mounting medium, microscope slides, and cover glasses.

Let us suppose that a thin ground section is to be prepared of a human mandibular molar tooth cut longitudinally in a mesiodistal plane. The coarse abrasive lathe wheel is attached to the lathe, water is directed onto the wheel, the tooth is held securely in the fingers, and its buccal surface is applied firmly to that flat surface of the rapidly rotating wheel. The tooth is ground down nearly to the level of the desired section.

The coarse wheel is now exchanged for a fine abrasive lathe wheel, and the cut surface of the tooth is ground again until the level of the desired section is reached.

At this point a piece of adhesive tape is wrapped around the wooden block in such a way that the sticky side of the tape is directed *outward*. The ground surface of the tooth is wiped dry and then is pressed onto the adhesive tape on one side of the wooden block. It will stick fast. With the block held securely in the fingers, the lingual surface of the tooth is applied to the coarse adhesive lathe wheel and the tooth is ground down to a thickness of about 0.5 mm. Then the coarse wheel is again exchanged for the fine-abrasive lathe wheel, and the grinding is continued until the section is as thin as desired.

The finished ground section is soaked off of the adhesive tape with ether and then dried for several minutes. Drying for too long will result in cracking. It is then mounted on a microscope slide. To do this, a drop of mounting medium is placed on the slide, the section is lifted with a camel's hair brush and placed on the drop, another drop of mounting medium is put on top of the section, and a cover glass is affixed for microscopic study.

The teeth used for ground sections should not be allowed to dry out after extraction, because drying makes the hard tissues brittle and the enamel may chip off in the process of grinding. Extracted teeth should be preserved in 10% formalin until used.

Precision equipment for making ground sections with much greater accuracy is available. The method described here is one in which equipment at hand in almost any laboratory is used. The technical literature contains a number of articles on the preparation of sections of undecalcified tissues.

PREPARATION OF FROZEN SECTIONS

Fixed soft tissues or fresh unfixed soft tissues may be cut into sections 10 to 15 μm thick by freezing the block of tissue with either liquid or solid carbon dioxide and cutting it on a freezing microtome. Frozen sections can be quickly prepared and are useful if the immediate examination of a specimen is required. Frozen sections are also useful when the tissue characteristics to be studied would be destroyed by the reagents used in paraffin embedding.

Details of the preparation of frozen sections can be obtained from books on microtechnique.

TYPES OF MICROSCOPY

A thin tissue section has the property to modify the color or intensity of light passing through it. The modified, light-containing information from the section is amplified through the lens system of a microscope and transmitted to the eye. Since the unstained tissues do not absorb or modify the light to a useful degree, tissue staining is utilized to induce differential absorption of light so that tissue components may be seen.

Many types of microscopes are used for the study of tissues. The most common is the bright-field microscope, which is a complex optical instrument that uses visible light. Modifications of this instrument have provided the phase-contrast, interference, dark-field, and polarizing microscopes. The optical systems that utilize invisible radiations include the ultraviolet microscope, roentgen-ray, and electron microscope. Each of these instruments has been a valuable tool in the study of oral tissues.

REFERENCES

Bodecker, C.F.: The Cape-Kitchin modification of the celloidin decalcifying method for dental enamel, J. Dent. Res. **16**:143, 1937.

Brewer, H.E., and Shellhamer, R.H.: Stained ground sections of teeth and bone, Stain Technol. **31**:111, 1956.

Davenport, H.A.: Histological and histochemical technics, Philadelphia, 1960, W.B. Saunders Co.

Fremlin, J.H., Mathieson, J., and Hardwick, J.L.: The grinding of thin sections of dental enamel, J. Dent. Res. **39**:1103, 1960.

Gatenby, J.B., and Beams, H.W., editors: The microtomist's vade-mecum (Bolles Lee), ed. 11, Philadelphia, 1950, The Blakiston Co.

Guyer, M.F.: Animal micrology, Chicago, 1953, University of Chicago Press.

Koehler, J.K.: Advanced techniques in biological electron microscopy, New York, 1973, Springer-Verlag.

Krajian, A.A., and Gradwohl, R.B.H.: Histopathological technic, ed. 2, St. Louis, 1952, The C.V. Mosby Co.

Mallory, F.B.: Pathological technique, Philadelphia, 1938, W.B. Saunders Co.

Morse, A.: Formic acid–sodium citrate decalcification and butyl alcohol dehydration of teeth and bones for sectioning in paraffin, J. Dent. Res. **24**:143, 1945.

Nikiforuk, G., and Sreebny, L.: Demineralization of hard tissues by organic chelating agents at neutral pH, J. Dent. Res. **32**:859, 1953.

Pearce, A.G.E.: Histochemistry, ed. 3, vol. 1, Baltimore, 1973, The Williams & Wilkins Co.

Sognnaes, R.F.: Preparation of thin serial ground sections of whole teeth and jaws and other highly calcified and brittle structures, Anat. Rec. **99**:134, 1947.

Sognnaes, R.F.: The organic elements of the enamel, J. Dent. Res. **29**:260, 1950.

Weber, D.F.: A simplified technique for the preparation of ground sections, J. Dent. Res. **43**:462, 1964.

Yaeger, J.A.: Methacrylate embedding and sectioning of calcified bone, Stain Technol. **33**:229, 1958.

Index

KALAMAZOO VALLEY COMMUNITY COLLEGE

Presented By

T. Hollowell

Shelby Foote

THE CIVIL WAR

WAR

A NARRATIVE

12

★ ★ ★

JAMES CROSSING
TO JOHNSONVILLE

40th Anniversary Edition

BY SHELBY FOOTE
AND THE EDITORS OF TIME-LIFE BOOKS,
ALEXANDRIA, VIRGINIA

All these were honoured in their generations,
and were the glory of their times.

There be of them,
that have left a name behind them,
that their praises might be reported.

And some there be, which have no memorial;
who are perished, as though they had never been;
and are become as though they had never been born;
and their children after them.

But these were merciful men,
whose righteousness hath not been forgotten.

With their seed shall continually remain
a good inheritance,
and their children are within the covenant.

Their seed standeth fast,
and their children for their sakes.

Their seed shall remain for ever,
and their glory shall not be blotted out.

Their bodies are buried in peace;
but their name liveth for evermore.

    ECCLESIASTICUS XLIV

Contents

★ ★ ★

★

Prologue

———✦———

In the summer of 1864, President Abraham Lincoln had been in a fight for his political life. As the Civil War had dragged on, his critics in the North had gotten bolder and more numerous. No longer were the Copperhead Democrats the only ones convinced he was not up to the challenges of his office in these trying times. A growing number of Republicans thought so as well. Many of the Jacobins in his party had already jumped ship, nominating in a special convention John C. Frémont. This had led some to doubt whether Lincoln would get the nod from his own party to stand for reëlection.

Lincoln had done what he could—undercutting troublemakers in his government such as Salmon Chase, renaming his Republican party the National Union party, choosing as a running mate the Tennessee Unionist Andrew Johnson, and, most crucially, giving the latest commander of the Union forces, Ulysses S. Grant, free rein in fighting the enemy. For Lincoln realized that his political future was tied to the fortunes of war, and he had placed his fate in the only Union general who seemed not only anxious to win at any cost but was also actually willing to fight Robert E. Lee. As the election drew nearer, however, Lincoln's resolve to stand off from as well as stand behind the dogged Grant was sorely tested.

Amid much ballyhoo, Grant had launched his new coordinated war effort with great energy and determination. He had put Phil Sheridan at the head of his cavalry, and Sheridan had given the rebel horseman Jeb Stuart a good run, which — at Yellow Tavern — cost the Confederate general his life. Grant had sent William Tecumseh Sherman off to take Atlanta, and his red-haired friend had rolled inexorably southward, forcing Joe Johnston to abandon position after position, then to contemplate abandoning the very city he was defending, then finally to be himself abandoned by Jefferson Davis. At sea, the famed raider C.S.S. *Alabama* had fallen spectacularly to the U.S.S. *Kearsarge*. And for forty days back in the spring and into the summer, Grant had fought on against Lee in the Virginia Wilderness when perhaps any other Union general, given his losses, would have retreated. And yet—

And yet the tide had not turned, or not turned enough, either militarily or politically. Grant had also sent Franz Sigel into the Shenandoah Valley, and his poor showing at New Market against John C. Breckinridge had forced Grant to replace him. The bold foray Grant had planned up the James River turned sour when P. G. T. Beauregard bottled up Union General Benjamin Butler at Bermuda Hundred. Nathan Bedford Forrest continued to grab headlines for his raids behind Sherman's line. Sherman, too, having reached Atlanta, seemed to stall for the rest of the summer. Finally, there was the horrendous body count, swelled by a costly stalemate at Cold Harbor. Lincoln had long thought he needed a general who could accept the grim arithmetic it would take for the North to win, and now that he had gotten what he wished for he was hard put not to question Grant's apparent heedlessness.

Not that the South was gaining ground; it was far too late for that. But both Lee and Davis hoped that by simply hanging on, by enduring, they could raise the price of victory too high even for Grant and cost Lincoln his reëlection, after which an acceptable peace might be reached with his successor. And, indeed, losing 100,000 men to gain a bloody stalemate would not keep Lincoln in the Oval Office. Thus the worried incumbent would pull out all the stops — exercising the first pocket veto, the suspension of habeus corpus, the culling of opponents from ballots and voting lists by questioning their loyalty, the furloughing of Union troops to go home and vote for their commander-in-chief —to win the election. Win Lincoln would, though his triumph was perhaps once again aided as much by the course of the war as by his political shenanigans.

For, come the fall elections, Sherman would have taken Atlanta from its new defender, John Bell Hood; the U.S. Navy would have captured the rebel cruiser *Florida* and sunk the ironclad North Carolina river raider *Albemarle;* Mobile Bay would have fallen to David Farragut; and Grant would have outmaneuvered Lee by laying siege to Petersburg, planning to choke off Richmond as he once had subdued Vicksburg. With Abe Lincoln back in the White House, with desperate but unofficial peace feelers once more being floated, with the Army of the Potomac and the Army of Northern Virginia locked in a most modern kind of trench warfare, even Robert E. Lee would admit, despite his occasionally renewed hopes, that it was now only a matter of time.

★ ★ ★

*Federal officers and soldiers
occupy Redan No. 5 – a strong
point in the Petersburg defenses
known as the Dimmock Line –
captured on June 15, 1864.*

O N E

1864 ★ ★ ★ ★ ★

Eastward, with Lee at last out-foxed, the blue tide ran swift and steady, apparently inexorable as it surged toward the gates of the capital close in his rear. But then, at the full, the outlying Richmond bulwarks held; Beauregard, as he had been wont to do from the outset — first at Sumter, three years back, then again two years ago at Corinth, and once more last year in Charleston harbor — made the most of still another "finest hour" by holding Petersburg against the longest odds ever faced by a major commander on either side in this lengthening, long-odds war.

Grant's crossing of James River went like clockwork, and the clock itself was enormous. Preceded in the withdrawal by Baldy Smith, whose corps took ship at White House Landing on June 13 for the roundabout journey to rejoin Butler at Bermuda Hundred, Hancock reached Wilcox Landing by noon of the following day, completing a thirty-mile hike from Cold Harbor to the north bank of the James, and began at once the ferrying operation that would put his corps on Windmill Point, across the way, by dawn of June 15. While he crossed, the engineers got to work on the pontoon bridge, two miles downriver, by which the other three corps of the Army of the Potomac were to march in order to reinforce Smith and Hancock in their convergence on Petersburg, the rail hub whose loss, combined with the loss of the Virginia Central — Hunter

★

With rail lines reaching Richmond, Norfolk, Tennessee, and the Carolinas, Petersburg was indepensable to the defense of the Confederate capital.

and Sheridan were presumed to be moving down that critical Shendandoah Valley supply line even now — would mean that Richmond's defenders, north as well as south of the James, would have to abandon the city for lack of subsistence, or else choose between starvation and surrender. In high spirits at the prospect, Grant was delighted to recover the mobility that had characterized the opening of the final phase of his Vicksburg campaign, which the current operation so much resembled. Now as then, he was crossing a river miles downstream from his objective in order to sever its lines of supply and come upon it from the rear. Whether it crumpled under a sudden assault, as he intended, or crumbled under a siege, which he hoped to avoid, the result would be the same; Richmond was doomed, if he could only achieve here in Virginia the concert of action he had enjoyed last year in Mississippi.

By way of ensuring that this would obtain, he did not tarry long on the north bank of the James, which he reached on the morning of June 14 to find the head of Hancock's column arriving and the engineers already hard at work corduroying approaches to the bridge the pontoniers would presently throw across the nearly half-mile width of river to Windmill Point. Instead, wanting to make certain that Butler understood his part in the double-

★

pronged maneuver, Grant got aboard a steamer for a fast ride up to Bermuda Hundred and a conference with the cock-eyed general. Butler not only understood; he was putting the final touches to the preliminary details, laying a pontoon bridge near Broadway Landing, where Smith would cross the Appomattox tonight for a quick descent on Petersburg next morning, and preparing to sink five stone-laden vessels in the channel of the James at Trent's Reach, within cannon range of his bottled-up right, to block the descent below that point of rebel gunboats which might otherwise make a suicidal attempt to disrupt the main crossing, some thirty winding miles downstream. Satisfied that no hitch was likely to develop in this direction, either from neglect or misconception, Grant prepared to return to Wilcox Landing for a follow-up meeting with Meade, but before he left he got off a wire to Halleck, who had opposed the movement from the outset in the belief that the scattered segments of both armies, Meade's and Butler's, would be exposed to piecemeal destruction by Lee while it was in progress. "Our forces will commence crossing the James today," Grant informed him. "The enemy show no signs yet of having brought troops to the south side of Richmond. I will have Petersburg secured, if possible, before they get there in much force. Our movement from Cold Harbor to the James River has been made with great celerity and so far without loss or accident."

The answer came next morning, not from Old Brains, who was not to be dissuaded from taking counsel of his fears, but from the highest authority of all:

> *Have just read your dispatch of 1 p.m. yesterday.*
> *I begin to see it. You will succeed. God bless you all.*
>
> *A. Lincoln*

★ ★ ★ **B**y that time Smith was over the Appomattox and moving directly on Petersburg, whose outer defenses lay within six miles of Broadway Landing. He had 16,000 men in his three infantry divisions, including one that joined him from City Point at daybreak — a Negro outfit under Brigadier General Edward Hincks, which had been left behind when the rest of the corps shifted northside for a share in the Cold Harbor nightmare — plus Kautz's 2400 wide-ranging troopers, over toward the City Point Railroad, where they covered the exposed southeast flank of the column on the march. Four miles from the river, after receiving long-range shots from rebel vedettes who scampered when threatened, the marchers came upon a fast-firing section of artillery posted atop an outlying hill with butternut infantry in support. Hincks, on the left, sent his unblooded soldiers forward at a run.

★

One gun got away, but they took the other, along with its crew, and staged a jubilation around the captured piece, elated at having made the most of a chance to discredit the doubts that had denied them a role in the heavy fighting two weeks ago. Baldy too was delighted, despite the delay, as he got the celebrants back into column, left and right, and resumed the march; for this was the route by which he believed Petersburg could have been taken in the first place, back in early May, and he had said as much, repeatedly though without avail, to Butler at the time. Another mile down the road, however, he came upon a sobering view, spirit-chilling despite the noonday heat, and called a halt for study and deployment.

What he saw, dead ahead down the tracks of the railroad, might well have given anyone pause, let alone a man who had just returned from playing a leading role in Grant's (and Lee's) Cold Harbor demonstration of what could happen to troops, whatever their numerical advantage, who delivered a hair-trigger all-out attack on a prepared position, however scantly it might be defended. Moreover, this one had been under construction and improvement not for two days, as had been the case beyond the Chickahominy, but for nearly two years, ever since August 1862, when Richmond's defenders learned that McClellan had wanted to make just such a southside thrust, as a sequel to *his* Peninsular "change of base," only to be overruled by Halleck, who had favored the maneuver no more then, when he had the veto, than he did now that he lacked any final say-so in the matter. Called the "Dimmock Line" for Captain Charles H. Dimmock, the engineer who laid them out, the Petersburg fortifications were ten miles in length, a half oval tied at its ends to the Appomattox above and below the town, and contained in all some 55 redans, square forts bristling with batteries and connected by six-foot breastworks, twenty feet thick at the base and rimmed by a continuous ditch, another six feet deep and fifteen wide. In front of this dusty moat, trees had been felled, their branches sharpened and interlaced to discourage attackers, and on beyond a line of rifle pits for skirmishers, who could fall back through narrow gaps in the abatis, the ground had been cleared for half a mile to afford the defenders an unobstructed field of fire that would have to be crossed, naked to whatever lead might fly, by whatever moved against them. Confronting the eastward bulge of this bristly, hard-shelled oval, Smith gulped and then got down to figuring how to crack it. First there was reconnoitering to be done; a risky business, and he did much of it himself, drawing sniper fire whenever he ventured out of the woods in which he concealed his three divisions while he searched for some apparently nonexistent weak point to assault.

Despite a superfluity of guns frowning from all those embrasures, there seemed to be a scarcity of infantry in the connecting works. Accordingly, he decided to try for a breakthrough with a succession of reinforced skirmish

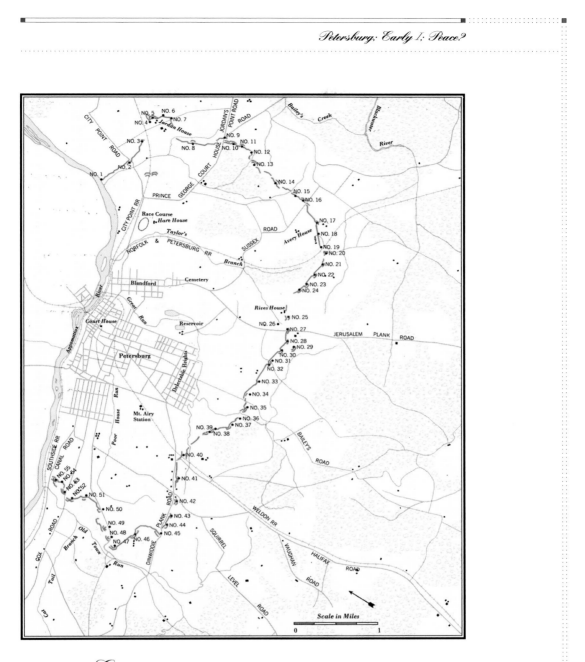

*Captain Charles H. Dimmock designed the defenses around
Petersburg. They stretched ten miles in length
and included 55 redans and breastworks six feet high.*

lines, strong enough to overwhelm the defenders when they came to grips, yet
not so thickly massed as to suffer unbearable losses in the course of their naked
advance across the slashings. All this took time, however. It was past 4 o'clock
when Smith wound up his reconnaissance and completed the formulation of his

★

plan. Aware that the defenders were in telegraphic contact with Richmond, from which reinforcements could be rushed by rail — the track distance was only twenty-three miles — he set 5 o'clock as the jump-off hour for a coördinated attack by elements from all three divisions, with every piece of Federal artillery firing its fastest to keep the heads of the defenders down while his troops were making their half-mile sprint from the woods, where they now were masked, to the long slow curve of breastworks in their front.

It was then that the first organic hitch developed. Unaware that an attack was pending (for the simple reason that no one had thought to inform him) the corps artillery chief had just sent all the horses off for water; which meant that there could be no support fire for the attackers until the teams returned to haul the guns into position along the western fringe of the woods. Angered, Baldy delayed the jump-off until 7. While he and his 18,000 waited, and the sun drew near the landline, word came that Hancock, after a similar hitch on Windmill Point this morning, was on the way but would not arrive till after dark. For a moment Smith considered another postponement; Hancock's was the largest corps in Meade's army, and the notion of more than doubling the Petersburg attack force to 40,000 was attractive. But the thought of Confederate reinforcements, perhaps racing southward in untold thousands even now, jam-packed into and onto every railway car available in this section of Virginia — plus the companion thought that Hancock outranked him and might therefore hog the glory — provoked a rejection of any further delay. The revised order stood, and at 7 o'clock the blue skirmishers stepped from the woods, supported by fire from the just-arrived guns, and started forward to where friendly shells were bursting over and around the rebel fortifications, half a mile ahead.

Once more Hincks and his green black troops showed the veterans how to do the thing in style. Swarming over the cleared ground and into the red after-glow of the sunset, they pursued the grayback skirmishers through the tangled abatis, across the ditch, and up and over the breastworks just beyond. Formidable as they had been to the eye, the fortifications collapsed at a touch; no less than seven of the individual

*B*lack troops under Brigadier General Edward Hincks overran five redans in the Petersburg defense line.

bastions fell within the hour, five of them to the jubilant Negro soldiers, who took twelve of the sixteen captured guns and better than half of the 300 prisoners. Astride and south of the railroad, the blue attackers occupied more than a mile of intrenchments, and Hincks, elated at the ease with which his men had bashed in the eastern nose of the rebel oval, wanted to continue the drive right into the streets of Petersburg, asking only that the other two divisions support him in the effort. Smith demurred. It was night now, crowding 9 o'clock, and his mind was on Lee, who was reported to have detached a considerable portion of his army for a crossing of the James that afternoon; they had probably arrived by now, in which case the Federals might be counterattacked at any moment by superior numbers of hornet-mad Confederate veterans. The thing to do, he told Hincks, was brace for the shock and prepare to hold the captured works until Hancock arrived to even or perhaps reverse the odds. Then they would see.

Hancock arrived something over an hour later; two of his three divisions, he said, were a mile behind him on the road from Prince George Court House. This had been a trying day for him and his dusty marchers, beginning at dawn, when he received orders to wait on Windmill Point for 60,000 rations supposedly on the way from Butler. He had no use for them, having brought his own, but he waited as ordered until 10.30 and then set out without them. That was the cause of the first delay, a matter of some five hours. The second, equally wasteful of time, was caused by an inadequate map, which misled him badly — with the result that the distance to Petersburg by the direct route, sixteen miles, was nearly doubled by the various countermarches he was obliged to make when he found that the roads on the ground ran in different directions from those inked on paper — and faulty instructions, which identified as his destination a point that later turned out to lie within the enemy lines. "I spent the best hours of the day," he would complain in his report, "marching by an incorrect map in search of a designated position which, as described, was not in existence."

Nor was that the worst of the oversights and errors that developed in the course of this long hot June 15, from which so much had been expected and of which some ten critical hours thus were thrown away. Approaching Prince George Court House about sunset, Hancock met a courier from Baldy Smith, who gave him a dispatch headed 4 p.m. and including the words: "If the II Corps can come up in time to make an assault tonight after dark, in the vicinity of Norfolk & Petersburg Railroad, I think we can be successful." This was the first he had heard that he and his 22,000 were intended to have any part in today's action; no one on Grant's staff had thought to tell Meade, who could scarcely be expected to pass along orders he himself had not received. Hancock hastened his march and rode ahead to join Smith at about 10.30, two miles east of Petersburg, only to find that the Vermonter had changed his mind about a night attack. He requested, rather, that Hancock relieve Hincks's troops — whether as a restful reward for all they had done today, or out of a continuing mistrust of their fighting qualities, he did not say — in occupation of the solid mile of rebel works they had taken when they charged into the sunset.

It was done, though Hincks continued to insist that he could march into Petersburg if his chief would only unleash and support him. Hancock rather agreed, though he declined to assume command, being unfamiliar with the ground and partly incapacitated by his Gettysburg wound, which had reopened under the strain of the fretful march. Smith — suffering too, as he said, "from the effects of bad water, and malaria brought from Cold Harbor" — was willing, even glad, to bide his time; his mind was still on all those probable grayback reinforcements coming down from Richmond in multi-thousand-man relays. The 40,000 Federals on hand would be about doubled tomorrow by the arrival of Burnside, who was over the James by now, and Warren, who had just begun to cross. Wilson and Wright would bring the total to roughly 100,000 the following morning; which would surely be enough for practically anything, Smith figured, especially since they had only to expand the gains already made today.

"Unless I misapprehend the topography," he wired Butler before turning in at midnight, "I hold the key to Petersburg."

★ ★ ★ **B**eauregard agreed that Baldy held the key. What was more, he also agreed with Hincks that the key was in the lock, that all the bluecoats had to do at this point was give the thing a turn and the gate would swing ajar. "Petersburg was clearly at the mercy of the Federal commander, who had all but captured it," he said later, looking back on that time of strain and near despair.

He had in all, this June 15, some 5400 troops in his department: 3200 with Bushrod Johnson, corking the bottle in which Butler was confined on Bermuda Hundred, and 2200 with Brigadier General Henry A. Wise at Petersburg. The rest — Hoke's division and the brigades of Ransom and Gracie; about 9000 in all — were beyond the James, detached to Lee or posted in the Richmond fortifications. Wise, it was true, had held his own last week in the "Battle of the Patients and the Penitents," which turned back a similar southside thrust, but the Creole identified this recent probe by Butler as no more than "a reconnaissance connected with Grant's future operations." Heavier blows were being prepared by a sterner commander, and he had been doing all he could for the past five days to persuade the War Department to return the rest of his little army to him before they landed. Smith had no sooner been spotted moving in transports up the James the day before, June 14, than Beauregard redoubled his efforts, insisting, now that the crisis he had predicted was at hand, that Hoke

★

and the others be sent without delay.
Next morning — today — with Smith
bearing ponderously down on him from
Broadway Landing and his detached
units still unreleased by Richmond, he
warned Bragg that even when these
were returned, as he was at last assured
they would be, he probably would have
to choose which of his two critical
southside positions to abandon, the
Howlett Line above the Appomattox or
the Dimmock Line below, if he was to
scrape together enough defenders to
make a fight for the other. While Wise
shifted his few troops into the eastern
nose of the intrenchments ringing Petersburg, thus to confront the enemy
approaching down the City Point Railroad, Beauregard put the case bluntly in
a wire to Richmond: "We must now

*Henry A. Wise's force of 2200 men
was charged with resisting the Federal
attack on Petersburg on June 15.*

elect between lines of Bermuda Hundred and Petersburg. We can not hold
both. Please answer at once." Evading the question, Bragg merely replied that
Hoke was on the way and should be used to the best advantage. Old Bory lost
patience entirely. "I did not ask your advice with regard to the movement of
troops," he wired back, "but wished to know preference between Petersburg
and lines across Bermuda Hundred Neck, for my guidance, as I fear my present
force may prove unequal to hold both."

Bragg made no reply at all to this, and while Wise and his 2200, outnumbered eight-to-one by the blue host assembling in front of their works,
made enough of a false show of strength to delay through the long afternoon an
assault that could scarcely fail, the Creole general fumed and fretted.

Smith's sunset attack was about as successful as had been expected,
though fortunately it was not pressed home; Hoke came up in time to assist in
work on the secondary defenses, to which Wise and his survivors had fallen back
when more than a mile of the main line caved in. Beauregard's strength was
now about 8000 for the close-up defense of the town, but this growth was incon-
siderable in the light of information that a second Federal column, as large as
the first, was approaching from Prince George Court House. Dawn would no
doubt bring a repetition of the sunset assault, which was sure to be as crumpling
since it could be made with twice the strength. Alone in the darkness, ignored
by his superiors, and convinced that Wise and Hoke were about to be swamped

★

unless they could be reinforced, the southside commander, who had joined them by then from his headquarters north of the Appomattox, notified Richmond that he had decided to risk uncorking Butler so as to reinforce Petersburg, even though this was likely to mean the loss of its vital rail and telegraph connections with the capital beyond the James. "I shall order Johnson to this point," he wired Bragg. "General Lee must look to the defenses of Drewry's Bluff and Bermuda Hundred, if practicable."

Notified of this development two hours past midnight, Lee reacted promptly. He had suspected from the outset that Grant would do as he had done; "I think the enemy must be preparing to move south of James River," he warned Davis at noon on June 14, before the first blue soldier crossed to Windmill Point. Still, that did not mean that he could act on the supposition. Responsible for the security of Richmond, he had his two remaining corps disposed along a north-south line from White Oak Swamp to Malvern Hill, where he covered the direct approach to the capital twelve miles in his rear, and he could not abandon or even weaken this line until he was certain that the Federals did not intend to come this way. Information that Smith was back at Bermuda Hundred, and then that he had crossed the Appomattox for an attack on Petersburg, was no real indication of what *Meade* would do; Smith was only returning to the command from which he had been detached two weeks ago. Nor was the report that a corps from the Army of the Potomac was on the march beyond the James conclusive evidence of what Grant had in mind for the rest of that army. Butler had reinforced Meade for the northside strike at Lee: so might Meade be reinforcing Butler for the southside strike at Beauregard — who, in point of fact, had yet to identify or take prisoners from any unit except Smith's; all he had really said, so far, was that he had an awesome number of bluecoats in his front, and that was by no means an unusual claim for any general to make, let alone the histrionic Creole.

However, when Lee was wakened at 2 o'clock in the morning to learn that the Howlett Line had been stripped of all but a skeleton force of skirmishers ("Cannot these lines be occupied by your troops?" Beauregard inquired. "The safety of our communications requires it") he no longer had any choice about what to do if he was to save the capital in his rear. A breakout by Butler, westward from Bermuda Hundred, would give the Federals control of the one railroad leading north from Petersburg, and that would have the same effect as if the three railroads leading south had been cut; Richmond would totter, for lack of food, and fall. Accordingly, Lee had Pickett's division on the march by 3 a.m. and told Anderson to follow promptly with one of his other two divisions, Field's, and direct the action against Butler, who almost certainly would have overrun the Howlett Line by the time he got there. Moreover, leaving instructions for A. P. Hill to continue shielding Richmond from a northside attack by

Meade — whose army, even with one corps detached, was still better than twice as large as the Army of Northern Virginia, depleted by Early's departure three days ago — Lee struck his tent at Riddell's Shop, while it still was dark, and mounted Traveller for the headquarters shift to Chaffin's Bluff, where Anderson's troops would cross by a pontoon bridge to recover the critical southside works Beauregard had abandoned the night before.

Sure enough, when Lee reached Chaffin's around 9.30 this June 16 and crossed the James behind Pickett, just ahead of Field, the nearby popping of rifles and the distant rumble of guns informed him, simultaneously, that Butler had indeed overrun the scantly manned Bermuda works, whose northern anchor was six miles downriver, and that Beauregard was fighting to hang onto Petersburg, a dozen miles to the south. Presently word came from Anderson that Butler's uncorked troops had advanced westward to Port Walthall Junction, where they were tearing up track and digging in to prevent the movement of reinforcements beyond that point, either by rail or turnpike. Lee replied that they must be driven off, and by nightfall they were, though only as far as the abandoned Howlett Line, which they held in reverse, firing west. All this time, Beauregard's guns had kept growling and messages from him ranged in tone from urgent to laconic, beginning with a cry for help — to which Lee replied, pointedly, that he could not strip the north bank of the James without evidence that more than one of Meade's corps had crossed — and winding up proudly, yet rather mild withal: "We may have force sufficient to hold Petersburg." In response to queries about Grant, whose whereabouts might indicate his intentions, Old Bory could only say at the end of the long day's fight: "No satisfactory information yet received of Grant's crossing James River. Hancock's and Smith's corps are however in our front."

Lee already knew this last. What he did not know, because Beauregard did not know it to pass it along to him, was that Burnside had been in front of Petersburg since midmorning (in fact, his was the corps responsible for such limited gains as the Federals made today) and that Warren was arriving even then, bringing the blue total to more than 75,000, with still another 25,000 on the way. Wilson, who had served Grant well in Sheridan's absence with the other two mounted divisions, was riding hard through the twilight from Windmill Point, and Wright would finish crossing the pontoon bridge by midnight with the final elements of Meade's army. Beauregard, whose strength had been raised in the course of the day to just over 14,000 by the arrival of Johnson from Bermuda Hundred and Ransom and Gracie from Richmond, might find the odds he had faced yesterday and today stretched unbearably tomorrow, despite the various oversights and hitches that had disrupted the Union effort south of the James for the past two days.

In all that time, hamstrung by conflicting orders and inadequate

Robert Ransom brought his troops to Petersburg, increasing the number of defenders along the newly-erected line just east of the city to about 15,000.

maps — and rendered cautious, moreover, by remembrance of Cold Harbor, fought two weeks ago tomorrow — the attackers had not managed to bring their preponderance of numbers to bear in a single concerted assault on the cracked and creaking Dimmock Line. Yet Grant, for one, was not inclined to be critical at this juncture. As he prepared for bed tonight in his tent at City Point, where he had transferred his headquarters the day before, he said with a smile, sitting half undressed on the edge of his cot: "I think it is pretty well, to get across a great river and come up here and attack Lee in the rear before he is ready for us."

So he said, and so it was; "pretty well," indeed. But June 17, even though all of Meade's army was over the James before it dawned and had been committed to some kind of action before it ended, turned out to be little different. Today, as yesterday, the pressure built numerically beyond what should have been the rebel breaking point — better than 80,000 opposed by fewer than 15,000 — yet was never brought decisively to bear. From the outset, things again went wrong: beginning with Warren, who came up the previous night. Instructed to extend the left beyond the Jerusalem Plank Road for a sunrise attack up that well-defined thoroughfare, he encountered skirmishers on the approach march and turned astride the Norfolk Railroad to drive them back, thus missing a chance (which neither he nor his superiors knew existed) to strike beyond the occupied portion of the Dimmock Line. If this had not happened, if Warren had brushed the skirmishers aside and continued his march as instructed, Beauregard later said, "I would have been compelled to evacuate Petersburg without much resistance." As it was, the conflict here at the south end of the line amounted to little more than an all-day long-range demonstration.

Northward along the center, where Burnside's and Hancock's corps were posted, the fighting was a good deal bloodier, although not much more productive in the end. One of Burnside's divisions started things off by seizing a critical hill, yet could not exploit the advantage because he failed to alert his other two divisions to move up quickly in support. The Confederates had time to shore up their crumbling defenses, both here and just to the north where Hancock's three divisions were lying idle; Hancock having been obliged by his reopened wound to turn the command over to Birney — a good man, but no Hancock — they too had failed to get the word, with the result that they were about as much out of things as were Wright's three divisions, one of which was used to bolster the fought-out Smith, inactive on the right, while the other two were sent in response to Butler's urgent plea for reinforcements to keep Lee from driving him back into the bottle he had popped out of yesterday. Wright went, but failed to arrive in time to do anything more than join the Bermuda Hundred soldiers in captivity. By midafternoon, Pickett and Field had retaken the Howlett Line from end to end; Butler was recorked, this time for good, and still more troops were reported to be on the march from Lee's position east of Richmond.

If they got there, if Petersburg was heavily reinforced, the Army of the Potomac would simply have exchanged one stalemate for another, twice the distance from the rebel capital and on the far side of a major river. There still was time to avoid this, however. None of Lee's veterans was yet across the Appomattox, and most of them were still beyond the James. With the railroad severed at Walthall Junction, even the closest were unlikely to reach the field by first light tomorrow; which left plenty of time for delivering the coördinated attack the Federals had been trying for all along, without success.

Happily, near sunset, at least a portion of the army recovered a measure of its old élan. Burnside and Birney, suddenly meshing gears, surged forward to seize another mile of works along the enemy center, together with a dozen guns and about 500 prisoners. A savage counterattack (by Gracie's brigade, it later developed, though at the time the force had seemed considerably larger) forestalled any rapid enlargement of the breakthrough, either in width or depth. Dusk deepened into darkness, and though the moon, only two nights short of the full, soon came out to flood the landscape with its golden light, Meade — like Smith before him, two dusks ago — declined to follow through by continuing the advance. Instead, he issued orders for a mass assault to be launched all along the line at the first wink of dawn.

Beauregard said afterwards that at this point, with his center pierced and Petersburg once more up for grabs, it seemed to him that "the last hour of the Confederacy had arrived." In fact, he had been expecting his patched-up line to crack all day, and he had begun at noon the laying out of a new defensive position, the better part of a mile in rear of the present one, to fall back on when the time came. He had no engineers, and indeed no reserves of any kind for digging; all he could do was mark the proposed line with white stakes, easily seen at night, and hope the old intrenchments would hold long enough for darkness to cover the withdrawal of his soldiers, who would do the digging when they got there. The old works, or what was left of them, did hold; or anyhow they nearly did, and Gracie's desperate counterattack delayed a farther blue advance until nightfall stopped the fighting. Old Bory ordered campfires lighted all along the front and sentinels posted well forward; then at midnight, behind this curtain of light and the fitful spatter of picket fire, the rest of his weary men fell back through the moon-drenched gloom to the site of their new line, which they then began to dig, using bayonets and tin cans for tools and getting what little sleep they could between shifts.

At 12.40 a.m. their commander got off his final dispatch of the day to Lee. "All quiet at present. I expect renewal of attack in morning. My troops are becoming much exhausted. Without immediate and strong reinforcements results may be unfavorable. Prisoners report Grant on the field with his whole army."

Lee now had a definite statement, the first in five days, not only that Meade's army was no longer in his front, but also that it was in Beauregard's, and he reacted accordingly. In point of fact, he had begun to act on this premise in response to a dispatch written six hours earlier, in which the southside commander informed him that increasing pressure along his "already much extended lines" would compel him to retire to a shorter line, midway between his original works and the vital rail hub in his rear. "This I shall hold as long as practicable," he added, "but without reinforcements I may have to evacuate the city very shortly." Petersburg's fate was Richmond's; Lee moved, as he had done two nights ago when the Creole stripped the Howlett Line, to forestall disaster — or anyhow to be in a better position to forestall it — by ordering Anderson's third division to proceed to Bermuda Neck and A. P. Hill to cross the James at Chaffin's Bluff and await instructions for a march in either direction, back north or farther south down the Petersburg Turnpike, depending on developments.

So much he had done already, and now that Beauregard's 1240 message was at hand, stating flatly that Grant was "on the field with his whole

Robert E. Lee ordered one of Richard Anderson's divisions to proceed to Bermuda Neck, where the Appomattox River, shown here, joined with the James River.

army," he followed through by telling Anderson to send his third division on to Petersburg at once and follow with the second. A. P. Hill would go as well, leaving one of his three divisions north of the Appomattox in case Richmond came under attack. This last seemed highly unlikely, however; for a report came in, about this time, that cavalry had ridden down the Peninsula the previous afternoon, as far as Wilcox Landing, and found that all four of Meade's corps had crossed to Windmill Point in the course of the past three days. Beauregard's information, gathered from prisoners, thus was confirmed beyond all doubt. It was now past 3.30 in the morning, June 18; Lee's whole army, except for one division left holding the Howlett Line against Butler — and of course Early, who made contact with Hunter at Lynchburg that same day — would be on the march for Petersburg within the hour.

Two staff officers arrived just then from beyond the Appomattox, sent by their chief to lend verbal weight to his written pleas for help. "Unless reinforcements are sent before forty-eight hours," one of them told Lee he had heard Old Bory declare, "God Almighty alone can save Petersburg and Richmond." Normally, Lee did not approve of such talk; it seemed to him tinged with irreverence. But this was no normal time. "I hope God Almighty will," he said.

★ ★ ★ For the first time since the crossing of the James, Meade's army gave him on schedule all he asked for. In line before dawn, the troops went forward before sunrise, under orders to take the Confederate works "at all costs." They took them, in fact, at practically no cost at all; for they were deserted, covered only by a handful of pickets who got off a shot or two, then scampered rearward or surrendered.

The result was about as disruptive to the attackers, however, as if they had met the stiffest kind of resistance. First, there was the confusion of calling a halt in the abandoned trenches, which had to be occupied for defense against a tricky counterstroke, and then there followed the testy business of groping about to locate the vanished rebels. All this took time. It was midmorning before they found them, nearly a mile to the west, and presently they had cause to wish they hadn't. Beauregard had established a new and shorter line, due south from the Appomattox to a connection with the old works beyond the Jerusalem Plank Road, and was dug in all along it, guns clustered thicker than ever. A noon assault, spearheaded by Birney, was bloodily repulsed: so bloodily and decisively, indeed, that old-timers among the survivors — who had encountered this kind of fire only too often throughout six weeks of crablike sidling from the Rapidan to the Chickahominy — sent back word that Old Bory had been reinforced: by Lee.

It was true. Anderson's lead division had arrived at 7.30 and the second marched in two hours later, followed at 11 o'clock by Lee himself, who

★

rode out to confer with Beauregard, now second-in-command, his lonely ordeal ended. As fast as the lean, dusty marchers came up they were put into line alongside the nearly fought-out defenders, some of whom tried to raise a feeble cheer of welcome, while others wept from exhaustion at the sudden release from tension. They were pleased to hear that A. P. Hill would also be up by nightfall to reduce the all-but-unbearable odds to the accustomed two-to-one, but as far as they were concerned the situation was stabilized already; they had considered their line unbreakable from the time the first of the First Corps veterans arrived to slide their rifles across the newly dug earth of the parapets and sight down them in the direction from which the Yankees would have to come when they attacked.

Across the way, the men who would be expected to do the coming flatly agreed. Remembering one Cold Harbor, they saw here the makings of another, and they wanted no part of it. The result, after the costly noon repulse, was a breakdown of the command system, so complete that Meade got hopping mad and retired, in effect, from any further participation in the effort. "I find it useless to appoint an hour to effect coöperation. . . . What additional orders to attack you require I cannot imagine," he complained in a message sent to all corps commanders. His solution, if it could be called such, was for them "to attack at all hazards and without reference to each other."

Under these circumstances, the army was spared another Cold Harbor only because its members, for the most part, declined to obey such orders as would have brought on a restaging of that fiasco. Hancock's troops had come up in high spirits, three days ago; "We knew that we had outmarched Lee's veterans and that our reward was at hand," one would recall. These expectations had died since then, however, along with a great many of the men who shared them. "Are you going to charge those works?" a cannoneer asked as a column of infantry passed his battery, headed for the front, and was told by a foot soldier: "No, we are not going to charge. We are going to run toward the Confederate earthworks and then we are going to run back. We have had enough of assaulting earthworks."

As the afternoon wore on, many declined to do even that much. Around 4 o'clock, for example, Birney massed a brigade for an all-out attack on the rebel center. He formed the troops in four lines, the front two made up of half a dozen veteran units, the rear two of a pair of outsized heavy-artillery regiments, 1st Massachusetts and 1st Maine. All four lines were under instructions to remain prone until the order came to rise and charge; but when it was given, the men in the front ranks continued to hug the ground, paying no attention to the shouts and exhortations of their saber-waving officers. They looked back and saw that the rear-rank heavies had risen and were preparing to go forward. "Lay down, you damn fools! You can't take them works!" they cried over their

A company of the 1st Maine Heavy Artillery carried this guidon. The regiment, outfitted as infantry, lost 74 percent of its members at Petersburg on June 18, 1864.

shoulders. For all their greenness, the Bay State troops knew sound advice when they heard it. They lay back down. But the Maine men were rugged. They stepped through and over the prone ranks of veterans and moved at the double against the enemy intrenchments, which broke into flame at their approach. None of them made it up to the clattering rebel line, and few of them made it back to their own. Of the 850 who went forward, 632 fell in less than half an hour. That was just over 74 percent, the severest loss suffered in a single engagement by any Union regiment in the whole course of the war.

This could not continue, nor did it. Before sunset Meade wired Grant that he believed nothing more could be accomplished here today. "Our men are tired," he informed his chief, "and the attacks have not been made with the vigor and force which characterized our fighting in the Wilderness; if they had been," he added, "I think we should have been more successful." Grant — who had maintained a curious hands-off attitude throughout the southside contest, even as he watched his well-laid plan being frustrated by inept staff work and the bone-deep disconsolation of the troops — invoked no ifs and leveled no reproaches. Declaring that he was "perfectly satisfied that all has been done that could be done," he agreed that the time had come to call a halt.

★

"Now we will rest the men," he said, "and use the spade for their protection until a new vein can be struck."

A new vein might be struck, in time, but not by the old army, which had suffered a further subtraction of 11,386 killed, wounded, or captured from its ranks since it crossed the James. That brought the grand total of Grant's losses, including Butler's, to nearly 75,000 men — more than Lee and Beauregard had had in both their armies at the start of the campaign. Of these, a precisely tabulated 66,315 were from the five corps under Meade (including Smith's, such time as it was with him) and that was only part of the basis for the statement by its historian, William Swinton, that at this juncture "the Army of the Potomac, shaken in its structure, its valor quenched in blood, and thousands of its ablest officers killed and wounded, was the Army of the Potomac no more."

Much the same thing could be said of the army in the Petersburg intrenchments. Though its valor was by no means "quenched," it was no longer the Army of Northern Virginia in the old aggressive sense, ready to lash out at the first glimpse of a chance to strike an unwary adversary; nor would it see again that part of the Old Dominion where its proudest victories had been won and from which it took its name. When Lee arrived that morning, hard on the heels of one corps and a few hours in advance of the other, Beauregard was in such a state of elation ("He was at last where I had, for the past three days, so anxiously hoped to see him," the Creole later wrote) that he proposed an all-out attack on the Union flank and rear, as soon as A. P. Hill came up. Lee rejected the notion out of hand, in the conviction that his troops were far too weary for any such exertion and that Hill's corps would be needed to extend the present line westward to cover the two remaining railroads, the Weldon and the Southside, upon which Richmond — and perhaps, for that matter, the Confederacy itself — depended for survival. He did not add, as he might have done, that he foresaw the need for conserving, not expending in futile counterstrokes, the life of every soldier he could muster if he was to maintain, through the months ahead, the stalemate he had achieved at the price of his old mobility. "We must destroy this army of Grant's before he gets to James River," he had told Early three weeks ago, in the course of the shift from the Totopotomoy. "If he gets there it will become a siege, and then it will be a mere question of time." It was not that yet; Richmond was not under direct pressure, north of the James, and Petersburg was no more than semi-beleaguered; but that too, he knew, was only a "question of time."

Grant agreed, knowing that the length of time in question would depend on the rate of his success in reaching around Lee's right for control of the two railroads in his rear. First, though, there was the need for making the hastily occupied Federal line secure against dislodgment. The following day, June 19, was a Sunday (it was also the summer solstice; *Kearsarge* and *Alabama* were engaged off Cherbourg, firing at each other across the narrowing circles they described in the choppy waters of the Channel, and Sherman was maneuvering, down in Georgia, for ground from which to launch his Kennesaw assault); Meade's troops kept busy constructing bomb-proofs and hauling up heavy guns and mortars that would make life edgy, not only for the grayback soldiers just across the way, but also for the civilians in Petersburg, whose downtown streets were so little distance away that the blue gun crews could hear its public clocks strike the hours when all but the pickets of both armies were rolled in blankets. Grant had it in mind, however, to try one more sudden lunge — a two-corps strike beyond the Jerusalem Plank Road — before settling down to "gradual approaches."

Warning orders went out Monday to Wright, whose three divisions would be reunited by bringing the detached two from Bermuda Hundred, and to Birney, whose corps would pull back out of line for the westward march, and on Tuesday, June 21, the movement got under way. Simultaneously, while still waiting for Sheridan to return from his failure to link up with Hunter near the Blue Ridge, Wilson, reinforced by Kautz, was sent on a wide-ranging strike at both the Petersburg & Weldon and the Southside railroads, with instructions to rip up sizeable stretches of both before returning. Grant had settled down at his City Point headquarters that afternoon to await the outcome of this double effort by half of Meade's infantry and all of the cavalry on hand, when "there appeared very suddenly before us," a staff colonel wrote his wife, "a long, lank-looking personage, dressed all in black and looking very much like a boss undertaker."

It was Lincoln. After sending his "I begin to see it" telegram to Grant on the 15th, he had gone up to Philadelphia for his speech next day at the Sanitary Fair; after which he returned to Washington, fidgeted through another three days while the Petersburg struggle mounted to climax, and finally, this morning, boarded a steamer for a cruise down the Potomac and a first-hand look at the war up the James. "I just thought I would jump aboard a boat and come down and see you," he said, after shaking hands all round. "I don't expect I can do any good, and in fact I'm afraid I may do harm, but I'll just put myself under your orders and if you find me doing anything wrong just send me right away."

Grant replied, not altogether jokingly, that he would do that, and the group settled down for talk. By way of reassurance as to the outcome of the campaign, which now had entered a new phase — one that opened with his army twice as far from the rebel capital as it had been the week before — the general took occasion to remark that his present course was certain to lead to victory. "You will never hear of me farther from Richmond than now, till I have taken it," he declared. "I am just as sure of going into Richmond as I am of any future event. It may take a long summer day, as they say in the rebel papers, but I will do it."

★

*Brigadier General James H. Wilson's Federal cavalrymen
rip up and burn railroad ties on the Southside Railroad
west of Petersburg on June 22, 1864.*

Lincoln was glad to hear that; but he had been watching the casualty lists, along with the public reaction they provoked. "I cannot pretend to advise," he said, somewhat hesitantly, "but I do sincerely hope that all may be accomplished with as little bloodshed as possible."

Aside from this, which was as close to an admonition as he came, he kept the conversation light. "The old fellow remained with us till the next day, and told stories all the time," the staff colonel informed his wife, adding: "On the whole he behaved very well."

One feature of the holiday was a horseback visit to Hincks's division, where news of Lincoln's coming gathered around him a throng of black soldiers ("grinning from ear to ear," the staffer wrote, "and displaying an amount of ivory terrible to behold") anxious for a chance to touch the Great Emancipator or his horse in passing. Tears in his eyes, he took off his hat in salute to them, and his voice broke when he thanked them for their cheers. This done, he rode back to City Point for the night, then reboarded the steamer next morning for an extension of his trip upriver to pay a courtesy call on Ben Butler, whose views on politics were as helpful, in their way, as were Grant's on army matters. He returned to Washington overnight, refreshed in spirit and apparently reinforced in the determination he had expressed a week ago at the Sanitary Fair: "We accepted this war for an object, a worthy object, and the war will end when that object is attained. Under God, I hope it never will until that time."

Helpful though the two-day outing was for Lincoln, by way of providing relaxation and lifting his morale, the events of that brief span around Petersburg had an altogether different effect on Grant, or at any rate on the troops involved in his intended probe around Lee's right. After moving up, as ordered, on the night of June 21, Wright and Birney (Hancock was still incapacitated, sloughing fragments of bone from the reopened wound in his thigh) lost contact as they advanced next morning through the woods just west of the Jerusalem Plank Road, under instructions to extend the Federal left to the Weldon Railroad. Suddenly, without warning, both were struck from within the gap created by their loss of contact. Lee had unleashed A. P. Hill, who attacked with his old fire and savagery, using one division to hold Wright's three in check while mauling Birney's three with the other two. The result was not only a repulse; it was also a humiliation. Though his loss in killed and wounded was comparatively light, no fewer than 1700 of Birney's men — including those in a six-gun battery of field artillery, who then stood by and watched their former weapons being used against their former comrades — surrendered rather than risk their lives in what he called "this most unfortunate and disgraceful affair." Hardest hit of all was Gibbon's division, which had crossed the Rapidan seven weeks ago with 6799 men and had suffered, including heavy reinforcements, a total of 7970 casualties, forty of them regimental commanders. Such losses, Gibbon declared in his formal report, "show why it is that troops, which at the commencement of the campaign were equal to almost any undertaking, became toward the end of it unfit for almost any."

Wilson, after a heartening beginning, fared even worse than the infantry in the end. Reinforced by Kautz to a strength of about 5000 horsemen and twelve guns, he struck and wrecked a section of the Weldon Railroad above Reams Station, nine miles south of Petersburg, then plunged on to administer the same treatment to the Southside and the Richmond & Danville, which crossed at Burkeville, fifty miles to the west. Near the Staunton River, eighty miles southwest of Petersburg, with close to sixty miles of track ripped up on the three roads, he turned and started back for his own lines, having been informed that they would have been extended by then to the Petersburg & Weldon. On the way there, he was harried by ever-increasing numbers of gray cavalry, and when he approached Reams Station he found it held, not by

★

Wright or Birney, who he had been told would be there, but by A. P. Hill. Moreover, the mounted rebels, pressing him by now from all directions, turned out to be members of Hampton's other two divisions, returned ahead of Sheridan from the fight at Trevilian Station. Outnumbered and all but surrounded, Wilson set fire to his wagons, spiked his artillery, and fled southward in considerable disorder to the Nottoway River, which he succeeded in putting between him and his pursuers for a getaway east and north. He had accomplished most of what he was sent out to do, but at a cruel cost, including 1500 of his troopers killed or captured, his entire train burned, and all twelve of his guns abandoned.

Grant had the news of these two near fiascos to absorb, and simultaneously there came word of still a third, one hundred air-line miles to the west, potentially far graver than anything that had happened close at hand. Wright and Birney at least had extended the Federal left beyond the Jerusalem Road, and Wilson and Kautz had played at least temporary havoc with no less than

Weary from a week of riding, fighting, and ripping up track, General August Kautz's cavalry division returns toward the Federal lines near Petersburg.

three of Lee's critical rail supply lines. But David Hunter, aside from his easy victory two weeks ago at Piedmont and a good deal of incidental burning of civilian property since, accomplished little more, in the end, than the creation of just such a military vacuum as Lee specialized in filling.

Descending on Lynchburg late in the day, June 17, Hunter found Breckinridge drawn up to meet him with less than half as many troops. He paused overnight, preparing to stage another Piedmont in the morning, only to find, when it broke, that Jubal Early had arrived by rail from Charlottesville to even the odds with three veteran divisions: whereupon Hunter (for lack of ammunition, he later explained) went over to the defensive and fell back that night, under cover of darkness, to the shelter of the Blue Ridge. Early came on after him, and Hunter decided that, under the circumstances, his best course would be to return to West Virginia without delay. For three days Early pursued him, with small profit, then gave it up and on June 22 — while A. P. Hill was mauling Birney, south of Petersburg — marched for Staunton and the head of the Shenandoah Valley, that classic route for Confederate invasion which Lee had used so effectively in the past to play on Halleck's and Lincoln's fears.

These last were likely to be enlarged just now, and not without cause. With Hunter removed from all tactical calculations, nothing blue stood between Early and the Potomac, and with the capital defenses stripped of their garrisons to provide reinforcements and replacements for Meade, little remained with which to contest a gray advance from the Potomac into Washington itself. Lincoln had come up the James this week for a first-hand look at the war, but now it began to appear that he needed only to have waited a few days in the White House for the war to come to him.

So much was possible; Halleck's worst fears as to the consequences of the southside shift for the failed assault might now be proved only too valid. But Grant was not given to intensive speculation on possible future disasters; he preferred to meet them when they came, having long since discovered that few of them ever did. Instead, in writing to Old Brains on June 23 he stressed his need for still more soldiers, as a way of forestalling requests (or, in Lincoln's case, orders) for detachments northward from those he had on hand. "The siege of Richmond bids fair to be tedious," he informed him, "and in consequence of the very extended lines we must have, a much larger force will be necessary than would be required in ordinary sieges against the same force that now opposes us." Two days later, in passing along the news that Hunter was indeed in full retreat, he added that Sheridan had at last returned, though with his horses too worn down to be of any help to Wilson, who was fighting his way back east against lengthening odds. "I shall try to give the army a few days' rest, which they now stand much in need of," Grant concluded, rather blandly.

★

After frightening Hunter's 18,000 away from Lynchburg, westward beyond the Blue Ridge, and enjoying a day's rest from the three-day Allegheny chase that followed, the 14,000 Confederates took up the march for Staunton via Lexington, where on June 25 part of the column filed past Stonewall Jackson's grave, heads uncovered, arms reversed, bands intoning a dirge with muted horns and muffled drums. This salute to the fallen hero was altogether fitting as an invocation of the spirit it was hoped would guide the resurrected Army of the Valley through the campaign about to be undertaken by his old Second Corps, now led by Jubal Early. "Strike as quick as you can," Lee had telegraphed a week ago, as soon as he learned that Meade's whole army was south of the James, "and, if circumstances authorize, carry out the original plan, or move upon Petersburg without delay."

The original plan, explained to Early on the eve of his departure from Cold Harbor, June 13, was for him to follow the slash at Hunter with a fast march down the Valley, then cross the Potomac near Harpers Ferry and head east and south, through western Maryland, for a menacing descent on the Federal capital itself. Lee's hope was that this would produce one of two highly desirable results. Either it would alarm Lincoln into ordering heavy detachments northward from the Army of the Potomac, which might give Richmond's defenders a chance to lash out at the weakened attackers and drive them back from the city's gates, or else it would provoke Grant into staging a desperate assault, Cold Harbor style, that would serve even better to bleed him down for being disposed of by the counterattack that would follow his repulse. Given his choice, Early stuck to the original plan. After driving Hunter beyond the mountains, which removed him from all immediate tactical calculations, the gray pursuers rested briefly, then passed for the last time in review by their great captain's grave in battered Lexington and continued on to Staunton, where their hike down the Valley Turnpike would begin.

Early got there next day, ahead of his troops, and reorganized the 10,000 foot soldiers into two corps while awaiting their arrival. By assigning Gordon's division to Breckinridge, who coupled it with his own, he gave the former Vice President a post befitting his dignity and put thirty-five-year-old Robert Rodes — a native of Lynchburg, which he had just helped to save from Hunter's firebrands, and a graduate and one-time professor at V.M.I., whose scorched ruins he viewed sadly, and no doubt angrily as well, after marching his veterans past that other V.M.I. professor's grave — in charge of the remaining corps, composed of his own and Dodson Ramseur's divisions; Ramseur, a North Carolinian, promoted to major general the day after his twenty-seventh birthday early this month, was the youngest West Pointer to achieve that rank in Lee's

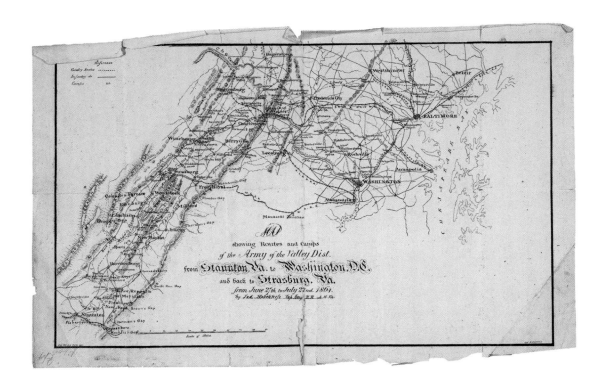

*This map by Jedediah Hotchkiss depicts the route
of Jubal A. Early's Second Corps from Staunton to
Washington, D.C., in late June and early July.*

army. The remaining 4000 effectives were cavalry and artillery, and these too
were included in the shakeup designed to promote efficiency in battle and on the
march. Robert Ransom, sent from Richmond for the purpose, was given command of the three mounted brigades ("buttermilk rangers," Early disaffectionately
styled these horsemen, riled by their failure to bring Hunter to bay the week
before) along with instructions to infuse some badly needed discipline into their
ranks. As for the long arm, it was not so much reshuffled as it was stripped by
weeding out the less serviceable guns and using only the best of teams to draw
the surviving forty, supplemented by ten lighter pieces the cavalry would bring
along. Recalling his predecessor Ewell's dictum, "The road to glory cannot be
followed with much baggage," Early stipulated that one four-horse "skillet wagon"
would have to suffice for transporting the cooking utensils for each 500 men, and
he even warned that "regimental and company officers must carry for themselves
such underclothing as they need for the present expedition." One major problem
remained unsolved: a lack of shoes for half the army. This would not matter

greatly in Virginia, but experience had shown that barefoot men suffered cruelly on the stony Maryland roads. Assured by the Quartermaster General that a shipment of shoes would overtake him before he crossed the Potomac, Early put the column in motion at first light June 28. Already beyond New Market two days later, some fifty miles down the turnpike, he informed Lee that his troops were "in fine condition and spirits, their health greatly improved. . . . If you can continue to threaten Grant," he added, "I hope to be able to do something for your relief and the success of our cause shortly. I shall lose no time."

True to his word, he reached Winchester on July 2, the Gettysburg anniversary, and there divided his army, sending one corps north, through Martinsburg, and the other east toward Harpers Ferry, where they were to converge two days later; Franz Sigel was at the former place with a force of about 5000, while the latter contained a garrison roughly half that size, and Early wanted them both, if possible, together with all their equipment and supplies. It was not possible. Sigel — who by now had been dubbed "The Flying Dutchman" — was too nimble for him, scuttling eastward to join the Ferry garrison before the rebel jaws could close and then taking sanctuary on Maryland Heights, which Early found too stout for storming when he came up on Independence Day. While one brigade maneuvered on Bolivar Heights to keep up the scare across the way, the rest of the Valley army settled down to feasting on the good things the Federals had left behind, here and at Martinsburg as well. Two days were spent preparing to cross the Potomac at Boteler's Ford, just upstream near Shepherdstown, and distributing the shipment of shoes that arrived on schedule from Richmond. On July 6 the crossing began in earnest; a third gray invasion was under way. No bands played "My Maryland," as before, but there was a chance for some of the veterans to revisit Sharpsburg, where they had fought McClellan, two Septembers back, from dawn to dusk along Antietam Creek. On they trudged, across South Mountain on July 8, breaking in their new shoes, and entered Frederick next morning in brilliant sunlight. East and southeast, beyond the glittering Monocacy River, the highway forked toward Baltimore and Washington, their goal.

Certain adjunctive matters had been or were being attended to by the time the infantry cleared Frederick. Coincident with the Potomac crossing, Imboden's cavalry had been sent westward, out the Baltimore & Ohio, to wreck a considerable stretch of that line and thus prevent a rapid return by Hunter's numerically superior force from beyond the Alleghenies, and simultaneously, by way of securing reparation for Hunter's recent excesses in the Old Dominion, a second mounted brigade, under Brigadier General John McCausland — another V.M.I. graduate and professor — was sent to Hagerstown with instructions to exact an assessment of $200,000, cash down, under penalty of otherwise having the torch put to its business district. En route, McCausland somehow dropped a digit, and the Hagerstown merchants, knowing a bargain when they saw one, were prompt in their payment of $20,000 for deliverance from the flames. No such arithmetical error was made at Frederick, where McCausland rejoined in time to see the full $200,000 demanded and paid in retaliation for what had been done, four weeks ago in Lexington, to Washington College and his alma mater. No sooner had he returned than the third brigade of horsemen, under Colonel Bradley Johnson, was detached. Hearing from Lee, in a sealed dispatch brought north by his son Robert, that a combined operation by naval elements and under-cover agents was planned for the liberation of 17,000 Confederate prisoners at Point Lookout, down Chesapeake Bay at the mouth of the Potomac, Early sent for Johnson — a native of Frederick, familiar with the region to be traversed — and told him to take his troopers eastward, cut telegraph wires and burn railroad bridges north and south of Baltimore in order to prevent the flow of information and reinforcements through that city when the gray main body closed on Washington, and then be at or near Point Lookout on the night of July 12, in time to assist in setting free what would amount to a full new corps for the Army of Northern Virginia. If things worked out just right, for them and for Early, the uncaged veterans might even return south armed with weapons taken from various arsenals, ordnance shops, and armories in the Federal capital, just over forty miles from Frederick, at the end of a two-day march down the broad turnpike.

Two days, that is, provided there was no delay en route: a battle, say, or even a sizeable skirmish, anything that would oblige a major portion of the army to deploy, engage, and then get back into march formation on the pike — always a time-consuming business, even for veterans such as these. And sure enough, Early had no sooner ridden southeast out of Frederick, down the spur track of the B & O toward its junction with the main line near the Monocacy, than he saw, drawn up to meet him on the far side of the river, with bridgeheads occupied to defend the crossings — the railroad itself and the two macadamized turnpikes, upstream and down — a considerable enemy force, perhaps as large as his own, with sunlight glinting from the polished tubes of guns emplaced from point to point along the line. Its disposition looked professional (which might signify that Grant had hurried reinforcements north from the Army of the Potomac, under orders from Lincoln to cover the threatened capital) but Early's first task, in any case, was to find out how to come to grips with this new blue assemblage and thereby learn its identity and size, preferably without a costly assault on one of the bridgeheads. McCausland promptly gave him the answer by plunging across a shallow ford, half a mile to the right of the Washington road, and launching a dismounted charge that overran a Federal battery. Counter-attacked in force, the troopers withdrew, remounted, and splashed back across the river. Though they were unable to hold the guns they had seized, they brought with them something far more valuable: the key to the enemy's un-doing. So Early thought at any rate.

★

By now it was noon, and he wasted no time in fitting the key to the lock. Rodes and Ramseur would feint respectively down the Baltimore pike and the railroad, while the main effort was being made downstream by Gordon, who would cross by the newly discovered ford for a flank assault, with Breckinridge in support. "No buttermilk rangers after you now, damn you!" Old Jube had shouted three weeks ago at Lynchburg, shaking his fist at the bluecoats as they backpedaled under pressure from his infantry, just off the cars from Charlottesville. He repeated this gesture today on the Monocacy, confident that victory was within his grasp whether the troops across the way were veterans, up from Petersburg, or hundred-day militia, hastily assembled from roundabout the Yankee capital and dropped in his path as a tub to the invading rebel whale.

They were both, but mostly they were veterans detached from the Army of the Potomac three days ago, on July 6, just as Early began crossing into Maryland. Warned by Halleck that Hunter had skittered westward, off the tactical margin of the map, and that Sigel too had removed his troops from contention with the 20,000 to 30,000 Confederates reported to be about to descend on Washington — which had nothing to defend it but militia, and not much of that — Grant loaded Ricketts' 4700-man VI Corps division onto transports bound

*The Federal complex at Point Lookout included
a hospital (lower left), stockades for prisoners (upper right),
and a beachfront command center (inset).*

for Baltimore, along with some 3000 of Sheridan's troopers, dismounted by the breakdown of their horses on the recent grueling raids beyond Burkeville and Louisa. Three days later, with Early across South Mountain and Washington approaching a state of panic, if not of siege, he not only followed through by ordering Wright to steam north in the wake of Ricketts with his other two divisions; he also informed Old Brains that he would be sending the XIX Corps, whose leading elements were due about now at Fortress Monroe, en route from New Orleans and the fiasco up Red River. This last came hard, badly needed as these far-western reinforcements were as a transfusion for Meade's bled-down army, straining to keep up the pressure south of the James. Yet Grant was willing to do even more, if need be, to meet the rapidly developing crisis north of the Potomac.

"If the President thinks it advisable that I should go to Washington in person," he wired Halleck that evening from City Point, while the last of Wright's men were filing aboard transports for the trip up Chesapeake Bay, "I can start in an hour after receiving notice, leaving everything here on the defensive."

Meantime Ricketts had landed at Baltimore, headquarters of Major General Lew Wallace's Middle Department, including Maryland, Delaware, and the Eastern Shore of Virginia. Wallace was not there, however. He had left two days ago, on July 5, after learning that the rebels were at Harpers Ferry in considerable strength, their outriders already on the loose in western Maryland as an indication of where they would be headed next. A former Illinois lawyer, now thirty-seven years old, he had been at the time of Shiloh the youngest major general in the Union army, but his showing there had soured Grant on him; the brilliant future predicted for him was blighted; he was shifted, in time, to this quiet backwater of the war. Quiet, that is, until an estimated 30,000 graybacks appeared this week on the banks of the Potomac, with nothing substantially blue between them and the national capital. Wallace said later that when he pondered the consequences of such a move by Early, "they grouped themselves into a kind of horrible schedule." If Washington fell, even temporarily, he foresaw the torch being put in rapid sequence to the Navy

Yankee Lew Wallace was again in the thick of the action when 30,000 rebels massed near Washington.

Yard, the Treasury, and the Quartermaster Depot, whose six acres of warehouses were stocked with $11,000,000 in equipment and supplies; "the war must halt, if not stop for good and all."

Accordingly, having decided to meet the danger near the rim of his department — though at considerable personal risk, for while he knew that Halleck was keeping tabs on him for Grant, watching sharply for some infraction that would justify dismissal, he could not inform his superiors of what he was about to do, since he was convinced that they would forbid it as too risky — he got aboard a train for Monocacy Junction, where the roads from nearby Frederick branched toward Baltimore and Washington. There he would assemble whatever troops he could lay hands on, from all quarters, and thus cover, from that one position, the approaches to both cities: not so much in hope of winning the resultant battle, he afterwards explained, as in hope of slowing the rebel advance by fighting the battle at all. Whatever the outcome, the delaying action on the Monocacy would perhaps afford the authorities time to brace for the approaching shock, not only by assembling all the available militia from roundabout states, but also by summoning from Grant, down in Virginia, a substantial number of battle-seasoned veterans to throw in the path of the invaders.

Sure enough, after managing to scrape together in two days, July 6-7, a piecemeal force of 2300 of all arms, he learned that this last had in fact been done, or at least was in the process of being done. Troops from the Army of the Potomac were debarking at Baltimore even then, hard-handed men in weathered blue who had taken the measure of Lee's touted veterans down the country and were no doubt willing and able to do the same up here. Greatly encouraged, Wallace sent for Ricketts to bring his division to Monocacy Junction without delay, leaving Sheridan's unhorsed troopers — more than a third of whom lacked arms as well as mounts — to man the Baltimore or Washington defenses, and thereby help, perhaps, to reduce the civilian panic reported to be swelling in both places. Ricketts arrived by rail next day, and none too soon; Early came over South Mountain that afternoon, July 8, and on into Frederick next morning. By noon he had his army moving by all the available roads down to the Monocacy, where Wallace had disposed his now 7000-man force to contest a crossing, posting Ricketts on the left, astride the Washington pike, where he figured the rebels would launch their main attack.

He figured right, but not right enough to forestall an end-on blow that soon resulted in a rout. Gordon struck from beyond the capital pike, not astride it, coming up from the ford downstream for an attack that Ricketts saw would roll up his line unless he effected a rapid change of front. He tried and nearly succeeded in getting his soldiers parallel to the turnpike, facing south, before they were hit. They gave ground, uncovering the unburnable iron railroad bridge for a crossing by Ramseur, who together with Breckinridge added the pressure that ended all resistance on this flank. Ricketts' two brigades, or what was left of them by now — the second, made up of veterans long known as "Milroy's weary boys," had been through this kind of thing before — scrambled northward for the Baltimore road, the designated avenue of retreat, and there lost all semblance of order in their haste to get out of range of the whooping rebels, one of whom afterwards called this hot little Battle of the Monocacy "the most exciting time I witnessed during the war."

By 4 o'clock it was over, and though Wallace (with 1880 casualties, including more than a thousand captured or otherwise missing, as compared to fewer than 700 killed or wounded on the other side) managed to piece together a rear guard not far east of the lost field, there was no real pursuit; Early did not want to be encumbered with more prisoners than he had already taken, more or less against his will. Nor did he want to move eastward, in the direction of Baltimore. His route was southeast, down the Washington pike, which Gordon's attack had cleared for his use in continuing the march begun that morning out of Frederick.

In any case he knew now, from interrogating captives with the canted VI Corps cross on the flat tops of their caps, that troops had arrived from the Army of the Potomac, and though he had whipped them rather easily — as well he might have expected to do, with the odds at two-to-one — he knew only too well that others were probably on the way, if indeed they were not already on hand in the capital defenses. If this was a source of satisfaction, knowing that he had fulfilled a considerable measure of Lee's purpose by obliging Grant to reduce the pressure on Petersburg and Richmond, it also recommended caution. Additional blue detachments might have arrived or be arriving from down the country in such numbers that his small army, cut off from the few available fords across the Potomac as he advanced, would be swamped and abolished. As it was, he had only to turn southwest; down the B & O to Point of Rocks, for a crossing that would gain him the security of the Virginia Piedmont, after which he could move south or west, unmolested, for a return to Lee or the Shenandoah Valley. Either course had its attractions, but Early dwelt on neither. He would move as he had intended from the outset, against Washington itself, and deal with events as they developed, knowing from past service under Jackson that audacity often brought its own rewards. Today was too far gone for resumption of the march, but he passed the word for his men to bed down for a good night's rest, here on the field where they had fought today, and be ready to move at "early dawn."

Sunday, July 10, was hot and dusty. By noon, the cumulative effect of all those twenty-mile hikes since the army left Staunton twelve days ago had begun to tell. Straggling increased as the day wore on, until finally the head of the column went into bivouac short of Rockville, just over twenty miles from the Monocacy and less than ten from the District of Columbia. Rear elements did not come up

★

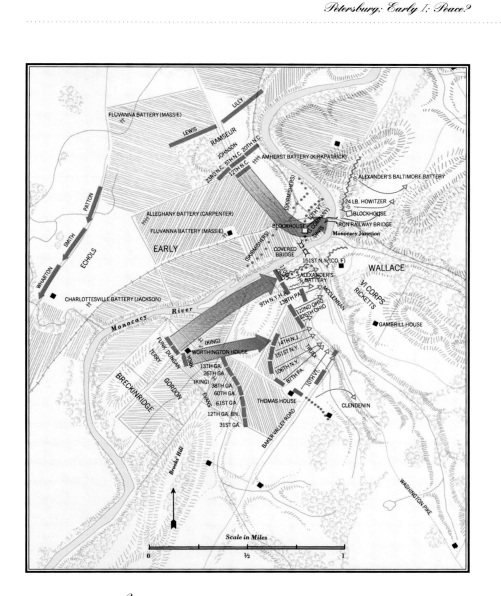

*At the Battle of Monocacy, the Confederates dislodged
Federal brigades from their positions on the Thomas farm and
drove skirmishers back across the iron railroad bridge.*

till after midnight, barely three hours before Early, hopeful of storming the Washington defenses before sundown, ordered the march resumed in the predawn darkness. Aware that he might be engaged in a race with reinforcements on the way there, he could afford to show his weary men no mercy, though he sought to encourage them, as he doubled the column on his lathered horse, with promises of rest and a high feast when the prize was won. Beyond Rockville, he had

McCausland's troopers hold to the main pike for a feint along the Tenallytown approaches, while the infantry forked left for Silver Spring, half a dozen miles from the heart of the city by way of the main-traveled Seventh Street Road.

Heat and dust continued to take their toll; "Our division was stretched out almost like skirmishers," one of Gordon's veterans, tottering white-faced with fatigue near the tail of the column, would recall. Then, close to 1 o'clock, the heavy, ground-thumping boom-bam-*boom* of loud explosions — guns: siege guns! — carried back from the front, where the head of the column had come within range of the outlying capital works.

Early rode fast toward the sound of firing, beyond the District line, and drew rein in time to watch his advance cavalry elements dismount and fan out to confront a large earthwork on rising ground to the right of the road, two miles below Silver Spring. Identified on the map as Fort Stevens, a major installation, it lay just over a thousand yards away, and when he studied it through his binoculars he saw a few figures on the parapet; by no means enough, it seemed to him, to indicate that the work was heavily, even adequately, manned. He had won his race with Grant. All he had to do, apparently, was bring up his men and put them in attack formation, then move forward and take it, along with much that lay beyond, including the Capitol itself, whose new dome he could see plainly in the distance, six miles south of where he stood.

Just now, though, his troops were in no condition for even the slightest exertion, whatever prize gleamed on the horizon. Diminished by cavalry detachments, by their losses in battle two days ago, and by stragglers who had fallen out of the column yesterday and today, they scarcely totaled 10,000 now, and of these no more than a third were fit for offensive action without a rest. All the same, he told Rodes, whose division was in the lead, to see what he could accomplish along those lines, and while Rodes did his best — which wasn't much; his men were leaden-legged, short of wind and spitting cotton — Early continued to study the objective just ahead. Beyond it, around 1.30, he saw a long low cloud of dust approaching from the rear, up the Seventh Street Road. Reinforcements, most likely; but how many? and what kind? Then he spotted them in his glass, the ones at the head of the fast-stepping column at any rate, and saw that they were dressed not in linen dusters and high-crowned hats, after the manner of home guardsmen or militia, but in the weathered blue tunics and kepis he had last encountered two days ago, when he found Ricketts' VI Corps veterans drawn up to meet him on the Monocacy.

Veterans they were, all right, and VI Corps veterans at that; Wright and the first of his other two divisions, the second relay of reinforcements ordered north from the Army of the Potomac, had begun debarking at the Sixth Street docks a little after noon and were summoned at once to the point of danger, out the Seventh Street Road. Grant himself might be on the way by now, moreover,

★

for Lincoln — under increasing pressure as the rebel column, having knocked Wallace out of its path, drew closer to Washington hour by hour — had responded approvingly to the general's offer to come up and take charge "in person," adding that it might be well if he brought still more of his soldiers along with him. "What I think," he told Grant, "is that you should provide to retain your hold where you are, certainly, and bring the rest with you personally and make a vigorous effort to destroy the enemy's force in this vicinity. I think there is really a fair chance to do this if the movement is prompt." In other words, hurry. But then, mindful once more of his resolution not to interfere in military matters, even with the graybacks practically at the gate, he closed by saying: "This is what I think, upon your suggestion, and it is not an order."

If he was jarred momentarily from his purpose — and, after all, the notion was Grant's in the first place; Lincoln merely concurred — it was small wonder, what with Hunter fled beyond recall up the Kanawah, Sigel holed up at Harpers Ferry, out of touch since July 4, and Washington panicked by rumors of Armageddon. Wallace, falling back down the Baltimore pike from his sudden drubbing on July 9, reported that Early had hit him with 20,000 of all arms, and though this was 10,000 fewer than Sigel had reported before the wire went dead

*B*efore Jubal Early could mount an attack on
scantly manned Fort Stevens, the northernmost fortification
of Washington, Federal reinforcements arrived.

★

in his direction, it still was 10,000 more than had been mustered, including War Department clerks and green militia, to man the capital defenses. Sheridan's dismounted troopers arrived about that time, a rather straggly lot who did less to bolster confidence here than their removal from Baltimore had done to provoke resentment there. When a group of that city's leading citizens telegraphed Lincoln that Sunday evening, July 10, protesting that they had been abandoned to their fate, he did what he could to reassure them. "Let us be vigilant, but keep cool," he replied. "I hope neither Baltimore nor Washington will be taken."

They remained disgruntled, wanting something more substantial. By next morning things looked better, however, at least in their direction. Returning with Ricketts, Wallace assured them that Early was headed for Washington, not Baltimore just yet. And even in the capital there was encouraging news to balance against reports that the rebel column had cleared Rockville soon after sunrise; Wright was expected hourly from Virginia with his other two divisions, and an advance detachment of 600 troops was already on hand from the XIX Corps, fine-looking men with skin tanned to mahogany by the Louisiana sun. Even Henry Halleck — who, according to an associate, had spent the past week "in a perfect maze, bewildered, without intelligent decision or self-reliance" — recovered his spirits enough to reply with acid humor to a telegram from an unattached brigadier at the Fifth Avenue Hotel, New York City, offering his services in the crisis now at hand. "We have five times as many generals here as we want," Old Brains informed him, "but are greatly in need of privates. Anyone volunteering in that capacity will be thankfully received." Then at noon the transports arrived at the Sixth Street docks (near which the Navy had a warship berthed with steam up, ready to whisk the President downriver in case the city fell); Wright's lead division came ashore and marched smartly through the heart of town to meet Early, who was reported to be approaching by way of Silver Spring. Presently the boom of guns from that direction made it clear how close the race had been, and was.

Lincoln, having ridden down to the docks to greet them from his carriage, also rode out the Seventh Street Road to watch them reinforce Fort Stevens; he may have been one of the figures — surely, if so, the tallest — Early saw etched against the sky when he focussed his binoculars on the parapet of the works just over a thousand yards ahead. Watching the dusty blue stream of veterans flow into position in the course of the next hour, Old Jube — or "Jubilee," as soldiers often styled him — knew there could be no successful assault by his weary men today. A good night's rest might make a difference, though, depending on how heavily the defenses had been reinforced by morning, either here or elsewhere along the thirty-seven miles of interconnected redans, forts, and palisades ringing the city and bristling with heavy guns at every point. What remained of daylight could be used for reconnaissance (and was; "Examination showed

★

what might have been expected," Early would report, "that every application of science and unlimited means had been used to render the fortifications around Washington as strong as possible") but the thing to do now, he saw, was put the troops into bivouac, then feed and get them bedded down, while he and his chief lieutenants planned for tomorrow. He and they had come too far, and Lee had risked too much, he felt, for the Army of the Valley to retire from the gates of the enemy capital without testing to see how stoutly they were hung.

Accordingly, he turned his horse and rode back toward Silver Spring, where his staff had set up headquarters, just beyond the District line, in the handsome country house of Francis P. Blair, who had decamped to avoid an awkward meeting with one-time friends among the invaders. A member of Andrew Jackson's "kitchen cabinet" and an adviser to most of the Presidents since, Old Man Blair had two sons in high Union places: Montgomery, Lincoln's Postmaster General, whose own home was only a short

"We have five times as many generals here as we want but are greatly in need of privates. Anyone volunteering in that capacity will be thankfully received."

— Henry Halleck

walk up the road, and Frank Junior, the former Missouri congressman, now a corps commander with Sherman.

Guards had been posted to protect the property; especially the wine cellar, which contributed to the festive spirit that opened the council of war with recollections by Breckinridge, as the toasts went round, of the good times he had had here in the days when he was Vice President under Buchanan. Someone remarked that tomorrow might give him the chance to revisit other scenes of former glory, such as the U.S. Senate, where he had presided until Lincoln's inauguration and then had sat as a member until he left, eight months later, to throw in with other Confederate-minded Kentuckians for secession. This brought up the question Early had called his lieutenants together to consider: Was an attack on Washington tomorrow worth the risk? Time was short and getting shorter; Hunter and Sigel could be expected to come up from the rear, eventually, and Grant was known to have sent what seemed to be most of a corps already. Doubtless other reinforcements were on the way, from other directions, and though the prize itself was the richest of all — perhaps even yielding foreign recognition, at long last, not to mention supplying the final straw that

★

*John Breckinridge,
pictured here in a
pre-war engraving,
entertained his fellow
rebel commanders with
stories of visits to the
Blair estate while he
was Vice President.*

might break the Federal home-front camel's back — was it worth the risk of losing
one fourth of Lee's army in the effort?

Early considered, with the help of his four division commanders, and
decided that it was. He would launch an assault at dawn, he told them, "unless
some information should be received before that time showing its impracticability."

Such information was not long in coming. The council of war had
scarcely ended when a courier arrived from Bradley Johnson, whose brigade was
still on its way to Point Lookout. After wrecking railroad bridges and tearing
down telegraph lines around Baltimore he had sent scouts into the city to confer
with Confederate agents, and from these he learned that not one but *two* Federal
corps, the VI and the XIX, were steaming up Chesapeake Bay and the Potomac
to bolster the Washington defenses. In the light of this intelligence that tomorrow
might find him outnumbered better than two-to-one by the bluecoats in the
capital intrenchments, Early countermanded his orders for a dawn assault. This
came hard. Just thirty days ago tomorrow he had received instructions from Lee
to attempt what he was on the verge of doing. Now though — as a result, he
perceived, of the victory Wallace had obliged him to win on the Monocacy, at
the cost of a twenty-hour delay — it began to appear that the verge was as close
as he was likely to get. Daylight would give him the chance to reconnoiter the
Union works and thus determine the weight of this new unwelcome information, but he could see already that an attack was probably beyond his means and
a good deal worse than risky.

Dawn broke, July 12, over a Washington in some ways even more
distraught than it had been the morning before, with the rebels bearing down
on its undermanned defenses. Overnight the shortage had been considerably

★

repaired; Wright's third division followed the second out the Seventh Street Road at dusk, and soon after dark the first of the two XIX Corps divisions landed. But as these 20,000 stalwarts arrived to join about the same number of militiamen, galvanized clerks, and dismounted cavalry in the outworks, so did a host of rumors, given unlimited opportunity for expansion by the fact that the city was cut off from all communication northward, either by rail or wire, newspapers or telegrams, speech or letters. Known secessionists did not trouble to mask broad smiles, implying that they knew secrets they weren't sharing. One that leaked out by hearsay was that Lee had given Meade the slip, down around Richmond, and was crossing the Potomac, close at hand, with an army of 100,000 firebrands yelling for vengeance for what had been done, these past three years, in the way of destruction to their homeland.

Lincoln rose early, despite a warning from Stanton that an assassination plot was afoot, and rode with Seward to visit several of the fortifications out on the rim of town, believing that the sight of him and the Secretary of State, unfled and on hand to face the crisis unperturbed, would help to reduce the panic in the streets through which their carriage passed. His main hope, now that he knew Grant would not be coming — "I think, on reflection, it would have a bad effect for me to leave here," the general had replied from City Point to the suggestion that he come north without delay — was in Horatio Wright, who had helped to drive these same gray veterans southward, down in Virginia, throughout the forty days of battle in May and June. Lincoln's belief was that the Connecticut general, now that he had the means, could do the same up here.

Wright rather thought so too. Taking Early's failure to attack this morning as a sign that the rebels were preparing to withdraw, probably after nightfall, he wanted to hit them before they got away unscathed. In particular he wanted to drive off their skirmishers, who had crept to within rifle range of Fort Stevens and were sniping at whatever showed above the parapet. However, when he requested permission, first of the fort commander and then of the district commander, Major Generals Alexander McCook and C. C. Augur — both of whom outranked him, although neither had seen any action for nearly a year, having been retired from field service as a result of their poor showings, respectively, at Chickamauga and Port Hudson — they declined, saying that they did not "consider it advisable to make any advance until our lines are better established."

By midafternoon this objection no longer applied; McCook, bearded in his command post deep in the bowels of the fort, agreed at last to permit a sortie by units from one of the VI Corps divisions. Wright started topside for a last-minute study of the terrain, and as he stepped out of the underground office he nearly bumped into Abraham Lincoln, who had returned from a cabinet meeting at the White House to continue his tour of the fortifications. Informed of what was about to be done, he expressed approval, and when the general

★

asked, rather casually, whether he would care to take a look at the field — "without for a moment supposing he would accept," Wright later explained — Lincoln replied that he would indeed. Six feet four, conspicuous in his frock coat and a stovepipe hat that added another eight inches to his height, he presently stood on the parapet, gazing intently at puffs of smoke from the rifles of snipers across the way. Horrified, wishing fervently that he could revoke his thoughtless invitation, Wright tried to persuade the President to retire; but Lincoln seemed not to hear him amid the twittering bullets, one of which struck and dropped an officer within three feet of him. From down below, a young staff captain — twenty-three-year-old Oliver Wendell Holmes, Junior, whose combat experience had long since taught him to take shelter whenever possible under fire — looked up at the lanky top-hatted civilian and called out to him, without recognition: "Get down, you damn fool, before you get shot!"

This got through. Lincoln not only heard and reacted with amusement to the irreverent admonition, he also obeyed it by climbing down and taking a seat in the shade, his back to the parapet, safe at last from the bullets that continued to twang and nicker overhead.

Relieved of the worst of his concerns, Wright turned now to the interrupted business of clearing his front. Deployment of the brigade assigned the task required more time than had been thought, however, with the result that it was close to 6 o'clock before the signal could be given to move out. The firing swelled, and Lincoln, popping up from time to time to peer over the parapet, had his first look at men reeling and falling in combat and being brought past him on stretchers, groaning or screaming from pain, leaking blood and calling on God or Mamma, in shock and out of fear. Presently the racket stepped up tremendously, and the brigade commander sent back for reinforcements, explaining that he had encountered, beyond the retiring screen of pickets, a full-fledged rebel line of battle. Supporting regiments moved up in the twilight and the attack resumed, though with small success against stiffened resistance. Gunflashes winked and twinkled along the slope ahead until about 10 o'clock, when they diminished fitfully and finally died away. The cost to Wright had been 280 killed and wounded in what one of his veterans called "a pretty and well-conducted little fight."

Across the way, the Confederates considered it something worse: especially at the outset, when it erupted in the midst of their preparations to depart. Early had needed no more than a cursory look at the enemy works that morning to confirm last night's report that they would be substantially reinforced by dawn. Permanently canceling the deferred assault, he ordered skirmishers deployed along a line that stretched for a mile to the left and a mile to the right of the Seventh Street Road to confront Forts Reno, Stevens, and De Russy, while behind this he had Rodes and Gordon form their divisions, in case the Federals tried a

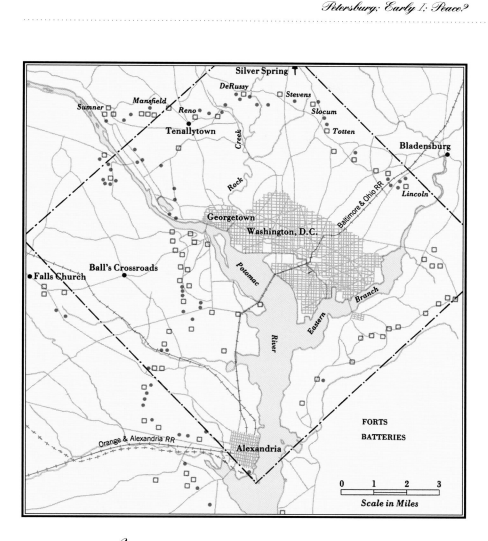

*A ring of forts (blue squares) and batteries (blue dots)
formed a 37-mile defensive perimeter around
the Federal capital in the summer of 1864.*

sortie, and sent word for McCausland to keep up the feint on the far right, astride the Georgetown pike. Here they would stay, bristling as if about to strike, until night came down to cover the withdrawal, back through Silver Spring to Rockville, then due west for a recrossing of the Potomac. Fortunately, the Yankees seemed content to remain within their works, and Early, having learned that the amphibious raid on Point Lookout had been called off because the prison authorities had been warned of it, had time to send a courier after Johnson, whose horsemen were beyond Baltimore by then, instructing him to turn back for the Confederate lines by whatever route seemed best now that the capture of Washington was no longer a part of the invasion plan. Preparations for the retirement were complete — were, in fact, about to be placed in execution — when Wright's attack exploded northward from Fort Stevens, flinging butternut skirmishers back on the main body, which then was struck by the rapid-firing Federals coming up in apparently endless numbers through the gathering dusk. The thing had the look of an all-out battle that would hold the Army of the Valley in position for slaughter tomorrow by preventing it from taking up its planned retreat tonight. Major Kyd Douglas, formerly of Jackson's staff and now of Early's, said quite frankly that he thought "we were gone up."

Presently though, to everyone's relief, the fireworks sputtered into darkness; the field grew still, except for the occasional jarring explosion of a shell from one of the outsized siege guns in the forts, and Early, resuming his preparations for withdrawal, summoned to headquarters Breckinridge and Gordon, whose divisions would respectively head and tail the column, for last-minute orders on the conduct of the march. They arrived to find him instructing Douglas to take charge of a rear-guard detail of 200 men and with them hold the present position until midnight, at which time he too was to pull out for Rockville: provided, of course, the bluecoats had not gotten wind of what was up, beforehand, and obliterated him.

When the handsome young Marylander left to assume this forlorn assignment, Early called after him, apparently in an attempt to lift his spirits: "Major, we haven't taken Washington, but we've scared Abe Lincoln like hell!"

Douglas stopped and turned. "Yes, General," he said, as if to set the record straight, "but this afternoon when that Yankee line moved out against us, I think some other people were scared blue as hell's brimstone."

"How about that, General?" Breckinridge broke in, smiling broadly beneath his broad mustache.

"That's true. But it won't appear in history," Early replied, thereby assuring the exchange a place in all the accounts that were to follow down the years.

It turned out there were no further losses, even for the rear-guard handful under Douglas, who took up the march on schedule without a parting shot being fired in his direction. He saw, as he went past it after midnight, that except for the depletion of its wine cellar and linen closets — all the bedclothes had been ripped into strips for bandages — Old Man Blair's mansion had suffered no damage from the occupation, but that his son Montgomery's house, just up the road, had been reduced to bricks and ashes by some vengeance-minded incendiary. Although the act perhaps was justified by Hunter's burning of Former Governor Letcher's home the month before, Early's regret that this had been done was increased when he learned that Bradley Johnson, off on his own, had also indulged in retaliation by setting fire to Governor A. W. Bradford's house near Baltimore. Such exactions, he knew, were unlikely to encourage pro-Confederate feelings, either here in Maryland or elsewhere. In any case, dawn of July 13 — thirty days, to the hour, since the re-created Army of the Valley pulled out of Cold Harbor, bound for Lynchburg and points north — found the column slogging through Rockville, where it turned left for Poolesville and the Potomac. At White's Ford by midnight, just upstream from Ball's Bluff and thirty miles from its starting point on the outskirts of Washington, the army crossed the river in good order next morning, still unmolested, to make camp near Leesburg for a much needed two-day rest; after which it shifted west, July 16, beyond the Blue Ridge. Back once more in the Lower Valley, within an easy day's march of Harpers Ferry, Early began preparing for further adventures designed to disrupt the plans of the Union high command.

This recent thirty-day excursion had accomplished a great deal in that direction, as well as much else of a positive nature, including the recovery of the grain-rich Shenandoah region from Hunter and Sigel, just in time for the harvesting of its richest crop in years, and the return from beyond the Potomac with a

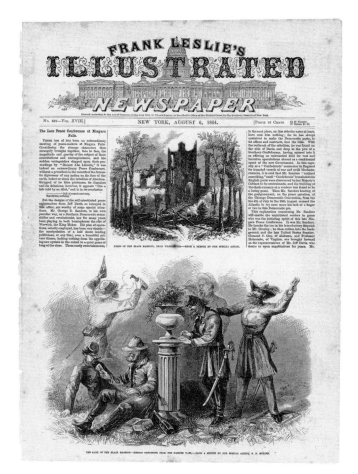

Frank Leslie's Illustrated Newspaper gave front-page coverage to the destruction of U.S. Postmaster General Montgomery Blair's estate near Silver Spring, Maryland.

large supply of commandeered horses and cattle, not to mention $220,000 in greenbacks for the hard-up Treasury and close to a thousand prisoners, most of them captured on the Monocacy, the one full-scale battle of the campaign. In fact, aside from his two main hopes — and hopes were all they were — that he could occupy Washington, even for a day, and that he could provoke Grant into making a suicidal assault on Lee's intrenchments, Early had accomplished everything that could have been expected of him. Best of all, he had obliged Grant to ease the pressure on Petersburg by sending large detachments north, and still had managed, despite the smallness of his force, if not to reverse the tide of the war, then anyhow to strike fear in the hearts of the citizens of Washington and Baltimore, both of which saw gray-clad infantry at closer range than any Federal had come, so far, to Richmond. This was much; yet there was more. For in the process Early had won the admiration not only of his fellow countrymen, whose spirits were lifted by the raid, but also of foreign observers, who still might somehow determine the outcome of this apparently otherwise endless conflict.

"The Confederacy is more formidable than ever," the London *Times* remarked when news of this latest rebel exploit crossed the ocean the following week. And closer at hand, on July 12 — even as Early and his veterans bristled along the rim of the northern capital, quite as if they were about to assail and overrun the ramparts in a screaming rush — the New York *World* asked its readers: "Who shall revive the withered hopes that bloomed on the opening of Grant's campaign?"

At White's Ford, Early's Confederates cross the Potomac on their return to the Shenandoah Valley after failing to capture the Federal capital in the summer of 1864.

★

———— ✧ ————

★ ★ ★ **W**ho indeed. The task was Lincoln's, as the national leader, but evidence piled higher every day that it would be his no longer than early March, when the outcome of the presidential election, less than four months off, was confirmed on the steps of the lately threatened Capitol. Despite setbacks, such as Cold Harbor, Petersburg, and this recent gray eruption on the near bank of the Potomac, he was convinced that he had found in U. S. Grant the man to win the war. But that was somewhat beside the point, which was whether or not the people could be persuaded, between now and November, to believe it too — and whether or not, believing it, they would agree that the prize was worth the additional blood, the additional money, the additional drawn-out anguish it was clearly going to cost. They, like Grant, would have to "face the arithmetic," and keep on facing it, to the indeterminate end.

One of the things that made this difficult was that the arithmetic kept changing, not only in the lengthening casualty lists, but also in the value fluctuations of what men carried in their wallets, a region where their threshold of pain was notoriously low. Gold opened the year at 152 on the New York market. By April it had risen to 175, by mid-June to 197, and by the end of that month to an astronomical 250. Reassurances from money men that the dollar was "settling down" brought the wry response that it was "settling down out of

sight." Sure enough, on July 11, as Early descended on Washington, gold soared to 285, reducing the value of the paper dollar to forty cents. Moreover, Lincoln faced this crisis without the help of the man who had advised him in such matters from the outset: Salmon Chase.

In late June, with the office of assistant treasurer of New York about to be vacated, the Secretary recommended a successor unacceptable to Senator Edwin D. Morgan of that state, who suggested three alternates for the post. "It will really oblige me if you will make a choice among these three," Lincoln wrote Chase, explaining the political ramifications of a tiff with Morgan at this time. Chase then requested a personal interview, which Lincoln refused "because the difficulty does not, in the main part, lie within the range of a conversation between you and me." In reaction to this snub, the Secretary went home and, as was his custom in such matters, "endeavored to seek God in prayer." So he wrote in his diary that night, adding: "Oh, for more faith and clearer sight! How stable is the City of God! How disordered the City of Man!" Mulling it over he reached a decision. His resignation was on the presidential desk next morning. "I shall regard it as a real relief if you think proper to accept it," he declared in a covering letter.

Lincoln read this fourth of the Ohioan's petulant resignations, and accepted it forthwith. "Of all I have said in commendation of your ability and fidelity, I have nothing to unsay," he replied, "and yet you and I have reached a point of mutual embarrassment in our official relationship which it seems cannot be overcome or longer sustained consistently with the public service." Ohio's Governor John Brough, who happened to be in town, went to the White House in an attempt to "close the breach," as he had done in one of the other instances of a threatened resignation, only to find that he could perform no such healing service here today. "You doctored the business up once," Lincoln told him, "but on the whole, Brough, I reckon you had better let it alone this time." Chase departed, still in something of a state of shock from the unexpected thunderclap, and retired to think things over, for a time, in the hills of his native New Hampshire.

A replacement was not far to seek. Next morning, July 1, when William Pitt Fessenden of Maine, chairman of the Senate Finance Committee, called on the President to recommend someone else for the Treasury post, Lincoln smiled and informed him that his nomination had just been sent for approval by his colleagues on the Hill. Fessenden's dismay was plain. "You must withdraw it. I cannot accept," he protested. His health was poor; Congress was to adjourn tomorrow, and he looked forward to a vacation away from the heat and bustle of the capital. "If you decline, you must do it in open day," Lincoln told him, "for I shall not recall the nomination."

Fessenden hurried over to the Senate in an attempt to block the move, only to find that he had been unanimously confirmed in about one minute. Regretfully, with congratulations pouring in from all quarters — even Chase's — he agreed to serve, at least through the adjournment. A soft-money man like his predecessor, he was sworn in on July 5, and it was observed that no appointment by the President, except perhaps the elevation of Grant four months before, had met with such widespread approval by the public and the press. "Men went about with smiling faces at the news," one paper noted.

Lincoln himself was not smiling by then. His trouble with Chase — whom he described as a man "never perfectly happy unless he is thoroughly miserable, and able to make everyone else just as uncomfortable as he is" — had been personal; Chase irked him and he got rid of him. But on the day after Fessenden's appointment he found himself in an even more irksome predicament, one that was susceptible to no such resolution because the men involved were not subject to dismissal; not by him, at any rate. On the morning of July 2, last day of the congressional session that was scheduled to adjourn at noon, Lincoln sat in the President's room at the Capitol, signing last-minute bills, including one that repealed the Fugitive Slave Law and another that struck the $300 commutation clause from the Draft Act. Both of these he signed gladly, along with others, but as he did so there was thrust upon him the so-called Wade-Davis bill, passed two months ago by the House and by the Senate within the hour. He set it aside to go on with the rest, and when an interested observer asked if he intended to sign it, he replied that the bill was "a matter of too much importance to be swallowed in that way."

He found it hard, in fact, to swallow the bill in any way at all, since what it represented was an attempt by Congress — more specifically, by the radicals in his party — to establish the premise that the legislative, not the executive, branch of government had the right and duty to define the terms for readmission to the Union by states now claiming to have left it; in other words, to set the tone of Reconstruction. Sponsored by Benjamin Wade in the Senate and Henry Winter Davis in the House, the bill proceeded from Senator Charles Sumner's thesis that secession, though of course not legally valid, nonetheless amounted to "State suicide," and it set forth certain requirements that would have to be met before the resurrected corpse could be readmitted to the family it had disgraced by putting a bullet through its head. Lincoln had done much the same thing in his Proclamation of Amnesty and Reconstruction, back in December, but this new bill, designed not so much to pave as to bar the path to reunion, was considerably more stringent. Where he had required that ten percent of the qualified voters take a loyalty oath, the Wade-Davis measure required a majority. In addition, all persons who had held state or Confederate offices, or who had voluntarily borne arms against the United States, were forbidden to vote for or serve as delegates to state constitutional conventions; the rebel debt was to be repudiated, and slavery outlawed, in

Thaddeus Stevens and other Jacobins passed the Wade-Davis bill pertaining to Reconstruction despite Lincoln's objections to it.

each instance. Moreover, this was no more than a precedent-setting first step; harsher requirements would come later, once the bill had established the fact that Congress, not the President, was the rightful agency to handle all matters pertaining to reconstruction of the South. Sumner and Zachariah Chandler in the Senate, Thaddeus Stevens and George W. Julian in the House — Jacobins all and accomplished haters, out for vengeance at any price — were strong in their support of the measure and were instrumental in ramming it through on this final day of the session.

Gideon Welles saw clearly enough what they were after, and put what he saw in his diary. "In getting up this law, it was as much an object of Mr. Henry Winter Davis and some others to pull down the Administration as to reconstruct the Union. I think they had the former more directly in view than the latter." Lincoln thought so, too, and was determined to keep it from happening, if he could only find a way to do so without bringing on the bitterest kind of fight inside his party.

The fact was, he had already found what he perceived might be the beginning of a way when he set the bill aside to go on signing others. Zachariah Chandler, who had asked him whether he intended to endorse it and had then been told that it was "too important to be swallowed in that way," warned him sternly, in reference to the pending election: "If it is vetoed, it will damage us fearfully in the Northwest. The important point is the one prohibiting slavery in the reconstructed states."

★

"That is the point on which I doubt the authority of Congress to act."

"It is no more than you have done yourself."

"I conceive that I may, in an emergency, do things on military grounds which cannot be done constitutionally by Congress," Lincoln replied, and Chandler stalked out, deeply chagrined.

His chagrin, and that of his fellow radicals, was converted to pure rage the following week — July 8; Early was crossing South Mountain to descend on Frederick — when Lincoln, having declined either to sign or to veto the bill, issued a public proclamation defending his action (or nonaction) on grounds that, while he was "fully satisfied" with some portions of the bill, he was "unprepared" to give his approval of certain others. "What an infamous proclamation!" Thaddeus Stevens protested. "The idea of pocketing a bill and then issuing a proclamation as to how far he will conform to it!"

By means of the "pocket veto," as the maneuver came to be called, Lincoln managed to avoid, at least for a season, being removed from all connection with setting the guidelines for Reconstruction; but he had not managed to avoid a fight. Indeed, according to proponents of the bill now lodged in limbo, he had precipitated one. Convinced, as one of them declared, that his proposed course was "timid and almost pro-slavery," they took up the challenge of his proclamation, which they defined as "a grave Executive usurpation," and responded in more than kind, early the following month in the New York *Tribune,* with what became known as the Wade-Davis Manifesto. Seeking "to check the encroachments of the Executive on the authority of Congress, and to require it to confine itself to its proper sphere," bluff Ben Wade and vehement Henry Davis charged that "a more studied outrage on the legislative authority of the people has never been perpetrated," and they warned that Lincoln "must understand that our support is of a cause and not of a man," especially not of a man who would connive to procure electoral votes at the cost of his country's welfare.

All this the manifesto set forth, along with much else of a highly personal nature from the pens of these Republican leaders, just three months before the presidential election. Lincoln declined to read or discuss it, not wanting to be provoked any worse than he was already, but he remarked in this connection: "To be wounded in the house of one's friends is perhaps the most grievous affliction that can befall a man."

Horace Greeley, editor of the paper in which the radical manifesto made its appearance, had been involved for the past month in an affair that added to Lincoln's difficulties in presenting himself as a man of war who longed for peace. Hearing privately in early July that Confederate emissaries were waiting on the Canadian side of Niagara Falls with full authority to arrange an armistice, Greeley referred the matter to the President and urged in a long, high-strung letter that he seize the opportunity this presented to end the fighting. "Confederates everywhere [are] for peace. So much is beyond doubt," he declared. "And therefore I venture to remind you that our bleeding, bankrupt, almost dying country also longs for peace — shudders at the prospect of fresh conscription, of further wholesale devastations, and of new rivers of human blood." Placed thus in the position of having to investigate this reported gleam of sunlight (which he suspected would prove to be moonshine) Lincoln was prompt with an answer. "If you can find any person anywhere professing to have any proposition of Jefferson Davis in writing, for peace, embracing the restoration of the Union and the abandonment of slavery, whatever else it embraces, say to him he may come to me with you." The editor, aware of the risk of ridicule, had not counted on being personally involved. He responded with a protest that the rebel agents "would decline to exhibit their credentials to me, much more to open their budget and give me their best terms." Lincoln replied: "I was not expecting you to send me a letter, but to bring me a man, or men."

"I venture to remind you that our bleeding, bankrupt, almost dying country also longs for peace — shudders at the prospect of fresh conscription, of further wholesale devastations, and of new rivers of human blood."

— Horace Greeley

He also told Greeley, in a message carried by John Hay, who was to accompany him on the mission, "I not only intend a sincere effort for peace, but I intend that you shall be a personal witness that it is made."

Being thus coerced, Greeley went with Hay to Niagara, where he discovered, amid the thunder and through the mist, what Lincoln had suspected from the start: that the "emissaries" not only had no authority to negotiate, either with him or with anyone else, but seemed to be in Canada for the purpose of influencing, by the rejection of their empty overtures, the upcoming elections in the North. He retreated hastily, though not in time to prevent a rash of Copperhead rumors that the President, through him, had scorned to entertain decent proposals for ending the bloodshed. Lincoln wanted to offset the effect of this by publishing his and Greeley's correspondence, omitting of course the editor's references to "our bleeding, bankrupt, almost dying country," as well as his gloomy prediction of a Democratic victory in November. Greeley said no; he would consent to no suppression; either print their exchange in full or not at all. Obliged thereby to let the matter drop, Lincoln explained to his cabinet that it was better

★

to withhold the letters, and abide the damaging propaganda, than "to subject the country to the consequences of their discouraging and injurious parts."

Simultaneously, in the opposite direction — down in Richmond itself — another peace feeler was in progress, put forth by Federal emissaries who had no more official sanction than their Confederate counterparts in Canada. Still, Lincoln had better hopes for this one, not so much because he believed that it would end the conflict, but rather, as he remarked, because he felt that it would "show the country I didn't fight shy of Greeley's Niagara business without a reason." What he wanted was for the northern public to become acquainted with Jefferson Davis's terms for an armistice, which he was sure would prove unacceptable to many voters who had been lured, in the absence of specifics, by the siren song of orators claiming that peace could be his for the asking, practically without rebel strings. Moreover, he got what he wanted, and he got it expressed in words as strong and specific as any he himself might have chosen for his purpose.

Colonel James F. Jaquess, a Methodist minister who had raised and led a regiment of Illinois volunteers, had become so increasingly shocked by the sight of fellow Christians killing each other wholesale — especially at Chickamauga, where he lost more than two hundred of his officers and men — that he obtained an extended leave of absence to see what he could do, on his own, to prepare the groundwork for negotiations. He had no success until he was joined in the effort by J. R. Gilmore, who enjoyed important Washington connections. A New York businessman, Gilmore had traveled widely in the South before the war, writing of his experiences under the pen name Edmund Kirke, and he managed to secure Lincoln's approval of an unofficial visit to Richmond by Jaquess and himself, under a flag of truce, for the purpose of talking with southern leaders about the possibility of arriving at terms that might lead to a formal armistice.

On Saturday, July 16, the two men were conducted past one of Ben Butler's outposts and were met between the lines by Judge Robert Ould, head of the Confederate commission for prisoner exchange. By nightfall they were lodged in the Spotswood Hotel, in the heart of the rebel capital, Jaquess wearing a long linen duster over his blue uniform. Next morning, amid the pealing of church bells, they conferred with Judah Benjamin, who promised to arrange a meeting for them that evening, here in his State Department office, with the President himself. They returned at the appointed time, and there — as Gilmore later described the encounter — at the table, alongside the plump and smiling Benjamin, "sat a spare, thin-featured man with iron-gray hair and beard, and a clear, gray eye full of life and vigor." Jefferson Davis rose and extended his hand. "I am glad to see you, gentlemen," he said. "You are very welcome to Richmond."

Although he neither mentioned the fact nor showed the strain it cost him, he had not been able to receive them earlier this Sunday because of the lengthy cabinet meeting that had resulted in the dismissal telegram Joe Johnston was reading now, on the outskirts of Atlanta. "His face was emaciated, and much wrinkled," Gilmore observed from across the table, "but his features were good, especially his eyes, though one of them bore a scar, apparently made by some sharp instrument. He wore a suit of grayish brown, evidently of foreign manufacture. . . . His manners were simple, easy and quite fascinating, and he threw an indescribable charm into his voice."

Jaquess opened the interview by saying that he had sought it in the hope that Davis, wanting peace as much as he did, might suggest some way to stop the fighting. "In a very simple way," the Mississippian replied. "Withdraw your armies from our territory, and peace will come of itself." When the colonel remarked that Lincoln's recent Proclamation of Amnesty perhaps afforded a basis for proceeding, Davis cut him short. "Amnesty, Sir, applies to criminals. We have committed no crime." Gilmore suggested that both sides lay down their arms, then let the issue be decided by a popular referendum. But Davis,

> *"I tried in all my power to avert this war. I saw it coming, and for twelve years I worked night and day to prevent it, but I could not.*
>
> — Jefferson Davis

thinking no doubt of the North's more than twenty millions and the South's less than ten, was having no part of that either. "That the *majority* shall decide it, you mean. We seceded to rid ourselves of the rule of the majority, and this would subject us to it again." It seemed to Gilmore that the dispute narrowed down to "Union or Disunion," and the Confederate President agreed, though he added that he preferred the terms "Independence or Subjugation." Despairing of semantics and the profitless exchange of opposite views that had brought on the war in the first place, the New Yorker made an appeal on personal grounds. "Can you, Mr Davis, as a Christian man, leave untried any means that may lead to peace?" Davis shook his head. "No, I cannot," he replied. "I desire peace as much as you do; I deplore bloodshed as much as you do." He spoke with fervor, but seemed to choose his words with care. "I tried in all my power to avert this war. I saw it coming, and for twelve years I worked night and day to prevent it, but I could not. And now it must go on till the last man of this generation falls in his tracks, and his children seize his musket and fight his battle, *unless you acknowledge our right to self-government . . .* We are fighting for Independence — and that, or extermination, we will have."

★

Additional matters were discussed or mentioned, including the military situation, which Davis saw as favorable to the South, and slavery, which he maintained was never "an essential element" in the contest, "only a means of bringing other conflicting elements to an earlier culmination." But always the talk came back to that one prerequisite. Whether it was called Self-Government or Disunion, all future discussion between the two parties would have to proceed from that beginning if there was to be any hope of ending the carnage they both deplored. The Confederate leader made this clear as he rose to see his visitors to the door, shook their hands, and spoke his final words. "Say to Mr Lincoln, from me, that I shall at any time be pleased to receive proposals for peace on the basis of our Independence. It will be useless to approach me with any other."

Whatever sadness he felt on hearing this evidence that the war was unlikely to end through negotiation, Lincoln perceived that the closing message, along with much that preceded it, would serve quite well to further his other purpose, which was to demonstrate his adversary's intransigence in the face of an earnest search for peace. He asked Gilmore, who had stopped by Washington on his return journey from Richmond, what he proposed to do with the transcript he had made of the interview. "Put a beginning and an end to it, Sir, on my way

Judah P. Benjamin (left) arranged a meeting of two northern peace seekers with Confederate President Jefferson Davis in Richmond.

home," the New Yorker said, "and hand it to the *Tribune*." Lincoln demurred. He had had enough of Horace Greeley for a while. "Can't you get it into the *Atlantic Monthly*? It would have less of a partisan look there." Gilmore was sure he could; but first, by way of counteracting what Lincoln called "Greeley's Niagara business," it was decided to release a shorter version in the Boston *Evening Transcript* the following week, while the full *Atlantic* text was being set in type and proofed for review by Lincoln. "Don't let it appear till I return the proof," he cautioned. "Some day all this will come out, but just now we must use discretion."

The *Transcript* piece appeared July 22, followed a month later by the one in the *Atlantic*, from which the President had deleted a few hundred words mainly having to do with terms he had found acceptable off the record. Both received much attention, especially the longer version. Indeed, so widely was it reprinted, at home and abroad, that another distinguished contributor — Oliver Wendell Holmes, whose son had lately cursed Lincoln off the parapet at Fort Stevens — soon told Gilmore that it had attracted more readers than any magazine article ever written.

Meantime (as always) Lincoln had kept busy with other problems, military as well as political. Often they overlapped, as in the case of facing up to the need for replacing the troops whose fall or discharge left gaps in the ranks of the two main armies: especially Meade's, which had a lower reënlistment quotient and had been further reduced, moreover, by detachments northward to shield Washington from attack by Early, still hovering nearby. On Sunday, July 17, while Jaquess and Gilmore talked in Richmond with Jefferson Davis — who had just put a message on the wire to Atlanta that presaged a step-up in the fighting there — Lincoln telegraphed Grant: "In your dispatch of yesterday to General Sherman I find the following, to wit: 'I shall make a desperate effort to get a position here which will hold the enemy without the necessity of so many men.' Pressed as we are by lapse of time, I am glad to hear you say this; and yet I do hope you may find a way that the effort shall not be desperate in the sense of a great loss of life." He sent this by way of preparation for a proclamation, issued next day, calling for 500,000 volunteers and ordering a draft to take place immediately after September 5 for any unfilled quotas.

This must surely be the last before November, he was saying, although there were already those who believed, despite the recent removal of the $300 exemption clause, that the results would not suffice even for the present. "We are not now receiving one half as many [troops] as we are discharging," Halleck complained to Grant the following day. "Volunteering has virtually ceased, and I do not anticipate much from the President's new call, which has the disadvantage of again postponing the draft for fifty days. Unless our government and people will come square up to the adoption of an efficient and thorough draft, we cannot supply the waste of our army."

★

Coming square up was easily said, but it left out factors that could not be ignored, including the reaction to this latest call for volunteers, which was seen as a velvet glove encasing the iron hand of a new draft. "Only half a million more! Oh that is nothing," one angry Wisconsin editor fumed, and followed through by saying: "Continue this Administration in power and we can all go to war, Canada, or to hell before 1868."

Now that the year moved into the dog days, with the fall elections looming just beyond, there was need for caution, if not in the military, then certainly in the political arena. Yet even caution might not serve, so portentous were the signs that a defeat was in the making. Frémont was something of a joke as an opponent, though not as a siphon for drawing off the Radical votes that would be needed if Lincoln was to prevail against the Democrats, who were scheduled to convene in Chicago in late August to adopt a platform and select a candidate for November. The platform would be strong for peace, and the

---

"Volunteering has virtually ceased, and I do not anticipate much from the President's new call . . ."

— Henry Halleck

---

candidate, it was believed, would be George McClellan: a formidable combination, one that might well snare both the anti-war and the soldier vote, not to mention the votes of the disaffected, likely to go to almost any rival of the present national leader. Indeed, the prospect so thoroughly alarmed a number of members of the Republican hierarchy that a secret call went out for a convention to meet in Cincinnati in September "to consider the state of the nation and to concentrate the Union strength on some one candidate who commands the confidence of the country, even by a new nomination if necessary."

For the present this was circulated privately, with the intention of bringing it out in the open when the time was ripe. In point of fact, however, the time seemed ripe enough already, to judge by the immediate response. Dissatisfaction with Lincoln had grown by now to include even close friends: Orville Browning, for example, who confessed he had long suspected that his fellow Illinoisan could not measure up to the task required. "I thought he might get through, as many a boy has got through college, without disgrace; but I fear he is a failure." Others agreeing were the eminent lawyer David Dudley Field, whose brother Lincoln had recently appointed to the Supreme Court, and Schuyler Colfax, Speaker of the House. Chase expressed interest in the supersession, of course, and Ben Butler lent encouragement from down on Bermuda

George McClellan, it was widely believed, would be the Democrats' candidate for the U.S. presidency and peace would be their platform in the election of 1864.

Hundred. Henry Davis was vehemently for it, but Wade and Sumner remained aloof for the time being, the former because he preferred to wait till after the Democratic convention, the latter because he thought it would make less trouble for the party if they gave Lincoln a chance to withdraw voluntarily. Many prominent editors favored the maneuver, including Parke Godwin of the New York *Evening Post* and Whitelaw Reid of the Cincinnati *Gazette*. But the most vociferous of them all was Horace Greeley, whose expression was cherubic but whose spirit had lately been strained beyond forbearance. "Mr Lincoln is already beaten," he declared. "He cannot be elected. And we must have another ticket to save us from overthrow."

Lincoln knew little or nothing yet of this plan by his friends and associates for a midstream swap, but he saw as clearly as they did that the drift was toward defeat and was likely to remain so unless some way could be found, between now and November, to turn the tide. A military victory would help, even one on a fairly modest scale — the more modest the better, in fact, so far as bloodshed was concerned — just so it encouraged the belief that things were looking up for one or another of the armies. But that was mainly up to Grant, locked in a stalemate below Richmond, and Sherman, apparently no better off in front of Atlanta. The other possibility was politics, Lincoln's field, and he was prepared to do all he could in that direction. His native Kentucky would be the first state to hold an election since his nomination; August 1 was the balloting date, and though only some county offices and an appellate judgeship were at stake, the contest was certain to be regarded as a bellwether for the rest, which were to follow in September. Consequently, he took off the gloves for this one. Declaring martial law, he suspended the writ of habeas corpus on July 5, continued the suspension through election day, and gave a free rein to Stephen Burbridge, who, having recently disposed of John Morgan at Cynthiana, proposed to move in a similar aggressive manner against all foes of the Administration throughout his Department of Kentucky. As a result, prominent Democrats were arrested wholesale for "disloyalty," and the name of their candidate for the judgeship was ordered stricken from the ballot on the same vague charge, obliging the survivors to make a last-minute substitute nomination for the post. Lincoln awaited the outcome with much interest, only to find on August 1 that all his pains had gone for nothing. The Democratic candidates swept the state.

There would be other contests; Maine, for instance, was coming up next, to be followed by Vermont. Although the snub just given him in his native state did not augur well for the result, he had no intention of doing anything less than his best to win in all of them, with the help of whatever devices he thought might help and despite the clamor of his critics, left and right, in his own party or the other. "The pilots on our western rivers steer from point to point, as they call it," he told a caller one of these days, "setting the course of the boat no farther than they can see. And that is all I propose to do in the great problems that are before us."

One such point now was Atlanta; or anyhow it seemed to him it might be. Events that followed hard on the rebel change of commanders there had brought the fighting to a pitch of intensity, throughout the last two weeks in July, that matched the savagery of the struggle here in the East before it subsided into stalemate. The same thing might happen there — for that seemed to be the pattern: alternate fury and exhaustion — but Lincoln kept peering in that direction, seeking a point to steer by in his effort to land the boat in his charge before it split and sank.

★   ★   ★

★

Shelby Foote

*Shaded from the sun by a plank
roof, a Federal sentry keeps
watch over a cleared field from
his seat on a captured Confederate
redoubt near Atlanta.*

Hood vs. Sherman; Mobile Bay; Memphis Raid; Atlanta Falls

1864 ★ ★ ★ ★ ★ "The appointment has but one meaning," the Richmond *Examiner* declared on July 19, in reference to Johnston's supersession down in Georgia the day before, "and that is to give battle to the foe." Because John Bell Hood, in contrast to his predecessor, was "young, dashing, and lucky," the rival *Whig* informed its readers that same day, "the army and the people all have confidence in his ability and inclination to fight, and will look to him to drive back Sherman and save Atlanta." Thus the two papers were in agreement on the matter, not only with each other, but also, for once, with the new western leader's red-haired adversary, who rarely subscribed to any journalist's opinion, North or South.

"I inferred that the change of commanders meant fight," Sherman remarked after conferring with subordinates who had known Hood in the days before the war. But he added, in contrast to the inference the two Confederate editors drew, five hundred miles away: "This was just what we wanted, viz., to fight in open ground, on anything like equal terms, instead of being forced to run up against prepared intrenchments."

He was about to get what he said he wanted. Hood — whose recent association with Johnston, he later explained, had made him "a still more ardent advocate of the Lee and Jackson school" — needed only one full day at his post

★

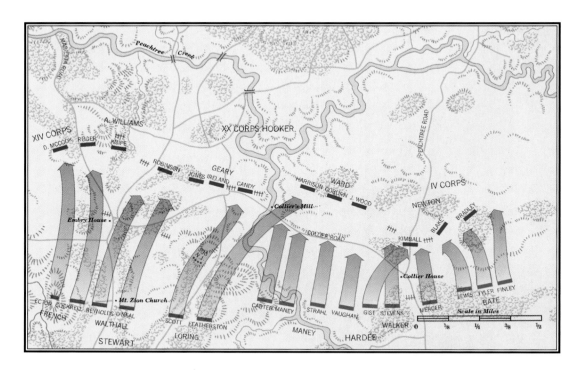

*By the time John Bell Hood's Confederates were ready
to attack on July 20, George Thomas had moved most
of his units to high ground along Peachtree Creek.*

before he resolved to go over to the offensive. By then, moreover, though he had
had to spend a good part of the time discovering where his own troops were, he
not only had decided to lash out at the encircling Federal host; he also had determined just when and where and how he would do so, with a minimal adjustment
of the lines now held by his three corps. Accordingly, on the evening of July 19,
he summoned Hardee and Stewart to headquarters along with Ben Cheatham,
his temporary successor as corps commander, and gave them face-to-face instructions for an attack to be launched soon after midday tomorrow in order to take
advantage of an opportunity Sherman was affording them, apparently out of
overweening contempt or unconcern, to accomplish his piecemeal destruction.

In the execution of what he termed "a general right wheel" from the
near bank of the Chattahoochee, with Thomas inching the pivot forward across
Peachtree Creek to close down on Atlanta from the north, and McPherson and
Schofield swinging wide to come in from the east along the Georgia Railroad, the
Ohioan had in effect divided his army and developed a better than two-mile gap
between the inner edges of its widespread wings. It was Hood's intention,

expressed in detail at his first council of war tonight on the outskirts of the city in his charge, not to plunge into but rather to preserve this gap, and thus keep the two blue wings divided while he crushed them in furious sequence, left and right.

Cheatham, with the help of Wheeler's troopers and some 5000 Georgia militia, would confront McPherson and Schofield from his present intrenchments east of Atlanta, taking care to mass artillery on his left and thus prevent the blue-coats in front from crossing the gap between them and Thomas, who meantime would be receiving the full attention of the other two corps. The Union-loyal Virginian's infantry strength was just above 50,000 — about the number Hood had in all — but the intention was to catch him half over Peachtree Creek, which he had begun to bridge today, and hit him before he could intrench or bring up reinforcements. Hardee on the right and Stewart on the left, disposed along a jump-off line roughly four miles north of the city, were to attack in echelon, east to west, each holding a division in reserve for immediate exploitation of any advantage that developed, "the effort to be to drive the enemy back to the creek, and then toward the river, into the narrow space formed by the river and creek." Once Thomas had been tamped into that watery pocket and ground up, the two gray corps would shift rapidly eastward to assist Cheatham in mangling Schofield and McPherson, with Wheeler's free-swinging horsemen standing by to carry out the roundup that would follow. Hood explained all this to his chief lieutenants "by direct interrogatory," having long since learned "that no measure is more important, upon the eve of battle, than to make certain in the presence of commanders that each thoroughly comprehends his orders."

His concern in this regard was not unfounded. Remembering, as he must have done, the Army of Tennessee's latest — and indeed, under Johnston, only — contemplated full-scale offensive at Cassville two months ago today, midway down the doleful road from Tunnel Hill to Atlanta, Hood knew only too well the dangers that lurked in tactical iotas. Nothing had come of the Cassville design, largely because of his own reaction to finding a misplaced blue column approaching his flank, and presently on July 20, with all his troops in position and the 1 o'clock jump-off hour at hand, there were signs that a repetition was in the making. Cheatham sent word before noon that he would have to shift his line southward to keep McPherson from overlapping his right, beyond the railroad. Hood could only approve, and issue simultaneous instructions for Hardee and Stewart to conform by sidestepping half a division-front to their right, thus to prevent too wide an interval from developing between them and Cheatham, through which Schofield might plunge when he came up alongside McPherson. Hardee then had a difficult choice to make. Sidestepping as ordered, he found the interval wider than Hood had supposed, which left him with the decision whether to continue the sidling movement, at the cost of delaying his jump-off, or go forward on schedule — it was 1 o'clock by now — with a mile-wide gap yawning empty on his right. He chose the former course, Stewart conforming on his left, and thus delayed the attack for better than two hours. Shortly after 3 o'clock he sent three of his four divisions plunging northward into the valley of Peachtree Creek.

George Thomas was there, in strength and largely braced. Though the attack achieved the desired surprise, those extra two hours had given him time, not only to get nearly all of his combat elements over the creek, but also to get started on the construction of intrenchments. Hardee struck them and rebounded as if from contact with a red-hot stove, followed by Stewart, who drove harder against the enemy right with no better luck. The Federals either stood firm or hurried reinforcements to shore up threatened portions of their line. Moreover, in the unexpected emergency, Thomas abandoned his accustomed role of Old Slow Trot. Urging his guns forward to "relieve the hitch," he used the point of his sword on the rumps of laggard battery horses, then crossed the stream to direct in person the close-up defense of the bridgehead. An Indiana officer judged the progress of the fighting by the way Old Tom fiddled with his short, thick, gray-shot whiskers. "When satisfied he smoothes them down; when troubled he works them all out of shape." They were badly tousled now, and presently, when he saw the attackers falling back from the blast of fire that met them, he moved even further out of character in the opposite direction. "Hurrah!" he shouted, and took off his hat and slammed it on the ground in pure exuberance. "His whiskers were soon in good shape again," the Hoosier captain noted.

They might have been worse ruffled shortly thereafter; Hardee was about to throw Cleburne's reserve division into the melee, and in fact had just summoned him forward, when an urgent dispatch from Hood directed that troops be sent at once to the far right, where Cheatham's flank was under heavy pressure from McPherson. Cleburne arrived after nightfall, in time to confront a piece of high, cleared ground known as Bald Hill, two miles east of Atlanta and a mile south of the Georgia Railroad; Wheeler's dismounted troopers, after being pushed back all morning, had managed to hang on there through most of the afternoon. Northward, the battle raged along Peachtree Creek, but with decreasing fury, until about 6 o'clock, when it sputtered out. At a cost of 2500 casualties suffered, and 1600 inflicted, Hood's plan for crushing first Thomas, then the other two Union armies, had failed because the Rock of Chickamauga declined, as usual, to be budged or flustered. The southern commander had only praise for Cheatham and Wheeler, who fought hard all day against long odds, and especially for Stewart, who, though his losses were close to two thirds of the Confederate total, "carried out his instructions to the letter." He put the blame for his lack of success on Hardee — his former senior, known since Shiloh as Old Reliable — whose corps, "although composed of the best troops in the army, virtually accomplished nothing"

Broken fences mark the graves of the dead at Peachtree Creek. Combined Union and Confederate casualties at the July 20 battle were 4100.

and in fact, as a comparison of casualties would show, "did nothing more than skirmish with the enemy."

So Hood would report afterwards, when he got around to distributing blame for the failure of his first offensive action; the Battle of Peachtree Creek, it was called, or "Hood's First Sortie." But that did not keep him from choosing Hardee to deliver the main effort, two days later, in what would be referred to as "Hood's Second Sortie" or the Battle of Atlanta.

While Cleburne struggled the following day to prevent a blue advance past Bald Hill — the fighting on this third anniversary of First Manassas, he said, was "the bitterest" of his life — Wheeler moved still farther to the right, another mile beyond the railroad, to forestall another Federal flanking effort. What he found instead was an invitation for just such a movement by the Confederate defenders. McPherson, apparently with his full attention drawn to the day-long contest with Cleburne, had his left flank "in the air," unprotected by cavalry and wide open to assault. Informed of the situation early that morning, Hood grasped eagerly at this chance to turn the tables on the attackers. It was one of the chief regrets of his career that he had missed Chancellorsville, having been on detached service with Longstreet around Suffolk while the Lee-Jackson masterpiece was being forged in the smoky, vine-choked Wilderness a hundred miles away. Now here was a God-given once-in-a-lifetime opportunity to stage a Chancellorsville of his own, down in the piny woods of Georgia, within a scant five days of his appointment to command the hard-luck Army of Tennessee.

In preparation for exploiting this advantage — and also because both ends of his present line were gravely threatened, Thomas having begun to build up pressure against the left about as heavy as McPherson had been exerting on the right — Hood directed that all three corps begin a withdrawal at nightfall to the works rimming the city in their rear, already laid out by Johnston the month before. These were to be held by Stewart and Cheatham, on the north and east, while Hardee marched south, then southeast, six miles down the McDonough Road to Cobb's Mill, where he would turn northeast and continue for the same distance up the Fayetteville Road to the Widow Parker's farm, south of the railroad about midway between Atlanta and Decatur. This would put his four divisions (including Cleburne's, which would join him on his way through town) in position for an all-out assault on McPherson's left rear. Though the route was as cir- cuitous and long as Stonewall's flanking march had been, fourteen months ago in Virginia, an early start this evening should enable Old Reliable to launch a dawn attack, and a dawn attack would give him a full day in which to accomplish McPherson's destruction, whereas Jackson had had only the few hours between sunset and dusk to serve Hooker in that fashion. Moreover, by way of increasing the blue confusion and distress, Wheeler's troopers, after serving as guides and outriders for the infantry column, would continue eastward to Decatur for a strike at McPherson's wagon train, known to be parked in the town square with all his reserve supplies and munitions. Hood explained further that once the flank attack got rolling he would send Cheatham forward to assail McPherson's front and keep Schofield from sending reinforcements to the hard-pressed Union left, while Stewart, around to the north, engaged Thomas for the same purpose.

Now, as before the Peachtree venture, he assembled a council of war to make certain that each of his lieutenants understood exactly what was required of him, and why. This was all the more advisable here, because of the greater complexity of what he was asking them to do. "To transfer after dark our entire line from the immediate presence of the enemy to another line around Atlanta, and to throw Hardee, the same night, entirely to the rear and flank of McPher- son — as Jackson was thrown, in a similar movement, at Chancellorsville and Second Manassas — and to initiate the offensive at daylight, required no small effort on the part of the men and officers. I hoped, however, that the assault would result not only in a general battle, but in a signal victory to our arms."

Such hope was furthered by the secrecy and speed of the nighttime withdrawal to Atlanta's "inner line," which Stewart and Cheatham then began improving with picks and shovels while Hardee set out on his march around the

Federal south flank. Almost at once the first hitch developed. Two miles to the east, confronting the enemy on Bald Hill, Cleburne had trouble breaking contact without giving away the movement or inviting an attack; it was crowding midnight before Hardee solved the problem by instructing him to leave his skirmishers in position and fall in behind W. H. T. Walker's men, marking time in rear of the other two divisions under Bate and George Maney, Cheatham's senior brigadier. Cleburne managed this by 1 a.m. of the projected day of battle — Friday, July 22 — but it was 3 o'clock in the morning before the final elements of the corps filed out of the unoccupied intrenchments south of town.

That was the first delay. Another was caused by the weariness of the marchers, still unrested from Wednesday's bloody work and Thursday's fitful skirmishing under the burning summer sun. Strung out on the single, narrow road, which had to be cleared from time to time when Wheeler's dusty horsemen clattered up or down it, the head of the column did not reach Cobb's Mill

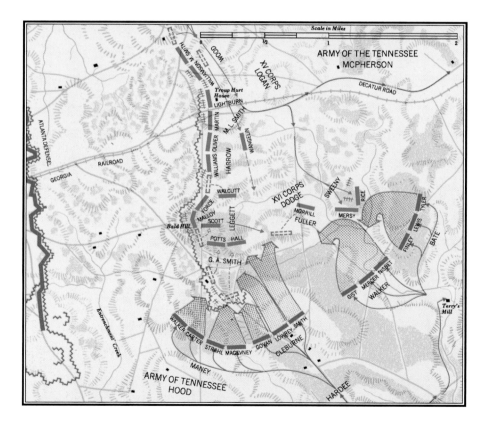

On July 22, Confederates attacked Federal troops closing in on Atlanta from the east. By nightfall, the opposing forces were back where they started.

until dawn, the supposed jump-off hour. Disgruntled, Hardee turned northeast for the Widow Parker's, another half dozen miles up the troop-choked road. It was close to noon by the time he got there, evidently unsuspected by the enemy in the woods across the way, and 12.30 before the corps was formed for assault, Maney and Cleburne on the left, astride the Flat Shoals Road, which ran northwest past Bald Hill, where McPherson's flank was anchored — Cleburne thus had nearly come full circle — and Walker and Bate on the right, on opposite sides of Sugar Creek, which also led northwest, directly into McPherson's rear. Old Reliable could take pride in being just where he was meant to be, in position to duplicate Jackson's famous end-on strike at Hooker, but he was also uncomfortably aware that he was more than six hours behind schedule.

This made him testy: as anyone near him could see in these final minutes before he gave the order to go forward. When Wheeler sent word that a sizeable column of blue troopers had passed this way a while ago, apparently headed southward on a raid, and requested permission to take out after them, Hardee was quick to say no; "We must attack, as we arranged, with all our force." So Wheeler, disappointed at being denied the chance to cross sabers with the intruders, set out eastward for Decatur and McPherson's unsuspecting and perhaps unguarded wagon train. Then Walker came to headquarters to report that he had discovered in his immediate front a giant brier patch, which he asked to be allowed to skirt when he advanced, despite the probable derangement of his line and the loss of still more time. Normally courteous, Hardee was emphatic in refusal. "No, sir!" he said roughly, not bothering to disguise his anger. "This movement has been delayed too long already. Go and obey my orders!"

Walker, a year younger at forty-seven than his chief, who had finished a year behind him at West Point — a veteran of the Seminole and Mexican wars, heavily bearded, with stern eyes, he was one of three West Pointers among the eight Confederate generals named Walker — then demonstrated a difficulty commanders risked with high-strung subordinates in this war, particularly on the southern side. He took offense at his fellow Georgian's tone, and he said as much to an aide who rode with him on the way back to his division. "Major, did you hear that?" he asked, fuming. The staffer admitted he had; "General Hardee forgot himself," he suggested. Walker was not to be put off, however. "I shall make him remember this insult. If I survive this battle, he shall answer me for it."

Just then an officer from Hardee's staff overtook them with the corps commander's regrets for "his hasty and discourteous language" and assurance that he would have "come in person to apologize, but that his presence was required elsewhere, and would do so at the first opportunity." So the envoy informed Walker, whose companion remarked soothingly, after they had ridden on: "Now that makes it all right." But Walker's blood was up. He was by no means satisfied. "No, it does not," he said hotly. "He must answer me for this."

As it turned out, no one on this earth was going to answer to W. H. T. Walker for anything. Ordered forward shortly thereafter, he and his three brigades clawed their way through the brier patch, hearing Maney's and Cleburne's attack explode on the left as it struck McPherson's flank, and then emerged from a stand of pines into what was to have been the Union rear, only to find a nearly mile-long triple line of bluecoats confronting them on ground that had been empty when it was reconnoitered, half an hour before. Walker had little chance to react to this discovery, however, for as he and his men emerged from the trees, sunlight glinting on his drawn saber and their rifles, a Federal picket took careful aim and shot him off his horse.

Hood, who had waited and watched impatiently for the past six hours in a high-sited observation post on the outskirts of Atlanta, was dismayed by what he saw no more than a mile away across the treetops. Plunging northwest, on the far left of the Confederate assault, Maney overlapped the Union flank and had to swing hard right as he went past it, which threw his division head-on against the enemy intrenchments facing west. This caused Hood to assume — and later charge — that Hardee's attack had been launched, not into the rear of the blue left flank, as directed, but against its front, with predictable results; Maney rebounded, then lunged forward again, and again rebounded. Beyond him, out of sight from Hood's lookout tower, Cleburne was doing better, having struck the Federals endwise, and was driving them headlong up the Flat Shoals Road, which ran just in rear of their works below Bald Hill. Still farther to the east, however, Bate and Walker's successor, Brigadier General Hugh Mercer, were having the hardest time of all. In this direction, the element of surprise was with the defenders, whose presence was as unexpected, here on the right, as the appearance of the attackers had been at the opposite end of the line.

Advancing westward yesterday and this morning, under instructions "not to extend any farther to the left" beyond the railroad, lest his troops be spread too thin, McPherson's front had contracted so much that he could detach one of his three corps, led by Major General Grenville M. Dodge, to carry out an order from Sherman to "destroy every rail and tie of the railroad, from Decatur up to your skirmish line." Dodge completed this assignment before midday and was moving up to take a position in support of Blair, whose corps was on the left, when he learned that a heavy force of graybacks was approaching from the southeast, up both banks of Sugar Creek. Under the circumstances, all he had to do was halt and face his two divisions to the left, still in march formation on an east-west road, to establish the triple line of defense whose existence Walker and Bate had not suspected until they emerged from the screen of pines and found it bristling in their front. If they had come up half an hour earlier they would have stepped into a military vacuum, with little or nothing between them and the rear of Blair and Logan, whose corps was on Blair's right, connecting McPherson

General Grenville M. Dodge led his Federal corps to carry out an order to "destroy every rail and tie of railroad from Decatur" to his skirmish line.

and Schofield. Now, instead, Walker was dead and Bate and Mercer were involved in a desperate fight that stopped them in their tracks, much as Maney had been stopped on the left, under different circumstances. Thus, of the four gray divisions involved in the attack from which so much had been expected, only Cleburne's was performing as intended. Yet he and his fellow Arkansans made the most of their advantage, including the killing of the commander of the Army of the Tennessee.

McPherson was not with his troops when Hardee's attack exploded on his flank. He was up in rear of Schofield's left, just over half a mile north of the railroad, conferring with Sherman in the yard of a two-story frame house that had been taken over for general headquarters, about midway of the line confronting Atlanta from the east. What he wanted was permission to open fire with a battery of long-range 32-pounders on a foundry whose tall smokestack he could see beyond the rebel works from a gun position he had selected and already had under construction on Bald Hill — or Leggett's Hill, as it was called on the Federal side, for Brigadier General Mortimer Leggett, whose division of Blair's corps occupied it. McPherson's notion was that if he could "knock down that foundry," along with other buildings inside Atlanta, he would hasten the fall of the city.

Moreover, he had personal reasons for wanting to accomplish this in the shortest possible time, since what he was counting on, in the way of reward, was a leave of absence that would permit him to go to Baltimore and marry a young lady to whom he had been engaged since his last leave, just after the fall of Vicksburg. He had tried his best to get away in March and April, but Sherman had been unwilling, protesting that there was too much to be done before the drive through Georgia opened in early May. So the thirty-five-year-old Ohioan

★

had had to bide his time; though only by the hardest. Just last week he had asked his friend Schofield when he supposed his prayers would be answered. "After the capture of Atlanta, I guess," Schofield replied, and McPherson had taken that as his preliminary objective, immediately preceding the real objective, which was Baltimore and a union that had little to do with the one he and more than a hundred thousand others would die fighting to preserve.

Sherman readily assented to the shelling of the city, and ordered it to begin as soon as the guns were in position. His first impression, on finding the rebel trenches empty in his front this morning, had been that Hood had evacuated Atlanta overnight; but that had lasted only until he relocated the enemy in occupation of the city's inner line, as bristly as ever, if not more so, and now he took the occasion of McPherson's midday visit to show him, on the headquarters map, his plan for shifting all three armies around to the west for the purpose of cutting Hood's remaining rail connections with Macon and Mobile, which would surely bring on the fall of Atlanta if the proposed bombardment failed. It was by then around 12.30, and as they talked, bent over the map, the sound of conflict suddenly swelled to a roar: particularly southward, where things had been quiet all morning. Sherman whipped out his pocket compass, trained it by earshot, and "became satisfied that the firing was too far to our left rear to be explained by known facts." McPherson quickly called for his horse and rode off to investigate, trailed by members of his staff.

Sherman stood and watched him go, curly bearded, six feet tall, with lights of laughter often twinkling in his eyes; "a very handsome man in every way," according to his chief, who thought of his fellow Ohioan as something more than a protégé or younger brother. He thought of him in fact as a successor — and not only to himself, as he would tell another friend that night. "I expected something to happen to Grant and me; either the rebels or the newspapers would kill us both, and I looked to McPherson as the man to follow us and finish the war."

From a ridge in rear of the road on which Dodge had been marching until he stopped and faced his two divisions left to meet the assault by Bate and Walker, McPherson could see that the situation here was less desperate than he had feared; Dodge was plainly holding his own, although the boom of guns from the east gave warning that a brigade he had posted at Decatur to guard the train in the cavalry's absence was also under attack. Sending the available members of his staff in both directions, with instructions for all units to stand firm at whatever cost, the army commander turned his attention westward to Blair's position, where the threat seemed gravest.

In point of fact it was graver than he knew. Cleburne by now had driven Blair's flank division back on Leggett, whose troops were fighting to hold the hill that bore his name, and numbers of enemy skirmishers had already worked

In this painting by James Taylor, Grenville Dodge aligns a division to block the movements of Confederates attempting to flank the main Federal line.

their way around in its rear to seize the wooded ground between there and Dodge's position. That was how it happened that McPherson, who had sent away all of his staff except an orderly, encountered graybacks while trotting along a road that led across to Leggett's Hill. Indeed, he was practically on top of one group of Confederates before he suspected they were there. An Arkansas captain, raising his sword as a signal for the two riders to surrender, was surprised by the young general's response ("He checked his horse slightly, raised his hat as politely as if he were saluting a lady, wheeled his horse's head directly to the right, and dashed off to the rear in a full gallop") but not for long. "Shoot him," the gray-clad officer told a corporal standing by, and the corporal did.

McPherson was bent over his mount's withers to keep from being swept from the saddle by the drooping limbs of trees along the road. He fell heavily to the ground, struck low in the back by a bullet that ranged upward through or near his heart. His companion, unhorsed and momentarily stunned by a low-hanging branch, recovered consciousness to find the general lying

beside him, clutching his breast in pain, and the butternut soldiers hurrying toward them. He bent over him and asked if he was hurt. "Oh, orderly, I am," McPherson said, and with that he put his face in the dust of the road, quivered briefly, and died. The orderly felt himself being snatched back and up by his revolver belt; "Git to the rear, you Yankee son of a bitch," he heard the rebel who had grabbed him say. Then the captain got there and stood looking down at the polished boots and buff gauntlets, the ornate sash about the waist, and the stars of a major general on both dead shoulders. "Who is this lying here?" he asked. The orderly had trouble answering. Sudden grief had constricted his throat and tears stood in his eyes. "Sir, it is General McPherson," he said. "You have killed the best man in our army."

Sherman's grief was as great, and a good deal more effusive. "I yield to no one but yourself the right to exceed me in lamentations for our dead hero," he presently wrote the Baltimore fiancée. "Though the cannon booms now, and the angry rattle of musketry tells me that I also will likely pay the same penalty, yet while life lasts I will delight in the memory of that bright particular star which has gone before to prepare the way for us more hardened sinners who must struggle to the end."

But that was later, when he could spare the time. Just now he responded to the news that McPherson's horse had come riderless out of the woods in back of Leggett's Hill by ordering John Logan, the senior corps commander, to take charge of the army and counterattack at once to recover the ground on which his chief might be lying wounded. Logan did so, and within the hour McPherson's body was brought to headquarters in an ambulance. Someone wrenched a door off its hinges and propped it on two chairs for a catafalque, and Sherman went on directing the battle from the room where his fellow Ohioan was laid out. Already he had sent a brigade from Schofield to support the one Dodge had defending Decatur from Wheeler's attack, but aside from this he sent no reinforcements to help resist the assault on his left flank and rear.

"I purposely allowed the Army of the Tennessee to fight this battle almost unaided," he later explained, partly because he wanted to leave to McPherson's veterans the honor of avenging his fall, and also because he believed that "if any assistance were rendered by either of the other armies, the Army of the Tennessee would be jealous."

Commander of the Army of the Tennessee, General James E. McPherson falls from his horse after being hit by a bullet fired by Confederate skirmishers.

His confidence in his old army — it had also once been Grant's, and had yet to come out loser when the smoke of battle cleared — was justified largely today because of Logan, who exercised his new command in style. Dubbed "Black Jack" by his soldiers, the former Illinois politician knew how to translate stump oratory into rousing military terms. Clutching his flop-brim hat in one hand so that his long raven hair streamed behind him in the wind, he spurred from point to embattled point and bellowed: "Will you hold this line with me? Will you hold this line?" The veterans showed they would. "Black Jack! Black Jack!" they chanted as they beat off attacks that soon were coming from all directions: particularly on Leggett's Hill, which Hood by now had ordered Cheatham to assault from the west while Cleburne kept up pressure from the south and east. Brigadier General Manning Force's brigade, menaced front and rear, was obliged at times to fight on alternate sides of its breastworks. At one critical point he called for a flag, and a young lieutenant, assuming from the look of things that the time had come to surrender, began a frantic search for a white handkerchief or shirt. "Damn you, sir!" Force shouted. "I don't want a flag of truce; I want the American flag!" Shot in the face shortly thereafter, he lost the use of his voice and fell back on conducting the hilltop defense with gestures, which were no less flamboyant and seemed to work as well.

The hill was held, though at a cost of ten guns — including the four McPherson had planned to use against Atlanta at long range — fifteen stands of colors, and better than a thousand prisoners, mostly from Blair's other division under Brigadier General Giles A. Smith (one of an even dozen Federal generals with that name, including one who spelled it Smyth) which had given way at the outset, badly rattled by Cleburne's unexpected flank-and-rear assault.

Although there were no other outright surprises, the issue continued to swing in doubt from time to time and place to place. Sherman watched with interest from his headquarters on the central ridge, and when Cheatham scored a breakthrough around 4 o'clock, just north of the railroad, he had Schofield mass the fire of several batteries to help restore Logan's punctured right. Word came then from Decatur that the two brigades of infantry had managed to keep Wheeler's troopers out of the town square, where the train was parked, and from Dodge that he was confident of holding against weakening attacks on the left rear. Mercurial as always, despite the tears that trickled into his stub red beard whenever he thought of McPherson laid out on his improvised bier inside the house, Sherman was in high spirits as a result of these reports, which reached him as he paced about the yard and watched the progress of the fighting in all directions.

Presently the headquarters came under long-range fire, obliging him and his attendants to take cover in an adjoining grove of trees. Sheltered behind one of these, he noticed a terrified soldier crouched nearby in back of another, moaning: "Lord, Lord, if I once get home," and: "Oh, I'll be killed!" Sherman grinned and picked up a handful of stones, which he then began to toss in that direction. Every pebble that struck the tree brought a howl or a groan from behind it. "That's hard firing, my man," he called to the unstrung soldier, who replied without opening his tight-shut eyes: "Hard? It's fearful! I think thirty shells have hit this tree while I was here." The fire subsided, and the general stepped into the open. "It's all over now; come out," he told the man, who emerged trembling. When he saw who had been taunting him, he took off running through the woods, pursued by the sound of Sherman's laughter.

From end to end, the Federal line was held or restored, except where Smith's unfortunates had been driven back across the lower slopes of Leggett's Hill, and though the fighting was sometimes hand-to-hand and desperate, on past sundown into twilight, there was by then no doubt that Hood's Second Sortie — aside, that is, from the capture of a dozen guns and an assortment of Union colors — had been no less a failure than his First, two days ago. It was, however, considerably more expensive; for this time the Confederate leader held almost nothing back, including the Georgia militia, which he used in a fruitless attack on Schofield that had no effect on the battle except to swell the list of southern casualties. In the end, Hood's loss was around 8000 killed, wounded, and missing, as compared to Sherman's 3700.

All next day the contending armies remained in position, licking their wounds, until Hardee withdrew unimpeded the following night into the Atlanta works. Saddened by the loss of Walker, who had called at headquarters on the eve of battle to assure him of his understanding and support, as well as by the news about McPherson — "No soldier fell in the enemy's ranks whose death caused me equal regret," he later said of his West Point friend and classmate — Hood was profoundly disappointed by the failure of his two sorties to accomplish the end for which they had been designed; but he was by no means so discouraged that he did not intend to attempt a third, if his adversary presented him with still another opportunity. He knew only too well how close he had come, except for the unlucky appearance of Dodge's corps in exactly the wrong place at the wrong time, to wrecking the encircling Union host entirely.

Frank Blair, for one, concurred in this belief. Hood's flanking movement, he afterwards declared, "was a very bold and a very brilliant one, and was very near being successful. The position taken up accidentally by [Dodge's] corps prevented the full force of the blow from falling where it was intended to fall. If my command had been driven from its position at the time that [Logan's] corps was forced back from its intrenchments, there must have been a general rout of all the troops of the Army of the Tennessee . . . and, possibly, the panic might have been communicated to the balance of the army."

Sherman was not much given to speculation on the might-have-beens

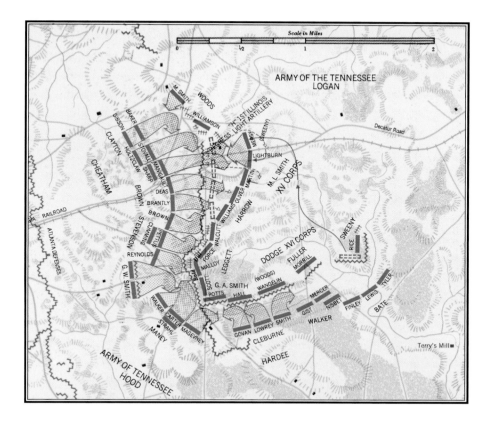

*During the afternoon of July 22, the Confederates
hurled themselves at the Union center pressing toward
Atlanta, but the Federal line stalled the onslaught.*

of combat, and in any case he no more agreed with this assessment than he did
with subsequent criticism that, in leaving Schofield and Thomas standing comparatively idle on the sidelines while Logan battled for survival, he had missed a
prime chance to break Atlanta's inner line, weakened as it was by the withdrawal
of a major portion of its defenders for the attack on his south flank. What he
mainly concluded, once the smoke had cleared, was that in staging two all-out
sorties in as many days — both of them not only unsuccessful but also highly
expensive in energy, blood, and ingenuity — Hood had shot his wad. And from
this Sherman concluded further that he was unlikely to be molested in his execution of the maneuver he had described to McPherson at their final interview;
that is, "to withdraw from the left flank and add to the right," thereby shifting
his whole force counterclockwise, around to the west of the city, in order to
probe for its rail supply lines to the south.

First, though, there was the problem of finding a permanent replacement for his fallen star, McPherson. On the face of it, Logan having performed spectacularly under worse than trying conditions, the solution should have been simple. But it turned out to be extremely complicated, involving the exacerbation of some tender feelings and, in the end, nothing less than the reorganization of the command structure of two of the three armies in his charge.

Thomas came promptly to headquarters to advise against keeping Logan at his temporary post. Although there was bad blood between them, dating back to Chattanooga, basically his objection was that Black Jack, like all the other corps and division leaders in the Army of the Tennessee — not one of them was a West Pointer, whereas two thirds of his own and half of Schofield's were Academy graduates — was a nonprofessional. "He is brave enough and a good officer," the Virginian admitted, "but if he had an army I am afraid he would edge over on both sides and annoy Schofield and me. Even as a corps commander he is given to edging out beyond his jurisdiction." Sherman agreed in principle that volunteers from civilian life, especially politicians, "looked to personal fame and glory as auxiliary and secondary to their political ambition. . . . I wanted to succeed in taking Atlanta," he later explained, "and needed commanders who were purely and technically soldiers, men who would obey orders and execute them promptly and on time." That ruled out Logan, along with Blair. Who then? he asked Thomas, who replied: "You cannot do better than put Howard in command of that army." Sherman protested that this would make Logan "terribly mad" and might also create "a rumpus among those volunteers," but then agreed. One-armed and two years younger even than McPherson, O. O. Howard, West Point '54, a Maine-born recent eastern import to the western theater, was then announced as the new commander of the army that had once been Sherman's own.

Returned to his corps, Logan managed to live with the burning aroused in his breast by this disappointment. But the same could not be said for Old Tom's ranking corps commander, the altogether professional Joe Hooker. Outraged at having been passed over in favor of the man he largely blamed for his defeat at Chancellorsville, Fighting Joe characterized the action as "an insult to my rank and services" and submitted at once a request to be relieved of his present duties. Thomas "approved and heartily recommended" acceptance of this application, which Sherman was quick to grant, remarking incidentally that the former commander of the Army of the Potomac had not even been considered for the post that now was Howard's, since "we on the spot did not rate his fighting qualities as high as he did." Hooker departed for an inactive assignment in the Northern Department, where he spent the rest of the war, further embittered by the news that his successor was Major General Henry W. Slocum, another enemy, who had been sent to Vicksburg on the eve of the present campaign to avoid personality clashes between them. Pending Slocum's arrival from Missis-sippi, Alpheus Williams would lead the corps as senior division commander, much as Major General David S. Stanley had succeeded to the command of Howard's corps, though on a permanent basis.

★ ★ ★ **B**y July 25, within five days of the Peachtree crossing, when work on it began, the railroad bridge over the Chattahoochee — 760 feet long and 90 high — was completed and track relaid to a forward base immediately in Thomas's rear. Sherman, his supplies replenished and generals reshuffled, was ready within another two days to begin the counterclockwise western slide designed to bring on the fall of Atlanta by severing its rail connection with the world outside. Already this had been accomplished up to the final step; for of the four lines in and out of the city all but one had been seized or wrecked by now, beginning with the Western & Atlantic, down which the Federals had been moving ever since they chevied Johnston out of Dalton. Then Schofield and McPherson had put the Georgia Railroad out of commission by dismantling it as they moved westward from Stone Mountain and Decatur. Of the remaining two — the Atlanta &

*U*nion soldiers examine a demolished shelter and an
abatis thrown up east of Atlanta by Confederates defending
the Georgia Railroad (shown at rear).

West Point and the Macon & Western, which shared the same track until they branched southwest and southeast at East Point, five miles south of the city — the former, connecting with Montgomery and Mobile, had been severely damaged the week before by Major General Lovell Rousseau, who raided southward through Alabama with 2500 troopers, practically unopposed, and tore up close to thirty miles of the line between Montgomery and Opelika, where it branched northeast for West Point and Atlanta. That left only the Macon road, connecting eastward with Savannah, for Hood's use in supplying his army and for Sherman to destroy. He began his large-scale semicircular maneuver to accomplish this on July 27, ordering Howard to swing north, then west — in rear of Schofield and Thomas, who would follow him in turn — for a southward march down the near bank of the Chattahoochee, which would serve as an artery for supplies, to descend as soon as possible on that one railroad still in operation out of a place that once had boasted of being "the turntable of the Confederacy."

Simultaneously, by way of putting two strings to his bow, he turned 10,000 horsemen loose on the same objective in an all-out double strike around both rebel flanks. Brigadier General Edward McCook, his division reinforced to a strength of 3500 by the addition of a brigade from Rousseau — who, it was hoped, had established the model for the current operation, over in Alabama the week before — would ride down the north bank of the Chattahoochee for a crossing at Campbelltown, under orders to proceed eastward and hit the Macon & Western at or below Jonesboro, just under twenty miles on the far side of Atlanta. This was also the goal of the second mounted column, 6500 strong, which would set out from Decatur under Stoneman, who had Garrard's division attached to his own for a southward lunge around the enemy right. Both columns were to start on July 27, the day the infantry slide began; Sherman expected them back within three days at the most. But when Stoneman asked permission to press on, once the railroad had been wrecked, to Macon and Andersonville for the purpose of freeing the prisoners held in their thousands at both places, he readily agreed to this hundred-mile extension of the raid, on condition that Garrard head back as soon as the Macon road was smashed, to work with McCook in covering the infantry's left wheel around Atlanta. The redhead's hopes were high, but not for long: mainly because of Joe Wheeler, who, though outnumbered three-to-two by the blue troopers, did not neglect this opportunity to deal with them in detail.

Right and left, at Campbelltown and Decatur, both of them closer to Jonesboro than they were to each other at the outset, the two columns took off on schedule, though not altogether in the manner Sherman intended. Stoneman's mind was fixed so firmly on his ultimate goal — Andersonville and its 30,000 inmates, whose liberation would be nothing less than the top cavalry exploit of the war — that he no longer had any discernible interest in the limited purpose for which the two-pronged strike had been conceived. Accordingly, without notifying anyone above him, he sent Garrard's 4300 troopers pounding due south to draw off the enemy horsemen while he and his 2200 rode east for Covington, which Garrard had raided five days ago during the Battle of Atlanta. In this he was successful; he reached Covington undetected and turned south, down the east bank of the Ocmulgee River, for Macon, the first of his two prison-camp objectives. Garrard meantime had been no less successful in carrying out his part of the revised design, which was to attract the attention of the rebels in his direction. On Snapfinger Creek that afternoon, barely ten miles out of Decatur, he ran into mounted graybacks whose number increased so rapidly overnight that at Flatrock Bridge next morning, another five miles down the road, he had to turn and ride hard, back to Decatur, to keep from losing every-thing he had. His nimbleness kept down his losses; yet even so these would have been much heavier if Wheeler, about to give chase with eight brigades — just over 6000 sabers in all — had not received word that McCook had crossed the Chattahoochee, en route for the Macon & Western, and that Stoneman was beyond the Ocmulgee, apparently headed for Macon itself. The Georgia-born Alabamian, two months short of his twenty-eighth birthday, left one brigade to keep up the pressure on Garrard and turned with the other seven to meet these rearward threats, sending three brigades to deal with Stoneman while he himself set out with the rest to intercept McCook.

  As it turned out, the interception came after, not before, McCook struck the railroad at Lovejoy Station, seven miles beyond Jonesboro. He got

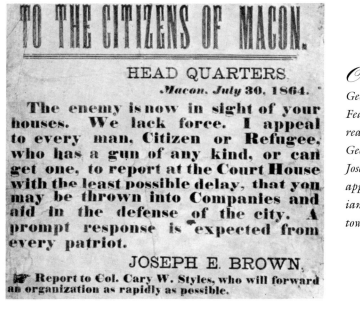

TO THE CITIZENS OF MACON.

HEAD QUARTERS.
Macon, July 30, 1864.

The enemy is now in sight of your houses. We lack force. I appeal to every man, Citizen or Refugee, who has a gun of any kind, or can get one, to report at the Court House with the least possible delay, that you may be thrown into Companies and aid in the defense of the city. A prompt response is expected from every patriot.

JOSEPH E. BROWN.

☞ Report to Col. Cary W. Styles, who will forward an organization as rapidly as possible.

On the day that George Stoneman's Federal raiders reached Macon, Georgia Governor Joseph Brown appealed to civilians to take up the town's defense.

there four hours ahead of Wheeler, which gave him time to burn the depot, tear up a mile and a half of track, and destroy a sizeable wagon train, along with its 800 mules, before the graybacks arrived to drive him off and pursue him all the way to the Chattahoochee. Overtaken at Newnan, due west on the West Point road, McCook lost 950 troopers killed and captured, along with his pack train and two guns, between there and the river, which he crossed to safety on July 30, reduced in strength by nearly a third and much the worse for wear.

By that time Stoneman had reached the outskirts of Macon, only to find it defended by local militia. While he engaged in a long-range duel across the Ocmulgee with these part-time soldiers, hoping to cover his search for a downstream ford, the three brigades sent after him by Wheeler came up in his rear. He tried for a getaway, back the way he had come, then found himself involved in a running fight that ended next day near Hillsboro, twenty-five miles to the north, when he was all but surrounded at a place called Sunshine Church. He chose one brigade to make a stand and told the other two to escape as best they could; which they did, while he and his chosen 700 were being overrun and rounded up. One of the two surviving brigades made it back to Decatur two days later, but the other, unable to turn west because of the swarm of rebels on that flank, was wrecked at Jug Tavern on August 3, thirty miles north of Covington. Stoneman and his captured fellow officers were in Macon by then, locked up with the unfortunates they had set out to liberate, and the enlisted men were in much the same position, though considerably worse off so far as the creature comforts were concerned, sixty miles to the southwest at Andersonville.

"On the whole," Sherman reported to Washington in one of the prize understatements of the war, "the cavalry raid is not deemed a success."

In plain fact, aside from McCook's fortuitous interception of the 800-mule train — the break in the track at Lovejoy's, for example, amounted to nothing worse than a two-day inconvenience, after which the Macon & Western was back in use from end to end — the raid not only failed to achieve its purpose, it was also a good deal harder on the raiders than on the raided. Sherman's true assessment was shown by what he did, on the return of his badly cut up horsemen, rather than by what he wrote in his report. Garrard's division, which had suffered least, was dismounted and used to occupy the intrenchments Schofield vacated when he began his swing around the city in Howard's wake, and the other two were reorganized, after a period of sorely needed rest and refitment, into units roughly half their former size. Not that Sherman expected much from them, offensively speaking, in the critical days ahead. "I now became satisfied," he said later, "that cavalry could not, or would not, make a sufficient lodgment on the railroad below Atlanta, and that nothing would suffice but for us to reach it with the main army."

But that turned out to be about as difficult an undertaking as the one assigned to Stoneman and McCook. For one thing — against all his expectations, which were founded on the belief that Hood by now had shot his wad — he had no sooner begun his counterclockwise wheel, shifting Howard around in rear of Schofield and Thomas to a position west of the city so that his right could be extended to reach the vital railway junction at East Point, than he was confronted with still a third sortie by his Confederate opponent, quite as savage as the other two.

All had gone well on the first day, July 27; Howard pulled out undeterred and took up the march, first north, then west along the near bank of Peachtree Creek. Riding south next morning in rear of Logan, whose corps was in the lead, Sherman and the new army commander came under fire from a masked battery as they approached the Lickskillet Road, which ran due east into Atlanta, three miles off. Howard did not like the look of things, and said so. "General Hood will attack me here," he told his companion, who scoffed at the notion: "I guess not. He will hardly try it again." But Howard remained persuaded

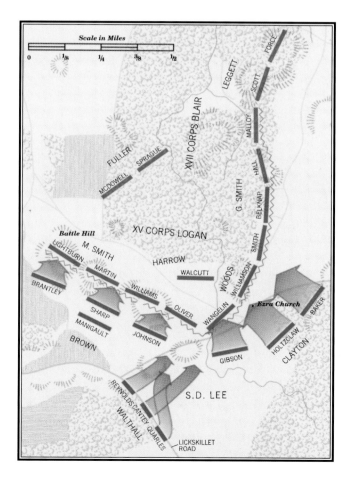

On July 28, John Bell Hood attempted to intercept a Federal encircling movement around the western side of Atlanta at the Battle of Ezra Church.

John Logan's XV Corps fires on Stephen D. Lee's Confederates, attempting to block the southwestward movement of the Federals near Ezra Church.

that he was about to be struck, explaining later that he based his conviction on previous acquaintance with the man who would do the striking; "I said that I had known Hood at West Point, and that he was indomitable."

Indomitable. Presented thus with a third chance to destroy an isolated portion of the enemy host, Hood had designed still another combined assault, once more after the manner of Lee and Jackson, to forestall this massive probe around his left. His old corps, now under Stephen D. Lee — the South Carolinian had been promoted to lieutenant general and brought from Alabama to take over from Cheatham — would march out the Lickskillet Road on the morning of July 28 to occupy a position from which it could block Howard's extension of the Union right and set him up for a flank attack by Stewart, who would bring his corps out the Sandtown Road that evening, a mile in Lee's rear, to circle the head of the stalled blue column and strike from the southwest at Howard's unguarded outer flank next morning. Hardee, reduced to three divisions, each of which received a brigade from the fallen Walker's broken-up division, would hold Atlanta's inner line against whatever pressure Schofield and Thomas might exert. Lee, who had assumed command only the day before, moved as ordered, determined to prove his mettle in this first test at his new post — two months short of his thirty-first birthday, he was six years younger than anyone else of his rank in the whole Confederacy — but found himself involved by midday, three miles out the Lickskillet Road, near a rural chapel known as Ezra Church, in a furious meeting engagement that left him no time for digging in or even getting set. So instead he took the offensive with all three of his divisions.

★

They were not enough: not nearly enough, as the thing developed. Howard, who was only two years older than Lee and no less anxious to prove his mettle, having also assumed command the day before, had foreseen the attack (or anyhow forefelt it, despite Sherman's scoff) and though there was no time for intrenching, once he had called a halt he had his lead corps throw up a rudimentary breastwork of logs and rails; so that when Lee's men charged — "with a terrifying yell," the one-armed commander would recall — they were "met steadily and repulsed." They fell back, then charged again, with the same result. Busily strengthening their improvised works between attacks, Logan's four divisions stood their ground, reinforced in the course of the struggle by others from Dodge and Blair, while Sherman rode back and alerted Thomas to be ready to send more. These last were unneeded, even though Hood by then had abandoned his plan for a double envelopment and instead told Stewart to go at once out the Lickskillet Road to Lee's assistance. Stewart added the weight of one division to the contest before sundown, without appreciable effect. "Each attack was less vigorous and had less chance than the one before it," a Union veteran was to note.

Alarmed by reports coming in all afternoon from west of Atlanta, Hood had Hardee turn his corps over to Cheatham, who had returned to his division, and proceed without delay to Ezra Church to take charge of the other two. Old Reliable arrived to find that the battle had sputtered out, and made no effort to revive it. Lee and Stewart between them had lost some 2500 killed and wounded — about the same number that had fallen along Peachtree Creek eight days ago — as compared to Howard's loss of a scant 700. Nor was that the worst of it, according to Hardee, who afterwards declared: "No action of the campaign probably did so much to demoralize and dishearten the troops engaged in it. . . ."

Sherman knew now that he had been wrong, these past five days, in thinking that Hood had shot his wad in the Battle of Atlanta. He would have been considerably closer to the truth, however, if he had reverted to this belief on the night that followed the Battle of Ezra Church. Moreover, there were Confederates in the still smoky woods, out beyond Howard's unbroken lines, who would have agreed with him; almost.

"Say, Johnny," one of Logan's soldiers called across the breastworks, into the outer darkness. "How many of you are there left?"

"Oh, about enough for another killing," some butternut replied.

This attitude on both sides, now that another month drew to a close, was reflected in their respective casualty lists. Including his cavalry subtractions, which were heavy, Sherman had lost in July about 8000 killed, wounded, and missing — roughly the number that fell in June, and better than a thousand fewer than fell in May. The over-all Federal total, from the outset back at Tunnel Hill, came to just under 25,000. Hood, on the other hand, had suffered 13,000 casualties in the course of his three sorties, which brought the Confederate total, including Johnston's, to 27,500. That was about the number Lee had lost during the same three-month span in Virginia, whereas Sherman had lost considerably fewer than half as many as Meade. Grant could well be proud of his western lieutenant, if and when he got around to comparing the cost, in men per mile, of the campaigns in Georgia and the Old Dominion, West and East.

Still, there was a good deal more to war than mere killing and maiming. "Lee's army will be your objective point," he had instructed Meade before the jump-off, only to have the eastern offensive wind up in a stalemate, a digging contest outside Petersburg. Similarly, he had told Sherman to "move against Johnston's army," and the red-haired Ohioan had done just that — so long as the army was Johnston's. But now that it was Hood's, and had come out swinging, a change set in: particularly after Ezra Church, the third of Hood's three roaring sorties. Lopsided as that victory had been for Sherman, it served warning that, in reaching for the railroad in his adversary's rear, his infantry might do no better than his cavalry had done, and indeed might suffer as severely in the process.

Inching southward all the following week he found rebel intrenchments bristling in his path. On August 5, having brought Schofield around in the wake of Howard, he reinforced him with a corps from Thomas and ordered the drive on the railroad resumed. Schofield tried, the following morning, but was soon involved in the toils of Utoy Creek and suffered a bloody repulse. It was then that the change in Sherman — or, rather, in his definition of his goal —

★

became complete. Formerly the Gate City had been no more than the anvil on which he intended to hammer the insurgent force to pieces. Now it became the end-all objective of his campaign. He would simply pound the anvil.

"I do not deem it prudent to extend any more to the right," he wired Halleck next day, "but will push forward daily by parallels, and make the inside of Atlanta too hot to be endured."

In line with McPherson's proposal at their farewell interview, he sent to Chattanooga for siege guns and began a long-range shelling of the city, firing over the heads of its defenders and into its business and residential districts. "Most of the people are gone; it is now simply a big fort," he informed his wife that week, and while this was by no means true at the time, it became increasingly the case with every passing day of the bombardment. "I can give you no idea of the excitement in Atlanta," a southern correspondent wrote. "Everybody seems to be hurrying off, especially the women. Wagons loaded with household furniture and everything else that can be packed upon them crowd every street, and women old and young and children innumerable are hurrying to and fro. Every train of cars is loaded to its utmost capacity. The excitement beats everything I ever saw, and I hope I may never witness such again." Presently, though the

Around Atlanta, the Confederates constructed
a formidable line of defense, at some points little more
than a mile from the center of the city.

destruction of property was great and the shelling continued day and night, the citizens learned to take shelter in underground bomb-proofs, as at Vicksburg the year before, and Hood said later that he never heard "one word from their lips expressive of dissatisfaction or willingness to surrender."

Sherman's reaction was to step up the rate of fire. "We can pick out almost any house in town," he boasted to Halleck. He was by nature "too impatient for a siege," he added, but "One thing is certain. Whether we get inside of Atlanta or not, it will be a used-up community when we are done with it."

His troops shared his ebullience, if not his impatience, finding much to admire in this notion of bloodless engagement at long range. "There goes the Atlanta Express!" they cheered as the big shells took off at fifteen-minute intervals over their and the rebel trenches. When one of the outsized guns developed the habit of dropping its projectiles short, they turned and shouted rearward through cupped hands: "Take her away! She slobbers at the mouth." Sherman moved among them, a reporter noted, with "no symptoms of heavy cares — his nose high, thin, and planted with a curve as vehement as the curl of a Malay cutlass — tall, slender, his quick movements denoting good muscle added to absolute leanness, not thinness." Uncle Billy, they called him, with an affection no blue-clad soldiers had shown for a commander, West or East, since Little Mac's departure from the war. What was more, unlike McClellan, he shared their life as well as their rations, though a staffer recorded that he was mostly "too busy to eat much. He ate hardtack, sweet potatoes, bacon, black coffee off a rough table, sitting on a cracker box, wearing a gray flannel shirt, a faded old blue blouse, and trousers he had worn since long before Chattanooga. He talked and smoked cigars incessantly, giving orders, dictating telegrams, bright and chipper."

Partly this was exuberance. Partly it was fret, which he often expressed or covered in such a manner. Either way, it was deadly: as was shown in a message he sent Howard, August 10, amid the roar of long-range guns. "Let us destroy Atlanta," he said, "and make it a desolation."

* * * * *Sherman's ebullience was heightened by news* that arrived next day, roundabout from Washington, of a great naval victory scored the week before by Farragut down in Mobile Bay. Long the target of various plans that had come to nothing until now — including Grant's, which went badly awry up the Red that spring with the near destruction of Banks's army and Porter's fleet — this last of the South's major Gulf of Mexico ports, second only to Wilmington as a haven for blockade runners, had been uppermost in Farragut's mind ever since the fall of New Orleans, more

than two years ago. He then solicited the Department for permission to steam booming into the bay before its defenses could be strengthened, only to be told that he and his sea-going vessels would continue to prowl the Mississippi until the big river was open from source to mouth. By the time this was accomplished, a year later at Port Hudson, both the admiral and his flagship *Hartford* were sorely in need of rest and repairs. However urgent its priority, the reduction of Mobile would have to await their return, respectively, from Hastings-on-Hudson, the Tennessee-born sailor's adoptive home, and the Brooklyn Navy Yard.

A Christmas visit to New York City was disrupted by an intelligence report that reached him amid the splendors of the Astor House, confirming his worst fears. Not only had Mobile's defenders greatly strengthened the forts guarding the entrance to the harbor; refugees now declared that they also were building a monster ironclad up the Alabama River, more formidable in armament and armor than any warship since the *Merrimac.* Farragut knew, from a study of what the latter had done in Hampton Roads before the *Monitor's* arrival — as well as from his own experience, near Vicksburg, when the *Arkansas* steamed murderously through the blue flotilla — just what damage one such vessel could do to any number of wooden ships. The answer, he saw, was to get back down there fast and, if possible, go up the river and destroy her before she was ready to engage; or else acquire some ironclads of his own, able to fight her on a give-and-take basis. In any case, after four months of rest and relaxation, he was galvanized into action. He went straight to Brooklyn and served notice that he expected the workmen to have the *Hartford* ready for sea by the evening of January 3. She was, and he dropped anchor at Pensacola two weeks later.

Off Mobile next day, January 18, he learned at first hand, not only that the rebel ironclad existed, as rumored, but also that she was now in the mouth of Dog River, up at the head of the bay. C.S.S. *Tennessee* was her name, and Admiral Franklin Buchanan, former commander of the *Merrimac-Virginia* and ranking man in the Confederate navy, was in charge; "Old Buck," Farragut called him, though at sixty-four Buchanan was only a year his senior and in fact had five years less service, having waited till he was fifteen to become a midshipman, which Farragut had done at the age of nine. Informed of a rumor that the ram was about to come down and attack the nine blockaders on station outside the bay, the Federal admiral braced his captains for the shock, and though he had small personal use for the new-fangled weapons ("If a shell strikes the side of the *Hartford,*" he explained, "it goes clean through. Unless somebody happens to be directly in the path, there is no damage excepting a couple of easily plugged holes. But when a shell makes its way into one of those damned tea-kettles, it can't get out again") he submitted an urgent request for at least a pair of monitors. "If I had them," he told Washington, "I should not hesitate to become the assailant instead of awaiting the attack."

Actually, though she had just completed the 150-mile downriver run from Selma, where she was built, there was little danger that the *Tennessee* would steam out into the Gulf. At this point, indeed, there was doubt that she could even make it into the bay, since she drew fourteen feet of water and the depth over Dog River Bar was barely ten. Ingenuity, plus three months of hard labor, solved the problem by installing "camels" — large floats attached to the hull below the water line — which lifted her enough to clear the bar with a good tide. By mid-May she was in Mobile Bay, and Farragut got his first distant glimpse of her from a gunboat cruising Mississippi Sound; "a formidable-looking thing," he pronounced her, though to one of his lieutenants "she looked like a great turtle."

Sixty-four-year-old Admiral Franklin Buchanan was commander of the rebel ironclad Tennessee.

More than 200 feet in length and just under 50 in the beam, she wore six-inch armor, backed by two solid feet of oak and pine, and carried six hard-hitting 6.4-and 7-inch Brooke rifles, one forward and one aft, mounted on pivots to fire through alternative ports, and two in each broadside. Her captain was Commander J. D. Johnston, an Alabama regular who had spent the past two years on duty in the bay, and her skeleton crew was filled out with volunteers from a Tennessee infantry regiment, inexperienced as sailors but proud to serve aboard a vessel named for their native state. Two drawbacks she had, both grave. One was that her engines, salvaged from a river steamboat, gave her a top speed of only six knots, which detracted from her maneuverability and greatly reduced her effectiveness as a ram. The other was that her steering chains led over, rather than under, her armored rear deck, and thus would be exposed to enemy fire. However, she also had one awesome feature new to warfare, described by her designer as "a hot water attachment to her boilers for repelling boarders, throwing one stream forward of the casemate and one abaft." What was more, with Buchanan directing events, there was every likelihood that the device would be brought into play; for he was a proud, determined man, with a fondness for close-quarter fighting and no stomach for avoiding dares.

"Everybody has taken it into their heads that one ship can whip a dozen," he wrote a friend while the ironclad was being readied for action, "and if the trial is not made, we who are in her are damned for life; consequently, the trial must be made. So goes the world."

Mobile's reliance was by no means all on the iron ram, however. In addition to three small paddle-wheel gunboats that completed the gray squadron — *Morgan* and *Gaines,* with six guns each, and *Selma* with four, all unarmored except for strips of plate around their boilers — three dry-land installations guarded the two entrances down at the far end of the thirty-mile-long bay. The first and least of these, Fort Powell, a six-gun earthwork on speck-sized Tower Island, a mile off Cedar Point, covered the approach from Mississippi Sound, off to the west, through Grant's Pass. Another was Fort Gaines, a pentagonal structure on the eastern tip of Dauphin Island, crowned with sixteen guns that commanded the western half of the main entrance, three miles wide, between there and Mobile Point, a long narrow spit of sand at whose extremity — the site of old Fort Bowyer, whose smoothbores had repelled the British fifty years ago — Fort Morgan, the stoutest and most elaborate of the three defensive works, reared its mass of dark red brick. This too was a five-sided structure, double-tiered and mounting no less than forty heavy guns in barbette and casemates, together with seven more in an exterior water battery on the beach in front of its northwest curtain. Both entrances had been narrowed by rebel contrivance, the one from the Sound by driving pilings from Cedar Point to Tower Island and from the northern end of Dauphin Island to within about half a mile of Fort Powell, the one from the Gulf by sinking others southeastward from Fort Gaines to within a mile of Mobile Point, while just in rear of the remaining gap a triple line of mines (called "torpedoes") had been strewn and anchored, barely out of sight below the surface, to within about two hundred yards of the western tip of the spit of land across the way. The eastern limit of this deadly underwater field was marked by a red buoy, fixed there for the guidance of blockade runners whose pilots could avoid sudden destruction by keeping to the right of it and steaming directly under the high-sited guns of Fort Morgan, almost within pistol range of those in the water battery on the beach.

Farragut planned to take that route, mainly because there seemed to be no other. Grant's Pass was too shallow for all but the lightest of his vessels, which would be no match for the iron ram once they entered the bay, and the combination of piles and mines denied him the use of any part of the main Gulf channel except that scant, gun-dominated 200-yard stretch just off the tip of Mobile Point. He was willing to take his chances there, as he had done in similar runs past Forts Jackson and St Philip and the towering bluffs at Vicksburg and Port Hudson, yet he did not enjoy the notion of getting inside the bay with the forts alive in his rear, his wooden ships crippled, and the *Tennessee* likely to pound or butt them into flotsam. Contemplating this, he saw more clearly than ever the need for ironclads of his own, and though four of these had been promised him by now, two from

*A*cross the entrance to Mobile Bay, the Confederates
placed a triple line of mines. The most common
Confederate mine was the Rains keg torpedo shown above.

the Atlantic squadron and two from the Mississippi, none had arrived by the time the squat metallic rebel monster steamed down the bay and dropped anchor behind Fort Morgan on May 20, intending either to await the entrance of the Union fleet or else run out and smash it in the Gulf. Farragut stormed at the delay, his patience stretched thin by the nonarrival of the monitors.

"I am tired of watching Buchanan," he wrote home in June, "and wish from the bottom of my heart that Buck would come out and try his hand upon us. The question has to be settled, iron versus wood, and there never was a better chance. . . . We are today ready to try anything that comes along, be it wood or iron, in reasonable quantities."

His plan was for the monitors to lead the way, holding to the right of the red buoy and providing an iron screen for the wooden ships as the two columns made their parallel runs past Fort Morgan, then going on to engage the ram in an all-out fight inside the bay, with such help as the multi-gunned sloops could provide. He would more or less ignore Fort Gaines while steaming in, not only because it was more than two miles off, but also because he planned to distract the attention of its gunners by having the army make a landing on the other end of Dauphin Island, then move east to invest the work from the landward side; after which Morgan would be served in the same fashion. But here too was a rub. The army, like the monitors, though promised, did not come. First there was Banks's drawn-out involvement up the Red, then a delay while Canby got the survivors back to New Orleans and in shape for the march to Mobile — which finally was cancelled when Grant was obliged to summon all but a handful to Virginia in late June, as replacements for Meade's heavy casualties. Canby visited the fleet in early July and agreed to send Major General

★

Gordon Granger with 2000 men in transports, admittedly a small force but quite as large as he felt he could afford.

Farragut had to be satisfied, and in any case his impatience was mainly with the monitors, which still had not arrived. By way of diversion from the heat and boredom, both of which were oppressive, he rehearsed the run past Fort Morgan, and the fight that was to follow inside the bay, on a wardroom table grooved with the points of the compass, maneuvering little boat-shaped wooden blocks carved for him by the *Hartford*'s carpenter. Meanwhile, Buchanan's inactivity puzzled and irked him more and more. "Now is the time," he declared in mid-July. "The sea is as calm as possible and everything propitious. . . . Still he remains behind the fort, and I suppose it will be the old story over again. If he won't visit me, I will have to visit him. I am all ready as soon as the soldiers arrive to stop up the back door of each fort."

He was not, of course, "all ready," nor would he be so until the monitors were on hand, the *Albemarle* having redemonstrated in April and May, at Plymouth and in the North Carolina Sound from which she took her name, what was likely to happen to his wooden ships if he had no ironclads of his own to stand between them and the *Tennessee.* Then on July 20 the first of the promised four arrived from the Atlantic coast; *Manhattan* she was called, wearing ten inches of armor on her revolving turret, which carried two 15-inch guns. Ten days later the *Chickasaw* put in from New Orleans, double-turreted with a pair of 11-inch guns in each, followed next day by her sister ship *Winnebago.* All were on hand by August 1 except the *Tecumseh,* en route from the Atlantic in the wake of her twin *Manhattan.* Farragut found the waiting even harder now that it was about to end; he improved the time by instructing his skippers in their duties, using the tabletop wooden blocks to show just where he expected their ships to be put in all eventualities. Meantime, as he had been doing for the past ten days, he continued to send out nightly boat crews, under cover of darkness and with muffled oars, to grapple for or sink as many as possible of the torpedoes anchored between the end of the line of pilings southeast of Dauphin Island and the red buoy just off Mobile Point. A number were so removed or destroyed, and the admiral was pleased to learn that many were found to be duds, their firing mechanisms having long been exposed to the corrosive effect of salt water.

Granger's 2000 soldiers arrived on August 2. They were taken around into Mississippi Sound the following night for a landing on the west end of Dauphin Island, and from there began working their way through heavy sand toward the back door of Fort Gaines. *Tecumseh* still had not appeared, but Farragut now was committed. "I can lose no more days," he declared. "I must go in day after tomorrow morning at daylight or a little later. It is a bad time, but when you do not take fortune at her offer you must take her as you can find her." Despite a heavy squall that evening, the grapplers went about their work in the mine field, undetected, and early next morning, August 4, the admiral took his fleet captains aboard the tender *Cowslip* for a closer look at the objective, cruising under the lee of Sand Island where the three monitors were anchored, ready to move out.

Returning, he went to his cabin, took out pen and paper, and composed a provisional farewell. "My dearest Wife: I write and leave this letter for you. I am going into Mobile Bay in the morning, if God is my leader, as I hope He is, and in Him I place my trust. . . . The Army landed last night, and are in full view of us this morning. The *Tecumseh* has not yet arrived."

Just then she did, steaming in from Pensacola to take position at the head of the iron column on the far side of Sand Island. The Union line of battle was complete. Asked at bedtime if he would consent to giving the men a glass of grog to nerve them up for the fight next morning, Farragut replied: "No, sir. I never found that I needed rum to enable me to do my duty. I will order two cups of good coffee to each man at 2 o'clock, and at 8 o'clock I will pipe all hands to breakfast in Mobile Bay."

Fog delayed the forming of the line past daybreak, the prearranged time for the start of the run, but a dawn breeze cleared the mist away by sunup, which came at 5.30 this Friday morning, August 5. As the four monitors began their movement eastward off the lee shore of Sand Island, in preparation for turning north beyond the line of pilings and the mine field — at which point the wooden column of seven heavy ships, each with a gunboat lashed to its port side for reserve power in case its boilers or engines were knocked out, would come up in their left rear for the dash past Mobile Point and the brick pentagon looming huge and black against the sunrise — Farragut was pleased to see that fortune had given him the two things he prayed for: a westerly wind to blow the smoke of battle away from the fleet and toward the fort, and a flood tide that would carry any pair of vessels on into the bay, even if both were disabled.

Captain James Alden's 2000-ton 24-gun *Brooklyn* led the way, given the honor because she was equipped with chase guns and an antitorpedo device called a cowcatcher. Then came Flag Captain Percival Drayton's *Hartford* with the admiral aboard, followed by the remaining five, *Richmond, Lackawanna, Monongahela, Ossipee,* and *Oneida,* each with its gunboat consort attached to the flank away from the fort and otherwise readied for action in accordance with instructions issued as far back as mid-July: "Strip your vessels and prepare for the conflict. Send down all superfluous spars and rigging. Trice up or remove the whiskers. Put up the splinter nets on the starboard side, and barricade the wheel and steersmen with sails and hammocks. Lay chains or sandbags on the deck over the machinery to resist a plunging fire. Hang the sheet chains over the side, or make any other arrangements for security that your ingenuity may suggest." As a result, according to a Confederate who studied the uncluttered ships from Mobile Point, "They appeared like prize fighters ready for the ring."

Buchanan, aboard the *Tennessee*, got word that they were coming at 5.45, shortly after they started his way. He hurried on deck in his drawers for a look at the Yankee vessels, iron and wood, and while he dressed passed orders for the ram and its three attendant gunboats to move westward and take up a position athwart the main channel, just in rear of the inner line of torpedoes, for crossing the Union T if the enemy warships — eighteen of them, mounting 199 guns, as compared to his own four with 22 — passed Fort Morgan in an attempt to enter the bay. Balding, clean-shaven like Farragut, with bright blue eyes and a hawk nose, the Marylander assembled the *Tennessee*'s officers and crew on her gun deck and made them a speech that managed to be at once brief and rambling. "Now, men, the enemy is coming, and I want you to do your duty," he began, and ended: "You shall not have it said when you leave this vessel that you were not near enough to the enemy, for I will meet them, and you can fight them alongside of their own ships. And if I fall, lay me on the side and go on with the fight."

Farragut came on deliberately in accordance with his plan, the flagship crossing the outer bar at 6.10 while the iron column up ahead was making its turn north into the channel. Ten minutes later the lead monitor *Tecumseh* fired the opening shot, a 15-inch shell packed with sixty pounds of powder and half a

Union Admiral David Glasgow Farragut's flagship, the Hartford, crossed the outer bar at Mobile Bay in the early morning hours of August 5, 1864.

bushel of cylindrical flathead bolts. It burst squarely over the fort, which did not reply until shortly after 7 o'clock, when the range to *Brooklyn,* leading the wooden column, had been closed to about a mile. Morgan's heaviest weapon was a 10-inch Columbiad, throwing a projectile less than half the weight of the one from *Tecumseh,* but the effect was altogether memorable for a young surgeon on the *Lackawanna,* midway down the line of high-masted vessels. "It is a curious sight to catch a single shot from so heavy a piece of ordnance," he later wrote. "First you see the puff of white smoke upon the distant ramparts, and then you see the shot coming, looking exactly as if some gigantic hand has thrown in play a ball toward you. By the time it is half way, you get the boom of the report, and then the howl of the missile, which apparently grows so rapidly in size that every green hand on board who can see it is certain that it will hit him between the eyes. Then, as it goes past with a shriek like a thousand devils, the inclination to do reverence is so strong that it is almost impossible to resist it."

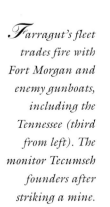

Farragut's fleet trades fire with Fort Morgan and enemy gunboats, including the Tennessee (third from left). The monitor Tecumseh founders after striking a mine.

Now the action became general, and by 7.30 the leading sloops, closing fast on the sluggish monitors, had their broadsides bearing fairly on the fort, whose gun crews were distracted by flying masonry, clouds of brickdust, and an avalanche of shells. Then two things happened, one in each of the tandem columns, for which Farragut had not planned while rehearsing the operation on the table in his cabin. Directly ahead of the flagship, *Brooklyn* had to slow to keep from overtaking the rear monitor *Chickasaw*. Presently, to the consternation of all astern, Alden stopped and began making signals: "The monitors are right ahead. We cannot go on without passing them. What shall we do?" While Farragut was testily replying, "Go ahead!" — and the guns of the fort and water battery, less than half a mile away, were stepping up their fire — Commander Tunis Craven of the *Tecumseh,* at the head of the iron column, reacted to a similar crisis in quite a different way, though it too involved a departure from instructions. Approaching the red buoy that marked the eastern limit of the mine field, he

saw the breakers off Mobile Point, just off his starboard bow, and said to his pilot, out of fear of running aground: "It is impossible that the admiral means us to go inside that buoy." He ordered a hard turn to port, which carried the *Tecumseh* to the left, not right, of the red marker. But not for long. A sudden, horrendous explosion against her bottom, square amidships — whether of one or more torpedoes was later disputed — shook and stopped the iron vessel, set her lurching from side to side, and sent water pouring down her turret as she wallowed in the waves.

All aboard her must have known the hurt was mortal, though no one guessed how short her agony would be. Craven and his pilot, for example, standing face to face at the foot of the ladder that led to the only escape hatch, staged a brief, courtly debate.

"Go ahead, Captain."

"After you, Pilot."

So they said; "But there was nothing after me," the pilot later testified. As he put his foot on the top rung of the ladder, *Tecumseh* and her captain dropped from under him.

Through a sight slit in the turret of *Manhattan,* next in line, an engineer watched the lead monitor vanish almost too abruptly for belief. "Her stern lifted high in the air with the propeller still revolving, and the ship pitched out of sight like an arrow twanged from the bow." With her went all but a score of her 114-man crew, including four who swam to Mobile Point and were taken captive, while the others who managed to wriggle out before she hit bottom were picked up by a boat from the *Hartford*'s consort, *Metacomet.*

Farragut sent the boat, though the fact was he had problems enough on his hands by then, including the apparent likelihood that such rescue work was about to be required in his own direction. *Brooklyn*'s untimely halt, practically under Morgan's guns, had thrown the wooden column into confusion; for when she stopped her bow yawed off to starboard, subtracting her broadside from the pounding the fort was taking, and what was worse she lay nearly athwart the channel, blocking the path of the other ships. Nor was that the end of the trouble she and her captain made. Alarmed by the sudden dive of the *Tecumseh* ("Sunk by a torpedo! Assassination in its worst form!" he would protest in his report) Alden spotted, just under his vessel's prow, "a row of suspicious-looking buoys" which he took to be floats attached to mines. He reacted by ordering *Brooklyn*'s engines reversed, and this brought her bearing down, stern foremost, on the *Hartford.* Farragut, who had climbed the mainmast rigging as far as the futtock shrouds for a view above the smoke — he was tied there with a rope passed round his body by a sailor, sent aloft by Drayton, lest a collision or a chance shot bring him crashing to the deck some twenty feet below — angrily hailed the approaching sloop, demanding to know the cause for such behavior, and got the reply: "Torpedoes ahead."

With spyglass in hand, Admiral David G. Farragut follows the action in Mobile Bay while tied to the rigging of his flagship Hartford to prevent a fall.

Like the *Brooklyn,* which took 59 hits in the course of the fight, *Hartford* was absorbing cruel punishment from the guns on Mobile Point: particularly from those in the water battery, whose fire was point-blank and deadly. Men were falling fast, their mangled bodies placed in a row on one side of the deck, while the wounded were sent below in numbers too great for the surgeons to handle. A rifled solid tore a gunner's head off; another took both legs off a sailor who threw up his arms as he fell, only to have them carried away by still another. Farragut looked back down the line, where the rest of his stalled vessels were being served in much the same fashion, and saw that it would not do. He either had to go forward or turn back. In his extremity, he said later, he called on God: "Shall I go on?" and received the answer from a commanding voice inside his head: "Go on." *Brooklyn* blocked the channel on the right, so he asked the pilot, directly above him in the maintop, whether there was enough water for the *Hartford* to pass her on the left. The pilot said there was, and the admiral, exultant, shouted down to Drayton on the quarterdeck: "I will take the lead!" Signaling "close order" to the ships astern, he had the *Metacomet* back her engines and the flagship go all forward. This turned her westward, clear of *Brooklyn,* which she passed as she moved out. Someone called up a reminder of Alden's warning, but Farragut, lashed to the rigging high above the smoke of battle, with Mobile Bay in full view before him, had no time or mind for caution. "Damn the torpedoes!" he cried. "Full speed ahead!"

Ahead he went, followed by the others, west of where the *Brooklyn* lay until she rejoined the column — and west, too, of the red buoy marking the eastern limit of the mine field. Though Farragut had been encouraged by the work of his nighttime grapplers, who not only had removed a considerable number of mines in the course of the past two weeks, but also reported a high percentage of duds among them, *Tecumseh* had just given an only-too-graphic demonstration of what might await him and all his warships, iron or wood, as a result of this sudden departure from his plan to avoid the doom-infested stretch of water the *Hartford* now was crossing. And sure enough, while she steamed ahead with all the speed her engines could provide, the men on deck — and, even worse, the ones cooped up below — could hear the knock and scrape of torpedo cases against her hull and the snap of primers designed to ignite the

"Damn the torpedoes!
Full speed ahead!"

— David Glasgow Farragut

charges that would blast her to the bottom. None did, either under the *Hartford* or any of the vessels in her wake, but the passage of Morgan became progressively more difficult as the lead sloops steamed out of range and left the tail of the column, along with the slow-moving monitors, to the less-divided attention of the cannoneers in the fort and on the beach. *Oneida,* which brought up the rear, took a 7-inch shell in the starboard boiler, scalding her firemen with escaping steam, and another that burst in the cabin, cutting both wheel ropes. Powerless and out of control, she too made it past, tugged along by her consort, only to emerge upon a scene of even worse destruction, just inside the bay.

Buchanan had succeeded in his design to cross the Union T; with the result that when Farragut ended his sprint across the mine field he found the *Tennessee* and the three rebel gunboats drawn up to receive him in line ahead, presenting their broadsides to the approaching column, whose return fire was limited to the vessels in the lead, and even these could bring only their bow guns into play. *Hartford*'s was promptly knocked out by a shot from *Selma,* smallest of the three, and this was followed by another that passed through the chain armor on the flagship's starboard bow, killing ten men, wounding five, and hurling bodies, or parts of bodies, aft and onto the decks of the *Metacomet,* lashed alongside.

Farragut kept coming, with *Brooklyn* and *Richmond* close astern, and managed to avoid an attempt by Buchanan to ram and sink him, meantime bringing his big Dahlgrens to bear on the gunboats, one of which then retired lamely toward Fort Morgan, taking water through a hole punched in her hull. This was the *Gaines;* she was out of the fight, and presently so were the others, *Morgan* and *Selma;* for *Hartford* and *Richmond* cast off their consorts to engage them and they fled. *Metacomet* led the chase, yawing twice to fire her bow gun, but then stopped firing to concentrate on speed. While *Morgan* made it to safety under the lee of Mobile Point, *Selma* kept running eastward across the shallows beyond the channel, still pursued despite the *Metacomet*'s deeper draft. Out on the bow of the northern vessel, a leadsman was already calling one foot less than the ship drew, but her captain, feeling the soft ooze of the bottom under her keel, refused to abandon the chase. "Call the man in," he told his exec. "He is only intimidating me with his soundings."

Persistence paid. Overtaken, *Selma* lost eight killed and seven wounded before she hauled down her flag. Westward, the *Gaines* burned briskly, set afire by her crew, who escaped in boats as she sank in shallow water. Only *Morgan* survived, anchored under the frown of the fort's guns to wait for nightfall, when she would steal around the margin of the bay to gain the greater safety of Mobile, inside Dog River Bar.

Left to fight alone, Buchanan steamed after the *Hartford* for a time, still hoping to ram and sink her, despite the agility she had shown in avoiding his first attempt, but soon perceived that her speed made the chase a waste of effort; whereupon he turned back and made for the other half-dozen sloops, advancing in closer order. *Tennessee* passed down the line of high-walled wooden men-of-war, mauling and being mauled. Two shots went through and through the *Brooklyn,* increasing her toll of killed and wounded to 54, but another pair flew high to miss the *Richmond*. Both ships delivered point-blank broadsides that had no effect whatever on the armored vessel as she bore down on *Lackawanna*, next in line, and *Monongahela*, which she struck a glancing blow, then swung round to send two shells crashing into the *Ossipee*. That left *Oneida*, whose bad luck now turned good, at least for the moment. Aboard the ram, defective primers spared the crippled ship a pounding; then one gun fired a delayed shot that cost the northern skipper an arm and the use of his 11-inch after pivot, which was raked. *Tennessee* turned hard aport in time to meet the three surviving monitors, just arriving, and exchanged volleys in passing that did no harm on either side. Then she proceeded to Fort Morgan and pulled up, out of range on the far side of the channel.

Farragut dropped anchor four miles inside the bay, and the rest of the blue flotilla, wood and iron, steamed up to join him, their crews already at work clearing away debris and swabbing the blood from decks, while belowdecks surgeons continued to ply their scalpels and cooks got busy in the galleys. It was 8.35; he was only a bit over half an hour behind schedule on last night's promise to "pipe all hands to breakfast in Mobile Bay" by 8 o'clock. All the same, despite the general elation at having completed another spectacular run past formidable works, rivaling those below New Orleans and at Vicksburg and Port Hudson, there was also a tempering sorrow over the loss of the *Tecumseh* and considerable apprehension, as well, from the fact that the murderous rebel iron ram was still afloat across the way.

Drayton promptly expressed this reservation to the admiral, who by now had come down from the flagship's rigging and stood on the poop. "What we have done has been well done, sir," he told him. "But it all counts for nothing so long as the *Tennessee* is there under the guns of Morgan." Farragut nodded. "I know it," he said, "and as soon as the people have had their breakfasts I am going for her."

As it turned out, there was no need for that, and no time for breakfast. At 8.50, fifteen minutes after *Hartford* anchored, there was a startled cry from aloft. "The ram is coming!" So she was, and presently those on deck saw her steaming directly for the fleet, apparently too impatient to wait for a fight in which she would have the help of the guns ashore. Farragut prepared for battle, remarking as he did so: "I did not think Old Buck was such a fool."

Fool or not, throughout the pause Buchanan had been unwilling to admit the fight was over, whatever the odds and no matter how far he had to go from Fort Morgan to renew it. Instrumental in the founding of the academy at Annapolis, he had served as its first superintendent and thought too highly of naval tradition to accept even tacit defeat while his ship remained in any condition to engage the enemy. "If he won't visit me, I will have to visit him," his adversary had remarked three weeks ago, and Buchanan felt much the same about the matter now as he gazed across three miles of water at the Yankee warships riding at anchor in the bay — his bay — quite as if there was no longer any question of their right to be there. Gazing, he drew the corners of his mouth down in a frown of disapproval, then turned to the *Tennessee*'s captain. "Follow them up, Johnston. We can't let them off that way." With that, the ram started forward: one six-gun vessel against a total of seventeen, three of them wearing armor heavier than her own, mounting 157 guns, practically all of them larger than any weapon in her casemate. That Buchanan was in no mood for advice was demonstrated, however, when one of his officers tried to call his attention to the odds. "Now I am in the humor, I will have it out," he said, and that was that. The ram continued on her way.

The monitors having proved unwieldy, Farragut's main reliance was on his wooden sloops, particularly the *Monongahela* and the *Lackawanna*, which were equipped with iron prows for ramming. Their orders were to run

★

As the Hartford grates past the Tennessee,
Federal sailors prepare to broadside the Confederate
ship by rushing to reload a Dahlgren.

the ram down, while the others pitched in to do her whatever damage they could manage with their guns. Accordingly, when the *Tennessee* came within range about 9.20, making hard for the flagship, *Monongahela* moved ahead at full speed and struck her amidships, a heavy blow that had no effect at all on the rebel vessel but cost the sloop her iron beak, torn off along with her cut-water. *Lackawanna* rammed in turn, with the result that an eight-foot section of her stem was crushed above and below the waterline. *Tennessee* lurched but held her course, and the two flagships collided nearly head on. "The port bow of the *Hartford* met the port bow of the ram," an officer aboard the Federal vessel later wrote, "and the ships grated against each other as they passed. The *Hartford* poured her whole port broadside against the ram, but the solid shot merely dented the side and bounded into the air. The ram tried to return the salute, but owing to defective primers only one gun was dis-charged. This sent a shell through the berth-deck, killing five men and wounding eight. The muzzle of the gun was so close to the *Hartford* that the powder blackened her side."

When the two ships parted Farragut jumped to the port quarter rail and held to the mizzen rigging while he leaned out to assess the damage, which was by no means as great as he had feared. Finding the perch to his liking he remained there, lashed to the rigging by friendly hands for the second time that day, and called for Drayton to give the *Tennessee* another thump as soon as possible. As the *Hartford* came about, however, she was struck on the starboard flank by the *Lackawanna,* which was also trying to get in position, crushing her planking on that side and upsetting one of the Dahlgrens. "Save the admiral! Save the admiral!" the cry went up, for it was thought at first that the flagship was sinking, so great was the confusion on her decks. Farragut untied himself, leaped down, and crossed to the starboard mizzen rigging, where he again leaned out to inspect the damage, which though severe did not extend to within two feet of the water. Again he ordered full speed ahead, only to find the *Lackawanna* once more looming on his starboard quarter. At this, one witness later said, "the admiral became a trifle excited." Forgetting that he had given the offending ship instructions to lead the ram attack, he turned to the communications officer on the bridge.

"Can you say 'For God's sake' by signal?"

"Yes, sir."

"Then say to the *Lackawanna,* 'For God's sake, get out of our way and anchor.'"

By now the ironclad had become the target for every ship that could get in position to give her a shot or a shove, including the double-turreted *Chickasaw,* which "hung close under our stern," the *Tennessee*'s pilot afterwards declared, "firing the two 11-inch guns in her forward turret like pocket pistols." Such punishment began to tell. Her flagstaff went and then her stack, giving the ram what one attacker called "a particularly shorn, stubby look" and greatly reducing the draft to her fires. Her steam went down, and then, as a sort of climax to her disablement, the monitor hard astern succeeded in cutting her rudder chain, exposed on the afterdeck, so that she would no longer mind her helm. Still she kept up the fight, exploiting her one advantage, which was that she could fire in any direction, surrounded as she was, without fear of hitting a friend or missing a foe. Presently, though, this too was reduced by shots that jammed half of her gunport shutters against the shield, thereby removing them from use. When this happened to the stern port, Buchanan sent for a machinist to unjam it, and while the man was at work on the cramped bolt, an 11-inch shell from the *Chickasaw* exploded against the edge of the cover just above him. "His remains had to be taken up with a shovel, placed in a bucket, and thrown overboard," a shipmate would recall. One of the steel splinters that flew inside the casemate struck Buchanan, breaking his left leg below the knee. "Well, Johnston," he said to the *Tennessee*'s captain as he was taken up to be carried down to the berth deck, "they've got me. You'll have to look out for her now. This is your fight, you know."

Johnston did what he could to sustain the contest with the rudderless, nearly steamless vessel, blind in most of her ports and taking heavy-caliber punches from two big sloops on each quarter and the monitor astern. Finally he went below and reported the situation to Buchanan. "Do the best you can, sir," the admiral told him, teeth gritted against the pain from the compound fracture of his leg, "and when all is done, surrender." Returning topside, the Alabamian found the battle going even worse. Unable to maneuver, the ram could not bring a single gun to bear on her tormentors; moreover, Johnston afterwards reported, "Shots were fairly raining upon the after end of the shield, which was now so thoroughly shattered that in a few moments it would have fallen and exposed the gun deck to a raking fire of shell and grape." He lowered the *Tennessee*'s ensign, in token of her capitulation, and when this did not slacken the encircling fire — it had been shot down before, then raised again on the handle of a rammer staff poked through the overhead grille of the smoky casemate — "I then decided, although with an almost bursting heart, to hoist the white flag."

At 10 o'clock the firing stopped, and presently Farragut sent an officer to demand the wounded admiral's sword, which then was handed over. *Tennessee*'s loss of two men killed and nine wounded brought the Confederate total for all four ships to 12 killed and 20 wounded. Union losses were 172 killed, more than half in the *Tecumseh,* and 170 wounded. Their respective totals, 32 and 342, were thus about in ratio of the strength of the two fleets, though in addition 243 rebel sailors were captured aboard *Selma* and the ironclad.

"The Almighty has smiled upon me once more. I am in Mobile Bay," Farragut wrote his wife that night, adding: "It was a hard fight, but Buck met his fate manfully. After we passed the forts, he came up in the ram to attack me. I made at him and ran him down, making all the others do the same. We butted and shot at him until he surrendered."

Westward across the bay, as he wrote, there was a burst of flame and a loud explosion off Cedar Point. The garrison of Fort Powell, taken under bombardment from the rear that afternoon by one of the big-gunned monitors at a range of 400 yards, had evacuated the place under cover of darkness and set a slow match to the magazine. Next morning the fleet dropped down and began shelling the eastern end of Dauphin Island, where Fort Gaines was under pressure from the landward side by Granger and his soldiers. This continued past nightfall, and the fort's commander asked for terms the following day, August 7. Told they were unconditional, he accepted and promptly surrendered his 818 men, together with all guns and stores. That left Fort Morgan; a much tougher proposition, as it turned out.

While the troops were being taken aboard transports for the shift to Mobile Point and a similar rear approach to the fortifications there, Farragut submitted under a flag of truce a note signed by himself and Granger, demanding the unconditional surrender of Fort Morgan "to prevent the unnecessary sacrifice of human life which must follow the opening of our batteries." The reply was brief and negative. "Sirs: I am prepared to sacrifice life, and will only surrender when I have no means of defense. . . . Respectfully, etc. *R. L. Page*, Brigadier General."

Approaching fifty-seven, Richard Page was a Virginian, a forty-year veteran of the Union and Confederate navies, who had transferred to the army five months ago when he assumed command of the outer defenses of Mobile Bay. His beard was white, his manner fiery; "Old Ramrod" and "Bombast Page" were two of his prewar nicknames, and if he bore a resemblance to R. E. Lee (both were born in 1807) it was no wonder. His mother had been Lee's father's sister.

Farragut's run past Morgan had come as a shock to its defenders, who fired close to 500 shots at the slow-moving Yankee column. "I do not see how I failed to sink the *Hartford*," Page said ruefully, shaking his head as the smoke cleared; "I do not see how I failed to sink her." Fort Powell's evacuation

*Gaping holes in the side of the brick lighthouse
at Fort Morgan attest to the strength of fire delivered
by the Federals during the battle in Mobile Bay.*

★

and the unresistant capitulation of Fort Gaines, neither of which had been done with his permission, angered and made him all the more determined to resist to the utmost the amphibious seige that got under way on August 9, shortly after he rejected unconditional surrender. Granger's men had been put ashore that morning on the bay side of Mobile Point, just over a mile to the east of the fort, and by nightfall — after they had performed the back-breaking labor of hauling guns and ammunition through shin-deep sand, which one of them said was "hot enough during the day for roasting potatoes" — took the east curtain and ramparts under fire with their batteries, while the sloops and ironclads, including the captured *Tennessee,* poured in shells and hotshot from the bay and Gulf. The fort shook under this combined pounding, but Page was no more of a mind to surrender now than he had been when he first declined the combined demand at midday.

For two weeks this continued, and throughout that time the pressure grew. Daily the troops drew closer on the landward side, increasing the number of weapons they brought to bear until at last there were 25 guns and 16 heavy mortars, their discharges echoed by those from the ships beyond and on both sides of the point. The climax came on August 22, when 3000 rounds were flung at the fort in the course of a twelve-hour bombardment, under whose cover the blue infantry extended its parallels to within reach of the glacis. All but two of the fort's guns were silenced and the citadel was burning; sharpshooters drew beads on anything that showed above the ramparts, and 80,000 pounds of powder had to be removed from the magazine and flooded, so close were the flames. Practically all that remained by now was wreckage and scorched debris. At 5 o'clock next morning two last shots were fired by the defenders, and one hour later the white flag went up. Farragut sent Drayton to arrange the formal surrender, which took place that afternoon amid the rubble. He had Buchanan's sword for a trophy, but he did not get Page's. The general and all his officers, displaying what Farragut called "childish spitefulness," had broken or thrown away their side arms just before the ceremony.

The admiral did get another 546 prisoners, however, which brought the total to better than 1700 on land and water — and he did get Mobile Bay, which after all was what he had come for. Blockade running might continue on the Atlantic coast, where Wilmington and Charleston still held out, but it was ended on the Gulf except for the sealed-off region west of the Mississippi, which in any case lay outside the constricting Anaconda coils. Mobile itself, thirty miles away at the head of the bay, was no part of Farragut's objective. Except as a port, it contributed little to the South's defense, and it was a port no longer. Moreover, Canby not only lacked the strength to expel the town's defenders; he could not have afforded to garrison it afterwards, so urgent were the calls for replacements for the men who had fallen in Georgia and, above all, in Virginia.

★

Best of all the immediate gains obtained from the naval battle, though, was the elation that followed, throughout the North, the announcement of the first substantial victory that had been scored, East or West, in the three months since the opening of Grant's spring offensive. Lincoln and his political supporters were pleased above all, perhaps, with the lift it seemed to give his chances for survival in the presidential contest, which by then was less than three months off.

★ ★ ★   *A*s usual, there was bad news with the good, and in this case the bad was double-barreled, concerning as it did a pair of highly spectacular reverses, one afloat and one ashore. In Washington on August 12, while the celebration of Farragut's week-old triumph over the *Tennessee* was still in progress at the Navy Department — word had come belatedly by wire from Ben Butler, who read of the bay battle in a Richmond paper smuggled through his Bermuda Hundred lines the day before — the telegraph line from coastal New Jersey began to chatter about a mysterious rebel cruiser at work off Sandy Hook. Yesterday she had taken seven prizes, and today she was adding six more to her list, which would reach a total of thirty U.S. merchant vessels within the week. It was as if the *Alabama,* eight weeks in her watery grave outside Cherbourg, had been raised, pumped out, and sped across the Atlantic to lay about her in a manner even more destructive than when she was in her prime. Quickly, all the available Federal warships within reach were ordered out to find and sink her at all costs. But who, or what, was she? Where had she come from? Who was her captain?

She was the *Tallahassee,* a former blockade runner, built up the Thames the year before and purchased that summer by the Confederates, who converted her into a raider by installing three guns and sent her out from Wilmington under Commander John T. Wood, a one-time Annapolis instructor, grandson of Zachary Taylor, aide to Jefferson Davis, and participant in a number of naval exploits, including the *Merrimac-Monitor* fight, New Bern, and the retaking of Plymouth. Setting out on the night of August 6 he showed the blockaders a clean pair of heels; for that was the ship's main virtue, speed. Twin stacked, with a 100-horsepower engine driving each of her two screws, she was 220 feet in length and only 24 in the beam, a combination that gave her a top speed of seventeen knots and had enabled her, on her shakedown cruise, to make the Dover-Calais crossing in seventy-seven minutes. Five mornings later, 500 miles up the Atlantic coast, *Tallahassee* encountered her first prize, the schooner *Sarah Boyce,* and before the day was over she ran down six more Union merchant vessels, ransoming the last to put all prisoners ashore. That was Thursday, August 11; "Pirate off Sandy Hook capturing and burning," the commandant of the Brooklyn Navy Yard wired Washington. Friday, off Long Island, she took six prizes, Saturday two, and Sunday — as if by way of resting on the Sabbath — one. By now she

★

was cruising the New England coast, and on Monday she took six ships, Tuesday five, and Wednesday three, rounding out a week that netted her thirty prizes, all burned or scuttled except seven that were ransomed to clear her crowded decks of captured passengers and crews. On August 18, running low on coal, she put into the neutral port of Halifax to refuel.

Under instructions from the Queen, and over ardent protests from the American consul, the Nova Scotia authorities gave Wood twenty-four hours to fill his bunkers, and when this did not suffice they granted him a twelve-hour extension. *Tallahassee* steamed out the following night in time to avoid half a dozen enemy warships that arrived next day, the vanguard of a fleet of thirteen ordered to Halifax as soon as the consul telegraphed word of the raider's presence in the harbor. She headed straight for Wilmington, taking so little chance on running out of coal that she only paused to seize one prize along the way, and arrived on the night of August 26 to speed and shoot her way through the blockade flotilla and drop anchor up the Cape Fear River, whose entrance was guarded by Fort Fisher. Her twenty-day cruise had cost the enemy 31 merchant vessels and had given Wood's fellow countrymen some welcome news to offset the bad from Mobile Bay, where Fort Morgan had fallen three days ago. They took pride in the fact that "this extemporaneous man-of-war," as Jefferson Davis called the *Tallahassee,* had "lit up the New England coast with her captures," and they could tell themselves, as well, that no matter what misfortunes befell their regular navy, outnumbered as it invariably was in combat, their irregular navy (so to speak) had won them the admiration of the world and was rapidly scouring the seas of Yankee shipping.

That was the first Federal reverse. The second, which occurred simultaneously ashore, was quite as spectacular and, if anything, even more "irregular" — as was often the case in operations involving Bedford Forrest. He had been given a free rein to conduct the defense of North Mississippi by Major General Dabney Maury, who succeeded to command of the Department of Alabama, Mississippi, and East Louisiana in late July, when Stephen Lee left to join Hood at Atlanta. "We must do the best we can with the little we have," Maury wrote from Meridian in early August, "and it is with no small satisfaction I reflect that of all the commanders of the Confederacy you are accustomed to accomplish the very greatest results with small means when left to your own untrammeled judgment. Upon that judgment I now rely."

Forrest took him at his word. "All that can be done shall be done," he replied, adding that since he lacked "the force to risk a general engagement" in resisting the next blue incursion, he would "resort to all other means."

Other means, in this case, included a raid on Memphis, the enemy's main base, under occupation for better than two years. Tactically, such a strike would be likely to disrupt the plans of the Federals for extending their conquest deep into Mississippi. Moreover, Forrest himself — a former alderman — would not only derive considerable personal satisfaction from returning to his home town, which no Confederate had entered, except as a spy or prisoner, since its fall in June of 1862; he would also be exacting vengeance for a battle fought the month before, near Tupelo, which was as close to a defeat as he had come so far in his career. Stephen Lee had been in command of the field, one week before his departure for Atlanta, but the memory rankled and Forrest was anxious to wipe it out or anyhow counterbalance it.

Hard on the heels of Brice's Crossroads in mid-June, when he received orders from Sherman "to make up a force and go out and follow Forrest to the

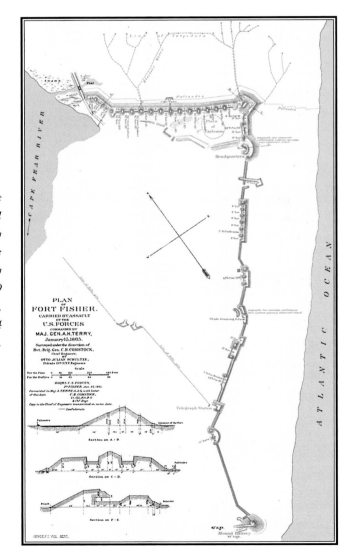

The L-shaped Fort Fisher guarded the approach to Confederate Wilmington, North Carolina, with 169 artillery pieces, including 44 heavy cannon.

death, if it costs 10,000 lives and breaks the Treasury," C. C. Washburn, the Memphis commander, assigned the task to A. J. Smith, reinforcing two of his divisions, just returned from their excursion up and down Red River, with Bouton's brigade of Negro infantry and Grierson's cavalry division, both of them recent graduates of the hard-knocks school the Wizard of the Saddle was conducting for his would-be conquerers down in Mississippi. On July 5 this column of 14,200 effectives, mounted and afoot, supported by six batteries of artillery and supplied with twenty days of rations — "a force ample to whip anything this side of Georgia," Washburn declared — set out southward from La Grange, fifty miles east of Memphis. Sherman's orders by then had been expanded; Smith and his gorilla-guerillas, who had polished their hard-handed skills in Louisiana under Banks, were to "pursue Forrest on foot, devastating the land over which he passed or may pass, and make him and the people of Tennessee and Mississippi realize that, although [he is] a bold, daring, and successful leader, he will bring ruin and misery on any country where he may pause or tarry. If we do not punish Forrest and the people now," the red-haired Ohioan wound up, "the whole effect of our past conquests will be lost."

Three days out, and just over fifty miles down the road, Smith showed that he took this admonition to heart by burning much of the town of Ripley, including the courthouse, two churches, the Odd Fellows Hall, and a number of homes. Next day, July 9, still mindful of his instructions to "punish Forrest and the people," he pressed on across the Tallahatchie and through New Albany, trailed by a swath of desolation ten miles wide.

Ahead lay Pontotoc, and beyond it Okolona, where Sooy Smith had come to grief five months before, checked almost as disastrously as Sturgis had been at nearby Brice's Crossroads, a month ago tomorrow. So far, only token opposition to the current march had developed, but at Pontotoc, which he cleared on July 11, this new Smith began to encounter stiffer resistance. Butternut troopers hung on the flanks of the column, as if to slow it down before it made contact with whatever was waiting to receive it up ahead, perhaps at Okolona. Smith would never know; for at dawn on July 13, well short of any ambush being laid for him there or south of there, he abruptly changed direction and struck out instead for Tupelo, fifteen miles to the east on the Mobile & Ohio, "his column well closed up, his wagon train well protected, and his flanks covered in an admirable manner."

So Forrest's scouts informed him at Okolona, where he was waiting — it was his forty-third birthday — for both Smith and Stephen Lee, who was on the way with 2000 troops and had ordered him not to commit his present force of about 6000 until these reinforcements got there to reduce the odds. Arriving from the south to find that the blue column had veered east, Lee took charge of pressing the pursuit. His urgency was based on reports from Dabney Maury, at Mobile, that Canby was preparing to march from New Orleans and attack the city from the landward side; Lee wanted Smith dealt with quickly so that the men he had brought to reinforce Forrest could be sent to Maury. "As soon as I fight I can send him 2000, possibly 3000," he explained in a dispatch to Bragg, though he added that this depended on whether the Mississippi invaders did or did not "succeed in delaying the battle." Smith was capable and canny, halting from time to time to beat off rearward threats while Grierson's horsemen rode on into Tupelo and began tearing up track above and below the town. All day the Federal infantry marched, then called a halt soon after nightfall at Harrisburg, two miles west of Tupelo, which had grown with the railroad and swallowed the older settlement as a suburb. Forrest came up presently in the darkness and "discovered the enemy strongly posted and prepared to give battle the next day."

Smith was at bay, and though his position was a stout one, nearly two miles long and skillfully laid out — flanks refused, rear well covered by cavalry, the line itself strengthened with fence rails, logs, timbers from torn-down houses, and bales of cotton — Forrest counted this a happy ending to an otherwise disappointing birthday. "One thing is certain," he told Lee; "the enemy cannot remain long where he is. He must come out, and when he does, all I ask or wish is to be turned loose with my command." No matter which way Smith headed when he emerged fretful and hungry, Forrest said, "I will be on all sides of him, attacking day and night. He shall not cook a meal or have a night's sleep, and I will wear his army to a frazzle before he gets out of the country."

Lee could see the beauty of that; but he had Mobile and Canby on his mind, together with the promises he had made to Bragg and Maury, and did not feel that he could afford the time it would take to deal with the penned-up bluecoats in this manner. There were better than 14,000 of them, veterans to a man, and though he had only about 8000 troops on hand he issued orders for an all-out assault next morning. Forrest would take the right and he the left. Together they would storm the Union works, making up for the disparity in numbers by the suddenness and ardor of their charge.

Ardor there was, and suddenness too, but these turned out to be the qualities that robbed Lee of what little chance he had for success in the first place. July 14 dawned hot and still, and the troops on line were vexed by delays in bringing several late-arriving units into position for the attack. Around 7.30, a Kentucky brigade near the center jumped the gun and started forward ahead of the others, who followed piecemeal, left and right, with the result that what was to have been a single, determined effort, all along the line, broke down from the outset into a series of individual lunges. Smith's veterans, snug behind their improvised breastworks, blasted each rebel unit as it advanced. "It was all gallantry and useless sacrifice," one Confederate was to say. To Smith, the disjointed attack "seemed to be a foot race to see who should reach us first. They were allowed to

★

Thirty-year-old Stephen D. Lee, the youngest lieutenant general in the Confederate army, brought 2000 reinforcements to aid Forrest in North Mississippi.

approach, yelling and howling like Comanches, to within canister range. . . . They would come forward and fall back, rally and forward again, with the like result. Their determination may be seen from the fact that their dead were found within thirty yards of our batteries." None got any closer, and after two hours of this Lee called a halt. He had lost 1326 killed and wounded and missing, Smith barely half that many, 674.

Skirmishing resumed next morning, but so fitfully and cautiously that it seemed to invite a counterattack. Smith instead clung fast to his position. He did, that is, until midday, when he was informed that much of the food in his train had spoiled in the Mississippi heat, leaving only one day's rations fit to eat, and that his reserve supply of artillery ammunition was down to about a hundred rounds per gun: whereupon he decided to withdraw northward, back in the direction he had set out from ten days ago, even though this meant leaving his more grievously wounded men behind in Tupelo. There followed the curious spectacle of a superior force retreating from a field on which it had inflicted nearly twice as many casualties as it suffered and being harassed on the march by a loser reduced to less than half the strength of the victor it was pursuing. In any case, after setting fire to what was left of Harrisburg, the Federals not only withdrew in good order and made excellent time on the dusty roads; they also succeeded, when they made camp at sunset on Town Creek, five miles north, in beating off a rebel attack and inflicting on Bedford Forrest, whom Lee had put in charge of the pursuit — and whom Smith had been told to "follow to the death" — his third serious gunshot wound of the war. The bullet struck him in the foot (the base of his right big toe, to be explicit) causing him so much pain that he had to relinquish the command, temporarily at least, and retire to a dressing station.

Smith kept going, unaware of this highly fortunate development, back through New Albany and across the Tallahatchie. Midway between there and La Grange he encountered a supply train sent to meet him. He kept going, despite this relief, and returned to his starting point on July 21, after sixteen round-trip days of marching and fighting. "I bring back everything in good order; nothing lost," he informed Washburn, who found the message so welcome a contrast to those received from other generals sent out after Forrest that he passed it along with pride to Sherman.

Far from proud, Sherman was downright critical, especially of the resultant fact that Forrest had been left to his own devices, which might well include a raid into Middle Tennessee and a strike against the blue supply lines running down into North Georgia. Engaged at the time in the Battle of Atlanta, Sherman replied that Smith was "to pursue and continue to follow Forrest. He must keep after him till recalled. . . . It is of vital importance that Forrest does not go to Tennessee." Smith returned to Memphis on July 23, miffed at this unappreciative reaction to his campaign, and began at once to prepare for a second outing, one that he hoped to improve beyond reproach.

This time the invasion column would number 18,000 of all arms, one quarter larger than before, and he would proceed by a different route — down the Mississippi Central, which he would repair as he advanced, thus solving the problem of supplies whose lack had obliged his recent withdrawal in mid-career. By August 2 the railroad was in running order down to the Tallahatchie, and Washburn notified Sherman that Smith's reorganized command, which he assured him could "whip the combined force of the enemy this side of Georgia and east of the Mississippi," would set out "as soon as possible. . . . Forrest's forces were near Okolona a week since," he added, saving the best news for last; "Chalmers in command. Forrest [has] not been able to resume command by reason of wound in fight with Smith. I have a report today that he died of lockjaw some days ago."

It was true that Chalmers was in nominal command, but not that Forrest was dead, either of lockjaw or of any other ailment, although a look at him was enough to show how the rumor got started. Troubled by a siege of boils even before he was wounded, "sick-looking, thin as a rail, cheekbones that stuck out like they were trying to come through the skin, skin so yellow it looked greenish, eyes blazing" — one witness saw him thus at Tupelo that week — he rode about the camps in a buggy, his injured foot propped on a rack atop the dashboard, waiting impatiently for it to heal enough for him to mount a horse and resume command of his two divisions. They were all that were left him now, about 5000 horsemen, after his casualties at Harrisburg and the departure of Stephen Lee, first for Mobile (where the reinforcements he took with him turned out not to be needed, Grant having ruled out Canby's attack by diverting his troops to Virginia) and then for Atlanta, to join Hood. Partly, too, Forrest's

★

haggard appearance was a result of the recent bloody repulse he had suffered in the assault on Smith. Even though he had advised against the attack, and was thereby absolved from blame for its failure, he was unaccustomed to sharing in a defeat and he burned with resentment over the useless loss of a thousand of his men, just at a time when they seemed likely to be needed most. Smith, he knew, was refitting in Memphis and would soon be returning to North Mississippi, stronger than before and with a better knowledge of the pitfalls.

Sure enough, by early August the new blue column of 18,000 effectives had moved out to Grand Junction and begun its advance down the Mississippi Central to Holly Springs, a day's march from the Tallahatchie. "We knew we couldn't fight General Smith's big fine army," a butternut artillery lieutenant would recall, "and we knew that we couldn't get any reinforcements anywhere, and we boys speculated about what Old Bedford was going to do."

Old Bedford wondered too, for a time. At first he thought Smith's movement down the railroad was a feint, designed to "draw my forces west and give him the start toward the prairies." Back in command — and in the saddle, though he only used one stirrup — he sent Chalmers's division over to cover the Mississippi Central, but kept Buford's around Okolona to oppose what he believed would be the main blue effort. He soon learned better. On August 8 Smith moved in strength from Holly Springs and forced a crossing of the Talla-hatchie, sending his cavalry ahead next day to occupy Oxford, twelve miles down the line. Forrest wired Chalmers to "contest every inch of ground," and set out at once for Oxford with Buford's division. Grierson fell back when he learned of this on August 10, and Smith remained at the river crossing, con-structing a bridge to ensure the rapid delivery of supplies when he continued his march south. It was then, in this driest season of the Mississippi year, that the rain began to fall. It fell and kept falling for a week, marking what became known thereafter in these parts as "the wet August."

Both sides were nearly im-mobilized by the deepening mud and washouts, but they sparred as best they could, in slow motion, and planned for the time ahead. On August 18, though

James R. Chalmers guarded the Mississippi Central while Confederate raiders pressed toward Memphis.

the weather still was rainy, Smith began inching southward; muddy or not, he had made up his mind to move, however slowly.

So by then had Forrest. At 5 o'clock that afternoon he assembled on the courthouse square at Oxford, after a rigorous "weeding out of sick men and sore-back and lame horses," close to 2000 troopers from two brigades and Morton's four-gun battery. In pelting rain and under a sky already dark with low-hanging clouds, the head of the column took up the march westward; Chalmers, left behind with the remaining 3000, had been told to put up such a show of resistance to the advancing Federals, who outnumbered him six to one, that Smith would not suspect for at least two days that nearly half of Forrest's command had left his front and was moving off to the west — in preparation for turning north around his flank, some were saying up and down the long gray column. "It got abroad in camp that we were going to Memphis," one rider later wrote. "That looked radical, but pleased us."

They knew they were right next morning, after a night march of twenty-five miles across swollen creeks and up and down long slippery hills, when they reached Panola and crossed the Tallahatchie, taking the route of the Mississippi & Tennessee Railroad, which ran north some sixty bee-line miles to Memphis. Four separate invasions they had repulsed in the past six months, three by pitched battle, one by sheer bluff, and now they were out to try their hand at turning back the fifth with a strike at the enemy's main base, close to a hundred miles in his rear. Radical, indeed. But Forrest knew what he would find when he got there; home-town operatives had kept him well informed. Washburn, under repeated urgings from Sherman to strengthen Smith to his utmost, had stripped the city's defense force to a minimum, and Fort Pickering, whose blufftop guns bore on the river and the city, but not on its landward approaches, offered little in the way of deterrent to an operation of this kind; Forrest did not intend to stay there any longer than it took his raiders to spread confusion among the defenders and alarm them into recalling Smith, who by now was skirmishing with Chalmers around Oxford, unaware that the man he was charged with following "to the death" had already rounded his flank and was about to set off an explosion deep in his rear.

Twenty miles the butternut column made that day, north from Panola to Senatobia, lighter by about two hundred troopers whose mounts had broken down before they reached the Tallahatchie and turned back, along with all but two of Morton's guns, whose teams were increased to ten horses each to haul them. The rain had stopped, as if on signal from the Wizard. All day the sun beamed down on roads and fields, but only enough, after eight days of saturation, to change the mud from slippery to sticky.

One mile north of Senatobia, which he cleared at first light, August 20, Forrest came upon Hickahala Creek, swollen to a width of sixty feet between its flooded banks; a formidable obstacle, but one for which he had planned by sending ahead a detachment to select a crossing point and chop down two trees on each bank, properly spaced, the stumps to be used for the support of a pair of cables woven from muscadine vines, which grew to unusual size and in great profusion in the bottoms. By the time the main body came up, the suspension cables had been stretched and were supported in midstream by an abandoned flatboat, which in turn was buoyed up by bundles of poles lashed to its sides. All that remained was for the span to be floored, and this was done with planks the troopers had ripped from gins and cabins on the approach march. In all, the crossing took less than an hour; but six miles north lay the Coldwater River, twice as wide. That took three, the work party having hurried ahead to construct another such grapevine bridge with the skill acquired while improvising the first. The heaviest loads it had to bear were the two guns, which were rolled across by hand, and several wagons loaded with unshucked corn for the horses, which were unloaded, trundled empty over the swaying rig, and then reloaded on the opposite bank. Forrest set the example by carrying the first armload, limping across on his injured foot, much to the admiration and amusement of his soldiers. "I never saw a command more like it was out for a holiday," one later wrote, while the general himself was to say: "I had to continually caution the men to keep quiet. They were making a regular corn shucking out of it."

Many of them, like him, were on their way home for the first time in years, and it was hard to contain the exuberance they were feeling at the prospect. Eight miles beyond the Coldwater by dark, Forrest called a rest halt at Hernando, where he had spent most of his young manhood, twenty-five miles from downtown Memphis. Near midnight the column pushed on, reduced to about 1500 sabers (so called, though for the most part they preferred shotguns and navy sixes) by the breakdown of another 200-odd horses, and stopped at 3 a.m. just short of the city limits, there to receive final instructions for the work ahead — work that was based on detailed information smuggled out by spies. One detachment under the general's brother, Captain William Forrest, would lead the way over Cane Creek Bridge and ride straight for the Gayoso House on Main Street, where Washburn's predecessor Stephen Hurlbut was quartered while awaiting reassignment; two other detachments, one of them under another brother, Lieutenant Colonel Jesse Forrest, would proceed similarly to capture Brigadier General R. P. Buckland, commander of the garrison, and Washburn himself, both of whom were living with their staffs in commandeered private residences. Two major generals and a brigadier would make a splendid haul and Forrest intended to have them, along with much else in the way of spoils assigned to still other detachments. Half an hour before dawn of this foggy Sunday morning, August 21, the head of the column entered the sleeping city whose papers had carried yesterday a special order from the department commander, prohibiting

During the Memphis raid, Bedford Forrest hoped
to capture two Union major generals and
a brigadier quartered there along with other spoils.

all "crying or selling of newspapers on Sunday between the hours of 9 a.m. and 5 p.m.," the better to preserve the peace and dignity of the Sabbath.

In some ways, the raid — the penetration itself — was anticlimactic. For example, all three Federal generals escaped capture, one because he slept elsewhere that night (just where became the subject of much scurrilous conjecture) and the other two because they were alerted in time to make a dash for safety under Fort Pickering's 97 guns, which Forrest had no intention of storming.

Buckland woke to a hammering, a spattering of gunfire some blocks off, and leaned out of his upstairs bedroom window to find a sentry knocking at the locked door of the house. He called down, still half asleep, to ask what was the matter.

"General, they are after you."

"Who are after me?"

"The rebels," he was told.

He had time to dress before hurrying to the fort. Not so Washburn, who had to make a run for it in his nightshirt through back alleys; so sudden was the appearance of the raiders at his gate, he barely had time to leave by the rear door as they entered by the front. By way of consolation, Jesse Forrest

★

captured two of his staff officers, along with his dress uniform and accouterments. Bill Forrest got even less when he clattered up Main Street to the Gayoso and, without pausing to dismount, rode his horse through the hotel doorway and into the lobby; Hurlbut, as aforesaid, had slept elsewhere and had only to lie low, wherever he was, to avoid capture. This he did, and survived to deliver himself of the best-remembered comment anyone made on either side in reference to the raid. "They removed me from command because I couldn't keep Forrest out of West Tennessee," he declared afterwards, "and now Washburn can't keep him out of his own bedroom."

By then enough blue units had rallied to bring on a number of vicious little skirmishes and fire-fights, resulting in a total of 35 Confederates and 80 Federals being killed or wounded, in addition to 116 defenders captured — many of them officers, rounded up in their night clothes at the Gayoso and elsewhere — along with some 200 horses. All this time, surprise reunions were in progress around town, despite the fact that recognition was not always easy: as, for example, in the case of a young raider who hailed his mother and sister from the gate of the family home, only to find that they had trouble identifying a tattered mud-spattered veteran as the boy they had kissed goodbye when he left three years ago, neatly turned out in well-pressed clothes for a war that would soon be won.

At 9 o'clock, satisfied that he had created enough disturbance to produce the effect he wanted, Forrest had the recall sounded and began the pre-arranged withdrawal. Beyond Cane Creek he paused to return, under a flag of truce, Washburn's uniform, which his brother Jesse proudly displayed as a trophy of the raid. (Whatever deficiencies he might show in other respects, Washburn knew how to return a courtesy. Some weeks later he sent Forrest, also under a flag of truce, a fine gray uniform made to measure by the cavalryman's own prewar Memphis tailor.) The column then took up the southward march, clearing Hernando that afternoon to ride back across the Tallahatchie and into Panola, late the following day. "If the enemy is falling back, pursue them hard," Forrest instructed Chalmers in a message taken cross-country by a courier who found him just below Oxford, still resisting Smith's advance.

That admonition — "pursue them hard" — was presently translated into action. Smith had entered Oxford that morning, but had no sooner done so than he began to backpedal in response to the news, brought forward under armed escort, that Forrest had raided Memphis the day before. Withdrawing, the Federals set fire to the courthouse, along with other public buildings and a number of private residences. "Where once stood a handsome little country town," an Illinois correspondent wrote, "now only remain the blackened skeletons of houses, and smouldering ruins." Smith's retrograde movement was hastened by a follow-up report next day, August 23, that the raiders were returning to Memphis for a second and heavier strike. The report was false (Forrest was still

★

at Panola, a hard two-day march to the south, resting his troopers from their 150-mile excursion through the Mississippi bottoms) but was almost as disruptive, in its effect, as if it had been true. Alarm bells rang; regulars and militiamen turned out — "eager for the fray," one of the latter said — and Washburn asked the naval commander to have a gunboat steam downriver, below Fort Pickering, to shell the southern approaches to the city. This was done, but with no more than pyrotechnical effect, since the raiders were only there by rumor, not in fact. "The whole town was stampeded," Washburn's inspector general declared, calling the reaction "the most disgraceful affair I have ever seen."

This too had its influence. Within another two days no part of A. J. Smith's command remained below the Tallahatchie, and so closely did Chalmers press him, in accordance with Forrest's instructions, that he soon abandoned close to a hundred miles of telegraph wire along the route from the river-crossing, all the way back to the outskirts of Memphis.

Washburn put the best possible interpretation on the outcome of the visit paid him by the raiders. "The whole Expedition was barren of spoils," he wrote his congressman brother Elihu. "They were in so great a hurry to get away that they carried off hardly anything. I lost two fine horses, which is about the biggest loss of anybody." So did Sherman tend to look on the bright side of the event. "If you get the chance," he wired Washburn on August 24, the day after the big stampede, "send word to Forrest that I admire his dash but not his judgment. The oftener he runs his head against Memphis the better."

There was much in that; Forrest's activities, these past four months, had been limited to North Mississippi and the southwest corner of Tennessee, with the result that he had been kept off Sherman's all-important supply line throughout this critical span. But it also rather missed the point that, with Memphis under cower and afflicted with a bad case of the shakes, the Wizard now was free to ride in practically any direction he or his superiors might choose: including Middle Tennessee, a region that nurtured a vital part of that supply line. The question was whether there was time enough, even if he were given his head at last, for Forrest's movement to be of much help to Hood in besieged Atlanta.

———————— ⟳ ————————

★ ★ ★   *E*ncouraged by Wheeler's recent victories over Stoneman and McCook, which he believed more or less disposed of the blue cavalry as a threat, Hood by then had thrown his own cavalry deep into the Union rear in North Georgia and East Tennessee, hoping, as he explained in a wire requesting the President's approval, that by severing Sherman's life line he would provoke him into rashness or oblige him to retreat. Davis readily concurred, having urged such a strike on Johnston, without success, from the outset to the time of his removal. He replied that he shared Hood's hope that this would "compel the enemy to attack you in position," but added, rather pointedly, and in a tone not unlike Lincoln's when cautioning Grant, down near Richmond the month before, on the heels of repulses even more costly than Hood had just suffered around Atlanta: "The loss consequent upon attacking him in his intrenchments requires you to avoid that if practicable."

Wheeler set out on August 10, taking with him some 4500 effectives from his eight brigades and leaving about the same number behind, including William Jackson's three-brigade division, to patrol and protect Hood's flanks and rear while he was gone. His itinerary for the following week, northward along the Western & Atlantic, resembled a synopsis, in reverse, of the Johnston-Sherman contest back in May. Marietta, Cassville, Calhoun, Resaca: all were hit on a five-day ride that saw the destruction of some thirty miles of track and the rebuilt bridge across the Etowah. On August 14, after detaching one brigade to escort his prisoners and captured livestock back to Atlanta, he began a two-day demonstration against Dalton, then continued north, around and beyond Chattanooga, to Loudon. He intended to cross the Tennessee River there, but found it in flood and had to continue upstream nearly to Knoxville, where he detached two more brigades to wreck the railway bridge at Strawberry Plains, then turned southwest, beyond the Holston and the Clinch, to descend on the Nashville & Chattanooga Railroad, which he broke in several places before he recrossed the Tennessee at Tuscumbia, Alabama, on September 10, his twenty-eighth birthday. At a total cost of 150 casualties on this month-long raid, in the course of which he "averaged 25 miles a day [and] swam or forded 27 rivers," Wheeler reported the seizure of "1000 horses and mules, 200 wagons, 600 prisoners, and 1700 head of beef cattle," and claimed that his command had "captured, killed, or wounded three times the greatest effective strength it has ever been able to carry into action."

As an exploit, even after allowing for the exaggeration common to most cavalry reports, this was much. In other respects, however, it amounted to little more than a prime example of how events could transform a tactical triumph into a strategic cipher. Although Wheeler accomplished practically everything he was sent out to do, and on a grander scale than had been intended, the only real effect of the raid was not on Sherman — whose work gangs were about as quick to repair damage to the railroads as the gray troopers had been to inflict it — but on Hood, who was deprived thereby of half his cavalry during the critical final stage of the contest for Atlanta; which, in point of fact, had ended before Wheeler recrossed the Tennessee. One further result of the raid, also negative, was that Hood at last was convinced, as he said later, "that no sufficiently effective number of cavalry could be assembled in the Confederacy to interrupt the enemy's line of supplies to an extent to compel him to retreat."

Sherman was no more provoked into rashness than he was into retreat, but Wheeler's absence did encourage him, despite the recent failure of such efforts, to venture still another cavalry strike at the Macon & Western, Hood's only remaining rail connection, whose rupture would oblige him to evacuate Atlanta for lack of supplies. Another persuasive factor was Judson Kilpatrick. Back in the saddle after a ten-week convalescence from the wound he had taken at Resaca, he seemed to Sherman just the man to lead the raid. Unlike Garrard — who, in Sherman's words, would flinch if he spotted "a horseman in the distance with a spyglass" — Little Kil had a reputation as a fighter, and though in the present instance he was advised "not to fight but to work," only boldness would assure success. Reinforced by two brigades from Garrard, the bandy-legged New Jerseyite took his division southeast out of Sandtown on the night of August 18, under instructions to "break up the Macon [rail]road about Jonesboro," twenty miles below Atlanta. He got there late the following day, unimpeded, and began at once to carry out Sherman's orders, passed on by Schofield: "Tell Kilpatrick he cannot tear up too much track nor twist too much iron. It may save this army the necessity of making a long, hazardous flank march."

First he set fire to the depot, then turned his attention to the road itself. But before he had ripped up more than a couple of miles of track he was attacked from the rear by a brigade of Texans from Jackson's division. Kilpatrick pressed on south, pursued by this and Jackson's other two brigades, but ran into infantry intrenched near Lovejoy Station and veered east, then north to reënter his own lines at Decatur. That was on August 22, and he proudly reported that he had done enough damage to Hood's life line to remove it from use for the next ten days. Sherman was delighted: but only overnight. Next morning, heavy-laden supply trains came puffing into Atlanta over tracks he had been

A Federal work crew stands before a rebuilt trestle bridge, bearing tracks of the Western & Atlantic, over the Chatta- hoochee River north of Atlanta.

assured were demolished. Told "not to fight but to work," Kilpatrick apparently had not done much of either, or else the rebel crews were as adept at repairs as their Union counterparts north of the city. In any case, Sherman said later, "I became more than ever convinced that cavalry could not or would not work hard enough to disable a railroad properly, and therefore resolved at once to proceed to the execution of my original plan."

This was the massive counterclockwise slide, the "grand left wheel around Atlanta," which he had designed to bring on the fall of the city by trans- ferring all but one of his seven infantry corps around to the south, astride its only rail connection with the outside world. Interrupted at Ezra Church in late July, the maneuver had been resumed only to stall again in the toils of Utoy Creek in early August. Since then, Sherman had sought by continuous long-range shelling, if not to convert the Gate City into "a desolation," as he had proposed two weeks ago, then in any case to reduce it to "a used-up community," and in this he had succeeded to a considerable extent, though not at a rate that matched his impatience, which was quickened by the spirit-lifting news of Farragut's triumph down in Mobile Bay. Now — Kilpatrick having failed, in Wheeler's absence, to spare him "the necessity of making a long, hazardous flank march" — he was ready to resume his ponderous shift.

Leaving Slocum's corps (formerly Hooker's) north of Atlanta, securely intrenched in a position from which to observe the reaction there and also protect the railway bridge across the Chattahoochee, he pulled all three armies rearward out the Sandtown Road on August 26 and started them south the following day in three wide-sweeping arcs, Howard and Schofield on the left and right, Thomas as usual in the center. Their respective objectives, all on the Macon Railroad, were Rough & Ready Station, four miles below East Point; Jonesboro, ten miles farther down the line; and a point about midway between the two. Thomas and Howard took off first, having longer routes to travel, and reached the inactive West Point Railroad next day at Red Oak and Fairburn, where they were to swing east. Then Schofield set out on his march, which was shorter but was presumably much riskier, since he would be a good deal closer to the rebels massed in and around Atlanta. As it turned out, however, he met with no more resistance than Howard and Thomas had done in the course of their wider sweeps; which was practically none at all. Welcome as this nonintervention was, Sherman also found it strange, particularly in contrast to his opponent's previous violent reaction to any attempt to move across his front or round his flank.

Hood's reaction, or nonaction, was stranger than any Federal supposed, being founded on a total misconception of what his adversary was up to. Not that his error had been illogically arrived at; it had not; but the logic, such as it was, was based insubstantially on hope. Suddenly, on August 26, after weeks of intensive shelling, the bombardment of Atlanta stopped as abruptly as a dropped watch, and when patrols went out at midday to investigate this unexpected silence — which somehow was even heavier with tension than the diurnal uproar that preceded it — they found the Union trenches empty and skirmishers posted rearguard-fashion along and on both sides of the road leading west to Sandtown and the Chattahoochee. Apparently a mass movement was in progress in that direction. Only on the north side of the city, in position to defend the indispensable railroad crossing and forward base, were the old works still occupied in strength. Hood's spirits took a leap at the news; for the brigade detached by Wheeler the week

Taking cover from the intensive shelling of their city, members of an Atlanta family and their slave tend to their chores in an underground structure.

before, up near Calhoun, had returned that morning with its haul of prisoners and cattle and a first-hand account of the extensive damage so far done to the Western & Atlantic, including the burning of the vital span across the Etowah. Wheeler himself, according to a report just in, was beyond Chattanooga with the rest of his command, preparing by now to cross the Tennessee River and descend on the blue supply line below Nashville. All this was bound to have its effect; Sherman must already be hurting for lack of food and ammunition. Indeed, there was testimony on hand that this was so. Six days ago, a woman whose home was inside Schofield's lines had appealed to one of his division commanders for rations, only to be refused. "No," she was told; "I would like to draw, myself. I have been living on short rations for seven days, and now that your people have torn up our railroad and stolen our beef cattle, we must live a damned sight shorter."

On such evidence as this, and out of his own sore need for a near miracle, Hood based his conclusion that Sherman, threatened with the specter of starvation by Wheeler's disruption of his life line, was in full retreat across the Chattahoochee with all of his corps but one, left temporarily in position north of the city to cover the withdrawal by rail of what remained of his sorely depleted stockpile of provisions.

Orders went out for Jackson to bring his overworked troopers in from the flanks and take up the pursuit toward Sandtown. Jackson did, beginning next day, but reported that the bluecoats seemed to him to be regrouping, not retreating. Hood rejected this assessment, preferring to believe that his cavalry simply lacked the strength to penetrate the Federal rear guard. So near the end of his military tether that he had nothing to fall back on but delusion, he held his three corps in the Atlanta intrenchments, which had been extended down to East Point, awaiting developments.

They were not long in coming. Sherman had Howard and Thomas spend a day astride the West Point Railroad, "breaking it up thoroughly," as he said, lest the rebels someday try to put it back in commission. His veterans were highly skilled at such work by now, and he later described how they went about it. "The track was heaved up in sections the length of a regiment, then separated rail by rail; bonfires were made of the ties and of fence rails on which the rails were heated, carried to trees or telegraph poles, wrapped around and left to cool." Not content with converting the rails into scrap iron — "Sherman neckties," the twisted loops were called — he then proceeded against the roadbed itself. "To be still more certain, we filled up many deep cuts with trees, brush, and earth, and commingled with them loaded shells, so arranged that they would explode on an attempt to haul out the bushes. The explosion of one such shell would have demoralized a gang of negroes, and thus would have prevented even the attempt to clear the road." Next morning, August 30, he started both armies east toward the headwaters of Flint River, which flowed south between the two converging railroads, the one he had just undone in his rear and the one ahead, whose loss would undo Hood.

Elated at the prospect of achieving this objective, he accompanied Thomas on the march, and as they approached the Flint that afternoon — still without encountering serious opposition, though the Macon road lay only a scant two miles beyond the river — he exulted to the Virginian riding beside him: "I have Atlanta as certainly as if it were in my hand!"

Hood by now had begun to emerge from his wishful three-day dream. Reports that Union infantry had appeared in strength on the West Point road the day before, above and below Fairburn and Red Oak, obliged him to concede that part at least of Sherman's host was headed for something other than the Chattahoochee River, and when follow-up dispatches informed him this morning that the same blue wrecking force was moving eastward, in the direction of the Macon road, he knew he had to act. All surplus goods were ordered packed for shipment out of the nearly beleaguered city, by whatever routes might be available when the time came, and Hardee was told to shift to Rough & Ready, bracing his corps for the defense of the rail supply line, there or farther down, while Lee moved out to take his place at East Point, under instructions to be ready for a march in either direction, southward to reinforce Hardee or back north to assist Stewart in the close-up defense of Atlanta, depending on which turned out to need him worst. Old Straight remained in the works that rimmed the city, not only because of Slocum's hovering menace, but also because Hood had revised — indeed, reversed — his estimate of the enemy's intentions. It seemed to him that Sherman was trying to draw him out of Atlanta with a strike at his supply line, say by half the Federal force, so that when he moved to meet this threat, the other half, concealed till then near the Chattahoochee, could swoop down and take the city. Hood's job, as he assessed it, was to avoid being lured out in such numbers that Atlanta would fall in their absence, its scantly manned intrenchments overrun, and yet at the same time to prevent the seizure or destruction of the Macon Railroad, whose loss would require him to give up the city for lack of subsistence.

Caught thus between the blue devil and the deep blue sea, Hood saw no choice, now that he had been shaken out of the dream that transformed his red-haired opponent from a destroyer into a deliverer, except to try to meet these separate dangers as they developed. All in all, outnumbered as he was, the situation was pretty much as Sherman was describing it to Thomas even now, a dozen-odd miles to the south: "I have Atlanta as certainly as if it were in my hand." What had the earmarks of a frothy boast — of a kind all too common in a war whose multi-thumbed commanders were often in need of reassurance, even if they had to express it themselves — was in fact merely a tactical assessment, somewhat florid but still a good deal more accurate than most.

Or maybe not. When Hood heard from Hardee, around midday, that the blue march seemed to be aimed at both Rough & Ready and Jonesboro, ten miles apart, he saw once more a chance to strike the enemy in detail. And having perceived this he was no less willing to undertake it than he had been three times before, in as many costly sorties. Now as then he improvised a slashing assault designed to subject a major portion of the Union host to destruction. His plan — refined to deal with a later, more specific report that Logan's corps had crossed the Flint that afternoon and gone into camp within cannon range of Jonesboro, supported only by Kilpatrick's horsemen, while the other two corps of the Army of the Tennessee remained on the west bank of the stream — was for Hardee to fall upon this exposed segment early next morning and "drive the enemy, at all hazards, into Flint River, in their rear." Moreover, when the rest of Howard's troops attempted to come to Logan's assistance they could be whipped in detail with help from Lee, whose corps would set out down the

---

"I have Atlanta as certainly as if it were in my hand!"

— William Tecumseh Sherman

---

railroad from East Point at the same time Hardee's moved from Rough & Ready on a night march that would put them in position for attack at first light, August 31. To make certain that his plan was understood, Hood wired both generals to leave their senior division commanders in charge of the march to Jonesboro and report to him in Atlanta, by rail, for the usual face-to-face instructions, which experience had shown were even more necessary than he had thought when he first took charge of the Army of Tennessee.

In Atlanta that night, at the council of war preceding this Fourth Sortie, Hood expanded his plan to include a follow-up attack September 1. After sharing in tomorrow's assault, which would drive the Federals away from the Macon road and back across the Flint, Lee was to return to Rough & Ready Station, where he would be joined by Stewart for an advance next morning, down the west bank of the river, that would strike the flank of the crippled bluecoats, held in position overnight by Hardee, and thus complete their destruction. This was in some ways less risky and in others riskier than Hood knew, believing as he did that only Howard's army was south of the city, which thus would be scantly protected from an assault by Thomas and Schofield. For that reason, Hood took what he believed was the post of gravest responsibility: Atlanta, whose defenses would be manned, through this critical time, only by Jackson's

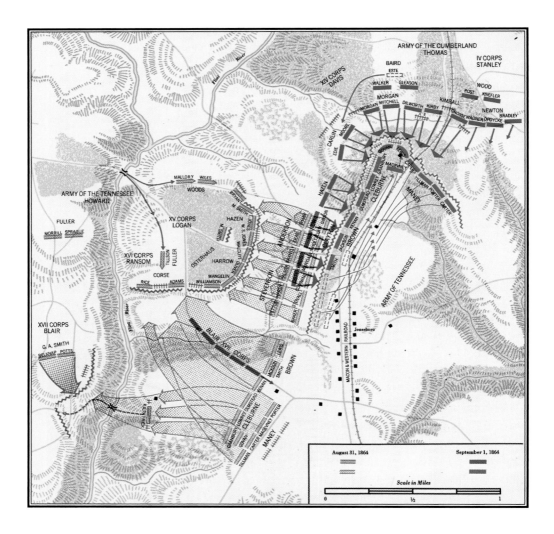

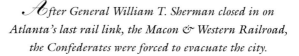

*After General William T. Sherman closed in on
Atlanta's last rail link, the Macon & Western Railroad,
the Confederates were forced to evacuate the city.*

dismounted troopers and units of the Georgia militia. It was late when the
council broke up and Hardee, who was put in charge of the attack, boarded a
switch engine for a fast ride to Jonesboro. He arrived before dawn, expecting to
find his and Lee's corps being posted for the assault at daybreak. Neither was
there; nor could he find anyone who could tell him where they were — Lee's,
which that general must have rejoined by now, or his own, which had set out
southward from Rough & Ready the night before.

Howard remained all morning in what he called a "saucy position," content to reinforce Logan's corps, intrenched on the east bank of the Flint, with a single division from Dodge, who was away recuperating from being struck on the forehead by a bullet the week before. He expected to be attacked by a rebel force that seemed to be gathering in Jonesboro, less than a mile across the way; that was why he kept most of his troops out of sight on the west side of the river, hoping, now that Logan's men had had plenty of time to strengthen their intrenchments, that the graybacks would come to him, rather than wait for him to storm their works along the railroad. But when nothing had come of this by the time the sun swung past the overhead, he decided he would have to prod them. He told Logan to move out at 3 o'clock. At 2.45, just as Black Jack's veterans were preparing to leave their trenches, long lines of butternut infantry came surging out of Jonesboro in far greater numbers than Howard had expected while trying to provoke them into making an attack.

Hardee was even tardier in launching Hood's Fourth Sortie than he had been in either of the other two committed to his charge, the first having opened two hours behind schedule, the second nearly seven, and this one more than nine. Yet here again the blame was hard to fix. Cleburne, left in corps command when Old Reliable went to Atlanta the night before, had found enemy units blocking his line of march and had had to detour widely around them, which delayed his arrival in Jonesboro until an hour after sunrise; while Lee, whose longer route was even worse obstructed, did not come up till well past noon. As a result, it was 2 o'clock before Hardee could get the two road-worn corps into jump-off positions and issue orders for the attack. These were for Cleburne to turn the enemy's right and for Lee to move against their front as soon as he heard Cleburne's batteries open.

Such a signal had often failed in the past, and now it did so here. Mistaking the clatter of skirmishers' rifles for the roar of battle, Lee started forward on his own and thus exposed his corps to the concentrated fire of the whole Union line, with demoralizing results. Cleburne then moved out, driving Kilpatrick's troopers promptly across the Flint, but found Logan's works too stoutly held for him to effect a lodgment without assistance. Hardee urged Lee to renew his stalled advance, only to be told that it was impossible; Howard was bringing reserves across the river to menace the shaken right. In reaction, Hardee called off the attack and ordered both Cleburne and Lee to take up defensive positions, saying later: "I now consider this a fortunate circumstance, for success against such odds could at best have only been partial and bloody, while defeat would have [meant] almost inevitable destruction to the army."

That ended the brief, disjointed Battle of Jonesboro; or half ended it, depending on what Howard would do now. Lee and Cleburne had suffered more than 1700 casualties between them, Logan and Kilpatrick less than a fourth as many, and these were the totals for this last day of August, as it turned out, since Howard did not press the issue. Late that night, in response to Hood's repeated summons, Hardee detached Lee's three divisions for the return march north, tomorrow's scheduled follow-up offensive down the west bank of the Flint having been ruled out by the failure of today's attempt to set up Howard for the kill.

What Hood now wanted Lee for, though, was to help Stewart hold Atlanta against the assault he expected Sherman to make next morning with the other two Federal armies, which he still thought were lurking northwest of the city. He presently learned better. Soon after dark, reports came in that bluecoats were across the Macon road in strength at Rough & Ready, as well as at several other points between there and Jonesboro. Lee not only confirmed this when he reached East Point at daylight, having managed to slip between the enemy columns in the darkness; he also identified them as belonging to Schofield and Thomas. This was a shock, and its meaning was all too clear. Atlanta was doomed. The only remaining question, now that Sherman had the bulk of his command astride the city's last rail supply line, squarely between Hardee and the other two corps, was whether the Army of Tennessee was doomed as well. Hood and his staff got to work at once on plans for the evacuation of Atlanta and the reunion, if possible, of his divided army, so that it could be saved to fight another day.

Such a reunion was not going to include Hardee's third of that army if Sherman had his way. Primarily he had undertaken this six-corps grand left

wheel as a railroad-wrecking expedition, designed to bring on the fall of Atlanta by severing its life line, but now that he saw in Hardee's isolation an opportunity to annihilate him, he extended its scope to achieve just that. Both Schofield and Thomas were told to move on Jonesboro without delay, there to combine their three corps with Howard's three — a total of more than 60,000, excluding cavalry — for an assault on Hardee's 12,500, still licking the wounds they had suffered in their repulse the day before. While this convergence was in progress Howard put the rest of Dodge's corps across the Flint, where Logan confronted the rebels in their works, and sent Blair to cut the railroad south of town and stand in the path of any escape in that direction.

Noon came and went, this hot September 1, still with no word from Thomas or Schofield, who were to attack the Confederates on their right while Howard clamped them in position from the front. Sherman fumed at the delay, knowing the graybacks were hard at work improving their intrenchments, and kept fuming right up to 3 o'clock, when the first of Slow Trot Thomas's two corps arrived, formerly John Palmer's but now under Jeff C. Davis, Palmer having departed in a huff after a squabble with Schofield, who he claimed had mishandled his troops in the Utoy Creek fiasco. The other Cumberland corps, David Stanley's, was nowhere in sight, and in fact did not turn up till after sundown, having got lost on its cross-country march, and Schofield moved so slowly from Rough & Ready, tearing up track as he went, that he arrived even later than Stanley. Combined with the detachment of Blair to close the southward escape hatch, the

Jefferson C. Davis's Federal XIV Corps charges the rebel line under Patrick Cleburne at Jonesboro, 15 miles south of Atlanta, on September 1, 1864.

nonappearance of these two corps reduced the size of the attacking force by half. But that still left Sherman with considerably better than twice the number he faced, and he also enjoyed the advantage of having Davis come down unexpectedly on the enemy right, which was bent back across the railroad north of town.

Davis was a driver, a hard-mannered regular who had come up through the ranks, thirty-six years old, with wavy hair and a bushy chin-beard, a long thin nose and the pale, flat eyes of a killer; which he was. Still a brigadier despite his lofty post and a war record dating back to Sumter, he had been denied promotion for the past two years because of the scandal attending his pistol slaying of Bull Nelson in Kentucky, long ago in '62, and he welcomed such assignments as this present one at Jonesboro, seeing in them opportunities to demonstrate a worth beyond the grade at which he had been stopped in his climb up the military ladder. He put his men in line astride the railroad — three divisions, containing as many troops as Hardee had in all — and sent them roaring down against the rebel flank at 4 o'clock. Cleburne's division was posted there, in trenches Lee had occupied the day before. Repulsed, Davis dropped back, regrouped quickly, and then came on again in a mass assault that went up and over the barricade to land in the midst of Brigadier General Dan Govan's veteran Arkansas brigade. Two batteries were overrun and Govan himself captured, along with more than half his men. "They're rolling them up like a sheet of paper!" Sherman cried, watching from an observation post on Howard's front.

But Granbury's Texans were next in line, and there the rolling stopped. Cleburne shored up his redrawn flank, massing fire on the lost salient, and Davis had all he could do to hold what he had won. Unwilling to risk a frontal assault by Howard, Sherman saw that what he needed now was added pressure on the weakened enemy right by Stanley, who was supposed to be coming up in rear of Davis. Angrily he turned to Thomas, demanding to know where Stanley was, and the heavy-set Virginian, who already had sent courier after courier in search of the errant corps, not only rode off in person to join the hunt, but also did so in a manner that later caused his red-haired superior to remark that this was "the only time during the campaign I can recall seeing General Thomas urge his horse into a gallop." Even so, the sun had set by the time Stanley turned up, and night fell before he could put his three divisions in attack formation. Darkness ended this second day of the Battle of Jonesboro, which cost Sherman 1275 casualties, mostly from Davis's corps, and Hardee just under 1000, two thirds of them captured in the assault that cracked his flank.

Disgruntled, Sherman bedded down, hopeful that tomorrow, with Schofield up alongside Stanley, he would complete the fate he planned for Hardee. He had trouble sleeping, he would recall, and soon after midnight, to add to his fret, "there arose toward Atlanta sounds of shells exploding, and other

As the Confederates prepared to evacuate Atlanta, they exploded an ordnance train and other stores, leveling the walls of the rolling mill in the background.

sounds like that of musketry." This was disturbing; Hood might well be doing to Slocum what he himself intended to do to Hardee. Yesterday he had instructed Thomas to have Slocum "feel forward to Atlanta, as boldly as he can," adding: "Assure him that we will fully occupy the attention of the rebel army outside of Atlanta." This last he had failed to do, except in part, and it seemed to him likely, from those rumblings twenty miles to the north, that he had thereby exposed Slocum to destruction by two thirds of Hood's command. Other listeners about the campfire disagreed, interpreting the muffled clatter as something other than battle, and Sherman decided to settle the issue by visiting a nearby farmhouse, where he had seen lights burning earlier in the evening. Shouts brought the farmer out into the yard in his nightshirt. Had he lived here long? He had. Had he heard such rumblings before? Indeed he had. That was the way it sounded when there was heavy fighting up around Atlanta.

★

The noise faded, then died away; which might have an even more gruesome meaning. Sherman returned to his campfire, still unable to sleep. Then at 4 o'clock it rose again, with the thump and crump and muttering finality of a massive coup de grâce. Again it died, this time for good. Dawn came, and with the dawn a new enigma. Thomas and Schofield moved as ordered, the latter on the left to sweep across the rebel rear — "We want to destroy the enemy," Sherman told them, anxious to be done with the work at hand — but found that Hardee had departed under cover of darkness and the distractive far-off rumblings from the north.

Atlanta Mayor James M. Calhoun asked Federal occupiers to protect noncombatants and private property.

Sherman took up the pursuit, southward down the railroad, still wondering what had happened deep in his rear. This was the hundred and twentieth day of the campaign, and while he was at Jonesboro another month had slipped into the past, costing him 7000 casualties and his adversary 7500: a total to date of 31,500 Federals and 35,000 Confederates, rough figures later precisely tabulated at 31,687 and 34,979 respectively. Close to 20,000 of the latter had been suffered by Hood in the nearly seven weeks since he took over from Johnston, while Sherman had lost just under 15,000 in that span.

Presently, as the six blue corps toiled southward down the railroad in search of Hardee's three vanished divisions, Schofield sent word that he took last night's drumfire rumblings from the direction of Atlanta to be the sound of Hood blowing up his unremovable stores, in preparation for evacuation. Two hours later, at 10.25, he followed this with a report that a Negro had just come into his lines declaring that the rebs were departing the city "in great confusion and disorder." Unconvinced, still troubled about "whether General Slocum had felt forward and become engaged in a real battle," Sherman kept up his pursuit of Hardee until he came upon him near Lovejoy Station, six miles down the line, his corps posted in newly dug intrenchments "as well constructed and as strong as if these Confederates had a week to prepare them." Such was his assessment after a tentative 4 o'clock probe was savagely repulsed. "I do not wish to waste lives by an assault," he warned Howard, explaining more fully to Thomas: "Until we hear from Atlanta the exact truth, I do not care about your pushing your men against breastworks." Still fretted by doubts about Slocum, he maintained his position of cautious observation through sunset into darkness. "Nothing positive from Atlanta," he informed Schofield within half an hour of midnight, "and that bothers me."

Finally, between then and sunup, September 3, a courier arrived with a dispatch from Slocum, who was not only safe but was safe inside Atlanta. Alerted by last night's racket, just across the way — it turned out to be the explosion of 81 carloads of ammunition, together with five locomotives, blown up in relays when they were found to be cut off from escape by the loss of the Macon road — he had felt his way forward at daylight to the city limits, where the commander of his lead division encountered a delegation of civilians. "Sir," their leader said with a formal bow. His name, it developed, was James M. Calhoun, and that was strangely fitting, even though no kinship connected him with the South Carolina original, John C. "The fortunes of war have placed the city of Atlanta in your hands. As mayor of the city I ask protection for noncombatants and private property." Slocum telegraphed the news to Washington: "General Sherman has taken Atlanta," and passed the word to his chief, approaching Lovejoy by then, that Hood had begun his withdrawal at 5 p.m. the day before, southward down the McDonough Road and well to the east of the Macon & Western, down which Howard and Thomas and Schofield were marching.

This meant that Hood had crossed their front and flank with Stewart and Lee and the Georgia militia, last night and yesterday, and by now had reunited his army in the intrenchments hard ahead at Lovejoy Station. Wise by hindsight, Sherman began to see that he had erred in going for Hardee, snug in his Jonesboro works, when he might have struck for the larger and more vulnerable prize in retreat on the McDonough Road beyond. Moreover, if he had been unable to pound the graybacks to pieces while he had them on the Atlanta anvil, there seemed little chance for success in such an effort now that they were free to maneuver as they chose. Such at last was the price he paid for having redefined his objective, not as the Army of Tennessee — "Break it up," Grant had charged him at the outset, before Dalton — but rather as the city that army had been tied to, until now.

In any case, he had it, and he was ready and anxious to take possession in person. "Atlanta is ours, and fairly won," he wired Halleck. "I shall not push much farther in this raid, but in a day or so will move to Atlanta and give my men some rest."

★   ★   ★

Shelby Foote

*Deep trenches like this one
concealed Union miners from
enemy view as they went in and out
of the long tunnel they dug under
rebel works at Petersburg.*

Crater; McClellan; Early II

1864 ★ ★ ★ ★ ★ ★ **S**locum's **wire,** received in Washington on the night of the day it was sent — "General Sherman has taken Atlanta" — ended a hot-weather span of anxiety even sorer than those that followed the two Bull Runs, back in the first two summers of the war. The prospect of stalemate, at this late stage, brought on a despondency as deep as outright defeat had done in those earlier times, when the national spirit displayed a resilience it had lost in the course of a summer that not only was bloody beyond all past imagining, but also saw Early within plain view of the Capitol dome and Democrats across the land anticipating a November sweep. Farragut's coup, down in Mobile Bay, provided no more than a glimmer of light, perfunctorily discerned before it guttered out in the gloom invoked by Sherman's reproduction, on the outskirts of Atlanta, of Grant's failure to take Richmond when he reached it the month before. Both wound up, apparently stalled, some twenty miles beyond their respective objectives, and by the end of August it had begun to appear that neither of them, having overshot the mark, was going to get back where he had been headed at the outset.

Nowhere, East or West or in between, was the disenchantment so complete as it was on the outskirts of Petersburg by then. Partly this was because of the high price paid to get there (Meade's casualties, exclusive of Butler's, were

★

more than twice as heavy as Sherman's, though the latter had traveled nearly twice as far by his zigzag route) and partly too because, time and again, the public's and the army's expectations had been lifted only to be dashed, more often than not amid charges of incredible blundering, all up and down the weak-linked chain of command. A case in point, supplementing the fiasco that attended the original attack from across the James, was an operation that came to be called "The Crater," which occurred in late July and marked a new high (or low) for mismanagement at or near the top, surpassing even Cold Harbor in that regard, if not in bloodshed.

Early that month, after the failure of his probe for the Weldon Railroad in late June, Grant asked Meade how he felt about undertaking a new offensive against Lee's center or around his flank. Faced as he was with the loss of Wright, whose corps was being detached just then to counter Early's drive on Washington, Meade replied that he was doubtful about the result of either a flank or a frontal effort, citing "the facility with which the enemy can interpose to check an onward movement." However, lest his chief suppose that he was altogether without aggressive instincts or intentions — which, in point of fact, he very nearly was by now — Meade did let fall that he had in progress a work designed to permit a thrust, not through or around, but *under* the Confederate intrenchments. Burnside was digging a mine.

The proposal had come from a regimental commander, Lieutenant Colonel Henry Pleasants, whose 48th Pennsylvania was made up largely of volunteers from the anthracite fields of Schuylkill County, one of whom he happened to hear remark, while peering through a firing slit at a rebel bastion some 150 uphill yards across the way: "We could blow that damned fort out of existence if we could run a mine shaft under it." Formerly a civil engineer engaged in railroad tunneling, Pleasants liked the notion and took a sketch of it to his division commander, Brigadier General Robert Potter, who passed it along to corps. Burnside told Pleasants to start digging, then went himself to Meade for approval and assistance. He got Meade's nod, apparently because the work at least would keep some bored men busy for a time, but not his help, his staff having advised that the project was impractical from the engineering point of view. No such tunnel could exceed 400 feet in length, the experts said, that being the limit at which fresh air could be provided without ventilation shafts, and this one was projected to extend for more than 500 feet from the gallery entrance to the powder chamber at its end.

Pleasants had been hard at work since June 25, the day Burnside told him to start burrowing into the steep west bank of an abandoned railway cut, directly in rear of his picket line and well hidden from enemy lookouts. By assigning his men to shifts so that the digging went forward round the clock, he managed to complete the tunnel within a month — though his miners later claimed they

*T*old by Meade's engineers no tunnel could exceed
400 feet, Colonel Henry Pleasants' miners
dug one 500-feet long to reach beyond rebel bastions.

could have done the job in less than half that time, if they had been given the proper tools. Not that Pleasants hadn't done his best in that regard. Denied any issue of special implements, such as picks, he contrived his own with the help of regimental blacksmiths, converted hardtack boxes into barrows for moving dirt, took over a wrecked sawmill to cut timbers and planks for shoring up the gallery walls and roof, and even borrowed a theodolite, all the way from Washington, when Meade's engineers declined to lend him one of theirs. Technical problems he solved in much the same improvisatory fashion, including some which these same close-fisted experts defined as prohibitive; ventilation, for example. Just inside the entrance he installed an airtight canvas door and beneath it ran a square wooden pipe along the floor of the shaft to the diggers at the end, extending it as they progressed. A fireplace near the sealed door sent heated air up its brush-masked chimney, creating a draft that drew the stale air from the far end of the tunnel and pulled in fresh air through the pipe, whose mouth was beyond the door. Working in the comparative comfort of a gallery five feet high, four feet wide at the bottom and two feet at the top — they had sweated and strained and wheezed and shivered through longer hours, with considerably less headroom and under far worse breathing conditions, back home in the Pennsylvania coal fields — the miners completed 511 feet of shaft by July 17.

This put them directly under the rebel outwork, whose defenders they could hear walking about, twenty feet above their heads, apparently unmindful of the malevolent, mole-like activity some half-dozen yards below the ground they stood on. Next day the soldier miners began digging laterally, right and left, to provide a powder chamber, 75 feet long, under the enemy bastion and the trenches on both flanks. By July 23 the pick and shovel work was done. After a four-day rest, Pleasants brought in 320 kegs of black powder, weighing 25 pounds each, and distributed this gritty four-ton mass among eight connected magazines, sandbagged to direct the explosion upward. When his requisition for insulated wire and a galvanic battery did not come through, he got hold of two fifty-foot fuzes, spliced them together, then secured one end to the monster charge and ran the other back down the gallery as far as it would reach; after which he replaced the earth of the final forty feet of tunnel, firmly tamped to provide a certain backstop. That was on July 28. All that remained was to put a match to the fuze, and get out before the boom.

Next afternoon, with the mine scheduled to be exploded early the following morning, Burnside assembled his division commanders to give them last-minute instructions for the assault that was to be launched through the resultant gap in the rebel works. Of these there were four, though only three of their divisions had done front-line duty so far in the campaign; the fourth, led by

The lead miner swings his pick as other Pennsylvanians pack soil in hardtack crates to cart out of the tunnel they were digging beneath Pegram's rebel battery.

Brigadier General Edward Ferrero, was composed of two all-Negro brigades whose service up to now had been confined to guarding trains and rearward installations, largely because of the continuing supposition — despite conflicting evidence, West and East — that black men simply were not up to combat. "Is not a Negro as good as a white man to stop a bullet?" someone asked Sherman about this time, over in Georgia. "Yes; and a sandbag is better," he replied. Like many eastern generals he believed that former slaves had their uses in war, but not as soldiers. Burnside felt otherwise, and what was more he backed up his contention by directing that Ferrero's division, which was not only the freshest but was also by now the largest of the four, would lead tomorrow's predawn charge. By way of preparation, he had had the two brigades spend the past week rehearsing the attack until every member knew just what he was to do, and how; that is, rush promptly forward, as soon as the mine was sprung, and expand the gap so that the other three divisions, coming up behind, could move unopposed across the Jerusalem Plank Road and onto the high ground immediately in rear of the blasted enemy intrenchments, which would give them a clear shot at Petersburg itself.

He was in high spirits, partly because the digging had gone so well and partly because Meade and Grant, catching a measure of his enthusiasm as the tunnel neared completion, had expanded the operation. Not only were Warren's and Baldy Smith's corps ordered to stand by for a share in exploiting the breakthrough — which was to be given close-up support by no less than 144 field pieces, mortars, and siege guns: more artillery, pound for pound, than had been massed by either side at Gettysburg — but Grant also sent Hancock's corps, along with two of Sheridan's divisions, to create a diversion, and if possible score an accompanying breakthrough, on the far side of the James. Hancock, who had returned to duty the week before, found the Confederates heavily reinforced in front of Richmond: as did Sheridan, who was worsted in a four-hour fight with Hampton on the day the fuze was laid to Pleasants's mine. Still, the feint served its purpose by drawing large numbers of graybacks away from the intended scene of the main effort, about midway down the five-mile rebel line below the Appomattox. Intelligence reported that five of Lee's eight infantry divisions were now at Bermuda Hundred or north of the James, leaving Beauregard with only three divisions, some 18,000 men in all, for the defense of the Petersburg rail hub. Moreover, there still was time for Hancock to return tomorrow — the day of Burnside's last-minute council of war — to lend still greater weight to the assault that would accompany the blasting of the undermanned enemy works before daylight next morning.

Burnside was happily passing this latest news along to his lieutenants when he was interrupted by a courier from army headquarters, bearing a message that had an effect not unlike the one expected, across the way, when the mine was sprung tomorrow. It contained an order from Meade, approved by

Grant, for the assault to be spearheaded not by Ferrero's well-rehearsed Negroes, but by one of the white divisions. This change, which landed like a bomb in the council chamber, was provoked by racism; racism in reverse. "If we put the colored troops in front and [the attack] should prove a failure," Grant would testify at the subsequent investigation, "it would then be said, and very properly, that we were shoving those people ahead to get killed because we did not care anything about them."

Stunned, Burnside tried to get the order rescinded, only to be told that it would stand; Meade was not about to give his Abolitionist critics this chance to bring him down with charges that he had exposed black recruits to slaughter in the forefront of a long-shot operation. By now the scheduled assault was less than twelve hours off, all but four of them hours of darkness, and the ruff-whiskered general, too shaken to decide which of his three unrehearsed white divisions should take the lead, had their commanders draw straws for the assignment. It fell to Brigadier General James H. Ledlie, a former heavy artilleryman, least experienced of the three. Potter and Brigadier General Orlando Willcox would attack in turn, behind Ledlie; Ferrero would bring up the rear.

As they departed to alert their troops, Burnside could find consolation only in reports that the Confederates — two South Carolina regiments, posted in support of the four-gun battery poised above the sealed-off powder chamber — seemed to have abandoned their former suspicion that they were about to be blown skyward. For a time last week they had tried countermining, without success, and when the underground digging stopped, July 23, so did their attempts at intersection. Apparently they too had experts who advised them that such a tunnel was impracticable; with the result that when the sound of picks and shovels stopped, down below, they decided that the Yanks had given up, probably after a disastrous cave-in or mounting losses from asphyxiation.

Eventually the troops were brought up in the darkness, groping their way over unfamiliar terrain to take up assigned positions for the jump-off: Ledlie's division out front, just in back of the ridge where the pickets were dug in, Potter's and Willcox's along the slope of the railway cut, and Ferrero's along its bottom, aggrieved at having been shunted to the rear. Elsewhere along the Union line the other corps stood by, including Hancock's, which had returned from its demonstration beyond the James. Shortly after 3 o'clock Pleasants entered the tunnel to light the fuze. The guns and mortars were laid, ammunition stacked and cannoneers at the ready, lanyards taut. Burnside had his watch out, observing the creep of its hands toward 3.30, the specified time for the springing of the mine. 3.30 finally came; but not the explosion. Half an hour went by, and still the night was black, unsplit by flame. Another half hour ticked past, bringing the first gray hint of dawn to the rearward sky, and though Pleasants had accepted his mine-boss sergeant's offer to go back into the tunnel and investigate the delay, there

Federal generals James Ledlie (left) and Edward Ferrero were censured after the Battle of the Crater for hiding in a bomb-proof during the attack.

still was no blast. Grant, losing patience, considered telling Burnside to forget the explosion and get on with his 15,000-man assault. Daylight grew, much faster now, and the flat eastern rim of earth was tinted rose, anticipating the bulge of the rising sun, by the time the sergeant and a lieutenant who had volunteered to join him — Harry Reese and Jacob Douty were their names — found that the fuze had burned out at the splice. They cut and relit it and scrambled for the tunnel entrance, a long 150 yards away, emerging just before 4.44, when the 8000-pound charge, twenty feet below the rebel works, erupted.

"A slight tremor of the earth for a second, then the rocking as of an earthquake," an awed captain would recall, "and, with a tremendous blast which rent the sleeping hills beyond, a vast column of earth and smoke shoots upward to a great height, its dark sides flashing out sparks of fire, hangs poised for a moment in mid-air, and then, hurtling down with a roaring sound, showers of stones, broken timbers and blackened human limbs, subsides — the gloomy pall of darkening smoke flushing to an angry crimson as it floats away to meet the morning sun." Another watcher of that burgeoning man-made cloud of dust and turmoil, a brigadier with Hancock, left an impression he never suspected would be repeated at the dawn of a far deadlier age of warfare, just over eighty years away: "Without form or shape, full of red flames and carried on a bed of lightning flashes, it mounted toward heaven with a detonation of thunder [and] spread out like an immense mushroom whose stem seemed to be of fire and its head of smoke."

Added to the uproar was the simultaneous crash of many cannon, fired by tense gunners as soon as they saw the ground begin to heave from the overdue explosion. Ledlie's men, caught thus between two shock waves, looked out and saw the rising mass of earth, torn from the hillside hard ahead, mount up and up until it seemed to hover directly above them, its topmost reaches glittering in the full light of the not-yet-risen sun. As the huge cluster started down, they recovered at least in part from their shock and reacted by breaking in panic for the rear. This was not too serious; their officers got them back in line within ten minutes and started them forward before the dust and smoke had cleared. But what happened next was serious indeed. In his dismay over the last-minute change in orders, Burnside had neglected to have the defensive tangle of obstacles cleared from in front of the parapets, with the result that the attack formation was broken

For more than a month their fighting had been confined to rifle pits and trenches, and now here at their feet was the biggest rifle pit in all the world.

up as soon as the troops set out. Instead of advancing on a broad front, as intended — a brigade in width, with the second brigade coming up in close support — they went forward through a hastily improvised ten-foot passway that not only delayed their start but also confined them to a meager file of wary individuals who advanced a scant one hundred yards, then stopped in awe of what they saw before them. Where the Confederate fort had stood there now was a monstrous crater, sixty feet across and nearly two hundred feet wide, ranging in depth from ten to thirty feet. All was silent down there on its rubbled floor except for the thin cries of the wounded — who, together with the killed, turned out to number 278 — mangled by the blast and buried to various depths by the debris.

As Ledlie's soldiers stood and gazed at this lurid moonscape, strewn with clods that ranged in size up to that of a small house, they not only forgot their instructions to fan out right and left in order to widen the breakthrough for the follow-up attack; they even forgot to keep moving. At last they did move, but not far. For more than a month their fighting had been confined to rifle pits and trenches, and now here at their feet was the biggest rifle pit in all the world. They leaped into it and busied themselves with helping the Carolinian survivors, many of whom, though badly dazed, had interesting things to say when they were uprooted and revived. Ledlie might have gotten his division

★

back in motion by exhortation or example, but he was not available just now. He was immured in a bomb-proof well behind the lines, swigging away at a bottle of rum he had cadged from a staff surgeon. It later developed that this had been his custom all along, in times of strain. In any case, there he remained throughout what was to have been a fast-moving go-for-broke assault on Petersburg, by way of the gap Henry Pleasants had blown in the rebel line.

That gap was already larger than any Federal knew. When the mine was sprung, the reaction of the graybacks right and left of the hoisted battery was the same as that of the intended attackers across the way. They too bolted rearward, panicked by the fury of the blast, and thus broadened the unmanned portion of their line to about 400 yards. What was more, it remained so for some time. The second and third blue waves rolled forward, paused in turn on the near rim of the crater, much as the first had done, and then, like it, swept down in search of cover amid the rubble at the bottom. By then, most of the bolted Confederates had returned to their posts on the flanks of the excavation, and Beauregard was bringing up reinforcements, along with all the artillery he could lay hands on.

They arrived, men and guns, at about the time Burnside's fourth wave started forward. Loosed at last (but without Ferrero; he had joined Ledlie in the bomb-proof, nearly a quarter-mile away) the Negro soldiers advanced in good order. "We looks like men a-marching on, We looks like men of war," they sang as they came up in the wake of the other three divisions, which were scarcely to be seen, having vanished quite literally into the earth. Disdaining the crater, they swung around it, in accordance with the maneuver they had rehearsed, and drove for the high ground beyond. However, now that the defenders had rallied and been reinforced, they not only failed to get there; they also lost a solid third of their number in the attempt — 1327 out of just under 4000. "Unsupported, subjected to a galling fire from batteries on the flanks, and from infantry fire in front and partly on the flank," a witness later wrote, "they broke up in disorder and fell back to the crater."

Conditions there were not much better. In some ways they were worse. Presently they were much worse in every way. More than 10,000 men, crowded hip to hip in a steep-walled pen less than a quarter-acre in extent, presented the gray cannoneers with a compact target they did not neglect. Counterbattery work by the massed Union guns was excellent, but the surviving rebel pieces, including hard-to-locate mortars, still delivered what one occupant of the crater termed "as heavy a fire of canister as was ever poured continuously upon a single objective point." The result was bedlam, a Bedlam in flames, and this got worse as the enemy infantry grew bolder, inching closer to the rim of the pit, where marksmanship would be about as superfluous as if the shots were directed into a barrel of paralyzed fish. Anticipating this, some bluecoats chose to run the gauntlet back to their own lines, while others preferred to remain and risk the prospect:

Advancing Confederates of the 12th Virginia clash at the edge of the Crater with the Union vanguard, including the black troops of Ferrero's division.

which was soon at hand. Around 9.30, with Grant's disgusted approval, Meade had cancelled the follow-up attack and told Burnside to withdraw his corps.

But that was easier said than carried out. Burnside by then had fallen into a state of euphoric despair, much as he had done at Fredericksburg twenty months ago, under similar circumstances, and delayed transmission of the order till after midday, apparently in hope of some miraculous deliverance. Shortly after noon, two brigades from Mahone's division — they had slipped away from Warren's front unseen — gained the lip of the crater, where they added rapid-fire rifle volleys to the horror down below, then followed up with a bayonet charge that shattered what little remained of blue resistance. Hundreds

surrendered, thousands fled, more hundreds fell, and the so-called Battle of the Crater was soon over. It had cost Burnside 3828 men, nearly half of them captured or missing, and losses elsewhere along Meade's line raised the Union total above 4000 for the day; Confederate casualties, mostly wounded, came to about one third that number. By nightfall, all that remained as evidence of this latest bizarre attempt to break Lee's line was a raw scar, about midway down its length below the Appomattox, which in time would green over and lose its jagged look, but would never really heal.

Nor would a new bitterness Southerners felt as a result of this affair. Not only had they been blown up while sleeping — "a mean trick," they declared — but for the first time, here in the Old Dominion, black soldiers had been thrown into the thick of a large-scale fight. That was something far worse than a trick; that was infamy, to Lee's men's way of thinking. And for this they cursed their enemy in cold blood. "Eyes gleamed, teeth clenched," a nurse who tended Mahone's wounded would recall, "as they showed me the locks of their muskets, to which blood and hair still clung, when, after firing, without waiting to reload, they had clenched the barrels and fought hand to hand." Privately — like the troopers who stormed Fort Pillow, out in the wilder West — they admitted to having bayoneted men in the act of surrender, and they were by no means ashamed of the act, considering their view of the provocation. It was noted that from this time forward there were no informal truces in the vicinity of the Crater. Sniping was venomous and continuous, dawn to dusk, along that portion of the line.

Ledlie (but not Ferrero, who was somehow overlooked in the cater-waul that followed) presently departed, condemned by a Court of Inquiry for his part in the mismanagement of what Grant pronounced "the saddest affair I have witnessed in this war." Burnside left even sooner, hard on the heels of a violent argument with Meade, an exchange of recriminations which a staff observer said "went far toward confirming one's belief in the wealth and flexibility of the English language as a medium of personal dispute." Meade wanted the ruff-whiskered general court-martialed for incompetence, but Grant, prefer-ring a quieter procedure, sent him home on leave. "He will never return whilst *I* am here," Meade fumed.

Nor did he. Resigning from the service, Ambrose Everett Burnside, forty years old, returned to his business pursuits in Rhode Island, where he not only prospered but also recovered the geniality he had lost in the course of a military career that required him to occupy positions he himself had testified he was unqualified to fill. In time he went into politics, serving three terms as governor, and would die well into his second term as a U.S. senator, twenty years after the war began.

★ ★ ★   **Tactically speaking, Lee no doubt regretted Burnside's departure**. He would miss him, much as he missed Mc-Clellan, now in retirement, and John Pope and Joe Hooker, who had been shunted to outlying regions where their ineptitudes would be less costly to the cause they served. This was not to say that mistakes came cheap from those commanders who remained near the violent center. Meade's losses for July, swollen by the botched attempt to score an explosive breakthrough near its end, totaled 6367, and he had scarcely an inch of ground to show for their subtraction. Yet Lee could take small comfort in the knowledge that his own were barely half that. In contrast to his custom in the old aggressive days, when a battle was generally followed by a Federal retreat, he now not only derived no positive gain for his losses; he was also far less able to replace them, so near was the Confederacy to the bottom of its manpower barrel. "There is the chill of murder about the casualties of this month," one of his brigadiers reported from the Petersburg intrenchments. Even such one-sided triumphs as the Crater were getting beyond his means, and much the same thing could be said of Early's recent foray to the gates of Washington, which, for all its success in frightening the authorities there, had failed to lure the Army of the Potomac into staging another Cold Harbor south of the James.

That was what Lee had wanted, and even expected. "It is so repugnant to Grant's principles and practice to send troops from him," he wrote Davis, "that I had hoped before resorting to it he would have preferred attacking me." Instead, Grant had detached two corps whose partial arrival discouraged Early from storming the capital defenses and obliged him to fall back across the Potomac. After a brief rest at Leesburg, in defiance of the superior blue force charged with pressing his pursuit, Old Jube returned to the lower Shenandoah Valley and continued to maneuver between Winchester and Harpers Ferry, Jackson style, as if about to move on Washington again. Before his adversaries managed to combine against him — they were drawn from four separate departments, with desk-bound Halleck more or less in charge by telegraph — he lashed out at George Crook near Kernstown, July 24, and after inflicting close to 1200 casualties, drove him all the way north across the Potomac. Following this, in specific retaliation for Hunter's burning of the homes of three prominent Virginians, Early sent two brigades of cavalry under John McCausland to Chambersburg, Pennsylvania, to demand of its merchants, under penalty of its destruction, $100,000 in gold or a cool half-million in greenbacks. When they refused, McCausland evacuated the 3000 inhabitants and set fire to the business district. That was on July 30, the day of the Crater, and by midnight two thirds of the town was in ashes, another casualty of a war that was growing harsher by the month.

Lee's acute concern for Early — whose foot-loose corps, though badly outnumbered, not only continued to disrupt the plans of the Union high command by bristling aggressively on both banks of the Potomac just upstream from Washington, but also served through this critical stretch of time as a covering force for the grain-rich Shenandoah region and the Virginia Central Railroad — was increased on August 4, five days after the Crater, by reports that Grant was loading another large detachment of troops aboard transports at City Point. "I fear that this force is intended to operate against General Early," Lee told Davis, "and when added to that already opposed to him, may be more than he can manage. Their object may be to drive him out of the Valley and complete the devastation they [had] commenced when they were ejected from it." In point of fact, next to provoking his adversary into making a headlong assault on his intrenchments, there was nothing Lee wanted more than just such a weakening of the pressure against them. However, there were limits beyond which a precarious balance would be lost; Early's defeat would mean the loss, as well, of the Shenandoah Valley and the Virginia Central, both necessary for the survival of the rest of the army, immobilized at Petersburg and Richmond. Lee conferred next day with the President and reached the conclusion that, whatever the risk to his thinly held works beyond the James, he would have to strengthen Early. Accordingly, on August 6 he ordered Richard Anderson to leave at once, with Kershaw's division of infantry and Fitz Lee's of cavalry, for Culpeper, where he would be in a position either to speed back to Richmond by rail, in case of an emergency there, or else to fall on the flank and rear of the Federals, just beyond the Blue Ridge, in case they advanced up the Valley.

As usual, Lee was right about Grant's intentions, though in this case they were more drastic than he knew. Not only did the Federal commander plan to "complete the destruction" begun by Hunter before Early drove him off; he already had directed that this was to be accomplished by a process of omnivorous consumption. When Early fell back in turn from Washington in mid-July, Grant told Halleck to see to it that he was pursued by "veterans, militiamen, men on horseback, and everything that can be got to follow," with specific instructions to "eat out Virginia clean and clear as far as they go, so that crows flying over it for the balance of this season will have to carry their own provender with them."

Nothing much had come of that, so far. The crows waxed fat on the Valley harvest, deep in Early's rear, while Halleck, convinced that all his doubts about Grant's movements since Cold Harbor had been confirmed by the events of the past month, fumbled his way through a pretense of directing the "pursuit" from his desk in Washington. "*Entre nous*," he wrote Sherman on July 16, "I fear Grant has made a fatal mistake in putting himself south of James River. He cannot now reach Richmond without taking Petersburg, which is strongly fortified, crossing the Appomattox, and recrossing the James. Moreover, by placing his army south of Richmond he opens the capital and the whole North to rebel raids. Lee can at any time detach 30,000 to 40,000 men without our knowing it till we are actually threatened. I hope we may yet have full success, but I find that many of Grant's general officers think the campaign already a failure." Old Brains was determined to play no active role in what he saw as a discredited operation, and Grant soon found there was little he himself could do from an even greater distance. One answer might be for him to go up the Potomac and take charge of the stalled pursuit, but the fact was he had problems enough on his hands at Petersburg just then, including Meade's immovability, Burnside's mine, and the presence of Ben Butler, who by virtue of his rank would assume command of all the forces south of the James if Grant went up the country.

★

Unable to get Butler transferred (though he tried — only to find that this was no time to risk offending a prominent hard-war Democrat who might retaliate by taking the stump against the Administration) Grant turned on his one-time favorite Baldy Smith, who by now, mainly because of what Rawlins called "his disposition to scatter the seeds of discontent throughout the army," had become as much of a thorn in Grant's side as he had been in his cock-eyed superior's all along. On July 19 he was relieved and Major General Edward Ord, in temporary command at Baltimore, was brought down to take charge of his three divisions. Similarly, when the dust of the Crater settled, Burnside was superseded by his long-time chief of staff, Major General John G. Parke. Both of these new corps commanders — Ord was forty-five, a West Pointer like Parke, who was thirty-six — had fought under Grant at Vicksburg, and he was pleased to have them with him, here in front of Petersburg, to help conduct another siege.

None of this improved conditions northwest of Washington, however, and on the last day of July, with the ashes of Chambersburg still warm in that direction, Grant went down the James to Fortress Monroe for a conference with Lincoln about the situation Early had created up the Potomac.

For weeks he had favored merging the separate departments around the capital under a single field commander, though when he suggested his

Ulysses S. Grant tried and failed to have the prominent hard-war Democrat Benjamin Butler, pictured here, transferred from his command at Petersburg.

classmate William Franklin for the post — Franklin was conveniently at hand in Philadelphia, home on leave from Louisiana — he was told that the Pennsylvanian "would not give satisfaction," apparently because of his old association with McClellan, which still rankled in certain congressional minds. Rebuffed, Grant then considered giving Meade the job, with Hancock as his successor in command of the Army of the Potomac, but then thought better of it and decided that David Hunter, with his demonstrated talent for destruction, was perhaps the best man for the assignment after all. By the time he got to Fortress Monroe on July 31, however, he had changed his mind again, and with the President's concurrence announced his decision next day in a telegram to Halleck: "I want Sheridan put in command of all the troops in the field, with instructions to put himself south of the enemy and follow him to the death."

Back in Washington, Lincoln saw the order two days later, and though he already had approved the policy announced, he was so taken with the message that he felt called upon to wire its author his congratulations — together with a warning. "This, I think, is exactly right as to how our forces should move," he replied, "but please look over the dispatches you may have received from here, even since you made that order, and discover, if you can, [whether] there is any idea in the head of anyone here of 'putting our army south of the enemy' or of 'following him to the death' in any direction. I repeat to you it will neither be done nor attempted unless you watch it every day and hour and force it."

This last was sound advice, and Grant reacted promptly despite his previous reluctance to leave the scene of his main effort. Delaying only long enough to compose a carefully worded note for Butler — "In my absence remain on the defensive," he told him, adding: "Please communicate with me by telegraph if anything occurs where you may wish my orders" — he was on his way down the James within two hours of reading Lincoln's message. In Washington next morning he visited neither the White House nor the War Department, but went instead to the railway station and caught a train for Monocacy Junction, where Hunter had gathered the better part of the 32,500-man force supposed to be in hot pursuit of Early. Grant arrived on August 5 to find him in a state of shock, brought on by having been harassed for more than a month by the rebels and his superiors, who had confused him with conflicting orders and unstrung his nerves with alarmist and misleading information. In any case, his jangled state facilitated the process of removal. Displaying what Grant later called "a patriotism none too common in the army," Hunter readily agreed not only to stand aside for Sheridan, whom he outranked, but also to step down for Crook, who took over his three divisions when he presently departed for more congenial duty in the capital.

Sheridan arrived on August 6, in time for a brief interview with Grant, who also gave him a letter of instructions. Two of his three cavalry divisions

Phil Sheridan, standing at far left, confers with four top subordinates – left to right, Wesley Merritt, George Crook, James W. Forsyth, and George A. Custer.

had been ordered up from Petersburg, and these, combined with the troops on hand, the Harpers Ferry garrison, and the rest of Emory's corps en route from Louisiana, would give him a total of just over **48,000** effectives: enough, Grant thought, to enable him to handle Jubal Early and any other problem likely to arise as he pressed south toward a reunion with Meade near Richmond, wrecking as he went. He would have to take preliminary time, of course, to acquaint himself with his new duties in an unfamiliar region, as well as to restore some tone to Hunter's winded, footsore men, now under Crook, and to Wright's disgruntled veterans, who had little patience with the mismanagement they had recently undergone. But Grant made it clear — despite protests from Stanton and Halleck, being registered in Washington even now, that the thirty-three-year-old cavalryman was too young for the command of three full corps of infantry — that he looked forward to hearing great things from this direction before long, when Sheridan began to carry out what was set forth in his instructions. "In pushing up the Shenandoah Valley, as it is expected you will have to do first or last," the letter read, "it is desirable that nothing should be left to invite the enemy to return. Take all provisions, forage, and stock wanted for the use of your command. Such as cannot be consumed, destroy. . . . Bear in mind, the object is to drive the enemy south, and to do this you want to keep him always in sight. Be guided in your course by the course he takes."

The interview was brief because Grant was in a hurry to get back down the coast before Lee reached into his bag of tricks and dangled something disastrously attractive in front of Butler's nose. Returning to Washington, he boarded the dispatch steamer that had brought him up Chesapeake Bay four days ago, and stepped ashore at City Point before sunrise, August 9.

His haste came close to costing him his life before the morning ended. Around noon he was sitting in front of his headquarters tent, which was pitched in the yard of a high-sited mansion overlooking the wharves and warehouses of the ordnance supply depot he had established near the confluence of the James and the Appomattox, when suddenly there was the roar of an explosion louder than anything heard in the region since the springing of Pleasants's mine, ten days back. "Such a rain of shot, shell, bullets, pieces of wood, iron bars and bolts, chains and missiles of every kind was never before witnessed. It was terrible — awful — terrific," a staffer wrote home. Grant agreed. "Every part of the yard used as my headquarters is filled with splinters and fragments of shell," he telegraphed Halleck before the smoke had cleared.

By then it was known that an ammunition barge had exploded, along with an undeterminable number of the 20,000 artillery projectiles on its deck and in its hold, though whether by accident or by sabotage was difficult to say, all aboard having died in the blast, which scattered parts of their bodies over a quarter-mile radius and flung more substantial chunks of wreckage twice that far. A canal boat moored alongside, for example, was loaded with cavalry saddles that went flying in every direction, one startled observer said, "like so many big-winged bats." These were nearly as deadly in their flight as the unexploded shells, and contributed to the loss of 43 dead and 126 injured along the docks, while others, killed or wounded on the periphery — including a headquarters orderly and three members of Grant's staff — nearly doubled both those figures. "The total number killed will never be known," an investigator admitted, though he guessed at "over 200," and it was not until the war ended that the cause of the disaster was established by the discovery of a report by a rebel agent named John Maxwell.

He had stolen through the Union lines the night before, bringing with him a "horological torpedo," as he called the device, a candle box packed with twelve pounds of black powder, a percussion cap, and a clockwork mechanism to set it off. Reaching City Point at daybreak — about the same time Grant arrived — he went down to the wharves to watch for a chance to plant his bomb. It came when he saw the captain of a low-riding ammunition vessel step ashore, apparently intent on business: whereupon the agent set the timer, sealed the box, and delivered it to a member of the crew, with a request from the skipper to "put it down below" till he returned. "The man took it without question," Maxwell declared, "while I went off a little distance." His luck held; for though, as he said, he was "terribly shocked by the explosion," which soon followed, he not only was uninjured by falling debris, he also made it back in safety to the Confederate lines, having accomplished overnight, with a dozen pounds of powder, more damage, both in lives and property, than the Federals had done ten days ago with four tons of the stuff, after a solid month of digging.

Fearful though the damage was — estimates ran to $2,000,000 and beyond — wrecked equipment could be repaired and lost supplies replaced. More alarming, in a different way, was an intelligence report, just in, that Lee had detached Anderson's entire First Corps three days ago, along with Fitz Lee's cavalry, to reinforce Early out in the Valley. If true (which it was not, except in part; Anderson had been detached, but only with Kershaw's, not all three of his infantry divisions) this would give Early close to 40,000 soldiers, veterans to a man; enough, in short, to enable him to overrun Sheridan's disaffected conglomeration for a second crossing of the Potomac, this time with better

In this Illustrated London News sketch, rebel General Richard Anderson's troops ford the Shenandoah to reinforce Early at Front Royal on August 21.

than twice the strength of the one that had wound up at the gates of Washington last month. As things stood now, Lincoln might or might not survive the November election, but with 40,000 graybacks on the outskirts of the capital, let alone inside it, there was little doubt which way the votes would go. And as the votes went, so went Grant — a hard-war man, unlikely to survive the inauguration of a soft-war President. Promptly he got off a warning to Little Phil that his adversary was being reinforced to an extent that would "put him nearer on an equality with you in numbers than I want to see." What was called for, under the circumstances, was caution: particularly on the part of a young general less than a week in command, whose total strategy up to now could be summarized in his watchword, "Smash 'em up!"

Caution he recommended; caution he got. Sheridan had begun an advance from Halltown, near Harpers Ferry, and had pressed on through Winchester, almost to Strasburg — just beyond which, after cannily fading back, Early had taken up a strong position at Fisher's Hill, inviting attack — when word came on August 14, via Washington, that Anderson was on the way from

"If it was his policy to produce the impression that he was too weak to fight me, he did not succeed, but if it was to convince me that he was not an energetic commander, his strategy was a complete success."

— Jubal Early

Richmond, if indeed he had not come up already, with reinforcements that would enable Early to go over to the offensive with close to twice his estimated present strength of better than 20,000 veterans. Little Phil, experiencing for the first time the loneliness of independent command, reacted with a discretion unsuspected in his makeup until now. "I should like very much to have your advice," he wrote Grant, rather plaintively, as he began a withdrawal that presently saw him back at Halltown, within comforting range of the big guns at Harpers Ferry.

Early too returned to his starting point in the Lower Valley, skirmishing with such enemy units as he could persuade to venture beyond reach of the heavy batteries in their rear, and resumed his harassment of the Baltimore & Ohio, threatening all the while to recross the Potomac for another march on the Yankee capital. He had 16,500 men, including detached cavalry, and when Kershaw and Fitz Lee joined him the total came to 23,000: about half the number his adversary enjoyed while backing away from a confrontation. The result was a scathing contempt which Old Jube did not bother to conceal, remarking then and later that Sheridan was not only "without enterprise" but also "possessed an excessive caution which amounted to timidity." As the stand-off continued, on through August and beyond, Early's confidence grew to overconfident proportions. "If it was his policy to produce the impression that he was too weak to fight me, he did not succeed," he said of Little Phil, "but if it was to convince me that he was not an energetic commander, his strategy was a complete success."

Grant meantime had not been long in finding that only one of Anderson's divisions had left the Richmond-Petersburg front; yet he still thought it best for Sheridan to delay his drive up the Valley until pressure from Meade obliged Lee to recall the reinforcements now with Early. Accordingly, he began at once to exert that pressure, first on one bank of the James, pulling the few Confederate reserves in that direction, then the other.

Hancock, with his own and one of Butler's corps, plus the remaining cavalry division, was ordered to repeat the northside maneuver he had attempted on the eve of the Crater. This began on August 14, the day Sheridan started to backtrack, and continued on the morrow, but with heavier casualties than before and even less success. Attacking at Deep Bottom Run with hopes of turning the Chaffin's Bluff defenses, Hancock found veterans, not reserves, in occupation of Richmond's outer works, and suffered a repulse. A renewal of the assault next day, just up the line, brought similar results until he called it off, confessing in his report that his men had not behaved well in the affair. His losses were just under 3000, more than three times Lee's, but Grant had him remain in position to distract his opponent's attention from a second offensive, off at the far end of the line.

Warren had the assignment, which was basically to repeat the late-June effort to get astride the Weldon Railroad a couple of miles southwest of where the present Union left overlapped the Jerusalem Plank Road. This time he succeeded. Moving with four divisions on the morning of August 18 he struck the railroad at Globe Tavern, four miles south of Petersburg, and quickly dispossessed the single brigade of cavalry posted in defense of the place while most of the gray infantry confronted Hancock on the far side of the James. Elated by their success, the attackers pushed north from the tavern, but soon found that holding the road was a good deal harder than breaking it had been. Beauregard counterattacked that afternoon, using such troops as he could scrape together, then more savagely next morning, when A. P. Hill came down with two of his divisions. Warren lost 2700 of his 16,000 men, captured in mass when two brigades were caught off balance in poorly aligned intrenchments, but managed to recover the ground by sundown. That night he fell back to a better position, just over a mile down the line, where he was reinforced for two more days of fighting before the Confederates were willing to admit that they could not dislodge him. His casualties for all four days came to 4500, while the

BATTLE OF THE WELDON RAIL-ROAD.
AUGUST 21ST 1864

*Colonel Frederick Winthrop's Federal gunners, astride
the Weldon Railroad, exchange fire with
Confederate batteries in the distance on August 21.*

rebel loss was only 1600 — plus of course the Weldon Railroad; or anyhow the final
stretch of track. Lee at once put teamsters to work hauling supplies in wagons by a
roundabout route from the new terminus at Stony Creek, twenty miles below
Petersburg and about half that distance beyond the limits of Federal destruction.

Grant was determined to lengthen this mule-drawn interval, if only to
keep up the pressure he hoped would bring Anderson back from the Valley, and
when Hancock recrossed the James on August 21 — the day Lee gave up trying
to drive Warren off the railroad — he received orders to proceed south with two
of his divisions, plus Gregg's troopers, for a follow-up strike at the vital supply
line near Reams Station, about five miles below Globe Tavern and ten above
Stony Creek. He reached his objective on August 23, and by the close of the following day had torn up three miles of track beyond it. That night, while resting
his wreckers for an extension of their work tomorrow, he learned that A. P. Hill

was moving in his direction. Arriving at noon, Little Powell drove in the blue cavalry so fast that the infantry had little time to get set. The main blow fell on three New York regiments, green troops lately assigned to Gibbon's division, some of whom fled, while most surrendered, and to Hancock's further outrage a reserve brigade, ordered into the resultant gap, "could neither be made to go forward nor fire." Before darkness ended the fighting, better than 2000 men here and elsewhere along the Union line chose prison over combat. Two more divisions were on the way as reinforcements, but Hancock decided not to wait for them and instead pulled out that night. He had lost 2750 killed or wounded or missing, along with nine guns, a dozen battle flags, and well over 3000 rifles abandoned on the field. Hill's loss was 720.

This came hard for Hancock — "Hancock the Superb," newsmen had called him ever since the Seven Days; *Hancock*, who had broken Pickett's Charge, stood firm amid the chaos of the Wilderness, and cracked the Bloody Angle at Spotsylvania — as well as for his veteran lieutenants, especially John Gibbon, former commander of the Iron Brigade, whose division had been considered one of the best in the whole army until it was bled down to skeleton proportions and then fleshed out with skulkers finally netted by the draft. Ashamed and angered, Gibbon submitted his resignation, then was persuaded to withdraw it, though he presently left both his division and the corps: the hard-driving II Corps, which had taken more than forty enemy colors before it lost one of its own, and then abandoned or surrendered twelve of these in a single day at Reams Station, August 25. After that, even Grant was obliged to admit that its three divisions were unfit for use on the offensive, now and for some time to come, and Hancock's adjutant later said of his chief's reaction to the blow: "The agony of that day never passed from that proud soldier, who for the first time, in spite of superhuman exertions and reckless exposure on his part, saw his lines broken and his guns taken."

Back at Petersburg next day, Hill was pleased but not correspondingly elated, having done this sort of thing many times before, under happier circumstances. Moreover, it was much the same for Lee, who saw deeper into the matter. A month ago, in a letter to one of his sons, he had said of Grant, with a touch of aspersion: "His talent and strategy consists in accumulating overwhelming numbers." Now he was faced with the product of that blunt, inelegant strategy — that "talent" — which included not only the loss of the final stretch of the Weldon Railroad, but also the necessity for extending his undermanned Petersburg works another two miles westward to match the resultant Federal extension beyond Globe Tavern.

Of the two problems thus posed for him, the first might seem more irksome at the moment, coming as it did at a time when the army's reserve supply of corn was near exhaustion; but the second was potentially the graver. For while there were other railroads to bring grain from coastal Georgia and the Carolinas — the Southside line, on this bank of the Appomattox, and the Richmond & Danville, coming down from beyond the James for an intersection at Burkeville — the accustomed influx of recruits from those and other regions had dwindled to a trickle. Lee could scarcely replace his losses, let alone avoid the thinning of a line already stretched just short of snapping. "Without some increase of our strength," he warned Seddon, even as Hill was moving against Hancock, "I cannot see how we are to escape the natural military consequences of the enemy's numerical superiority."

Ten days later he reviewed the situation in a letter to the President, stressing "the importance of immediate and vigorous measures to increase the strength of our armies. . . . The necessity is now great," he said, "and will soon be augmented by the results of the coming draft in the United States. As matters now stand, we have no troops disposable to meet movements of the enemy or to strike where opportunity presents, without taking them from the trenches and exposing some important point. The enemy's position enables him to move his troops to the right or left without our knowledge, until he has reached the point at which he aims, and we are then compelled to hurry our men to meet him, incurring the risk of being too late to check his progress and the additional risk of the advantage he may derive from their absence. This was fully illustrated in the late demonstration north of James River, which called troops from our lines here, who if present might have prevented the occupation of the Weldon Railroad."

Across the way, at City Point, admonitions flowed in the opposite direction. Halleck warned Grant in mid-August that draft riots were likely to occur at any time in New York and Pennsylvania, as well as in Indiana and Kentucky: in which case he would be called upon, as Meade had been last summer, to furnish troops to put them down. Anticipating such troubles between now and the election in November, Old Brains suggested it might be well for the army to avoid commitment to any operation it could not discontinue on short notice. "Are not the appearances such that we ought to take in sail and prepare the ship for a storm?" he asked.

Grant thought not, and said so. Such police work should be left for the various governors to handle with militia, which should be called out now for the purpose. "If we are to draw troops from the field to keep the loyal states in harness," he declared, "it will prove difficult to suppress the rebellion in the disloyal states." Besides, he added, to ease the pressure on Lee at Petersburg and Richmond would be to allow him to reinforce Hood at Atlanta, just as he had reinforced Bragg at Chickamauga a year ago this month, and that "would insure the defeat of Sherman." In short, Grant had no intention of relaxing his effort on either bank of the James, whatever civilian troubles might develop up the country in his rear.

Lincoln read this reply on August 17 and promptly telegraphed approval. "I have seen your dispatch expressing your unwillingness to break your hold where you are. Neither am I willing. Hold on with a bulldog grip, and chew and choke as much as possible."

Scanning the words at his headquarters overlooking City Point, Grant laughed aloud — a thing he seldom did — and when staffers came over to see what had amused him so, passed them the message to read. "The President has more nerve than any of his advisers," he said.

⋆ ⋆ ⋆ Nerve was one thing, hope another, and Lincoln was fast running out of that: not so much because of the current military situation — though in point of fact this was glum enough, on the face of it, with Meade and Sherman apparently stalled outside Petersburg and Atlanta, Forrest rampant in Memphis, and the *Tallahassee* about to light up the New England coast with burning merchantmen — as in regard to his own political survival, which was seen on all sides as unlikely, especially in view of what had happened this month in his native Kentucky despite some highly irregular efforts to forestall defeat for a party that soon was still worse split by the Wade-Davis Manifesto. Six days after his chew-and-choke message to Grant, and six days before the Democrats were scheduled to convene in Chicago to nominate his November opponent — a time, he would say, "when as yet we had no adversary, and seemed to have no friends" — Lincoln sat in his office reading the morning mail. Thurlow Weed, an expert on such matters, recently had informed him that his reëlection was impossible, the electorate being "wild for peace." Now there came a letter from Henry J. Raymond, editor of the friendly *New York Times* and chairman of the Republican National Executive Committee, who said much the same thing.

"I feel compelled to drop you a line," he wrote, "concerning the political condition of the country as it strikes me. I am in active correspondence with your staunchest friends in every state, and from them all I hear but one report. The tide is setting strongly against us." Oliver Morton, Simon Cameron, and Elihu Washburne had respectively warned the New Yorker that Indiana, Pennsylvania, and Illinois were probably lost by now. Moreover, he told Lincoln, he was convinced that his own state "would go 50,000 against us tomorrow. And so of the rest. Nothing but the most resolute action on the part of the government and its friends can save the country from falling into hostile hands. . . . In some way or other the suspicion is widely diffused that we can have peace with Union if we would. It is idle to reason with this belief — still more idle to denounce it. It can only be expelled by some authoritative act, at once bold enough to fix attention and distinct enough to defy incredulity and challenge respect."

⋆

What Raymond had in mind was another peace commission, armed with terms whose rejection by Richmond would "unite the North as nothing since the firing on Fort Sumter has hitherto done." Lincoln knew only too well how little was apt to come of this, having tried it twice in the past month, and was correspondingly depressed. If this was all that could save the election he was whipped already. Sadly he took a sheet of paper from his desk and composed a memorandum.

> *Executive Mansion*
>
> *Washington, Aug. 23, 1864*
>
> *This morning, as for some days past, it seems exceedingly probable that this Administration will not be reëlected. Then it will be my duty to so coöperate with the President-elect as to save the Union between the election and the inauguration; as he will have secured his election on such ground that he cannot possibly save it afterwards.*
>
> *A. LINCOLN*

He folded the sheet, glued it shut, and took it with him to the midday cabinet meeting, where, without so much as a hint as to the subject covered, he had each member sign it on the back, in blind attestation to whatever it might contain — a strange procedure but a necessary precaution, since to tell them what was in the memorandum would be to risk increasing the odds against his reëlection by having it spread all over Washington, by sundown, that he himself had predicted his defeat. "In this peculiar fashion," his two secretaries later explained, "he pledged himself and the Administration" (so far, at least, as the pledge was binding: which was mainly on himself, since he alone knew the words behind the seal) "to accept loyally the anticipated verdict of the people against him, and to do their utmost to save the Union in the brief remainder of his term of office."

Not that he did not intend to do all he could, despite the odds, in the eleven weeks between now and the day the issue would be settled. Treading softly where he felt he must, and firmly where he didn't, he attended to such iotas as recommending in advance to field commanders that Indiana soldiers, who were required by law to be present to cast their ballots, be given furloughs in October to go home and offset the pacifist vote in their state election, considered important as a forecast of what to expect across the nation in November and as an influence on those whose main concern was that their choice be a winner. Besides, he foresaw trouble for his opponents once they came out in the open, where he had spent the past four years, a target for whatever mud was flung. The old Democratic rift, which had made him President in the first place, was even wider than it had been four years ago, except that now the burning issue was the war itself, not just slavery, which many said had caused it, and Lincoln expected the rift to widen further when a platform was adopted and a candidate named to stand on it. The front runner was Major General George B. McClellan, who was expected to attract the soldier vote, although numbers of Democrats were saying they would accept no candidate "with the smell of war on his garments." Either way, as Lincoln saw the outcome, platform and man were likely to be mismatched, with the result that half the opposition would be disappointed with one or the other, perhaps to the extent of bolting or abstaining when election day came round. "They must nominate a Peace Democrat on a war platform, or a War Democrat on a peace platform," he told a friend who left that weekend for the convention in his home state, "and I personally can't say I care much which they do."

He was right. Convening in Chicago on August 29, in a new pine Wigwam like the one set up for the Republicans in 1860, the Democrats heard New York's Governor Horatio Seymour establish the tone in a keynote speech delivered on taking the gavel as permanent chairman. "The Administration cannot save the Union. We can. Mr Lincoln views many things above the Union. We put the Union first of all. He thinks a proclamation more than peace. We think the blood of our people more precious than edicts of the President." After this, the assembly got down to adopting a platform framed in part by Clement L. Vallandigham, the nation's leading Copperhead and chairman of the Resolutions Committee, who had returned last year from presidential banishment, first beyond the rebel lines, then back by way of Canada, to run unsuccessfully for governor of Ohio. The former congressman's hand was most apparent in the peace plank, which resolved: "That this convention does explicitly declare, as the sense of the American people, that after four years of failure to restore the Union by the experiment of war . . . justice, humanity, liberty, and the public welfare demand that immediate efforts be made for a cessation of hostilities, with a view to an ultimate convention of the States, or other peaceable means, to the end that at the earliest practicable moment peace may be restored on the basis of the Federal Union of the States."

The stress here, as in Seymour's keynote speech, was on achieving peace through restoration of the Union, not "at any price," as was claimed by hostile critics. Vallandigham had emphasized this on the eve of the convention,

★

saying: "Whoever charges that I want to stop this war in order that there may be Southern independence charges that which is false, and lies in his teeth, and lies in his throat!" But presently the nominee himself lent strength to the charge by repudiating the plank in question. It was McClellan, as expected; he was chosen by acclaim on the first ballot, with Congressman George H. Pendleton of Ohio, long an advocate of negotiated peace, as his running mate. Ten days after his nomination — a delay that prompted a Republican wit to remark in the interim that Little Mac was "about as slow in getting up on the platform as he was in taking Richmond" — he tendered the notification committee his letter of acceptance. "I could not look in the face of my gallant comrades of the army and navy who have survived so many bloody battles," he declared, "and tell them that their labors and the sacrifices of so many of our slain and wounded brethren have been in vain, that we had abandoned that Union for which we have so often periled our lives. A vast majority of our people, whether in the army and navy or at home, would, as I would, hail with unbounded joy the permanent restoration of peace, on the basis of the Union under the Constitution, without the effusion of another drop of blood. But no peace can be permanent without Union."

Thus McClellan sought to deal with the dilemma Lincoln had foreseen, and wound up infuriating the faction that admired what he rejected: as Lincoln also had foreseen. But that was not as important by then as it had seemed the week before, when the charge that the "experiment of war" had

As this political cartoon shows, hard-war Democratic presidential nominee George B. McClellan sought to whitewash the party's pro-peace platform.

★

been a failure, East and West, was one that could perhaps be contested but could scarcely be refuted in the face of evidence from practically every front. Aside from Farragut's coup in Mobile Bay — seen now as rather a one-man show, with the credit all his own — incredible casualties had produced only stalemates or reverses, whether out in North Mississippi, down around Richmond and Atlanta, or up in the Shenandoah Valley. United in their anticipation of victory at the polls in November, whatever internal troubles racked the party, the Democrats adjourned on August 31, having wound up their business in jig time. Then two days later fate intervened, or seemed to. Slocum's wire reached Washington on September 2, followed next day by Sherman's own: "Atlanta is ours, and fairly won."

Church bells rang across the land as they had not rung since the fall of Vicksburg, fourteen months ago. "Sherman and Farragut have knocked the bottom out of the Chicago platform," Seward exulted, and Lincoln promptly tendered "national thanks" to the general and the admiral, issuing at the same time a Proclamation of Thanksgiving and Prayer, to be offered in all churches the following Sunday, for "the glorious achievements" of the army and the navy at Atlanta and in Mobile Bay. Grant too rejoiced, and telegraphed Sherman next day: "In honor of your great victory, I have ordered a salute to be fired with *shotted* guns from every battery bearing upon the enemy." Within earshot of that cannonade, the editor of the Richmond *Examiner* spoke of "disaster at Atlanta in the very nick of time when a victory alone could save the party of Lincoln from irretrievable ruin. . . . It will obscure the prospect of peace, late so bright. It will also diffuse gloom over the South."

Gladdened by congratulations from all sides, including some from political associates who he knew had been about to desert what they had thought was a sinking ship, Lincoln enjoyed the taste of victory so well that it made him hungry for still more. "Sheridan and Early are facing each other at a deadlock," he wired Grant on September 12. "Could we not pick up a regiment here and there, to the number of say ten thousand men, and quietly but suddenly concentrate them at Sheridan's camp and enable him to make a strike? This is but a suggestion." A suggestion was enough. Grant replied next day that he had been intending for a week "to see Sheridan and arrange what was necessary to enable him to start Early out of the Valley. It seems to me it can successfully be done." Content to have Meade in charge while he was gone — Butler was conveniently on leave — he set out the following day on his second trip up the Potomac in six weeks. Once more without stopping in Washington, he reached Sheridan's headquarters near Harpers Ferry on September 16.

"That's Grant," a veteran sergeant told a comrade, pointing him out. "I hate to see that old cuss around. When that old cuss is around there's sure to be a big fight on hand."

This applied even more to the present visit than to most, since Grant had in his pocket a plan for a campaign to drive Early all the way to Richmond, destroying first the Shenandoah Valley and then the Virginia Central Railroad in his wake. However, he was not long in finding that Little Phil had plans of his own which he was anxious to place in execution, having received from a spy in Winchester, just that morning, word that the time was ripe for an advance. A Quaker schoolteacher, Rebecca Wright by name, had smuggled out a note, wrapped in tinfoil and cached in the mouth of a Negro messenger, informing him that Anderson had left the Valley two days ago, with Kershaw's division and three batteries of artillery, recalled by Lee to help meet the stepped-up pressure from Meade on both sides of the James. What was more, Early — encouraged, as Lee had been in withdrawing the reinforcements, by his opponent's apparent quiescence under cover of the guns on Bolivar Heights for the past month — had posted three of his four infantry divisions in scattered positions above Winchester, toward the Potomac, to promote the fear that he was about to take the offensive with many more troops than the 18,000 or so which Sheridan now knew were all he had. Sheridan's plan was to use his field force of 40,000 not merely to drive Early from the Valley but to annihilate him by attacking his lone division at Winchester, then moving over or around it to cut off the escape of the rest up the Valley Turnpike.

Grant heard the ebullient young general out, and finding him "so clear and so positive in his views, and so confident of success," said nothing about the plan that remained in his pocket. Instead — today was Friday — he asked if the whole blue force could be ready to move by Tuesday. Sheridan replied that, subject to Grant's approval, he intended to take up the march before daybreak Monday, September 19. Grant thought this over, then nodded and issued his briefest order of the war: "Go in."

He left next morning, and though he still avoided Washington he managed a side excursion to Burlington, New Jersey, where his wife had taken a house after coming East. That night and part of Sunday he spent with her and the children, then returned to City Point on Monday, hoping for news of the Valley offensive, which had been scheduled to open that morning. Delayed by breakdowns, Sheridan's wire did not arrive till the following day, but when it did it more than justified the buildup of suspense.

Headed "Winchester, 7.30 p.m." — itself a confirmation of success — the telegram read: "I have the honor to report that I attacked the forces of General Early on the Berryville pike at the crossing of Opequon Creek, and after a most stubborn and sanguinary engagement, which lasted from early in the morning until 5 o'clock in the evening, completely defeated him." There followed a list of their losses, including "2500 prisoners, five pieces of artillery, nine army flags, and most of their wounded," but a companion message, written in

*A Negro messenger conveys Sheridan's request
for aid to southern Quaker Rebecca Wright, who sends
word of a rebel withdrawal from the Shenandoah.*

greater heat by his chief of staff, better caught the public's fancy, being quoted in all the papers: "We have just sent them whirling through Winchester, and we are after them tomorrow. This army behaved splendidly."

Actually, there had been a good deal more to it than that. For one thing, Sheridan's loss was considerably heavier than Early's — just over 5000 killed, wounded, or missing, as compared to just under 4000 — and for another, despite his achievement of surprise at the outset, he had come close to getting whipped before he got rolling. On the approach march, against orders, Wright brought his corps train along, old-army style, which so clogged the Berryville Pike in his rear that Emory was unable to cross Opequon Creek in time to join the dawn assault on Ramseur's division and Fitz Lee's troopers, posted three miles east of Winchester. Ramseur alternately held his position and withdrew slowly, in good order, and thus not only gave Early time to call in his other three infantry divisions, six to ten miles north of town, but also enabled him to launch a counterattack by Gordon and Rodes when Emory came up around midmorning, led onto the field by Sheridan himself, who, in a rage at the delay, had ordered Wright's wagons flung into ditches to clear the pike. Here fell Robert Rodes, the tall blond Virginia-born Alabamian who had led Jackson's flank attack at Chancellorsville, thirty-five years old and a veteran of all the army's major battles, from First Manassas on. Shot from his horse while directing the charge into the breach between Emory and Wright, he did not live to see it healed by the latter's reserves when they arrived. Emory, badly shaken — he had finished at West Point in 1831, the year Sheridan was born — had to be reinforced by Crook, whose two divisions had been intended for use in a flanking effort to block the path of a Confederate escape. Still, as the fight continued the weight of numbers told. Early, with some 14,000 men on hand, gave ground steadily all afternoon, under pressure from Sheridan's 38,000, and finally, about 5 o'clock, fell back through the streets of the town and retreated up the Valley Turnpike, which Fitz Lee's horsemen managed to keep open although Fitz himself had had to retire from the conflict, pinked in the thigh by a stray bullet. The battle — called Third Winchester by the defenders and Opequon Creek by the attackers — was over. Early did not stop till he reached Fisher's Hill, beyond Strasburg, twenty miles to the south, where Sheridan had ended his advance the month before, preceding his withdrawal to Harpers Ferry.

Grant's response next day was threefold. Wiring Stanton a recommendation that

Rebel General Robert E. Rodes (inset) lies mortally wounded in a horse drawn ambulance as his division retreats down Berryvike Pike at Winchester.

Sheridan be rewarded with a promotion to regular-army brigadier (which was promptly conferred) he also ordered the firing of a hundred-gun celebration salute in front of Richmond, just as he had done two weeks ago in Sherman's honor, and telegraphed Sheridan his congratulations for "your great victory," adding: "If practicable, push your success and make all you can of it."

Sheridan — whose 5018 casualties, though more than a thousand heavier than Early's 3921, had cost him only an eighth of his command, whereas Early had lost a solid fourth — intended to do just that. Late next day, with a force that was now three times the size of the one he was pursuing, he called a halt near Strasburg, advancing two corps across Cedar Creek and holding the third in reserve while he went forward to study the rebel positions, two miles beyond the town. He found it quite as formidable as it had been six weeks ago, when he had declined to test its strength.

Massanutton Mountain, looming dead ahead between the sun-glinted forks of the Shenandoah, divided the Valley into two smaller valleys: Luray on the left, beyond Front Royal, and what remained of the main valley on the right, narrowed at this point to a width of about four miles between the North Fork of the Shenandoah River and Little North Mountain, a spur of the Alleghenies. His flanks anchored east and west on the river and the mountain, Early also enjoyed the advantage of high ground overlooking a boggy stream called Tumbling Run, which the Federals would have to cross, under fire from massed artillery and small arms, if they were to attack him from the front. Down to fewer than 10,000 effectives as a result of his battle losses and the need for detaching two of Fitz Lee's three brigades to hold the midway notch in Massanutton (lest Sheridan send part of his superior force up the Luray Valley for a crossing there to get astride the turnpike at New Market, twenty miles in the Confederate rear) Early had to dismount troops from his other cavalry division, under Lunsford Lomax — most of whom had arrived too late for yesterday's fight, having been involved in railroad wrecking around Martinsburg, some fifty miles to the north — to man the western extension of his four-mile line to the lower slopes of Little North Mountain. Although the Winchester defeat had gone far toward disabusing him of the notion that his opponent "possessed an excessive caution which amounted to timidity," he had confidence in the natural strength of his position on Fisher's Hill, as well as in the veterans who held it, and believed that the bluecoats had little choice except to come at him head-on, in which case they were sure to be repulsed.

He was mistaken: grievously mistaken, as it turned out. Sheridan intended to approach him only in part from the front, using Wright's three and Emory's two divisions to fix him in place while Crook's two, kept hidden in reserve, made a flanking march, under cover of Little North Mountain, for a surprise descent on the Confederate left — where Early, expecting an assault on his right center, had posted his least dependable troops. All next day this

This James E. Taylor sketch depicts the climactic Union charge at Fisher's Hill from his vantage point on a bluff overlooking Tumbling Run and the rebel intrenchments.

misconception was encouraged by the sight of heavy blue columns filing through Strasburg, down toward Tumbling Run. Moreover, here as at Winchester two days ago, Little Phil intended to do more than merely whip or wreck his adversary; he planned to bag him entirely, and with this in mind he detached two of his three cavalry divisions, under Torbert, for a fast ride up Luray Valley and across Massanutton Mountain, through the midway notch in its knife-edge crest, to get control of the Valley Turnpike at New Market and thus prevent the escape of such gray fugitives as managed to slip through the net he would fling over Fisher's Hill tomorrow.

Crook set out before dawn, September 22, marching with flags and guidons trailed to keep them from being spotted by butternut lookouts while he rounded the wooded upper slopes of Little North Mountain, beyond the rebel left. Wright and Emory began their frontal demonstration after sunup, banging away with all their guns and bristling along Tumbling Run, as if about to splash across at any moment. This was a drawn-out business, continuing well past midday, since Crook's West Virginians — so-called because that was where they had done most of their fighting until now, though in fact they were in large part from Ohio, with a sprinkling of Pennsylvanians and New Yorkers thrown in to leaven or "easternize" the lump — had a long hard way to travel, much of it uphill. Finally at 4 o'clock, twelve hours after they set out, they struck.

"Flanked! Outflanked!" the cry went up on Early's left as the dismounted horsemen he had scorned from the outset, calling them buttermilk rangers and worse, fled before the onslaught of Crook, whose two divisions came whooping down the mountainside to strike them flank and rear. Eastward along Fisher's Hill, where the defenders had begun to remark that Sheridan must have lost his nerve and called off the attack he had been threatening all day, the confusion spread when Wright's corps joined the melee, advancing division by division across Tumbling Run as the gray line crumbled unit by unit from the shattered left. Fearful of being trapped in the angle between river and run, they too bolted, leaving the teamless cannoneers to slow the blue advance while they themselves took off, first down the rearward slope, then southward up the turnpike.

"Forward! Forward everything!" Sheridan yelled, coursing the field on his black charger and gesturing with his flat-topped hat for emphasis. "Don't stop! Go on!" he shouted as his infantry overran and captured twelve of the guns on Fisher's Hill.

Anticipating "results still more pregnant," he counted on Averell, whose division he presently launched in pursuit of the rebels fleeing through the twilight, to complete the Cannae he had had in mind when he sent Torbert with two divisions up the Luray Valley for a crossing of Massanutton to cut off Early's retreat at New Market. Alas, both cavalry generals failed him utterly in the crunch. Torbert came upon Fitz Lee's two brigades, posted in defense of a narrow gorge twelve miles beyond Front Royal, and decided there was nothing to be gained from being reckless. He withdrew without attempting a dislodgment. Sheridan was "astonished and chagrined" when he heard of this next morning. But his anger at Torbert was mild compared to what came over him when he learned that Averell had put his troopers into bivouac the night before to spare them the risk of attacking Early's rear guard in the darkness. Enraged, Little Phil fired off a message informing the cavalryman that he expected "resolution and actual fighting, with necessary casualties, before you retire. There must be no more backing and filling," he fumed, and when Averell did no better today, despite this blistering, he relieved him of command and sent him forthwith back to West Virginia, "there to await orders from these headquarters or higher authority."

By that time Early had cleared New Market, and though Sheridan kept up the pursuit beyond Harrisonburg, where the graybacks turned off eastward around the head of Massanutton to find shelter near one of the Blue Ridge passes a dozen miles southeast of Staunton, he had to be content with what he had won at Fisher's Hill and picked up along the turnpike afterwards. This included four additional guns, which brought the total to sixteen, and more than a thousand prisoners. Early's over-all loss, in the battle and on the retreat, was about 1400 killed, wounded, and missing; Sheridan's came to 528.

Gratifying as the comparison was, another was even more so. When Sheridan took over Hunter's frazzled command at Monocacy eight weeks ago, the rebs were bristling along the upper Potomac, as if their descent on Washington the month before had been no more than a rehearsal for a heavier blow. Now they were a hundred miles from that river, and it seemed doubtful they would ever return to its banks, so complete had been his triumph this past week, first near Winchester and then, three days later, at Fisher's Hill. "Better still," Grant replied to his protégé's announcement of the second of these victories, "it wipes out much of the stain upon our arms by previous disasters in that locality. May your good work continue is now the prayer of all loyal men."

Exultation flared among Lincoln supporters, whose number had grown considerably in the course of the three-week September span that opened with news of Atlanta's fall and closed with this pair of Shenandoah victories to balance the tally East and West. The candidate himself was in "a more gleeful humor," friends testified after visits to the White House. "Jordan has been a hard road to travel," he told one caller, "but I feel now that, notwithstanding the enemies I have made and the faults I have committed, I'll be dumped on the right side of that stream."

Abrupt though it was, he had cause for this change in mood from gloom to glee. Within two weeks of his August 23 pledge-prediction,

A reluctant Salmon P. Chase, shown above, took to the stump in Ohio for Lincoln's reëlection in 1864.

countersigned blindly by the cabinet as a prelude to defeat, the news from Sherman down in Georgia produced a scurry by disaffected Jacobins to get back aboard the bandwagon: especially after the mid-September elections in Maine and Vermont showed the party not only holding its own, contrary to pre-Atlanta expectations, but also registering a slight gain. These straws in the wind grew more substantial with the announcement of Sheridan's triumphal march up the Valley. Salmon Chase paid his respects at the White House, then left to take the stump in Ohio, Vallandigham's stamping ground, while Horace Greeley, privately declaring that he intended to "fight like a savage in this campaign — I hate McClellan," he explained — announced that the *Tribune* would "henceforth fly the banner of Abraham Lincoln for President." Even Ben Wade and Henry Davis, whose early-August manifesto had sought to check what they called his "encroachments," took to the stump, like Chase, in support of the very monster they had spent the past two months attacking, though they maintained a measure of consistency by spending so much of their time excoriating the Democratic nominee that they had little left for praise in the other direction. "To save the nation," Wade told a colleague in explanation of his support for a leader he despised, "I am doing all for *him* that I could possibly do for a better man."

Meantime Lincoln, no doubt as amused as he was gratified by these political somersaults, did not neglect the particulars incident to victory and available to the candidate in office. Patronage and contracts were awarded to those who could do most for the party, and a binding promise went to James Gordon Bennett that he would be appointed Minister to France in exchange for his support in the New York *Herald*. There remained the thorny problem of Frémont, whose continuation in the race threatened to siphon off a critical number of die-hard radical voters. These had long been calling for the removal of Montgomery Blair, whose presence in the cabinet they considered an affront, and though Lincoln, aware that his compliance would be interpreted as an act of desperation, had resisted their demand for the Postmaster General's removal, now that Atlanta had turned the tide he felt willing to be persuaded: provided, that is, he got something commensurate in exchange.

The something in this case was Frémont's withdrawal, and he got it without having to drop the pretense of unwillingness he had kept up all along. "The President was most reluctant to come to terms, *but came,*" Zachariah Chandler informed his wife after serving as go-between in the bargain. On September 22 — by coincidence, the day Sheridan hustled Early off Fisher's Hill — Frémont renounced his candidacy. "The union of the Republican Party has become a paramount necessity," he explained in his announcement of withdrawal, but he added, by way of a backhand lick in parting: "In respect to Mr Lincoln I continue to hold exactly the sentiments contained in my letter of acceptance. I consider that his administration has been politically, militarily, and financially a failure, and that its necessary continuance is a cause of regret for the country."

Blair's head rolled next day. "My dear Sir," Lincoln wrote him: "You have generously said to me more than once that whenever your resignation could be a relief to me it was at my disposal. The time has come." There followed compliments and thanks, if not regrets. Blair saw clearly enough that he was in fact "a peace offering to Frémont and his friends." The thought rankled. "The President has, I think, given himself, and me too, an unnecessary mortification in this matter," he wrote his wife before clearing out his desk, "but then I am not the best judge and I am sure he acts from the best motives." A good party

man, like all the Blairs, he soon was out wooing voters for the chief who had let him go when bargain time came round.

While this high-level politicking was in progress up the country, Grant tried another pendulum strike at opposite ends of Lee's line, first north then south of the James. Encouraged by news from the Valley, which seemed to show what determination could accomplish, he was also provoked by a mid-September coup the rebel cavalry scored at his expense. On Coggins Point, six miles downriver from his headquarters, a large herd of cattle awaited slaughter for Meade's army; or so it was thought until a rustling operation, dubbed "Hampton's Cattle Raid," caused the beef to wind up in stomachs unaccustomed to such fare. Hampton set out with three brigades on a wide swing around the Union left, September 14, and reached his objective before dawn two days later. Two brigades fought a holding action, hard in the Federal rear, while the third rounded up the animals on Coggins Point; then all three turned drovers and rode back into their own lines next day with just over 300 prisoners and just under 2500 beeves, at a cost of fewer than 60 casualties. Lee's veterans were feasting on Yankee beef by the time Grant returned from his Harpers Ferry conference with Sheridan to find that in his absence, and to his outrage, the graybacks had foraged profitably half a dozen miles in rear of City Point. Determined to avenge this indignity — and aware, as well, that the year was about to move into the final month before the national election, still without the main eastern army having chalked up a gain to compare with

Cracking whips, Confederate cavalry-men led by Wade Hampton drive home 2486 head of cattle meant for Meade's army in what became known as "Hampton's Cattle Raid."

those scored recently in Georgia and the nearby Shenandoah Valley — he told Meade to proceed with another of those sequential right-left strikes, such as he had attempted twice in the past month, designed to throw Lee off balance and overrun at least a portion of his works.

Both times before, the initial attack north of the James had been made by Hancock, but his corps by now was practically *hors de combat* as a result of these and other efforts there and elsewhere. So this time the assignment went to Butler. Presumably refreshed by his recent leave, the Massachusetts general drew up a plan whereby 20,000 men from Kautz's cavalry and the two corps of infantry under Ord and David Birney — successors to the disgruntled and departed Baldy Smith and Quincy Gillmore — crossed the river on the night of September 28 for a double-pronged assault on Forts Harrison and Gilmer, works that were part of Richmond's outer line, down near the James, and covered Lee's critical Chaffin's Bluff defenses. Ord, coming up on schedule through a heavy morning fog, launched an all-out attack which quickly overran the first of these, a mile beyond the river, along with its surprised and meager garrison, though at the cost of a crippling wound that caused him to be carried off the field. Alerted by the racket, just over a mile away, the defenders of Fort Gilmer were ready when Birney struck. Repulsed, he drew back and struck again, with help from Ord, only to find that the place had been reinforced from Richmond, where the tocsin still was sounding. Grant arrived that afternoon to order still a third assault, which was also unsuccessful, and the effort here was abandoned in favor of bracing Fort Harrison against Lee's expected attempt to retake it. This came next day, September 30, when two gray divisions and part of a third, 10,000 men in all, came over from Petersburg under Richard Anderson to make three desperate attacks, all of which failed. Butler's loss for the two days was 3327 of all arms. Lee's was about 2000; plus the fort.

This last was no great deprivation. Lee promptly drew a retrenchment in rear of Fort Harrison, still beyond small-arms range of Chaffin's Bluff, that resulted in a stronger line than the one laid out before. Still, Ben Butler had provided northern journalists with an item fit for crowing over, and best of all — potentially at least — Lee once more had been decoyed into stripping that portion of his defenses where the main blue effort was about to land, off beyond the far end of the long curve of intrenchments south of the James.

Warren and Parke, with two divisions each and Gregg's cavalry in support, set out westward from Globe Tavern while Butler's assault on the forts was in progress. Their mission was to cut, and if possible hold, both the Boydton Plank Road and the Southside Railroad, the two remaining arteries whose severance would bring on the collapse of Petersburg. They were stopped next day along Vaughan Road, less than halfway to the first of these objectives, by Hampton, who skirmished with Warren's column at Poplar Springs Church.

★

*Union troops attacking Fort Harrison on
September 29 faced this forbidding parapet with its
enclosed guns fronted by rows of sharpened stakes.*

Moving west to meet the threat with two divisions from the Petersburg defenses — already weakened by the detachment of Anderson for the attempt to retake Fort Harrison that same day — A. P. Hill encountered Parke at nearby Peebles Farm. Badly shot up, Parke managed to hang on until Warren sent reinforcements to help him hold his ground along Squirrel Level Road, where both corps dug in at nightfall. That was the limit of their lateral advance, and it cost them 2889 casualties, all told, as compared to about 900 for Hill and Hampton. With scarcely a pause for rest, the Federals got busy with picks and shovels, constructing a line of intrenchments from their new position, back east to Globe Tavern, two miles away on the Weldon Railroad. Lee, of course, was obliged to conform, extending once more the length of line his dwindling army had to cover to keep its flank from being turned.

By ordinary standards, Grant's gain in this third of his pendulum strikes at the Richmond-Petersburg defenses — a rather useless rebel earthwork, one mile north of the James, plus a brief stretch of country road, two miles beyond the previous western limit of his line — was incommensurate with his loss of just over 6000 men, a solid half of them captives already on their way to finish out the struggle in Deep South prison camps, as compared to just under 3000 for Lee, most of them wounded and soon to return to the gray ranks. But with the presidential contest barely five weeks off, this was no ordinary juncture.

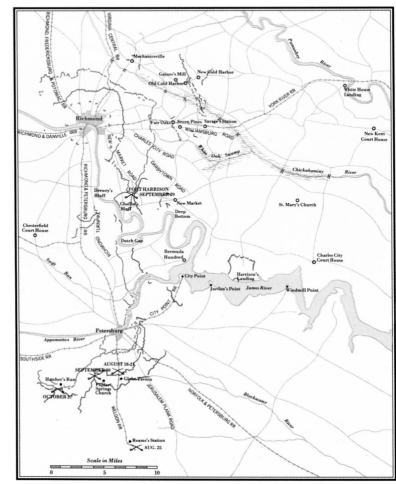

In late summer Grant hammered at Lee, whose lines were stretched to the breaking point, with coordinated attacks at Fort Harrison, the Weldon Railroad, and Hatcher's Run.

Ordinary standards did not apply. What did apply was that Lincoln supporters now had something they could point to, down around the Confederate seat of government itself, which seemed to indicate, along with recent developments in Atlanta and the Shenandoah Valley, that the war was by no means the failure it had been pronounced by the opposition in Chicago, five weeks back.

In recognition of this, Democrats lately had shifted their emphasis from the conduct to the nature of the war; "The Constitution as it is, the Union as it was," was now their cry. How effective this would prove was not yet known, for all its satisfying ring. But the evidence from Pennsylvania, Ohio, and Indiana, all of which held their state and congressional elections on October 11, was far from encouraging to those who were out of power and wanted in. With help from Sherman, who at Lincoln's urging not only granted furloughs wholesale to members of the twenty-nine Hoosier regiments in his army down in Georgia, but also sent

John A. Logan and Frank Blair with them on electioneering duty, all three states registered gains for the Union ticket, both in Congress and at home.

"There is not, now, the slightest uncertainty about the reëlection of Mr Lincoln. The only question is, by what popular and what electoral majority?" Chase had told a friend in Ohio the week before, and once the ballots were tallied in these three states — all considered spheres of Copperhead influence — *Harper's Weekly* was quick to agree with the former Treasury head's assessment: "The October elections show that unless all human foresight fails, the election of Abraham Lincoln and Andrew Johnson is assured."

Neither of these nominees campaigned openly, any more than McClellan or Pendleton did, but their supporters around the country — men of various and sometimes awesome talents, such as the stout-lunged New Orleans orator, who "when he got fairly warmed up," one listener declared, "spoke so loud it was quite impossible to hear him" — more than made up for this traditional inactivity, which was designed to match the dignity of offices too lofty to be sought. Behind the scenes, other friends were active, too; especially those on the Union executive committee, responsible for funding the campaign. Cabinet members were assessed $250 each for the party coffers, and a levy of five percent was taken from the salaries of underlings in the War, Treasury, and Post Office departments. Gideon Welles alone refused to go along with this, pronouncing the collectors "a set of harpies and adventurers [who] pocket a large portion of the money extorted," and though workers in the Brooklyn Navy Yard "walked the plank in scores" for demonstrating support or sympathy for the opposition, Welles was by no means as active in this regard as Edwin Stanton, who at a swoop fired thirty War Department clerks for the same cause, including one whose sole offense was that he let it be known he had placed a bet on Little Mac. Such methods had produced excellent results in the recent state elections, held four weeks, to the day, before the national finale, scheduled for November 8, when still better returns were not only hoped for but expected, as the result of yet a third Sheridan-Early confrontation, providentially staged within three weeks of that all-important first Tuesday following the first Monday in November.

After Fisher's Hill, Sheridan's progress southward up the Valley — described by a VI Corps veteran as "a grand triumphal pursuit of a routed enemy" — ended at Mount Crawford, beyond the loom of Massanutton, where he gave his three infantry corps some rest while the cavalry raided Staunton and Waynesboro, a day's march ahead on the Virginia Central. Grant wanted the whole force, horse and foot, to move in that direction and down that railroad for a junction with Meade, wrecking Lee's northside supply lines as it went. "Keep on," he wired, "and your good work will cause the fall of Richmond." But Sheridan, with Hunter's unhappy example before him — not to mention that of bluff John Pope, who had tried such a movement two years ago, only to wind up riding herd on Indians out in Minnesota — replied that, even though Early had been eliminated as a deterrent, this was "impracticable with my present means of transportation. . . . I think that the best policy will be to let the burning of the crops in the Valley be the end of this campaign, and let some of this army go elsewhere." Lured by the notion of bringing Wright's hard-hitting corps back down the coast to Petersburg, Grant agreed that Sheridan would do well to make a return march down the Valley, scorching and smashing left and right to ensure that this classic "avenue of invasion" would no longer furnish subsistence even for those who lived there, let alone for Lee's army around Richmond. "Carry off stock of all descriptions, and negroes, so as to prevent further planting," he reminded Little Phil, elaborating on previous instructions. "If this war is to last another year we want the Shenandoah Valley to remain a barren waste."

He knew his man. Beginning the countermarch October 6, Sheridan reported the following night from Woodstock, forty miles away, that he had "destroyed over 2000 barns filled with wheat, hay, and farming implements; over 70 mills filled with flour and wheat; have driven in front of the army over 4000 head of stock, and have killed and issued to the troops not less than 3000 sheep. . . . Tomorrow I will continue the destruction of wheat, forage, &c. down to Fisher's Hill. When this is completed the Valley, from Winchester up to Staunton, 92 miles, will have but little in it for man or beast." Others attested to his proficiency in destruction, which continued round the clock. "The atmosphere, from horizon to horizon, has been black with the smoke of a hundred conflagrations," a correspondent wrote, "and at night a gleam brighter and more lurid than sunset has shot from every verge. . . . The completeness of the devastation is awful. Hundreds of nearly starving people are going north. Our trains are crowded with them. They line the wayside. Hundreds more are coming." They had little choice, a staff captain noted, having been "left so stripped of food that I cannot imagine how they escaped starvation."

To hurt the people, the land itself was hurt, and the resultant exodus was both heavy and long-lasting. A full year later, an English traveler found the Valley standing empty as a moor.

By now, although Early was being careful to maintain a respectful distance with his twice-defeated, twice-diminished infantry, butternut cavalry was snapping at the heels of the blue column, and Sheridan took this as continuing evidence of the timidity his own cavalry had shown, just over two weeks ago, after Fisher's Hill. Approaching that place from the opposite direction, October 9, he gave Torbert a specific order: "Either whip the enemy or get whipped yourself," then climbed nearby Round Hill for a panoramic view of the result. It was not long in coming. After crossing Tom's Brook, five miles short of Strasburg, Torbert had Merritt and Custer whirl their divisions around and charge the two pressing close in their rear under Lomax and Tom Rosser, who had recently arrived from

Union General George Armstrong Custer oversees the withdrawal of his cavalry from the havoc they wreaked in the Shenandoah Valley on October 7.

Richmond with his brigade. Startled, the gray troopers stood for a time, exchanging saber slashes till their flanks gave way, then panicked and fled southward up the pike, pursued by the whooping Federals, who captured eleven of the dozen rebel guns in the course of a ten-mile chase to Woodstock and beyond, along with some 300 graybacks on fagged horses. "The Woodstock Races," the victors dubbed the affair, taking their cue from the Buckland Races, staged at Custer's expense by Jeb Stuart, a year ago this month, on the far side of the Blue Ridge. His temper cooled, his spirits lifted, Sheridan passed through Strasburg and crossed Cedar Creek next morning to put Crook's and Emory's corps in bivouac on the high ground, while Wright prepared his three divisions for an eastward march through Ashby's Gap, as agreed upon beforehand, to rejoin Grant at Petersburg.

They set out two days later, on October 12: only hours, as it developed, before Early reappeared on Fisher's Hill, five miles to the south. He had been reinforced from Richmond, not only by Rosser's cavalry brigade, but also by Kershaw's infantry division, which had been with him last month until it was recalled by Lee on the eve of the Federal strike at Winchester. Aware of these acquisitions, Sheridan was not disturbed, knowing as he did that they barely lifted Early's strength to half his own. If Old Jubal was in search of a third drubbing, he would be happy to oblige him when the time came.

All the same, he recalled the three VI Corps divisions from Ashby's Gap next day, deferring their departure until the situation cleared, and set about making his Cedar Creek position secure against attack while he determined his next move. Amid these labors, which included preparations for a horseback raid to break up the railroad around Charlottesville, he was summoned to Washington by Halleck for a strategy conference, October 16. He left that morning to catch a train at Front Royal, and when he got there he was handed a telegram from Wright, whom he had left in command on Cedar Creek, quoting a message just intercepted from a rebel signal station on Massanutton Mountain: "Be ready to move as soon as my forces join you, and we will crush Sheridan." The signature was *Longstreet;* which was news in itself, if the message was valid. Little Phil considered it "a ruse," however, designed to frighten him out of the Valley, and he declined to be frightened. Besides, he had confidence in Wright, who assured him: "I shall hold on here until the enemy's movements are developed, and shall only fear an attack on my right, which I shall make every preparation for guarding against and resisting." Aside from calling off the Charlottesville raid, Sheridan did not change his plans. Boarding the train for Washington, he advised Wright: "Look well to your ground and be well prepared. Get up everything that can be spared," he added, and promised to return within two days, "if not sooner."

He was right in assuming the intercepted dispatch was a plant, and right as well about its purpose. But he was altogether wrong if he thought his twice-whipped adversary did not intend to try something far more drastic if the invoked ghost of Old Peter failed to frighten him away. In point of fact, so thoroughly had the bluecoats scorched the country in his rear, Early believed he had no choice except "to move back for want of provisions and forage, or attack the enemy in his position with the hope of driving him from it." Another reason, despite his usual crusty disregard for the opinions of others in or out of the army, was that he had a reputation to retrieve; "To General Sheridan, care of General Early," cynics had chalked on the tubes of guns sent from Richmond to replace the 21 pieces he had lost in battle this past month, exclusive of the eleven abandoned by the cavalry last week in its panicky flight from Tom's Brook to Woodstock.

Admittedly, with the blue force nearly twice his size, securely in position on high ground, its front covered by a boggy creek and one flank anchored on the Shenandoah, the odds against a successful assault were long. But his predecessor Jackson, in command of these same troops, had taught him how far audacity could go toward evening such odds, and Lee himself, in a letter that followed the sending of reinforcements, had just told him: "I have weakened myself very much to strengthen you. It was done with the expectation of enabling you to gain such success that you could return the troops if not rejoin me yourself. I know you have endeavored to gain that success, and believe you have done all in your power to assure it. You must not be discouraged, but continue to try. I rely upon your judgment and ability, and the hearty coöperation of your officers and men still to secure it. With your united force it can be accomplished."

Sustained and appealed to thus, Early was "determined to attack." But how, against such odds, could he do so with any real hope of success? Crippled as he was by arthritis, which aged him beyond his not quite forty-eight years and prohibited mountain climbing, he sent John Gordon, his senior division commander since the fall of Rodes, and Major Jedediah Hotchkiss, a staff cartographer inherited from Jackson, atop Massanutton to study the enemy position, which lay spread out below them, facing southwest along Cedar Creek. Crook's two divisions were nearest, on the Federal left, then Emory's two, beyond the turnpike, and finally Wright's three, on the distant right, where most of the blue cavalry was posted, obviously in expectation that if an attack was made it would come from that direction. Hotchkiss had discovered and recommended the route for the movement around Hooker's flank at Chancellorsville, but what he and Gordon saw from their high perch this bright fall morning, October 18, was an opportunity for an end-on strike that might outdo even Stonewall's masterpiece. A night march around the steep north face of Massanutton, following a crossing of the Shenandoah near Fisher's Hill, would permit a recrossing of the river beyond its confluence with Cedar Creek, and this in turn would place the flanking column in direct confrontation with the unsuspecting Union left, which could be assaulted at first light in preparation for further assaults on Emory and Wright, once Crook's position had been overrun. Gordon, in fact, was so confident of success that when he came down off the mountain to urge the adoption of the plan, he offered to take all responsibility for any failure that occurred.

Early had never been one to avoid responsibility, nor did he delay approval of the plan. He would march tonight and strike at dawn, he announced at a council of war called that afternoon. Gordon would be in charge of the turning column made up of his own and the divisions of Ramseur and Rodes, the latter now commanded by its senior brigadier John Pegram, recently recovered from the leg wound he had taken in the Wilderness. Kershaw would move through Strasburg, also under cover of darkness, and attack on the right of the Valley pike, crossing lower Cedar Creek to join the flanking effort as soon as he heard Gordon open fire, and Brigadier General Gabriel Wharton — successor to Breckinridge, who had been recalled to eastern Virginia on the eve of Fisher's Hill — would advance along and to the left of the turnpike, accompanied by Rosser's troopers, to menace and fix the Federals in position on the far side of the creek while the massed Second Corps, with Kershaw's help, struck their flank and drove them north across his front. Rosser then would take up the pursuit, as would Lomax, whose horsemen were to come upon the field by a roundabout march through Front Royal in order to cut off the blue retreat this side of Winchester, fifteen miles beyond Middletown, which was close in the Union rear.

The plan was elaborate, involving a convergence by three columns, but it seemed pat enough to Early and his lieutenants, who went straight from the meeting to prepare for the various night marches designed to yield revenge for the two defeats they had recently suffered, here in the Valley from which their army took its name. The first of these — Third Winchester — had occurred exactly a month ago tomorrow, and this made them and their butternut veterans all the more eager to get started on the observance of that anniversary.

Aided by the light of a moon only three nights past the full, Gordon's column set out shortly after dark, the men of all three divisions having left their cooking utensils and even their canteens behind to avoid any give-away clink of unnecessary metal, and was in position in the shadows close to Bowman's Ford before daybreak, half a mile beyond the confluence of Cedar Creek and the river, prepared to splash across on signal. Similarly, accompanied by Early and his

Dressed as farmers, rebel General John Gordon (foreground) and Major Jedediah Hotchkiss pretend to cut cornstalks as they reconnoiter Sheridan's forces.

★

staff, as well as by most of the army's guns, Kershaw moved undetected around Strasburg to the near bank of the creek, across which he could see low-burnt campfires glowing in the darkness. Wharton followed, turning off to the left of the macadamized pike, preceded by Rosser, whose troopers rode at a walk to muffle the sound of hoofbeats on the stony ground. At 4.30, after an hour's wait on the creekbank, Early told Kershaw to go ahead and cross. He did, and while he was getting his men back into column on the other side, the boom of Rosser's horse artillery came from well upstream, along with the rattling clatter of picket fire nearby on the right, where Gordon was fording the Shenandoah just off the unalerted Union flank. The surprise was complete, if not quite over-whelming at the outset. "As we emerged from a thicket into the open," one of Kershaw's South Carolinians later wrote, "we could see the enemy in great commotion. But soon the works were filled with half-dressed troops, and they opened a galling fire upon us."

Kershaw charged, and as he did so, racing uphill through the spreading dawn, Gordon struck the left rear of the hastily formed blue line, which promptly broke. Elated (for these were Crook's men, the so-called West Virginians who had flanked them unceremoniously off Fisher's Hill four weeks ago) the Confederates surged forward on a broad front across the turnpike, pursuing and taking prisoners by the hundreds. With only a bit more time for getting set, Emory's corps fared little better, its unbraced ranks plowed by shells from rebel batteries massed on a hill beyond the creek. Fugitives from the four routed divisions fled northward through Wright's camps, in rear of which his Potomac veterans were falling in for battle. By now the sun was rising, alternately bright and pale as drifts of smoke blew past it, and the graybacks — joined at this stage by Wharton, who had been left with nothing in his front — came on yelling as they drove Wright's troops northeast across the open fields, first to a second and then to still a third position nearly two miles in rear of Middletown, where Jackson had captured Banks's wagon train in May of '62. This seemed to some a comparable achievement, while others went further afield in search of a parallel triumph. "The sun of Middletown! The sun of Middletown!" Early kept exclaiming, as if to say he had found his Austerlitz.

It was now past 9 o'clock, and he was delighted that within a scant four hours he had driven seven infantry divisions from the field with only five of his own, taking in the process more than 1300 prisoners, 18 guns, and an uncounted number of flags.

He was delighted; but he was also satisfied, it seemed. "Well, Gordon, this is glory enough for one day," he declared on meeting the Georgian near the front soon afterward. They stood looking across the fields at the Yankees reduced to stick men in the distance. "This is the 19th," he went on. "Precisely one month ago today we were going in the opposite direction." Gordon too was

★

happy, but his thoughts were on the immediate future, not the past. "It is very well so far, General," he replied, "but we have one more blow to strike, and then there will not be left an organized company of infantry in Sheridan's army." His chief demurred. "No use in that. They will all go, directly." The Georgian was doubtful, and said so, indicating the bluecoats on the horizon. "This is the VI Corps, General. It will not go unless we drive it from the field." Once more Early shook his head. "Yes, it will go directly," he insisted as he continued to wait for the whipped Federals to withdraw.

Gordon said no more just then, but he later wrote: "My heart went into my boots." He was remembering "that fatal halt on the first day at Gettysburg," as well as Old Jube's daylong refusal, back in May, to let him strike Grant's unguarded flank in the Wilderness, which he believed had cost the Army of Northern Virginia the greatest of all its victories.

His heart might have sunk still deeper if he had known what was happening, across the way, while he and his chief stood talking. Sheridan had just arrived and was reassembling his scattered army for an all-out counterattack. True to his promise to return from the capital in two days, "if not sooner," he had slept last night in Winchester and had heard the guns of Cedar Creek, some fifteen miles away, while still in bed this morning. Dismissing the cannonade as "irregular and fitful" — most likely a reconnaissance-in-force by one of Wright's brigades — he tried to get back to sleep, without success. At breakfast, the guns still were muttering in the distance, faint but insistent, and he ordered his staff and cavalry escort to saddle up without delay. On the way out of town, he noticed "many women at the doors and windows of the houses, who kept shaking their skirts at us and who were otherwise markedly insolent in their demeanor." It occurred to him that they "were in rapture over some good news," mysteriously received, "while I as yet was utterly in ignorance of the actual situation." What was more, the sound of firing seemed to be moving to meet him; an ominous development. But it was not until he crossed Mill Creek, beyond Kernstown, and reached the crest of a low hill on the far side, that he and his staff and escort saw their worst fears confirmed by "the appalling spectacle of a panic-stricken army."

His first notion was to rally what was left of his command, here if not still farther back toward Winchester, for a last-ditch stand against the rebel force, which might or might not include Longstreet and his famed First Corps. With this in mind, Little Phil ordered his staff and escort to form a straggler line along the crest of the hill: all, that is, except two aides and a score of troopers, who would proceed with him toward Cedar Creek to find out what had happened.

In the course of the twelve-mile ride — "Sheridan's Ride," it came to be called — his purpose changed. Partly this was because of his aggressive nature, which reasserted itself, and partly it was the result of encountering groups of men along the roadside boiling coffee. That did not seem to indicate demoralization;

★

On October 19, slipping back down the Shenandoah Valley after defeat at Fisher's Hill, Early's rebels launched a surprise attack on the Federal camp at Cedar Creek.

nor did the cheers they gave when they saw him coming up the turnpike. "As he galloped on," one of the two aides later wrote, "his features gradually grew set, as though carved in stone, and the same dull red glint I had seen in his piercing black eyes when, on other occasions, the battle was going against us, was there now." Grimness then gave way to animation. He began to lift his little flat-topped hat in jaunty salute, rather as if in congratulation for a victory, despite the contradictory evidence. "The army's whipped!" an unstrung infantry colonel informed him, only to be told: "You are, but the army isn't." He put the spurs to Rienzi — an undersized, bandy-legged man, perched high on the pounding big black horse he had named for the town in Mississippi where he acquired him two years ago — and called out to the retreaters, "About face, boys! We are going back to our camps. We are going to lick them out of their boots!" He kept saying that, shouting the words at the upturned faces along the pike. "We are going to get a twist on those fellows. We are going to lick them out of their boots!"

And did just that: but not with the haste his breakneck manner had implied. Arriving about 10.30 he found Crook's corps disintegrated and Emory's not much better off, though most of it at least was still on hand. Wright's, however, was holding firm in its third position, a couple of miles northwest of Middletown, its line extended southeast across the turnpike by Merritt's and Custer's horsemen. Sheridan got to work at once, concentrating on getting Emory's troops, together with a trickle of retreaters who were returning in response to the exhortations he had shouted as he passed them on the pike, regrouped to support Wright in his resistance to the expected third assault by Early's whooping graybacks. Nor was he unmindful, even at this stage, of the fruits a sudden counterstroke might yield. "Tell General Emory if they attack him again to go after them, and to follow them up, and to sock it to them, and to give them the devil. We'll have all those camps and cannon back again." Emory got the message, and reacted with a sort of fervid resignation. "We might as well whip them today," he said. "If we don't, we shall have to do it tomorrow. Sheridan will get it out of us sometime."

Noon came and went, then 1 o'clock, then 2, and Little Phil continued to withhold his hand: as did Early, across the way.

At 3 o'clock, having at last persuaded his chief to let him undertake a limited attack, Gordon probed the Federal position beyond Middletown, but was easily repulsed. Still Sheridan held back, his numbers growing rapidly as more and more blue fugitives returned from their flight down the turnpike. Finally, after interrogating prisoners to make certain Longstreet was not there, he gave orders for a general advance at 4 o'clock. At first, though their ranks were thinned by looters prowling the Yankee camps in search of food and booty, the graybacks refused to budge. But then one of Emory's brigades found a weak spot in the rebel line, and before it could be reinforced Custer struck with his whole division, launching an all-out mounted charge that sundered the Confederate force and sent the two parts reeling back on Cedar Creek. "Run! Go after them!" Sheridan cried. "We've got the God-damnedest twist on them you ever saw! "

Early did what he could; which, at that stage, wasn't much. For the past four hours — hearing nothing from Lomax, whose roundabout march with half the cavalry later turned out to have been blocked near Front Royal by Torbert's third division — he had watched the steady build-up across the way, aware that this, combined with the rearward leakage from his idle ranks, restored the odds to about what they had been at daybreak, when he enjoyed the lost advantage of surprise. Increasingly apprehensive, he withdrew his captured guns beyond Cedar Creek for quick removal in a crisis, and started his nearly two thousand prisoners on their long trek south to Staunton. All this time, the vaunted "sun

★

of Middletown" was declining, and the nearer it drew to the peaks of the Alleghenies the clearer he saw that the Federals not only had no intention of quitting
their third position, in which they had little trouble fending off a belated
feeling-out by Gordon, but were in fact preparing to launch a massive counterstroke. When it came, as it did at straight-up 4 o'clock, Early managed to withstand
the pressure, left and center, until Emory drove a wedge between two of Gordon's
brigades, opening a gap into which Custer flung his rapid-firing troopers;
whereupon the Georgian's veterans, foreseeing disaster, began a scurry for the
crossings in their rear. Rapidly the panic spread to the divisions of Kershaw and
Ramseur, next in line. Dodson Ramseur — a major general at twenty-seven, the
youngest West Point graduate to attain that rank in the Confederate army —
tried his best to stay the rout, appealing from horseback to his men, but took a
bullet through both lungs and was left to die in enemy hands next day, near
Sheridan's reclaimed Belle Grove headquarters, where he fell.

By then there would be no uncaptured rebels within twenty miles;
Sheridan, having spared his hand until he felt that victory was clearly within reach,
exploited the break for all he was worth. "It took less time to drive the enemy

*Ramseur (inset), a division commander
in the Army of the Valley, collapses into
the arms of an aide after being shot while
attempting to rally his Confederates.*

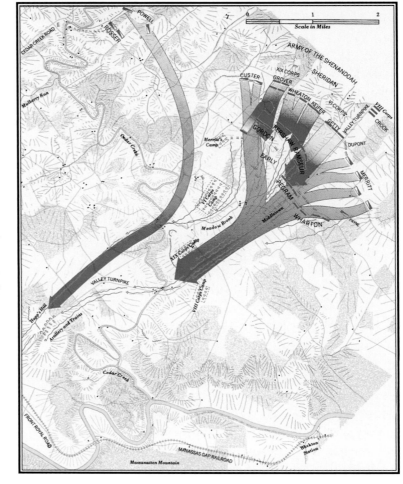

Midday at Cedar Creek, retreating Yankees formed a defensive line and Early called his spent troops to a halt. Sheridan counterattacked and routed the rebels.

from the field than it had for them to take it," according to Merritt, whose division clashed with Rosser's and overran the Confederate far left. Early pulled in Wharton and Pegram to brace the center, under assault from the VI Corps, but only succeeded in delaying Wright's advance. Rearward, meantime, a flying column of Union cavalry wrecked the bridge at Spangler's Mill, just west of Strasburg, with the result that the three miles of turnpike between there and the crossing at Cedar Creek were crowded with artillery and vehicles of all kinds, trapped and at the mercy of the pursuers. Little Phil thus recovered all the guns lost that morning, together with 25 of his adversary's, which enabled him to report that he had taken no less than 43 pieces at one swoop, though he neglected to mention that 18 of them were his own, recaptured in the confusion of the gray retreat.

      Early fell back to Fisher's Hill in the twilight, intending to make a stand there in the morning, but soon saw that it would not do. Though his casualties were only a bit over half as heavy as Sheridan's this day — 2910, as compared to 5665 — his army, routed for the third time in thirty days, was in no condition for further resistance to an enemy twice its size. He took up the march for New Market before daylight, fighting off Custer's and Merritt's horsemen, who snapped at his heels all the way. Summing it up afterwards, Old Jube remarked sadly: "The Yankees got whipped. We got scared."

No explanation could shield him now, however, from the blame about to be heaped upon his head by his own people; blame that outweighed the praise that had come his way, three months ago, when he hovered defiantly on the outskirts of the northern capital. Indeed, the brightness of that midsummer exploit only served to deepen, by contrast, the shadows that gathered in this dark autumn of the Confederacy, which some were already saying would be its last. In the past thirty days Early had fought three full-scale battles, and all three had turned out to be full-scale routs. It mattered little to his critics that he had obliged Grant to lessen the pressure on Lee by detaching a veteran corps from Meade and rerouting another, on its way by sea to reinforce him, in order to meet Jubal's threat, first on the far and then on the near side of the Potomac. Nor did it matter that in the course of his follow-up campaign in the Valley, where he was outnumbered roughly three-to-one from start to finish, he inflicted a total of 16,592 casualties on his adversary — the equivalent of still another blue corps, by Sheridan's own count, and about as many combat troops as he himself had been able to scrape together for any one of those several confrontations — at a cost of less than 10,000 of his own. What mattered in the public's estimation was that, here on the field of Stonewall Jackson's glory, Early had been whipped three times running, each time more soundly than before. Tart of tongue, intolerant of the shortcomings of others since the outset of the war, the former Commonwealth's Attorney of Franklin County now found himself accused of ineptness, inefficiency, incompetence, even drunkenness and cowardice, in the journals and in public and private talk, here in his native Virginia as well as elsewhere in the South.

It was otherwise for Sheridan, whose praises now were being sung throughout the North. "With great pleasure," Lincoln wrote him, three days after Cedar Creek, "I tender to you and your brave army the thanks of the nation and my own personal admiration and gratitude for the month's operations in the Shenandoah Valley, and especially for the splendid work of October 19." The following evening, shortly before midnight, he was awakened by Assistant Secretary of War Charles A. Dana, who had just arrived from Washington to present him with the most prized of all rewards: his commission as a major general in the regular army, together with a commendation from the Adjutant General's office citing him "for the personal gallantry, military skill, and just confidence in the courage and patriotism of his troops . . . whereby, under the blessing of Providence, his routed army was reorganized, a great national disaster averted, and a brilliant

★

victory achieved." Riding through the camps with Little Phil next morning, October 25, Dana thought he had never seen a general so popular with all ranks: not even Sherman or Pap Thomas — maybe not even McClellan in his heyday.

Grant by then was ready to try still another of his pendulum swings at Lee. After ordering a second hundred-gun salute fired with shotted guns in honor of his protégé's third victory in the Valley, he wrote his wife: "I hope we will have one here before a great while to celebrate," and put his staff to work at once on plans for the heaviest strike, so far, at the Richmond-Petersburg defenses. Butler would feint north of the James, with the same number as before, but this time the lunge around the enemy right would be made by no less than 43,000 troops from Hancock, Warren, and Parke, on the theory that what two corps had failed to achieve, just under a month ago, might be accomplished now by three.

On October 27, with Butler already over the river, demonstrating for all he was worth at Fair Oaks, the companion blow was launched. As a further diversion, Parke was to hit the western end of the gray line, just east of Hatcher's Run, while Hancock and Warren swung wide around that stream to cross the Boydton Plank Road and then press north to get astride the Southside Railroad. Alas, no part of this flanking effort went well, and most parts went very badly indeed. Parke encountered stiff resistance and was stalled, and though Hancock made it to his initial objective on schedule, he had to stop and wait for Warren, who was delayed by difficult terrain. While Hancock waited Hill and Hampton struck him flank and front, attacking with about half of the 23,000 effectives Lee had kept south of the river, and forced him to withdraw that night, nearly out of ammunition and altogether out of patience. Meantime Warren turned east, under orders from Grant to help Parke envelop the Hatcher's Run defenses, but was unable to cross the creek; so he too withdrew. None of the three corps in this direction, Parke's or Hancock's or Warren's, had carried out its part of a plan whose only tangible result was the loss of 1758 men — plus the confirmation of Hancock's resolution to seek duty elsewhere; which he would do the following month, suffering as much from recent damage to his pride as from the continuing discomfort of his Gettysburg wound. North of the James, where Lee was not deceived by his gyrations around Fair Oaks, Butler lost 1103 killed, wounded, and missing, as compared to a Confederate loss of 451 there and perhaps twice that number in the opposite direction, along the Boydton Road and Hatcher's Run.

All lines remained the same, north and south of the river, as both armies prepared to go into winter quarters. No more discouraged by this latest failure than he had been by those others outside Petersburg and Richmond, Grant maintained what Lincoln called his "bulldog grip," prepared to "chew and choke" as long as need be. He could fail practically any number of times, and only needed to succeed but once. "I will work this thing out all right yet," he told his wife in a home letter.

In any case, this late-October affair down around Richmond went practically unnoticed by a public still absorbed in the recent Shenandoah drama, finding it restorative of the romantic, picture-book aspect so long missing from the war. "The nation rings with praises of Phil Sheridan," the Chicago *Tribune* noted, three days after the famous ride that saved the day at Cedar Creek and prompted black Rienzi's master to change his name to Winchester in commemoration of the exploit. Various poets tried their hand at the subject, including Herman Melville, but the one who caught the public's fancy best was T. Buchanan Read in a ballad titled "Sheridan's Ride."

> *Hurrah! Hurrah for Sher-i-dan!*
> *Hurrah! Hurrah for horse and man!*

its refrain went. Availing himself of a poetic license which the general he praised sometimes employed in his reports, Read doubled the distance of the gallop, eliminated all stops along the way, and had Rienzi himself announce the nick-of-time arrival to the troops:

> *"I have brought you Sheridan, all the way*
> *From Winchester, down to save the day."*

Widely read and recited, the piece made a fine recruiting and electioneering appeal, especially when delivered by professionals such as James E. Murdoch, a retired actor and celebrated "reader," whose declamation of the poem at a theater in Cincinnati on November 1, just one week before the presidential contest was to be settled at the polls, threw the crowd into a frenzy of approval for the war and for the men who fought and ran it.

★　★　★

*Attacking from the prairie
south of Westport, Missouri, Sidney
D. Jackson's brigade overruns
Union militia near the Big Blue
River on October 22, 1864.*

FOUR

Price Raid; "Florida"; Cushing; Forrest Raids Mid-Tennessee

1864 ★ ★ ★ ★ ★ ★

Elsewhere — not only in the embattled heartland of the South, but also in places as far afield as Kansas, Vermont, and Brazil — both sides undertook desperate measures, throughout the critical two-month span that opened with the fall of Atlanta, in attempts to influence militarily the early-November political decision that perhaps would begin to end the war itself, come Inauguration Day. For example:

Aside from an abortive Union gunboat probe down White River in late June, which was turned back at Clarendon before the flotilla could enter the Arkansas to help patrol that line of Federal occupation, there had been no significant clash of arms in the Transmississippi since Frederick Steele retired from Camden in late April and Banks and Porter abandoned in May their effort to ascend the Red. Since then, Kirby Smith had seemed content to rest on his laurels, clinging precariously to what was left of Texas, Louisiana, Arkansas, and the Indian Territory — "Kirby-Smithdom," this vast but empty stretch of the continent was called — and resisted all efforts by Richmond and homesick subordinates to persuade him to go over to the offensive, either toward New Orleans or St Louis. Discontent to have so many good troops standing idle, even against such odds as here obtained, the authorities instructed him in mid-July to prepare Richard Taylor's corps, along with "such other infantry as can be spared," for a

★

prompt movement across the Mississippi to assist in the defense of Atlanta and Mobile. Smith passed the order to Taylor, who had been sulking in Natchitoches for the past six weeks, his hurt feelings, if not his animosity toward his chief, somewhat relieved by a promotion to lieutenant general as a reward for his repulse of Banks. Eager to shake the dust of Kirby-Smithdom from his feet, Taylor looked into the possibility of a crossing, either by ferries or by the employment of what would have been the longest pontoon bridge in history, but replied in the end that it couldn't be done, since the Federals, getting wind of the project, had stationed ironclads at twelve-mile intervals all the way from Vicksburg past the mouth of the Red, with gunboats on constant patrol between them, day and night. "A bird, if dressed in Confederate gray, would find it difficult to fly across the river," a reconnoitering cavalryman declared.

Regretfully, for he was as anxious to get rid of Taylor as Taylor was to be quits with him, Smith informed his superiors in Virginia that the shift could not be made. By then the year had moved into August, and Richmond's answer solved at least a part of his problem by dusting the gadfly Taylor off his back. Stephen Lee having been sent to Georgia to head a corps under Hood, the

Price not only had discouraged the sending of more troops east across the Mississippi, he had even provoked a drain in the opposite direction . . .

Kentucky-born Louisianian (and presidential brother-in-law) was ordered to replace him in command of the Department of Alabama, Mississippi, and Eastern Louisiana, temporarily under Maury at Mobile. On a moonless night, within a week of receiving the order on August 22, Taylor crossed the river in a dugout canoe, swimming his mare alongside, and set out eastward for his new headquarters in Meridian. Before he reached it, Smith — or, more specifically, Sterling Price — had placed an alternate plan in execution, back in the Transmississippi, by launching 12,000 horsemen northward into Missouri.

Originally designed to draw attention away from the downriver crossing, the operation was now to be undertaken for its own sake: first against St Louis, where government warehouses bulged with the goods of war, then westward along the near bank of the Missouri River to the capital, Jefferson City — whose occupation, however brief, would refurbish the somewhat tarnished star representing the state on the Confederate battle flag — then finally back south "through Kansas and the Indian Territory, sweeping that country of its mules, horses, cattle, and military supplies." So Price was told by Smith in his instructions for the raid, which was also to serve the double-barreled purpose of discouraging the departure of still more bluecoats to lengthen the odds against Hood and Lee, east of the Mississippi, and of attracting recruits to the gray column as it swept through regions whose voters were about to get their chance, as the case was being put to them in the campaign already under way, to "throw off the yoke of oppression." Mounted on Bucephalus, a warhorse as gray as its rider and stockily built to withstand his two hundred and ninety dead-weight pounds, Old Pap left Camden on August 28 and was joined next day at Princeton by the divisions of Marmaduke and Fagan, who rode with him across the Arkansas River at Dardanelle on September 2, midway between Little Rock and Fort Smith, neither of whose blue garrisons ventured out to challenge the invaders. At Pocahontas on the 13th, up near the Missouri line, Jo Shelby added his division to the column, now 12,000 strong, with fourteen guns, though only about two thirds of the troopers were adequately armed — a deficiency Price intended to repair when he encountered opposition. On September 19, the day before his fifty-fifth birthday, he crossed into his home state, headed for Ironton, eighty miles to the north, terminus of the railroad running south out of St Louis, another eighty miles away. At nearby Pilot Knob there was a Union fort, Fort Davidson, with a garrison of about one thousand men and seven guns, and he had chosen this as his first prize of the campaign, to be followed by those other, larger prizes, north and west.

Assembling his three divisions at Fredericktown on the 25th — a day's ride east of Pilot Knob, which he intended to move against tomorrow — he received news from St Louis that was both good and bad, from different points of view. Department Commander William Rosecrans, on learning in early September that the graybacks had crossed the Arkansas in strength, wired Halleck a request that A. J. Smith's two veteran divisions, then aboard transports at Cairo on their way to rejoin Sherman after service up the Red and in North Mississippi, be sent instead to help defend Missouri against this new incursion. Old Brains complied by ordering Smith upriver at once to "operate against Price & Co." This meant that one purpose of the raid had been achieved before the first blow landed; Price not only had discouraged the sending of more troops east across the Mississippi, he had even provoked a drain in the opposite direction, though at the cost of lengthening the odds against fulfilling his other objectives, including the strike at goods-rich St Louis, whose defenses now were manned by Smith's 8000 gorilla-guerillas, in addition to its regular complement. In any case, after sending a brigade to rip up track on the railroad above Ironton and thus prevent the sudden arrival of reinforcements, he completed his plans for the reduction of Fort Davidson, twenty miles west of Fredericktown, and had it invested by nightfall the following day. He badly wanted its thousand-man garrison and their arms: especially those seven guns, whose addition would increase by half the firepower of the artillery he had brought along for blasting a path through his beloved Missouri.

Brigadier General Thomas Ewing, commander of the District of St Louis — Sherman's brother-in-law and author, too, of last year's infamous Order 11, which emptied Missouri's western counties of civilians in an attempt to ferret out guerillas whose bloody work grew bloodier in reaction to the hardships thus imposed on their women and children — had come down to the fort on an inspection trip, only to have the railroad cut in his rear, and decided not to abandon the place under threat from ten times the number he had for its defense. Accordingly, when a rebel delegation came forward under a flag of truce that night, demanding surrender, he sent it back with a defiant challenge, and when the demand was repeated a few hours later he did the same thing, adding that he would fire on the next white flag that approached his works. These were extremely stout, heptagonal in shape, with earthen walls nine feet tall and ten feet thick, surrounded by a dry moat as deep as the walls were high. Next day, September 27, they were tested in a furious six-hour fight that cost the attackers 1500 casualties, half again more than the total number of defenders, who lost 200. Falling back at dark, Old Pap's troopers began the construction of scaling ladders to use when they renewed the assault at dawn, and Ewing, knowing the fort could not hold out past then — and that he himself, as the author of Order 11, was unlikely to survive capture — assembled a council of war to decide whether to surrender or risk attempting a getaway. The vote was for the latter; which succeeded. Under cover of darkness the blue garrison built a drawbridge, draped it with canvas to muffle the sound of boots and hoofs, and withdrew undetected through a gap in the gray lines, leaving behind a slow fuze laid to the powder magazine. Slogging along in a column of twos, Ewing and his 800 survivors were well out the road to Rolla, seventy miles northwest, when the magazine blew with a great eruption of flame that gave the investors their first hint the fort was empty.

Marmaduke and Shelby, furious over their losses and fairly itching to fit Ewing for a noose, wanted to take out after him at once, but their fellow Missourian Price, already regretting a fruitless three-day interlude which had deprived him of more than a tenth of his command and netted him nothing but rubble and spiked guns, was unwilling to use up still more time on a project that he suspected had already cost him whatever chance there had been for surprising Rosecrans in St Louis. Sure enough, after following the Iron Mountain Railroad to within thirty miles of the city, he found its garrison reinforced to a strength reportedly greater than his own. So he turned west, as planned — though he had not intended to do so empty-handed — up the south bank of the Missouri, wrecking bridges and culverts along the Pacific Railroad as he proceeded, first across the Gasconade River and then the Osage, which he cleared on October 6 to put his raiders within easy reach of Jefferson City.

★

But this too was untakable, he decided upon learning that its defenses were manned by bluecoats drawn from beyond the river despite a flurry of apprehension caused there the week before by a ruthless attack on Centralia, fifty miles north of the capital, by a force of about 200 butternut guerillas under William Anderson, who bore and lived up to the nickname "Bloody Bill." A former lieutenant in William C. Quantrill's gang, of Lawrence and Fort Baxter fame, he had quarreled with his chief in Texas and returned to his old stomping ground, near the Missouri-Kansas border, along with other disaffected members

Led by William Anderson (top left), rebel guerillas Frank James (top right), Jesse James (bottom left), and George Todd joined the raid on Centralia.

Looking to the West. 151

Topeka farmer Samuel J. Rider joined the 4000 Kansas militiamen "invading" Missouri to fight "Pap" Price's rebels and sketch the action beginning October 21.

of the band, including George Todd and David Pool, as well as Frank James and his seventeen-year-old brother Jesse. Clattering into Centralia at midday, September 27 — the day of the Fort Davidson assault, one hundred and fifty miles southeast at Pilot Knob — they held up a stagecoach and an arriving train, killed two dozen unarmed soldiers aboard on furlough, along with two civilians who tried to hide valuables in their boots, and left hurriedly, with $3000 in greenbacks from the express car, when three troops of Union cavalry unexpectedly appeared and gave chase. Three miles out of town, the guerillas turned on their pursuers, who numbered 147, and shot dead or cut the throats of all but 23 who managed to escape on fast horses. "From this time forward I ask no quarter and give none," Anderson had announced on the square in Centralia, and then proceeded to prove he meant it, first in town and then out on the prairie.

Price's decision to forgo a strike at Jefferson City, the main political objective of his raid, was based on more than information that the capital had been reinforced, not only from beyond the Missouri, but also from scattered posts on this side of the river, including Springfield and Rolla. He learned too, while skirmishing on the outskirts after crossing the Moreau, that Rosecrans, supposedly left holding the bag in St Louis, had sent Smith's 8000 infantry westward in his wake, along with 7000 troopers under Major General Alfred Pleasonton, who had served the better part of a year as cavalry commander in the Army of the Potomac until Grant replaced him with Sheridan, back in March, and sent him west to share Old Rosy's exile. Price was aware that any prolonged attempt to break through the capital defenses was likely to be interrupted by the arrival of Pleasonton and Smith, now toiling along the demolished Pacific Railroad with a combined strength greater than his own. Moreover, scouts coming in from the Kansas border, a hundred and forty miles in the opposite direction, reported that more than 20,000 regulars and militia were being assembled there for his reception by the department commander, Major General Samuel R. Curtis, his old Pea Ridge adversary. The thing to do, he reasoned, was get there fast, before Curtis got organized or Smith and Pleasonton came up in his rear to make the fight for Kansas City a two-front affair. Accordingly, he turned his back on the state capitol, plainly visible on its hill beyond the treetops, and continued his march another forty miles upriver to Boonville, which he reached October 9. Riding due west for Lexington, sixty-odd miles away — the scene of his one unassisted victory, back in the first September of the war, hard on the heels of the triumph he had shared with Ben McCulloch at Wilson's Creek — he put Marmaduke's division in the lead and had Shelby strike out left and right at Sedalia and Glasgow, both of which were taken on the 15th, together with their garrisons, while Fagan covered the rear, on the lookout for Pleasonton's horsemen, who were known to have reached Jefferson City four days ago. Four days later at Waverly, his home town on the south bank of the Missouri, twenty miles short of Lexington, Shelby encountered a force of Coloradans and Kansans under Major General James Blunt, brought in from the plains by Curtis and sent forward to delay the approach of the raiders. Here were fired the opening shots of what turned out to be a week-long running skirmish, covering more than a hundred miles of the border region, with several pauses for full-scale engagements along the way.

Shelby drove Blunt back through Lexington, October 20, and on across the Little Blue next day, fighting house-to-house through Independence to the Big Blue, just beyond. Curtis had established a line of works along the opposite bank, manned by 4000 regulars and an equal number of Kansas militia, some 16,000 of whom had come forward in the current emergency, though only about one fourth of them were willing to cross into Missouri, the remainder having called a halt at the state line, half a dozen miles to the west. His plan was to hang on there, securely intrenched, till Pleasonton came up in Price's rear, then go over to the offensive, east and west, against the graybacks trapped between the Big and Little Blues. It did not work out quite that way: partly because of the timid militia, skulking rearward on home ground, but mainly because of black-plumed Jo Shelby. While Marmaduke and Fagan took the bluecoats under fire from across the river on the morning of the 22d, Shelby splashed his three brigades across an upstream ford to flank the defenders out of their works and throw them into retreat on Westport, immediately south of Kansas City and within two miles of the state line. As a result, when Pleasonton arrived that night he found Curtis's intrenchments bristling in his path, occupied by the butternut invaders he had been trailing ever since he left St Louis, three weeks back.

Confronted east and west by forces that totaled three times his own, Old Pap took stock and pondered his next move. Staffers advised that this be south without delay, while the long road home lay open for a withdrawal in good order. But he was urged by Shelby, whose blood was up, to take advantage of a position which, though not without obvious dangers, fairly glittered with Napoleonic possibilities. Using one division to hold Pleasonton in check on the far side of the Big Blue, he could move with the other two against Curtis at nearby Westport, then turn, having disposed of the Kansan and his green militia, to crush Pleasonton and thus cap the raid with a stunning double victory; after which, according to Shelby, he could proceed at his leisure, rounding up Federal garrisons and Confederate recruits, as intended from the outset, on the final leg of his march back across the Arkansas. Price liked the notion, partly for its own glittering sake, partly because of the chance it gave him to put a gainful end to a campaign that so far had profited his country and his reputation next to nothing. Accordingly, after lodging Marmaduke's two brigades in the Union intrenchments overlooking the Big Blue, he ordered Fagan and Shelby to prepare their six for the attack on Curtis, whose troops were deployed along Brush Creek below Westport, at daybreak tomorrow, October 23.

Pleasonton, having posted his four brigades for a dawn assault on the former Union works across the river west of town, spent the night in

General Samuel R. Curtis, above, led the Union forces, mostly militia, against Sterling Price in Missouri.

Independence. A graduate of West Point and the hard-knocks school of combat in the East — including Brandy Station, where he had taken Jeb Stuart's measure on the eve of Gettysburg — he intended to do to Price tomorrow what Price had done to Curtis today; that is, dispossess him of those works. Even though no blue infantry was at hand (A. J. Smith's two divisions had turned south at Lexington, under orders from Rosecrans to head off a rebel swerve in that direction, and thus were removed from all possible contact with the raiders, now or later) the forty-year-old cavalryman was satisfied he could do the job on his own, and with this in mind had his cannoneers keep heaving shells across the Blue to discourage the intrenched defenders from getting much sleep till after midnight, a scant five hours before he planned to strike them.

By that time Curtis was planning to strike them too, despite his mistrust of the balky militia that comprised about four fifths of his command. Persuaded by Blunt — as Price had been by Shelby — that a victory was within his reach if he would only grasp it, the fifty-nine-year-old department head reversed his previous decision to fall back on Fort Leavenworth, twenty-five miles north on the Missouri, and agreed instead, under pressure from Blunt and others at a council of war in the Gillis House that night in Kansas City, to go over to the offensive in the morning. Down along Brush Creek all this while, his green recruits were kept awake by the boom of Pleasonton's guns on the far side of the river and by the nerve-jarring crump of shells on the near bank, close in their rear. "I'd rather hear the baby cry," one married volunteer remarked. Presently the guns left off, but he continued to fret, confiding in a friend that he expected to be killed in tomorrow's contest, and found small comfort in assurances that the future life was superior to this one. "Well, I don't know about that," he said, still worried.

His chances for survival were better than he knew. Next day's battle, though numerically the largest ever fought in the Transmississippi — out of 40,000 Federals and Confederates on the field, close to 30,000 were engaged, as compared to just under 27,000 at Pea Ridge, the next largest, and only about half that many at Wilson's Creek — was neither as hotly contested nor as bloody as both sides had expected when they lay down to sleep the night before. Fagan and Shelby went forward as ordered, shortly after daybreak, and threw Curtis's greenhorns into skittery retreat, much as Shelby had predicted and Curtis, who watched the action through a spyglass from the roof of a convenient farmhouse, had feared. But not for long. Thrown back on Westport and the Kansas line, the militiamen and regulars, outnumbering the attackers better than two to one, not only rallied and held their own against renewed assaults by the yelling graybacks, but even, in response to a horseback appeal from their commander, who came down off his roof to ride among them, began massing for a counterattack to recover the lost ground along the creek. Whereupon, in this moment of crisis — it was now about midmorning — Price was informed that Pleasonton had broken

Marmaduke's line on the near bank of the Big Blue and was approaching his right rear, threatening to come between the raiders and their train, parked southward on the road he had been persuaded not to take the night before.

Enraged to find the dawn attack deferred to await his arrival from Independence, Pleasonton had begun his day with on-the-spot dismissals of two brigade commanders — "You're an ambulance soldier and belong in the rear," he told one of the brigadiers, shaking a cowhide whip in his face quite as if he meant to use it — and peremptory orders for their successors to throw everything they had against Byram's Ford, a strongly defended crossing on the rebel right. He did this on the theory that the enemy would least expect a major effort there, and the result was all he hoped for. When the dismounted horsemen splashed across the ford, through the abatis on the opposite bank, then up and over the intrenchments on the ridge beyond, he followed with a third brigade to deepen and widen the breakthrough, while the fourth came on behind. Marmaduke's rattled defenders, turned suddenly out of their works by twice their number, fled rearward across the prairie that stretched to the Kansas line, unobstructed except by the trees along Brush Creek, where Price's effort against Curtis was in crisis.

Pleasonton reined in his horse to watch them flee, and as he did he stabbed the air with one hand, pointing at the sticklike figures, running or wavering, near and far. "Rebels! Rebels! Rebels!" he shouted at his troopers, who had stopped, much as he himself had done, to watch this flight across the rolling tableland. "Fire! Fire, you damned asses!" he kept shouting.

There was not much time for that, however. Faced with the threat of annihilation on the open prairie, Price disengaged Fagan, pulled him back alongside Marmaduke's reassembled fugitives, and used them both to cover the withdrawal of his train, southward down the road on which it had been parked for ready accessibility or a sudden getaway. Shelby — as was only fair, since he was the one who had talked his chief into this predicament in the first place — was charged with stalling the blue pursuit, at least until the wagons and guns and the other two divisions, remounted to make the best possible time, escaped the closing jaws of the trap and got a decent head start down the road to Little Santa Fe, a dozen miles below on the Kansas border. Hemmed in as he was on three sides (and grievously outnumbered; Curtis and Pleasonton had just over 20,000 infantry and cavalry engaged from first to last — less than three quarters of their total force — while Price had only about 9000 — all that he had arms for) this was no easy task; but Shelby managed it in style, cutting his way out with a mounted charge in the final stage, near sunset, to join the gray column grinding its way south in the darkness.

Too ponderous for even heavy-hocked Bucephalus to bear his weight for long, Price rode in a carriage on the retreat, depressed by the knowledge that Westport — sometimes disproportionately referred to as "the Gettysburg of the Transmississippi," though in point of fact it was fought for no real purpose and

★

*This heroic mural by famed illustrator N. C. Wyeth
depicts the clash of Confederate and Union
cavalry during the Battle of Westport on October 23.*

settled nothing — had merely added another repulse to his long list of reverses, east and west of the Mississippi River. Fortunately it was not a costly one, however. Neither commander filed a casualty report, but their losses seem not to have reached a thousand men on either side.

A heavier defeat, with heavier losses, came two days later, fifty miles beyond Little Santa Fe, soon after the raiders crossed the Marais des Cygnes, which flowed eastward into Missouri and the Osage. They had made good time, marching day and night through wet and blustery weather, but Pleasonton and Curtis dogged their heels, eager to close in for the kill. Swinging west to take advantage of better roads leading south beyond the Kansas line, Price halted Marmaduke on the far bank of the tributary river — mostly referred to hereabouts as the Mary Dayson — in hope of delaying his pursuers at that point. This the Missouri West Pointer did, briefly at least, and then fell back to a similar position on Mine Creek, three miles below, where Fagan had been deployed to support the rear-guard effort with ten of the column's fourteen pieces of artillery. Here on that same morning, October 25, occurred the first and last full-scale engagement between regulars, Federal and Confederate, to be fought on Kansas soil. The first Price knew of its outcome was when he saw troops from both divisions come stumbling toward him in disorder, pursued by whooping bluecoats, mounted and afoot. All ten guns were lost in the rout, along with close to a thousand prisoners, including Marmaduke himself, Brigadier General William Cabell — Old Pap's only other West Pointer, in charge of one of Fagan's Arkansas brigades — and four colonels. Hit in the arm and thrown from his horse, Marmaduke was taken single-handedly by James Dunlavy, an Iowa private, who marched his muddy, dejected captive directly to army headquarters. "How much longer have you to serve?" the department commander asked. Told, "Eight months, sir," Curtis turned to his adjutant: "Give Private Dunlavy a furlough for eight months." The Iowa soldier left for home next day, taking with him the long-haired rebel general's saber for a souvenir of the war that was now behind him, and Marmaduke and Cabell were soon on their way to northern prison camps, the war behind them too.

Once more Price called Shelby back to contest a further advance by the exultant Federals, who were delayed in following up their victory by an argument that broke out between Curtis and Pleasonton as to whether the latter's prisoners were to be sent to Leavenworth or St Louis and thus be credited to Curtis or to Rosecrans. While Shelby fought successive rear-guard actions on the Little Osage and the Marmiton, Price reassembled the other two divisions and pressed on south with the train. Beyond the Little Osage the road forked, one branch leading to Fort Scott, six miles south across the Marmiton, the other back southeast into Missouri. Formerly the fort had been on Old Pap's list of trophies to be picked up on this final leg of the raid, but now he had neither the time nor the strength to move against it. After pausing to lighten the train by burning some 400 wagons, together with the excess artillery ammunition — excess because only one four-gun battery remained — he took the left-hand fork and set out on a forced march of just over sixty miles to Carthage, down near the southwest corner of his home state. Although most of the blue pursuers stopped for food and a night's sleep at Fort Scott, and though Shelby managed to keep the rest from overtaking the train and its escort, still the night-long day-long night-long trek, ending at Carthage on the morning of the 27th, was an experience not soon forgotten by those who made it. "I don't know that a longer march graces history; a fatal day for horse flesh," one weary raider noted in his journal at its close.

Price rewarded their efforts with a full day's rest, then resumed the march next morning, hoping to reach and cross the Arkansas River, still more than a hundred miles away, without having to stop for another time- and man-killing fight for survival. His hope was not fulfilled. At Newtonia that afternoon, twenty-odd miles beyond Carthage, the Federals came up in his rear and obliged him to turn and form ranks for a battle no one knew was to be the last ever fought between regular forces west of the Mississippi.

★

Back at Fort Scott two days ago, the Kansas militia and two of the Missouri cavalry brigades had retired from the chase — as had Pleasonton himself, after falling sick — but Curtis, with his regulars and Blunt's plainsmen still on hand, as well as Pleasonton's other two brigades, was determined to overtake the still-outnumbered raiders before they escaped. Here at Newtonia he got his chance; along with cause to regret it. Spotting dust clouds south of town, Blunt thought Price was attempting a getaway and galloped hard around his flank to cut him off, only to be cut off himself by Shelby, who handled him roughly until other blue units broke through to cover his withdrawal. The fighting sputtered out at sundown, with little or no advantage on either side, and Price took up his march southward, unpursued, while Curtis waited for Blunt to lick his wounds.

Reduced by casualties and desertions, badly worn by a thousand miles of marching, and even lower in spirits than he was on food and ammunition . . . Price was in no condition to risk another heavy engagement . . .

"I must be permitted to say that I consider him the best cavalry officer I ever saw," Old Pap wrote gratefully of Shelby in his report of the campaign: an opinion echoed and enlarged upon by Pleasonton years later, when he said flatly that the Missourian was "the best cavalry general of the South."

Curtis rested briefly, then proceeded, no longer in direct pursuit of Price, who veered southwest beyond Newtonia, but rather by a shorter route, due south across the Arkansas line, in hope of intercepting the raiders when they swung back east to recross the Arkansas River between Fort Smith and Little Rock; probably at Dardanelle, he figured, where they had crossed on their way north eight weeks ago. Hurrying from Pea Ridge to the relief of Fayetteville, which was reported under attack by a detachment from the rebel main body at Cane Hill, just under twenty miles southwest, the Kansan supposed that his cut-off tactics had succeeded. When he reached Fayetteville on November 4, however, he not only found the attackers gone, he also learned that his adversary was moving en masse in the opposite direction, away from the trap contrived for his destruction. Reduced by casualties and desertions, badly worn by a thousand miles of marching, and even lower in spirits than he was on food and ammunition — which was low indeed — Price was in no condition to risk another heavy engagement, and to avoid one he had decided not to attempt a march east of Fort Smith, whose garrison would be added to the force that would surely intercept him before he made it across the river in that direction. Instead, he would move on west, toward Tahlequah in the Indian Territory, for an upstream crossing of the Arkansas twenty-odd miles beyond the border. Curtis followed as far as a north-bank settlement called Webber's Falls, November 8, only to find that the raiders, assisted by friendly Choctaws, had destroyed all the available boats on reaching the south bank the day before. So he pronounced the campaign at an end, fired a 24-gun salute in celebration, the booms reverberating hollowly across the empty plains, and turned back toward Kansas, glad to be done with an opponent who, as he declared in closing his report, had "entered Missouri feasting and furnishing his troops on the rich products and abundant spoils of the Missouri Valley, but crossed the Arkansas destitute, disarmed, disorganized, and avoiding starvation by eating raw corn and slippery-elm bark."

Worse things were said of Price by his own soldiers in the course of their detour through the wintry territorial wilds. "God damn Old Pap!" was among the milder exclamations on the march, and afterwards there was to be a formal inquiry into charges of "glaring mismanagement and distressing mental and physical military incapacity." One trooper wrote that his unit subsisted for four days on parched acorns, while another told how he and his comrades butchered and devoured a fat pony along the way. A cold wind cut through their rags, freezing the water in their canteens, and coyotes laughed from the darkness beyond their campfires, a terrifying sound to men too weak from hunger or dysentery to keep up with the column. Even so, hundreds fell out in the course of this last long stage of the raid, south through Indian country, down across the Red into Texas, and finally back east to Laynesport, Arkansas, which they reached on December 2, still a hundred miles west of Camden, which Price had left just over three months ago. Though he put the case as best he could in his report — "I marched 1434 miles; fought 43 battles and skirmishes; captured and paroled over 3000 Federal officers and men . . . [and] do not think I go beyond the truth when I state that I destroyed in the late expedition to Missouri property to the amount of $10,000,000 in value" — his claim that his own losses totaled fewer than a thousand men, in and out of combat, scarcely tallied with the fact that he returned with only 6000, including recruits, or barely half the number who had ridden northward with him in September.

★ ★ ★ *W*hatever the true figures were, in men or money, and however great the disruption had been along the Missouri River and the Kansas border, this last campaign in the Transmississippi had no more effect on the outcome of the national conflict than did a much smaller, briefer effort made at the same time, up near the Canadian border, against St Albans, a Vermont town of about 5000 souls. This too was a raid designed to bring home to voters remote from the cockpit

★

*On October 15, Bennett Young led a group of
rebels down from Canada on a raid of St Albans, Vermont,
whose placid Main Street is pictured here.*

of war — Westport and St Albans were both just under a thousand miles from
Charleston — some first-hand notion of the hardships involved in a struggle
they were about to decide whether to continue or conclude: with the difference
that the New England blow was struck primarily at what was reputed to be a
New Englander's tenderest spot, his wallet.

First Lieutenant Bennett Young, a twenty-one-year-old Kentuckian
who had ridden with Morgan, reconnoitered St Albans on a visit from Canada, fifteen miles away, and returned on the evening of October 18 with twenty followers,
most of them escaped or exchanged prisoners like himself. Arriving in twos and
threes to avoid suspicion, they checked into various hotels and boarding houses,
then assembled at 3 o'clock the following afternoon in the town square, where
they removed their overcoats to reveal that each wore a gray uniform and a pair
of navy sixes. At first, when Young announced that the place was under formal
occupation and ordered all inhabitants to gather in the square, the townspeople

thought they were being treated to some kind of joke or masquerade, but when the raiders began discharging pistols in the direction of those who were slow to obey the lieutenant's order, they knew better. Meantime, three-man details proceeded to the three banks and gathered up all the cash on hand, though not before outraged citizens began to shoot at them from second-story windows. In the skirmish that ensued, one townsman was killed, three invaders were wounded, and several buildings around the square were set aflame with four-ounce bottles of Greek fire, brought along to be flung as incendiary grenades.

Back in Canada not long after nightfall, once more in civilian dress, Young and his men counted the take from this farthest north of all Confederate army operations. It came to just over $200,000; none of which ever found its way to Richmond, as originally intended, being used instead to finance other disruptions in other Federal regions that had not felt the hand of war till now.

★ ★ ★ *A*float as ashore, throughout this critical span of politics and war, there were desperate acts by desperate men intent on winning a reputation before it was too late. Commander Napoleon Collins, for example, a fifty-year-old Pennsylvanian with thirty years of arduous but undistinguished service, learned while coaling at Santa Cruz de Tenerife in mid-September that the rebel cruiser *Florida* had been there for the same purpose the month before; reports attending her departure, August 4, were that her next intended port of call was Bahia, just around the eastern hump of South America, some 1500 nautical miles away. His orders, as captain of the U.S.S. *Wachusett* — a sister ship of the *Kearsarge* — were to intercept and sink her, much as Winslow had sunk the *Alabama* three months ago off Cherbourg, and he wasted no time in clearing the Canaries for Brazil. Arriving in early October he did not find the prize he sought in Bahia harbor; nor, despite her six-week head start and her reputed greater speed, had she been there. Apparently the Santa Cruz report was false, or else she had been terribly busy on the way. Then two days later, shortly after dark, October 4, a trim, low-lying sloop of war put into All Saints Bay, and when Collins dispatched a longboat to look her over he found to his delight that the report had been true after all. The twin-stacked handsome vessel, riding at anchor no more than a long stone's throw off his starboard flank, was indeed the *Florida,* one of the first and now the last of the famed Confederate raiders that had practically driven Federal shipping from the Atlantic.

Since her escape from Mobile Bay in January of the previous year, *Florida* had burned or ransomed 37 prizes, and to these could be added 23

The C.S.S. Florida *(center), second only to the* Alabama
*in damage done to the Union navy, takes on coal
and provisions off Funchal, Madeira, in early 1864.*

more, taken by merchantmen she had captured and converted into privateers, thereby raising her total to within half a dozen of the *Alabama*'s record 66. Most of the time she had been in Commander John Maffitt's charge, but since the beginning of the current year, Maffitt having fallen ill, she had been under her present skipper, Lieutenant Charles M. Morris. Her most recent prize was taken a week ago, and Collins had it very much in mind to see that she took no more. Employing Winslow's tactics, he sent Morris next day, through the U.S. consul at Bahia, a formal invitation to a duel outside the three-mile limit. But Morris not only declined the challenge, he even declined to receive the message, addressed as it was to "the sloop *Florida*," quite as if he and his ship were nationless. He would leave when he saw fit, he said, having been granted an extension of the two-day layover allowed by international law, and would be pleased to engage the *Wachusett* if he chanced to meet her on the open sea. Collins absorbed the failure of this appeal to "honor," which had worked so well for Winslow against Semmes, then fell back on a secondary plan, rasher than the first and having nothing whatever to do with

★

honor. Tomorrow night would be the *Florida*'s third in Bahia harbor, and he was determined, regardless of the security guaranteed by her presence in a neutral port, that it would be her last.

Suspecting nothing, Morris coöperated fully in the execution of the plan now being laid for his undoing. He had had the shot withdrawn from his guns, as required by law before entering the harbor, and assured the port authorities — who seemed disturbed by the thought of what he (not Collins, with whose government their own had long-standing diplomatic relations) might do in the present edgy situation — that he would commit no hostile act, in violation of their neutrality, against the enemy vessel anchored off his flank. This done, he let his steam go down, hauled his fires, and gave the port and starboard watches turnabout shore leave while off duty. On the night of October 6 he went ashore himself, with several of his officers, to attend the opera and get a good night's sleep in a hotel, leaving his first lieutenant aboard in charge of half the crew. Long before dawn next morning he was awakened by the concierge, who informed him that his ship was under attack by the *Wachusett* in the harbor down below.

Collins had planned carefully and with all the boldness his given name implied. Slipping his cables in the deadest hour of night, he backed quietly to give himself space in which to pick up speed for a ram that would send the raider to the bottom, then paused to build up a full head of steam before starting his run on the stroke of 3 o'clock. His intention was to bear straight down on the sitting vessel and thus inflict a wound that would leave her smashed beyond repair; but *Wachusett* went a bit off course and struck instead a glancing blow that crushed the bulwarks along the rebel's starboard quarter and carried away her mizzenmast and main yard. Convinced that he had inflicted mortal damage, Collins was backing out to let his adversary sink, when there was a spatter of small arms fire from the wreckage on her deck. He replied in kind and added the boom of two big Dahlgrens for emphasis, later saying: "The *Florida* fired first." As he withdrew, however, he saw that the raider was by no means as badly hurt as he had thought. Accordingly, he changed his plan in mid-career and decided to take her alive. Guns reloaded, he stopped engines at a range of one hundred yards and called out a demand for the sloop's immediate surrender before he blew her out of the water.

Aboard the crippled *Florida*, with no steam in her boilers, no shot in her guns, and only a leave-blown skeleton crew on hand, the lieutenant left in charge had little choice except to yield, though he did so under protest at this hostile action in a neutral port. Collins promptly attached a hawser to the captive vessel and proceeded to tow her out to sea, fired on ineffectively by the guns of a harbor fort and pursued by a Brazilian corvette which he soon outdistanced. Morris arrived from the hotel in time to see the two sloops leave the bay in this

★

tandem fashion, *Wachusett* in front and his own battered *Florida* in ignominious
tow, and though he too protested this "barbarous and piratical act," they were
by then beyond recall on the high seas, bound for Norfolk.

After a stopover in the West Indies, Napoleon Collins brought the
two warships into Hampton Roads on November 12, both under their own
power. There he received a welcome as enthusiastic as the one that had greeted
his former squadron commander, Captain Charles Wilkes — also at one time
skipper of the *Wachusett* — following his removal, three years ago, of Mason
and Slidell from the British steamer *Trent*. Seward, on learning of what had
happened in Bahia harbor, was only too aware that the two cases were uncomfortably similar, except that this was an even more flagrant violation of international
law. Like the two Confederate envoys, the *Florida* was likely to prove an elephant
on the State Department's hands, and he began to regret that Collins had not
sunk her outright instead of bringing her in, since there could be little doubt
that the courts would order her returned intact to the neutral port where he had
seized her. "I wish she was at the bottom of the sea," the Secretary was afterwards reported to have remarked in discussing the affair with David Porter,

*Slipping through a pitch black Brazilian harbor,
the Federal sloop Wachusett rams the Florida, crushing
her bulwarks and snapping her mizzenmast.*

★

recently transferred from duty on the Mississippi to command the North Atlantic Blockading Squadron. "Do you mean it?" Porter asked, and Seward replied: "I do, from my soul." The admiral returned to his headquarters in Hampton Roads and ordered the captive sloop moved to Newport News and anchored, as an act of poetic justice, near the spot where the *Merrimac* had sunk the *Cumberland*. In the course of the shift, the raider collided with a transport, losing her jibboom and figurehead and being severely raked along one side. She began leaking rather badly, and though her pumps were put to work, suddenly and mysteriously in the early-morning hours of November 28 she foundered and went to the bottom, nine fathoms down. Or maybe not so mysteriously after all; Porter subsequently confided that he had put an engineer aboard with orders to "open her sea cock before midnight, and do not leave that engine room until the water is up to your chin."

This might or might not account for her loss (for with Porter as an unsupported witness, no set of facts was ever certain) but in any case Seward's task in responding to the formal Brazilian protest, which arrived next month, was greatly simplified. "You have justly expected that the President would disavow and regret the proceedings at Bahia," he replied, adding that the captain of the *Wachusett* would be suspended from duty and court-martialed. As for the rebel sloop, there could be no question of returning her, due to "an unforeseen accident which casts no responsibility upon the United States." All the same, a U.S. gunboat was to put into All Saints Bay on the Emperor's birthday, two years later, and fire a 21-gun salute as the *amende honorable* for this offense against the peace and dignity of Brazil. Collins himself was tried within six months, as Seward promised, and despite his plea that "the capture of the *Florida* was for the public good," was sentenced to be dismissed from the service. Gideon Welles, much pleased with the commander's response to a situation that had worked out well in the end, promptly set the verdict aside, restored the Pennsylvanian to duty, and afterwards promoted him to captain. Like Charles Wilkes, he would be a rear admiral before he died, a decade later.

Welles's pleasure was considerably diminished, however, by reports that followed hard on the heels of Collins's exploit, indicating that this was by no means the end of rebel depredations against Federal shipping on the sea lanes of the world. By coincidence, on October 8 — the day after the *Florida* was taken under tow in Bahia harbor — the Clyde-built steamer *Sea King,* a fast sailer with a lifting screw, an iron frame, and six-inch planking of East India teak, left London bound for Madeira, which she reached ten days later to rendezvous with a Liverpool-based tender bearing guns and ammunition and James I. Waddell, a forty-year-old former U.S. Navy lieutenant who had gone over to the Confederacy, with equal rank in its infant navy, when his native North Carolina left the Union. He took over at once as captain of the *Sea King,* supervised the transfer and installation of her armament, formally commissioned her as the C.S.S. *Shenandoah,* and set out two days later, October 20, on a cruise designed to continue the *Alabama-Florida* tradition.

In point of fact, his mission was to extend that tradition into regions where his country's flag had never flown. Like the raid on St Albans, staged the day before he left Madeira, and the recent 31-prize sortie by the *Tallahassee,* to Halifax and back, *Shenandoah*'s maiden effort was designed as a blow at the pocketbooks of New England, although Waddell had no intention of sailing her anywhere near that rocky shore. "The enemy's distant whaling grounds have not been visited by us," Secretary Mallory had noted in an August letter of instructions. "This commerce constitutes one of his reliable sources of national wealth no less than one of his best schools for seamen, and we must strike it, if possible."

Nothing in the new captain's orders precluded the taking of prizes en route to the field of his prime endeavor. He took six — two brigs, two barks, a schooner, and a clipper — between the day he left Madeira and November 12, the day the captive *Florida* steamed into Hampton Roads. Three more he took —

James I. Waddell (inset) left the U.S. Navy to join the Confederates and captain the Shenandoah (below), which has just set aflame nine Yankee whalers.

another schooner and two barks, bringing the total to nine in as many weeks — in the course of a stormy year-end voyage around the Cape of Good Hope to Hobson's Bay, Australia, where the *Shenandoah* stopped to refit before setting out again, northward through the Sea of Japan and into the North Pacific, to take up a position for intercepting Yankee whaling fleets bound for Oahu with the product of their labors in the Arctic Ocean and the Bering Sea. A whaler filled with sperm oil, Waddell had been told, would give a lovely light when set afire.

★ ★ ★ **C**ruisers were and would remain a high-seas problem, mainly viewed through a murk of inaccurate reports. But there were other problems the Union navy considered far more pressing, especially through this critical season of decision, because they were closer to home and the November voters. One was blockade-runners; or, more strictly speaking, the discontent they fostered. Although by now only three out of four were getting through the cordon off the Carolina coast, as compared to twice that ratio two years back, there was general agreement that they could never really be stopped until their remaining ports were sealed from the landward side. Meantime, sleek and sneaky, they kept weary captains and their crews on station in all weathers, remote from combat and promotion and contributing for the most part nothing but their boredom to a war they felt could be quickly won if only they were free to bring their guns to bear where they would count. Another problem was rebel ironclads, built and building, which threatened not only to upset plans for future amphibious gains, but also to undo gains already made.

A prime example of this last, now that the *Merrimac-Virginia,* the *Arkansas,* and the *Tennessee* had been disposed of, was the achievement of the *Albemarle* in reclaiming the region around the Sound whose name she bore. Since mid-April, when she retook Plymouth and blocked ascent of the Roanoke toward Petersburg and Richmond, a stalemate advantageous to the Confederacy had obtained there, and though the commander of the half-dozen Federal vessels lying off the mouth of the river had devised a number of highly imaginative plans for her discomfort — including one that involved the use of stretchers for lugging hundred-pound torpedoes across the intervening swamps, to be planted and exploded alongside the Plymouth dock where she was moored — none had worked, so vigilant were the graybacks in protecting this one weapon whose loss would mean the loss of everything within range of her hard-hitting rifles, all up and down the river she patrolled. Not since early May, when she tried it and came uncomfortably close to being sunk or captured for her pains, had the ironclad ventured out to engage the fleet, but neither could the Union ships invite destruction by steaming up to engage her at close quarters within the confines of that narrow stream. It was clear, however, that something had to be done about her

★

before long: for there were reports that two more rams were under construction up the river, one of them in the very cornfield where she herself had taken shape. One *Albemarle* was fearful enough to contemplate, even from a respectful distance. A flotilla of three, churning down into the Sound, was quite unthinkable.

The answer came from Lieutenant William B. Cushing, who presented two plans for getting rid of the iron menace. One involved the use of India-rubber boats, to be packed across the swamps to within easy reach of the objective, then inflated for use by a hundred-man assault force that would board the ram under cover of darkness, overpower her crew, and take her down to join the fleet at the mouth of the river, eight miles off. Plan Two, also a night operation, called for the boarding party to move all the way by water in a pair of

After the war, thirty-year-old William B. Cushing would become the youngest commander in the U.S. Navy.

light-draft steamers, each armed with a bow howitzer and a long spar tipped with a torpedo, to be used to sink the rebel warship if the attempt to seize her failed. He submitted his proposal in July, and when the Hampton Roads authorities chose the second plan and passed it on to Washington — where Welles approved it too, though with misgivings, since it seemed likely to cost the service one of its most promising young officers, not to mention the volunteers he proposed to take along — he left at once for New York, his home state, to purchase "suitable vessels" for the undertaking up the Roanoke.

No one who knew or knew of Cushing, and he was well known by now on both sides of the line, would have been surprised, once they learned that he was the author of the plan, at the amount of risk and verve its execution would require. Wisconsin-born, the son of a widowed schoolteacher, and not yet twenty-two — the age at which his brother Alonzo had died on Cemetery Ridge the year before, a West Pointer commanding one of the badly shot-up batteries that helped turn Pickett's Charge — he already had won four official commendations for similar exploits he had devised and carried out in the course of the past three years. Perhaps this was compensatory daring; he had been at Annapolis until midway through his senior year in 1861, when he was permitted to resign and thus avoid dismissal for unruly conduct and a lack of what the authorities called "aptitude for the naval service." He volunteered as an acting master's mate, in reaction to Sumter, and was restored to the rank of midshipman within six months.

"Where there is danger in the battle, there will I be," he informed a kinsman at the time, "for I will gain a name in this war." By now he had done so, and had won promotion to lieutenant, first junior, then senior grade, as well as those four commendations signed by Welles. None of this was enough; he wanted more; nothing less, indeed, than the highest of all military honors. "Cousin George," he wrote as he left New York in mid-October to keep his appointment with the *Albemarle* near Plymouth, "I am going to have a vote of thanks from Congress, or six feet of pine box by the next time you hear from me."

He had secured two open launches originally built for picket duty, screw-propelled vessels thirty feet long and narrow in the beam, of shallow draft and with low-pressure engines for quiet running, his notion being that one could stand by to provide covering fire and to pick up survivors if the other was sunk in the assault. As it turned out, this duplication was useful much sooner than he had expected; for one was lost in a Chesapeake storm on the way down, and he decided to go ahead with a single boat rather than wait for a replacement. Steaming in through Hatteras Inlet — whose bar no Union monitor could cross to ascend the Roanoke and engage the homemade iron ram — he joined the fleet riding at anchor fifty miles up Albemarle Sound. Two days he spent reconnoitering and drilling his volunteer crew, including fourteen men in the launch

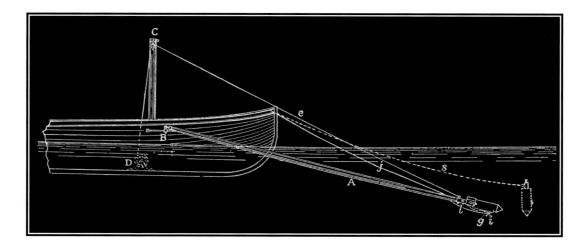

William Cushing designed this weapon – a buoyant torpedo on a wooden spar, lowered by a pulley, released and triggered by tugs on the lanyards – to destroy the C.S.S. Albemarle.

★

with him and another twelve in a towed cutter, the latter group to be used to silence rebel lookouts posted aboard the wreck of the *Southfield,* sunk in April a mile downstream from the dock where the *Albemarle* was moored. Soon after moonset, October 26, Cushing began his eight-mile run, the cutter in tow, only to be challenged just beyond the mouth of the river by Federal pickets who nearly opened fire when they heard the launch approaching. He turned back, warned by this apparent mishap that the expedition would have failed, and next day had a carpenter box-in the engine to muffle its sound, then set out again the following night, having added a tarpaulin to reduce the noise still more.

This time all went well on the run upriver. A rainstorm afforded such good additional cover that the launch chugged past the grounded *Southfield* undetected, thus enabling Cushing to keep the cutter with him in hope of using its dozen occupants to help overpower the crew of the ram when he went aboard. But that was not to be. Challenged by a sentry as he drew within hailing distance of the wharf, he changed his plan in mid-career; "Ahead fast!" he called out, and cast the cutter loose with orders to return downriver and deal with the pickets on the *Southfield.*

As he approached the ram, a signal fire blazed up ashore and he saw by its light that the ironclad was surrounded by a pen of logs chained in position to shield her from just such an attack as he was about to make. Hailed by a sailor on her deck, he replied with a shot from his howitzer and ran within pistol range for a better look at the problem. The logs were placed too far out for him to reach the ram with the torpedo attached to the tip of its fourteen-foot spar, although closer inspection showed that they perhaps were slimy enough for the launch to slide onto or even over them if it struck hard, at a direct angle. (Getting off or out was of course another matter, but that was no part of the plan as he had revised it.) He came about, under heavy fire from the enemy ship and shore, and picked up speed for the attempt. The launch struck and mounted and slithered across the encircling pen of logs, and Cushing found himself looking into the muzzle of one of the big rifles on the *Albemarle,* which he later described as looming before him like a "dark mountain of iron."

Then came the hardest part. To control and produce the explosion he had three lines tied to his wrists: one to raise or lower the long spar goose-necked to the bow of the launch, another to arm the torpedo by dropping it into a vertical position, and a third to activate the firing mechanism. All three required the coolness and precision of a surgeon performing a delicate operation, since too sudden a pull on any one of the lines would result in a malfunction. In this case, moreover, the surgeon was grievously distracted, having lost the tail of his coat to a blast of buckshot and the sole of one shoe to a bullet. Working as calmly under fire as he had done while rehearsing the performance in the quiet of his quarters, Cushing maneuvered the spar and swung the torpedo under the

Pro-Union sight-seers celebrate atop the casemate of the Albemarle after she was sunk in a shallow stretch of the Roanoke River outside Plymouth, North Carolina.

overhang of the ram's iron deck to probe for a vital spot before he released the firing pin. As he did so, the big rifle boomed, ten feet ahead, and hurled its charge of grape across the bow and into the stern of the stranded launch, which then was swamped by the descent of a mass of water raised by the explosion, nearly strangling all aboard. "Abandon ship!" the lieutenant cried, removing his shoes and shucking off his coat to go over the side.

The river was cold, its surface lashed by fire from the shore and the now rapidly sinking ram, whose captain would later testify that the hole blown in her hull was "big enough to drive a wagon through." Cushing struck out for the opposite bank, intent on escape, and as he did, heard one of his crew, close behind him, give "a great gurgling yell" as he went down. Ceasing fire, the Confederates came out in boats to look for survivors; Cushing heard them call his name, but continued to go with the current, paddling hard to keep afloat until he made it to shallow water, half a mile below. Exhausted, he lay in the mud till daylight, then crept ashore to take cover in the swamp. Later he found an unguarded bateau, and at nightfall began a stealthy trip downstream.

"Ship ahoy! Send a boat!" the crew of a Union patrol ship heard someone call from the darkness of the mouth of the river before dawn. An

★

armed detail sent to investigate presently returned with Cushing and the news that he had sunk the *Albemarle*. Cheers went up, as did rockets, fired to inform the other ships of the triumph scored two nights ago, and before long the weary lieutenant, who had been reported lost with all his crew, was sipping brandy in the captain's cabin. A few days later he was with Porter at Hampton Roads. "I have the honor to report, sir, that the rebel ironclad is at the bottom of the Roanoke River."

By then Plymouth, untenable without the protection of the ram, was back in Federal hands, having been evacuated after its works were taken under bombardment by the fleet on October 31. Upriver, the two unfinished ironclads were burned in their stocks when the whole region passed from rebel occupation. Cushing was promptly rewarded with a promotion to lieutenant commander, along with the thanks of Congress, upon Lincoln's recommendation, for having displayed what Porter called "heroic enterprise seldom equaled and never excelled." Much was expected of him in his future career, and he gave every sign of fulfilling those expectations. Before he was thirty, six years after the conflict ended, he would become the youngest full commander in the U.S. Navy. But that was as far as he went. He died at the age of thirty-two in a government

★

asylum for the insane, thereby provoking much discussion as to whether heroism and madness, like genius and tuberculosis, were related — and, if so, had insanity been at the root of his exploits? or had the strain of performing them, or even of having performed them, been more than a sane man could bear? In any case Farragut himself, in a subsequent conversation with Welles, stated flatly that "young Cushing was the hero of the war."

★ ★ ★ **W**estward to the Mississippi and north to the Ohio, Confederates did what they could to offset the loss of Atlanta by harassing the supply lines that sustained its Federal occupation. John Morgan was not one of these, for two sufficient reasons. One was that his command had by no means recovered from its unauthorized early-summer excursion into Kentucky, which had cost him half of his "terrible men," along with at least as great a portion of what remained of a reputation already diminished by the collapse of his Ohio raid the year before. The other was that he was dead — shot down in a less-than-minor skirmish on September 4, two days after Atlanta fell and nine months short of his fortieth birthday.

Informed that a blue column had set out from Knoxville for a strike at Saltville and the Southwest Virginia lead mines, he left Abingdon on September 1 and two days later reached Greeneville, Tennessee, where he prepared to confront the raiders when they emerged from Bull's Gap tomorrow or the next day. Down to about 2000 men, he deployed them fanwise to the west, covering three of the four roads in that direction, and retired for the night in the finest house in town, which as usual meant that its owner had Confederate sympathies. Greeneville, like many such places in East Tennessee, was a town with divided loyalties; Longstreet had wintered here, awaiting orders to rejoin Lee, and Andrew Johnson had been its mayor in the course of his rise from tailor to Lincoln's running mate in the campaign now in progress. Around sunup, after a rainy night, Morgan was wakened this Sunday morning by rifle fire, spattering in the streets below his bedroom window, and by a staff captain who brought word that the Union advance guard had arrived by the untended road. He pulled on his trousers and boots and went out by a rear door in an attempt to reach the stable and his horse, but was cut off and had to turn back, taking shelter in a scuppernong arbor that screened the walkway from the house.

"That's him! That's Morgan, over there among the grape vines!" a woman called from across the street to the soldiers pressing their search for the raider.

★

The body of John Hunt Morgan, who was shot dead after surrendering and tossed in a ditch by Union troopers, was reclaimed after battle and returned to his widow.

"Don't shoot; I surrender," Morgan cried.

"Surrender and be God damned — I know you," a blue trooper replied as he raised and fired his carbine at a range of twenty feet.

"Oh God," Morgan groaned, shot through the breast, and collapsed among the rain-wet vines, too soon dead to hear what followed.

"I've killed the damned horse thief!" the trooper shouted, and he and his friends tore down an intervening fence in their haste to get at Morgan's body, which they threw across a horse for a jubilant parade around the town before they flung it, stripped to a pair of drawers, into a muddy roadside ditch. Two captured members of the general's staff were allowed to wash and dress the corpse in the house where he had slept the night before, and others, returning after the enemy withdrew, reclaimed the body and sent it back to Abingdon, where his widow — the former Mattie Ready, pregnant with the daughter he would never see — had it removed to a vault in Richmond, to await the time when it could be returned in peace to the Bluegrass region he had loved and raided. That was the end of John Hunt Morgan.

It was otherwise with Forrest. Not only was he still very much alive, he now also had a department commander who would use him for something more than repelling Memphis-based raids into North Mississippi; would use him, indeed, on raids of his own against Sherman's life line up in Middle Tennessee. One of Richard Taylor's first acts, on assuming command at Meridian in early September, was to notify his presidential brother-in-law of this intention, while summoning the cavalryman to headquarters for instructions. Davis approved, and Forrest arrived by rail on September 5, "a tall, stalwart man, with grayish hair, mild countenance, and slow and homely of speech."

★

Taylor saw him thus for the first time, two weeks after his Memphis strike — three days after Atlanta fell and the day after Morgan died — though he knew him, of course, by reputation: nothing in which had prepared him for the Wizard's initial reaction to the news that he was to be sent at last "to worry Sherman's communications north of the Tennessee River." Forrest responded more with caution than with elation, inquiring about the route prescribed, the problem of subsistence, his possible lines of retreat in case of a check, and much else of that nature. "I began to think he had no stomach for the work," Taylor later wrote. But this was in fact his introduction to the Forrest method; for presently, he noted, "having isolated the chances of success from causes of failure with the care of a chemist experimenting in his laboratory," the Tennessean rose and brought the conference to an end with an abrupt transformation of manner. "In a dozen sharp sentences he told his wants, said he would leave a staff officer to bring up his supplies, asked for an engine to take him back north to meet his troops, informed me he would march with the dawn, and hoped to give an account of himself in Tennessee."

That was how Taylor would recall the parting, but here again he misconstrued the method. Far from marching "with the dawn," Forrest took ten days to get ready before he set out from below Tupelo with everything in order, plans all laid and instructions clearly understood by subordinates charged with carrying them out. Chief among these was Abraham Buford, in command of his own two brigades and one from Chalmers, who would remain behind to patrol the region around Memphis. Eight guns rolled with the column, which left on September 16 with just over 3500 effectives, anticipating a meeting near the Tennessee River with nearly a thousand Alabama troopers under William Johnson, who had shown his mettle at Brice's Crossroads back in June.

At Tuscumbia on the 20th Forrest also met someone he had not expected: Joe Wheeler. The diminutive Georgian was recrossing the river to wind up his long raid through East and Middle Tennessee, begun on August 10. Although the destruction he had wrought was about as extensive as he claimed to Hood, he neglected to add

Richard Taylor (above) was surprised by Bedford Forrest's caution when ordered to "worry Sherman."

that Sherman's road gangs had repaired the damage about as fast as it was in-flicted, often appearing on the scene before the twisted rails were cool. Moreover, there was something else the young West Pointer did not include in his report, and this was the condition of his command. Grievously diminished (for he tallied only his combat losses, which were barely a twentieth of the total suffered in the course of his six-week ride from Atlanta, up to Strawberry Plains near Knoxville, then back into North Alabama) the survivors were scarecrow examples of what could happen to troopers off on their own behind enemy lines. Originally 4500 strong — the number Forrest would have when Johnson joined tomorrow — they now counted fewer than 2000. A good many of the missing were stragglers whose mounts had broken down, and Forrest wrote Taylor that night, amid preparations for crossing the river next day: "I hope to be instrumental in gathering them up."

Fording his horsemen and floating his guns and wagons across on flatboats, he camped the following night on the north bank of the river, five miles west of Florence, which he passed through next morning, September 22, on the way to his main objective, the Tennessee & Alabama Railroad, just over forty miles to the east. One of Sherman's two main supply lines, running from Nashville through Columbia and Pulaski to Decatur, where it joined the Memphis & Charleston to connect with Chattanooga and Atlanta, its nearest point was Athens, and that was where Forrest was headed. He got there after sunset on the 23d to begin his investment of the town and its adjoining fort, a ditched and palisaded work a quarter-mile in circumference, occupied by a force of 600 infantry and considered impregnable to assault: as indeed perhaps it was, although no one would ever know. Soon after daybreak John Morton opened fire with his eight guns, "casting almost every shell inside the works," according to the garrison commander. Before long, Forrest halted fire to send in a white-flag note demanding "immediate and unconditional surrender." The Federal declined, but then unwisely consented to a parley, in the course of which Forrest pulled his customary trick of exposing troops and guns in tripli-cate, thereby convincing his adversary that he was besieged by a host of 15,000 of all arms, with no less than two dozen cannon. Capitulation came in time for the graybacks to give their full attention to a relief column that arrived from Decatur to take part in a brief skirmish before joining the surrender. Reduction of two nearby railway blockhouses raised the day's bag to 1300 prisoners, two pieces of artillery, 300 horses, and a mountain of supplies and equipment, in-cluding two locomotives captured with their cars in Athens. Forrest put the torch to the stores and installations, issued the horses to those of his men who needed them, smashed the rolling stock, and sent the prisoners back through Florence for removal south. Then he took up the march northward along the railroad, wrecking as he went.

Halfway to the Tennessee line next morning, September 25, he came upon the Sulphur Branch railway trestle, 72 feet high and 300 long, guarded by a double-casemated blockhouse at each end and a large fortress-stockade with a garrison of about one thousand men. Surrender declined, Morton opened fire and kept it up for two cruel hours, slamming in 800 rounds that left the fort's interior "perforated with shell, and the dead lying thick along the works." So Forrest would report, adding that a repeated demand for surrender was promptly accepted. This time the yield was 973 bluecoats, two more guns, another 300 horses, and a quantity of stores. Again he sent his prisoners rearward, together with the captured guns and four of his own, so greatly had the bombardment reduced his supply of artillery ammunition, and after setting fire to the two blockhouses, the buildings in the fort, and the long trestle they had been designed to shield, rode on north to the Elk River, which he reached next day, about midway between Athens and Pulaski. Here too there was a blockhouse at each end of a bridge even longer than the trestle at Sulphur Branch; but they were unmanned, abandoned by a commander who had heard from below how little protection they afforded, either to the installations they overlooked or to the garrisons they contained. Forrest burned them, along with the Elk River span, and pushed on to Richland Creek, seven miles beyond the Tennessee line and the same distance from Pulaski. Here there was a 200-foot-long truss bridge, stoutly built to take the weight of heavy-laden supply trains. The raiders crossed and sent it up in flames.

Now the character of the expedition changed. "Enemy concentrating heavily against me," Forrest notified Taylor the following night, September 27, from the vicinity of Pulaski. Touched where he was tender, Sherman had reacted hard and fast, sending George Thomas himself from Atlanta with two divisions to take charge in Middle Tennessee, with instructions for "the whole resources" of the region, including Kentucky and North Alabama, to be "turned against Forrest . . . until he is disposed of." Other divisions were on the way by rail and river from Memphis and Chattanooga, and Rosecrans had been urged to return A. J. Smith's gorillas from Missouri. As a result, fully 30,000 reinforcements were converging by now from all directions upon Pulaski, where Lovell Rousseau, arriving from Nashville to meet the threat, already had more men in its fortifications than were in the gray column on its outskirts. "Press Forrest to the death," Thomas wired ahead, "keeping your troops well in hand and holding them to the work. I do not think that we shall ever have a better chance than this."

The chance was not as good as the blue Virginian thought: not yet at any rate. Though he kept his Pulaski defenders "well in hand," Rousseau found the raiders gone from his front next morning. Forrest had built up his campfires the night before, and leaving them burning had pulled out. Having done what he could, at least for the present, to cripple the Tennessee & Alabama, he now was moving toward that other, more vital supply line, the Nashville &

Chattanooga, fifty miles to the east. He was obliged, however, to do it no more than superficial damage, learning from scouts when he got beyond Fayetteville on the 29th that the Chattanooga road was heavily protected by reinforcements hurried up it from Georgia and down it from Kentucky. He contented himself with detaching a fifty-man detail to tear up wires and track around Tullahoma, then confused the regathering Federals still more by splitting his force in two. Buford turned south with his division and Morton's four remaining guns, under orders to return to the Tennessee River by way of Huntsville, which he was to capture if possible, and tear up track on the Memphis & Charleston, between there and Decatur, before recrossing. Forrest himself, with the other two

---

"Press Forrest to the death, keeping your troops well in hand and holding them to the work. I do not think that we shall ever have a better chance than this."

— George Thomas

---

brigades, turned northwest through Lewisburg, then north across Duck River, passing near his Chapel Hill birthplace on the last day of September to descend once more, at high noon of the following day, on the already hard-hit Tennessee & Alabama near Spring Hill, ten miles north of Columbia and about four times that distance above Pulaski, which he had left four days ago.

He turned south, ripping up track, capturing three more block-houses — mainly by bluff, since Buford had the guns — firing bridges, and smashing culverts all the way to Columbia, which he bypassed on October 2 to avoid the delay of a gunless fight with the bluecoats in its works. The time had come to get out, and Forrest, as one of his troopers said, was "pretty good on a git." Taking off southwest away from what remained of the Tennessee & Alabama, he moved by country roads through Lawrenceburg, where he camped on the night of the 3d, and crossed the Alabama line the next day to return to Florence on October 5, one day less than two weeks after he left it. Buford was there ahead of him, having found Huntsville too stoutly garrisoned to be taken, and though the Tennessee was swollen past fording he had managed to get his men and guns across in relays on three rickety ferries, swimming the horses alongside. Now it was Forrest's turn.

A slow and risky business, with the enemy reported close astern, the piecemeal crossing took two full days, and was only accomplished, a veteran would recall, with "considerable disregard of the third commandment." Fretted

*Abraham Buford, ordered by Forrest to capture
Huntsville, Alabama, found the town, pictured here at
mid-century, too stoutly garrisoned to attack.*

and tired, the general was in the last boat to leave. While helping to pole against the swift-running current he noticed a lieutenant standing in the bow and taking no part in the work. "Why don't you take hold of an oar or pole and help get this boat across?" The lieutenant replied that, as an officer, he did not feel "called on to do that kind of work" while private soldiers were available to perform it. Astounded by this implied reproach — for he himself was as hard at work as anyone aboard — Forrest slapped the young man sprawling into the river, then held out the long pole and hauled him back over the gunwale, saying: "Now, damn you, get hold of the oars and go to work! If I knock you out of the boat again I'll let you drown." Another passenger observed that the douched lieutenant "made an excellent hand for the balance of the trip."

In the two weeks spent south of Nashville, within the great bend of the Tennessee, Forrest had captured 2360 of the enemy and killed or wounded an estimated thousand more, at a cost to himself of 340 casualties, only 47 of whom were killed. He had destroyed eleven blockhouses, together with the extensive trestles and bridges they were meant to guard, and had taken seven U.S. guns, 800 horses, and more than 2000 rifles, all of which he brought out with him, in addition to fifty captured wagons loaded with spoils too valuable for burning. Best of all, he had wrecked the Tennessee & Alabama so thoroughly that even the skilled blue work crews would need six full weeks to put it back in operation.

Indeed, Taylor was so encouraged by this Middle Tennessee expedition that he promptly authorized another, to be aimed this time at Johnsonville, terminus of the newly extended Nashville & Northwestern Railroad, by which supplies, unloaded from steamboats and barges on the Tennessee, were sent to Sherman by way of Nashville, seventy-five miles due east. A blow at this riverport depot, whose yards and warehouses were crowded with stores awaiting transfer, would go far toward increasing the Union supply problem down in Georgia, and Forrest spent only a week resting and refitting his weary troopers, summoning Chalmers to join him en route, and adding a pair of long-range Parrotts to Morton's two batteries, before he took off again for Johnsonville, a hundred miles north of Corinth, to which he had returned on October 9.

Much was expected of this follow-up strike, even though the first — successful as it had been, within its geographic limitations — had failed to achieve its major purpose, which was to make Sherman turn loose of Atlanta for lack of subsistence for his army of occupation. Not only did the red-haired Ohioan by then have ample stockpiles of supplies, he also had the scarcely interrupted use of the Nashville & Chattanooga line, having repaired within twelve hours the limited damage inflicted near Tullahoma by the fifty-man detail Forrest had detached when he turned north beyond Fayetteville. If the raid had been made a month or six weeks earlier, while the Federals were fighting outside Atlanta, opposed by an aggressive foe and with both overworked railroads barely able to meet their daily subsistence needs, the result might have been different. Even so, Forrest with only 4500 troopers had managed to disrupt Sherman's supply arrangements, as well as the troop dispositions in his rear, and had brought him to the exasperated conclusion, expressed to Grant on October 9, that it would be "a physical impossibility to protect the roads, now that Hood, Forrest, Wheeler, and the whole batch of devils are turned loose without home or habitation."

★ ★ ★

*U*nder orders to evacuate
Atlanta, many civilians departed
by train, leaving behind at the railroad yard, shown here, the wagons
they used to carry their possessions.

Hood-Davis; Lincoln Reëlected

1864 ★ ★ ★ ★ ★

First there had been the fret of **verbal contention.** Drawing back from Jonesboro, as he said, "to enjoy a short period of rest and to think well over the next step required in the progress of events," Sherman announced on September 8 that "the city of Atlanta, being exclusively required for warlike purposes, will at once be evacuated by all except the armies of the United States." He foresaw charges of inhumanity, perhaps from friends as well as foes, but he was determined neither to feed the citizens nor to "see them starve under our eyes. . . . If the people raise a howl against my barbarity or cruelty," he told Halleck, "I will answer that war is war and not popularity-seeking."

Sure enough, when Mayor Calhoun protested that the suffering of the sick and aged, turned out homeless with winter coming on, would be "appalling and heart-rending," Sherman replied that while he gave "full credit to your statement of the distress that will be occasioned," he would not revoke his orders for immediate resettlement. "They were not designed to meet the humanities of the case, but to prepare for the future struggle. . . . You cannot qualify war in harsher terms than I will. War is cruelty, and you cannot refine it. . . . You might as well appeal against the thunder storm as against these terrible hardships of war. . . . Now you must go," he said in closing, "and take with you your

★

Heading into exile, residents of Atlanta load their belongings onto wagons provided by the Union army after General Sherman gave them ten days to leave the city.

old and feeble, feed and nurse them, and build for them, in more quiet places, proper habitations to shield them against the weather until the mad passions of men cool down and allow the Union and peace once more to settle over your old homes at Atlanta. Yours in haste."

Hood attacked as usual, head down and full tilt, in response to a suggestion for a truce to permit the removal southward, through the lines, of the unhappy remnant of the city's population. He had, he said, no choice except to accede, but he added: "Permit me to say that the unprecedented measure you propose transcends, in studied and ingenious cruelty, all acts ever

★

brought to my attention in the dark history of war. In the name of God and humanity, I protest."

"In the name of common sense," Sherman fired back, "I ask you not to appeal to a just God in such a sacrilegious manner. You who, in the midst of peace and prosperity, have plunged a nation into war — dark and cruel war — who dared and badgered us to battle, insulted our flag, seized our arsenals and forts." There followed an arm-long list of Confederate outrages, ending: "Talk thus to the marines, but not to me, who have seen these things. . . . If we must be enemies, let us be men and fight it out as we propose to do, and not deal in

★

such hypocritical appeals to God and humanity. God will judge us in due time, and he will pronounce whether it be more humane to fight with a town full of women and the families of a brave people at our backs, or to remove them to places of safety among their own friends."

For two more days, though both agreed that "this discussion by two soldiers is out of place and profitless," the exchange continued, breathy but bloodless, before a ten-day truce was agreed on and the exodus began. Union troops escorted the refugees, with such clothes and bedding as they could carry, as far as Rough & Ready, where Hood's men took them in charge and saw them south across the fifteen-mile railroad gap to Lovejoy Station, within the rebel lines. Sherman was glad to see them go, and truth to tell had rather enjoyed the preceding altercation, which he saw as a sort of literary exercise, beneficial to his spleen, and in which he was convinced he had once more gotten the best of his opponent. But in other respects, having little or nothing to do with verbal fencing, he was far less satisfied, and a good deal more perturbed.

On September 8, the day he ordered Calhoun and his people to depart, he also issued a congratulatory order proclaiming to his soldiers that their capture of Atlanta "completed the grand task which has been assigned us by our Government." This was untrue. Welcome as the fall of the city was at this critical time — he was convinced, for one thing, that it assured Lincoln's reëlection, and for another he could present it, quite literally, to his troops as a crowning reward for four solid months of combat — his real objective, agreed on before-hand and identified by Grant in specific instructions, was the Army of Tennessee; he had been told to "break it up," and Atlanta had been intended merely to serve as the anvil upon which the rebel force was to be fixed and pounded till it shattered. That had been, and was, his true "grand task." Not only was Hood's army still in existence, it was relatively intact, containing close to 35,000 effectives, even with Wheeler gone for the past month; whereas Sherman's own, though twice as strong as Hood's at the time of occupation, started dwindling from the wholesale loss of veterans whose three-year enlistments ran out about the time the truce began. Subtractions from the top were even heavier in proportion. Schofield had to return for a time to Knoxville to attend to neglected adminis-trative matters in his department, and Dodge, wounded soon after he received a promotion to major general, took off on sick leave, never to return; his corps was broken up to help fill the gaps in Howard's other two, whose commanders, Logan and Blair — "political soldiers," Sherman scornfully styled them — had been given leaves of absence to stump for Lincoln in their critical home states. Presently even George Thomas was gone, along with two of his nine infantry divisions, sent back to Tennessee when the news came down that Forrest was on the rampage there, scooping up rear-guard detachments and providing the rail repair gangs with more work than they could handle in a hurry.

Various possibilities obtained, even so, including a march on Macon, Selma, or Mobile; but what the army needed most just now was rest and refitment, a brief period in which to digest its gains and shake its diminished self together, while its leader pondered in tranquillity his next move. Fortified Atlanta seemed an excellent place for this, although the situation afforded little room for error. "I've got my wedge pretty deep," Sherman remarked in this connection, "and must look out I don't get my fingers pinched." One drawback was that the interlude surrendered the initiative to Hood, who had shown in the past that he would be quick to grasp it, however stunned his troops might be as a result of their recent failures, including the loss of the city in their charge. Wheeler's damage to the supply line running back to Chattanooga had long since been repaired, but it seemed likely that his chief would strike there again, this time in heavier force; perhaps, indeed, with all he had.

Sherman ordered Union General George Thomas, at left, back to Tennessee along with two infantry divisions to run down Bedford Forrest, once more on the rampage.

This was in fact what Hood intended, if only because he felt he had no other choice. Determined to do *something*, yet lacking the strength to mount a siege or risk another large-scale confrontation on the outskirts of Atlanta, he had begun to prepare for a rearward strike while exchanging verbal shafts with his opponent inside the city. First he asked Richmond for reinforcements, and was told: "Every effort [has been] made to bring forward reserves, militia, and detailed men for the purpose. . . . No other resource remains." This denial had been expected, but it was promptly followed by another that had not. By gubernatorial proclamation on September 10, one week after Atlanta's fall, Joe Brown withdrew the Georgia militia beyond Confederate reach, granting blanket furloughs for his "pets," as they were called, "to return to their homes and look for a time after other important interests," by which he meant the tending of their farms.

Discouraged but not dissuaded by this lengthening of the numerical odds, Hood held to his plan for a move northward, requesting of the government that the 30,000 Andersonville inmates, ninety miles in his rear, be transferred beyond reach of the Federals in his front and thus permit him to shift his base from Lovejoy Station, on the Macon & Western, to Palmetto on the Atlanta & West Point; that is, from south of the city to southwest. This, he explained in outlining his proposed campaign, would open the way for him to recross the Chattahoochee, west of Marietta, for a descent on the blue supply line north of the river. Sherman most likely would follow to protect his communications, leaving a strong garrison to hold Atlanta; in which event Hood would be able to fight him with a far better chance of winning than if he tried to engage him hereabouts, with the odds at two-to-one. If, on the other hand, Sherman responded to the shift by moving against Augusta, Mobile, or some other point to the east or south, Hood would return and attack his rear. In any case, whatever risk was involved in his proposal, he was convinced that this was the time to act, since "Sherman is weaker now than he will be in the future, and I as strong as I can expect to be."

Richmond, approving this conditional raid-in-force, ordered the transfer of all able-bodied prisoners from Andersonville, near Americus, to stockades down in Florida. This began on September 21, by which time Hood had completed his twenty-mile shift due west to Palmetto, about the same distance southwest of Atlanta, and had his subordinates hard at work on preparations for the march north around Sherman's flank. They were still at it, four days later — September 25, a rainy Sunday that turned the red dust of their camps to mire — when Jefferson Davis arrived for a council of war.

He came for other purposes as well, including the need — even direr now than at the time of his other western trips, in early winter and late fall of the past two years, when Bragg had been the general in trouble — "to arouse all classes to united and desperate resistance." Outwardly at least, Davis himself never quailed or wavered under adversity, Stephen Mallory would testify after working

★

close to him throughout the war. "He could listen to the announcement of defeat while expecting victory, or to a foreign dispatch destructive to hopes widely cherished, or to whispers that old friends were becoming cold or hostile, without exhibiting the slightest evidence of feeling beyond a change of color. Under such circumstances, his language temperate and bland, his voice calm and gentle, and his whole person at rest, he presented rather the appearance of a man, wearied and worn by care and labor, listening to something he knew all about, than of one receiving ruinous disclosures." But this reaction was by no means characteristic of the high-strung people, in or out of uniform, to and for whom he was responsible as Commander in Chief and Chief Executive: and it was especially uncharacteristic now that the Federal penetration of the heartland had regional leaders of the caliber of Brown and Aleck Stephens crying havoc and talking of calling the dogs of war to kennel. Leaving Richmond five days ago, the day after Early's defeat at Winchester provided a companion setback in the eastern theater, Davis remarked to a friend: "The first effect of disaster is always to spread a deeper gloom than is due to the occasion." Then he set out for Georgia, as he had done twice before, in an attempt to dispel or at any rate lighten the gloom that had gathered and deepened there since the fall of Atlanta, three weeks back.

Army morale was a linked concern. Addressing himself to this on the day of his arrival at Palmetto, he attempted to lift the spirits of the troops with a speech delivered extemporaneously to Cheatham's Tennesseans, who flocked to meet him at the station. "Be of good cheer," he told them, "for within a short while your faces will be turned homeward and your feet pressing the soil of Tennessee."

Shouts of approval greeted this extension of the plan Hood had proposed; but other responses had a different tone. "Johnston! Give us Johnston!" Davis heard men cry or mutter from the ranks, and though he made no reply to this, it pointed up another problem he had come west to examine at first hand — the question of possible changes in the structure of command. Hardee, for example, had recently repeated his request for a transfer that would free him from further service under Hood, who blamed him for the collapse of two of his three Atlanta sorties, as well as for his failure to whip the enemy at Jonesboro, which had brought on the fall of the city. So Hood said, at any rate, wiring Richmond: "It is of the utmost importance that Hardee should be relieved at once. He commands the best troops in this army. I must have another commander." One or the other clearly had to go. Now at Palmetto, in tandem interviews, Davis heard the two generals out, recriminations abounding, and arrived at a decision that pleased them both: Hood by replacing Hardee with Cheatham, his senior division commander, and Hardee by ordering him to proceed at once to Charleston, where he would head the Department of South Carolina, Georgia, and Florida.

★

That was Beauregard's old bailiwick, and he was there even now, conducting a rather superfluous inspection of the coastal defenses. But there would be no overlapping of duties when Hardee arrived, since Davis planned for the Creole to be gone by then, summoned west as the solution to another command problem in the Army of Tennessee, this one at the very top. In mid-September, just before he left Richmond, he had received from Samuel French, who led a division in Stewart's corps, a private communication reminiscent of the famous round-robin letter that reached him after Chickamauga. This one was signed only by French, though it was written, he said, at the request of several high-ranking friends "in regard to a feeling of depression more or less apparent in parts of this army." His suggestion — or theirs, for the tone of the letter was strangely indirect — was that the President "send one or two intelligent officers here to visit the different divisions and brigades to ascertain if that spirit of confidence so necessary for success has or has not been impaired within the past month or two." Hood was not mentioned by name or position, as Bragg had been in the earlier document, but he was clearly responsible for conditions in a command which he had assumed "within the past month or two" and from which, the letter implied, he ought to be removed. This, combined with the public outcry over the loss of Atlanta, was part of what prompted the President's visit, and even before he set out he had arrived at a tentative solution to the problem by inviting Beauregard to go along. Old Bory was down in Charleston at the time, and Davis could not wait for him. He did, however, ask R. E. Lee to find out whether the Louisianian would be willing to return to duty in the West. Frustrated by subservience to Lee for the three months since Petersburg came under formal siege, Beauregard replied that he would "obey with alacrity" any

Jeff Davis left Richmond for the Georgia front to attend a council of war with his new commander, to rally support among the citizenry, and to shore up army morale.

such order for a transfer, and Davis wired from Palmetto for the Creole to meet him in Augusta on his way back in early October.

Beauregard, receiving the summons, assumed that he was about to return, as Hood's successor, to command of the army that had been taken from him more than two years ago, after Shiloh and the evacuation of Corinth. In this he was mistaken: though not entirely. Davis had it in mind to put him in charge not only of Hood's but also of Taylor's department, the whole to be known as the Military Division of the West, containing all of Alabama and Mississippi, together with major parts of Georgia and Louisiana and most of Tennessee. Assigned primarily in an advisory capacity, he would exercise direct control of troops only when he was actually with them — and only then, in Davis's words, "whenever in your judgment the interests of your command render it expedient." This was the position in which Johnston had fretted so fearfully last year; "a political device," a later observer was to term its creation, "designed to silence the critics of Hood, satisfy the friends of Beauregard, and save face for the Administration." That was accurate enough, as far as it went, but for Davis the arrangement had two other pragmatic virtues. One was that Hood's accustomed rashness might be tempered, if not controlled, by the presence of an experienced superior close at hand, and the other was that there was no room left for Joe Johnston, whose return Davis was convinced would result in a retreat down the length of the Florida peninsula. In any case, Beauregard was highly acceptable to the generals Davis talked with at Palmetto, including Hood, and he was determined to offer him the post when they met in Augusta the following week.

Mainly, though, the presidential visit was concerned with the strategy Hood had evolved for drawing the blue army north by striking at its supply line beyond the Chattahoochee, where he would take up a strong defensive position inviting a disadvantageous attack. Now in discussion this was expanded and improved. If Sherman appeared too strong even then, or if Hood, as Davis put it, "should not find the spirit of his army such as to justify him in offering battle" at that point, he was to fall back down the Coosa River and through the mountains to Gadsden, Alabama, where he would establish a new base, supplied by the railroad from Selma to Blue Mountain, and there "fight a conclusive battle" on terrain even more advantageous to the defender; Sherman, drawn far from his own base back in Georgia, might then be annihilated. If, on the other hand, the Ohioan declined battle on those terms and returned to Atlanta, Hood would follow, and when Sherman, his supply line cut, moved from there, Hood would still pursue: either northward, across the Tennessee — which would undo the Federal gains of the past four months and open the way for a Confederate march on Nashville — or south or east, through Selma or Montgomery to the Gulf or through Macon or Augusta to the Atlantic, in which case the Union rear could be assaulted. That was the expanded plan, designed to cover all contingencies, as

Hood and the Commander in Chief developed it over the course of the three-day visit. Then on the evening of September 27 Davis took his leave.

In Macon next morning, at a benefit for the impoverished Atlanta refugees, he took up the spirit-lifting task he had begun at Palmetto when he told the Tennessee soldiers their faces would soon turn homeward. "What though misfortune has befallen our arms from Decatur to Jonesboro," he declared, "our cause is not lost. Sherman cannot keep up his long line of communications; retreat sooner or later he must. And when that day comes, the fate that befell the army of the French Empire in its retreat from Moscow will be re-enacted. Our cavalry and our people will harass and destroy his army, as did the Cossacks that of Napoleon, and the Yankee general, like him, will escape with only a bodyguard. . . .

"Let no one despond," he said in closing, and repeated the words the following day in Montgomery, speaking at the Capitol where he had been inaugurated forty-three months ago. "There be some men," he told the Alabamians,

---

"Our cause is not lost. Sherman cannot keep up his long line of communications; retreat sooner or later he must. And when that day comes, the fate that befell the army of the French Empire in its retreat from Moscow will be re-enacted."

— Jefferson Davis

---

in support of his advice against despondence, "who when they look at the sun can only see a speck upon it. I am of a more sanguine temperament perhaps, but I have striven to behold our affairs with a cool and candid temperance of heart, and, applying to them the most rigid test, am more confident the longer I behold the progress of the war. . . . We should marvel and thank God for the great achievements which have crowned our efforts."

Closeted that night with Richard Taylor, who had transferred his headquarters from Meridian to Selma, he was glad to learn the particulars of Forrest's current raid into Middle Tennessee, but disappointed to be told that any hopes he retained for securing reinforcements from beyond the Mississippi were quite groundless, not only because the situation there would not permit it, but also because of the gunboats Taylor had had to dodge, even at night in a small boat, when he returned. Davis was able to counter this with news that Hood had begun today a crossing of the Chattahoochee near Campbelltown, twenty miles southwest of Atlanta, for his strike at the Federal life line. Taylor

★

was pleased to hear it, remarking that the maneuver would no doubt "cripple [Sherman] for a time and delay his projected movements." Whatever enthusiasm surged up in him on hearing of this new offensive was certainly well contained. Moreover: "At the same time," he later wrote of the exchange, "I did not disguise my conviction that the best we could hope for was to protract the struggle until spring. It was for statesmen, not soldiers, to deal with the future."

This was chilling in its implications, coming as it did from a friend and kinsman whose opinion he respected and whose experience covered all three major theaters of the war, but Davis refused to be daunted; like Nelson off Copenhagen, putting the telescope to his blind eye, he declined to see these specks upon the Confederate sun. The two men parted to meet no more in the course of a conflict Taylor believed was drawing to a close, and Davis resumed his journey eastward from Montgomery next day, joined en route by Hardee for the scheduled meeting with Beauregard in Augusta on October 2, the President's second Sunday away from Richmond. Old Bory's spirits took a drop when he learned that he was to occupy an advisory rather than a fighting post, but they soon revived at the prospect of conferring with Hood on plans for reversing the western tide of battle. In the end, he was as pleased as Hardee was with his new assignment, and both generals sat on the rostrum with their chief the following day at a patriotic rally. "We must beat Sherman; we must march into Tennessee," Davis told the Augustans. "There we will draw from 20,000 to 30,000 to our standard, and, so strengthened, we must push the enemy back to the banks of the Ohio and thus give the peace party of the North an accretion no puny editorial can give."

Such was the high point of his last speech in Georgia, and having made it he presented the two generals to the crowd. Beauregard, who had fired the first gun of the war, was cheered for saying that he "hoped to live to fire the last," and Hardee, a native son, drew loud applause when he reported that Hood had recently told him "he intended to lay his claws upon the state road in rear of Sherman, and, having once fixed them there, it was not his intention to let them loose their hold."

Next day, October 4 — by which time the three speakers had reached or were moving toward their separate destinations: Beauregard west, Hardee east, and Davis north to the South Carolina capital — Hood had carried out at least the first part of this program. Completing his crossing of the Chattahoochee before September ended, he struck the Western & Atlantic at Big Shanty and Acworth, capturing their garrisons, and now was on the march for Allatoona, the principal Union supply base near the Etowah. Best of all, Sherman had taken the bait and was hurrying northward from Atlanta with most of his army, apparently eager for the showdown battle this gray maneuver had been fashioned to provoke. While the opening stage of the raid was in progress, and even as Hood's troops were tearing up some nine miles of track around Big Shanty, Davis delivered in Columbia the last in his current series of addresses designed to lift the spirits of a citizenry depressed by the events of the past two months.

"South Carolina has struggled nobly in the war, and suffered many sacrifices," he declared, beginning as usual with praise for the people of the state in which he spoke. "But if there be any who feel that our cause is in danger, that final success may not crown our efforts, that we are not stronger today than when we began this struggle, that we are not able to continue the supplies to our armies and our people, let all such read a contradiction in the smiling face of our land and in the teeming evidences of plenty which everywhere greet the eye. Let them go to those places where brave men are standing in front of the foe, and there receive the assurance that we shall have final success and that every man who does not live to see his country free will see a freeman's grave." He himself was on his way back from such a visit, and he had been reassured by what he saw. "I have just returned from that army from which we have had the saddest accounts — the Army of Tennessee — and I am able to bear you words of good cheer. That army has increased in strength since the fall of Atlanta. It has risen in tone; its march is onward, its face looking to the front. So far as I am able to judge, General Hood's strategy has been good and his conduct has been gallant. His eye is now fixed upon a point far beyond that where he was assailed by the enemy. He hopes soon to have his hand upon Sherman's line of communications, and to fix it where he can hold it. And if but a half — nay, one fourth — of the men to whom the service has a right will give him their strength, I see no chance for Sherman to escape from a defeat or a disgraceful retreat. I therefore hope, in view of all the contingencies of the war, that within thirty days that army which has so boastfully taken up its winter quarters in the heart of the Confederacy will be in search of a crossing of the Tennessee River." Having claimed as much, he pressed on and claimed more. "I believe it is in the power of the men of the Confederacy to plant our banners on the banks of the Ohio, where we shall say to the Yankee: 'Be quiet, or we shall teach you another lesson.'"

So he said, bowing low to the applause that followed, and after a day's rest — badly needed, since two weeks of travel on the buckled strap-iron of a variety of railroads amounted to a form of torture rivaling the rack — ended his fifteen-day absence from Richmond on the morning of October 6.

The warm bright pleasant weather of Virginia's early fall belied the strain its capital was under; Fort Harrison had toppled just one week ago, creating a dent in the city's defenses north of the James, and the fight next day at Peebles Farm, though tactically a victory, had obliged Lee to extend his already thin-stretched Petersburg lines another two miles west. For Davis, however, any day that brought him back to his family was an occasion for rejoicing. And rejoice he did: especially over its newest member, three-month-old Varina Anne. Born in late

June, while the guns were roaring on Kennesaw and Jubal Early was heading north from Lynchburg, she would in time be referred to as the "Daughter of the Confederacy," but to her father she was "Winnie," already his pet name for her mother, or "Pie-Cake," which her sister and brothers presently shortened to "Pie." He was glad to be back with her and the others, Maggie, Little Jeff, Billy, and his wife, who was pleased, despite her distress at the wear he showed, to hear how well the trip had gone in regard to his efforts to lift the flagging morale of the people with predictions of great success for Hood — whose troops were moving northward even now — and "defeat or a disgraceful retreat" for Sherman.

★ ★ ★ *G*rant, for one, disagreed with this assessment of the situation in North Georgia. Informed of Davis's late-September prediction that the fate that crumpled Napoleon in Russia now awaited Sherman outside Atlanta, he thought it over briefly, then inquired: "Who is to furnish the snow for this Moscow retreat?"

  Afterwards, Sherman took this one step further, professing to have been delighted that the rebel leader's "vainglorious boasts" had in effect presented "the full key to his future designs" to those whom they were intended to undo; "To be forewarned was to be forearmed," he explained. But that was written later, when he seemed to have taken what he called "full advantage of the occasion." Davis in fact had said very little more in his recent impromptu speeches, including his proposal "to plant our banners on the banks of the Ohio," than he (and, indeed, many other Confederate spokesmen) had expressed on previous tours undertaken to lift spirits that had sagged under the burden of defeat. As for Hood's reported promise to "lay his claws" on the railroad north of Atlanta, they were already fixed there by the time Sherman heard from his spies or read in the papers of what Davis or Hardee was supposed to have said — days after Hood's whole army was across the Chattahoochee in his rear. Besides, the red-haired Ohioan was far too busy by then, attempting to deal with this newly developed threat to his life line, to conjecture much about what Hood might or might not have in mind as a next step.

  Leaving Slocum's corps to hold Atlanta, he began recrossing the Chattahoochee with the other five — some 65,000 of all arms, exclusive of the two divisions sent back to Tennessee with Thomas the week before — when he discovered on October 3 that Hood, after crossing in force near Campbelltown, was moving north through Powder Springs, apparently with the intention of getting astride the Western & Atlantic somewhere around or beyond Marietta. Sherman rushed a division from Howard north by rail, under Brigadier General John M. Corse, to cover Rome in case the graybacks veered in that direction, but by the time he got the last of his men over the river next day he learned that the rebs had taken Big Shanty and Acworth, along with their garrisons, and had

★

Ordered up by Sherman, General John M. Corse (right) arrived by rail with his troops to bolster the Union garrison only hours before the Battle of Allatoona Pass.

torn up nine miles of track on their way to seize his main supply base at Allatoona, which they would reach tomorrow. He got a message through for Corse to shift his troops by rail from Rome to Allatoona, reinforcing its defenders, and to hang on there till the rest of the army joined him.

Corse complied, but only by the hardest. When Sherman climbed Kennesaw next morning, October 5, he could see the Confederate main body encamped to the west around Lost Mountain, his own men at work repairing the railroad past Big Shanty, just ahead, and gunsmoke lazing up from Allatoona Pass, a dozen air-line miles to the north, where Corse was making his fight. Hood had detached Stewart's corps for the Acworth strike, and Stewart, before heading back to rejoin Hood last night, had in turn detached French's division to extend the destruction to the Etowah. "General Sherman says hold fast; we are coming," the Kennesaw signal station wigwagged Allatoona over the heads of the attackers. Corse — a twenty-nine-year-old Iowan who had spent two years at West Point before returning home to study law and run for public office, only to lose the election and enter the army, as was said, "to relieve the pain of political defeat" — had arrived, although with less than half of his division, in time to receive a white-flag note in which French allowed him five minutes "to avoid a needless effusion of blood" by surrendering unconditionally. He declined, replying: "We are prepared for the 'needless effusion of blood' whenever it is agreeable to you."

The engagement that followed was as savage as might have been expected from this exchange. Corse had just under 2000 men, French just over 3000, and their respective losses were 706 and 799 killed, wounded, or captured. After two of the three redoubts had fallen, Corse withdrew his survivors to the third, near the head of the pass, and kept up the resistance, despite a painful face

★

wound and the loss of more than a third of his command. By 4 o'clock, having intercepted wigwag messages that help was on the way from the 60,000 Federals in his rear, French decided to pull out before darkness and Sherman overtook him. Corse was exultant: so much so that when Sherman, still on Kennesaw, inquired by flag as to his condition the following day, he signaled back: "I am short a cheekbone and an ear, but am able to whip all hell yet."

Such was the stuff of which legends were made, including this one of the so-called Battle of Allatoona Pass. "Hold the fort, for I am coming," journalists quoted Sherman as having wigwagged from the top of Kennesaw, and that became the title of P. P. Bliss's revival hymn, inspired by the resolute valor Corse and his chief had shown in defending a position of such great natural strength that the latter had chosen not to risk an attack when he found it looming across his southward path in May. French, moreover, got clean away, long before any blue relief arrived, and when Sherman encountered the high-strung young Iowa brigadier a few days later he was surprised to find on his cheek only a small bandage, removal of which revealed no more than a scratch where the bullet had nicked him in passing, and no apparent damage to the ear he had claimed was lost. Sherman laughed. "Corse, they came damned near missing you, didn't they?" he said.

He laughed, yet the fact was he found small occasion for humor in the present situation. Hood withdrew his reunited army westward beyond Lost Mountain to New Hope Church and Dallas. There he stopped, or anyhow paused. Sherman, however, had no intention of reëntering that tangled wasteland, even though this meant leaving the initiative to an adversary who had just shown that he would use it to full advantage and now seemed about to do as much again. Sure enough, when the sun came up on October 7 the graybacks had disappeared. Wiring Slocum that they had "gone off south," Sherman warned that they might be doubling back for a surprise attack on Atlanta, and when he discovered later in the day that they were actually headed north, he charged that Hood was an eccentric: "I cannot guess his movements as I could those of Johnston, who was a sensible man and only did sensible things."

Delayed by an all-day rain next day, he did not reach Allatoona until October 9, when he heard from scouts that the butternut column was on the march for Rome. But that was not true either, it turned out. Crossing the Coosa River west of Rome, then moving fast up the right bank of the Oostanaula, Hood struck Resaca on October 12 and wrecked a dozen miles of railroad between there and Dalton, where he captured the thousand-man garrison next day and then ripped up another five miles of track on his way to Tunnel Hill, where the contest for North Georgia had begun five months ago. When Sherman moved against him from Rome and Kingston, he fell back through Snake Creek Gap to a position near LaFayette, some twenty miles south of where

★

Bragg and Rosecrans had clashed about this time last year at Chickamauga, and there took up a defensive stance, both flanks stoutly anchored and a clear field of fire to his front. Sherman came on after him from Resaca, reaching LaFayette on October 17. By the time he got his troops arrayed for battle, however, Hood was gone again — vanished westward, across the Alabama line, into even more rugged terrain where Sherman would be obliged to risk defeat a long way from his base. Exasperated, the red-head complained bitterly that everything his adversary had done for the past three weeks was "inexplicable by any common-sense theory." Recalling Jefferson Davis's boast of Hood's intentions: "Damn him," he said testily of the latter. "If he will go to the Ohio River I will give him rations. . . . Let him go north. My business is down South."

Whether this last was to be the case or not was strictly up to the general-in-chief, and that was the main cause of Sherman's irritability through this difficult and uncertain time, even more than the loss of much of the railroad in his rear. The railroad could be rebuilt — would in fact be back in use within ten days — but Hood's evident ability to smash it, more or less at will, might have an adverse influence on the decision Grant had been pondering for the past month, ever since Sherman first made it clear what he meant when he said that his business was "down South."

Back in early May, at the start of his campaign to "knock Jos. Johnston," a staffer had asked what he planned to do at its end; "Salt water," he replied, flicking the ash from his cigar. Mobile and the Gulf had been what he meant, but thanks to Farragut there was not much left in that direction worth the march. He now had a different body of water in mind, rimming a different coast. In brief, his proposal — first made on September 20, while the refugee truce was still in effect below Atlanta — was that the navy secure and provision a base for him on the Atlantic seaboard — probably Savannah, since that was the closest port — and his army would "sweep the whole state of Georgia" on its way there. Such a march, he told Grant, would be "more than fatal to the possibility of Southern independence. They may stand the fall of Richmond, but not of all Georgia," he declared, and added a jocular, upbeat flourish to close his plea: "If you can whip Lee and I can march to the Atlantic, I think Uncle Abe will give us a twenty days' leave of absence to see the young folks."

Grant had doubts. With its attention fixed on Wilmington, the last major port still open to blockade runners, the navy would not willingly divert its strength to a secondary target more than two hundred miles down the coast; besides which, the mounting of such an effort would take months, and previous attempts against Charleston had shown there was little assurance of success, even if every ironclad in the fleet was employed in the attack. His main objection, however, was the continued existence of Hood's army. Speaking in Georgia, Alabama, and South Carolina, hard on the heels of Sherman's proposal, Jefferson

★

Davis announced plans for a northward campaign that might well succeed if Sherman marched eastward and thus removed from Hood's path the one force that could stop him. Grant said as much, opposing the expedition on both counts, but Sherman replied that he did not really need for the navy to take Savannah before he got there; all he wanted was for supply ships to be standing by, ready to steam in after he reduced the city from the landward side. As for Hood, Thomas was on the way to Nashville even now with two divisions which he would combine with troops already there and others on the way; "Why will it not do to leave Tennessee to the forces which Thomas has, and the reserves soon to come to Nashville, and for me to destroy Atlanta and march across Georgia to Savannah or Charleston, breaking roads and doing irreparable damage? We cannot remain on the defensive."

A supply train with General Sherman's army makes slow progress during the pursuit of Hood's Confederate forces into Alabama in October 1864.

That was written October 1. By the time the message reached City Point, Forrest had rampaged through Middle Tennessee, smashing installations within thirty miles of Nashville, and Hood was across the Chattahoochee, ripping up track on the Western & Atlantic thirty miles north of Atlanta. Grant saw these strikes as confirmation of his objection to Sherman's departure, but Sherman

took them as proof of his contention that he was wasting time by remaining where he was; that it was, in fact, as he insisted on October 9, "a physical impossibility to protect the roads, now that Hood, Forrest, Wheeler, and the whole batch of devils are turned loose. . . . By attempting to hold the roads, we will lose a thousand men each month and will gain no result." Having said as much, he returned to his plea

> *"The utter destruction of its roads, houses, and people will cripple their military resources. . . . I can make this march, and make Georgia howl!"*
>
> — William Tecumseh Sherman

that he himself be "turned loose" to make for the coast. This time, noting that he had some 8000 head of cattle on hand, as well as 3,000,000 rations of bread, and expected to find "plenty of forage in the interior of the state," he went into logistical details of the expedition. "I propose that we break up the railroad from Chattanooga forward, and that we strike out with our wagons for Milledgeville, Millen, and Savannah. Until we can repopulate Georgia, it is useless for us to occupy it; but the utter destruction of its roads, houses, and people will cripple their military resources. . . . I can make this march, and make Georgia howl!"

Hood by then had retired westward, but soon he was on the go again, about to throw another punch at the railroad forty miles farther north. Even before it landed, Sherman predicted that it would be successful and renewed his appeal to be spared the patchwork soldiering that would follow, urging Grant to let him "send back all my wounded and unserviceable men, and with my effective army move through Georgia, smashing things to the sea. Hood may turn into Tennessee and Kentucky," he admitted, "but I believe he will be forced to follow me." In any case, Thomas could handle him, he said, and best of all, "instead of being on the defensive, I will be on the offensive. Instead of my guessing at what he means to do, he will have to guess at my plans. The difference in war would be fully 25 percent. . . . Answer quick, as I know we will not have the telegraph long."

Grant's reply next day, October 12 — the day Hood landed astride the railroad at Resaca — was encouraging. "On reflection I think better of your proposition," he wired back. "It will be much better to go south than to be forced to come north." He suggested that the move be made with "every wagon, horse, mule, and hoof of stock, as well as the Negroes," and that plenty of spare weapons be taken along to "put them in the hands of Negro men," who could serve as otherwise unobtainable reinforcements on the march. All the same, his approval was only tentative, not final, and Sherman continued to fume, irked in front by Hood and from the rear by Grant.

The former got away westward again, through Snake Creek and Ship's gaps, to a position just below LaFayette, which he abandoned at the approach of the blue army, and fell back down the valley of the Chattooga River, across the Alabama line. "It was clear to me that he had no intention to meet us in open battle," Sherman later wrote, "and the lightness and celerity of his army convinced me that I could not possibly catch him on a stern-chase." Angry at being drawn in the direction he least wanted to go — and resentful, above all, at the mounting proof of his error in having turned back to Atlanta, when the city fell to Slocum in his rear, instead of pressing after Hood to achieve the true purpose of his campaign — the red-head called a halt at Gaylesville, thirty miles short of Gadsden, and there continued to fret and fume as October wore away, still with no definite go-ahead from the general-in-chief. Evidence of his snappishness appeared in a telegram he sent a cavalry brigadier, posted at Calhoun on rear-guard duty, when he heard that a sniper had taken pot shots at cars along the newly repaired Western & Atlantic: "Cannot you send over about Fairmont and Adairsville, burn ten or twelve houses of known secessionists, kill a few at random, and let them know that it will be repeated every time a train is fired on from Resaca to Kingston?"

★ ★ ★ **A**cross the way at Gadsden, while Sherman thus was breathing fire and threatening random slaughter, Hood's troubles were not so much with his superior, Beauregard, as they were with his subordinates, who he felt had let him down. Drawn up for combat near LaFayette the week before, he had "expected that a forward movement of one hundred miles would reinspirit the officers and men to a degree to impart to them confidence, enthusiasm, and hope of victory," but when he took a vote at a council of war, assembled on the eve of what he intended as an all-out effort to whip Sherman, "the opinion was unanimous that although the army was much improved in spirit, it was not in a condition to risk battle against the numbers reported." Disappointed, he withdrew down the Chattooga Valley

★

and the Coosa River to Gadsden for a meeting on October 21 with Beauregard, who had formally assumed command of the new Military Division of the West only four days ago. To the Creole's great surprise, Hood presented for his approval a broad-scale plan, conceived en route, for "marching into Tennessee, with a hope to establish our line eventually in Kentucky."

"Broad-scale" was perhaps not word enough; spread-eagle was more like it. But knowing as he did that time was on the side of the Union — that delay would enable Thomas to complete his build-up in Tennessee and combine with Sherman to corner and crush the fugitive gray army, wherever it might turn — Hood was determined to extend and enlarge the flea-bite offensive by which he had managed, ever since he left Palmetto three weeks back, to keep his adversaries edgy and off-balance. A northward march, into or past the mouth of the Federal lion, was admittedly a risky undertaking, but he was of the Lee-Jackson school, whose primary tenet was that the smaller force must take the longest chances, and moreover he had before him the example of Bragg, who by just such a maneuver after the fall of Corinth, two years ago, had reversed the gloomy situation in this same theater by dispersing the superior enemy combinations then being assembled to bring on his destruction.

His plan, he said, was to cross the Tennessee River at Guntersville, which would place him within reach of Sherman's single-strand rail supply line in the delicate Stevenson-Bridgeport area, and move promptly on Nashville, smashing Thomas's scattered detachments on the way. Possessed of the Tennessee capital, he would resupply his army from its stores, thicken his ranks with volunteers drawn to his banner, and move on through Kentucky to the Ohio, where he would be in a position to threaten Cincinnati and receive still more recruits from the Bluegrass. If Sherman followed, as expected, Hood would then be strong enough to whip him; after which he would either send reinforcements to beleaguered Richmond or else take his whole command across the Cumberlands to come up in rear of the blue host outside Petersburg. Or if Sherman did not follow, but instead took off southward for the Gulf or eastward for the Atlantic, Hood explained that he would move by the interior lines for an attack on Grant "at least two weeks before he, Sherman, could render him assistance." Such a shift, he said, winding up in a blaze of glory, "would defeat Grant and allow General Lee, in command of our combined armies, to march upon Washington or turn upon and annihilate Sherman."

Old Bory was amazed, partly by the bold sweep of the plan, which seemed to him as practicable as it was entrancing, and partly by the shock of recognition, occasioned by its resemblance to the half-dozen or so which he himself had submitted to friends and superiors over the course of the past three years, invariably without their being adopted. One difference was that he had always insisted on heavy reinforcement at the outset, whereas Hood proposed to

★

strike with what he had. If this seemed rash, Beauregard could see that it might well be a virtue in the present crisis, not only because no reinforcements were available, but also because it would save time, and time was of the essence in a situation depending largely on how rapidly the invaders moved — especially against Thomas, who must not be given a chance to pull his scattered forces together for the protection of the capital in his care. In any case, approval was little more than a formality; Hood had informed the government two days ago that he intended to cross the Tennessee, and only yesterday had wired ahead to Richard Taylor, whose department he had entered for the crossing: "I will move tomorrow for Guntersville."

Beauregard did not withhold his blessing, though after much discussion he insisted that Wheeler's cavalry, which had rejoined the army near Rome ten days ago, be left behind to operate against Sherman's communications and attack his rear if he set out

On October 21, Pierre G. T. Beauregard took command of the C.S.A.'s new Military Division of the West.

south or east, through otherwise undefended regions between Atlanta and the Gulf or the Atlantic. Hood readily agreed to this subtraction when the Creole added that Forrest would join him on the march, replacing Wheeler, as soon as he and his troopers returned from their current raid on Johnsonville; which, incidentally, would add to the Federal confusion Hood hoped to provoke when he moved on Nashville.

Word went out to the camps that the shift northward would begin at daylight, and their commander later recalled that the news was greeted with "that genuine Confederate shout so familiar to every Southern soldier." By this he meant the rebel yell, the loudest of which no doubt came from the bivouacs of the Tennesseans. Davis had told them four weeks ago that their feet would soon be pressing native soil, and now they whooped with delight at finding the promise about to be kept.

It was kept, although by no means as promptly as they and Beauregard expected when they parted at Gadsden next morning. Guntersville, thirty-odd miles northwest, turned out to be crowded with bluecoats, and Hood decided to veer west for a crossing at Decatur, just over forty miles downriver. However, when he drew close to there on October 26, after four days on the march, he found that it, like Guntersville, was too stoutly garrisoned to be stormed without heavier losses than he felt he could afford; so he pressed on for Courtland, twenty miles beyond Decatur, which he bypassed the following day. It was not until then that Beauregard, who had been off making supply arrangements and was miffed at not having been informed of the change in route, caught up with the column some fifty miles west of its original objective. He was aggrieved not only because the detour had ruled out the disruptive strike at Stevenson, now clearly beyond range of the butternut marchers, but also because of the loss of time, which Sherman and Thomas would surely use to their advantage. He had said from the start that celerity was Hood's best hope for success in this

Informed by his engineers that they did not have enough pontoons to bridge the rain-swollen Tennessee at Courtland, Hood decided to push on and use the partly demolished railway span at Tuscumbia . . .

long-odds undertaking; yet five whole days had already been spent in search of a crossing that still had not been reached. Nor was that the worst of it. Informed by his engineers that they did not have enough pontoons to bridge the rain-swollen Tennessee at Courtland, Hood decided to push on and use the partly demolished railway span at Tuscumbia, another twenty-five miles downstream and well over eighty from Guntersville, where he had intended to ford the river a week ago. At Tuscumbia on the last day of October, he further alarmed his superior by announcing that he lacked sufficient provisions for the march that would follow the crossing, as well as shoes for his men and the horses in Jackson's two slim brigades, which were all the cavalry he would have until Forrest returned from Johnsonville, more than a hundred miles downriver to the north.

Taylor had unwelcome news for them in that regard as well. Unmindful of the need for haste, he had waited till Hood drew near Decatur on the 26th to send a courier summoning Forrest, who had left five days ago, and even then had told him to complete his mission before heading back. Hood took this, then and later, as evidence that he had done well to shift his infantry westward in search of a crossing, since this reduced the gap between it and the cavalry he was obliged

★

to wait for anyhow. Moreover, while he marked time at Tuscumbia, doing what he could to repair his supply deficiencies and giving his men some well-earned rest through the first fine days of November, word came back that the delay had perhaps been worth the vexation after all, adding as it did a highly colorful chapter to the legend surrounding the Wizard of the Saddle.

After reaching the Tennessee River near the Kentucky line on October 28, thirty miles north of Johnsonville, Forrest converted a portion of his 3500 troopers into literal horse marines and put them aboard two Union vessels, the gunboat *Undine* and the transport *Venus,* which he captured by posting batteries at both ends of a five-mile stretch of river to prevent their escape when he took them under fire with other guns along the bank. For three days, November 1-3, while this improvised two-boat navy molested traffic and drew attention north-ward, he led his horsemen south, up the west bank of the swollen Tennessee, to carry out the devastation that was the purpose of his raid. Well before midday November 4, after losing the *Venus* in an engagement with two gunboats and burning the eight-gun *Undine* to prevent her recapture, the two divisions were directly opposite Johnsonville, masked from view by trees and brush.

While Morton was sneaking his guns into position, under orders to open fire at 2 o'clock, Forrest examined with his binoculars the unsuspecting target on the far side of the half-mile-wide river. Three gunboats, eleven trans-ports, and eighteen barges were moored at the wharves, aswarm with workers unloading stores, and beyond them, spread out around a stockade fortress on high ground, warehouses bulged with supplies and acres of open storage were piled ten feet high with goods of every description, covered with tarpaulins to protect them from the weather. Two freight trains were being made up for the run to Nashville, just under eighty miles away, and neither the soldiers at work nor the officers scattered among them seemed aware that they were in any more danger now than they had been at any time since the base — named for the military governor who was Lincoln's running mate in the election only four days off — was put in operation, six months back.

Promptly at 2 o'clock they found out better. Morton having syn-chronized the watches of his chiefs of section, all ten pieces went off with an enormous bang that seemed to come from a single heavy cannon. For nearly an hour, after this introductory clap of thunder out of a cloudless sky, their fire was concentrated on the gunboats, the most dangerous enemy weapon, and when these were abandoned by their crews, who left them to burn and sink with the transports and barges they had been ordered to protect, the rebel artillerists shifted their attention to the landward installations, including the hilltop fortress whose unpracticed cannoneers replied wildly, blinded by smoke from riverside sheds and warehouses that had been set afire by sparks from the burning wharves and exploding vessels down below. Soon all those acres of high-piled stores were a mass of flames, and the exultant rebel gunners chose individual targets of opportunity, neglected until now. Perhaps the most spectacular of these was a warehouse on high ground, which, when struck and set afire, turned out to be stocked with several hundred barrels of whiskey that burst from the heat and sent a crackling blue-flame river of bourbon pouring down the hillside. Tantalized by the combined aroma of burnt liquor, roasting coffee beans, and frizzled bacon, wafted to them through a reek of gunsmoke, Morton's hungry veterans howled with delight and regret as they kept heaving shells into the holocaust they had created across the way. Forrest himself took a hand in the fun, directing the fire of one piece. "Elevate the breech of that gun a little lower!" he shouted, and the crew had little trouble understanding this unorthodox correction of the range. Within two hours all of Johnsonville was ablaze, resulting in a scene that "beggared description," according to one Federal who confined himself to the comment that it was "awfully sublime."

It was also awfully expensive. The base commander later put his loss at $2,200,000, taking the burned-out steamers and barges into account, but not the three sunken gunboats — four, including the *Undine*, subtracted during the naval phase of the raid, along with three more transports and three barges, mounting a total of 32 guns. Forrest's estimate of $6,700,000 included all of these, and probably came closer to the truth. His own loss, over-all, was two men killed and nine wounded, plus two guns lost when the *Venus* was recaptured.

Retiring southward by the glare of flames still visible when he made camp six miles away, he encountered in the course of the next few days a series of couriers from Beauregard, all bearing orders for him to report at once to Hood, who was waiting at Tuscumbia for the outriders he would need on his march north. Forrest did what he could to hurry, but the going was slow through the muddy Tennessee bottoms, especially for the artillery. Even with sixteen horses to each piece, spelled by oxen impressed from farms adjoining the worst stretches along the way, he could see that he would need

A reluctant Ulysses S. Grant, above, finally gave Sherman the go-ahead to march through Georgia.

more than a week to reach Hood in Northwest Alabama.

Beauregard's distress at this development was matched by opposite reactions up the Coosa and beyond the Tennessee. Not only did the delay give Thomas added time to prepare for the blow Hood's drawn-out march had warned him was about to land; it also prompted Sherman to send still more reinforcements to Nashville, even while putting the final touches to his plan for making Georgia howl by slogging roughshod across it to the sea.

Grant by now had assented unconditionally to the expedition, though not until he recovered from a last-minute fit of qualms brought on by the news that Hood was headed north. Sherman at Gaylesville had not known that the gray army had left Gadsden, thirty miles away, until it turned up near Decatur, ninety miles to the west, on October 26. His reaction, once Hood's departure had ruled out a confrontation near the Alabama-Georgia line, was to send Stanley's corps to strengthen Thomas, and when he learned that Hood was still in motion westward, apparently intending to force a crossing at Tuscumbia, he also detached Schofield's one-corps Army of the Ohio and directed that A. J. Smith's divisions return at once from Missouri to join in the defense of Middle Tennessee. Between them, Stanley, Schofield, and Smith had close to 40,000 men, and these, added to those already on hand — including more than half of Sherman's cavalry, sent back earlier; sizeable garrisons at Murfreesboro, Chattanooga, Athens, and Florence; and recruits coming down from Kentucky and Ohio, in response to Forrest's early October penetration of the region below Nashville — would give Thomas about twice as many troops as Hood could bring against him. Surely that was ample, even though most of them were badly scattered, others were green, and some had not arrived. Best of all, however, from Sherman's point of view, this new arrangement provided a massive antidote for dealing with Grant's reawakened fears as to what might happen if Old Pap was left to face the invasion threat alone.

"Do you not think it advisable, now that Hood has gone so far north, to entirely ruin him before starting on your proposed campaign?" Grant inquired on November 1, and added, rather more firmly: "If you see a chance of destroying Hood's army, attend to that first, and make your other move secondary."

This, of all things, was the one Sherman wanted least to hear, and in his reply he marshaled his previous arguments in redoubled opposition. "No single army can catch Hood," he declared, "and I am convinced that the best results will follow from our defeating Jeff. Davis's cherished plan of making me leave Georgia by maneuvering." Edgy and apprehensive, fearing a negative reaction, he followed this with a second, more emphatic plea, before there was time for an answer to the first. "If I turn back, the whole effect of my campaign will be lost. By my movements I have thrown Beauregard (Hood) well to the west, and Thomas will have ample time and sufficient troops to hold him. . . . I am clearly of opinion that the best results will follow my contemplated movement through Georgia."

To his great relief, Grant wired back on November 2 that he was finally persuaded that Thomas would "be able to take care of Hood and destroy him." Moreover, he added, echoing his lieutenant's words in closing, "I really do not see that you can withdraw from where you are to follow Hood without giving up all we have gained in territory. I say, then, go as you propose."

Here at last was the go-ahead Sherman had been seeking all along, and now that he had it he moved fast, as if in fear that it might be revoked. Trains that had been shuttling between Chattanooga and Atlanta for the past two months, heavy-laden coming down and empty going back, now made their runs the other way around, returning all but the supplies he would take along in wagons when he set out for the sea with his four remaining corps, two from what was left of the Army of the Cumberland, under Slocum, and two from his

The Union Army of the Tennessee, heading back to Georgia, crosses the Coosa River on a pontoon bridge after Sherman broke off his pursuit of Hood in Alabama.

old Army of the Tennessee, under Howard. They numbered better than 60,000 of all arms, including a single division of cavalry under Kilpatrick. He saw this mainly as an infantry operation, much like the one against Meridian last year, and had ordered the rest of his troopers back to Nashville for reorganization under James Wilson, who had recently been promoted to major general and sent by Grant to see what he could do about the poor showing western horsemen had been making ever since the start of the campaign. Sherman might have taken him along, a welcome addition on a march into the unknown, except that Thomas would most likely need him worse. Besides, he said, "I know that Kilpatrick is a hell of a damned fool, but I want just that sort of a man to command my cavalry on this expedition."

In "high feather," as he nearly always was when he was busy, he reëstablished headquarters at Kingston, the main-line railroad junction on the Etowah east of Rome, and there, with trains grinding north and rattling south at all hours of the day and night, supervised the final runs before the Western & Atlantic was closed down and its several depot garrisons withdrawn to become part of Major General J. B. Steedman's command at Chattanooga, on call for service under Thomas against Hood. His own army seemed to Sherman in splendid condition, fattened by veterans returning from thirty-day reënlistment furloughs, yet trimmed for hard use by evacuating all who were judged by surgeons not to be in shape for the 300-mile cross-Georgia march.

On Sunday, November 6, he took time out to compose a farewell letter to Grant, a general statement of his intention, as he put it, "to act in such a manner against the material resources of the South as utterly to negative Davis' boasted threat." While he wrote, paymasters were active in all the camps, seeing to it that the soldiers would be in an appreciative frame of mind to support the Administration in the election two days off. "If we can march a well-appointed army right through his territory, it is a demonstration to the world, foreign and domestic, that we have a power which Davis cannot resist. This may not be war, but rather statesmanship. Nevertheless it is overwhelming to my mind that there are thousands of people abroad and in the South who reason thus: If the North can march an army right through the South, it is proof positive that the North can prevail."

He would set out, he told his chief, hard on the heels of Lincoln's reëlection — "which is assured" — and would thereby have the advantage of the confusion, not to say consternation, that event would provoke in the breasts of secessionists whose heartland he would be despoiling. What he would do after he reached Savannah he would decide when he got there and got back in touch with City Point. Meantime, he said, "I will not attempt to send couriers back, but trust to the Richmond papers to keep you well advised."

★

Grant — observing with hard-won equanimity the unusual spectacle of the two main western armies, blue and gray, already more than two hundred miles apart, about to take off in opposite directions — replied next day: "Great good luck go with you. I believe you will be eminently successful, and at worst can only make a march less fruitful than is hoped for."

In Richmond that same day, November 7 — election eve beyond the Potomac — Congress was welcomed back into session by a message from the Chief Executive, who had continued in Virginia the efforts made on his Georgia trip to lift spirits depressed by the outcome of the Hood-Sherman contest for Atlanta. Indeed, Davis went further here today in his denial that the South could be defeated, no matter what calamities attended her resistance to the force that would deny her independence.

After speaking of "the delusion fondly cherished [by the enemy] that the capture of Atlanta and Richmond would, if effected, end the war by the overthrow of our government and the submission of our people," he said flatly: "If the campaign against Richmond had resulted in success instead of failure, if the valor of [Lee's] army, under the leadership of its accomplished commander, had resisted in vain the overwhelming masses which were, on the contrary, decisively repulsed — if we had been compelled to evacuate Richmond as well as Atlanta — the Confederacy would have remained as erect and defiant as ever. Nothing could have been changed in the purpose of its government, in the indomitable valor of its troops, or in the unquenchable spirit of its people. The baffled and disappointed foe would in vain have scanned the reports of your proceedings, at some new legislative seat, for any indication that progress had been made in his gigantic task of conquering a free people." And having said as much he said still more in that regard. "There are no vital points on the preservation of which the continued existence of the Confederacy depends. There is no military success of the enemy which can accomplish its destruction. Not the fall of Richmond, nor Wilmington, nor Charleston, nor Savannah, nor Mobile, nor of all combined, can save the enemy from the constant and exhaustive drain of blood and treasure which must continue until he shall discover that no peace is attainable unless based on the recognition of our indefeasible rights."

He spoke at length of other matters, including foreign relations and finances — neither of them a pleasant subject for any Confederate — and referred, near the end, to the unlikelihood of being able to treat for peace with enemy leaders "until the delusion of their ability to conquer us is dispelled." Only then did he expect to encounter "that willingness to negotiate which is now confined to our side." Meantime, he told the assembled representatives, the South's one recourse lay in self-reliance. "Let us, then, resolutely continue to devote our united and unimpaired energies to the defense of our homes, our lives, and our liberties. This is the true path to peace. Let us tread it with confidence in the assured result."

Nowhere in the course of the long message did he mention tomorrow's election in the North, although the outcome was no less vital in the South — where still more battles would be fought if the hard-war Union party won — than it was throughout the region where the ballots would be cast. For one thing, any favorable reference to McClellan by Jefferson Davis would cost the Pennsylvanian votes he could ill afford now that Atlanta's fall and Frémont's withdrawal had transformed him, practically overnight, from odds-on favorite to underdog in the presidential race. In point of fact, much of the suspense had gone out of the contest, it being generally conceded by all but the most partisan of Democrats, caught up in the hypnotic fury of the campaign, that Little Mac had only the slimmest of chances.

Lincoln himself seemed gravely doubtful the following evening, however, when he crossed the White House grounds, soggy from a daylong wintry rain, to a side door of the War Department and climbed the stairs to the telegraph office, where returns were beginning to come in from around the country. These showed him leading in Massachusetts and Indiana, as well as in Baltimore and Philadelphia, and the trend continued despite some other dispatches that had McClellan ahead in Delaware and New Jersey. By midnight, though the storm delayed results from distant states, it was fairly clear that the turbulent campaign would end in Lincoln's reëlection.

Earlier he had said, "It is strange that I, who am not a vindictive man, should always, except once, have been before the people in canvasses marked by great bitterness. When I came to Congress it was a quiet time, but always, except that, the contests in which I have been prominent have been marked with great rancor." Now he lapsed into a darkly reminiscent mood, telling of that other election night, four years ago in Springfield, and a strange experience he had when he came home, utterly worn out, to rest for a time on a horsehair sofa in the parlor before going up to bed. Across the room, he saw himself reflected in a mirror hung on the wall above a bureau, almost at full length, murky, and with two faces, one nearly superimposed upon the other. Perplexed, somewhat alarmed, he got up to study the illusion at close range, only to have it vanish. When he lay down again it reappeared, plainer than before, and he could see that one face was paler than the other. Again he rose; again the double image disappeared. Later he told his wife about the phenomenon, and almost at once had cause — for both their sakes — to wish he hadn't. She took it as a sign, she said, that he would be reëlected

★

By 2 a.m. on November 8, Lincoln knew he had won reëlection, but he did not yet know he had won with 55 percent of the popular vote and 212 out of 233 electoral ballots.

four years later, but that the pallor of the second face indicated that he would not live through the second term.

The gloom this cast was presently dispelled by further reports that put all of New England and most of the Middle West firmly in his column. Around 2 o'clock, word came that serenaders, complete with a band, had assembled on the White House lawn to celebrate a victory whose incidentals would not be known for days. These would show that, out of some four million votes cast this Tuesday, Lincoln received 2,203,831 — just over 55 percent — as compared to his opponent's 1,797,019. Including those of Nevada, whose admission to the Union had been hurried through, eight days ago, so that its three votes could tip the scales if needed, he would receive 212 electoral votes and McClellan only the 21 from Delaware, New Jersey, and Kentucky. Yet the contest had been a good deal closer than these figures indicated. Connecticut, for example, was carried by a mere 2000 votes and New York by fewer than 7000, both as a result of military ballots, which went overwhelmingly for Lincoln, here as elsewhere. Without these two states, plus four others whose soldier voters swung the

★

balance — Pennsylvania, Illinois, Maryland, and Indiana — he would have lost the election. Moreover, even in victory there were disappointments. New York City and Detroit went Democratic by majorities that ran close to three to one, and McClellan not only won the President's native state, Kentucky, he also carried Sangamon County, Illinois, and all the counties on its border. Lincoln could say to his serenaders before turning in that night, "I give thanks to the Almighty for this evidence of the people's resolution to stand by free government and the rights of humanity," but there was also the sobering realization, which would come with the full returns, that only five percent less than half the voters in the nation had opposed with their ballots his continuance as their leader.

Still, regardless of its outcome, he found consolation in two aspects of the bitter political struggle through which the country had just passed, and he mentioned both, two nights later, in responding to another group of serenaders. One was that the contest, for all "its incidental and undesirable strife," had demonstrated to the world "that a people's government can sustain a national election in the midst of a great civil war." This was much, but the other aspect

Union troops, such as these Pennsylvania absentees voting at army headquarters on the James River, gave Lincoln enough swing votes to win reëlection.

was more complex, involving as it did the providence of an example distant generations could look back on when they came to be tested in their turn. "The strife of the election is but human nature practically applied to the facts of the case," he told the upturned faces on the lawn below the window from which he spoke. "What has occurred in this case must ever recur in similar cases. Human nature will not change. In any future great national trial, compared with the men of this, we shall have as weak and as strong, as silly and as wise, as bad and as good. Let us therefore study the incidents of this, as philosophy to learn wisdom from, and none of them as wrongs to be revenged."

Even so, a cruel paradox obtained. McClellan the loser was soon off on a European tour, a vacation that would keep him out of the country for six months, whereas Lincoln now more than ever, despite the stimulus of victory at the polls, could repeat what he had said two years before, in another time of trial: "I am like the starling in Sterne's story. 'I can't get out.'"

He had this to live with, as well as the memory of that double-image reflection in the mirror back in Springfield: both of which no doubt contributed, along with much else, to the nighttime restlessness a member of the White House guard observed as he walked the long second-story corridor, to and fro, past the door of the bedroom where the President lay sleeping. "I could hear his deep breathing," the sentry would recall. "Sometimes, after a day of unusual anxiety, I have heard him moan in his sleep. It gave me a curious sensation. While the expression of Mr Lincoln's face was always sad when he was quiet, it gave one the assurance of calm. He never seemed to doubt the wisdom of an action when he had once decided on it. And so when he was in a way defenseless in his sleep, it made me feel the pity that would almost have been an impertinence when he was awake. I would stand there and listen until a sort of panic stole over me. If he felt the weight of things so heavily, how much worse the situation of the country must be than any of us realized! At last I would walk softly away, feeling as if I had been listening at a keyhole."

★　★　★

Epilogue

Outfoxed by Grant, Robert E. Lee had rushed the majority of his troops to the outskirts of Richmond, where he expected to see the 16,000 Federal troops that on June 15, 1864, arrived instead at Petersburg. There only Beauregard's 3000 or so rebels were on hand to defend the place. Luckily for Lee, the Union soldiers, commanded by Baldy Smith, were combat weary, with the slaughter at Cold Harbor fresh in their memories, and were slow to attack. When expected reinforcements got lost on the way and failed to arrive, Smith called off the fight with victory virtually within reach. Beauregard was reinforced, and the war went on as repeated assaults were beaten back, and both sides settled in for a siege.

A siege was the one thing Lee had said his army could not afford. While Grant touted the siege as a reprise of Vicksburg, it was clearly going to be drawn out, which was something Lincoln, facing a rough reëlection campaign, could have done without. Yet six weeks of constant combat had crippled both armies, so now they burrowed into the ground, creating a labyrinth of trenches exposed to mortar bombardment and plagued by hunger and diseases, but free of the kind of assaults that at Cold Harbor had given even Grant pause. The President himself was relieved to find Grant more mindful of his casualty lists, and Lincoln told him so in a communication careful to make its point without appearing meddlesome.

Yet soldiers did die in sieges, and it was a kind of fighting hard on the nerves. Sometimes it even led to desperate gambles, like the one a regiment of Pennsylvania coal miners persuaded General Burnside to let them venture in late July. They wanted to dig a tunnel underneath Confederate lines, pack it with explosives, and blow a hole wide enough to march an army through. With Grant's approval, they did precisely that, blasting a huge crater in the Petersburg defenses that sent rebels reeling back in terror. Half an hour passed, however, before the Federals followed up, and even then they rushed into the hole rather than around it, forgot to bring ladders, and milled about long enough for both sides to realize they were trapped. The regrouped Confederates commenced a deadly fire, and when the Union attackers surrendered later that afternoon, many white troops were taken captive while black soldiers were shot down, bayoneted, or bludgeoned to death by the hundreds.

By then, back in Washington, Lincoln had more to worry about than Grant's huge casualty figures. Jubal Early and his troops had marched to within sight of the capital, a fact to which the President could personally attest. He had gone out to see them with his own eyes, and the Confederates had fired on him, without realizing at whom they were shooting, before Union soldiers ordered the tall onlooker to get down off the ramparts, without realizing to whom they were giving orders. Grant sent reinforcements, and Early soon withdrew, but by then Lincoln had more to worry about than a possible raid on the Federal District. Radicals in Congress, critical of his emphasis on reconstituting the Union, passed a harsh reconstruction bill, which Lincoln — mindful of the damage it would do to his chances for reëlection — killed by exercising the first pocket veto in history.

In fact, that summer and fall, Lincoln used all the powers that a President in wartime can garner to assure him of victory in November, a lot of them more questionable than putting a bill he disliked in his pocket and forgetting it. Whatever else he did, however, Lincoln kept a close eye on developments in the war, for it was there he would rise or fall. He cheered Farragut's taking of Mobile Bay and William Cushing's long-odds but successful plan to destroy the troublesome *Albemarle*. He fretted over Bedford Forrest's nettlesome raids in Tennessee. And he exulted in Sherman's defeat of Hood down in Georgia and the subsequent fall of Atlanta.

Lincoln won the election, despite all that the disenchanted Radicals, the carping Copperheads, and Lee and Davis could do to prevent it. As he prepared for his second inaugural, the siege would continue in Petersburg. For some ten months, the two sides would face each other across a landscape that foreshadowed the evils of modern trench warfare as reproduced half a century later by the belligerents of World War I. An ebullient Sherman would take off from Atlanta on a march to the sea, and then a march through the Carolinas, burning as he went, in yet another presaging of the twentieth century's "total" warfare. And Hood, anxious to redeem his failure at Atlanta, would fight a battle near Franklin, Tennessee, as horrible as Cold Harbor before going on to suffer the near total destruction of his army on the outskirts of Nashville, the last great battle of the war.

★   ★   ★

★

Picture Credits

of Wisconsin. **190:** Map by William L. Hezlep. **193:** Drawing by Alfred R. Waud, from *Battles and Leaders of the Civil War,* vol. 4, published by The Century Co., New York, 1887. **196:** Drawing by James E. Taylor, The Western Reserve Historical Society, Cleveland, photographed by Michael McCormick. **199:** Map by Walter W. Roberts and William L. Hezlep. **201:** Museum of the Confederacy, Richmond, Va.—Sketch by James E. Taylor, The Western Reserve Historical Society, Cleveland, photographed by Michael McCormick. **202:** Map by Walter W. Roberts and William L. Hezlep. **206-208:** Painting by Benjamin Mileham, Kansas State Historical Society. **213:** State Historical Society of Missouri, Columbia; Jackson County Historical Society, Independence. Mo.—Tilgman 138, Western History Collections, University of Oklahoma; George Hart Collection. **214:** Watercolor by Samuel J. Reader, Kansas State Historical Society. **216:** Zenda, Inc. **219:** Mural by N. C. Wyeth, Missouri State Capitol, Jefferson City, Mo., photographed by Jack A. Savage. **223:** St. Albans Historical Society. **225:** National Archives, Record Group No. 45. **227:** American Heritage Picture Collection. **229:** U.S. Navy Photo, Courtesy of

The Mariners' Museum, Newport News, Va.—Museum of the Confederacy, Richmond, Va. **231:** Courtesy Paul de Haan. **232:** Culver Pictures, Inc. **234, 235:** © Collection of The New-York Historical Society, New York. **237:** Zenda, Inc. **238:** From *The Photographic History of the Civil War,* vol. 10, Reivew of Reviews Co., New York, 1912. **242:** Courtesy Harry M. Rhett, Jr., Huntsville, Ala., photographed by George Flemming. **244-246:** Library of Congress, Neg. No. 381-72-2712. **248, 249:** Courtesy Frank and Marie-Thérèse Wood Print Collections, Alexandria, Va. **251:** Painting by Alexander Lawrie, West Point Museum Collections, U. S. Military Academy, photographed by Henry Groskinsky. **254:** Zenda, Inc. **260:** Courtesy State Historical Society of Iowa, Des Moines. **263:** Courtesy Frank and Marie-Thérèse Wood Print Collections, Alexandria, Va. **264:** Zenda, Inc. **267:** National Archives, Neg. No. 111-B-5176. **270:** Zenda, Inc. **272, 273:** Courtesy Frank and Marie-Thérèse Wood Print Collections, Alexandria, Va. **277:** Courtesy James Mellon Collection. **278:** Drawing by William Waud, Library of Congress.

Index

Numerals in italics indicate an illustration of the subject mentioned.

A